Achieving sustainable cultivation of apples

It is widely recognised that agriculture is a significant contributor to global warming and climate change. Agriculture needs to reduce its environmental impact and adapt to current climate change whilst still feeding a growing population, i.e. become more 'climate-smart'. Burleigh Dodds Science Publishing is playing its part in achieving this by bringing together key research on making the production of the world's most important crops and livestock products more sustainable.

Based on extensive research, our publications specifically target the challenge of climate-smart agriculture. In this way we are using 'smart publishing' to help achieve climate-smart agriculture.

Burleigh Dodds Science Publishing is an independent and innovative publisher delivering high quality customer-focused agricultural science content in both print and online formats for the academic and research communities. Our aim is to build a foundation of knowledge on which researchers can build to meet the challenge of climate-smart agriculture.

For more information about Burleigh Dodds Science Publishing simply call us on +44 (0) 1223 839365, email info@bdspublishing.com or alternatively please visit our website at www.bdspublishing.com.

Related titles:

Achieving sustainable cultivation of tomatoes
Print (ISBN 978-1-78676-040-1) Online (ISBN 978-1-78676-042-5, 978-1-78676-043-2)

Chapters are available individually from our online bookshop: https://shop.bdspublishing.com

climate•SMART•publishing
IN AGRICULTURAL SCIENCE

BURLEIGH DODDS SERIES IN AGRICULTURAL SCIENCE

NUMBER 018

Achieving sustainable cultivation of apples

Edited by Professor Kate Evans, Washington State University, USA

burleigh dodds
SCIENCE PUBLISHING

Published by Burleigh Dodds Science Publishing Limited
82 High Street, Sawston, Cambridge CB22 3HJ, UK
www.bdspublishing.com

Burleigh Dodds Science Publishing, 1518 Walnut Street, Suite 900, Philadelphia, PA 19102-3406, USA

First published 2017 by Burleigh Dodds Science Publishing Limited

Library of Congress Control Number: 2017936825

British Library Cataloguing in Publication Data
A catalogue record for this book is available from the British Library

ISBN 978-1-78676-032-6 (print)
ISBN 978-1-78676-035-7 (online)
ISBN 978-1-78676-034-0 (online)
ISSN 2059-6936 (print)
ISSN 2059-6944 (online)

Typeset by Deanta Global Publishing Services, Chennai, India
Printed by Lightning Source

Contents

Series list

Title	Series number
Achieving sustainable cultivation of maize - Vol 1 From improved varieties to local applications Edited by: Dr Dave Watson, CGIAR Maize Research Program Manager, CIMMYT, Mexico	001
Achieving sustainable cultivation of maize - Vol 2 Cultivation techniques, pest and disease control Edited by: Dr Dave Watson, CGIAR Maize Research Program Manager, CIMMYT, Mexico	002
Achieving sustainable cultivation of rice - Vol 1 Breeding for higher yield and quality Edited by: Prof. Takuji Sasaki, Tokyo University of Agriculture, Japan	003
Achieving sustainable cultivation of rice - Vol 2 Cultivation, pest and disease management Edited by: Prof. Takuji Sasaki, Tokyo University of Agriculture, Japan	004
Achieving sustainable cultivation of wheat - Vol 1 Breeding, quality traits, pests and diseases Edited by: Prof. Peter Langridge, The University of Adelaide, Australia	005
Achieving sustainable cultivation of wheat - Vol 2 Cultivation techniques Edited by: Prof. Peter Langridge, The University of Adelaide, Australia	006
Achieving sustainable cultivation of tomatoes Edited by: Dr Autar Mattoo, USDA-ARS, USA & Prof. Avtar Handa, Purdue University, USA	007
Achieving sustainable production of milk - Vol 1 Milk composition, genetics and breeding Edited by: Dr Nico van Belzen, International Dairy Federation (IDF), Belgium	008
Achieving sustainable production of milk - Vol 2 Safety, quality and sustainability Edited by: Dr Nico van Belzen, International Dairy Federation (IDF), Belgium	009
Achieving sustainable production of milk - Vol 3 Dairy herd management and welfare Edited by: Prof. John Webster, University of Bristol, UK	010
Ensuring safety and quality in the production of beef - Vol 1 Safety Edited by: Prof. Gary Acuff, Texas A&M University, USA & Prof.James Dickson, Iowa State University, USA	011
Ensuring safety and quality in the production of beef - Vol 2 Quality Edited by: Prof. Michael Dikeman, Kansas State University, USA	012
Achieving sustainable production of poultry meat - Vol 1 Safety, quality and sustainability Edited by: Prof. Steven C. Ricke, University of Arkansas, USA	013
Achieving sustainable production of poultry meat - Vol 2 Breeding and nutrition Edited by: Prof. Todd Applegate, University of Georgia, USA	014
Achieving sustainable production of poultry meat - Vol 3 Health and welfare Edited by: Prof. Todd Applegate, University of Georgia, USA	015

Acknowledgements

We wish to acknowledge the following for their help in reviewing particular chapters:

- Chapter 3: Emeritus Professor Ian Warrington Massey University, New Zealand; and Dr Ben Van Hooijdonk, Plant and Food Research, New Zealand
- Chapter 9: Dr Nigel Swarts, University of Tasmania, Australia
- Chapter 15: Emeritus Professor Alan Biggs, West Virginia University, USA

Introduction

The apple is an iconic fruit recognized in most countries of the world and is produced in many. China is the biggest producer by far, producing around 40% of the world's 63 million tons of apples, almost ten times that of the second largest producer, the United States. Production levels and advances in storage technologies have resulted in the availability of apples year round in many markets. As a fruit with a high content of flavonoids and dietary fibre, apple has been linked to many health benefits. Indeed, apple has achieved what most other fruits have not in that it has successfully made its way into the fast food industry and can be frequently found, typically sliced, accompanying a considerably less healthy burger.

We have been cultivating apples for thousands of years, but the last century has seen the biggest changes in our production systems. Like other crops, we have moved into intensive agriculture. Dwarfing rootstocks for apple have enabled intensive, more efficient plantings of almost-two-dimensional trees before most other tree fruits. Such high-density plantings are moving apple production towards precision horticulture and provide orchard structure that fits mechanization and automation, rapidly becoming a requirement as labour shortage is the principal challenge currently facing sustainability of apple in many areas of the world. Our ability to optimize cultural practices through integrated pest management, biocontrol, and measured irrigation and nutrition, increases the sustainability of the orchard systems. Growers and pack-houses worldwide are striving to achieve Global GAP certification of farm assurance leading to Good Agricultural Practice status; apple precision horticulture is perfectly aligned.

Apple cultivation can be challenging in many ways; growers need the orchard to remain in the ground often for decades to recoup their very high initial investments. In addition, they must develop and execute careful management plans. However, while complex, working with a composite tree brings a level of flexibility and purposefully intensive management that is not present in some other production systems. As opposed to most annual crop producers, skilled apple growers can literally 'build' the tree to their requirements by carefully choosing a rootstock to suit their soil conditions and fitness for their preferred production system, and a scion to suit their climate and the available market. Once the tree is planted, the growers can train it into a structure that is desired by themand manipulate the leaf canopy and crop load to achieve the optimum balance of fruit quality and yield. Of course, none of this is easy, so perhaps the key skill for a grower is the ability to forecast 10–20 years ahead and prepare for change. Will it be hotter and drier, or will our precipitation patterns change dramatically? What will new technology bring and will it fit the current orchard training system? Will the variety still be profitable? And of course, there is the constant threat of new exotic pests. While apple growers using conventional farming practices confront these challenges, those utilizing organic systems must follow appropriate, and often quite different, practices. Nonetheless many are doing so successfully.

More growers are moving towards the physical protection of their crop using netting, initially for mitigation of hail but now also to filter potentially damaging levels of light that can damage fruit and stress vegetative growth, and in some areas, are also being considered for pest protection. Such investments in infrastructure add to the already expensive establishment costs, so obviously the grower has to take into account the market price and typical grade-out of the variety to recover their initial costs and generate a revenue stream.

Ultimately, in my view, it is consumer demand that will sustain apple production as long as we can produce and deliver quality pieces of fruit consistently throughout the year. One thing that sets apple apart from most other fruit crops is the range of fruit types that are in demand, in terms of both internal eating quality and, of course, appearance. Another advantage for apple is that new varieties are clearly named and can easily be distinguished, providing the retailer with a marketing incentive and ensuring consumers can be confident that the ones they chose to purchase again have the characteristics that they prefer. We can offer the consumer more than simply fresh fruit by extending the sliced apple market, developing high-quality juices, sauces, dried apple and other lightly processed products.

Worldwide, consumers prefer different fruit flavour and texture attributes. Breeders are challenged to combine those diverse consumer-preferred attributes with varieties that also enable producer- and environmentally-friendly production practices. Continued breeding and selection to feed the demand for new improved eating experiences as well as to develop varieties that will face the challenges of a changing climate is essential for long-term sustainability. Cultivated *Malus* are only a few generations removed from wild progenitors compared to many other crop species, so we have plenty of opportunities for genetic gain to develop superior new cultivars. Further, we have widely diverse genetic resources available to us and, with new DNA-based tools, breeders are becoming better enabled to access this diversity and attempt the introgression of new traits.

The authors and topics in the book have been carefully chosen to provide the reader with an introduction to apple as a crop species plus present updates from leading international scientists on multiple aspects of apple production. Each chapter contributes to the ultimate aim of achieving sustainable cultivation, whether the word sustainable is interpreted as meaning economically viable, environmentally sound and/or socially acceptable, or the more general definition to endure or last indefinitely. Specific chapters on environmental impact assessment, economics of production and consumer trends in apple sales add to the bigger picture of the current position of global apple sustainability. Our goal was not just to provide a current review, but also to provide a useful resource for further information in terms of publications, research groups and conference series. Apple cultivation is certainly heading in a more sustainable direction. Advances over recent years have been dramatic. This book is a unique, timely and comprehensive review of these scientific advances and their potential implementation by apple producers worldwide.

Part 1 Plant physiology and breeding

The first group of chapters looks at advances in understanding apple physiology and breeding. Chapter 1 highlights the critical importance of genetic diversity. Widely used apple cultivars exhibit low resistance to pathogens and low tolerance to climatic threats. Apple breeding programmes therefore need to develop new cultivars which are more resilient in the face of key threats, while also selecting for other important qualities. Previously, it was difficult to use wild apple species for breeding because typically they have undesirable fruit, non-uniform ripening times, and other traits that are not amenable to commercial apple production. Apple progenitor species, *M. sieversii*, *M. orientalis*,

M. sylvestris, *M. prunifolia*, and *M. baccata* offer desirable variation, and are more amenable to use in breeding programmes because they are diploid and often have fruit traits more similar to *M.* × *domestica* than many of the more distant wild *Malus* species. The chapter explores the opportunities and challenges of using progenitor species of *Malus* × *domestica* in traditional breeding programmes.

Key challenges include understanding *Malus* diversity, capturing and maintaining this diversity, and ensuring that researchers and breeders have access to living apple accessions. The chapter highlights the role of genebanks, such as the USDA-ARS National Plant Germplasm System. Genebanks provide breeders access to *Malus* wild species and local cultivars (which may also have desirable traits). The USDA-ARS National Plant Germplasm System, for example, maintains an apple collection with 3070 unique grafted trees representing *M.* × *domestica* cultivars and 33 *Malus* species, and 15 hybrid species that are available to breeding programmes.

It is important to note that many wild *Malus* species are, however, poorly represented in genebank collections, and phytosanitary restrictions and treaty agreements may limit access to wild materials, particularly at the international level. Ongoing coordinated efforts among genebanks to compare collections at the phenotypic and genetic levels as well as confirm cultivar identities will facilitate use of genebank collections in the future. These efforts will enable future researchers specializing in apple improvement to provide consumers with higher quality fruit that can be produced more sustainably. Finally, new technologies that make use of marker-assisted selection, genomic selection, genetic engineering, genome-wide association mapping, high-throughput genotyping and/or rapid-cycling plants are paving the way toward the increased use of wild *Malus* species in apple breeding.

Chapter 2 provides an overview of the development of genome sequencing technology, reviews the process of sequencing the apple genome and then considers how this information can be employed, both to develop new and better varieties of apple and, in the shorter term, to improve current horticultural practices. In the future, knowledge of specific genes can be used to improve vegetative growth and development, biotic and abiotic disease tolerance or resistance, pre-harvest stages of fruit development, postharvest fruit storage, shelf-life and organoleptic traits.

The next group of chapters reviews the latest research in understanding apple tree growth and its implications for cultivation practice, starting with rootstocks and planting systems. Chapter 3 discusses the importance of rootstocks, with an emphasis on the mechanisms and morphological effects of dwarfing. This chapter outlines the key attributes of widely used dwarfing and semi-dwarfing rootstocks including relatively recent releases that confer resistance to economically important pests and diseases and tolerance to abiotic conditions of drought, low temperature and waterlogging. These traits significantly widen the potential sites and regions that can potentially support modern apple systems.

Understanding how rootstocks and scions interact to manifest in various growth attributes of size, architecture and precocity (earliness of flowering and bearing) has been a major focus of research. Understanding of growth attributes also enables managers to design growing systems that allow the tree to fill the allocated canopy space to optimise light capture early in the life of the orchard (and prevent excessive shading after establishment) in order to produce better yields. The chapter summarises recent research on the manipulation of tree growth and development, with an emphasis on the use of plant growth regulators, followed by a discussion of the effects of different planting systems.

Chapter 4 provides a detailed discussion of the practices which affect canopy development, dormancy release, bud break and vegetative growth. As an example, the chapter explores how application of branching agents has become standard nursery practice while notching has gained acceptance for branch induction in older trees, particularly in combination with cytokinin. Under conditions of inadequate winter chilling, bud break can be manipulated by cultural practices or application of chemical rest-breaking agents. Understanding how scion cultivars interact with rootstocks has driven studies on training and pruning, underpinned by knowledge of architectural traits and tree physiology. The chapter ends by presenting two case studies which show the importance of post-planting apple tree management, and illustrate how it can interact with choice of rootstock and planting system to determine yield, including artificial spur extinction and chemical thinning for crop load control.

Chapter 5 looks at recent research on the biology and horticultural aspects of flowering with a particular focus on pollination. Fruit production begins with the transition of a bud from vegetative to floral state. The bud differentiates, overwinters, and emerges as a flower the following spring. Flowers are then pollinated, fertilized, and the fruit grows firstly by cell division and later by cell enlargement. All of these processes are vital to the development of high-quality fruit. Sub-optimal environmental, biological or cultural conditions during any of these stages can reduce both productivity and fruit quality. The chapter discusses the biological processes and genetic controls of these developmental stages. It also highlights some of the key environmental effects and how these processes can be manipulated by cultural management.

As with earlier processes, understanding the ways in which apple fruit grow and develop is crucial for achieving sustainable apple cultivation. Chapter 6 examines how apples grow and ripen, what we know about the factors that support or limit growth, why fruit abscise and how growers can manipulate fruit growth and abscission to optimize cropping. Amongst other topics, the chapter explores in detail seasonal growth patterns, the chemical composition of apples in different seasons, the role of hormones in abscission, and seasonal ripening patterns as well as competition within the flower/fruit cluster. The chapter suggests ways to model optimal crop load and suggests future trends for research in this area.

The foundations of a productive and healthy orchard are the rootstocks that provide anchorage, water and nutrients essential to the above-ground portions of the trees. Utilization of composite trees has increased the efficiency of breeding productive apple trees by dividing the selection of scion traits and rootstock traits into two genetically (and functionally) different specimens which are then brought together through grafting. As part of the tree, the rootstock influences many factors in addition to tree size, particularly productivity, fruit quality, pest resistance and stress tolerance. Understanding how scion properties are modulated by rootstocks allows targeting of traits that may be selected to improve whole tree performance by improving rootstock performance. Chapter 7 examines apple breeding methods and explores how rootstocks affect scion traits, before addressing the impact of rootstocks on disease and pest resistance.

Breeding and selecting new, improved cultivars by exploiting natural genetic diversity is an essential route forward in ensuring sustained and sustainable apple production. There can be no doubt that breeders have an expanding set of tools at their disposal to achieve these goals with the advent of marker-assisted (and DNA-informed) breeding. Chapter 8 discusses the promise of marker-assisted breeding in apple and reviews the advances that have been made, particularly over the last two decades, in practical application of genetic markers to breeding programmes, current tools, and the impacts of this technology. A case study of genetic marker application in the Washington State University apple breeding programme is also described. The chapter suggests what the near future may hold for marker-assisted breeding in apple. The rapid increase in genomic data will present its own challenges in terms of managing decision-making although the result will be enhanced efficiency, accuracy and pace of new apple cultivar development.

Part 2 Cultivation techniques

The next group of chapters reviews improvements in cultivation techniques. Chapter 9 provides a summary of recent innovations in orchard training system design, pruning techniques, thinning, plant growth regulators and fruit finishing with the aim of identifying more sustainable practices. New training systems like the 'bi-axis' are described to achieve planar, vertical or angle canopies, as well as mechanical and other pruning and thinning techniques to minimize labour requirements. It is now possible to purchase specific typologies of tree that are more compatible with training techniques which will further improve production. New uses of plant growth regulators to manipulate tree growth and fruit finish are also analysed.

Chapter 10 explores the emerging issues that will influence future approaches to soil and nutrient management in apple orchards. These include climate change and variability, the degradation of soil and water resources, and the future availability and cost of fertilizers. The chapter then reviews strategies for precision nutrient and water management. These include fertigation and targeted foliar application as well as improved types of fertilizer and technologies for sensing nutrient limitations. Improved irrigation scheduling and conservation irrigation techniques, including partial and deficit irrigation, are also discussed. Finally, the chapter considers the prospects for alternative production systems and production areas given the changing climate.

As Chapter 11 points out, sustaining large-scale commercial apple production depends on the availability of a large, seasonal and suitably skilled workforce. Costs and overall productivity could be improved by mechanization of apple production operations. Chapter 11 begins by reviewing the levels of mechanization that may be introduced into apple production and then considers a variety of mechanization solutions for each of the four key operations in apple production: training and pruning, thinning, pest and disease control, and harvesting. In each case, the chapter first considers those solutions which employ lower levels of mechanization, and then moves up to solutions which involve a higher level of mechanization, including automation and robotics. The advantages and disadvantages of each solution and directions for further development are also discussed.

Use of fungicides to control disease in postharvest apples is becoming less acceptable to many consumers, and the search for alternative approaches that are more sustainable in controlling postharvest diseases of fruits has intensified during the last three decades. Chapter 12 describes various alternative approaches to synthetic fungicides for controlling postharvest diseases of apples that are potentially safer for human health and the environment. The main focus has been on exploring natural products and substances generally regarded as safe (GRAS) which have anti-fungal properties. The chapter also reviews physical treatments (such as heat treatment, irradiation or the use of controlled atmospheres), biological control and natural fruit resistance, either by induction in harvested fruit or through breeding programmes. When used in combination, these techniques can now rival fungicide treatments in effectiveness.

Postharvest handling of apples, as for most horticultural products, is largely concerned with the maintenance of product quality after harvest. Chapter 13 outlines recent advances in understanding the various factors affecting fruit maturation and ripening. It reviews harvest, handling and grading operations, and discusses postharvest storage technologies including refrigeration, controlled atmosphere (CA) storage and the use of 1-Methylcyclopropene (1-MCP). It also considers a range of postharvest treatments of apples and the physiological disorders to which apples are subject.

Part 3 Diseases and pests

Fungal plant pathogens cause significant economic losses in the field and during storage which decreases fruit quality. Several pathogens also produce mycotoxins that are harmful to human health. Pathogens include *Alternaria mali* (Alternaria leaf spot), *Colletotrichum* species (bitter rot), *Glomerella cingulata* (Glomerella leaf spot), *Gymnosporangium juniperi-virginianae* (cedar apple rust), *Podosphaera leucotricha* (powdery mildew), *Botryosphaeria dothidea* (white rot), *Schizathyrium pomi* (flyspeck) and, in particular, *Venturia inequalis* (apple scab). There have been a number of advances in biological, chemical and cultural methods of control which help to combat these diseases. Chapter 14 reviews these developments together with a discussion of the main pre- and postharvest apple pathogens and methods of integrated disease management.

Like fungal diseases, viruses and virus-like agents are a recurring problem in apple production. Viruses include so-called 'latent viruses', apple mosaic virus, flat apple disease and viroids. They are graft-transmissible, and are readily disseminated through the use of infected budwood. Since many of these agents do not have arthropod or nematode vectors, the most effective control strategy is to develop orchards with clean, virus-tested planting stock. As Chapter 15 states, the implementation of quarantine and certification programmes globally is essential to safeguard apple production and minimize the risk of spreading disease by removing virus-infected plants from the supply chain. Procedures for virus testing and elimination are constantly being revised to incorporate the best and most appropriate technologies, including developments in high-throughput sequencing techniques.

As Chapter 16 indicates, bacterial diseases remain an ongoing challenge to the sustainability of apple production. Fire blight, for example, is a serious economic threat to apple production. Chapter 16 uses it as an example to present the key issues and challenges which bacterial diseases raise for apple production. The chapter reviews the history of fire blight and its spread around the globe, economic impact, symptoms, disease biology and management. The chapter also includes a case study on current efforts to develop apple cultivars which are resistant to fire blight disease. Other apple diseases caused by bacteria, including blister spot, crown gall and hairy root, bacterial blister bark and apple proliferation are also considered.

Management of apple pests is a necessary and challenging part of crop production. Chapter 17 begins by covering key pests of apple such as the codling moth. The chapter then considers indirect pests, which may be tolerated in moderate densities, and are therefore more amenable to other control tactics, especially biological control. The chapter suggests that integrated pest management (IPM) tactics which control key pests without disrupting biological control of secondary pests are the best way forward for sustainable IPM in apple production. After considering the tools and tactics available as part of a sustainable IPM programme, the chapter considers the challenges facing IPM, such as pesticide resistance and invasive species, which require innovative solutions for future IPM systems.

Disease- and pest-resistant cultivars can significantly contribute to sustainable and resilient cultivation of apples. Chapter 18 reviews the development of apple varieties resistant to a number of important diseases, including apple scab, powdery mildew, fire blight, nectria canker and Marssonina apple blotch. The chapter also consider the development of pest-resistant apple varieties, before discussing DNA-based selection techniques for developing resistance and the mechanisms on which resistance depends.

Part 4 Sustainability

A viable business is a key element of sustainability. Chapter 19 reviews and compares costs in apple production worldwide. The chapter analyses the economics of apple production around the world, including in countries such as Germany, Italy, South Africa, Switzerland and Chile. A detailed case study of apple production in Washington State, USA, is also provided. The chapter includes a comparison of apple production costs by variety across the United States, Germany, Italy, and South Africa. Given the importance of international markets for major apple producing countries in the world, the chapter considers apple production in relation to global trends in production and international trade. It discusses structural changes in the apple industry around the world, as well as trends in production systems and technologies.

As Chapter 20 indicates, the world of apple marketing presents many complex challenges. Consumers have become increasingly disparate and selective in what they expect from the products they buy. They remain concerned about the intrinsic qualities of apples, such as colour, taste and texture, but continue to add new concerns such as social responsibility and environmental sustainability. The chapter reviews topics such as the influence of suppliers and retailers on apple sales, global forces affecting apple demand, recent trends in apple consumption, factors affecting consumer demand for apples and the challenges this creates in marketing apples.

As Chapter 21 suggests, the environmental impact of apple production can be managed through cultivation practices and inputs. Pesticides typically have the greatest environmental impact, but energy use from cultivation techniques (including the costs of nutrient, irrigation and pesticide inputs) is also substantial. This chapter shows that apple production sustainability has improved in the area of pest management and decreased for resource use. It demonstrates that high-density apple orchards require higher levels of infrastructure and resource inputs than older systems, and argues that enhancing biocontrol of pests and lengthening the usable life of equipment and infrastructure are important strategies for reducing the environmental impact of modern production systems.

The final chapter in the book, Chapter 22, reviews key research areas for those involved in organic apple production. It covers the development of suitable varieties and rootstocks, soil fertility management, and strategies for maintaining apple tree health.

The latter include weed and disease control as well as methods for regulating crop set and tree growth. A number of case studies are used to show how research can have real practical impacts, for example by adding to the organic farmer's toolbox of disease control strategies.

Part 1

Plant physiology and breeding

Chapter 1

Ensuring the genetic diversity of apples

Gayle M. Volk, USDA-ARS-CARR National Laboratory for Genetic Resources Preservation, USA

1. Introduction: the importance of apples
2. Threats to production and the need for new varieties
3. Sources of genetic diversity
4. Understanding *Malus* diversity
5. Capturing, maintaining and accessing genetic diversity
6. Future trends: new technologies to facilitate enhanced and sustainable crop production
7. Where to look for further information
8. References

1 Introduction: the importance of apples

Apples are a healthy and nutritious part of the human diet. They provide antioxidants, polyphenolics, vitamins and fibre. Consumption of apple products has been shown to be associated with a reduced risk of major diseases, including cancer, cardiovascular disease and asthma (Boyer, 2004; Hyson, 2011). Apple consumption may also improve human responses to Alzheimer's disease, cognitive decline due to ageing, diabetes, weight management and bone health (Hyson, 2011).

Apple cultivars have been propagated by grafting since the time of early Greek civilizations (Mudge, 2009). In fact, 'Herefordshire Pearmain' and 'White Winter Pearmain' are apple cultivars that are available today and that have been clonally propagated ever since they were described in the 1200s (Smith, 1971). Many of the early European apple cultivars were valued as fruit used in hard cider production (Leforestier, 2015; Luby, 2003). By the 1700s, apple cultivars that were sought for fresh consumption were widely grown (Luby, 2003). Some of the fresh consumption apple cultivars that are grown today have been around for over a century, including Cox's Orange Pippin (1825), McIntosh (1811), Red Delicious (1880) and Rome Beauty (1848) (Smith, 1971). Many of these older varieties are vulnerable to abiotic and biotic threats.

This chapter reviews the threats to apple production and the resulting need for new varieties of apple. It considers the advantages of using progenitor species of *Malus × domestica* in traditional breeding programmes, and then presents and addresses three challenges to ensuring the genetic diversity of apples: understanding *Malus* diversity,

http://dx.doi.org/10.19103/AS.2016.0017.01

capturing and maintaining this diversity, and ensuring that researchers and breeders have access to living apple accessions.

2 Threats to production and the need for new varieties

Apple orchards are particularly susceptible to threats as most apple-growing regions are dominated by the production of only a few cultivars (Hampson, 2003). These cultivars are propagated clonally by grafting, resulting in genetically identical orchard trees. This lack of genetic diversity in orchards makes entire orchards susceptible if new diseases or pathogens become threats to non-tolerant cultivars. The concern regarding orchards of identical cultivars is further magnified by the fact that many of the popular cultivars share a common ancestry. A study of worldwide genetic diversity among cultivars in apple breeding programmes revealed that 'Cox's Orange Pippin', 'Golden Delicious', 'Red Delicious', 'Jonathan' and 'McIntosh' are the most common parents of new cultivars (Noiton, 1996). The status effective numbers (the equivalent number of panmictic-mated, unrelated, non-inbred genotypes) were 8–10 for the mainstream cultivars and those with resistance to apple scab (*Venturia inaequalis*) (Noiton, 1996). Consequently, different cultivars may not exhibit a wide range of tolerance to threats, resulting in similar levels of susceptibility across orchards that are thought to be diverse. Orchards are long-term investments, with a relatively low turnover in cultivars or rootstocks over the decades (Volk, 2015a). The high cost of planting and relatively slow return on investment results in low orchard turnover and therefore a slow rate of adoption of new varieties that may provide resistance to biotic threats (Volk, 2015a).

Pathogens continue to threaten orchards. Some apple cultivars that carried a single gene resistance to a disease, such as apple scab, are now showing signs of susceptibility (Parisi, 2004). Through mutation and selection, the apple scab pathogen is overcoming the resistance in apple cultivars that carry the most common resistance gene, *Rvi6 (Vf)*, which was introgressed from *Malus floribunda* 821 (Crosby, 1992; Soriano, 2009). Durable cultivars with new forms of resistance, ideally in pyramided form, are needed so that orchards remain resistant to pathogens over time (Gessler, 1994; Patocchi, 2009; Soriano, 2009).

Apple orchards are also threatened by the effects of changing climate. Warm winters may not result in adequate chilling hours for budbreak in the spring (Petri, 2004). Alternatively, warm spring weather may cause trees to flower early, and spring frosts could easily damage the crop if they occur during bloom or when fruit has set on the trees (Eccel, 2009).

There is a need for new apple varieties that can resist pathogens and climatic threat. These new cultivars would ideally have pyramided forms of disease resistance, desirable fruit quality and storage characteristics, and minimal need for pesticide applications. In addition, these new cultivars would flower at times to avoid spring frost damage, and have foliar and fruit architectures that are amenable to high-density production and mechanical harvesting (Volk, 2015a).

3 Sources of genetic diversity

3.1 Wild apple species

Wild apple species offer genetic variation for desirable traits for improvement: disease and insect resistance, fruit quality, growth attributes and physiological characters (Table 1;

Table 1 Crop wild relatives (excluding progenitor species) of apple (*Malus*) that have been identified to potentially have pathogen resistance useful to breeding programs

Trait	Pathogen	*Malus* species source	References
Crown gall resistance	*Rhizobium radiobacter*	*Malus sieboldii*	Moriya, 2010
Fire blight resistance	*Erwinia amylovora* (Burill)	*M. x robusta* 5	Broggini, 2014; Giovanni, 2014; Peil, 2007
Powdery mildew resistance	*P. leucotricha* (Ell. Et Everh.) Salm.	*M. coronaria*	Sestras, 2011
Powdery mildew resistance	*P. leucotricha* (Ell. Et Everh.) Salm.	*M. niedzwetzkyana*	Sestras, 2011
Powdery mildew resistance	*P. leucotricha* (Ell. Et Everh.) Salm.	*M. robusta*	Dunemann, 2007; Markussen, 1995
Powdery mildew resistance	*P. leucotricha* (Ell. Et Everh.) Salm.	*M. zumi*	Sestras, 2011
Scab resistance	*Venturia inaequalis* (Cke.) Wint.	*M. atrosanguinea* 804	Crosby, 1992
Scab resistance	*V. inaequalis* (Cke.) Wint.	*M. baccata*	Erdin, 2006; Gygax, 2004; Kumar, 2012
Scab resistance	*V. inaequalis* (Cke.) Wint.	*M. coronaria*	Sestras, 2011
Scab resistance	*V. inaequalis* (Cke.) Wint.	*M. floribunda* 821	Belfanti, 2004; Bénaouf, 2000; Crosby, 1992; Gessler, 2012
Scab resistance	*V. inaequalis* (Cke.) Wint.	*M. micromalus*	Crosby, 1992; Patocchi, 2005
Scab resistance	*V. inaequalis* (Cke.) Wint.	*M. niedzwetzkyana*	Sestras, 2011
Scab resistance	*V. inaequalis* (Cke.) Wint.	*M. pumila*	Crosby, 1992
Scab resistance	*V. inaequalis* (Cke.) Wint.	*M. zumi*	Sestras, 2011

Brozynska, 2015; Hokanson, 1997; Luby, 2001). Wild *Malus* species are native throughout the northern hemisphere in North America, Europe and Asia (Fig. 1). The species centre of diversity is believed to be China, where at least 31 species of *Malus* have been identified in the wild (Zhi-Qin, 1999). Most *Malus* species are considered to be temperate, with winter chilling requirements that must be satisfied for fruit bud formation. However, there are some wild *Malus* species (and cultivars) that are not winter hardy, including *M. doumeri* (Robinson, 2001).

Taxonomic sections of the genus *Malus* have been previously described based on phenotypic traits (Table 2; Phipps, 1990; Robinson, 2001; Yu, 1979). More recently, chloroplast-based sequence analyses have been performed to identify genetic relationships among *Malus* species. The phylogenies based on chloroplast sequence

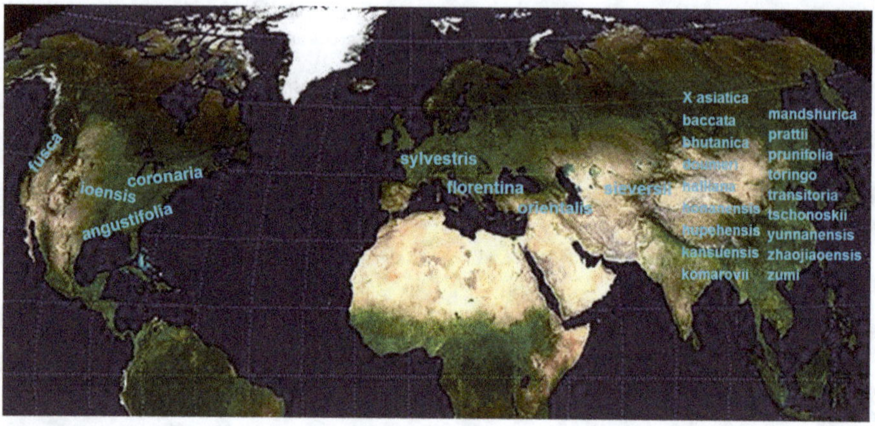

Figure 1 Distribution of wild *Malus* species throughout the world.

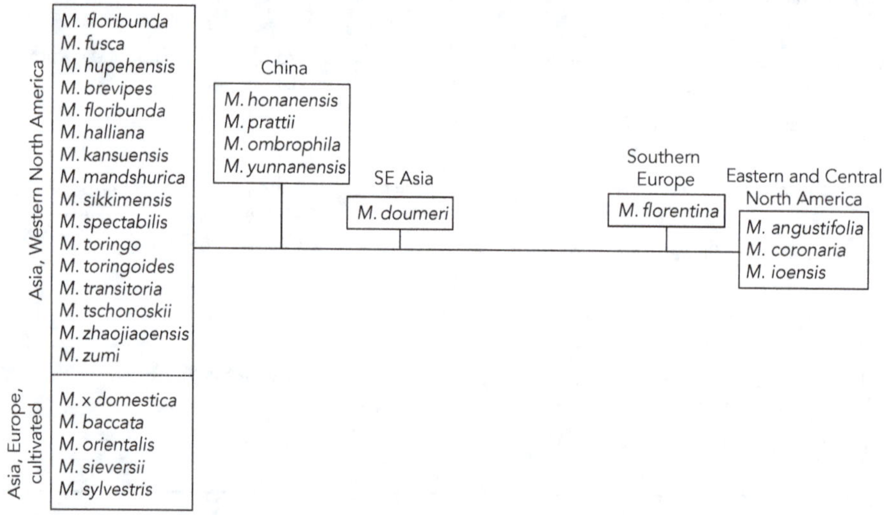

Figure 2 Diagram depicting relationship among clades of apple species, summarized from Volk (2015b).

did not align with the previously described *Malus* sections. The 47 chloroplast genome sequences compared by Nikiforova et al. (2013) identified five differentiated clades, with only the Chloromeles section aligning with the previous definition. Volk et al. (2015b) sequenced four regions of the chloroplast from 412 individuals representing 30 *Malus* species, and found that only three clades of species could be differentiated (Fig. 2; Table 2). Further assessments of the taxonomic relationships among *Malus* species will require the use of both chloroplast and nuclear data for multiple individual representatives per species (Volk, 2015b).

Table 2 Taxonomic sections or clades of Malus based on Robinson (2001), Nikiforova (2013) and Volk (2015)

Robinson, 2001	Nikiforova, 2013	Volk, 2015
Eumalus	**A**	**A**
M. asiatica	*M. asiatica*	*M. brevipes*
M. baccata	*M. baccata*	*M. baccata*
M. floribunda	*M. fusca*	*M. fusca*
M. halliana	*M. halliana*	*M. halliana*
M. hupehensis	*M. hupehensis*	*M. hupehensis*
M. mandshurica	*M. mandshurica*	*M. mandshurica*
M. orientalis	*M. orientalis*	*M. orientalis*
M. sieversii	*M. sieversii*	*M. sieversii*
M. spectabilis	*M. pumila*	*M. spectabilis*
M. zhaojiaoensis	*M. kirghisorum*	*M. zhaojiaoensis*
M. prunifolia		*M. prunifolia*
M. micromalus	**B**	*M. sikkimensis*
M. sylvestris	*M. sylvestris*	*M. sylvestris*
	M. domestica	*M. domestica*
		M. toringo
	C	*M. toringoides*
	M. sargentii	*M. tschonoskii*
	M. prunifolia	*M. zumi*
Sorbomalus	*M. floribunda*	*M. floribunda*
M. transitoria	*M. transitoria*	*M. transitoria*
M. florentina	*M. zhaojiaoensis*	
M. fusca	*M. sieboldii*	
M. ombrophila	*M. spectabilis*	
M. sargentii	*M. rockii*	
M. sieboldii	*M. sikkimensis*	
M. toringo		
M. toringoides	**D**	**B**
M. kansuensis	*M. kansuensis*	*M. kansuensis*
M. honanensis	*M. honanensis*	*M. honanensis*
M. prattii	*M. prattii*	*M. prattii*
M. yunnanensis	*M. yunnanensis*	*M. yunnanensis*
		M. omprophila
Chloromeles	**E**	**C**
M. angustifolia	*M. angustifolia*	*M. angustifolia*
M. ioensis	*M. ioensis*	*M. ioensis*
M. coronaria		*M. coronaria*
Eriolobus		
M. trilobata		
Docyniopsis		
M. doumeri		
M. tschonoskii		

Wild species may offer sources of resistance or desirable traits to overcome abiotic and biotic threats (Table 1; Hokanson, 1997; Luby, 2001). The challenge to using wild species in breeding programmes is that the wild species usually have undesirable fruit, non-uniform ripening times and other traits that are not amenable to commercial apple production. Desirable traits are difficult to introgress from wild species using traditional breeding methods because of the long juvenile phases and possible linkages between desirable and undesirable traits.

3.2 Progenitor species of *Malus* × *domestica*

Traditional breeding programmes may be able to make use of the progenitor species of *M.* × *domestica* more easily than more distant wild species of *Malus*. Given the closer relationship between the progenitor species and the cultivated apple, these species may be more easily introgressed into cultivars, thus possibly reducing the time to new cultivar development (Cornille, 2012). Progenitor species could be desirable in breeding programmes because they are usually diploid and often have larger and more desirable fruit characteristics than the more distantly related species. The primary progenitor species of *M.* × *domestica* are likely *M. sieversii*, *M. sylvestris*, *M. orientalis*, *M. baccata* and *M. prunifolia* (Fig. 3; Cornille, 2014; Volk, 2015a).

Malus × *domestica* was domesticated thousands of years ago. Progenitor species *M. sieversii* served as the source species. Other western species, including *M. orientalis* and *M. sylvestris*, were hybridized with *M. sieversii* as seedlings and were transported from the region of the Tien Shan Mountains in Central Asia through the Middle East and into Europe. Alternative domestication processes may have occurred as *M. sieversii* was transported across China, where it may have hybridized with *M. baccata* and *M. prunifolia* (Luby, 2003). Molecular evidence has supported the primary contribution of *M. sieversii* as a progenitor of the domesticated apple in Europe, with *M. sylvestris* serving as a secondary contributor (Cornille, 2012, 2014).

Some accessions representing *M.* × *domestica* progenitor species exhibit desirable traits. For example, *M. baccata* species representatives have resistance to apple scab, Botryosphaeria canker (*Botryosphaeria* (Sawada) Shang), fire blight (*Erwinia amylovora* (Burrill) Winslow et al.) and powdery mildew (*Podosphaera leucotricha* (Ellis and Everh.) E.S. Salmon), in addition to cold hardiness and resistance to waterlogging (Dunemann, 2009; Hokanson, 2001; Le Roux, 2012; Luby, 2002; Wan, 2011; Yang, 2011; Zhi-Qin, 1999). Similarly, *M. sieversii* accessions have resistance to apple scab, Botryosphaeria canker,

Figure 3 Images of fruits and flowers of *Malus* progenitor species: (A) *M. sieversii*, (B) *M. orientalis*, (C) *M. sylvestris*, (D) *M. baccata* and (E) *M. prunifolia*. Flower images provided by C.T. Chao.

fire blight, powdery mildew, woolly apple aphid (*Eriosoma lanigerum* Hausmann), light brown apple moth (*Epiphyas postvittana* Walker), apple leaf curling midge (*Dasyneura mali* Kieff.), Phytophthora, western white root rot (*Rosellinia nacatrix* Prill.), violet root rot (*Helicobasidium mompa* Tanaka), apple maggot (*Rhaoletis pomonella* Walsh), blue mould (*Penicillium expansum* link) and apple replant disease (Bus, 2005; Forsline, 2004; Hokanson, 2001; Isutsa, 2000; Kumar, 2010; Luby, 2001, 2002; Norelli, 2014; Zhi-Qin, 1999). *M. sieversii* also exhibits cold hardiness, late flowering, variation in water use efficiency, short juvenility, resistance to sunburn, some variants of fruit quality and drought tolerance (Bassett, 2011; Khan, 2014; Luby, 2001; Maguylo, 2014; Yan, 2008). Progenitor species *M. sylvestris* has resistance to blue mould (Jurick, 2011). Accessions of *M. orientalis* have been documented to have resistance to apple scab, cedar apple rust (*Gymnosporangium juniperi-virginianae* Schwein.), fire blight, sooty blotch (*Peltaster fructicola* Johnson, Sutton and Hodges, *Geastrumia polystigmatus* Batista and M. A. Farr, *Leptodontium elatus* (F. Mangenot) de Hoog, *Gloeodes pomigena* (Schwein) Colby), and flyspeck (*Schizothyrium pomi* (Mont. and Fr.) Arx; Höfer, 2013; Volk, 2008). *M. orientalis* is also later blooming, adapted to a wide variety of habitats, and can be stored for extended lengths of time (Höfer, 2013). *M. prunifolia* has exhibited resistance to apple scab, Botryosphaeria canker and fire blight (Fischer, 1999; Hokanson, 2001; Le Roux, 2012; Wan, 2011; Zhi-Qin, 1999).

4 Understanding *Malus* diversity

Taxonomic relationships among *Malus* species have been difficult to discern. Molecular marker and sequence data have been used for species differentiation and to determine genomic relationships among *Malus* species (Forte, 2002; Nikiforova, 2013; Robinson, 2001; Velasco, 2010; Volk 2015b). Much less is known about the diversity of *Malus* species in the wild. Overall, species have a weak spatial genetic structure, likely due to the long-distance dispersal of pollen and seeds (Cornille, 2013b). Localized, regional structures among some subsets of populations have been identified (Richards, 2009b; Volk, 2008).

Assessments using microsatellite markers have shown that populations sampled over one thousand kilometres in Kazakhstan were significantly differentiated. Two primary clusters of individuals were identified based on the microsatellite markers. The southwestern *M. sieversii* collection sites were more admixed and diverse than the northern sites. There were higher levels of differentiation among half-sib seedlings than among sites, likely due to the self-incompatible, outcrossing nature of the species (Richards, 2009b). *Malus sieversii* from the Xinjiang Province of China exhibited significant differences among populations, with low levels of gene exchange (Zhang, 2007).

Malus orientalis populations in Turkey, Georgia, Armenia, and Southern Russia were genetically distinguishable and exhibited high levels of diversity (Volk, 2008). *M. orientalis* from Russia was more homogeneous, whereas the populations sampled from Turkey were highly diverse and exhibited disease resistance. Some pockets of novelty were identified, which could possibly be attributed to local hybridization with domesticated cultivars (Volk, 2008).

Wild *Malus sylvestris* is distributed throughout Europe in a manner that has been partially dictated by high-disturbance events, including winds, fire, and floods. Analyses of the population structure of Rhine Valley revealed, similar to other wild apple species, a weak population structure, despite the clumped settings (Schnitzler, 2014). Further analyses of the population structure of *M. sylvestris* in Europe revealed that there were three differentiated populations in Europe: (1) a western population from France to Norway, (2) a northeastern population near the Carpathian Mountains, and (3) a southeastern population along the Balkan Peninsula (Cornille, 2013a). Genetic assessments of these populations suggest that *M. sylvestris* populations retreated to glacial refugia in the Iberian Peninsula and the Balkan region during the times of the ice ages (Cornille, 2013a).

Malus diversity can also be assessed at the cultivar level. Apple cultivars have been vegetatively propagated for centuries, and many of these ancient cultivars remain available as either local cultivars or within genebanks (Guarino, 2006; Pereira-Lorenzo, 2007; Urrestarazu, 2012). Cultivar diversity assessments within genebanks reveal genetic relationships (parentage, sports, synonyms, mislabelling) among cultivars (Gross, 2012; van Treuren, 2010). Due to the high level of allelic diversity in *Malus × domestica*, relatively few markers can be used to differentiate distinct apple cultivars (aside from sports), which makes for straightforward cultivar identification of the diploid cultivars (Foroni, 2012; Moriya, 2011). Collections with genetic fingerprints available also serve as standards for the identification of unknown cultivars (Routson, 2009).

Dozens of genebanks worldwide maintain *Malus* collections including both local cultivars and those with international economic significance. Regional genebank collaborations as well as those between individual genebanks have performed collection comparisons that have revealed novel local varieties (Gao, 2015; Patzak, 2012; Zhang, 2012). These local varieties may have novel alleles for key traits that are not commonly available in the cultivated genepools (Leforestier, 2015; Tartarini, 2004). A large-scale comparison among dessert apples in European genebanks identified strong historical processes of cultivar migration and selection as well as adaptive responses to diverse environments. These findings demonstrate the limited population structure, yet extensive variation among apple cultivars (Urestarazu, 2016).

The past decade has brought about many advances in understanding the genetic diversity in *Malus*; however, there is much to be done. Diversity assessments have not been performed for many *Malus* species. In addition, correlations among unique habitat or environmental niches (or selection pressures) and genetic adaptations may reveal desirable new genes that could be used in breeding programmes. Access to phenotypic, genotypic and sequence data for individuals representing wild populations of *Malus* species will facilitate progress in this area. The use of genetic marker and sequence data to compare individuals within and between collections will greatly aid in the understanding of collection composition, overlaps and vulnerabilities. The use of novel local cultivars with desirable phenotypes in breeding programmes may generate new cultivars that make apple production more sustainable in the future.

5 Capturing, maintaining and accessing genetic diversity

5.1 Capturing and maintaining the diversity

Natural settings with wild apple trees are at risk of being lost due to the encroachment of humans and the presence of livestock (Forsline, 2003; Fowler, 2004; Williams, 2005). Forest settings may be destroyed as habitations expand and lands are cleared for alternative uses (Forsline, 2003). Trees within boundaries of parks or forest preserves may have some level of protection. Obtaining access to protected areas may be a challenge.

The role of genebanks for species conservation has grown as the natural habitats of species continue to be threatened (Esquinas-Alcázar, 2005; Forsline, 2003). Representatives of most *Malus* wild species are available in national genebanks; however, in many cases, the diversity of wild species is poorly represented in *ex situ* genebanks (Vincent, 2013). Many national genebanks and arboreta include *Malus* species, crab apples or domesticated apples in their collections. These genebanks may be limited in size or scope because of the expense of maintaining collections of living trees. In some cases, collections may focus on ornamental crab apple collections, whereas others may focus on historic cultivars, crop wild relatives or species diversity. Vincent (2013) reported that 55% of the *M.* × *domestica* crop wild relatives are represented by fewer than 50 *ex situ* accessions.

The USDA-ARS National Plant Germplasm System (NPGS) maintains a large apple collection as part of the Plant Genetic Resources Unit in Geneva, New York. This collection currently has 3070 unique grafted trees in the field representing *M.* × *domestica* cultivars

Figure 4 USDA-ARS National Plant Germplasm System apple collection in Geneva, New York. (A) Grafted apple cultivar orchard and (B) own-rooted seedling orchard of *M. sieversii* trees.

Figure 5 Images of *Malus* × *domestica*: (A) seeds, bar: 1 mm; (B) flower; (C) dormant budwood for use in cryopreservation and (D) grafted apple bud after cryopreservation.

and 33 *Malus* species, and 15 hybrid species. The collection also has an orchard that has been planted with own-root seedling trees grown from seeds of *Malus* species collected in the wild (Fig. 4). In addition, there is a planting of trees that are the progeny of crosses between 'Gala' and seven *Malus sieversii* accessions (GMAL 4335, GMAL 4448, GMAL 4331, GMAL 4333, GMAL 4327, GMAL 4455 and GMAL 4334; Volk, 2015a, Thomas Chao, pers. comm.). These populations segregate for key traits such as resistance to blue mould, fire blight, apple scab, oblique banded leaf roller (*Choristoneura rosaceana* (Harris)) and codling moth resistance as well as water use efficiency (Wisniewski, 2016). Quantitative trait loci and linked markers have been identified for blue mould resistance traits (Norelli, 2014).

The NPGS apple collection was established as part of a national network of clonal repositories in the 1980s (Postman, 2006). Budwood for grafting trees representing cultivars was originally donated from breeders, arboreta and other genebank collections. In addition, a series of collection trips were performed between 1989 and 2015 to acquire seeds from wild *Malus* species (Forsline, 2003; Volk, 2015a; Thomas Chao, pers. comm.). The NPGS apple collection is actively maintained in the field and propagules are distributed for research purposes. A seed collection, composed primarily of wild-collected seeds from plant explorations, is maintained at −18°C. In addition, 2052 of the trees in the field collection are backed up as dormant buds at cryogenic temperatures as a long-term secure base collection at the National Laboratory for Genetic Resources Preservation in Fort Collins, Colorado (Fig. 5; Volk, 2016).

The apple collection is maintained using a complementary conservation approach that includes grafted trees for the permanent collection, seedling trees in evaluation orchards, seeds for conserving the diversity of wild species and a cryogenic back-up of the vegetatively propagated accessions (Volk, 2010b; Walters, 2008). Grafted trees in the field are available for phenotypic and genetic evaluations, breeding and budwood distribution. Seeds in long-term storage conserve the diversity of wild species in an efficient manner, and cryogenic storage of vegetatively propagated collections serves as a secure back-up if trees in the field are lost (Forsline, 1998; Towill, 2004; Volk, 2010b).

In the 1990s, a core collection of the field apple collection was proposed as a smaller subset of NPGS apple accessions that was believed to capture much of the diversity represented in the collection (Gross, 2013; Hokanson, 1998). Microsatellite markers have since been used to assess how well the original core collection captured the genetic diversity, and additional accessions were suggested for inclusion in the core collection based on these results (Gross, 2013). In addition, core collections have been proposed for *M. sieversii* and *M. orientalis* species collections within the NPGS (Richards, 2009a; Volk, 2005, 2009).

5.2 Access to the diversity

Genebank collections provide access to living apple accessions (in either tree or seed form) for use in taxonomy, physiology, molecular and evolutionary biology, horticulture and breeding programmes. Available accessions in the NPGS apple collection are distributed both within the United States and internationally using the appropriate permits (Volk, 2015a). Genebank accessions are known to be most valuable when standardized characterization and evaluation data are available (Rubenstein, 2006; Volk, 2010a). Genebank customers are most readily served by the use of descriptors that are relevant to their communities. They also benefit from collections that have materials whose taxonomic identities are correct and cultivar identifications have been verified (Evans, 2010). Descriptor sets that are selected to correctly identify wild species may be different than those that are used to evaluate materials for potential use in breeding programmes (Phipps, 1990; Schmitz, 2013; Yu, 1979).

Passport, phenotypic and molecular data for NPGS accessions are publicly available on the GRIN-Global database (http://www.ars-grin.gov/npgs/gringlobal/webpages/publicwebsite.html). Future database interoperability should provide seamless data access on *Malus* collections from genomic database portals (such as the Genome Database for Rosaceae, GDR). Efforts to compare national and regional collections of *Malus* genetic resources will be facilitated by the collection of phenotypic and genetic data in standardized formats (Kellerhals, 2012; Schmitz, 2013; Wegrzyn, 2012).

6 Future trends: new technologies to facilitate enhanced and sustainable crop production

Traditional apple breeding is a complex process that involves many years of crossing, field evaluations, multi-site testing and consumer trials to ensure that new varieties meet the expectations of both producers and consumers (Evans, 2013). The multiple generations of modified backcrosses that were previously necessary to introgress genes of interest from a wild species into a desirable background made it an arduous process to use wild species in breeding programmes.

New technologies that make use of marker-assisted selection, genomic selection, genetic engineering, genome-wide association mapping, high-throughput genotyping and/or rapid-cycling plants are paving the way towards the increased use of wild *Malus* species in apple breeding (Bianco, 2014; Broggini, 2014; Evans, 2013; Flachowsky, 2011; Kumar, 2012a,b, 2013; Troggio, 2012), thus facilitating the movement of desirable alleles from *Malus* species into the cultivated apple and reducing deleterious effects on fruit quality, quantity, and production traits (Kumar, 2010). These techniques become feasible as markers linked to key traits, trait-associated breeding values and quantitative traits become better understood (Gardner, 2014; Kouassi, 2009; Longhi, 2013; Troggio, 2012). The process of identifying allelic variation in *Malus* wild species will be aided by an improved understanding of the diversity available within wild species and the characterization of the key genes that influence desirable traits.

Technological advances in the area of whole-genome sequencing will soon enable the genomic sequencing of entire apple collections (Velasco, 2010). This will allow users to determine the genetic relationships among accessions within apple genebanks. The use of fine resolution genomic data in breeding programmes focused on wild species will be dependent upon access to high-quality phenotypic evaluation data that allows users to link desirable traits to their genetic control (Volk, 2015a). New genomic assessment tools that easily accommodate polyploid accessions will also facilitate the analyses of polyploid cultivars and species (Bianco, 2014).

Access to wild *Malus* species will be critical for advancements in future plant breeding programmes. Wild *Malus* species are most readily available to the breeding community from genebanks. Many species are, however, poorly represented in genebank collections, and phytosanitary restrictions and treaty agreements may limit access to wild materials, particularly at the international level (Bretting, 2007; Varaprasad, 2010). Ongoing coordinated efforts among genebanks to compare collections at the phenotypic and genetic levels as well as confirm cultivar identities will facilitate the use of genebank collections in the future (Kellerhals, 2012; Volk, 2011). These efforts will enable future researchers specializing in apple improvement to provide consumers with higher quality fruit that can be produced more sustainably.

7 Where to look for further information

There are many challenges in conserving and utilizing apple genetic resources. With limited funding and high costs associated with genebanks and breeding programmes, researchers must work together to capitalize on their strengths. Key findings and results relating to conservation and breeding are shared at many conferences including the Rosaceae Genomics Conference and the Eucarpia Plant Breeding Symposium, in additional to national and international horticultural and genomics conferences and topic-specific symposia.

The Rosaceae Genomics Conference is organized by a committee that meets in the form of the Rosaceae International Genomics Initiative (RosIGI; https://www.rosaceae.org/community/rosgi). In the United States, the U.S. Rosaceae Genomics, Genetics and Breeding

Executive Committee (RosExec; https://www.rosaceae.org/community/us_rosexec), is comprised of scientists specializing in Rosaceous crops who represent academia, government and industry programmes. This group meets quarterly to serve as a communication and coordination focal point in for Rosaceae genomics, genetics and breeding in the United States. In Europe, the *Malus/Pyrus* European Cooperative Programme for Plant Genetic Resources (http://www.ecpgr.cgiar.org/working-groups/maluspyrus/) has working groups with meetings to ensure long-term conservation and facilitate the utilization of plant genetic resources in Europe. In the United States, the Apple Crop Germplasm Committee (CGC; http://www.ars-grin.gov/npgs/cgclist.html#Apple) is comprised of NPGS users who provide collection management recommendations to the NPGS apple collection curator. Apple CGC members published an apple crop vulnerability report in 2015 that summarizes the status of apple production and conservation in the United States (Volk, 2015a). Rosaceae passport, taxonomic, genomic, genetic and phenotypic data are available in English in multiple databases, including, but not limited to, the GDR (Jung, 2014), GRIN-Global, NCBI (http://www.ncbi.nlm.nih.gov/gquery/gquery.fcgi), GBIF (http://www.gbif.org/), Brogdale National Fruit Collection Database (http://www.nationalfruitcollection.org.uk/index.php) and Genesys (https://www.genesys-pgr.org/welcome).

8 References

Bassett, C. L., Glenn, D. M., Forsline, P. L., Misniewski, M. E. and Farrell, R. E., Jr. (2011), 'Characterizing water use efficiency and water deficit responses in apple (*Malus × domestica* Borkh. and *Malus sieversii* Ledeb) M. Roem', *HortScience* 46, 1079–84.

Belfanti, E., Silfverberg-Dilworth, E., Tartarini, S., Patocchi, A., Barbieri, M., Zhu, J., Vinatzer, B. A., Gianfranceschi, L., Gessler, C. and Sansavini, S. (2004), 'The HcrVf2 gene from a wild apple confers scab resistance to a transgenic cultivated variety', *Proc. Nat. Acad. Sci. U.S.A.* 101, 886–90.

Bénaouf, G. and Parisi, L. (2000), 'Genetics of host-pathogen relationships between *Venturia inaequalis* races 6 and 7 and *Malus* species', *Phytopathology* 90, 236–42.

Bianco, L., Cestaro, A., Sargent, D. J., Banchi, E., Derdak, S., Di Guardo, M., Salvi, S., Jansen, J., Viola, R., Gut, I., Laurens, F., Chagné, D., Velasco, R., van de Weg, E. and Troggio, M. (2014), 'Development and validation of a 20K single nucleotide polymorphism (SNP) whole genome genotyping array for apple (*Malus × domestica* Borkh)', *PLoS ONE* 9, e110377.

Boyer, J. and Liu, R. H. (2004), 'Apple phytochemicals and their health benefits', *Nutrition J.* 3, 5. doi: 10.1186-1475-2891-3-5.

Bretting, P. K. (2007), 'The U.S. National Plant Germplasm System in an era of shifting international norms for germplasm exchange', *Acta Hort.* 760, 55–60.

Broggini, G. A. L., Wöhner, T., Fahrentrapp, J., Kost, T. D., Flachowsky, H., Peil, A., Hanke, M.-V., Richter, K., Patocchi, A. and Gessler, C. (2014), 'Engineering fire blight resistance into the apple cultivar "Gala" using the *FB_MR5* CC-NBS-LRR resistance gene of *Malus* x *robusta* 5', *Plant Biotechnol. J.* 12, 728–33.

Brozynska, M., Furtado, A. and Henry, R. J. (2015), 'Genomics of crop wild relatives: expanding the gene pool for crop improvement', *Plant Biotechnol. J.* doi: 10.1111/pbi.12454.

Bus, V. G. M., Laurens, F. N. D., Van De Weg, W. E., Rusholme, R. L., Rikkerink, E. H. A., Gardiner, S. E., Bassett, H. C. M., Kodde, L. P. and Plummer, K. M. (2005), 'The *Vh8* locus of a new gene-for-gene interaction between *Venturia inaequalis* and the wild apple *Malus sieversii* is closely linked to the Vh2 locus in *Malus pumlia* R12740–7A', *New Phytol.* 166, 1035–49.

Cornille, A., Giraud, T., Bellard, C., Tellier, A., Le Cam, B., Smulders, M. J. M., Kleinschmit, J., Roldan-Ruiz, I. and Gladieux, P. (2013a), 'Postglacial recolonization history of the European crabapple (*Malus sylvestris* Mill.), a wild contributor to the domesticated apple', *Mol. Ecol.* 22, 2249–63.

Cornille, A., Giraud, T., Smulders, M. J. M., Roldán-Ruiz, I. and Gladieux, P. (2014), 'The domestication and evolutionary ecology of apples', *Trends Genet.* 30, 57–65.
Cornille, A., Gladieux, P. and Giraud, T. (2013b), 'Crop-to-wild gene flow and spatial genetic structure in the closest wild relatives of the cultivated apple', *Evolut. Applic.* 6, 737–48.
Cornille, A., Gladieux, P., Smulders, M. J. M., Roldán-Ruiz, I., Laurens, F., Le Cam, B., Nersesyan, A., Clavel, J., Olonova, M., Feugey, L., Gabrielyan, I., Zhang, X.-G., Tenaillon, M. I. and Giraud, T. (2012), 'New insight into the history of domesticated apple: Secondary contribution of the European wild apple to the genome of cultivated varieties', *PLoS Genet.* 8, e1002703.
Crosby, J. A., Janick, J., Pecknold, P. C., Korban, S. S., O'Connon, P. A., Ries, S. M., Goffreda, J. and Voordeckers, A. (1992), 'Breeding apples for scab resistance: 1945–90', *Fruit Var. J.* 46, 145–66.
Dunemann, F., Peil, A., Urbanietz, A. and Garcia-Libreros, T. (2007), 'Mapping of the apple powdery mildew resistance gene *Pl1* and its genetic association with an NZBS-LRR candidate resistance gene', *Plant Breed.* 126, 476–81.
Dunemann, F. and Schuster, M. (2009), 'Genetic characterization and mapping of the major powdery mildew resistance gene *Plbj* from *Malus baccata* jackii', *Acta Hort.* 814, 791–7.
Eccel, E., Rea, R., Caffarra, A. and Crisci, A. (2009), 'Risk of spring frost to apple production under future climate scenarios: the role of phenological acclimation', *Int. J. Biometeorol.* 53, 273–86.
Erdin, N., Tartarini, S., Broggini, G. A. L., Gennari, F., Sansavini, S., Gessler, C. and Patocchi, A. (2006), 'Mapping of the apple scab resistance gene *Vb*', *Genome* 49, 1238–45.
Esquinas-Alcázar, J. E. (2005), 'Protecting crop genetic diversity for food security: political ethical and technical challenges', *Nature Rev.* 6, 946–53.
Evans, K. (2013), 'Apple breeding in the Pacific Northwest', *Acta Hort.* 976, 75–8.
Evans, K. M., Patocchi, A., Rezzonico, F., Mathis, F., Durel, C. E., Fernández- Fernández, F., Boudichevskaia, A., Dunemann, F., Stankiewicz-Kosyl, M., Gianfranceschi, L., Komjanc, M., Lateur, M., Madduri, M., Noordijk, Y. and van de Weg, E. (2011), 'Genotyping of pedigreed apple breeding material with a genome-covering set of SSRs: trueness-to-type of cultivars and their parentages', *Mol. Breeding* 28, 535–47.
Fischer, M. and Fischer, C. (1999), 'Evaluation of *Malus* species and cultivars and the fruit genebank Dresden-Pillnitz and its use for apple resistance breeding', *Genet. Resour. Crop Evol.* 46, 235–41.
Flachowsky, H., Le Roux, P.-M., Peil, A., Patocchi, A., Richter, K. and Hanke, M.-V. (2011), 'Application of a high-speed breeding technology to apple (*Malus × domestica*) based on transgenic early flowering plants and marker-assisted selection', *New Phytol.* 192, 364–77.
Foroni, I., Baptista, C., Monteiro, L., Lopes, M. S., Mendonça, D., Melo, M., Carvalho, C., Monjardino, P., Lopes, D. J. H. and Machado, A. (2012), 'The use of microsatellites to analyze relationships and to decipher homonyms and synonyms in Azorean apples (*Malus × domestica* Borkh.)', *Plant Syst. Evol.* 298, 1297–313.
Forsline, P. L. and Aldwinckle, H. S. (2004), 'Evaluation of *Malus sieversii* seedling populations for disease resistance and horticultural traits', *Acta Hort.* 663, 529–34.
Forsline, P. L., Aldwinckle, H. S., Dickson, E. E., Luby, J. J. and Hokanson, S. C. (2003), 'Collection, maintenance, characterization, and utilization of wild apples of Central Asia', *Hort. Rev.* 29, 1–61.
Forsline, P. L., Towill, L. E., Waddell, J. W., Stushnoff, C., Lamboy, W. F. and McFerson, J. R. (1998), 'Recovery and longevity of cryopreserved dormant apple buds', *J. Amer. Soc. Hort. Sci.* 123, 365–70.
Forte, A. V., Ignatov, A. N., Ponomarenko, V. V., Dorokhov, D. B. and Savelyev, N. I. (2002), 'Phylogeny of the *Malus* (apple tree) species, inferred from the morphological traits and molecular DNA analysis', *Russian J. Genet.* 38, 1150–60.
Fowler, C. and Hodgkin, T. (2004), 'Plant genetic resources for food and agriculture: Assessing global availability', *Annu. Rev. Environ. Resour.* 29, 143–79.
Gao, Y., Liu, F., Wang, K., Wang, D., Gong, X., Liu, L., Richards, C. M., Henk, A. D. and Volk, G. M. (2015), 'Genetic diversity of *Malus* cultivars and wild relatives in the Chinese National Repository of Apple Germplasm Resources', *Tree Genet. Genom.* 11(5), 1–9.

Gardner, K. M., Brown, P., Cooke, T. F., Cann, S., Costa, F., Bustamante, C., Velasco, R., Troggio, M. and Myles, S. (2014), 'Fast and cost-effective genetic mapping in apple using next-generation sequencing', *G3: Genes| Genomes| Genetics* 4, 1681–7.

Gessler, C. and Blaise, P. (1994), 'Differential resistance in apple against scab and its use in breeding and in orchard planting strategies to control the disease', in H. Schmidt and M. Kellerhals (eds), *Progress in Temperate Fruit Breeding*, Kluwer Academic Pub., A. A. Dordrech, the Netherlands, 99–104.

Gessler, C. and Pertot, I. (2012), 'Vf scab resistance of *Malus*', *Trees* 26, 95–108.

Giovanni, A., Broggini, L., Wöhner, T., Fahrentrapp, J., Kost, T. D., Flachowsky, H., Peil, A., Hanke, M.-V., Richter, K., Patocchi, A. and Gessler, C. (2014), 'Engineering fire blight resistance into the apple cultivar "Gala" using the *FB-MR5* CC-NBS-LRR resistance gene of *Malus × robusta* 5', *Plant Biotechnol. J.* 12, 728–33.

Gross, B. L., Volk, G. M., Richards, C. M., Forsline, P. L., Fazio, G. and Chao, C. T. (2012), 'Identification of "duplicate" accessions within the USDA-ARS National Plant Germplasm System *Malus* collection', *J. Amer. Soc. Hort. Sci.* 137, 333–42.

Gross, B. L., Volk, G. M., Richards, C. M., Reeves, P. A., Henk, A. D., Forsline, P. L., Szewc-McFadden, A., Fazio, G. and Chao, C. T. (2013), 'Diversity captured in the USDA-ARS National Plant Germplasm System apple core collection', *J. Amer. Soc. Hort. Sci.* 138, 375–81.

Guarino, C., Santoro, S., De Simone, L., Lain, O., Cipriana, G. and Testolin, R. (2005), 'Genetic diversity in a collection of ancient cultivars of apple (*Malus × domestica Borkh.*) as revealed by SSR-based fingerprinting', *J. Hort. Sci. Biotechnol.* 81, 39–44.

Gygax, M., Gianfranceschi, L., Liebhard, R., Kellerhals, M., Gessler, C. and Patocchi, A. (2004), 'Molecular markers linked to the apple scab resistance gene *Vbj* derived from *Malus baccata jackii*', *Theor. Appl. Genet.* 109, 1702–9.

Hampson, C. R. (2003), 'Characteristics of important commercial apple cultivars', in D. C. Ferree and I. J. Warrington (eds), *Apples: Botany, Production and Uses*, CAB International, Wallingford, UK, 61–89.

Höfer, M., Flachowsky, H., Hanke, M.-V., Semënov, V., Šlâvas, A., Bandurko, I., Sorokin, A. and Alexanian, S. (2013), 'Assessment of phenotypic variation of *Malus orientalis* in the North Caucasus region', *Genet. Resour. Crop Evol.* 60, 1463–77.

Hokanson, S. C., Lamboy, W. F., Szewc-McFadden, A. K. and McFerson, J. R. (2001), 'Microsatellite (SSR) variation in a collection of *Malus* (apple) species and hybrids', *Euphytica* 118, 281–94.

Hokanson, S. C., McFerson, J. R., Forsline, P. L., Lamboy, W. F., Luby, J. J., Djangaliev, A. D. and Aldwinckle, H. S. (1997), 'Collecting and managing wild *Malus* germplasm in its center of diversity', *HortScience* 32, 173–6.

Hokanson, S. C., Szewc-McFadden, A. K., Lamboy, W. F. and McFerson, J. R. (1998), 'Microsatellite (SSR) markers reveal genetic identities, genetic diversity and relationships in a *Malus x domestica* Borkh. core subset collection', *Theor. Appl. Genet.* 97, 671–83.

Hyson, D. A. (2011), 'A comprehensive review of apples and apple components and their relationship to human health', *Adv. Nutr.* 2, 408–20,

Isutsa, D. and Merwin, I. A. (2000), '*Malus* germplasm varies in resistance or tolerance to apple replant disease in a mixture of New York orchard soils', *HortScience* 35, 262–8.

Jurick, W. M., Janisiewicz, W. J., Saftner, R. A., Vico, I., Gaskins, V. L., Park, E., Forsline, L. P., Fazio, G. and Conway, W. S. (2011), 'Identification of wild apple germplasm (*Malus* spp.) accessions with resistance to the postharvest decay pathogens *Penicillium expansum* and *Colletotrichum acutatum*', *Plant Breeding* 130, 481–6.

Jung, S., Ficklin, S. P., Lee, T., Cheng, C.-H., Blenda, A., Zheng, P., Yu, J., Bobarely, A., Cho, I., Ru, S., Evans, K., Peace, C., Abbott, A. B., Mueller, L. A., Olmstead, M. A. and Main, D. (2014), 'The Genome Database for Rosaceae (GDR): year 10 update', *Nucleic Acid Res.* 42, D1237–44.

Kellerhals, M., Szalatnay, D., Hunziker, K., Duffy, B., Nybom, H., Ahmadi-Afzadi, M., Höfer, M., Richter, K. and Lateur, M. (2012), 'European pome fruit genetic resources evaluated for disease resistance', *Trees* 26, 179–89.

Khan, M. A., Olsen, K. M., Sovero, V., Kushad, M. M. and Korban, S. S. (2014), 'Fruit quality traits have played critical roles in domestication of the apple', *Plant Genome* 7(3), 1–18.

Kouassi, A. B., Durel, C.-E., Costa, F., Tartarini, S., van de Weg, E., Evans, K., Fernandez-Fernandez, F., Govan, C., Boudichevskaja, A., Dunemann, F., Antofie, A., Lateur, M., Stankiewicz-Kosyl, M., Soska, A., Tomala, K., Lewandowski, M., Rutkovski, K., Zurawicz, E., Guerra, W. and Laurens, F. (2009), 'Estimation of genetic parameters and prediction of breeding values for apple fruit-quality traits using pedigreed plant material in Europe', *Tree Genet. Genomes* 5, 659–72.

Kumar, S., Bink, M. C. A. M., Volz, R. K., Bus, V. G. M. and Chagné, D. (2012a), 'Towards genomic selection in apple (*Malus × domestica* Borkh.) breeding programmes: Prospects, challenges and strategies', *Tree Genet. Genomes* 8, 1–14.

Kumar, S., Chagné, D., Bink, M. C. A. M., Bolz, R. K., Whitworth, C. and Carlisle, C. (2012b), 'Genomic selection for fruit quality traits in apple (*Malus × domestica* Borkh.)', *PLoS ONE* 7, e36674.

Kumar, S., Garrick, D. J., Bink, M. C. A. M., Whitworth, C., Chagné, D. and Volz, R. K. (2013), 'Novel genomic approaches unravel genetic architecture of complex traits in apple', *BMC Genomics* 14, 393.

Kumar, S., Volz, R. K., Alspach, P. A. and Bus, V. G. M. (2010), 'Development of a recurrent apple breeding programme in New Zealand: a synthesis of results, and a proposed revised breeding strategy', *Euphytica*, 173, 207–22.

Leforestier, D., Ravon, E., Muranty, H., Cornille, A., Lemaire, C., Giraud, T., Durel, C. E. and Branca, A. (2015), 'Genomic basis of the differences between cider and dessert apple varieties', *Evol. Appl.* 8, 650–61.

Le Roux, P.-M., Flachowsky, H., Hanke, M.-V., Gassler, C. and Patocchi, A. (2012), 'Use of a transgenic early flowering approach in apple (*Malus × domestica* Borkh.) to introgress fire blight resistance from cultivar Evereste', *Mol. Breeding* 30, 857–74.

Longhi, S., Hamblin, M. T., Trainotti, L., Peace, C. P., Velasco, R. and Costa, F. (2013), 'A candidate gene based approach validates Md-PG1 as the main responsible for a QTL impacting fruit texture in apple (*Malus × domestica* Borkh.)', *BMC Plant Biol.* 13, 37.

Luby, J. J. (2003), 'Taxonomic classification and brief history', in D. C. Ferree and I. J. Warrington (eds), *Apples: Botany, Production and Uses*, CAB International, Wallingford, UK, 1–14.

Luby, J. J., Alspach, P. A., Bus, V. G. M. and Oraguzie, N. C. (2002), 'Field resistance to fire blight in a diverse apple (*Malus* sp.) germplasm collection', *J. Am. Soc. Hort. Sci.* 127, 245–53.

Luby, J., Forsline, P., Aldwinckle, H., Bus, V. and Geibel, M. (2001), 'Silk road apples – Collection, evaluation, and utilization of *Malus sieversii* from Central Asia', *HortScience* 36, 225–31.

Maguylo, K. and Bassett, C. (2014), 'Phenotyping M. sieversii collections from Kazakhstan for leaf traits and tree architecture', *Acta Hort.* 1058, 335–41.

Markussen, T., Krüger, J., Schmidt, H. and Dunemann, F. (1995), 'Identification of PCR-based markers linked to the powdery-mildew-resistance gene *Pl1* from *Malus robusta* in cultivated apple', *Plant Breed.* 114, 530–4.

Moriya, S., Iwanami, H., Okada, K., Yamamoto, T. and Abe, K. (2011), 'A practical method for apple cultivar identification and parent-offspring analysis using simple sequence repeat markers', *Euphytica* 177, 135–50.

Moriya, S., Iwanami, H., Takahashi, S., Kotoda, N., Suzaki, K., Yamamoto, T. and Abe, K. (2010), 'Genetic mapping of the crown gall resistance gene of the wild apple *Malus sieboldii', Tree Genet Genomes* 6, 195–203.

Mudge, K., Janick, J., Scofield, S. and Goldschmidt, E. E. (2009), 'A history of grafting', *Hort. Rev.* 35, 437–93.

Nikiforova, S. V., Cavalieri, D., Velasco, R. and Goremykin, V. (2013), 'Phylogenetic analysis of 47 chloroplast genomes clarifies the contribution of wild species to the domesticated apple maternal line', *Mol. Biol.* Evol. 30, 1751–60.

Noiton, D. A. M. and Alspach, P. A. (1996), 'Founding clones, inbreeding, coancestry, and status number of modern apple cultivars', *J. Amer. Soc. Hort. Sci.* 121, 773–82.

Norelli, J. L., Wisniewski, M. and Droby, S. (2014), 'Identification of a QTL for postharvest disease resistance to *Penicillium expansum* in *Malus sieversii*', *Acta Hort.* 1053, 199–203.

Parisi, L., Fouillet, V., Schouten, H. J., Groenwold, R., Laurens, F., Didelot, F., Evans, K., Fischer, C., Gennari, F., Kemp, H., Lateur, M., Patocchi, A., Thissen, J. and Tsipouridis, C. (2004), 'Variability of the pathogenicity of *Venturia inaequalis* in Europe', *Acta Hort.* 663, 107–13.

Patocchi, A., Frei, A., Frey, J. E. and Kellerhals, M. (2009), 'Towards improvement of marker assisted selection of apple scab resistance cultivars: *Venturia inaequalis* virulence surveys and standardization of molecular marker alleles associated with resistance genes', *Mol. Breed.* 24, 337–47.

Patocchi, A., Walser, M., Tartarini, S., Broggini, G. A. L., Gennari, F., Sansavini, S. and Gessler, C. (2005), 'Identification by genome scanning approach (GSA) of a microsatellite tightly associated with the apple scab resistance gene *Vm*', *Genome* 48, 630–6.

Patzak, J., Paprštein, F., Henychová, A. and Sedlák, J. (2012), 'Comparison of genetic diversity structure analyses of SSR molecular marker data within apple (*Malus × domestica*) genetic resources', *Genome* 55, 647–65.

Peil, A., Garcia-Libreros, T., Richter, K., Trognitz, F. C., Hanke, M.-V. and Flachowsky, H. (2007), 'Strong evidence for a fire blight resistance gene of *Malus robusta* located on linkage group 3', *Plant Breed.* 126, 470–6.

Pereira-Lorenzo, S., Ramos-Cabrer, A. M. and Díaz-Hernández, M. B. (2007), 'Evaluation of genetic identity and variation of local apple cultivars (*Malus* × *domestica* Borkh.) from Spain using microsatellite markers', *Genet. Resour. Crop Evol.* 54, 405–20.

Petri, J. L. and Leite, G. B. (2004), 'Consequences of insufficient winter chilling on apple tree bud-break', *Acta Hort.* 662, 53–60.

Phipps, J. B., Robertson, K. R., Smith, P. G. and Rohrer, J. R. (1990), 'A checklist of the subfamily Maloidease (Rosaceae)', *Can. J. Bot.* 68, 2209–69.

Postman, J., Hummer, K., Stover, E., Krueger, R., Forsline, P., Grauke, L. J., Zee, F., Ayala-Silva, T. and Irish, B. (2006), 'Fruit and nut genebanks in the U.S. National Plant Germplasm System', *HortScience* 41, 1188–94.

Richards, C. M., Volk, G. M., Reeves, P. A., Reilley, A. A., Forsline, P. L. and Aldwinckle, H. S. (2009a), 'Selection of stratified core sets representing wild apple (*Malus sieversii*)', *J. Amer. Soc. Hort. Sci.* 134, 228–35.

Richards, C. M., Volk, G. M., Reilley, A. A., Henk, A. D., Lockwood, D. R., Reeves, P. A. and Forsline, P. L. (2009b), 'Genetic diversity and population structure in *Malus sieversii*, a wild progenitor species of domesticated apple', *Tree Genet. Genome.* 5, 339–47.

Robinson, J. P., Harris, S. A. and Juniper, B. E. (2001), Taxonomy of the genus *Malus* Mill. (Rosaceae) with emphasis on the cultivated apple, *Malus domestica* Borkh.', *Plant Syst. Evol.* 226, 35–58.

Routson, K. J., Reilley, A. A., Henk, A. D. and Volk, G. M. (2009), 'Identification of historic apple trees in the southwestern United States and implications for conservation', *HortScience* 44, 589–94.

Rubenstein, K. D., Smale, M. and Widrlechner, M. P. (2006), 'Demand for genetic resources and the U.S. National Plant Germplasm System', *Crop Sci.* 46, 1021–31.

Schmitz, C. A., Clark, M. D., Luby, J. J., Guan,Y., Evans, K., Orcheski, B., Brown, S., Verma, S. and Peace, C. (2013), 'Fruit texture phenotypes of the RosBREED U.S. apple reference germplasm set', *HortScience* 48, 296–303.

Schnitzler, A., Arnold, C., Cornille, A., Bachmann, O. and Schnitzler, C. (2014), 'Wild European apple (*Malus sylvestris* (L.) Mill.) population dynamics: Insight from genetics and ecology in the Rhine Valley. Priorities for a future conservation Programme', *PLoS ONE.* 9, e96596.

Sestras, A. F., Pamfil, D., Dan, C., Bolboaca, S. D., Jäntschi, L. and Sestras, R. E. (2011), 'Possibilities to improve apple scab (*Venturia inaequalis* (Cke.) Wint.) and powdery mildew [*Podosphaera leucotricha* (Ell. Et Everh.) Salm.] resistance on apple by increasing genetic diversity using potentials of wild species', *Aust. J. Crop Sci.* 5, 748–55.

Smith, M. W. G. (1971), National apple register of the United Kingdom. London Ministry of Agriculture, Fisheries and Food. 2002 reprint by Castlepoint Press, Colvend, UK.

Soriano, J. M., Joshi, S. G., van Kaauwen, M., Voordijk, Y., Groenwold, R., Henken, B., van de Weg, W. E. and Schouten, H. J. (2009), 'Identification and mapping of the novel apple scab resistance gene Vd3', *Tree Genet. Genomes* 5, 475–82.

Tartarini, S., Gennari, F., Pratesi, D., Palazzetti, C., Sansavini, S., Parisi, L., Fouillet, A., Fouillet, V. and Durel, C. E. (2004), 'Characterisation and genetic mapping of a major scab resistance gene from the old Italian apple cultivar "Durello di Forli"', *Acta Hort.* 6631, 129–33.

Towill, L. E., Forsline, P. L., Walters, C., Waddell, J. W. and Laufmann, J. (2004), 'Cryopreservation of *Malus* germplasm using a winter vegetative bud method: Results from 1915 accessions', *CryoLetters* 25, 323–34.

Troggio, M., Gleave, A., Salvi, S., Chagné, D., Cestaro, A., Kumar, S., Crowhurst, R. N. and Gardiner, S. E. (2012), 'Apple, from genome to breeding', *Tree Genet. Genomes* 8, 509–29.

van Treuren, R., Kemp, H., Ernsting, G., Jongejans, B., Houtman, H. and Visser, L. (2010), 'Microsatellite genotyping of apple (*Malus* × *domestica* Borkh.) genetic resources in the Netherlands: application in collection management and variety identification', *Genet. Resour. Crop Evol.* 57, 853–65.

Urrestarazu, J., Denancé, C., Ravon, E., Guyader, A., Guisnel, R., Feugey, L., Poncet, P., Lateur, M., Houben, P., Ordidge, M., Fernandez-Fernandez, F., Evans, K. M., Paprstein, F., Sedlak, J., Nybom, H., Garkava-Gustavsson, L., Miranda, C., Gassmann, J., Kellerhals, M., Suprun, I., Pikunova, A. V., Krasova, N. G., Torutaeva, E., Dondini, L., Tartarini, S., Laurens, F. and Durel, C.-E. (2016), 'Analysis of the genetic diversity and structure across a wide range of germplasm reveals prominent gene flow in apple at the European level', *BMC Plant Biol.* In press.

Urrestarazu, J., Miranda, C., Santesteban, L. G. and Royo, J. B. (2012), 'Genetic diversity and structure of local apple cultivars from Northeastern Spain assessed by microsatellite markers', *Tree Genet. Genome.* 8, 1163–80.

Varaprasad, K. S. and Sivaraj, N. (2010), 'Plant genetic resources conservation and use in light of recent policy developments', *Electronic J. Plant Breed.* 1, 1276–93.

Velasco, R., Zharkikh, A., Affourtit, J., Dhingra, A., Cestaro, A., Kalyanaraman, A., et al. (2010), 'The genome of the domesticated apple (*Malus* × *domestica* Borkh.)', *Nature Genet.* 42, 883–9.

Vincent, H., Wiersema, J., Kell, S., Fielder, H., Dobbie, S., Castañeda-Álvarez, N. P., Guarino, L., Eastwood, R., León, B. and Maxted, N. (2013), 'A prioritized crop wild relative inventory to help underpin global food security', *Biological Conserv.* 167, 265–75.

Volk, G. M. (2010a), 'Advantages for the use of standardized phenotyping in databases', *HortScience* 45, 1310–13.

Volk, G. M., Chao, C. T., Norelli, J., Brown, S. K., Fazio, G., Peace, C., McFerson, J., Zhong, G.-Y. and Bretting, P. (2015a), 'The vulnerability of US apple (*Malus*) genetic resources', *Genet. Resour. Crop Evol.* 62, 765–94.

Volk, G. M., Henk, A. D., Baldo, A., Fazio, G., Chao, C. T. and Richards, C. M. (2015b), 'Chloroplast heterogeneity and historical admixture within the genus *Malus*', *Am. J. Bot.* 102, 1198–208.

Volk, G. M., Jenderek, M. and Chao, C. T. (2016), 'Prioritization of *Malus* accessions for collection cryopreservation at the USDA-ARS National Center for Genetic Resources Preservation', *Acta Hort.* In press.

Volk, G. M. and Richards, C. M. (2011), 'Integration of georeferencing, habitat, sampling, and genetic data for documentation of wild plant genetic resources', *HortScience* 46, 1446–9.

Volk, G. M., Richards, C. M. and Forsline, P. L. (2010b), 'A comprehensive approach toward conserving *Malus* germplasm', *Acta Hort.* 859, 177–82.

Volk, G. M., Richards, C. M., Henk, A. D., Reilley, A. A., Reeves, P. A., Forsline, P. L. and Aldwinckle, H. S. (2009), 'Capturing the diversity of wild *Malus orientalis* from Georgia, Armenia, Russia, and Turkey', *J. Amer. Soc. Hort. Sci.* 134, 453–9.

Volk, G. M., Richards, C. M., Reilley, A. A., Henk, A. D., Forsline, P. L. and Aldwinckle, H. S. (2005), 'Ex situ conservation of vegetatively propagated species: development of a seed-based core collection for *Malus sieversii*', *J. Amer. Soc. Hort. Sci.* 130, 203–10.

Volk, G. M., Richards, C. M., Reilley, A. A., Henk, A. D., Reeves, P. A., Forsline, P. L. and Aldwinckle, H. S. (2008), 'Genetic diversity and disease resistance of wild *Malus orientalis* from Turkey and Southern Russia', *J. Amer. Soc. Hort. Sci.* 133, 383–9.

Walters, C., Volk, G. M. and Richards, C. M. (2008), 'Genebanks in the post-genomic age: Emerging roles and anticipated uses', *Biodiversity*, 9, 68–71.

Wan, Y., Li, D., Zhao, Z., Mei, L., Han, M., Schwaninger, H. and Fazio, G. (2011), 'The distribution of wild apple germplasm in northwest China and its potential application for apple rootstock breeding', *Acta Hort.* 903, 123–41.

Wegrzyn, J. L., Main, D., Figueroa, B., Choi, M., Yu, J., Neale, D. B., Jung, S., Lee, T., Stanton, M., Zhang, P., Ficklin, S., Cho, I., Peace, C., Evans, K., Volk, G., Oraguzie, N., Chen, C., Olmstead, M., Gmitter, G. Jr. and Abbott, A. G. (2012), 'Uniform standards for genome databases in forest and fruit trees', *Tree Genet. Genomes.* doi: 10.1007/s11295-012-0494-7.

Williams, K. A. (2005), 'An overview of the U.S. National Plant Germplasm System's exploration program', *HortScience* 40, 297–301.

Wisniewski, M., Norelli, J., Droby, S., Levin, E., Ballester, A.-R. and Abdelfattah, A. (2016), 'Genomic tools for developing markers for postharvest disease resistance in Rosaceae fruit crops', *Acta Hort.* In press.

Yan, G., Long, H., Song, W. and Chen, R. (2008), 'Genetic polymorphism of *Malus sieversii* populations in Xinjiang, China', *Genet. Resour. Crop Evol.* 55, 171–81.

Yang, W., Liu, X.-D., Chi, X.-J., Wu, C.-A., Li, Y.-Z., Song, L.-L., Liu, X.-M., Wang, Y.-F., Wang, F.-W., Zhang, C., Liu, Y., Zong, J.-M. and Li, H.-Y. (2011), 'Dwarf apple MbDREB1 enhances plant tolerance to low temperature, drought, and salt stress via both ABA-dependent and ABA-independent pathways', *Planta* 233, 219–29.

Yu, T. C. (1988), 'Pome fruits, 1. *Malus*', in C.-K. K. Shu, F. L. Hsueh, T. P. Chang and J. X. Lei (eds), *Taxonomy of Fruit Trees in China*, Muhammad Ali Society, Karachi, Pakistan, 87–122.

Zhang, C., Chen, X., He, T., Liu, X., Feng, T. and Yuan, Z. (2007), 'Genetic structure of *Malus sieversii* population from Xinjiang, China, revealed by SSR markers', *J. Genet. Genomics* 34, 947–55.

Zhang, Q., Li, J., Zhao, Y., Korban, S. S. and Han, Y. (2012), 'Evaluation of genetic diversity in Chinese wild apple species along with apple cultivars using SSR markers', *Plant Mol. Biol. Rep.* 30, 539–46.

Zhi-Qin, Z. (1999), 'The apple genetic resources in China: The wild species and their distributions, informative characteristics and utilization', *Genet. Resour. Crop Evol.* 46, 599–609.

Chapter 2

The apple genome – harbinger of innovation for sustainable apple production

Amit Dhingra, Washington State University, USA

1 Introduction

The road to sequencing of the apple genome was paved with multiple developments, starting from the endeavour to sequence the human genome (Lander et al., 2001). A spate of technological innovations, primarily in the area of high-throughput and cost-effective sequence acquisition, ushered in the current era of genomics where draft genomes are generated and released on a regular basis (Heather and Chain, 2016). This technological innovation has now enabled several large-scale genera-wide genome-sequencing projects that aim to use population-wide variation to associate important traits with genomic regions or genes (Goodwin et al., 2016).

The primary premise to sequencing any genome is to understand the genetic blueprint of a species. Having the knowledge of the genome can then enable the improvement of the species by devising practice-based or pharmacological or agrochemical interventions. In humans and animals, there is a predominance of development of pharmacological approaches while in crops the primary approach for improvement has been via traditional breeding, with DNA-informed breeding gaining ground in recent years.

Apple (*Malus × domestica* Borkh.) belongs to the Rosaceae family, which is an ensemble of some of the most diverse, economically important, nutritious, and culturally and historically relevant plant species (Jackson, 2003). These include pear (*Pyrus communis*), peach (*Prunus persica*), almond (*Prunus dulcis*), strawberry (*Fragaria × ananassa*) and

http://dx.doi.org/10.19103/AS.2016.0017.02

several more (Hummer and Janick, 2009). Breeding programmes across the globe are aimed at improving these crops for agronomic and consumer traits. In the tree fruit crops, the long duration between subsequent generations necessitates maintenance of all the siblings from a cross over several years. While DNA-informed breeding does not reduce the time between generations, it can certainly enable the early elimination of undesirable seedlings saving resources for the programme. In order to perform DNA-informed breeding, it is important to establish associations between traits and genomic regions or genes, but is not a straightforward task. Combined with horticultural and biotechnological approaches to reduce the generation times, DNA-informed breeding can enable the development of superior cultivars in an economically efficient way (van Nocker and Gardiner, 2014).

While there are several 'omics' studies encompassing transcriptomics, proteomics and metabolomics approaches that have preceded the release of the apple genome, subsequent 'omics' studies are bound to benefit from an annotated apple genome. Several biological processes have been studied using techniques such as cDNA-amplified fragment length polymorphism, suppression subtractive hybridization, differential display and EST-based microarrays (Schaeffer et al., 2016; Dal Cin et al., 2007, 2009; Breyne et al., 2003; Degenhardt et al., 2005; Janssen et al., 2008; Schaffer et al., 2007; Troggio et al., 2012). As the apple reference genome sequence becomes more accurate, all these studies can now be augmented to expand our understanding of apple biology and help develop technologies and products with direct impact on sustainability of apple production in both the short and longer term.

2 Brief overview of genome sequencing technology development

The very first genome of a free-living organism to be sequenced was that of the bacterium *Haemophilus influenzae* using the classical Sanger sequencing methodology (Fleischmann et al., 1995). The same methodology was used to sequence the genome of *Arabidopsis thaliana* representing the first angiosperm to be sequenced (Arabidopsis_Genome_Initiative, 2000). Since then, several plant and other eukaryotic genomes continue to be sequenced using hybrid or Sanger and next-generation sequencing platforms, although recently, most genomes are being sequenced using next-generation sequencing technologies. These genomes can be accessed and analysed using diARK 2.0, an online genome repository (Hammesfahr et al., 2011) or GOLD – Genomes OnLine Database (Mukherjee et al., 2016).

As mentioned before, most genomes sequenced prior to the early 2000s used Sanger's dideoxy termination method (Sanger et al., 1977). This low-throughput approach is time consuming and expensive; however, it currently remains the gold standard for its accuracy in DNA base-calling. Development of automated capillary sequencing combined with Sanger sequencing enabled the initiation of the human genome project (Collins et al., 2003; Smith et al., 1986). Sanger method-based sequencing necessitates capture of the genome as fragments ranging in 150–250 kilo bases (kb) in bacterial artificial chromosome (BAC) libraries. Thereafter, the BAC clones are fingerprinted usually with restriction enzymes to identify those that are unique, which are then sequenced using Sanger's dideoxy-termination sequencing method (Sanger et al., 1977). Despite the use of robotics, the cost to sequence the 3 billion base pair (bp) human genome using the Sanger method was US$ 2.7 billion. The DNA read length obtained with Sanger sequencing ranges from 500 bp to 1000 bp and the unsurpassable accuracy of DNA base-calling ensures that the genome is reconstructed to best possible approximation.

Next generation or second generation sequencing platforms, with Sanger sequencing being the first generation platform, first appeared on the scene in the early 2000s (Margulies et al., 2005; Heather and Chain, 2016), performing massively parallel sequencing reactions which generate billions of reads ranging from 35 bp to 250 bp in length depending on the platform. This is an exponential increase in throughput that does not rely on preparing and sequencing of BAC libraries. Each next-generation sequencing platform uses a unique set of chemical reactions to detect DNA bases and each one also has a unique set of issues associated with DNA base detection leading to errors in the final output and consequently challenges with the final assembly (Ambardar et al., 2016). Recently, third-generation sequencing platforms, such as PacBio sequencing, have also become available, and they are based on sequencing by synthesis, sequencing by degradation or direct physical examination of the DNA molecule and can read long stretches of DNA running into a few kilo bases, but with moderate DNA bp quality (Schadt et al., 2010). Some of these challenges are overcome by the capacity of next-generation sequencing to generate enough data that each DNA nucleotide in the final assembled sequence is represented multiple times by the sequencing reads. The number of times a DNA base is represented by sequencing reads is called depth of coverage. Usually, it is represented as an average coverage of the whole genome where the total number of reads are multiplied by the length of the reads and divided by the total genome size (Sims et al., 2014). Higher depth of coverage allows for base-call correction of reads from all platforms using specific error-correction software (Marçais et al., 2015; Marinier et al., 2015; Salmela and Rivals, 2014), which continue to be improved, and thus enhancing our ability to accurately assemble next-generation sequencing data (Park and Kim, 2016).

3 Sequencing of the apple genome and lessons learned

The draft genome of the domesticated apple (*Malus* × *domestica* L.) was unveiled in 2010 by an Italy-led consortium (Velasco et al., 2010). The 'Golden Delicious' genotype was chosen for sequencing since it is an economically important cultivar in Europe but also because 'Golden Delicious' is in the pedigree of many commercially relevant cultivars (Salvi et al., 2014). While the goal to sequence the genome of apple was to develop improved tools for breeding programs, the origins of 'Golden Delicious' itself remain ambiguous. Anecdotally, 'Golden Delicious' was proposed to be derived from a cross between 'Golden Reinette' and 'Grimes Golden'; in fact, 'Grimes Golden' was recently confirmed as one of the parents (Salvi et al., 2014). The available anecdotal information states that 'Golden Delicious' originated as a chance seedling, initially known as 'Mullins Seedling', in West Virginia, USA, in the early 1900s. Stark Nursery started selling the trees around 1914, and it soon became a household variety in the United States. In addition, this humble product of a chance plant hybridization event became the messenger of hope for the Western European fruit industry in 1945 under the Marshall Plan as millions of trees of cloned 'Golden Delicious' reached Europe after World War II (Juniper, 2007).

Sequencing of the apple genome used two sources of 'Golden Delicious'. The majority of the data were generated from DNA derived from young leaves of a two-year-old heterozygous, allotetraploid, typical cultivated accession of 'Golden Delicious', grown in the greenhouse at Fondazione Edmund Mach–Istituto Agrario di San Michele all 'Adige, Italy. The group in Italy that led the sequencing of the heterozygous 'Golden Delicious' genome was able to utilize their expertise in assembling a heterozygous genome using their prior experience with the grape genome, which was assembled using the assembly algorithms available at that time (Velasco et al., 2010). The overall data amounting to ~75× of the estimated genome size were generated using Sanger sequencing of BAC and Fosmid libraries and 454 sequencing. The genome was assembled using a custom iterative assembly workflow previously used for grape genome assembly (Zharkikh et al., 2008). Funding for the Italian apple genome project was provided by the research office of the Provincia Autonoma di Trento. The second source used was a doubled haploid derivative of 'Golden Delicious' developed by the French National Institute for Agricultural Research (Lespinasse et al., 1998). The data from the second genotype was generated as part of a USDA NIFA, Washington State University and Washington Tree Fruit Research Commission funded project in the Dhingra Genomics Lab (Dhingra, 2013). The doubled haploid sequence enabled accurate haplotype phase determination and thus assembly of the genome sequence; it is estimated that 80% of the 3.3 million SNPs identified from the heterozygous apple genome were represented in the doubled haploid dataset (Troggio et al., 2012). Since the release of the apple genome in 2010, Apple genome 2.0, a new assembly has been released and is accessible via Genomic Database for Rosaceae (Jung et al., 2014). The *Malus* × *domestica* genome v2.0 assembly is a derivative of the original v1.0 assembly, representing four pseudo haplotypes corresponding to the four different haplotypes present in the apple genome. This assembly was derived from the *Malus* × *domestica* assembly v1.0 by removing 34,882 problematic contigs. Another hybrid assembly, which has been released recently, combines 76 Gb of Illumina HiSeq data representing ~102 × genome coverage and 21.7 Gb of PacBio data representing ~29 × genome coverage, which approximates to ~90% of the estimated genome size (Li et al., 2016). A stand-alone assembly of the doubled haploid genome version 1.0 was developed using 2.8 × genome coverage as 454 unpaired data, 75 × genome coverage as Illumina paired end data and 1.3 × genome coverage as PacBio data. The assembly was performed using CLC Genomics workbench 6.0.1. This assembly is available from the Genome Portal at Washington State University (Dhingra, 2013).

Of all the plant genomes sequenced at the time, apple had the highest number of predicted genes (57,386). Of these putative genes, nearly 11,444 were specific to apple which speaks to the unique biology of this species (Velasco et al., 2010). Comparative analysis across seven genomes for plastid-targeted genes revealed that apple had the highest number of apple-specific plastid-targeted genes (Schaeffer et al., 2014). Both reports point to the polyploid nature of the species that may have contributed to some inflation in predicted number of genes. Further, comparative analysis with other genomes has shed light on various aspects of the unique biology of this important member of the plant kingdom and its ancestry. It was surmised that the most likely direct ancestor of cultivated *Malus* × *domestica* is the central Asian species *Malus sieversii*. Further, it was determined that a genome duplication around 50 million years ago was responsible for the 17 chromosome karyotype in both apples and pears. This number was derived from loss of a chromosome from an 18-chromosome ancestor which was itself derived from an autopolyploidization event of a 9-chromosome progenitor. This 9-chromosome ancestor is most likely *Gillenia*, which has fruit resembling the apple core without the flesh (Evans and Campbell, 2002; Troggio et al., 2012).

Both apple and pear belong to the Pyreae tribe within the Rosaceae family and are characterized by a unique type of fruit termed as pome, where the bulk of the fruit tissue is derived from the receptacle, the tissue at the base of the floral organs. It was revealed that several of the genes in the chromosomal duplication regions were related to fruit development. Analysis revealed something very interesting about the MADS box II family of transcription factor genes which control floral organ identity and development in other plant species (Irish and Litt, 2005). The MADS Box II gene family was present in greater diversity in apples compared to other species, indicating that these genes may have a role to play in the development of the pome fruit. Most plant species transport their photosynthesis-derived carbohydrates as sucrose; however apples, like only a few other plant species, transport their carbohydrates as sorbitol. The apple genome also possesses an expanded set of the sorbitol transporter genes (Velasco et al., 2010; Giovannoni, 2010).

4 Harvesting the genome for improvement of apple

The ultimate goal of establishing the genome of any species is to be able to use it as a reference for performing studies that provide information regarding gene-trait relationships for developing molecular markers which can be utilized in breeding superior apple varieties. Prior to the establishment of such associations, there is a need to genotype a large set of accessions to sample the available genotypic differences that exist across a species or genera. Genotyping can be accomplished in two ways. One such approach that has been used previously in human genomics studies is the development of a SNParray (Clark et al., 2005). For apple, three SNParrays with 8K and 20K SNPs and another recent one with 480K SNPs have been developed (Chagné et al., 2012; Bianco et al., 2014, 2016). However, extensive studies in human genomics have revealed that there is an inherent ascertainment bias while using SNParrays (Clark et al., 2005; Albrechtsen et al., 2010) unless one intends to utilize the arrays only for the population that were initially genotyped. It is difficult to expect such a scenario in breeding programmes which are constantly trying to infuse novel sources of genetics to introgress exotic and economically important consumer and agronomic traits. A recent study evaluated multiple genotyping approaches to compare five closely related cultivars of sweet cherry. Except for the SNParray, all other approaches were able to identify the differences in the five genotypes (Hewitt et al., 2017). While the ascertainment bias can be compensated for, the choice of genotyping-calling algorithms can have a substantial impact on the outcome and such corrections are not applicable across different studies (Albrechtsen et al., 2010). SNParrays may have a limited shelf life unless this limitation can be overcome by developing programme-specific SNParrays that represent all the genotypic diversity available in each breeding programme or collaborative group of programmes. Of course, with nearly 10,000 documented apple cultivars, besides the wild germplasm which also number in thousands (Janick et al., 1996), producing community-wide arrays would appear to be an improbable task. The second approach which is based on genotyping-by-sequencing (GBS) addresses such a limitation (Myles et al., 2009). GBS is the application of next-generation sequencing to identify SNPs in order to perform genotyping experiments which include genome-wide associate studies (GWAS) (Poland and Rife, 2012). GBS is performed on a reduced but similar representation of the genomes from the samples which is achieved by using restriction enzymes, PCR or shearing of the genomic DNA. GBS enables the comparison of similar loci across multiple samples of a family or population (He et al., 2014; Kim et al., 2016). In the GBS approach, sequences are used directly for both detecting and scoring SNPs, thereby completely bypassing the SNP marker assay development stage required for developing SNParrays (Deschamps et al., 2012). This advantage allows for the application of GBS to any plant system and circumvents the issue of the ascertainment bias discussed in the case of SNParrays. There are several software that allow for performing genotyping studies using GBS data (Perea et al., 2016; Catchen et al., 2013). As the cost of sequencing continues to drop, GBS seems to be emerging as the preferred approach for genotyping and performing association studies. With that, the emphasis is moving towards efficient experimental design to determine genotype-trait associations (Myles et al., 2009).

Some progress has already been made in determining trait-loci associations; however, as the genotyping platforms expand, it is expected that a better understanding of traits and underlying loci will become available. Agronomic traits in perennial crops represent complex biological questions that have not been previously researched in model plant systems. Availability of the apple genome – combined with other genetic resources, means to establish gene-trait relationships, protocols and techniques to reduce generation time, and biotechnology tools such as genome editing – should enable development of varieties that incorporate resilience to abiotic stresses such as water limitations and temperature fluctuations and to biotic stresses such as microbial diseases both above ground and soil-borne, and that can be grown sustainably. In particular, genome editing using CRISPR/Cas9, TAL effector nucleases (TALENs), zinc-finger nucleases (ZFN) and gene repair oligonucleotides (GRONs) offer an unprecedented opportunity to make efficient and targeted changes to elite germplasm, saving the enormous amounts of resources that go into developing a new variety in perennial crops (Schaeffer and Nakata, 2015; Weeks et al., 2016).

Enhancement of photosynthetic efficiency could have a substantial impact on sustainability of apple production. Recent developments in this area in other species could be incorporated into new apple varieties. A transcription factor, HYR, which enhances most photosynthesis-related genes with a concomitant increase in resilience to environmental stresses resulting in high yields in rice could be introgressed into apples (Ambavaram et al., 2014). Another recent report has demonstrated the enhancement of radiation use efficiency of a plant by enabling accelerated change in non-photochemical quenching, increasing productivity by up to 20% (Kromdijk et al., 2016). While this was accomplished using a transgenic approach, use of genome editing can now enable deleting, silencing or adding genomic fragments to rapidly improve agricultural crops (Ishii and Araki, 2016).

There are areas of apple production which were previously thought to be beyond the realm of genetics. Having a genome sequence is expected to allow for the identification of genomic loci that regulate highly complex traits such as tree architecture or fruit abscission. This knowledge can be applied towards developing custom rootstocks to control vigor or branching in scion varieties. Also the force required to separate fruit from the tree can be modulated, reducing damage to the fruit during harvest. These traits can be complemented with emerging orchard mechanization efforts so that both the genetics and the machines can be customized for most efficient production. Automation translates into precision agriculture which in turn enhances the sustainability factor in any agricultural operation (Zude-Sasse et al., 2016).

5 Utility of the apple genome in resolving issues in the apple industry

While on the surface there appears to be a long way to go before apple genome sequence information can assist apple growers worldwide in producing better fruit for the consumer, opportunities for application are already available. Of course the development of new and superior varieties using DNA-informed breeding will have an impact in the long term but genome sequence data is also being used to identify horticultural or agrochemical solutions in the short term. Knowledge of specific genes can be combined with their expression and function in the context of important biological processes that impact vegetative growth and development, biotic and abiotic disease tolerance or resistance, pre-harvest stages of fruit development, post-harvest fruit storage, shelf life and organoleptic traits.

Indeed such approaches are already being employed in plants, as illustrated by a chemical intervention in wheat where plant permeable analogues of trehalose-6-phosphate enhanced yields by up to 20% and increased stress resilience (Griffiths et al., 2016). Another recent genome-based study highlighted the lack of specificity in currently used phytohormones and herbicides and how the use of small molecules can fine-tune ethylene response in plants (Le Deunff et al., 2016). In the Dhingra lab, we have demonstrated delay of ripening in pear by targeting a non-ethylene pathway where pre-harvest application of glycine betaine can delay harvest by a few days and also improve storage of the fruit in controlled atmosphere conditions (Dhingra and Schaeffer, 2014). Carefully designed studies that combine phenomics and transcriptomics can enable the identification of genes that then may be targeted by already existing or newly designed compounds to precisely modulate the physiology. This chemogenomics approach has been successfully employed in drug discovery (Bredel and Jacoby, 2004). A similar approach in plants can be very useful as it circumvents the need for generation of transgenics as well as enabling direct application to food crops (Stokes and McCourt, 2014). One such example is pears, a close relative to apple; there is a well-defined conditioning process to ripen fruit which requires incubating the fruit in the cold for 15 to 90 days, depending on the genotype, or treating fruit with ethylene (Sugar and Basile, 2013). If pears are treated with the ethylene receptor blocker 1-methyl cyclopropene after harvest, cold conditioning or exogenous application of ethylene fails to ripen the fruit (Chiriboga et al., 2013; Watkins, 2006). Using the cold conditioning model of pear ripening and RNAseq to identify metabolic pathways that are involved in conditioning and ripening, we devised a chemical intervention strategy to overcome the limitation in ripening imposed by 1-methycyclopropene (Dhingra and Hendrickson, 2017). As such transcriptomics studies based on real-time and well-defined phenotyping expand, it is expected that additional agrochemicals will be identified to address short-term needs of the growers. In the long term, information regarding genetic targets of agrochemicals can be utilized in developing molecular markers for raising new varieties through breeding or mutagenesis.

6 The future of genomics research in apple

Since the release of the Arabidopsis genome (Arabidopsis_Genome_Initiative, 2000), several complex plant genomes have been sequenced and are available in the public domain. The next phase seems to focus on resequencing multiple genotypes of a given species as the cost of sequencing continues to plummet. One can only conjecture at this point what the future of genomics and genetics in a complex crop like apple will be. Even though all the cultivated varieties will be sequenced, will it be feasible to utilize genomics in unravelling the random output of recombination in a breeding population? Technology platforms that allow for microdissection of whole or parts of a chromosome followed by sequencing of those chromosomes already exist (Seifertova et al., 2013). There are still challenges associated with sequencing and assembly of large genomic fragments although third- and fourth-generation sequencing platforms promise to sequence fragments of several million bases directly in fixed cells (Ke et al., 2016). These approaches may seem in their infancy right now; however, as the field advances, these tools may allow us to map every recombination event in apple seedlings. It will of course require gargantuan levels of data handling and analysis capacity. With these new approaches, will linkage maps become obsolete? One expects that the ongoing studies will also establish a large number of relationships between desirable traits and associated genomic loci. As in the case of the promise of personalized medicine for individuals (Lu et al., 2014), will it be feasible to generate customized genotypes of apples or other perennial crops that can be produced in any environment both efficiently and sustainably? With the prior knowledge of desirable genomic loci and their underlying sequences, and the recent ability to perform targeted mutagenesis and rewrite the DNA language using genome editing (Georges and Ray, 2017), will there be a need to develop new breeding populations?

7 Conclusion

In conclusion, having access to the genome or the underlying algorithm of any plant species presents boundless possibilities. The development of new apple varieties for sustainable production is often considered only from the perspective of the Western world. The technological developments broadly covered in this overview have the potential to impact apple production strategies and the dependent industries worldwide. Advances in genetics can also facilitate local and sustainable production of apples and contribute feeding the burgeoning population.

It is ironic that the genome of the forbidden fruit that is believed to have been the cause of populating the world, will now lead to its sustainable production across the globe to feed those very people.

8 Where to look for further information

The apple genome continues to be harnessed for useful information to address various aspects of the complex and unique biology that this member of Rosaceae represents. This information along with the genome sequences can be accessed under the accession numbers PRJNA28845 and PRJNA249071 from the National Center for Biotechnology

Information (NCBI, 2017). The apple genome and other genomic resources can be visualized and investigated at the Genome Database for Rosaceae (Jung et al., 2014). The doubled haploid apple genome can be accessed via the Genome Portal at Washington State University (Dhingra, 2013). The annual Plant and Animal Genome Conference (PAG, 2017) is expected to continue to provide latest information as the apple genome bears fruits of knowledge that will contribute to sustainable production of apple in the years to come.

9 References

Albrechtsen, A., Nielsen, F. C. and Nielsen, R. 2010. Ascertainment biases in SNP chips affect measures of population divergence. *Molecular Biology and Evolution* 27, 2534–47.

Ambardar, S., Gupta, R., Trakroo, D., Lal, R. and Vakhlu, J. 2016. High Throughput Sequencing: An Overview of Sequencing Chemistry. *Indian Journal of Microbiology* 56, 394–404.

Ambavaram, M. M., Basu, S., Krishnan, A., et al. 2014. Coordinated regulation of photosynthesis in rice increases yield and tolerance to environmental stress. *Nature Communications* 5, 5302.

Arabidopsis_Genome_Initiative, 2000. Analysis of the genome sequence of the flowering plant Arabidopsis thaliana. *Nature* 408, 796–815.

Bianco, L., Cestaro, A., Linsmith, G., et al. 2016. Development and validation of the Axiom®Apple480K SNP genotyping array. *The Plant Journal* 86, 62–74.

Bianco, L., Cestaro, A., Sargent, D. J. et al. 2014. Development and Validation of a 20K Single Nucleotide Polymorphism (SNP) Whole Genome Genotyping Array for Apple (Malus x domestica Borkh). *Plos One* 9. https://doi.org/10.1371/journal.pone.0110377.

Bredel, M. and Jacoby, E. 2004. Chemogenomics: an emerging strategy for rapid target and drug discovery. *Nature Reviews Genetics* 5, 262–75.

Breyne, P., Dreesen, R., Cannoot, B. et al. 2003. Quantitative cDNA-AFLP analysis for genome-wide expression studies. *Molecular Genetics and Genomics* 269, 173–9.

Catchen, J., Hohenlohe, P. A., Bassham, S., Amores, A. and Cresko, W. A. 2013. Stacks: An analysis tool set for population genomics. *Molecular Ecology* 22, 3124–40.

Chagné, D., Crowhurst, R. N., Troggio, M. et al. 2012. Genome-wide SNP detection, validation, and development of an 8K SNP array for apple. *Plos One* 7, e31745. https://doi.org/10.1371/journal.pone.0031745.

Chiriboga, M.-A., Saladie, M., Gine Bordonaba, J., Recasens, I., Garcia-Mas, J. and Larrigaudiere, C. 2013. Effect of cold storage and 1-MCP treatment on ethylene perception, signalling and synthesis: Influence on the development of the evergreen behaviour in 'Conference' pears. *Postharvest Biology and Technology* 86, 212–20.

Clark, A. G., Hubisz, M. J., Bustamante, C. D., Williamson, S. H. and Nielsen, R. 2005. Ascertainment bias in studies of human genome-wide polymorphism. *Genome Research* 15, 1496–502.

Collins, F. S., Morgan, M. and Patrinos, A. 2003. The Human Genome Project: Lessons from Large-Scale Biology. *Science* 300, 286–90.

Dal Cin, V., Barbaro, E., Danesin, M., Murayama, H., Velasco, R. and Ramina, A. 2009. Fruitlet abscission: A cDNA-AFLP approach to study genes differentially expressed during shedding of immature fruits reveals the involvement of a putative auxin hydrogen symporter in apple (Malus domestica L. Borkh). *Gene* 442, 26–36.

Dal Cin, V., Danesin, M., Botton, A., Boschetti, A., Dorigoni, A. and Ramina, A. 2007. Fruit load and elevation affect ethylene biosynthesis and action in apple fruit (Malus domestica L. Borkh) during development, maturation and ripening. *Plant Cell and Environment* 30, 1480–5.

Degenhardt, J., Al-Masri, A. N., Kurkcuoglu, S., Szankowski, I. and Gau, A. E. 2005. Characterization by suppression subtractive hybridization of transcripts that are differentially expressed in leaves of apple scab-resistant and susceptible cultivars of Malus domestica. *Molecular Genetics and Genomics* 273, 326–35.

Deschamps, S., Llaca, V. and May, G. D. 2012. Genotyping-by-Sequencing in Plants. *Biology* 1, 460–83.

Dhingra, A. 2013. Pre-publication Release of Rosaceae Genome Information. In. https://genomics.wsu.edu/research/ Washington State University. (2016).

Dhingra, A. and Hendrickson, C. 2017. Control of ripening and senescence in pre-harvest and post-harvest plants and plant materials by manipulating alternative oxidase activity. https://www.google.com/patents/US9591847

Dhingra, A. and Schaeffer, S. 2014. Use of photosynthetic pigment stabilizing agents to regulate ripening and quality in fruits and vegetables. In: https://www.google.com/patents/US8901039.

Evans, R. C. and Campbell, C. S. (eds), 2002. The origin of the apple subfamily (Maloideae; Rosaceae) is clarified by DNA sequence data from duplicated GBSSI genes. *American Journal of Botany* 89, 1478–84.

Fleischmann, R., Adams, M., White, O. et al. 1995. Whole-genome random sequencing and assembly of Haemophilus influenzae Rd. *Science* 269, 496–512.

Georges, F. and Ray, H. 2017. Genome editing of crops: a renewed opportunity for food security. *GM Crops Food* 1–12.

Giovannoni, J. 2010. Harvesting the apple genome. *Nature Genetics* 42, 822–3.

Goodwin, S., Mcpherson, J. D. and Mccombie, W. R. 2016. Coming of age: ten years of next-generation sequencing technologies. *Nature Reviews Genetics* 17, 333–51.

Griffiths, C. A., Sagar, R., Geng, Y. et al. 2016. Chemical intervention in plant sugar signalling increases yield and resilience. *Nature* 540, 574–8.

Hammesfahr, B., Odronitz, F., Hellkamp, M. and Kollmar, M. 2011. diArk 2.0 provides detailed analyses of the ever increasing eukaryotic genome sequencing data. *BMC Research Notes* 4, 338.

He, J., Zhao, X., Laroche, A., Lu, Z.-X., Liu, H. and Li, Z. 2014. Genotyping-by-sequencing (GBS), an ultimate marker-assisted selection (MAS) tool to accelerate plant breeding. *Frontiers in Plant Science* 5.

Heather, J. M. and Chain, B. 2016. The sequence of sequencers: The history of sequencing DNA. *Genomics* 107, 1–8.

Hewitt, S., Kilian, B., Hari, R., Koepke, T., Sharpe, R., & Dhingra, A. (2017). Evaluation of multiple approaches to identify genome-wide polymorphisms in closely related genotypes of sweet cherry (*Prunus avium* L.). Computational and Structural Biotechnology Journal. doi:http://dx.doi.org/10.1016/j.csbj.2017.03.002

Hummer, K. E. and Janick, J. 2009. Rosaceae: Taxonomy, Economic Importance, Genomics. In: Folta, K. M. and Gardiner, S. E. (eds), *Genetics and Genomics of Rosaceae*. New York, NY: Springer New York, 1–17.

Irish, V. F. and Litt, A. 2005. Flower development and evolution: gene duplication, diversification and redeployment. *Current Opinion in Genetics and Development* 15, 454–60.

Ishii, T. and Araki, M. 2016. Consumer acceptance of food crops developed by genome editing. *Plant Cell Reports* 35, 1507–18.

Jackson, J. E. 2003. *The Biology of Apples and Pears*. Cambridge: Cambridge University Press.

Janick, J., Cummins, J., Brown, S. and Hemmat, M. 1996. Apples. In: Janick, J. and Moore, N. (eds), *Fruit breeding, vol I—tree and tropical fruits*. New York: Wiley, 1–66.

Janssen, B. J., Thodey, K., Schaffer, R. J. et al. 2008. Global gene expression analysis of apple fruit development from the floral bud to ripe fruit. *BMC Plant Biology* 8, 1–16.

Jung, S., Ficklin, S. P., Lee, T. et al. 2014. The Genome Database for Rosaceae (GDR): year 10 update. *Nucleic Acids Research* 42, D1237–44.

Juniper, B. 2007. The Mysterious Year of the Sweet Apple. In. *American Scientist*. Sigma Xi, *The Scientific Research Society*, 44(95).

Ke, R., Mignardi, M., Hauling, T. and Nilsson, M. 2016. Fourth Generation of Next-Generation Sequencing Technologies: Promise and Consequences. *Human Mutation* 37, 1363–7.

Kim, C., Guo, H., Kong, W., Chandnani, R., Shuang, L. S. and Paterson, A. H. 2016. Application of genotyping by sequencing technology to a variety of crop breeding programs. *Plant Science* 242, 14–22.

Kromdijk, J., Glowacka, K., Leonelli, L. et al. 2016. Improving photosynthesis and crop productivity by accelerating recovery from photoprotection. *Science* 354, 857–61.

Lander, E. S., Consortium, I. H. G. S., Linton, L. M. et al. 2001. Initial sequencing and analysis of the human genome. *Nature* 409, 860–921.

Le Deunff, E., Lecourt, J. and Malagoli, P. 2016. Fine-tuning of root elongation by ethylene: a tool to study dynamic structure-function relationships between root architecture and nitrate absorption. *Annals of Botany* 118, 607–20.

Lespinasse, Y., Bouvier, L., Djulbic, M. and Chevreau, E. Haploidy in apple and pear, 1998: International Society for Horticultural Science (ISHS), Leuven, Belgium, pp. 49–54.

Li, X. W., Kui, L., Zhang, J. et al. 2016. Improved hybrid de novo genome assembly of domesticated apple (Malus x domestica). *GigaScience* 5.

Lu, Y. F., Goldstein, D. B., Angrist, M. and Cavalleri, G. 2014. Personalized medicine and human genetic diversity. *Cold Spring Harbor Perspectives in Medicine* 4, a008581.

Marçais, G., Yorke, J. A. and Zimin, A. 2015. QuorUM: An Error Corrector for Illumina Reads. *Plos One* 10, e0130821. https://doi.org/10.1371/journal.pone.0130821.

Margulies, M., Egholm, M., Altman, W. E. et al. 2005. Genome sequencing in microfabricated high-density picolitre reactors. *Nature* 437, 376–80.

Marinier, E., Brown, D. G. and Mcconkey, B. J. 2015. Pollux: platform independent error correction of single and mixed genomes. *BMC Bioinformatics* 16, 10.

Mukherjee, S., Stamatis, D., Bertsch, J. et al. 2017. Genomes OnLine Database (GOLD) v.6: data updates and feature enhancements. *Nucleic Acids Research*. 45 (D1), D446-D456. DOI: https://doi.org/10.1093/nar/gkw992.

Myles, S., Peiffer, J., Brown, P. J. et al. 2009. Association mapping: critical considerations shift from genotyping to experimental design. *The Plant Cell* 21, 2194–202.

NCBI, 2017. Malus domestica. In. https://www.ncbi.nlm.nih.gov/genome/?term=apple, Apple Genome Data.

PAG, 2017. Plant and Animal Genome Conference.

Park, S. T. and Kim, J. 2016. Trends in Next-Generation Sequencing and a New Era for Whole Genome Sequencing. *International Neurourology Journal* 20, S76–83.

Perea, C., De La Hoz, J. F., Cruz, D. F. et al. 2016. Bioinformatic analysis of genotype by sequencing (GBS) data with NGSEP. *BMC Genomics* 17, 498.

Poland, J. A. and Rife, T. W. 2012. Genotyping-by-Sequencing for Plant Breeding and Genetics. *Plant Genome* 5, 92–102.

Salmela, L. and Rivals, E. 2014. LoRDEC: accurate and efficient long read error correction. *Bioinformatics* 30, 3506–14.

Salvi, S., Micheletti, D., Magnago, P. et al. 2014. One-step reconstruction of multi-generation pedigree networks in apple (Malus × domestica Borkh.) and the parentage of Golden Delicious. *Molecular Breeding* 34, 511–24.

Sanger, F., Nicklen, S. and Coulson, A. R. 1977. DNA sequencing with chain-terminating inhibitors. *Proceedings of the National Academy of Sciences of the United States of America* 74, 5463–7.

Schadt, E. E., Turner, S. and Kasarskis, A. 2010. A window into third-generation sequencing. *Human Molecular Genetics* 19, R227–R40.

Schaeffer, S., Harper, A., Raja, R., Jaiswal, P. and Dhingra, A. 2014. Comparative analysis of predicted plastid-targeted proteomes of sequenced higher plant genomes. *Plos One* 9, e112870.

Schaeffer, S., Hendrickson, C., Fox, R. and Dhingra, A. 2016. Identification of Differentially Expressed Genes between 'Honeycrisp' and 'Golden Delicious' Apple Fruit Tissues Reveal Candidates for Crop Improvement. *Horticulturae* 2, 11.

Schaeffer, S. M. and Nakata, P. A. 2015. CRISPR/Cas9-mediated genome editing and gene replacement in plants: Transitioning from lab to field. *Plant Science* 240, 130–42.

Schaffer, R. J., Friel, E. N., Souleyre, E. J. et al. 2007. A genomics approach reveals that aroma production in apple is controlled by ethylene predominantly at the final step in each biosynthetic pathway. *Plant Physiology* 144, 1899–912.

Seifertova, E., Zimmerman, L. B., Gilchrist, M. J. et al. 2013. Efficient high-throughput sequencing of a laser microdissected chromosome arm. *BMC Genomics* 14, 357.

Sims, D., Sudbery, I., Ilott, N. E., Heger, A. and Ponting, C. P. 2014. Sequencing depth and coverage: key considerations in genomic analyses. *Nature Reviews Genetics* 15, 121–32.

Smith, L. M., Sanders, J. Z., Kaiser, R. J. et al. 1986. Fluorescence detection in automated DNA sequence analysis. *Nature* 321, 674–9.

Stokes, M. E. and Mccourt, P. 2014. Towards personalized agriculture: what chemical genomics can bring to plant biotechnology. *Frontiers in Plant Science* 5, 344. https://doi.org/10.3389/fpls.2014.00344.

Sugar, D. and Basile, S. R. 2013. Integrated ethylene and temperature conditioning for induction of ripening capacity in 'Anjou' and 'Comice' pears. *Postharvest Biology and Technology* 83, 9–16.

Troggio, M., Gleave, A., Salvi, S. et al. 2012. Apple, from genome to breeding. *Tree Genetics and Genomes* 8, 509–29.

Van Nocker, S. and Gardiner, S. E. 2014. Breeding better cultivars, faster: applications of new technologies for the rapid deployment of superior horticultural tree crops. *Horticulture Research* 1, 14022.

Velasco, R., Zharkikh, A., Affourtit, J. et al. 2010. The genome of the domesticated apple (Malus x domestica Borkh.). *Nature Genetics* 42, 833–9.

Watkins, C. B. 2006. The use of 1-methylcyclopropene (1-MCP) on fruits and vegetables. *Biotechnology Advances* 24, 389–409.

Weeks, D. P., Spalding, M. H. and Yang, B. 2016. Use of designer nucleases for targeted gene and genome editing in plants. *Plant Biotechnology Journal* 14, 483–95.

Zharkikh, A., Troggio, M., Pruss, D. et al. 2008. Sequencing and assembly of highly heterozygous genome of *Vitis vinifera* L. cv Pinot Noir: problems and solutions. *Journal of Biotechnology* 136, 38–43.

Zude-Sasse, M., Fountas, S., Gemtos, T. A. and Abu-Khalaf, N. 2016. Applications of precision agriculture in horticultural crops. *European Journal of Horticultural Science* 81, 78–90.

Chapter 3

Advances in understanding apple tree growth: rootstocks and planting systems

Dugald C. Close and Sally A. Bound, University of Tasmania, Australia

1 Introduction

The apple is an important fruit in many cultures; indeed in Australia the phrase of British origin 'an apple a day keeps the doctor away' is often stated in reference to the health benefits of eating apples regularly. The cultivated apple probably originated from natives of northern Asia (Siberian, *Pyrus prunifolia;* Chinese, *Pyrus spectabilis;* and cherry-crab, *Pyrus baccata* (Anon. 1898)) and is referred to in historical texts dating back as far as AD 973 (Hogg 1851). Modern large-scale production occurs predominantly in China, Europe, United States, New Zealand and South Africa. Recent research and development has been conducted with the objective to provide the market with consistent quality of consumer-desired fruit, that is, firm (Wills et al. 1980), juicy (Feliciano et al. 2010), high-coloured fruit (Kootstra et al. 1994) with sugar–acid balance typical of the cultivar (Asrey et al. 2008). Recent foci for advancing our understanding of apple tree growth towards the objective of consistently providing the market with consumer-desired fruit include optimizing yield produced per unit orchard area and optimizing quality or 'marketable yield'. Breeding programmes have released many new cultivars that have improved and/ or distinct characteristics desired by the consumer, in addition to semi-dwarfing and

http://dx.doi.org/10.19103/AS.2016.0017.03

dwarfing rootstocks. Understanding how rootstocks and scions interact to manifest in various growth attributes of size, architecture and precocity (earliness of flowering and bearing) has been a major focus. Understanding of growth attributes enables managers to design growing systems that allow the tree to fill the allocated canopy space to optimize light capture early in the life of the orchard (and prevent excessive shading after establishment) in order to produce early yields to pay for investments. How tree growth attributes are manipulated through plant management and the use of plant bioregulators (PBRs) to manipulate developmental processes to induce branching and manage vigour is an important aspect for optimizing tree growth. Key manipulations of tree training and pruning to arrive at a desired tree structure that optimizes within canopy light distribution has been an area of significant R&D investment to provide desired fruit quality, particularly size, firmness, colour and sugar–acid levels. Another important area of study has been the manipulation of bud break through either cultural application or application of PBRs to overcome inadequate winter chilling. Finally, in order to inform how best to optimize total light interception and therefore yield, planting systems research that studies interactions of within and between row spacing of a given rootstock/scion combination trained to various tree structure configurations has been investigated.

This chapter begins with a review of the critically important aspects of tree growth and development. It then discusses rootstocks, focusing on the mechanisms and morphological effects of dwarfing. It provides a brief introduction to the manipulation of tree growth and development, with an emphasis on the use of plant growth regulators. However, more detailed discussion of the methods of managing tree growth and development is reserved for Chapter 3b. The chapter ends with a discussion of planting systems.

2 Tree growth and the development and propagation of apple

The tree structure, comprising roots and primary stem, or trunk, with lateral branches, arises from cell division within apical meristems. Apical meristems are small cell clusters that ensure self-perpetuation and that give rise to plant organs (see review by Steeves and Sussex 1989). Apical meristems originate from the radicle (embryonic root) and epicotyl (shoot) that emerge from the seed (Bewley and Black 1978). Cells arising from meristems subsequently differentiate into specialized ones to form tissues of distinct function. The fundamental unit, the phytomer, comprises an internode and a node with one or more leaves that contain lateral vegetative and/or floral meristems or buds. Phytomers are repeated, termed reiteration, to form growth units (Hollender and Dardick 2015). Architectural development arises through rhythmic growth and monopodial branching until the apical meristem differentiates into an inflorescence (Crabbe 1984).

Apple trees flower autonomously, with floral initiation dependent on vegetative development of the previous season (Wilkie et al. 2008). Attainment of a critical node number is a requirement before the vegetative bud transitions to floral (Fulford 1965). Floral initiation occurs with changes in the shoot apical meristem, after coinciding with the cessation of shoot growth (Hirst and Feree 1995). It was reported that buds of diameter >3.1 mm were generally floral, but this only correctly classified 65% of buds (Hirst and Feree 1995). Floral differentiation occurs in resting buds in terminal or axillary positions on one-year-old wood (long shoots and short shoots, called 'spurs'); on two-year-old wood; and on older wood. The spur growth unit is called a 'bourse' and includes an axillary vegetative unit that develops without any rest period. Wilkie et al. (2008) provides a well-considered review of the regulation of floral induction. Whilst floral buds initiate growth in spring, vegetative buds can initiate growth or remain dormant depending on environmental conditions. Vegetative growth can be sylleptic, within the same season as bud formation, or proleptic, where the bud undergoes a period of dormancy.

Apical dominance is the inhibition of lateral bud outgrowth by the shoot and branch apical meristems. Subordinate buds can be released by branch bending, where the apical shoot is reorientated below subordinate buds, or by heading-back of the shoot. Research since Thimann and Skoog (1933) has shown that apical dominance is affected, at least in part, by an auxin concentration gradient arising from the apical bud that represses cytokinin synthesis at the nodes and its transport to dormant buds. Apical dominance affects branch angle, branch length and diameter (Wilson 2000). The hypotheses and known mechanisms relating to these influences are reviewed by Hollender and Dardick (2015). In combination, these give rise to the five types of apple trees as described by 'Lespinasse's typology' (Lespinasse and Delort 1986). These are further described in 'Training and pruning' section of 'Advances in understanding apple tree growth: the manipulation of tree growth and development'.

Apples are propagated vegetatively to ensure fruit is 'true to type' as this is not the case when propagated from seed. Propagation by cuttings is generally not reliable or successful so that grafting to form a composite tree is the commercial approach. This is achieved by harvesting 'scion' wood from the mother trees. This is grafted onto a rootstock that confers particular preferred traits, for example dwarfing, water logging, pest and/or disease resistance in addition to shortening the juvenile (vegetative only) growing phase of the tree. See Wertheim and Webster (2003) for a review of apple propagation.

3 Dwarfing rootstocks: history of selection and key morphological effects

3.1 A brief history of selection programmes and releases

The East Malling Research Station was established in 1913 in Kent, England and released the Malling (M; M1-M16 between 1912 and 1914, M17-M24 in 1924, M25 in 1952, M26 in 1959, M27 in 1975 and M116 in 2001) and Malling/Merton (MM; with the John Innes Institute in the 1950s) dwarfing and semi-dwarfing apple rootstock series (http://www.emr.ac.uk/projects/rootstock-research-east-malling-history/; accessed 23 December 2015). It should be noted that M9 is a rootstock selected from a chance seedling in France in the 1800s (Carrière 1897, cited in Foster et al. 2014) that was renamed from the previous 'Jaune de Metz'. Various clones of M9 selected and currently used include M9 NAKB 337, a virus-free release from the Netherlands, M9 EMLA, a virus-free release from East Malling/Long Ashton, Pajam 1 and 2 from France (http://extension.psu.edu/plants/tree-fruit/commercial-tree-fruit-production/cultivars-and-rootstocks/apple-rootstocks; accessed 3 February 2016) and the Belgian Nic29 (http://www.anfic.com.au/shop/item/nic-29; accessed 11 April 2016). The Geneva (G) apple rootstock breeding program was established in 1968 with the objective of developing dwarfing rootstocks tolerant to fire blight (*Erwinia amylovora*), crown rot (*Phytophthora* spp.) and woolly apple aphid (http://www.ars.usda.gov/SP2UserFiles/Place/80600500/Gennaro/

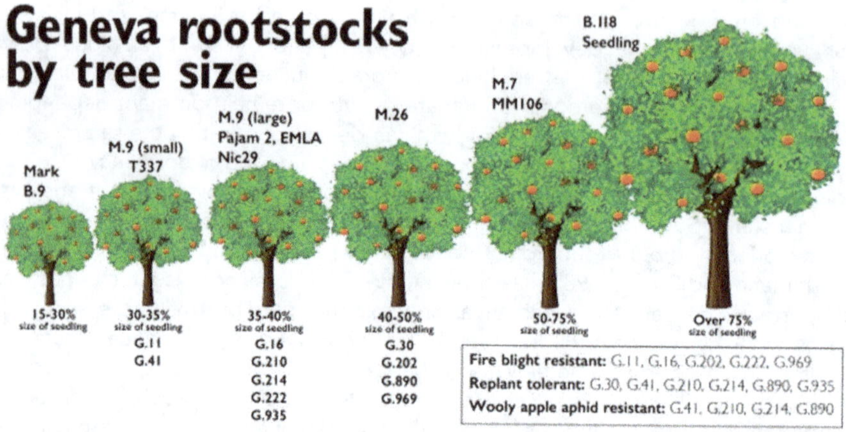

Figure 1 Relative dwarfism of the predominant M, MM and G rootstocks. The Center for Technology Licensing at Cornell University is acknowledged for provision of the GENEVA® Apple Rootstocks Comparison Chart.

GeneraldescriptionofTHEGENEVABreedingprogram.pdf; accessed 22 December 2015). The G16 and G41 (both similar in vigour to that of M9), G202 (similar to M.26) and G935 (slightly larger than M.26) rootstocks are key releases from the breeding programme. Budagovsky 9 was released from the Soviet Union, and appears to be resistant to collar rot and fire blight and is very cold-hardy. Relative dwarfism of the predominant M, MM and G rootstocks is depicted in Fig. 1. More recent foci for rootstock breeding programmes have been tolerance to environmental stress, particularly to drought (Webster 2001) and fire blight, phytophthora, apple replant disorder, woolly apple aphid, winter hardiness and waterlogging (Robinson 2011). As there are too many other releases to mention, this summary is aimed to cover the contemporary and commonly used rootstocks only (see Webster 2001 for further reading).

3.2 Morphological effects of dwarfing rootstocks and interstems

Two key traits induced in the apple scion by dwarfing rootstocks are: 1. precocity (flowering and fruiting early in the life of the tree) and increased flowering intensity and 2. reduction in mature tree dry mass and volume. The second trait is underpinned by the fact that a particular shoot:root ratio remains consistent for a given cultivar at maturity, independent of the vigour of the rootstock. This is speculated to be based on integration of signalling of auxin from shoots translocated to roots and effects on root growth and cytokinin synthesis that in turn are transported to shoots which subsequently affect shoot growth (Lockard and Schneider 1981). The two key traits of precocity and reduction in mature tree dry mass and volume induced by dwarfing rootstocks enable high-density plantings that can increase fruit yield per unit land area in the early years after establishment (expanded upon in later sections). It has been shown that rootstocks *per se* do not alter the overall growth habit (typology as discussed in a later section of this chapter) of the tree canopy architecture. Nor does rootstock directly influence apple fruit size, as reported by Al-Hinai and Roper (2004) comparing M26, Ottawa 3, M9 Pajam 1 and Vineland-605; though all other factors are equal, it is believed that G11 produces larger fruit than M9 does (N Dallabetta and G Fazio pers. comm.). Rather, fruit size is a response of key factors such as light interception, training system, climate, crop load and floral bud type. According to Schechter et al. (1990), rootstock also does not affect the timing of bud-break, full bloom or full canopy establishment in apples.

Is dwarfing mediated by the scion or the rootstock?

A major difficulty in understanding rootstock effects arises from successive cycles of shoot growth, flowering and fruiting that lead to cumulative and compound effects on tree structure and function (Webster 1995). In a six-year field study of rootstock effects on apple scions of different growth habits, Tworkoski and Miller (2007) showed that scion had more influence than rootstock on monthly growth rate, although M9 and seedling rootstock had the lowest and greatest growth in terms of tree height and diameter, respectively. However, growth measured as tree height or node number is not a true measure of whole tree dwarfing in contrast to measurement of total tree dry mass. Vyvyan (1955) showed that total tree dry mass (root and scion) was smaller for M9 scions budded onto M12 rootstocks when compared with M12 scions budded onto M12 rootstocks. Combined with unpublished results of van Hooijdonk et al., these findings indicate that the scion is a more powerful regulator of whole tree vigour than the root system – the physiology of this is discussed in the next section.

Rootstock effects on precocity

Dwarfing rootstocks increase the probability in young trees of axillary flowering, of terminal flowering and of return bloom compared to ungrafted own rooted trees of the same age derived from tissue culture (studied in the absence of pruning; Costes and García-Villanueva (2007)). It should be noted that micro-propagation via tissue culture is a confounding factor of the study, given that the practice can increase juvenility and delay flowering (Webster et al. 1985). Seleznyova et al. (2008) demonstrated that treatments on M9 rootstock developed on average 0.52 and 0.17 extension growth units from vegetative and floral shoots, respectively, compared with 0.77 extension growth units of vegetative shoots in MM106 rootstocks. They concluded that a major effect of dwarfing rootstocks is reduced vigour of annual shoots that is amplified by the relative early transition to flowering compared with semi-dwarfing rootstocks.

Rootstock effects on growth and tree volume

Dwarfing rootstocks impose earlier cessation of annual shoot growth that gives rise to shorter shoots due mostly to a reduced number of neoformed nodes (i.e. nodes not preformed within the bud). So, shorter internodes and reduction in the number of nodes on vegetative shoots effect dwarfing rather than the popular belief of reduced mean internode length (Seleznyova et al. 2003). For instance Seleznyova et al. (2003) found that the mean internode length for populations of shoots with a given number of nodes was the same for 'Royal Gala' scions grafted to either M9 or MM106 rootstocks. Whilst mean internode length was lower on average for treatments on M9 rootstock, this was related to differences in node number distribution. Costes and García-Villanueva (2007) demonstrated that dwarfing rootstocks reduced the number of proleptic shoots developed and that proportionally there were more medium axillaries and spurs than long shoots. These occurred after the induction of flowering in the second year of growth at axillary and terminal positions. Dwarfing rootstocks increase the probability of sympodial shoot development during early tree development that consequently reduces tree vigour and limits development of structure (Seleznyova et al. 2008). Monopodial axillary annual shoots are replaced with less vigorous sympodial axillary annual shoots, and monopodial annual shoots are of reduced vigour.

Van Hooijdonk et al. (2011) reported that, whilst dwarfing was expressed in the first year of development, at the end of the first growing season, primary shoot length and node number were the same for 'Royal Gala'/M9 as for a grafted self-rooted 'Royal Gala' rootstock control. Although no difference in primary shoot node number was reported by Seleznyova et al. (2008), other studies found that primary shoot growth was less on M9 by the end of the first growing season compared with more vigorous rootstocks (Costes et al. 2001; van Hooijdonk et al. 2010), as was total tree dry mass (the most rigorous measure of whole tree growth; van Hooijdonk et al. 2010, 2011). For M9 in some growing environments, fewer sylleptic shoots formed on primary shoots in summer and there was less extension of secondary shoots in autumn (see Fig. 2 and 3). Critically these architectural characteristics arising from dwarfing rootstocks provide favourable light relations in the lower, inner canopy of conically trained trees, relative to semi-dwarfing or non-dwarfing rootstocks (Tustin et al. 2001), in addition to perhaps allowing more time for starch accumulation and floral bud development (Swarbrick 1929), though this has not specifically been addressed in the contemporary literature.

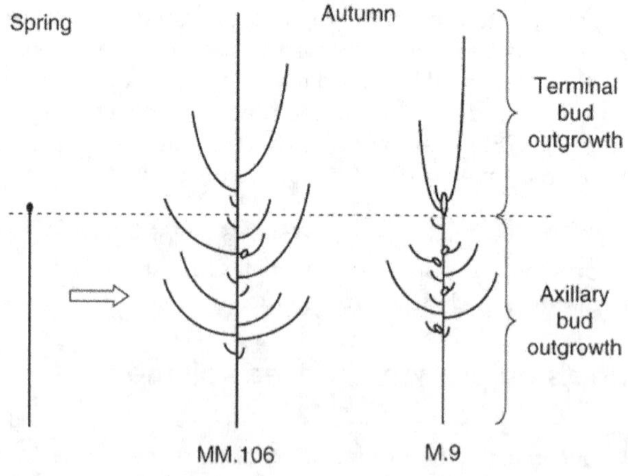

Figure 2 Schematic representation of rootstock effects on structural development of young 'Royal Gala' apple trees. Second-year outgrowths from axillary and terminal buds are separated by a dashed line. Ovals represent bourses; and lines represent vegetative growth units. Boundaries between consecutive growth units are not shown. The drawing does not show the exact measured values but rather represents the conclusion that the trees on dwarfing rootstocks developed fewer extension axillary annual shoots, very few of which had two extension growth units, while the outgrowth from the terminal buds was not so much affected (from Seleznyova et al. 2008). Oxford University Press is acknowledged for provision of this figure.

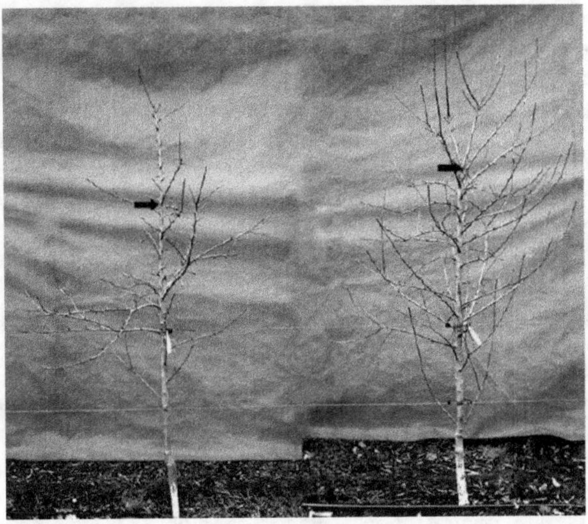

Figure 3 Examples of 'Royal Gala' apple trees from treatments with M.9 (left) and MM.106 (right) rootstocks and no interstocks at the end of the third year of growth. The photographs of the two trees were taken from the same distance. The arrows show limits of the first year of growth. The photographs clearly show the higher number of long branches developed from the first-year section of the trunk for treatment with MM.106 rootstock (right). Hence the overall shape of the tree and the branch density within the canopy envelope were affected by the rootstock treatments (from Seleznyova et al. 2008). Oxford University Press is acknowledged for provision of this figure.

4 Dwarfing rootstocks: mechanisms and example application

4.1 Mechanisms of precocity and dwarfing

It was postulated by Seleznyova et al. (2008) that dwarfing in the second spring after grafting of the compound tree is partly due to flowering (and sink competition for carbohydrate and other resources) of scion on M9. Foster et al. (2014) found higher expression of the FT/TFL1 (flowering) gene family in M9 and M27 relative to M793. These genes promote flowering, accelerate the transition from juvenile to mature and regulate cycles of seasonal growth and termination. Celton et al. (2009) first reported DW1 that was required for dwarfing to be expressed, but found that additional genes were involved in the expression of dwarfing. Fazio et al. (2014) reported that the dwarfing effect of M9 is related to two main quantitative trait loci (QTL) that were necessary for the dwarfing phenotype to be expressed. Foster et al. (2015) further discovered that the two QTLs (Dw1 and Dw2) were carried by over 41 rootstock accessions that confer a range of effects on scion growth and Harrison et al. (2016) have recently found a third QTL associated with dwarfing that may be considered *Dw3*. Thus the mechanism of dwarfing could relate to independent genetic factors. Tustin and van Hooijdonk (2013) emphasize that, whilst traits for dwarfing and precocity can co-occur, they may be unrelated as they can be separately expressed in pear. Vattiprolu et al. (2012) found that 'Royal Gala' scion grafted onto M9 rootstock had fewer flowers, relative to winter bud number, than 'Royal Gala' scion grafted to 'Royal Gala' rootstock, emphasizing that dwarfing and precocity can co-occur but not necessarily be linked. This unexpected result also emphasizes the need for the use of appropriate controls in rootstock–scion research so that incorrect conclusions are not arrived at.

Atkinson et al. (2000) found that the reduced leaf population of dwarfing rootstocks had reduced stomatal conductance relative to that on vigorous rootstocks. However it may be that these observations were not related to rootstock treatments as they were confounded, given that tree leaf canopy size and root morphology were not estimated and accounted for in the one-year-old potted plant material used for the study. It was observed that the amount of functional xylem tissue in the graft and scion increased with rootstock vigour (Atkinson et al. 2003). This led Atkinson et al. (2003) to speculate that water flow restriction and reduced basipetal transport of indole-3-acetic acid (IAA) across the graft is consistent with different degrees of dwarfism. Though observed differences in functional xylem tissue were initially thought to be related to restricted hydraulic conductivity, differences were not apparent when root mass was accounted for (Atkinson et al. 2003). Additionally hydraulic limitations could be compensated for by increases in the diameter of the graft union observed to develop on dwarfing rootstocks, potentially due to auxin accumulation (Atkinson et al. 2003). In contrast, Lockard and Schneider (1981) demonstrated that bark of M26 grafted into 'Gravenstein' scion on M111 rootstock induced dwarfing. Phloem appears key to the dwarfing mechanism of dwarfing rootstocks.

Perhaps the most definitive evidence to date is the reports of van Hooijdonk et al. (2010; 2011). Van Hooijdonk et al. (2010) found that 1-N-naphthylphthalamic acid (NPA), an inhibitor of auxin transport when applied to the graft union of relatively vigorous rootstocks, had similar results on the 'Gala' scion, in terms of dwarfing response, to that of 'Gala' on M9 (see Fig. 4). That is, both had a higher proportion of primary and secondary shoots that terminated early and a low number of secondary shoots that formed. Van Hooijdonk et al. (2010; 2011) hypothesized that limited transport of IAA from scion to roots may differentially affect the synthesis of root-derived cytokinins (Kamboj et al. 1999). Van Hooijdonk et al. (2011) found that relatively low extension growth in autumn of scions on M9 coincided with less gibberellin 19 (GA19) in xylem sap that may have resulted in less shoot extension growth. Consistent with the earlier hypothesis of Lockard and Schneider (1981), van Hooijdonk et al. (2010) showed that dwarfing rootstocks appear to decrease the basipetal transport of IAA and therefore reduce the amount of some root-produced cytokinin and GA forms transported to the scion. Decreased cytokinin was speculated to limit sylleptic branching and decreased GA was thought to reduce the duration of shoot extension growth. This was refined by van Hooijdonk et al. (2011) to add that M9 rootstocks restrict biosynthesis and supply of cytokinin and GA to the scion.

Some elegant, although yet to be published, work by van Hooijdonk et al. (2011)used reciprocal grafting of rootstocks, including M9 and M793, in order to investigate influences of rootstock and scion on dwarfing outcomes. As expected, M9 scion on M9 rootstock was very dwarfed (as estimated by tree height one growing season post grafting) relative to M793 scion grafted to M793 rootstock. However M793 scion height growth was almost identical whether grafted to M9 rootstock or M793 rootstock. In contrast the hydraulic limitation hypothesis would predict greater growth of M793 scion on M793 rootstock relative to that on M9 rootstock. Additionally M9 scion growth was similar, whether grafted to M793 or to M9 rootstock, and root dry mass of M9 rootstock increased when grafted with a vigorous scion! Clearly this suggests that the mechanism of dwarfing is via the M9

Figure 4 Architecture of 'Royal Gala' apple scions at the end of their first growing season after grafting onto M.9, MM.106, M.793 and 'Royal Gala' (left to right, respectively) rootstocks, and the same combinations after application of the auxin transport inhibitor 1-N-naphthylphthalamic acid (NPA) applied to the rootstock stem (Dr Ben van Hooijdonk is acknowledged for provision of photos that formed part of his PhD thesis 'The physiological basis of vigour control by apple rootstocks – an unresolved paradigm' Massey University, 2009).

shoot and not via M9 roots per se, at least during the first season of growth post grafting, and this is consistent with the earlier work of Vyvyan (1955).

Tworkoski and Fazio (2015) have recently hypothesized a possible role of specific conductivity of hormones, nutrients, metabolites and/or water in relation to dwarfing. However, if conductivity of nutrients was part of the dwarfing mechanism in apple, we might expect deficiencies relative to vigorous rootstocks which are not typically observed. Recent work has shown that the ability to accumulate key plant nutrients was not well correlated with dwarfing or initial tree performance, with the exception of a close association between leaf phosphorus and initial tree vigour (Neilsen and Havipson 2014), whilst Tagliavini et al. (1992) concluded that decreased concentrations of N, Ca and Mg in relatively vigorous rootstocks were due to dilution across larger leaf areas. What is consistently observed is early shoot termination of scions on M9 rootstock (van Hooijdonk et al. 2010; Foster et al. 2014). Given the current lack of certainty in the literature and the critical importance of dwarfing rootstocks for apple production systems, going forward the mechanisms that underpin rootstock-induced dwarfing and precocity of apple will no doubt involve active research in hormone signalling and genetic by environmental influences.

4.2 Rootstocks for tolerance to soil water limitation

There has been recent focus on the use of seedling rootstocks, used in conjunction with M9 or M27 interstocks (Di Vaio et al. 2009), for use in production systems on dry and/or windy sites. This is because of the relatively poor anchoring and shallow rooting of some dwarfing rootstocks, rendering trees prone to nutrient and water stress if not properly managed. As expected, scion growth was reduced with dwarfing rootstocks by 80% (M27) and 50% (M9), and increased interstock length (20 cm compared to 10 cm) further decreased vigour. Bauerle et al. (2011) showed greater tolerance to limited soil water availability of a high vigour rootstock underpinned by greater growth, 28% lower current season xylem ring width and fewer embolisms. This was suggested to be simply due to greater biomass investment in roots, relative to dwarfing rootstocks, that enabled access to water via a greater soil volume. Unfortunately Bauerle et al. (2011) did not assess pre-dawn stem water potentials which could have confirmed less severe plant water deficit in the case of scions grafted to vigorous, relative to dwarfing, rootstocks. In an alternative approach to tolerating water availability limitation, Li et al. (2015) reported that planting dwarfing rootstocks at 15 cm, relative to 5 or 10 cm, depth resulted in increased fine root biomass in soil layers to 40 cm which had benefits associated with higher fruit yield and improved quality; they concluded that fine root growth was promoted at a planting depth of 15 cm. In relation to direct comparisons of rootstocks of distinct vigour pot trials, Tworkoski et al. (2016) have indicated that, whilst M9 produced higher levels of abascisic acid (ABA) that may provide superior stomatal control and confer short-term drought resistance, MM111 rootstocks may be more tolerant of drought over longer terms due to the development of a more extensive root system. Clearly, as water for irrigation becomes more scarce, increased research effort will be required to enable optimized water use efficiency in orchard production. This topic is covered in subsequent chapters.

5 An introduction to manipulating tree growth and development

In intensive apple production, a key objective is to maximize yield in the early years without impacting on fruiting potential when the tree is mature. In order to achieve this goal, the framework of the tree needs to be established early, and once the tree structure is developed, a balance should be maintained between growth and fruiting; trees also need to be contained within their allocated space to prevent crowding and excessive shading. Planting density influences pruning requirements; if trees are too close, then more containment pruning is required, and this may invigorate the tree, which in turn results in greater shading. Site characteristics need to be matched with scion and rootstock when determining planting density as these can all influence tree vigour.

Management of tree growth and structure can be achieved through cultural methods such as pruning, root restriction, girdling and training techniques such as limb spreading or tying down or limiting water or nutrients. However plant bioregulators (PBRs) can be used effectively to manipulate tree growth which may be less labour intensive than cultural methods. Bioregulators act by promoting, inhibiting or otherwise affecting the biological and biochemical processes in plants (Miller 1988), and can assist orchardists to maximize the genetic potential of their trees. However it is important to note that a plant cannot be

Table 1 Common bioregulators used in apple tree canopy development, rest-breaking and vigour control

Common name	Active ingredient	Brand name/s	Activity
Armobreak	Fatty amine polymer	Akso	Dormancy release
BA	6-Benzyladenine	CyLex MaxCel BAPsol	Lateral branching; Post-bloom thinning
Chlormequate chloride (CCC)	(2-Chloroethyl) trimethyl ammonium chloride	Cycocel	Gibberellin biosynthesis inhibitor
Cyclanilide	1-((2,4-Dichlorophenyl) carbamoyl) cyclopropanecarboxylic acid	Tiberon	Branching agent
Daminozide	2,2-Dimethylhydrazide	Alar	Suppresses shoot growth; Controls preharvest drop; Enhances apple colour; Promotes flower initiation
DNBP	Dinitro-butyl-phenol		Dormancy release
DNOC	Dinitro-ortho-cresol		Dormancy release; Flower thinner
Ethephon	(2-Chloroethyl) phosphonic acid	Ethrel	Growth control; Flower/fruit thinner; Flower initiation
Gibberellins (GAs)	GA_3, GA_4, GA_7		Promotes shoot elongation; Russet reduction; Reduces return bloom
Hydrogen cyanamide		Dormex	Dormancy release
Paclobutrazol; PP333	(2RS,3RS)-1-(4-Chlorophenyl)-4,4-dimethyl-2-(1,2,4-triazol- 1-yl)-pentan-3-ol	Cultar® Payback™	Growth inhibitor
Prohexadione-calcium	3-Oxido-4-propionyl-5-oxo-3-cyclohexene-carboxylate	Regalis® Apogee® Kudos	Growth inhibitor
Promalin	6-Benzyladenine plus GA_{4+7},	Promalin Cytolin	Promotes lateral branching; Improves fruit shape
Thidiazuron (TDZ)	N-Phenyl-N'-1,2,3-thidiazol-5-ylurea		Increases bud break; Reduces blind wood; Promotes shoot growth
Waiken	Methyl esters of fatty acids	Waiken®	Enhances dormancy release

forced to do anything that it is genetically incapable of doing. In addition to being useful tools for managing tree growth and structure, PBRs can also be valuable aids in releasing buds from dormancy, regulating crop load, improving fruit quality (shape, size, colour, skin finish), preventing pre-harvest fruit drop or altering fruit maturity.

Bioregulators can be plant hormones or synthetic analogues. The common categories of bioregulators include auxins, ABA, cytokinins, GAs and ethylene; however, other more recently discovered compounds that can regulate plant growth and development include brassinosteroids, jasmonates, oligosaccharins, polyamines, strigalactones and salicylates. Over the last few decades there have been numerous chemicals evaluated for their impact on apple tree growth and development, but because of inconsistency of results in their evaluation only few reach the commercial registration stage. The action and efficacy of a PBR is dependent on a range of factors, including product concentration, time of application and cultivar; environmental conditions can also influence the physiological response of the plant. Miller (1988) discusses the factors that influence PBR response in depth and Bukovac (1997) describes the performance of PBRs in relation to the leaf surface. Some common bioregulators used in apple production are shown in Table 1.

A detailed discussion of the methods of manipulating tree growth and development, including the application of PBRs and selected cultural practices for canopy/tree structure development, management of dormancy release and bud break, and vegetative growth control, is provided in chapter 'Advances in understanding apple tree growth: the manipulation of tree growth and development'.

6 Planting systems

If there is one key principle or 'golden rule' to optimizing marketable yield, and therefore profit, it is that that both light interception and within-canopy light distribution must be optimized for conversion to photosynthate and production of economic yield at the per unit orchard area level because light falling on alleyways does not contribute to fruit production (Jackson 1980)! Seminal work by Palmer (1989) has clearly demonstrated this across cultivars, regions and countries. Whilst this principle is straightforward, achieving it in practice is not, particularly given the practical limitations of access requirements of current tractors and other machinery. The orchard production system can be defined as the combination of cultivar, rootstock, tree spacing and training/pruning regime (Robinson et al. 1991). The integration of these with other variables such as tree quality, support system, tree arrangement, soil and climate and management ability has been termed the 'orchard systems puzzle' by Barritt (1992 cited in Hoying and Robinson 2000). Whilst many advances have been made in apple production, the general upper limit of 60% light interception of mature orchards noted in 1980 by Jackson has not increased today.

Generally north–south tree row orientation is ideal as this maximizes light interception and minimizes seasonal effects on interception so that interception is primarily determined by changes in canopy size (Jackson 1980; Palmer 1989). Thus the relationship between tree height and inter row width is critical for optimizing light interception, with early modelling of 'solid conical' trees in rows indicating in theory that tree height:row spacing of 1:1 could intercept over 80% of available light (Jackson and Palmer 1972). The global trend of increased cost of labour, coupled with a lack of automation, led to systems using small trees that could be accessed without ladders or 'pedestrian' orchards. Decreased tree height has not always been matched with decreased alley width, which has sometimes resulted in reduced light interception. However, disadvantages associated with trees pruned for pedestrian orchards, including invigoration and less efficient harvest index associated with 'heading' (described in Chapter 3b), have driven the development of modern tall systems. Labour efficiency is now gained in such systems via using platforms to elevate people to fruit height as opposed to vice versa! Ratios of tree height–row spacing of 0.8–0.9:1 optimize light interception (Robinson 2011). Ratios of 1.0–1.1:1.0 may be more appropriate for open, porous, contemporary tall spindle and fruiting wall systems (Robinson pers. comm. 2014).

Tree density dictates light interception during orchard establishment; for example, tree densities below 500 trees/ha can take 7–12 years before developing a full canopy (Robinson 2011). Ideal density and optimizing the early filling of the space allocated to a tree depends on vigour that is in turn dependent on the scion/rootstock combination and how this responds to the climate and soils of the site. Thus 'there is no (one) correct planting system for every grower' (Hoying and Robinson 2000). Whilst the early precocity conferred by dwarfing rootstock systems was as important for generating yield as light interception in the early years of establishment, relatively low light interception of some systems using dwarfing rootstocks limited production in subsequent years where density was not appropriately matched to scion/rootstock vigour. For example, Tustin et al. (2001) reported that yields from year 4 onwards were lower on 'Mark' dwarf rootstock at 4 x 2 m spacing relative to that on MM106 at 5 x 3 m and M26 at 4 x 2 m due to the limited canopy growth. Tustin et al. (2001) also noted that quality in terms of adequate colour development in the 'Fuji' investigated was not as high in the MM106 system due to the shading influence of vegetative growth. This is consistent with the contemporary work investigating architectural effects of dwarfing rootstocks, outlined in Section 3 of this chapter.

Tree planting density has steadily increased since the 2000s to densities around 3000 trees/ha as the best benefit–cost of rapidly filling the canopy and delivering high, early yields without overinvesting in trees to provide return on investment for growers. Wagenmakers (1991) used model simulations to demonstrate that a theoretical optimum for light interception is 3584 trees/ha. Orchard tree densities began to increase initially based on Vertical Axis structure, and later, V or Y Trellis and spindle tree structures on dwarfing rootstocks. This had the dual advantage of improved early and mature light interceptions, given that tree density and cumulative yield is linear up to around year 5, following which it becomes curvilinear (Robinson et al. 2007). However, in the ultra-high density range (4000+ trees/ha), tree cost becomes a pragmatic limiting factor (case study this chapter). Economic and risk analyses indicate that densities of 2500–3300 trees/ha are optimal in NY state (Robinson et al. 2007). Thus whilst upper limits of light interception by mature orchards has not significantly changed since 1980, the rapidity with which this is achieved has improved dramatically with the use of modern scion/rootstock combinations at densities between 2500 and 3300 trees/ha.

The principles of matching cultivar to rootstock vigour to site climate conditions for maximizing early canopy development and high early yields and the trade-off with tree numbers are well demonstrated by a study of intensive growing systems undertaken at Grove Research Station in the apple-growing district of the Huon Valley in south-east Tasmania, Australia. This will be discussed in chapter 'Advances in understanding apple tree growth: the manipulation of tree growth and development', after the detailed discussion of tree growth management.

7 Summary

This chapter outlined the key attributes of widely used dwarfing and semi-dwarfing rootstocks, including relatively recent releases that confer resistance to economically important pests and diseases and tolerance to abiotic conditions of drought, low temperature and waterlogging. These traits significantly widen the potential sites and regions that can potentially support modern apple systems. The key trait of dwarfing rootstocks (restricted tree size) can similarly be effected using interstocks. Dwarfing is coupled with general architectural effects such as a reduced number of axillary shoots that are floral more often or, if vegetative, of lower vigour and proportionately more medium axillaries and spurs than long shoots. Critically these traits provide favourable light relations for the inner and lower canopy regions to support fruitfulness and fruit quality. Restricted tree size is thought to be due to restricted hydraulic flow of water and/or altered hormone transport. The former thinking is based on work with significant methodological issues. The latter, more methodologically rigorous work, indicates that restricted tree size is likely to involve decreased rates of basipetal IAA transport late season, which in turn reduces some forms of root-produced cytokinin and gibberellic acid that are transported to the scion. The other key trait of precocity is driven by activity of flowering genes beginning in the first year after grafting onwards.

The management of tree growth and development is a vital factor in maximizing apple tree yield, and plant bioregulators are valuable tools for manipulating tree growth and development, and can assist orchardists to maximize the genetic potential of their trees. More detailed discussion of this topic is provided in chapter 'Advances in understanding apple tree growth: the manipulation of tree growth and development'.

8 Future trends in research

Continued research into the mechanisms that underpin the physiological dwarfing of rootstocks in concert with the ongoing activities of breeding programmes will lead to improved and better matched cultivars and rootstocks. This will importantly enable better matching of rootstock–scion combinations to micro and meso sites to current and new growing regions and further offers the opportunity for making incremental improvements to tree structure and growing systems.

Plant bioregulators will always have a place in fruit production, although several of the PBRs available today have a limited life due to their high cost and toxicity. The focus for development of new PBRs is on safety as well as efficacy; for commercial registration, new products need to have low toxicity and reduced environmental risk. Overall, future investment and ongoing use of PBRs are likely to drop, given the trend of increasing consumer concern over the use of 'chemicals' in fruit production, bringing governments and/or retailers to regulate and/or specify compliance by growers.

9 Where to look for further information

- Rootstocks: See http://www.goodfruit.com/rootstocks-under-trial/ (accessed 3 February 2016) for an excellent summary of recent trials comparing the performance of various commercially used dwarfing rootstocks in Washington State, USA.
- John E. Jackson's review article 'Light interception and utilisation by orchard systems' (1980, *Horticultural Reviews* 2: 208–67) and Alan N. Lakso's book chapter (Lakso, A. N. (1994), 'Apple', in *Handbook of Environmental Physiology of Fruit Crops. Vol. I: Temperate Crops*, B. Schaffer and P. C. Andersen (Eds), pp. 3–42.) are essential reading to those interested in the fundamental principles of light interception and distribution, yield, quality and growing systems.

- Terence L. Robinson's book chapter (2003), 'Apple-orchard planting systems', in *Apples: Botany, Production and Uses*, Ferree, D. C. and Warrington, I. J. (Eds), pp. 345–407. CABI Publishing, United Kingdom.' Provides a comprehensive review on planting systems.
- Plant bioregulators: review article by Stephen S. Miller 'Plant bioregulators in apple and pear culture' (1988, *Horticultural Reviews* 10, 309–401) gives a good background to the use of PBRs in apple.
- The International Society for Horticultural Science (ISHS: http://www.ishs.org/) organizes the International Horticulture Congress (IHC), the premier global horticultural event, which covers all aspects of horticulture and runs every four years. Additionally, chapters and branches of the ISHS organize individual symposia that run every four years – the most relevant to this chapter being 'Integrating Canopy, Rootstock and Environmental Physiology in Orchard Systems' and 'Plant Bioregulators in Fruit Production'.
- The ISHS publishes *Acta Horticulturae*, a journal of refereed papers that captures the proceedings of each symposium. This journal, as well as other relevant journals, is cited in the references section of this chapter.

10 Acknowledgements

Drs Ben van Hooijdonk and Stuart Tustin are gratefully acknowledged for many insightful comments that significantly improved an earlier version of this chapter.

11 References

Al-Hinai, Y. K. and Roper, T. R. (2004), 'Rootstock effects on growth and quality of "Gala" apples', *Hort. Sci.*, 39(6), 1231–3.

Anon (1898), 'The apple'. The Kansas State Horticultural Society. State Capitol, Topeka, Kansas, USA.

Asrey, R., Patel, V. B., Singh, S. K. and Sagar, V. R. (2008), 'Factors affecting fruit maturity and maturity standards – a review', *J. Food Sci. Tech.*, 45(5), 381–90.

Atkinson, C. J., Else, M. A., Taylor, L. and Dover, C. J. (2003), 'Root and stem hydraulic conductivity as determinants of growth potential in grafted trees of apple (*Malus pumila* Mill.)', *J. Exp. Bot.*, 54(385), 1221–9.

Atkinson, C. J., Policarpo, M., Webster, A. D. and Kingswell, G. (2000), 'Drought tolerance of clonal Malus determined from measurements of stomatal conductance and leaf water potential', *Tree Physiol.*, 20(8), 557–63.

Bauerle, T. L., Centinari, M. and Bauerle, W. L. (2011), 'Shifts in xylem vessel diameter and embolisms in grafted apple trees of differing rootstock growth potential in response to drought', *Planta*, 234, 1045–54.

Bewley, J. D. and Black, M. (1978), 'Physiology and biochemistry of seeds in relation to germination', vol 1: development, germination and growth. Springer-Verlag, New York, USA.

Bukovac, M. J. (1997), 'The plant surface: a defining factor in performance of plant bioregulators', *Acta Hort.*, 463, 23–8.

Carrière, E. A. (1897), 'Un nouveau sujet pour greffer les pommes', *Rev. Horti.*, 436–7.

Celton, J. M., Tustin, D. S., Changé, D. and Gardiner, S. E. (2009), 'Construction of a dense genetic linkage map for apple rootstocks using SSRs developed from *Malus* ESTs and *Pyrus* genomic sequences', *Tree Gen. Genom.*, 5, 93–107.

Costes, E. and García-Villanueva, E. (2007), 'Clarifying the effects of dwarfing rootstock on vegetative and reproductive growth during tree development: A study on apple trees', *Ann. Bot.*, 100, 347–57.

Costes, E., Salles, J. C. and García-Villanueva, E. (2001), 'Growth and branching pattern along the main axis of two apple cultivars grafted on two different rootstocks', *Acta Hort.*, 557, 131–8.

Crabbè, J. (1984), 'Vegetative vigor control over location and fate of flower buds in fruit trees', *Acta Hort.*, 149, 55–63.

Di Vaio, C., Cirillo, C., Buccheri, M. and Limongelli, F. (2009), 'Effect of interstock (M.9 and M.27) on vegetative growth and yield of apple trees (cv "Annurca")', *Sci. Hort.*, 119(3), 270–4.

Feliciano, R. P., Antunes, C., Ramos, A., Serra, A. T., Figueira, M. E., Duarte, C. M. M., de Carvalho, A. and Bronze, M. R. (2010), 'Characterisation of traditional and exotic apple varieties from Portugal. Part 1 – Nutritional, phytochemical and sensory evaluation', *J. Funct. Foods*, 2(1), 35–45.

Foster, T. M., Celton, J-M., Chagné, D., Tustin, D. S. and Gardiner, S. E. (2015), 'Two quantitative trait loci, *Dw1* and *Dw2*, are primarily responsible for rootstock-induced dwarfing in apple', *Hort. Res.*, 2, 15001; doi:10.1038/hortres.2015.1.

Foster, T. M., Watson, A. E., van Hooijdonk, B. M. and Schaffer, R. J. (2014), 'Key flowering genes including FT-like genes are upregulated in the vasculature of apple dwarfing rootstocks', *Tree Gen. Genom.*, 10, 189–202.

Fulford, R. M. (1965), 'The morphogenesis of apple buds. I. The activity of the apical meristem', *Ann. Bot.*, 29, 167–82.

Harrison, N., Harrison, R. J., Barber-Perez, N., Cascant-lopez, E., Cobo-Medina, M., Lipska, M., Conde-Ruíz, R., Brain, P., Gregory, P. J. and Fernández-Fernández, F., (2016), 'A new three-locus model for rootstock-induced dwarfing in apple revealed by genetic mapping of root bark percentage', *J. Expt. Bot.*, 67(6), 1871–81.

Hirst, P. M. and Feree, D. C. (1995), 'Rootstock effects on the flowering of delicious apple. 1. Bud Development', *J. Am. Soc. Hort. Soc.*, 120, 1010–17.

Hogg, R. (1851), 'British pomology; the history, description, classification and synonymes of the fruit and fruit trees of Great Britain'. London; Printed by William Ford, 28 Russell Court, Brydges Street, Covent Garden.

Hollender, C. A. and Dardick, C. (2015), 'Molecular basis of angiosperm tree architecture', *New Phytol.*, 206, 541–56.

Hoying, S. A. and Robinson, T. L. (2000), 'The apple orchard planting systems puzzle', *Acta Hort.*, 513, 257–60.

Jackson, J. E. (1980), 'Light interception and utilization by orchard systems', *Hort. Rev.*, 1, 208–67.

Jackson, J. E. and Palmer, J. W. (1972), 'Interception of light by model hedgerow orchards in relation to latitude, time of year and hedgerow configuration and orientation', *J. Appl. Ecol.*, 9(2), 341–57.

Kamboj, J. S., Browning, G., Blake, P. S., Quinlan, J. and Weaver, D. A. (1999), 'GC-MS-SIM analysis of abscisic acid and indole-3-acetic acid in shoot bark of apple rootstocks', *J. Plant Growth Reg.*, 28, 21–7.

Kootstra, A., Dong, Y. H. and Mitra, D. (1994), 'Apple skin colours a feast for consumer eyes', *Orchardist N.Z.*, 67(2), 40–3.

Lespinasse, J. M. and Delort, J. F. (1986), 'Apple tree management in vertical axis: appraisal after ten years of experiments', *Acta Hort.*, 160, 139–55.

Li, B., Wang, J., Ren, X., Bao, L., Zhang, L., Zhang, L., Han, M. and Zhang, D. (2015), 'Root growth, yield and fruit quality of "Red Fuji" apple trees in relation to planting depth of dwarfing interstock on the Loess Plateau', *Eur. J. Hort. Sci.*, 80(3), 109–16.

Lockard, F. G. and Schneider, G. W. (1981), 'Stock and scion relationships and the dwarfing mechanism in apple', *Hort. Rev.*, 3, 315–75.

Miller, S. S. (1988), 'Plant bioregulators in apple and pear culture', *Hort. Rev.*, 10, 309–401.

Neilsen, G. and Havipson, C. (2014), '"Honeycrisp" apple leaf and fruit nutrient concentration is affected by rootstock during establishment', *J. Am. Pom. Soc.*, 68(4), 178–89.

Palmer, J. W. (1989), 'The effects of row orientation, tree height, time of year and latitude on light interception and distribution in model apple hedgerow canopies', *J. Hort. Sci.*, 64(2), 137–45.

Robinson, T. L. (2011), 'Advances in apple culture worldwide', *Rev. Bras. Frutic.*, Special Issue, 37–47.

Robinson, T. L., Hoying, S. A., DeMaree, A., Lungerman, K. and Fargione, M. (2007), 'The evolution towards more competitive apple orchard systems in New York', *N.Y. Fruit Q.*, 15(1), 3–9.

Robinson, T. L., Lakso, A. N. and Carpenter, S. G. (1991), 'Canopy development, yield and fruit quality of "Empire" and "Delicious" apple trees grown in four orchard production systems for ten years', *J. Am. Soc. Hort. Sci.*, 116(2), 179–87.

Schechter, I., Elfing, D. C. and Proctor, J. T. A. (1990), 'Apple tree canopy development and photosynthesis as affected by rootstock', *Can. J. Bot.*, 69, 295–300.

Seleznyova, A. N., Thorp, T. G., White, M., Tustin, S. and Costes, E. (2003), 'Application of architectural analysis and AMAPmod methodology to study dwarfing phenomenon: the branch structure of 'Royal Gala' apple grafted on dwarfing and non-dwarfing rootstock/interstock combinations', *Ann. Bot.*, 91(6), 665–72.

Seleznyova, A. N., Tustin, D. S. and Thorp, T. G. (2008), 'Apple dwarfing rootstocks and interstocks affect the type of growth units produced during the annual growth cycle: precocious transition to flowering affects the composition and vigour of annual shoots', *Ann. Bot.*, 101, 679–87.

Steeves, T. A. and Sussex, I. M. (1989), 'Patterns in Plant Development', Cambridge: Cambridge University Press.

Swarbrick, T. (1929), 'Factors governing flower bud formation. VII. The seasonal elongation of apple varieties on some vegetative rootstocks and its possible relation to fruit bud formation', *J. Pom. Hort. Sci.*, 7, 100–29.

Tagliavini, M., Schdellari, M., Marangoni, B., Bastianel, A., Franzin, F. and Zamborlini, M. (1992), 'Leaf mineral-composition of apple tree – sampling date and effects of cultivar and rootstock', *J. Plant Nutr.*, 15(5), 605–19.

Thimann, K. V. and Skoog, F. (1933), 'Studies on the growth hormone of plants: III. The inhibiting action of the growth substance on bud development', *Proc. Nat. Acad. Sci. USA*, 19, 714–16.

Tustin, D. S. and van Hooijdonk, B. M. (2013), 'Roots, fruits and the pieces in between: a review of fruit production systems towards a step-change increase in productivity', *Aspects Appl. Biol.*, 119, 109–18.

Tustin, D. S., Cashmore, W. M. and Bensley, R. B. (2001), 'Pomological and physiological characteristics of slender pyramid central leader apple (*Malus domestica*) planting systems grown on intermediate vigour, semi-dwarfing and dwarfing rootstocks', *N.Z. J. Crop Hort. Sci.*, 29, 195–208.

Tworkoski, T. and Fazio, G. (2015), 'Effects of size-controlling apple rootstocks on growth, abscisic acid, and hydraulic conductivity of scion of different vigour', *Int. J. Fruit Sci.*, 15, 369–81.

Tworkoski, T. and Miller, S. (2007), 'Rootstock effect on growth of apple scions with different growth habits', *Sci. Hort.*, 111, 335–43.

Tworkoski, T., Fazio, G. and Glenn, D. M. (2016), 'Apple rootstock resistance to drought', *Sci. Hort.*, 204, 70–8.

Van Hooijdonk, B. M., Woolley, D. J. and Warrington, I. J. (2011), 'Rootstocks modify scion architecture, endogenous hormones and root growth of newly grafted 'Royal Gala' apple trees', *J. Am. Soc. Hort. Sci.*, 136(2), 93–102.

Van Hooijdonk, B. M., Woolley, D. J., Warrington, I. J. and Tustin, D. S. (2010), 'Initial alteration of scion architecture by dwarfing apple rootstocks may involve shoot-root-shoot signalling by auxin, gibberellin and cytokinin', *J. Hort. Sci. Biotech.*, 85(1), 59–65.

Vyvyan, M. C. (1955), 'Interrelation of scion and rootstock in fruit trees I. Weights and relative weights of young trees formed by the reciprocal unions, as scion and rootstock, of three apple rootstock varieties: M.IX, M.IV, and M.XII', *Ann. Bot.*, 19(3), 401–23.

Wagenmakers, P. S. (1991), 'Simulation of light distribution in dense orchard systems', *Agric. Forest Meteorol.*, 57, 13–25.

Webster, A. D. (1995), 'Rootstock and interstock effects on deciduous fruit tree vigour, precocity, and yield productivity', *N.Z. J. Crop Hort. Sci.*, 23(4), 373–82.

Webster, A. D. (2001), 'Rootstocks for temperate fruit crops: Current uses, future potential and alternative strategies', *Acta Hort.*, 557, 25–34.

Webster, A. D., Oehl, H. V., Jackson, J. E. and Jones, O. P. (1985), 'The orchard establishment, growth and precocity of four micropropagated apple scion cultivars', *J. Hort. Sci.*, 60(2) 169–80.

Wertheim, S. J. and Webster, A. D. (2003), 'Propagation and nursery tree quality', in *Apples: Botany, Production and Uses*, Ferree, D. C. and Warrington, I. J. (Eds), pp. 126–36. Google Books.

Wilkie, J. D., Sedgley, M. and Olesen, T. (2008), 'Regulation of floral initiation in horticultural trees', *J. Exp. Bot.*, 59(12), 3215–28.

Wills, R. B.H., Bambridge, P. A. and Scott, K. J. (1980), 'Use of flesh firmness and other objective tests to determine consumer acceptability of Delicious apples', *Aust. J. Exp. Agr.*, 20(103), 252–6.

Wilson, B. F. (2000), 'Apical control of branch growth and angle in woody plants', *Am. J. Bot.*, 87, 601–7.

Chapter 4

Advances in understanding apple tree growth: the manipulation of tree growth and development

Dugald C. Close and Sally A. Bound, University of Tasmania, Australia

1 Introduction

As noted in the chapter 'Advances in understanding apple tree growth: rootstocks and planting systems', apple tree management strategies after planting are an important factor in apple tree development and growth, and their effects must be understood in order to provide a complete picture of the factors determining yield.

How tree growth attributes are manipulated through plant management and the use of plant bioregulators (PBRs) to manipulate developmental processes to induce branching and manage vigour is an important aspect for optimizing tree growth. Key manipulations of tree training and pruning to arrive at a desired tree structure that optimizes within-canopy light distribution has been an area of significant R&D investment to provide desired fruit quality, particularly, size, firmness, colour and sugar–acid levels. Another important area of study has been the manipulation of bud break through either cultural application or application of PBRs to overcome inadequate winter chilling. Finally, in order to inform how

http://dx.doi.org/10.19103/AS.2016.0017.29

best to optimize total light interception and therefore yield, planting systems research that studies interactions of within and between row spacing of a given rootstock/scion combination trained to various tree structure configurations has been investigated.

This chapter will provide detailed discussion of the practices which affect canopy development, dormancy release, bud break and vegetative growth. It also reviews the effects of training and pruning techniques. The chapter ends by presenting two case studies which show the importance of post-planting apple tree management, and illustrates how it can interact with choice of rootstock and planting system to determine yield.

2 Canopy development

Prior to 1970 most nurseries produced single shoot whips that were headed at planting in the orchard and then grown vigorously for several years to develop a large, strong branch framework (Robinson 2011). This practice of heading one-year-old wood normally results in clusters of branches near the cut with narrow crotch angles. Heading discourages early fruit production, which is one of the aims of intensive orchards, by promoting vigorous shoot growth and removing potential fruiting wood (Ouellette and Young 1994). Other cultural techniques used to induce lateral branching and stimulate canopy development include limb spreading and leaf removal, but these methods produce either fewer branches than can be achieved through PBR application or short branches (Greene and Autio 1994). There is also variation in cultivars, with some cultivars branching naturally, providing nutrition and water availability are adequate (e.g. 'Royal Gala'), while others are sparser branching (e.g. 'Fuji'). With the advent of high-density intensive orchard systems, it has become essential to have techniques for effective stimulation of branch development that avoid pruning, thus shortening the time needed to develop a fruitful canopy (Elfving 2015).

2.1 Plant bioregulator (PBR) application

Cytokinins such as 6-benzyladenine (BA) were developed during the 1960s. A proprietary mix of BA and GA_{4+7} (Promalin), first registered in the late 1970s for improving the size and shape of apples, was shown to have potential as a branching agent (Jones 1967; Forshey 1982; Greene 1983; Looney 1983). The combination of BA and gibberellins (GAs) is effective as BA stimulates bud growth, while the GAs stimulate shoot elongation. Williams and Billingsley (1970), Elfving (1985), Joosse (1986), Greene and Miller (1988), Koen et al. (1989) and Olwell and Andrews (1992) have all demonstrated the ability of the combination of BA and GA found in Promalin to improve the plant structure of several fruit tree species, including apples, by stimulating branching and increasing primary branch crotch angles. Miller (1989) suggests that 250 mg L^{-1} Promalin is necessary to obtain a good branching response. Koen et al. (1989) reported differences in the response of spur and non-spur bearing 'Delicious' to BA+GA_{4+7}, with good branching obtained in spur-bearing trees but no response in non-spur-bearing trees. Examining concentrations from 100-500 mg L^{-1}, they also reported that, while higher concentrations increased branch number, both branch and leader length were reduced and average branch angle was decreased. They consequently recommended a rate no higher than 100 mg L^{-1}. However, Miller (1983) obtained very broad crotch angles on young 'Delicious' trees with 400 mg

L^{-1} Promalin. Cultivar differences were reported by Miller (1989) who found that some cultivars respond to 100 mg L^{-1}, while others require 500 mg L^{-1}.

Application of branching agents containing BA or BA plus GA_{4+7} are now standard nursery practice to stimulate branching, enabling the development of feathered trees possessing a good scaffold structure, and thus allowing significant production within two years after planting (Petracek et al. 2003). However the rates of BA+GA_{4+7} (Promalin) required to stimulate branching can thin developing fruit and inhibit flower bud formation (Greene et al. 1990). Promalin can also be mixed with paint or lanolin and applied directly to the buds at bud swell to induce bud development. The uptake and efficacy of BA is influenced by temperature, being more effective when followed by a three-hour period of at least 18°C (Robinson and Sazo 2014).

Cyclanilide, an auxin inhibitor, is a useful tool for branch induction, particularly in nursery stock. When used alone it blocks apical dominance and enhances lateral branching in young trees (Elfving and Visser 2005). According to these authors, it produces a significant number of well-developed feathers with good crotch angles and, unlike BA and GA_{4+7}, promotes branching without deformity or long-term carry-over effects. In comparative studies of cyclanilide, BA and BA+GA_{4+7} in New York State, Washington State and Chile, Robinson and Sazo (2014) report that cyclanilide is a potent branching agent but it can have negative effects on tree growth, including reduction in terminal shoot elongations and reduction in tree height, in some regions. They found that BA can also reduce leader growth rate and final tree height, but not to the extent of cyclanilide, and concluded that BA was an effective branching agent but requires multiple applications (up to five sprays) to produce five or more feathers. They also reported that, while BA+GA_{4+7} stimulated leader growth rate over the course of the summer resulting in improved final tree height, it produced feathers with more acute angles. It should be noted that an increase in shoot dry mass as a result of PBR application results in a corresponding reduction in root dry mass as a result of reallocation of resources. Doric et al. (2015) confirm the stimulating effect of GA_{4+7} on shoot growth and its counteraction of the growth slowing effect of BA (due to ethylene evolution). Robinson (2011) reports that cyclanilide, BA and BA+GA_{4+7} are all useful tools for induction of lateral branching in both the nursery and orchard. Another PBR that exhibits cytokinin-like activity is thidiazuron (TDZ), which has been used to stimulate budbreak in blind wood of young 'Scifresh' trees in the field, and to enhance the axillary budbreak of newly planted rootstock stems in stool beds (Dayatilake et al. 2011).

2.2 Notching and branching

One cultural technique that has been successfully used to stimulate branching is the practice of notching. This technique is suitable for trees of all ages, and can also be combined with PBR application.

Notching involves removal of a narrow strip of bark directly above a bud with the aim of blocking the downward flow of growth-regulating compounds from the apical meristem, thus releasing the bud from apical dominance. Although it is well known that notching can stimulate branching, there is conflicting advice on the optimal time that notching should be performed and hence results have been inconsistent. Greene and Autio (1994) speculated that inconsistent results were probably due to a lack of understanding of the variables affecting bud growth after notching and identified factors influencing the response of young trees to notching. They found that the amount of bark removed was important; notching procedures removing <2mm of bark are less effective as the wound may heal and apical dominance may be reimposed before the bud has grown sufficiently to overcome apical dominance. Performing notching on large buds on vigorous wood between green tip and 1-cm green (half-inch green) stages of bud development gave >80% success in development of buds into lateral shoots (Greene and Autio 1994). A major advantage of notching is that buds can be stimulated to grow where they are required, it is not influenced by wood age and there is no effect on fruit set and flower initiation, and so, notching can be used on fruiting trees. However McArtney and Obermiller (2015) suggest that notching may create infection sites for pathogenic bacteria, and so, care needs to be taken for maintaining hygiene.

Clements et al. (2010) reported that both notching and Promalin ($BA+GA_{4+7}$) increased the number of laterals and did not affect leader growth, and suggested that a combination of both is a good option for improving branching in poorly branched first leaf trees. The benefits of this practice for older trees were confirmed by McArtney and Obermiller (2015). These authors reported that notching of 2-year-old wood on 5-year-old 'Granny Smith' trees stimulated 59% of the buds to resume growth, while spraying BA directly into the notch increased bud break to >90%.

3 Dormancy release and bud break

Deciduous fruit trees, such as apple, enter a dormant period during the winter and this is characterized by two stages: (i) endo-dormancy (rest phase) where dormancy is due to endogenous inhibition within the buds and (ii) eco-dormancy (quiescent phase after endo dormancy is overcome by low-temperature degradation of inhibitors) where dormancy persists due to unfavourable environmental conditions for growth. The speed and uniformity of bud-break is an indication of the completion of endo-dormancy. Duration of the endo-dormancy period is controlled by the length of time of exposure of the buds to low temperatures – this chilling requirement is genetically determined and varies amongst cultivars. In endo-dormancy, the depth of dormancy of individual buds can vary, and hence, some buds will end their endo-dormant state before others (Saure 1985; Erez 1995). According to Erez (1995) terminal buds generally have a lower chilling requirement than lateral buds, and floral buds generally require less exposure to chilling than vegetative buds. Insufficient chill accumulation over the dormant winter period can result in physiological damage to trees and non-synchronous bud burst and flowering (Bound and Miller 2006; Thomas et al. 2012). Multiple chill accumulation models have been developed, but Erez and Lavee (1971) recommended that chilling be measured by weighted chilling hours to account for the varying efficiency of different temperatures in inducing dormancy release. The model developed by Richardson et al. (1974) to calculate chill units has adopted this concept. According to Darbyshire (2014), the Dynamic chill model (Fishman et al. 1987) is currently the best practice model. It calculates chill in units known as 'chill portions' based on hourly temperatures, and captures temperature–winter chill relationships which are lacking in other models. In the Dynamic model, effective winter chill temperatures follow a bell shape with an optimum chilling temperature at 6°C, tapering to zero at −2°C and 14°C (Darbyshire 2014); high temperatures act to negate previously accumulated chill and moderate temperatures can enhance chill accumulation. As an example, the chill requirement for apples ('Golden Delicious') is 50.0 chill portions (UC-Davis Fruit & Nut Research and Information Center 2011).

While many apple-growing regions across Europe and the Northern United States have marked seasonal transitions, with adequate winter chill resulting in even dormancy release and rapid bud break in spring, the climate across most Australian, South American and South African apple-growing regions is characterized by gradual transitions from one season to another, and often indistinct and occasionally warm winters. Worldwide, there is also a shift in the production of temperate tree fruits to warmer areas with insufficient winter chill for cultivars with a high chill requirement (e.g. Brazil, South Africa). Trees grown under these conditions come out of winter dormancy slowly and bud break is erratic, consequently foliar development is sporadic, and some cultivars flower over an extended period of 6–8 weeks, which can have major implications on fruit size, quality and maturity (Bound and Jones 2004). This problem is also exacerbated by climate change, with many regions becoming marginal production areas for temperate tree crops (Rai et al. 2015).

Under conditions of inadequate winter chilling, bud break can be manipulated by cultural practices or application of chemical rest-breaking agents.

3.1 Cultural practices

Timing of management practices such as irrigation and fertilization can modify the amount of chill hours required; late season pruning, irrigation and fertilizer application have been shown to promote new growth, thus prolonging growth late into the season and increasing the chilling hours required to complete rest (Walser et al. 1981; Melke 2015). Saure (1985) also describes delayed bud-break compared with other shoot types in late-growing water-shoots. As early as 1885, Müller-Thurgau (cited in Saure 1985) reported that withholding water supply to the roots was a common practice to promote early inception of dormancy and to shorten its duration. Melke (2015) suggests that techniques such as regulated deficit irrigation or partial root drying that limit soil water availability, hence arresting shoot growth earlier in the season, could be worth examining with respect to control of dormancy and subsequent regulation of bud break. Early leaf removal can also reduce the rest period (Saure 1985; Melke 2015), although Diaz et al. (1987) reported that chemical defoliation with copper sulphate was more effective than hand defoliation, with chemical defoliation resulting in earlier flowering and reducing the length of the flowering period. It should be noted, though, that this can have implications for accumulation of carbohydrate reserves during the post-harvest period (Breen et al. unpub.). Application of nitrogen during autumn after termination of growth has also been reported to promote early bud break (Terblanche and Strydom 1973; Terblanche et al. 1979); however, this is not recommended in cold climates as susceptibility to freeze injury is increased with late nitrogen application.

Another technique to overcome chilling requirements in warm regions is the use of evaporative cooling which can be achieved through the use of overhead irrigation sprinklers. During the rest period, high maximum bud temperatures can be counteracted by the use of overhead sprinklers during the day to cool the buds, enhancing both floral and vegetative bud break (Erez 1995; Rai et al. 2015).

3.2 Chemical application

Application of chemical rest-breakers to compensate for insufficient chilling and to induce synchronous bud break has become a widespread practice in many apple-growing regions, particularly those that are marginal in terms of winter chill. However rest-breakers are more effective when a significant proportion of the chilling requirement of the plant has already been satisfied – Petri et al. (2014) indicate that 50% or more of the chilling requirement should already have been met, while Erez (1995) states that chemicals will not compensate for more than approximately 30% of the chilling requirement. With any chemical rest-breaker, time of application greatly influences the effectiveness of the chemical (Dostal, cited in Saure 1985; Petri 1987; Bound and Jones 2004; Bound and Miller 2006) and with appropriate timing many rest-breakers can also be used to manipulate the time and length of the flowering period (Petri 1987; Bound and Miller 2006).

The first chemical recommended for breaking dormancy of apple trees was mineral oil (Black 1936). Black also reported that the effectiveness of oil sprays is cultivar dependent. The mode of action of oils is described by Melke (2015) as creation of oxygen starvation through coating of the branch/buds – as oxygen starvation develops, the products of anaerobic respiration are implicated in breaking dormancy.

The respiration uncouplers DNBP (dinitro-butyl-phenol) and DNOC (dinitro-ortho-cresol) gained acceptance as rest-breakers in the 1940s. Either alone or combined with mineral oil, DNOC was applied commercially in many parts of the world until the compound was banned in the 1980s. Thiourea was found to have a strong effect on breaking vegetative bud dormancy, particularly in combination with potassium nitrate (KNO_3) and oil-DNOC, but because of human toxicity it has been banned from use (Erez 1995).

Potassium nitrate has been reported to have a rest-breaking effect in a range of tree crops, but even at a high concentration of 10%, the effect is mild (Erez 1995). The fatty amine polymer Armobreak is reported to enhance cuticle penetration, carrying other chemicals with it (Erez 1995). Armobreak, combined with KNO_3, condenses the bloom period without advancement, and Rai et al. (2015) indicate that the combination of KNO_3 and Armobreak can improve efficacy by 20–30% compared with either chemical alone. Chemicals such as Armobreak and KNO_3 are useful in combination with other rest-breakers to enhance their effect. Erez (1995) suggests that, by lowering the concentration of rest-breaking agents such as hydrogen cyanamide (HCN), it may reduce the phytotoxicity and certainly the cost.

Hydrogen cyanamide has become the leading rest-breaking chemical in apple production since the 1980s. Erez (1995) reported contradicting results on damage to flower buds with HCN, and postulated that the main cause for the varied results was in the level of endo-dormancy of the buds. Bud damage and reduced fruit set have been reported following the application of HCN close to budburst (Bound and Jones 2004). Erez (1995) suggested that, as resistance to HCN declines rapidly upon release from the endo-dormant state, in order to avoid damage to flower buds, it is essential to avoid late application unless it is evident that dormancy is not released. In spite of these issues, HCN is an effective rest-breaker and when used carefully gives good results, with the added benefit that both time of flowering and the length of the flowering period can be manipulated with HCN (Bound and Jones 2004). The recommended application rate for HCN is 2–3%, but Rai et al. (2015) report that low concentrations of 0.5–1% in combination with 3–4% winter oil is now the commercial recommendation in South Africa.

Due to its toxicity considerable care needs to be taken when applying HCN, hence the need for a replacement has been flagged (Bound and Miller 2006). There is now a ban on the use of HCN in some European countries and it is likely to be prohibited in other countries in the future.

Efforts to develop a safer chemical that would penetrate the bark of dormant trees to stimulate dormancy break have led to the development of Waiken® (an emulsified vegetable oil compound; SST Australia Pty Ltd). According to the product label, Waiken® induces a period of dormancy after which floral bud burst is uniform. However, as this is an oil-based product, it is possible that it creates oxygen starvation in a process similar to that described by Melke (2015). Bud burst and flowering can be manipulated with application timing, and it can be safely applied up until bud-swell/greentip stage without lasting phytotoxic damage (Bound and Miller 2006). As it can be applied late in the season, compared with HCN which needs to be applied 40–30 days before estimated bud-burst, Waiken® provides a greater window of opportunity for application as a rest-breaker.

Erger® is another product developed as a rest-breaker to potentially replace HCN. According to Petri et al. (2014), this compound is a penetrating surfactant consisting of inorganic nitrogen, mono-, di- and polysaccharides, proteins, enzymes and citric acid radicals. The recommendation is to mix Erger with a nitrogen source (calcium nitrate), and it is claimed that this allows the nitrogen source to penetrate the outer bark layer and stimulate bud movement. Petrie et al. (2014) reported that Erger with calcium nitrate produced similar results in inducing budbreak to HCN mixed with mineral oil; however, this study was only over one season.

There is a lack of information on the mechanism of how many rest-breakers work; however, work in grapes suggests that a respiratory stress may be involved in the release of buds from dormancy (Vergara et al. 2012). This is consistent with the dormancy-breaking mechanism caused by oxygen starvation when applying oils (Melke 2015). According to Zhuang et al. (2015), antioxidant systems and redox systems, such as reactive oxygen species (ROS), have been reported to play a role in regulating bud dormancy release.

4 Managing vegetative growth

In many apple cultivars, particularly older less floriferous cultivars, there is an inverse relationship between vegetative growth and flowering. In managing canopy vigour, the aim is to achieve 'calm' trees with a good balance between vegetative growth and cropping (Robinson 2011). Byers and Yoder (1999) reported that in 'Stayman' apple trees on seedling rootstock, 75% of shoot growth occurred in the first 30 days after bloom; however, growth varies depending on cultivar and age, and growth in younger trees will often extend over the entire season (van Hooijdonk, pers. com.). Growth is normally encouraged in younger trees to achieve rapid canopy fill; however, if trees have filled their allocated space and continue to grow vigorously, then growth control is recommended. Shoots will terminate naturally in trees on dwarfing rootstocks with a well-managed canopy, and so, no intervention is required, but where trees are growing vigorously and suppression of growth is required, an understanding of the proportion of growth in each flush and how many flushes occur throughout the season is important.

A range of methods are available to mitigate excessive tree vigour, including rootstock, tree training, trunk scoring, crop load, soil fertility management and root pruning. While plant bioregulators play an important role as they are convenient, can be applied as needed and the effect is not irreversible (Greene 2015), the practices of trunk girdling and root pruning can be useful tools in vigour management in both young and established orchards.

4.1 Cultural practices

Trunk scoring and girdling

Trunk scoring and girdling are commonly used to reduce tree vigour, and involve cutting through the bark and phloem to temporarily disrupt movement of photosynthates through the phloem. Scoring is simply the practice of cutting through the bark to the cambium, while girdling or ringing involves removal of a strip of bark, and is normally performed with a double-bladed knife. The wider the strip the more prolonged the effect, but the risk of damage to the tree is also increased (Goren et al. 2004), particularly where a complete ring is removed which may result in tree death if the tree is unable to form new tissue to bridge the gap (Ferree and Schupp 2003). Several studies reported by Greene and Lord (1983) and Autio and Greene (1992) on a range of cultivars show that phloem interruption by girdling can reduce terminal growth, trunk cross-sectional area, shoot diameter and a number of shoots. Samud (1998) also reported a reduction in trunk diameter, branches per tree and shoot length in several cultivars following girdling treatments. An increase in flower number, fruit set and yield has also been reported following girdling (Samud 1998; Goren et al. 2004), although it can reduce fruit size (Goren et al. 2004).

Care needs to be taken to avoid damage to the cambium as healing depends on cambial activity for regeneration of the vascular connection. The degree of vigour control is determined by timing as well as the severity of the treatments. As the healing process is longer when girdling is performed during dormancy, the recommendation is to girdle during periods of intensive growth when the cambium is active (Goren et al. 2004). According to Ferree and Schupp (2003), the effects of girdling can last for more than 4 weeks compared with scoring where trees often recover within 2 weeks. Scoring is preferred for younger trees.

While a continuous cut is made around the trunk with scoring, it is preferable when girdling to overlap the cut in a spiral pattern or use a double 'C' in which cuts are made on either side of the trunk at different levels 5–10 cm apart, with approx. 2-cm overlap. Girdling knives are available from 2–6 mm cutting width, and to avoid ringbarking the tree, a strip no wider than 6 mm is recommended. For large mature trees, chain saws can also be used.

Root pruning

The effectiveness of root pruning in reducing tree growth can be observed in the ancient art of bonsai where it is used to assist in dwarfing of the tree as well as to stimulate new root growth (Geisler and Feree 1984). In the 1800s, Rivers (1866) advocated autumn root pruning of vigorous trees, particularly those growing in rich soils, by digging a trench around the tree about 45 cm from the stem. However, he stipulated that it is important for the soil to be moist. In mature trees he suggested that the trench should be 1 m from the stem and only two-thirds of the roots should be pruned the first season, leaving the remainder to anchor the tree. According to Wilton (1999), natural dwarfing of trees on vigorous rootstocks is often seen where orchards are planted on shallow soils which restrict root growth.

Root pruning reduces the root:shoot ratio and the plant reacts by directing more assimilates to the root system. Hence growth is redistributed in favour of the roots and the relative shoot growth is reduced (Maggs 1964, 1965; Geisler and Feree, 1984). In discussing the response of plants to root pruning, Geisler and Feree (1984) suggest that the mechanisms underpinning the effects of root pruning on growth are complex and may involve water stress induced by limited water absorption, reduced mineral absorption and assimilation, and a reduction in hormone synthesis, particularly cytokinins which are produced in the root tips. The distance of the cut from the trunk is critical – if too few roots are cut, excessive root development is induced, leading to increased shoot growth; however, cutting too close to the trunk may result in tree death, particularly if insufficient water is available (Geisler and Ferree 1984).

Time of pruning will influence the response, and Schupp and Feree (1987, 1990) demonstrated a reduction in shoot length following root pruning at the dormant or full-bloom stage, but Schupp (1990) advised that the best time to root prune is from bloom until two weeks after bloom. Both Feree (1992) and Khan et al. (1998) found that the growth-retarding effect of root pruning lasted for more than one year.

In discussing vigour control methods, Wilton (2011) recommends use of a tractor-drawn cutting blade for root pruning, beginning as the orchard approaches full canopy. The cut should be made 30–50 cm out from the trunk to a depth of 30–40 cm to sever the surface feeder roots. The amount of vegetative growth control can be adjusted by adjusting the distance from the cuts to the trunk (Schupp 1990). Normally only one side of the tree is done during the dormant or pre-bloom period; however, if this gives inadequate vigour control, a second root prune can be done to the other side of the tree during late spring/early summer. One drawback is that a tractor of at least 40–50 horsepower is required to pull the cutting blade (Schupp 1990), and so, this practice may not be suitable for all orchardists.

While many authors have reported success in reducing tree growth with root pruning, Miller (1995) reported a lack of response with root pruning in young trees and suggested that this may have been due to (i) the cut being too far from the trunk to effectively reduce root volume; (ii) irrigation and/or use of a mechanical tree planter confining the bulk of the root system to the centre of the tree row, thus limiting the amount of roots cut; and (iii) deep fertile soils encouraged downward rather than lateral root development, especially in conjunction with trickle irrigation. Crop load can also influence the degree of response following root pruning; Schupp (1990) reported up to 40% reduction in vegetative growth in bearing trees, but only 14% in non-bearing ones.

In their review of the literature, Geisler and Feree (1984) concluded that root pruning has potential as a size-controlling mechanism for woody plants, but the interaction of factors such as time and degree of pruning, environmental conditions, soil moisture and crop load with the plant response to root pruning do not allow a commercially predictive response. Khan et al. (1998) concluded that root pruning is an effective method of reducing tree size in an ultra-high-density planting system. It is likely that, with careful consideration of soil fertility, water availability and distance of the cut from the trunk, root pruning can be an effective technique for controlling shoot growth in growing systems that result in vigour requiring management. Ideally the density of rootstock and scion combinations can be matched to site conditions in order to minimize shoot and/or root pruning to enable maximization of the harvest index as outlined in the Training and Pruning section of this chapter.

4.2 Use of plant bioregulators

The first effective PBR reported for growth control in apples was daminozide (Batjer et al. 1964). Registered as Alar, this PBR was a successful commercial strategy for controlling vegetative growth until registration was cancelled in 1989 (Byers 1993; Unrath 1999; Greene 2010).

Ethephon can be used to control growth, but because it causes fruit abscission at rates that retard growth, its use is often limited to non-bearing trees. Prior to the loss of daminozide, a tank-mix of daminozide and ethephon was effective for controlling growth without fruit abscission in bearing trees (Unrath 1999). Multiple low weekly doses of ethephon applied in the first 45 days after full bloom have been shown to control tree growth without causing fruit abscission (Byers 1993), but as ethephon is known to cause fruit russetting (Jones et al. 1991), this strategy needs to be treated with caution.

Growth retardants that are inhibitors of GA biosynthesis can be grouped into three different chemical types according to their structural features and mode of action (Rademacher 1995): (1) onium compounds – these compounds possess a positively charged ammonium, phosphonium or sulfonium moiety; (2) compounds with a nitrogen-containing heterocycle – these are pyrimidine derivatives or possess a triazole moiety and (3) acylcyclohexanediones and related compounds. Rademacher (1995) provides a detailed description of the parts of the GA biosynthesis pathways affected by each compound type.

Several triazole gibberellin biosynthesis inhibitors were tested in the 1980s and as a result paclobutrazol (PP-333) was registered for use as a growth retardant for apple trees (Greene 2010). Paclobutrazol is metabolized slowly within the tree, with effects of a single application lasting up to 4 years (Miller 1988). Nickell (1994) has suggested that, by reducing excessive terminal growth, tree efficiency is increased, and hence, under some conditions the yield of paclobutrazol-treated trees can be increased due to increase in light to the fruiting spurs. High rates of paclobutrazol resulting in excessive reduction in shoot growth can flatten fruit (Miller 1988; Deyton et al. 2014); however, at rates that are effective for acceptable shoot growth control, there is no impact on fruit quality. There is considerable discussion on the most efficient method of application, including foliar spray, trunk injection, trunk spray/paint or soil drench around the base of the tree. Miller (1988) discusses the extensive research undertaken on paclobutrazol and the different methods of application. Young stems and shoot tips are the most active uptake sites (Quinlan and Richardson 1986). Sequential low concentration foliar sprays are more effective in controlling shoot growth than a single high concentration application (Quinlan and Richardson 1986; Shearing and Jones 1986) – this is in keeping with young shoots and leaves being more active uptake sites than mature leaves. Application during flowering can cause abscission or damage to the developing fruitlet (Lever 1986). The effectiveness of soil-applied paclobutrazol is impacted by factors such as tree size, rooting pattern, water availability and distribution in the soil, and soil physical properties (Lever 1986). According to Lever (1986), paclobutrazol is of low water solubility and tends to bind reversibly to soil, and also to wood in the tree's vascular system; this constitutes a reservoir which releases the chemical over time to give continued growth control.

Although paclobutrazol is labelled for use only on ornamental trees in the United States (Deyton et al. 2014) due to its persistence in the tree and potential residue problems (Miller 1988), it is still successfully used in several countries, including Australia, to reduce vegetative growth in apple trees.

Chlormequat chloride (Cycocel) is an onium-type compound which inhibits GA production and has been used for controlling canopy growth and initiating flower buds in pear orchards in Europe, but it is only moderately effective on apple (Miller 1988); it can also persist in the tree in an unmetabolized form for up to six months (Smit et al. 2005).

Due to increased environmental concerns, its use has been prohibited in Denmark since 2002 (Wang et al. 2014).

The gibberellin biosynthesis inhibitor prohexadione-calcium (Pro-Ca) was identified in the early 1990s – this chemical is an acylcyclohexanedione. Extensive testing led to registration as Apogee in the USA/Canada and Regalis in other parts of the world. Pro-Ca has a similar mode of action to daminozide (Rademacher 1995). Suppression of shoot growth with Pro-Ca is short-lived (Byers and Yoder 1999; Cline 2000; Greene 2015), and hence for effective control throughout the season, multiple applications are recommended (Cline 2000; Greene 2015). However, this short-term effect and lack of persistence can make Pro-Ca a flexible tool in developing growth management strategies to fit specific situations (Evans et al. 1999; Greene 2010).

There is a reasonable consensus in the literature that an initial Pro-Ca should be applied early in the season, and Evans et al. (1999) reported that applying Pro-Ca before new shoot growth reaches 10 cm provides the most effective control of vegetative growth. Greene (2010) suggests application as soon as sufficient leaf area has emerged for absorption to obtain maximum benefit for growth control as it can take 10 days for the compound to take effect. Cline (2000) also recommends application as extension growth is beginning followed by 3–4 sprays spaced 2 weeks apart. According to Unrath (1999), response to a single application of Pro-Ca lasts 3–4 weeks, and he suggests that multiple applications at 2–3-week intervals are appropriate.

Early work with Pro-Ca indicated that it could reduce fruit size and increase fruit set (Greene 1999), and this effect has been observed in multiple studies (Cline et al. 2008). In discussing work by Greene, Warner (2016) reports a correlation between increasing concentration of Pro-Ca and increased fruit set, and Greene has suggested that Pro-Ca does not increase initial fruit set, but rather the reduction in vegetative growth retards June (December in southern hemisphere) drop. Undertaking trials on Pro-Ca in Spain, Medjdoub et al. (2004) found that flower initiation was not affected. The negative effects of Pro-Ca on fruit size are most likely a direct result of the increased fruit set and crop density (Cline et al. 2008); however, standardization of crop loads in studies such as these would enable clarification of the impact of the growth retardant on fruit size.

As discussed above, like all PBRs, Pro-Ca can have negative side effects such as increased fruit set and reduction in fruit drop, leading to increased difficulty in thinning; however, these problems may potentially be overcome with more aggressive thinning programmes. One major advantage of Pro-Ca is that it reduces secondary fireblight infections (Cline 2000). Fruit from Pro-Ca treated trees has improved red colour; however, this is most likely a secondary effect due to increased light within the tree.

5 Training and pruning

Fundamental principles that underpin practices of training and pruning for any particular tree structure include:

- Buds need exposure to no less than 30% incident light for flower initiation and development (Palmer 1989).
- Leaves exposed to uninterrupted light are efficient sources of photosynthates relative to shaded leaves.

- Spur leaves provide the majority of photosynthate for fruit development and quality attributes such as size and sugars (e.g. Tustin et al. 1988), acids and secondary metabolites, relative to leaves on vegetative shoots.
- Fruit require exposure to light (at least 40%; Heinicke 1966) in order to develop adequate colour.
- Harvest index, the proportion of annual biomass produced as fruit relative to total annual tree biomass, should be maximized to prevent wasted investment of carbon into non-harvested structures.

Therefore a key objective at the tree structure level is to reduce investment in branches and vegetative shoots that self-shade, as light distribution within the canopy is more important than total light interception with regard to fruit quality, irrespective of the training system (Wagenmakers 1991; Sansavini and Corelli 1992). In the context of the above principles, consider that the interior of a tree that is not trained or pruned receives light intensity–only 6–30% that of full sunlight (Looney 1968)!

5.1 Types of training and pruning

Severe dormant (winter) pruning decreases flowering and yield (Barden et al. 1989) because of the direct removal of carbohydrate stores and other resources, and by limiting root development and cytokinin production. Further, heading or shortening of the leader or branches leads to local, or entire canopy, invigoration, particularly of the tree apex, because of the loss of apical dominance. Hence, when a dominant branch must be pruned, the approach of 'renewal pruning', where the entire branch is removed and replaced with a new branch, is the preferred approach. Heading back during summer can be effective for improving fruit quality (but can also decrease quality through sun-scald effects) and flower bud development, relative to winter pruning (Marini and Barden 1982), although Bound and Summers (2001) reported reduced fruit weight, length/diameter ratio and total soluble solids concentration in summer pruned 'Fuji' trees. However the aim is to use tree structures that require minimal pruning, given that this represents a potential cost to harvest index as outlined in the following subsections. Given that training and pruning are the 'annual control mechanism of canopy properties needed to optimise physiological function' (Tustin and van Hooijdonk 2013), and that growers are constantly training workers on how to do this to suit their desired system, it is little wonder that this issue is one of constant discussion and conjecture.

Historically fruit trees were minimally trained and pruned, perhaps into 'goblet' or 'vase' structures as the understanding of the importance of canopy light distribution started to develop. These structures relied on structural branches creating a 'scaffold' that resulted in delayed development of the mature tree and relative late return on investment for growers. Long-term studies by Bernhard, Lespinasse et al. (see Lauri et al. 2009) investigated the architecture, in relation to fruiting, of a range of cultivars. The consequent Lespinasse's Typology describes four main cultivar-related growth, branching and fruiting habits of trees in the absence of training and pruning (Lespinasse and Delort 1986): Type I exhibits upright scaffolds and lateral fruiting spurs and a tendency to alternative bearing (e.g. 'McIntosh' and 'Starkrimson') to Type IV of a tip-bearing habit, globular, weeping crown and regular bearing (e.g. 'Granny Smith'), with Types II and III intermediate between these two extremes (see Fig. 1). In 1986 Lespinasse and Delort showed that branches of 30–120° had balance between fruiting and vigour with optimum branch angle between 30 and

45°. This led to the use of branch 'spreaders' that were placed between branches and the main trunk with the aim of 30–45°angle from horizontal. With the increased understanding and availability of dwarfing rootstocks, tree structures have evolved to better suit higher-density planting systems (higher efficiency in light capture and management operations) and to facilitate early fruiting.

The Central leader system of Heinicke (1975 as cited in Robinson et al. 1991) utilized semi-dwarfing rootstocks and intermediate tree densities (240–500 trees/ha). The structure involved a conical shape of 3–4 tiers of branches and a height of 4–5 m. The conical shape was preferable to goblet or vase tree structures for ensuring adequate light distribution to lower portions of the tree for both fruit yield and quality. However upper limbs often outgrew and shaded lower limbs, and the lost resources as a result of pruning these led to lost potential yield as Gvozdenović et al. (2008) showed in their study comparing bending versus heading back of various cultivars trained to the Central Leader structure. The central leader system could often lead to major tree restructuring requirements (Warrington et al. 1995).

The slender spindle (Wertheim 1978) involved a central leader with only one permanent tier of four to five scaffold branches above which only small fruiting branches were developed. Limbs were spread or tied down in the second and third seasons. After the fifth year, larger branches were removed from the top of the tree and lower scaffold branches were cut back to weak side branches in order to contain the tree to its allocated space. Larger pruning cuts were made in winter, whilst summer pruning involved removal of upright shoots and cutting back some fruiting branches to the first apple in the top of the tree to maintain conical shape. Trees were maintained at desired height by heading back central leaders to side branches. Wertheim (1978) discussed cultivar-specific modifications, consistent with the principles of the work of Lauri and Corelli (2014) described in the next section.

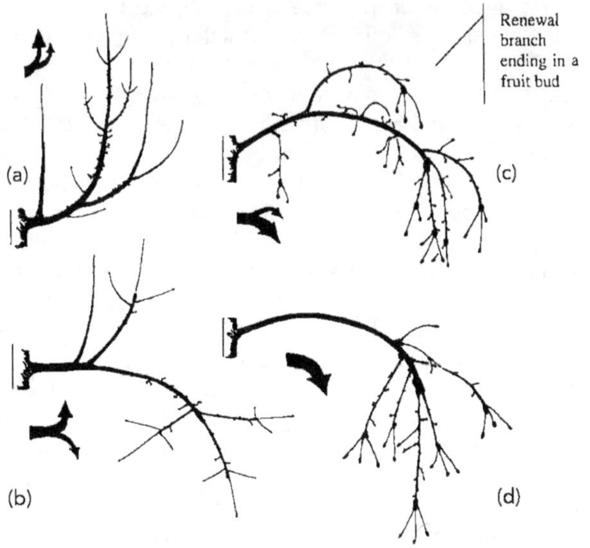

Figure 1 Silhouette of a four-year-old fruiting branch from 'Reine des Reinettes' (a), 'Golden Delicious' (b), 'Granny Smith' (c) and 'Rome Beauty' (d). Arrows indicate vegetative growth distribution (from Lauri and Lespinasse 2000). *Acta Horticulturae* is acknowledged for provision of this figure.

The 'vertical axis system' was developed by Lespinasse et al. in 1975 based on the natural branching and fruiting habits of the cultivars and on the relationship between position within the canopy and fruit quality. The structure involves a central leader with regularly distributed fruiting branches and is conical in shape. Two principles were key to the structure: 1. terminal buds of the central axis and fruiting branches are of key importance to optimal vegetative/fruit balance and 2. limbs aim to have terminally located 'zone A' of optimal light interception for strong vegetative growth and predominantly 'zone b' the prime fruiting zone with good light and moderate shoot growth and minimal 'zone c' where light is inadequate for the required fruit quality. This structure, however, maintained fruiting buds on one-year-old wood which form generally small fruit of poor quality and, although supporting very productive yields, was labour intensive as the renewal pruning of branches to maintain tree shape led to excessive vigour.

V or Geneva Y trellis systems (known as the Tatura Trellis in Australia) use trellis to support two single-plane canopies at 60° that partly cover the alley ways, increasing intercepted light to over 70% (Robinson and Lakso 1991; Robinson 2000). Whilst Y trellises optimize theoretical light interception, they require detailed pruning and can produce excessive annual growth that shades spurs (Palmer and Warrington 2000), and so, in terms of cost–benefit do not compare favourably with comparably simpler systems such as the tall spindle.

Tustin et al. (1990), Lespinasse et al. (1975) and Lauri and Lespinasse (2000) developed the Slender Pyramid and Solaxe structures, respectively, where the central axis is bent rather than headed and branches are simplified and tied down if not attaining desired angles from the weight of fruit (see Fig. 2) on semi-vigorous rootstocks such as MM106. Pruning during establishment did not involve heading and was minimal, focussed on removal of excessive structure on the top of the tree via early renewal pruning in summer and not in winter. This encouraged rapid early growth and balanced trees. The Slender Pyramid aimed for the initial eight limbs in the basal tier of the tree managed down to four from years 4–6.

A significant issue with the Solaxe system was excessive flowering in the 3rd and 4th leaf. Different cultivars demonstrated different responses to this: (a) at the spur level the ability to produce a fruit bud after a fruit bud on two successive years, that is, the 'bourse-over-bourse' phenomenon (Lauri et al. 1995) and (b) at the branch level where cultivars do not produce bourse-over-bourse there were two cases: spur fruiting can be synchronized, leading to alternative bearing, or desynchronized. Natural spur extinction was described by Lauri et al. (1997) where there was a positive correlation between the annual proportion of physiological extinction of spurs (their death), which would otherwise flower year on year, and the ability of the cultivar to fruit annually. This led to the practice of removing flower clusters in shaded parts of the tree and on one-year-old wood, given the inefficiencies of chemical or hand-thinning flowers and fruit on these positions in the canopy. The thinning of a portion of flowering spurs was termed 'artificial spur extinction' (ASE). This led to improved canopy porosity to light and increased light distribution to fruiting spurs compared to vegetative shoots, resulting in a 20% increase of within-canopy irradiance (estimated by silhouette to total area ratio) relative to the standard Solaxe tree structure and increased the leaf area of fruiting shoots by approximately 15% (Willaume et al. 2004). This also avoids the need for summer pruning with heading cuts to get light to spurs and leads to faster achievement of balance between vegetative and fruit growth. Improved light relations, in addition to crop load management of ASE-managed trees, results in greater fruit size and colour (Lauri et al. 2007; Tustin et al. 2011, 2012). The critical

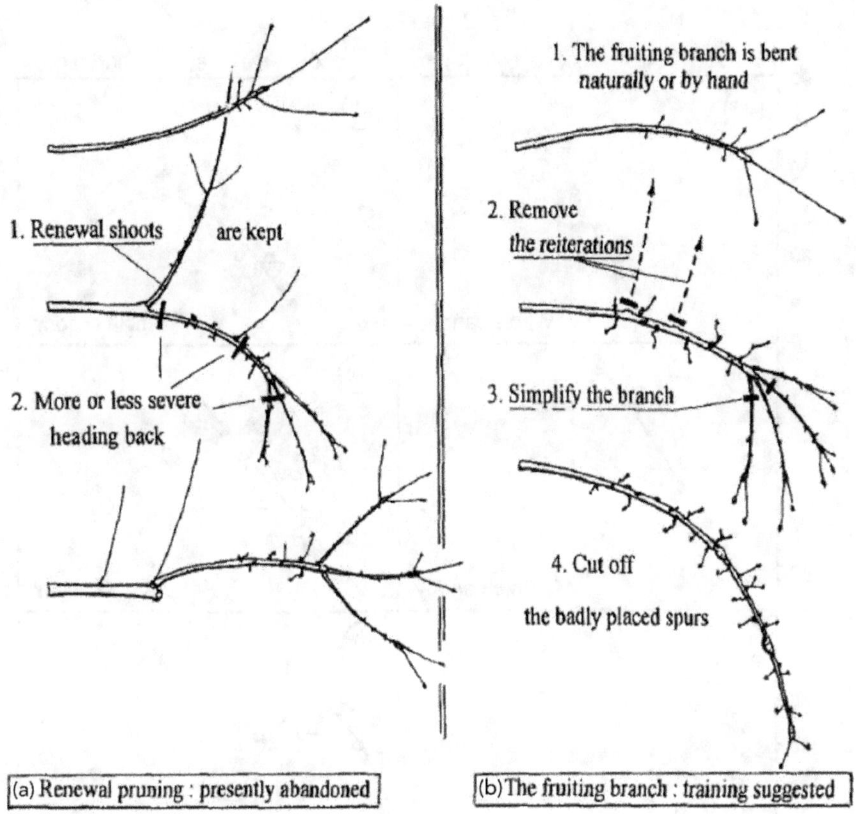

Figure 2 Pruning concept developed by Lauri and Lespinasse (2000) associated with the Solaxe tree structure. From previously practiced heading back (a) to bending, simplification and removal of spurs where spur density is too high and/or poorly positioned to receive light. *Acta Horticulturae* is acknowledged for provision of this Figure.

importance of light interception by the spur canopy has been elegantly shown by Wünsche et al. (1996) and Wünsche and Lakso (2000). Yield is linearly related to light interception by the spur canopy, yet not necessarily light interception by the whole canopy (see Fig. 3). This is because extension shoots utilize carbohydrate production of their own leaves until a critical leaf area (10–12 leaves; Johnson and Lakso 1986) is reached before exporting carbohydrate to other sinks. See the second case study of this chapter for further details on ASE in practice.

Modern tall slender spindle trees (Robinson et al. 2011, 2014) on dwarfing rootstocks are not 'headed' during tree establishment (as opposed to early approaches to this tree structure which led to invigorated upper branches and shaded lower canopy) to encourage rapid early growth. The central axis is grown to 3.3 m and then sometimes bent (similar to solaxe). Nursery trees have 10–15 'feathers' that are tied down at or below horizontal in the second year. Early fruiting slows vegetative growth. In addition to early fruiting, the reduced extension growth ensures that individual trees do not exceed their allocated

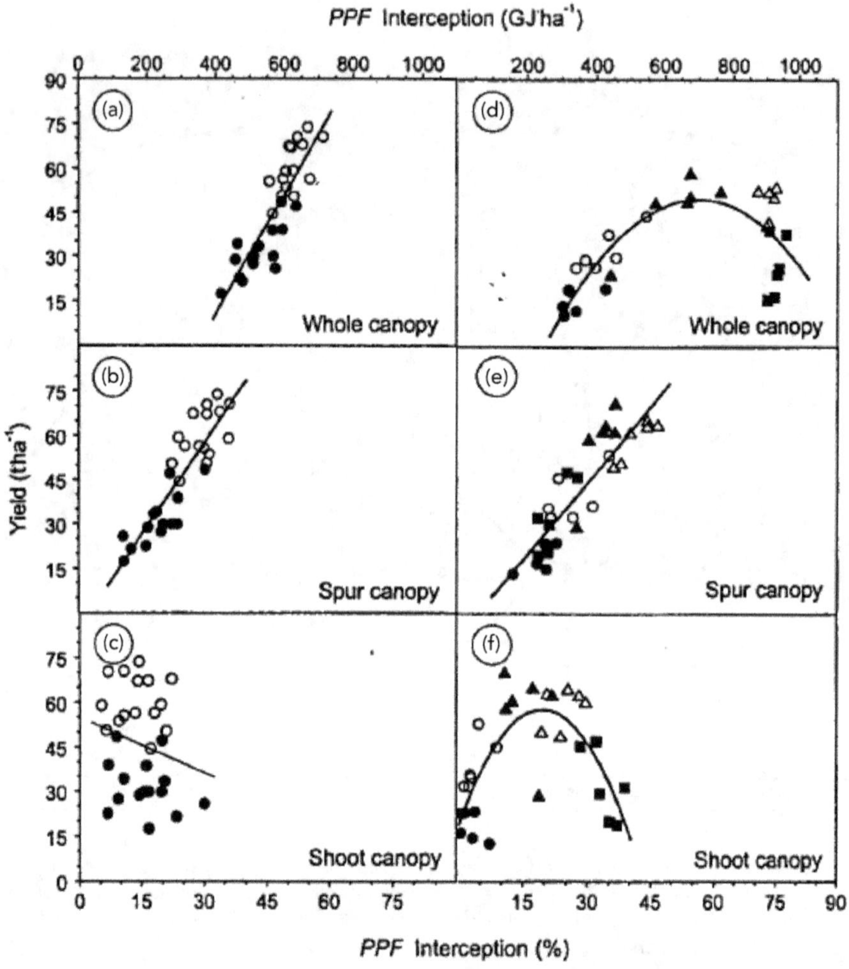

Figure 3 The relationship between fruit yield and mean whole canopy, spur canopy and extension shoot canopy light interception (average energy intercepted over 2 weeks and average percentage of available light intercepted daily at 3–5 and 10–12 weeks after full bloom) in several apple production systems (Wünsche, J. N. and Lakso, A. N. (2000)).

space in high-density systems and that their narrow canopies ensure light penetration to the lower canopy and fruitful buds that develop fruit of adequate colour development. Renewal pruning of dominant branches (>2 cm diameter or when lateral branches are more than 30% of the diameter of the main trunk in high-density plantings) with acute crotch angles are cut back to the main trunk. Replacement branches are often flat and bend under the weight of fruit, thus maintaining tree shape and ideal light distribution. The mature structure is narrow with no permanent scaffold branches and a simple canopy (similar to super spindle) of 10–15 small fruiting branches. This results in low ongoing requirement for pruning, therefore maximizing the harvest index. However potential issues with this structure include: 1. overcropping in 2nd and 3rd years preventing filling of allocated canopy space and potentially inducing biennial bearing in the absence of effective crop load management (Robinson et al. 2011) and 2. shading of lower branches as trees age which has been widely observed in Italy (Dallabetta et al. 2014) without sufficient effective limb renewal.

'Click' pruning (Dallabetta et al. 2014) has been applied to contain trees within their allocated area, though this represents greater labour requirement (10–20%) than the previously employed 'long pruning' technique (Mészáros et al. 2015). 'Click' pruning involves heading back the leader and the scaffold branches on new wood at the 2nd and 3rd bud positions. Lateral branches that have a terminal vegetative bud are vigorous or 4 years or older are cut back to a stub so that all fruiting branches are 2, 3 or 4 years old. Fruit yield and quality have been reported to be similar (Mészáros et al. 2015). Dallabetta et al. (2014) reported no significant increases in fruit size on trees of statistically similar crop load (N Dallabetta pers comm 2016) but suggested that improved homogeneity may reduce the number of picks required. Light summer pruning was similarly effective (Mészáros et al. 2015).

Robinson et al. (2014) described preliminary investigations into mechanization of dormant pruning via annual side-wall hedging, coupled with corrective pruning every third year, of 'non-invigorated' Tall Spindle structured trees minimally pruned during establishment. The careful matching of rootstock and scion structured into tall spindle trees, along with technology-assisted (GPS) alignment of rows and adaptation of existing 'cutting bar' technologies, will likely become the most labour efficient of contemporary tree structure systems. It should be noted that this is not the view of some other leading pomologists whose counter argument is that systems and tree structures optimized to maximize yield and quality will have greater impact on profitability relative to marginal cost savings in pruning, which may be to the detriment of yield and quality. Only time will tell!

5.2 Tailoring training and pruning to rootstocks and cultivars

Lauri and Corelli Grappadelli (2014) emphasized that cultivars of distinct typology are specific in how they respond to training and pruning: 1. 'plastic' cultivars, where training and pruning increase fruit quality and regularity of bearing and 2. 'non-plastic' cultivars, where training and pruning improve fruit quality in the high yielding year of an alternate bearing cycle without a consistent effect on the regularity of bearing there after Lauri et al. (2011) argued that training and pruning need to be based on the cultivar specific architecture. Type IV 'Granny Smith' requires little manipulation, whereas Type II 'Galaxy Gala' needs detailed bending to control growth *and* shape *in addition to* pruning to thin out reiterations and ASE to improve light distribution within the canopy. Key architectural features to consider include, at the branch level: occurrence of reiteration, fruiting response to bending and fruiting on one-year-old wood, and at the spur level: bourse-over-bourse phenomenon, natural spur extinction and bourse volume (Lauri et al. 2011). It should be noted that a high vigour rootstock increases alternative bearing of standard and tip-bearing cultivars, whereas it reduces the alternative bearing of spur-type cultivars (Barritt et al. 1997).

6 Case study 1: intensive growing systems in Tasmania, Australia

6.1 Introduction

Tasmania, Australia, is situated around 42°C latitude and enjoys long summer daylength, clear sky days and consistent chill unit accumulation over winter. These conditions support an industry that, at its peak in 1964, supplied 8.9 M boxes (160 000 T) of apples before Australia lost preferential market access after the negotiation by Britain of the European Free Trade Agreements in 1974. The high cost of labour and relatively expensive freight costs to move fruit to the Australian mainland port of Melbourne before international shipping led to the establishment of a demonstration trial of an intensive growing system in 2000 (Jotic and Oakford 2004). In addition to considering some of the training and pruning techniques discussed above, this case study demonstrates the principles outlined in planting systems in chapter 'Advances in understanding apple tree growth: rootstocks and planting systems'. It also focuses on a growing region that is relatively poorly represented in the literature.

6.2 Trial design and pruning

Given their economic value and contrasting growing and cropping characteristics, the cultivars 'Gala', 'Fuji', 'Pink Lady' and 'Sundowner' were compared. A range of dwarfing and semi-dwarfing rootstocks and interstems were planted at five densities as per Table 1.

Table 1 Density, cultivar and rootstock combinations investigated in a study of intensive growing systems in Grove, Tasmania

Tree density/ha	1904	2857	3333	4444	6666
Spacing (m)	3.5 x 1.5	3.5 x 1.0	3.0 x 1.0	3.0 x 0.75	3.0 x 0.5
Cultivar/rootstock					
'Gala'			M9	M9	M9
'Fuji'	MM106	MM106	M9/MM106	M9/MM106	M9/MM106
'Pink Lady'	MM102	MM102	M9/MM102	M9/MM102	M9/MM102
'Sundowner'			M26	M26	M26

Table 2 Feather development in Fuji 'Ogura'/MM102 one-year-old nursery trees planted as rods in September 2000 with treatments of scoring ('v blade'), notching and Cytolin applied in late October 2000 (adapted from Jotic and Oakford 2004)

Treatment	Ave. no. feathers in 0.7–1.1 m zone	Labour cost cents/tree (in year 2000)
Control	5.4	–
Cytolin (15mL/L)	5.7	4.5
Scored at 1.1m	8.9	1.6
Scored at 1.1m + Cytolin (15 mL/L)	9.5	6.1
Bud notched 0.7-1.1m zone	10.9	28.0
Bud notched + cytolin (15mL/L) in 0.7-1.1m zone	10.6	32.5

Tree structure was trained and pruned to slender pyramid supported by trellis. Additional comparisons were trialled: feathered versus rod one-year-old nursery trees; heading at 1.1 m versus minimal pruning and; Cytolin, scoring, notching and combined treatments of one-year-old rods to promote feathering after planting. Costs of trees, trellis and labour were tracked in order to enable comparison of yield and quality with input costs for cost–benefit analysis.

Minimal pruning, limited to laterals below 0.7 m on the trunk or those in competition with the centre leader, occurred in the first season. At the end of the dormant period, primary laterals were tied down to just below horizontal to reduce vigour and promote

Table 3 Yield (T/ha) in the second season after planting of feathered versus non-feathered ('rod') 'Pink Lady' trees from the nursery (adapted from Jotic and Oakford 2004)

Spacing (m)/Density per ha	Feathered	M9/MM106	M9/MM102	M26
3.0 x 0.75/4444	No	34.0	30.2	28.8
	Yes	43.8	41.7	33.4
3.0 x 1.0/3333	No	29.1	26.4	25.1
	Yes	37.6	31.5	31.9

Table 4 Cumulative yield (T/ha) after second and third season (adapted from Jotic and Oakford 2004)

Tree spacing (m)	Density (/ha)	M9/MM106	M9/MM102	M26	M9	MM102	MM106
'Fuji'							
3.0 x 0.5	6666	92.2	64.8	83.1	74.7		
3.0 x 0.75	4444	76.2	49.2	72.9	62.4	56.3	
3.0 x 1.0	3333	63.0	44.0	65.5		29.7	
3.5 x 1.0	2857					49.9	88.5
3.5 x 1.5	1904					48.2	62.1
'Gala'							
3.0 x 0.5	6666	83.0	75.2	84.5	88.6		
3.0 x 0.75	4444	76.2	67.5	77.3			
3.0 x 1.0	3333	63.5	53.8	62.4			
3.5 x 1.0	2857					45.2	75.5
3.5 x 1.5	1904					42.0	54.6
'Pink Lady'							
3.0 x 0.5	6666	83.5	80.1	82.8	92.2		
3.0 x 0.75	4444	76.5	70.8	74.1			
3.0 x 1.0	3333	70.6	63.9	77.8			
3.5 x 1.0	2857						87.4
3.5 x 1.5	1904						

fruitfulness. Minimal pruning for shoot thinning and laterals of narrow crotch angles occurred over years two and three.

6.3 Results

The major costs in the first season in descending order were trees, land preparation and trellis. In the second season, branch tying down and pruning and in the third season hand thinning were the major input costs.

6.4 Discussion

This case study (results from Jotic and Oakford 2004) has demonstrated the following:

1. Scoring was an inexpensive and effective method of promoting shoot development. Bud notching was similarly effective but relatively expensive, whilst Cytolin was relatively ineffective;
2. Feathered nursery trees, non-headed after establishment, consistently had significantly elevated yields and
3. Many cultivar/rootstock and density combinations produced yields of 30 T/ha and some 40 T/ha, in the second season that, except for 'Fuji', were generally exceeded in year three.

However by year three the highest yields were consistently observed on M9 rootstocks or interstocks. This emphasizes the value of architecture and precocity of dwarfing rootstocks. It should be noted that the positive effects of dwarfing rootstocks, at appropriate densities, would be expected to continue their trajectory of improved yield relative to semi-dwarfing rootstocks, due to the effects on scion architecture, and consequent within-canopy light relations, as well as precociousness. Shortcomings of this study, however, were the lack of fruit quality assessment and that it was limited to three seasons. It might have been expected that marketable yield would be higher on dwarfing rootstocks due to better colour development. Further, the results of this study need to be considered within the context that it was a demonstration trial for growers and not replicated to an extent that allowed statistical analysis.

With regard to tree expenses and quality versus early yield and cost–benefit, a particularly instructive comparison is that feathered 'Gala' on M9/MM106 at 3333 trees/ha yielded only 16% less crop over two seasons relative to rods at 6666 trees/ha at double the tree cost. Moderate high density of 2500–3500 well-feathered trees/ha that are minimally pruned and not headed represents the 'sweet spot' of not overinvesting in tree costs whilst maximizing tree canopy development and early yields. This conclusion is in line with current recommendations from Robinson et al. (2014) and emphasizes the importance of the key principles of early canopy development for light interception and within canopy light distribution effected by rootstock–scion combinations. These principles are clearly overriding factors, given the distinct soil and climatic conditions between Tasmania and New York State.

A significant cost in year 2 and the major cost in year 3 of this study were labour inputs for hand thinning following ineffective chemical thinning. The next case study describes recent findings on the technique of ASE, which allows effective crop load management with minimal hand thinning.

7 Case study 2: comparison of artificial spur extinction and chemical thinning for crop load control

7.1 Introduction

Artificial spur extinction (ASE) is a crop load management method using bud-thinning techniques to precisely define where and how much fruit is set on each limb of the tree. To maximize the benefits of any crop load management technology, including ASE, it is important to optimize tree canopy structure through pruning and tree training to achieve maximum light interception by the orchard canopy. This can be achieved by thinning out of limbs to establish a framework of 6–7 limbs per metre of tree height and tying down of upright limbs to reduce tree vigour. Spur extinction (the removal of unwanted buds) is performed immediately prior to bud break, with inferior quality buds and those in poor canopy positions being removed first; remaining buds are then spaced along the limb to obtain the desired bud density. Bud (fruit) density is a ratio of X fruit per unit cross-sectional area at the base of the fruiting limb and is typically in the range of four to six buds or fruit cm^{-2} limb cross-sectional area (LCSA). Hence ASE-managed trees commence spring growth in an already significantly 'crop-thinned' state, carrying fewer but stronger flower buds than conventionally managed trees (Fig. 4). This reduces competition for tree resources, enabling optimization of crop load and maximizing fruit quality potential. Hand thinning is simplified because spacing, position and the number of clusters are already set by the bud-thinning process – all that is required is to break up fruit bunches and remove fruit with defects. It is important to note that ASE will not compensate for poor pruning techniques leading to poor tree structure.

Figure 4 Comparison of 'Scifresh' trees with and without spur extinction. Limb numbers are similar in both trees (Courtesy of Tustin et al., Plant & Food Research Ltd, NZ).

A five-year programme examining the implementation of ASE on 'Pink Lady' and 'Gala' across three growing regions in Australia from 2009 to 2014 demonstrated the benefits of ASE over conventional crop-loading practices with yields approaching 100 T/ha of high-quality fruit obtained with ASE management (Close 2014). This programme compared precise crop load management using ASE with conventional hand thinning after final fruit drop in late November/early December (dependent on site); no chemical thinning was used. While, in theory, application of ASE to trees should eliminate the need for chemical thinning, growers asked for a comparison of ASE and chemical thinning for crop load management, and hence, a study was undertaken on the 'Royal Gala' site in Tasmania in the final year of the programme.

7.2 Trial design and treatments

Treatments were ASE or conventional hand thinning plus or minus a chemical-thinning regime. Floral buds were thinned to 5 buds cm^{-2} LCSA on the ASE trees in late August just prior to bud break. Chemical thinning was undertaken during the flowering period, with 10 ppm naphthalene acetic acid (NAA) applied at full bloom (FB). Unfortunately continuing severe weather conditions did not permit application of a post-bloom thinner as originally planned. All trees were hand-thinned 9 weeks after FB, following natural fruit drop, to 5 fruit cm^{-2} LCSA. Fruit was harvested at normal commercial harvest and a subsample of fruit selected randomly from each tree for laboratory assessment of fruit quality.

7.3 Results

Crop load management method influenced the number of fruit set per floral bud (Fig. 5). Conventional trees had significantly fewer buds setting 3, 4, 5 or 6 fruit per bud than ASE

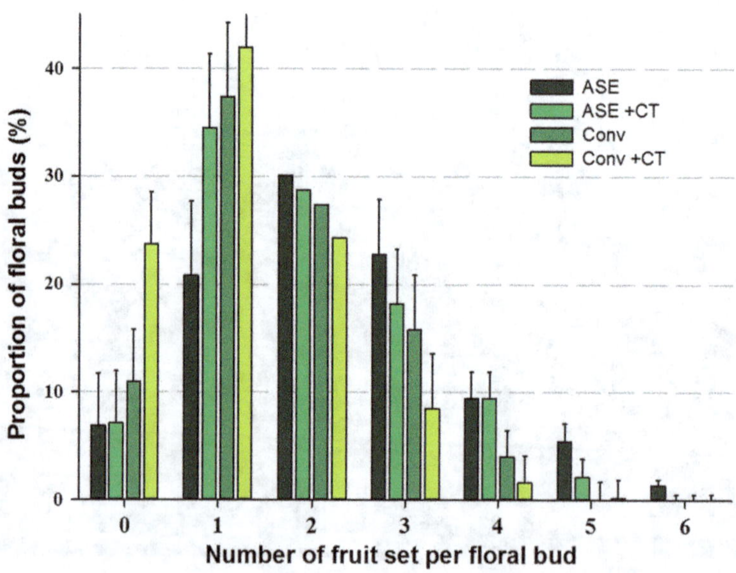

Figure 5 Effect of crop load management method on the number of fruit set per floral bud in 'Royal Gala'. ASE = Artificial Spur Extinction, CT = chemical thinning.

trees. Chemically thinned trees had a higher proportion of buds setting one fruit than trees without chemical thinning. Conventional plus chemical thinning resulted in the highest number of buds not setting any fruit.

Fruit set was significantly higher in ASE trees than in conventional ones (Table 5), with a twofold increase in the number of fruit set per 100 blossom clusters in ASE trees compared with traditional ones. Chemical thinning had no additional effect on fruit set per 100 blossom clusters on ASE trees but did reduce fruit set in conventionally managed trees. The percentage of clusters setting fruit was ~15% higher in ASE trees than in conventional ones, and chemical thinning had no additional effect on clusters set.

Crop load at harvest was the same for both ASE and conventionally managed trees, determined on a branch level and a whole tree basis (Table 5). Conventionally managed trees had more limbs than ASE trees, and this was reflected in the total limb cross-sectional area; however, this did not result in an increase in total fruit cm^{-2} trunk CSA.

ASE-managed trees produced heavier fruit than conventionally managed ones (Table 6), but fruit weight was the same in all ASE trees, regardless of chemical thinning. Fruit from ASE-managed trees was more mature than the one from conventional trees, with higher (yellower) background colour and greater conversion of starch (higher starch pattern index).

Table 5 Effect of crop load management method on fruit set, final crop load and total limb area in 'Royal Gala' trees. ASE = artificial spur extinction; CT = chemical thinning; Conv = conventional; LCSA = limb cross-sectional area

	No. fruit set per 100 blossom clusters	% clusters set	Crop load at harvest (fruit cm^{-2} LCSA)	Total fruit cm^{-2} trunk CSA)	Total limb area (cm^2)
ASE	205 c	93 b	4.3	7.1	30.9 a
ASE + CT	174 bc	89 b	4.4	6.8	30.8 a
Conv	137 b	80 a	4.4	8.8	42.9 b
Conv + CT	94 a	78 a	4.6	6.6	43.1 b

Comparisons within each column followed by different letters are significantly different at P=0.05 using the LSD test.

Table 6 Effect of crop management method and chemical thinning on fruit weight, skin background colour, and starch pattern index of 'Royal Gala' fruit. ASE = artificial spur extinction; CT = chemical thinning

	Mean fruit weight (g)	Skin background colour	Starch pattern index
ASE	152 b	5.51 bc	3.72 b
ASE + CT	150 b	5.67 c	3.54 b
Conv	135 a	5.24 a	2.97 a
Conv + CT	141 ab	5.36 ab	3.66 b

Comparisons within each column and main effect followed by different letters are significantly different at P = 0.05 using the LSD test.

7.4 Discussion

This study demonstrates that ASE is a viable alternative to chemical thinning as a crop load management tool, and confirms the postulation by Tustin et al. (2012) that bud (spur) extinction could replace chemical thinning. Although chemical thinning is a valuable crop load management practice that is widely used in apple production, it has limitations due to its inherent unpredictability resulting from complex interactions between the plant and the environment; this unpredictability normally results in the need for extensive hand thinning. However, as hand thinning is usually delayed until natural fruit drop is complete, there is considerable wastage of the tree's resources arising from the delay in achieving target fruit numbers.

A major advantage of ASE is the early removal of excess floral buds, allowing a reduction of floral bud density without the loss of commercial yield (Breen et al. 2014). This removal of competitive floral/fruitlet sinks prior to bud break and flowering when competition for resources is high improves fruit size, as illustrated in this study with a 9.4% increase in mean fruit weight in ASE trees for the same crop density set by post-drop hand thinning; this is confirmed by Robinson et al. (2014) and van Hooijdonk et al. (2014). Bud position is optimized in ASE, fruit is well spaced and light distribution into the canopy is enhanced.

Hand thinning is still necessary in ASE trees, but is simplified compared with chemically thinned trees, as bud number and spacing is already set; hence, all that is required is to break up bunches and remove damaged fruit. As observed in this work, several studies have demonstrated that reducing floral bud density through ASE increases the proportion of floral buds that set fruit and also increases the number of fruit set on individual buds (Tustin et al. 2012; Breen et al. 2014; Close 2014; van Hooijdonk et al. 2014). Hence in the event that not all buds set fruit, the desired crop load can still be readily achieved by leaving two fruit, rather than singles, on some bud sites. As fruit buds in ASE-managed trees are all well positioned, carrying multiple fruit on some buds does not appear to detract from fruit quality.

Maturity of fruit in ASE-managed trees is usually more even (background colour and SPI). The commercial outcome of this is that a higher percentage of fruit can be picked at the first harvest where multiple harvests are required, or in many cases all fruit can be picked at a single time, removing the need for multiple harvests.

The lack of effect of chemical thinning in further reducing crop load on ASE-managed trees demonstrates that ASE is a suitable tool for managing crop load without additional chemical thinning. There is also the added advantage that the achievement of target crop loads is no longer dependent on unpredictable weather conditions during the flowering and post-bloom periods when thinners are normally applied. A move away from chemical thinning, combined with simplified hand thinning and more even maturity, will reduce both time and cost to the grower.

This case study has demonstrated that application of ASE for crop load management is a viable alternative to chemical thinning, and is beneficial to growers in providing predictable fruit set and crop load responses. However implementation of ASE and a move away from chemical thinning will require a paradigm shift for growers.

8 Summary

Application of branching agents containing synthetic cytokinins and gibberellins has become standard nursery practice to stimulate branching. Notching has gained acceptance for branch induction in older trees, particularly in combination with cytokinin. Under conditions of inadequate winter chilling, bud break can be manipulated by cultural practices or application of chemical rest-breaking agents. It has been suggested that respiratory stress may be involved in the release of buds from dormancy, but there is still a lack of information on the mechanisms of how many rest-breakers work. While bioregulators are valuable tools, they cannot compensate for poor tree management.

How scion cultivars interact with rootstocks has driven studies on training and pruning, underpinned by knowledge of architectural traits and tree physiology, in order to achieve the fundamental principles of the following:

- Buds need exposure to no less than 30% incident light for adequate flower initiation and development (Palmer 1989).
- Leaves exposed to uninterrupted light are efficient sources of photosynthates relative to shaded leaves.
- Spur leaves provide a major quantity of photosynthate for fruit development (and quality attributes such as size and sugars (e.g. Tustin et al. 1988), acids and secondary metabolites), relative to leaves on vegetative shoots.
- Fruit require exposure to light (at least 40%; Heinicke 1966) in order to develop adequate colour.
- Harvest index, the proportion of biomass produced as fruit relative to other tree biomass, should be maximized to prevent wasted investment of carbon into non-harvested structures.

Modern tree structures, on dwarfing rootstocks, such as the tall slender spindle, are characterized by the following: minimal pruning during establishment, no heading back of the apex, branches are tied down to below horizontal, no scaffold branches are developed and the number and position of floral buds can be managed by ASE. Modern tree structures are carefully developed to suit the desired planting system: the combination of cultivar, rootstock, tree spacing and training/pruning regime for a given site. The golden principle of the growing system is to maximize light interception per unit area as this is directly proportional to marketable yield, assuming tree structures allow good within-canopy light distribution. Whilst contemporary best practice still achieves only around 60% light interception of mature orchards, a level that has not improved significantly for many decades, significant improvement in the rate at which a full canopy is achieved is a feature of modern growing systems. This is through finding the 'sweet spot' of tree density that balances early canopy development and fruit yield with the cost of trees. This sweet spot is considered to be within the range of 2500–3300 trees/ha in general across different soils, climates and even hemispheres.

Advancing our understanding of apple tree growth has led to significant incremental improvement in yield, fruit quality and the efficiency of achieving these, so that consumers can enjoy an affordable, quality eating experience in many parts of the world.

9 Future trends in research

Palmer et al. (2002) calculate that 169 T/ha is theoretically possible if systems can be designed that intercept 90% of incident light. Tustin and van Hooijdonk (2013) suggest that a more radical simplification of tree canopies may be possible along the lines of the

Bibaum (Musacchi 2008; Dorigoni et al. 2011) structure of two axes along the row with relatively (compared to Slender Spindle) short laterals that can be easily managed in terms of vigour distribution, using current technologies and techniques. The Bibaum® structure can increase harvest index above that of the Slender Spindle, improves light penetration (Dallabetta et al. unpub.) and reduces the costs of trees required per ha. This could shift to multiple axis trees to further simplify secondary branch structure to short shoots and spurs subtended off vertically positioned fruiting shoots arising from horizontal cordons (Tustin and van Hooijdonk 2013; Tustin 2014) that would reasonably be best suited to Type I or spur-type cultivars. The planar canopy that minimizes vegetative shoots further improves light distribution and enables reduction of row spacing to 1.5–2.0 m (Tustin and van Hooijdonk 2013). Certainly mechanical technology is available for 1.5 m alleys; French viticulture appellations have required between row spacings as low as 1 m for centuries! Current issues with multiple axis trees include managing for uniform vigour of the uprights, although this is achieved in the conceptually similar 'Upright Fruiting Offshoots' or UFO cherry tree structure (Whiting 2011) where rootstock–cultivar and site are carefully matched and tree management during establishment is sufficiently skilled. Planar architecture has added benefits in terms of chemical spray efficiency and, most importantly for countries with high labour costs, relatively high harvest efficiency (Ampatzidis and Whiting 2013).

Probably the greatest opportunity for step-change in quality and yield of apple production, underpinned by advances in our understanding of tree growth, is that afforded by multidisciplinary research that aligns systems with new and emerging technologies of mechanization, automation, remote sensing and robotics. The complex architectures of tree crops have limited investment and discovery in these areas up till now, particularly when one considers the successful development and deployment of these technologies in crops of relatively simple architectures, such as wheat for example.

Finally the consolidating and modelling of experimental data will enable the development of scenario testing models that can then be developed into web-based applications or 'apps' that will provide a level of decision support to horticulturalists, advisors and growers, something that has never before been available. This will lead to further efficiency gains, not only in terms of production, but also in terms of efficiency of inputs such as water and nutrients, which will minimize off-site impacts, water, carbon and nitrogen 'footprints' which are increasingly of interest to retailers and consumers.

10 Where to look for further information

- John E. Jackson's review article 'Light i.nterception and utilisation by orchard systems' (1980, *Horticultural Reviews* 2: 208–67) and Alan N. Lakso's book chapter (Lakso, A. N. (1994), 'Apple', in *Handbook of Environmental Physiology of Fruit Crops. Vol. I: Temperate Crops*, B. Schaffer and P. C. Andersen (Eds), pp. 3–42) are essential reading to those interested in the fundamental principles of light interception and distribution, yield, quality and growing systems.
- Plant bioregulators: review article by Stephen S. Miller 'Plant bioregulators in apple and pear culture' (1988, *Horticultural Reviews* 10: 309–401) gives a good background to the use of PBRs in apple.

- For background information on dormancy, refer to Saure, M. M. (1985), 'Dormancy release in deciduous fruit trees', *Horticultural Reviews* 7: 239–300.
- The International Society for Horticultural Science (ISHS: http://www.ishs.org/) organizes the International Horticulture Congress (IHC), the premier global horticultural event, which covers all aspects of horticulture and runs every four years. Additionally, chapters and branches of the ISHS organize individual symposia that run every four years – the most relevant to this chapter being 'Integrating Canopy, Rootstock and Environmental Physiology in Orchard Systems' and 'Plant Bioregulators in Fruit Production'.
- The ISHS publishes *Acta Horticulturae*, a journal of refereed papers, which captures the proceedings of each symposium. This journal, as well as other relevant journals, is cited in the references section of this chapter.

11 Acknowledgements

Drs Ben van Hooijdonk and Stuart Tustin are gratefully acknowledged for many insightful comments that significantly improved an earlier version of this chapter.

12 References

Ampatzidis, Y. G. and Whiting, M. D. (2013), 'Training system affects sweet cherry harvest efficiency', *HortSci*, 48(5), 547–55.

Autio, W. R. and Greene, D. W. (1992), 'Ringing and scoring to control growth of apple trees: results from 1991', *Massachusetts Cooperative Extension Service Fruit Notes*, 57(3), 1–5.

Barden, J. A., DelValle, T. B. G. and Myers, S. C. (1989), 'Growth and fruiting of "Delicious" apple trees as affected by severity and season of pruning', *J. Am. Soc. Hort. Sci.*, 114, 184–6.

Barritt, B. H., Konishi, B. S. and Dilley, M. A. (1997), 'Tree size, yield and biennial bearing relationships with 40 apple rootstocks and three scion cultivars', *Acta Hort.*, 451, 105–12.

Batjer, L. P., Williams, M. W. and Martin, G. C. (1964), 'Effects of N-dimethylaminosuccinamic acid (B9) on vegetative and fruit characteristics of apples, pears and sweet cherries', *Proc. Amer. Soc. Hort. Sci.*, 85, 11–19.

Black, M. W. (1936), 'Some physiological effects of oil spray on deciduous fruit trees', *J. Pomol.*, 14, 175–202.

Bound, S. A. and Jones, K. M. (2004), 'Hydrogen cyanamide impacts on flowering, crop load and fruit quality of red 'Fuji' apple (*Malus domestica*)', *N.Z. J. Crop Hort. Sci.*, 32, 227–34.

Bound, S. A. and Miller, P. (2006), 'Effects of Waiken on flowering and spring growth in apple', *Acta Hort.*, 727, 167–74.

Bound, S. A. and Summers, C. R. (2001), 'The effect of pruning level and timing on fruit quality in red "Fuji" apple', *Acta Hort.*, 557, 295–302.

Breen, K. C., van Hooijdonk, B. M., Tustin, D. S., Wilkie, J. D., Bound, S. A., Middleton, S. G. and Close, D. C. (2014), 'Changes in fruit set of 'Gala' apple in response to environment and artificial spur extinction', *Acta Hort.*, 1058, 77–83.

Byers, R. E. (1993), 'Controlling growth of bearing apple trees with ethephon', *HortSci.*, 28(11), 1103–5.

Byers, R. E. and Yoder, K. S (1999), 'Prohexadione-calcium inhibits apple, but not peach, tree growth, but has little influence on apple fruit thinning or quality', *HortSci.*, 34(7), 1205–9.

Clements, J. M., Autio, W. R. and Cowgill, W. P. (2010), 'Using notching vs. notching with or without BA application to induce branching in non-feathered, first-leaf apple trees', *Fruit Notes*, 75, 7–11.

Cline, J. (2000), 'Plant growth regulators (PGRs) for apples: understanding and using them to your advantage', *Elgin Grower Meeting*, 23 March 2000.

Cline, J. A., Embree, C. G., Hebb, J. and Nichols, D. S. (2008), 'Performance of prohexadione-calcium on shoot growth and fruit quality of apple — Effect of spray surfactants', *Can. J. Pl. Sci.*, 88, 165–74.

Close, D. C. (2014), 'Productivity, Irrigation, Pests and Soils (PIPS) Orchard Productivity Program', *Horticulture Innovation Australia Final Report AP09031*.

Dallabetta, N., Forno, F., Mattedi, L., Giordan, M. and Wehrens, R. (2014), 'The implication of different pruning methods on apple training systems', *Ag. & Forestry*, 60(4), 173–9.

Darbyshire, R. (2014), 'Winter chill and fruit trees', Research Fact Sheet, Primary Industries Climate Challeges Centre. http://www.hin.com.au/__data/assets/pdf_file/0004/18265/Horticulture-winter-chill-fact-sheet-FINAL.pdf; accessed 21 July 2016.

Dayatilake, G. A., Palmer, J. W. and Tustin, D. S. (2011), 'Enhancing budbreak and axillary shoot development of 'M.9' apple rootstock during the early period of stool-bed establishment', *N.Z. J. Crop Hort. Sci.*, 39(4), 263–70.

Deyton, D. E., Sams, C. E. and Milne, C. G. (2014), 'Effect of foliar applied paclobutrazol on morphology and physiology of apple trees', *Acta Hort.*, 1042, 105–11.

Diaz, D. H., Alvarez, A., Sandoval, J. (1987), 'Cultural and chemical practices to induce uniform bud break of peach and apple under warm climates in Mexico', *Acta Hort.*, 199, 129–36.

Doric, M., Keserovic, Z., Magazin, N. and Milic, B. (2015), 'The effects of BA and BA+GA_{4+7} on the main shoot dynamics and the feather formation in two-year-old 'knip-boom' apple trees', *Proceedings 50th Croatian and 10th International Symposium on Agriculture*, Opatija, Croatia, pp. 555–9.

Dorigoni, A., Lezzer, P., Dallabetta, N., Serra, S. and Musacchi, S. (2011), 'Bi-axis: an alternative to slender spindle for apple orchards', *Acta Hort.*, 309, 581–8.

Elfving, D. (2015), 'Plant bioregulators', Washington State University Horticultural Resources. http://hort.tfrec.wsu.edu/pages/Plant_Bioregulators (accessed 5 January 2016).

Elfving, D. C and Visser, D. B. (2005), 'Cyclanilide induces lateral branching in apple trees', *HortSci.*, 40(1), 119–22.

Elfving, D. C. (1985), 'Comparison of cytokinin and apical-dominance inhibiting growth regulators for lateral branch induction in nursery and orchard apple trees', *J. Hort. Sci.*, 60(4), 447–54.

Erez, A. (1995), 'Means to compensate for insufficient chilling to improve bloom and leafing', *Acta Hort.*, 395, 81–95.

Erez, A. and Lavee, S. (1971), 'The effect of climatic conditions on dormancy development of peach buds. I. Temperature', *J. Amer. Soc. Hort. Sci.*, 96, 711–14.

Evans, J. R., Evans, R. R., Regusci, C. L. and Rademacher, W. (1999), 'Mode of action, metabolism and uptake of BAS 125W, prohexadione-calcium', *HortSci.*, 34, 1200–1.

Feree, D. C. (1992). 'Time of root pruning influence growth, fruit size, biennial bearing and yield of Jonathan apple', *J. Am. Soc. Hort. Sci.*, 117, 198–202.

Ferree, D. C. and Schupp, J. R. (2003). 'Pruning and training physiology', in *Apples: Botany, Production And Uses*, D. C. Feree. and I. J. Warrington (eds), Chapter 14, pp. 319–44. CABI International, Oxfordshire, UK.

Fishman, S., Erez, A. and Couvillon, G. A. (1987), 'The temperature dependence of dormancy breaking in plants – mathematical analysis of a two-step model involving a cooperative transition, *J. Theor. Biol.*, 124, 473–83.

Forshey, C. G. (1982), 'Branching responses of young apple trees to applications of 6-benzylaminopurine and gibberellin A_{4+7}', *J. Amer. Soc. Hort Sci.*, 107, 538–41.

Geisler, D. and Feree, D. C. (1984), 'Response of plants to root pruning', in *Horticultural Reviews*, J. Janick (ed.), Vol. 6, John Wiley & Sons Inc., Hoboken, NJ, pp. 155–88.

Goren, R., Huberman, M. and Goldschmidt, E. E. (2004), 'Girdling: physiological and horticultural aspects', in *Horticultural Reviews*, J. Janick (ed.), Vol. 30, pp. 1–36. John Wiley & Sons Inc., Hoboken, NJ.

Greene, D. W. (2015), 'Plant growth regulators in tree manipulation', Washington State University Fruit School, November 17–18, 2015.

Greene, D. W. (2010), 'The development and use of plant bioregulators in tree fruit production', *Acta Hort.*, 884, 31–40.

Greene, D. W. (1999), 'Tree growth management and fruit quality of apple trees treated with prohexadione-calcium (BAS 125)', *HortSci.*, 34, 1209–12.

Greene, D. W. (1983), 'Use of Promalin to increase branching of young trees', *Fruit Notes*, 48(2), 20–2.

Greene, D. W. and Autio, W. R. (1994), 'Notching techniques increase branching of young apple trees', *J. Amer. Soc. Hort. Sci.*, 119, 678–82.

Greene, D. W, Autio, W. R. and Miller, P. (1990), 'Thinning activity of benzyladenine on several apple cultivars', *J. Amer. Soc. Hort. Sci.*, 115, 394–400.

Greene, D. W. and Lord, W. J. (1983). 'Effects of dormant pruning, summer pruning, scoring and growth regulators on growth, yield and fruit quality of 'Delicious' and 'Cortland' apple trees', *J. Am. Soc. Hort. Sci.*, 108, 590–5.

Greene, D. W. and Miller, P. (1988), 'Effects of growth regulator sprays and notching on growth and branching of "Starking Delicious" apple trees', *J. Amer. Soc. Hort. Sci.*, 113, 18–23.

Gvozdenović, D., Magazin, N. and Lazić, S. (2008), 'Influence of different central leader training treatments on apple tree growth and productivity', *Acta Hort.*, 772, 381–4.

Heinicke, D. R. (1966), 'Characteristics of McIntosh and Red Delicious apples as influenced by exposure to sunlight during the growing season', *Proc. Am. Soc. Hort. Sci.*, 89, 10–13.

Johnson, R. S. and Lakso, A. N. (1986), 'Carbon balance model of a growing apple shoot. II. Simulated effects of light and temperature on long and short shoots', *J. Am. Soc. Hort. Sci.*, 111(2), 164–9.

Jones, O. P. (1967), 'Effect of benzyladenine on isolated apple shoots'. *Nature*, 215, 1514–15.

Jones, K. M., Koen, T. B., Bound, S. A. and Oakford, M. J. (1991), 'Some reservations in thinning "Fuji" apples with naphthalene acetic acid (NAA) and ethephon', *N.Z. J. Crop Hort. Sci.*, 19, 225–8.

Joosse, M. L. (1986), 'Growth regulators and tree quality: more branching and more profit!', *Fruitteelt*, 76(22), 663–4.

Jotic, P. and Oakford, M. (2004), 'Intensive apple growing systems', Horticulture Australia Limited, Final Report AP 98022.

Khan, Z. U., McNeil, D. L. and Samud, A. (1998). 'Root pruning reduces the vegetative and reproductive growth of apple trees growing under an ultra high density planting system', *Sci. Hort.*, 77, 165–76.

Koen, T. B., Jones, K. M. and Oakford, M. J. (1989), 'Promoting branching in young trees of apple cv. Red Delicious using growth regulators', *J. Hort. Sci.*, 64(5), 521–5.

Lauri, P-É. and Corelli Grappadelli, L. (2014), 'Tree architecture, flowering and fruiting – thoughts on training, pruning and ecophysiology', *Acta Hort.*, 1058, 291–8.

Lauri, P-É., Costes, E., Regnard, J. L., Brun, L., Simon, S., Monney, P. and Sinoquet, H. (2009), 'Does knowledge on fruit tree architecture and its implications for orchard management improve horticultural sustainability? An overview of recent advances in the apple', *Acta Hort.*, 817, 243–50.

Lauri, P-É. Crété, X. and Ferré, G. (2007), 'Centrifugal training in apple – Appraisal of a two-year experiment on cv. 'Galaxy' in southeast France', *Acta Hort.*, 732, 391–6.

Lauri, P-É., Hucbourg, B., Ramonguilhem, M. and Méry, D. (2011), 'An architectural-based tree training and pruning – identification of key features in the apple', *Acta Hort.*, 903, 589–96.

Lauri, P-É. and Lespinasse, J. M. (2000), 'The vertical axis and solaxe systems in France', *Acta Hort.*, 513, 287–96.

Lauri, P-É., Térouanne, É. and Lespinasse, J. M. (1997), 'Relationship between the early development of apple fruiting branches and the regularity of bearing – An approach to the strategies of various cultivars', *J. Hort. Sci. Biotech.*, 72(4), 519–30.

Lauri, P-É., Térouanne, É., Lespinasse, J. M., Regnard, J. L. and Kelner, J. J. (1995), 'Genotypic differences in the axillary bud growth and fruiting pattern of apple fruiting branches over several years – an approach to regulation of fruit bearing', *Sci. Hort.*, 64, 265–81.

Lespinasse, J. M. and Delort, J. F. (1986), 'Apple tree management in vertical axis: appraisal after ten years of experiments', *Acta Hort.*, 160, 139–55.

Lever, B. G. (1986), 'Cultar – a technical review', *Acta Hort.*, 179, 459–66.

Looney, N. E. (1983), 'Growth regulator usage in apple and pear production', in *Plant Growth Regulating Chemicals,* L. G. Nickell (ed.), Vol. 1, pp. 1–26. CRC Press, Boca Raton, Florida.

Looney, N. E. (1968), 'Light regimes within standard sized apple trees as determined spectrophotometrically', *Proc. Am. Soc. Hort. Sci.*, 93, 1–6.

Maggs, D. H. (1964). 'Growth rates in relation to assimilate supply and demand. I. Leaves and roots as limiting regions', *J. Exp. Bot.*, 15, 574–83.

Maggs, D. H. (1965). 'Growth rates in relation to assimilate supply and demand. II. The effect of particular leaves and growing regions in determining the dry matter distribution in young apple trees', *J. Exp. Bot.*, 16, 387–404.

Marini, R. P. and Barden, J. A. (1982), 'Yield, fruit size and quality of 3 apple cultivars as influenced by summer or dormant pruning', J. *Am. Soc. Hort. Sci.*, 107(3), 474–9.

McArtney, S. and Obermiller, J. D. (2015), 'Effect of notching, 6-benzyladenine, and 6-benzyladenine plus gibberellin A_4+A_7 on budbreak and shoot development from paradormant buds on the leader of young apple trees', *HortTech.*, 25(2), 233–7.

Medjdoub, R., Val, J. and Blanco, A. (2004), 'Prohexadione-Ca inhibits vegetative growth of "Smoothee Golden Delicious" apple trees', *Sci. Hort.*, 101, 243–53.

Melke, A. (2015), 'the physiology of chilling temperature requirements for dormancy release and bud break in temperate fruit trees grown at mild winter tropical climate', *J. Pl. Studies*, 4(2), 110–56.

Mészáros, M., Sus, J., Laňar, L. and Náměstek, J. (2015), 'Evaluation of slender spindle form in young 'Topaz' apple orchard', *Sci. Agri. Bohemica*, 46(4), 167–71.

Miller, S. S. (1988), 'Plant bioregulators in apple and pear culture', *Hort. Rev.*, 10, 309–401.

Miller, S. S. (1989), 'Effect of 500 ppm 6BA and spray volume on shoot development in bearing Redchief Delicious', *Abbott Laboratories, Chemical and Agricultural Division, Report No 9074.*

Miller, S. S. (1995). 'Root pruning and trunk scoring have limited effect on young bearing apple trees', *HortSci.*, 30(5), 981–4.

Miller, P. (1983), 'Cytolin for branching of young Delicious apple trees', *Report No. 134/F1–1 Horticultural Research Institute, Knoxfield.*

Musacchi, S. (2008), 'Bibaum®: a new training system for pear orchards', *Acta Hort.*, 800, 763–8.

Nickell, L. G (1994), 'Chapter 1: Plant growth regulators in agriculture and horticulture', in P. Hedin (ed.), *Bioregulators for Crop Protection And Pest Control*, American Chemical Society, Washington, DC, pp. 1–14.

Ouellette, D. R. and Young, E. (1994), 'Branch inducement in apple stoolbed shoots by summer leaf removal and tipping', *HortSci.*, 29(12), 1478–80.

Palmer, J. W. (1989), 'The effects of row orientation, tree height, time of year and latitude on light interception and distribution in model apple hedgerow canopies', *J. Hort. Sci.*, 64(2), 137–45.

Palmer, J. W. and Warrington, I. J. (2000), 'Underlying principles of successful apple planting systems', *Acta Hort.*, 513, 357–63.

Palmer, J. W., Wunsche, J. N., Meland, M. and Hann, A. (2002), 'Annual dry matter production by three apple cultivars at four within-row spacings in New Zealand', *J. Hort. Sci. Biotech.*, 77, 712–17.

Petracek, P. D., Silverman, F. P. and Greene, D. W. (2003), 'A history of commercial plant growth regulators in apple production', *HortSci.*, 38(5), 937–42.

Petri, J. L. (1987), 'Breaking dormancy of apple trees with chemicals', *Acta Hort.* 199: 109–17.

Petri, J. L., Leite, G. B., Couto, M., Gabardo. G. C. and Haverroth, F. J. (2014), 'Chemical induction of budbreak: new generation products to replace hydrogen cyanamide', *Acta Hort.*, 1042, 159–66.

Quinlan, J. D. and Richardson, P. J. (1986), 'Uptake and translocation of paclobutrazol and implications for orchard use', *Acta Hort.*, 179, 443–51.

Rademacher, W. (1995), 'Growth retardants: biochemical features and applications in horticulture', *Acta Hort.*, 394, 57–73.

Rai, R., Joshi, S., Roy, S., Singh, O., Samir, M. and Chandra, A. (2015), 'Implications of Changing Climate on Productivity of Temperate Fruit Crops with Special Reference to Apple', *J. Hort.* 2, 135. doi:10.4172/2376-035401000135.

Richardson, E. A., Seeley, S. D. and Walker, D. R. (1974), 'A model for estimating the completion of rest for 'Redhaven' and 'Elberta' peach trees', *HortSci.*, 9, 331–2.

Rivers, T. (1866). '*The miniature fruit garden*'. 13th English ed. Orange Judd & Company, New York. Downloaded 18 August 2016 from: https://archive.org/details/miniaturefruitga01rive.

Robinson, T. L. (2011), 'Advances in apple culture worldwide', *Rev. Bras. Frutic.*, Special Issue, 37–47.

Robinson, T. L. (2000), 'V-shaped apple planting systems', *Acta Hort.*, 513, 337–47.

Robinson, T. L., Dominguez, L. I. and Acosta, F. (2014), 'Pruning strategy effects fruit size, yield and biennial bearing of Gala and Honeycrisp apples', *N.Y. Fruit Quarterly*, 22, 27–32.

Robinson, T. L., Hoying, S. A., Sazo, M. M., Dominguez, L. I. and Fachinello, J. C. (2014), 'Yield fruit quality and mechanisation of the tall spindle apple production system', *Acta Hortic.*, 1058, 95–103.

Robinson, T. L., Hoying, S. A. and Reginato, G. H. (2011), 'The tall spindle planting system: principles and performance', *Acta Hort.*, 903, 571–9.

Robinson, T. L. and Lakso, A. N. (1991), 'Bases of yield and production efficiency in apple orchard systems', *J. Am. Soc. Hort. Sci.*, 116(2), 188–94.

Robinson, T. L., Lakso, A. N. and Carpenter, S. G. (1991), 'Canopy development, yield and fruit quality of "Empire" and "Delicious" apple trees grown in four orchard production systems for ten years', *J. Amer. Soc. Hort. Sci.*, 116(2), 179–87.

Robinson, T. L. and Sazo, M. M. (2014), 'Effect of Promalin, benzyladenine and cyclanilide on lateral branching of apple trees in the nursery', *Acta Hort.*, 1042, 293–302.

Samud, A. (1998), 'Effect of girdling and bridge grafting on apple tree growth, yield, apple quality and biochemical composition for apple trees growing under moderate or ultra-high density conditions. *PhD Thesis, Lincoln University*. https://researcharchive.lincoln.ac.nz/handle/10182/1910; accessed 21 July 2016.

Sansavini, S. and Corelli, L. (1992), 'Canopy efficiency of apple as affected by microclimatic factors and tree structure', *Acta Hort.*, 322, 69–77.

Saure, M. C. (1985), 'Dormancy release in deciduous fruit trees', *Hort. Rev.*, 7, 239–300.

Schupp, J. R. (1990). 'Root pruning advantages', *Fruit Grower*, March 1990, pp. 24–5.

Schupp, J. R. and Feree, D. C. (1987). 'Effect of root pruning at different stages in growth and fruiting of apple trees', *HortSci.*, 22, 387–90.

Schupp, J. R. and Feree, D. C. (1990). 'Influence of time of root pruning on growth, net photosynthesis, and transpiration of young apple trees', *Sci. Hort.*, 42, 299–306.

Smit, M., Meintjes, J. J., Jacobs, G., Stassen, P. J. C. and Theron, K. I. (2005), 'Shoot growth control of pear trees (Pyrus communis L.) with prohexadione calcium', *Sci. Hort.*, 106(4): 515–29.

Terblanche, J. H. and Strydom, D. K. (1973), 'Effects of autumnal nitrogen nutrition, urea sprays and a winter rest-breaking spray on budbreak and blossoming of young 'Golden Delicious' apple trees grown in sand culture', *Decid. Fruit Grower*, 23, 8–14.

Terblanche, P. J. C., Hesebeck, I. and Strydom, D. K. (1979), 'Effects of autumn nitrogen application and a winter ret-breaking spray on the growth, development and composition of young 'Golden Delicious' apple trees grown in sand culture', *Sci. Hort.*, 10, 37–48.

Thomas, D., Hayman, P. and James, P. (2012), 'Understanding and managing the risks and opportunities from climate change on cherry production. South Australian Research and Development Institute (Sustainable Systems)', *DAFF project R1#087*.

Tustin, D. S., Dayatilake, G. A., Henroid, R. E., Breen, K. C. and Oliver, M. (2011), 'Changes in fruiting behaviour and vegetative development of "Scifresh" apple in response to artificial spur extinction using centrifugal training', *Acta Hort.*, 903, 603–10.

Tustin, D. S., Hirst, P. M. and Warrington, I. J. (1988), 'Influence of orientation and position of fruiting laterals on canopy light penetration, yield and fruit quality of "Granny Smith" apple', *J. Am. Soc. Hort. Sci.*, 113(5), 693–9.

Tustin, D. S. and van Hooijdonk, B. M. (2013), 'Roots, fruits and the pieces in between: a review of fruit production systems towards a step-change increase in productivity', *Aspects Appl. Biol.*, 119, 109–18.

Tustin, D. S., Dayatilake, G. A., Breen, K. C. and Oliver, M. J. (2012), 'Fruit set responses to changes in floral bud load – a new concept for crop load regulation', *Acta Hort.*, 932, 195–202.

Tustin, D. S., Hirst, P. M., Cashmore, W. M., Warrington, I. J. and Stanley, C. J. (1990), 'The principles and practices of training slender pyramid trees for high intensity production', *Compact Fruit Tree*, 23, 83–92.

Unrath, C. R. (1999), 'Prohexadione-Ca: a promising chemical for controlling vegetative growth of apples', *HortSci.*, 34, 1197–200.

Van Hooijdonk, B. M., Tustin, D. S., Oliver, M. J., Breen, K. C. and Dayatilake, G. A. (2014), 'Modification of canopy architecture imposed by artificial spur extinction promotes reliable cropping behaviour and enhances fruit quality of "Scilate" apple trees', *Acta Hort.*, 1058, 63–70.

Vergara, R., Rubio, S. and Perez, F. J. (2012), 'Hypoxia and hydrogen cyanamide induce bud-break and up-regulate hypoxic responsive genes (HRG) and VvFT in grapevine-buds', *Pl. Mol. Biol.*, 79, 171–8.

Wagenmakers, P. S. (1991), 'Simulation of light distribution in dense orchard systems', *Ag and For Met*, 57, 13–25.

Walser, M. N., Walker, D. R., and Seeley, S. D. (1981), 'Effects of temperature fall defoliation and gibberellic acid on the rest period of peach leaf buds', *J. Amer. Soc. Hort. Sci.*, 106, 91–4.

Wang, Y., Travers, S., Bertelsen, M. G., Thorup-Kristensen, K., Petersen, K. K. and Liu, F. (2014), 'Effect of root pruning and irrigation regimes on pear tree: growth, yield and yield components', *Hort Sci. (Prague)* 41(1), 34–43.

Warner, G. (2016), 'Getting the most out of Apogee', *Good Fruit Grower*, 14 January 2016. http://www.goodfruit.com/getting-the-most-out-of-apogee/ (accessed 11 February 2016)

Warrington, I. J., Stanley, C. J., Julian, J. F., Tustin, D. S., Hirst, P. M. and Cashmore, W. (1995), 'Pruning strategies for restructuring top-dominant central leader 'Granny Smith' apple trees', *N. Z. J. Crop Hort. Sci.*, 23, 315–22.

Wertheim, S. J. (1978), 'Pruning of slender spindle type trees', *Acta Hort.*, 65, 173–80.

Whiting, M. D. (2011), 'The U.F.O. Architecture: A novel system for high efficiency sweet cherry orchards', https://www.researchgate.net/publication/267295355_The_UFO_Architecture_A_Novel_System_for_High_Efficiency_Sweet_Cherry_Orchards (accessed 24 February 2016).

Willaume, M., Lauri, P-É. and Sinoquet, H. (2004), 'Light interception in apple trees influenced by canopy architecture manipulation', *Trees*, 18, 705–13.

Williams, M. W. and Billingsley, H. D. (1970), 'Increasing the number and crotch angles of primary branches of apple trees with cytokinins and gibberellic acid', *J. Amer. Soc. Hort. Sci.*, 95, 649–51.

Wilton, J. (2011), 'Growing for your market', *Australian Fruitgrower*, May 2011. Downloaded 18 August 2016 from: http://apal.org.au/wp-content/uploads/2013/07/au-fruitgrower-2011–04.pdf.

Wilton, J. (1999), 'Vigour control hints', *The Orchardist*, 72(7), 14–18.

Wünsche, J. N. and Lakso, A. N. (2000), 'Relationship between leaf area and light interception by spur and extension shoot leaves and apple orchard productivity', *HortSci.*, 35(7), 1202–6.

Wünsche, J. N., Lakso, A. N., Robinson, T. L., Lenz, F. and Denning, S. S. (1996), 'The bases of productivity in apple production systems: the role of light interception by different shoot types', *J. Am. Soc. Hort. Sci.*, 121(5), 886–93.

Zhuang, W., Gao, Z., Wen, L., Huo, X., Cai, B. and Zhang, Z. (2015), 'Metabolic changes upon flower bud break in Japanese apricot are enhanced by exogenous GA4', *Hort. Res.*, 2, 15046. doi:10.1038/hortres.2015.46.

Chapter 5

Advances in understanding flowering and pollination in apple trees

Peter M. Hirst, Purdue University, USA

1 Introduction

Understanding flower development is a major area of research in the plant sciences. Not only are flowers, seeds and fruit economically important, but the transition from vegetative to reproductive growth is one of the key developmental processes in plant biology. The process of producing a fruit begins with the transition of a bud from a vegetative to a floral state. In apple, this occurs early in the summer, with differentiation of the floral bud often occurring ~100 days after bloom, although both earlier and later differentiation can take place. Obviously, the signal driving this transition must occur sometime earlier, but the exact nature and timing of the signal in apple are unknown. In recent years, considerable progress has been made in elucidating some of the key genetic regulators of the flowering process. Inadequate flowering can be a severe limitation to fruit production, and understanding the floral induction process is a necessary step to the development of precision orchard management strategies to ensure regular cropping, and reduce biennial bearing.

From an ecological, evolutionary and survival point of view, plants produce fruit to facilitate seed dispersal. Mechanisms to ensure outcrossing and heterozygosity increase genetic diversity to provide an ecological advantage, although not a commercial advantage. Ecologically speaking, the optimum situation is for trees to produce many small fruit containing the maximum number of viable seeds, thus comprising the next

http://dx.doi.org/10.19103/AS.2016.0017.04

generation. So while trees are adapted to produce high-quality seeds, for commercial production we are more concerned about high-quality fruit. Orchards are typically comprised of clonally propagated trees with little genetic diversity, a far cry from the natural state. In this sense, the objective of commercial fruit growers is diametrically opposed to the natural characteristics of the trees arising from thousands of years of natural selection. The requirement for outcrossing has important implications for commercial fruit production. Growing orchard blocks of a single cultivar is only feasible if pollinizer trees are interspersed throughout the planting. More commonly, commercial orchards consist of a number of cultivars (as few as 5 up to 20 or more) for both pollination and management reasons. In general, most apple cultivars will effectively cross-pollinate most other cultivars provided there is overlap in flowering times. The exception is triploid cultivars (such as 'Jonagold', 'Mutsu' and 'Winesap') which do not produce viable pollen. Bees or other pollinators deposit compatible pollen on the stigmatic surface, the pollen is recognized, then germinates and grows through the stigma, down the style, through the micropyle and into the ovary where it fertilizes the ovary producing an embryo, and also fuses with polar nuclei to form the endosperm. Seed set is typically a requirement for fruit set and fruit development.

Sustainable and consistent fruit production depends on adequate flowers being produced and then successful pollination and fertilization of those flowers. The sustainable production methods discussed elsewhere in this book also contribute to maintaining adequate populations of pollinators and optimum fruit development and fruit quality.

2 Flowering: introduction and biology

2.1 Introduction to flowering

The transition of a bud from vegetative to flowering is one of the pivotal developmental steps in plant biology, and especially relevant to horticulture. The control of this transition has been intensely studied by floriculturists with the goal of manipulating flowering time so that flowering is synchronized with major holidays and marketing windows (Latimer and Whipker, 2012; Matta and Botto, 2009). However, the transition of buds from vegetative to a floral state is also critical for the production of fruit crops, such as apple.

Endogenous control of flowering in apple has been the subject of numerous research studies over the last 100 years or so, but our understanding remains imprecise. Following Chan and Cain's famous paper published in 1967, many considered that the role of seeds in inhibiting flowering completely explained the basis for floral induction and that the case was closed. Their work was carried out on 'Spencer Seedless', which is capable of setting fruit either with or without seeds, depending on whether pollination and fertilization has occurred. The theory was that seeds are the source of inhibition of flowering in the bourse bud. Previously it was thought that fruit inhibited subsequent flowering due to nutritional competition. However, more recent evidence with different genotypes have questioned the earlier hypothesis. For example, with the very biennial cultivar 'Honeycrisp', fruit weight on the spur was more closely related to flower development than seed number (Elsysy and Hirst, unpublished data). I liken this conundrum to completing a jigsaw puzzle. When only a small proportion of the pieces are on the table, it is very difficult to determine where they fit. As more and more pieces of the flowering puzzle are understood, the overall picture starts to come more into focus. Recent advances in understanding some of the genetic controls of flowering in apple have added a number of previously unknown pieces, but the overall picture has yet to come into clear focus. A recent integrative model for flowering in apple has been suggested that may prove to be a major advance in our understanding (Xing et al., 2015).

2.2 Biology of flowering

Apples produce vegetative buds and mixed buds containing both flowers and leaves. They are borne terminally on short shoots (spurs) or laterally on long shoots. Flowering spurs are typically borne on two year and older wood, but the largest flowers and best quality fruit is from 2- to 3-year-old spurs (Volz et al., 1994). Some apple cultivars also produce flowers within buds located in leaf axils of 1-year-old shoots. *Flower induction* results when meristems receive a signal and change from a vegetative to a floral status (Buban and Faust, 1982). This occurs at a molecular level and no morphological sign of induction is visible until *flower bud differentiation* occurs.

There has been some variation in the reported descriptions of the stages of bud differentiation from vegetative to highly differentiated floral buds (Foster et al., 2003, Hoover et al., 2004). However, given the differences in cultivars, rootstocks, climate and tree management, perhaps the similarity in the descriptions of the process is more striking. For example, Hoover et al. (2004) outlined five stages of development: Stage 1, apical meristem flat (vegetative); Stage 2, apical meristem domed (floral commitment); Stage 3, bractlets differentiating on the terminal floral meristem; Stage 4, sepals clearly differentiated on the terminal floral meristem and bractlets differentiating on lateral floral meristems; Stage 5, sepals clearly differentiated on lateral floral meristems. By comparison, Foster et al. (2003) classified the stages as Stage 0, flattened, narrow meristems (vegetative); Stage 1, flattened broad meristems (floral commitment); Stage 2, meristem domed; Stage 3, first lateral floral meristem visible; Stage 4, bract subtending apical meristem visible; Stage 5, sepals visible on terminal meristem; Stage 6, lateral meristem initiate sepals; Stage 7, sepals visible on all lateral meristems. The most obvious point of difference is what constitutes the first visible sign of floral commitment, broadening of the apex (Foster et al., 2003) or doming of the apex (Hirst and Ferree, 1995b; Hoover et al., 2004; McArtney et al., 2001). It seems likely that broadening of the apex is the first visible sign of floral commitment but doming of the apex is probably more definitive and easier to recognize (Fig. 1).

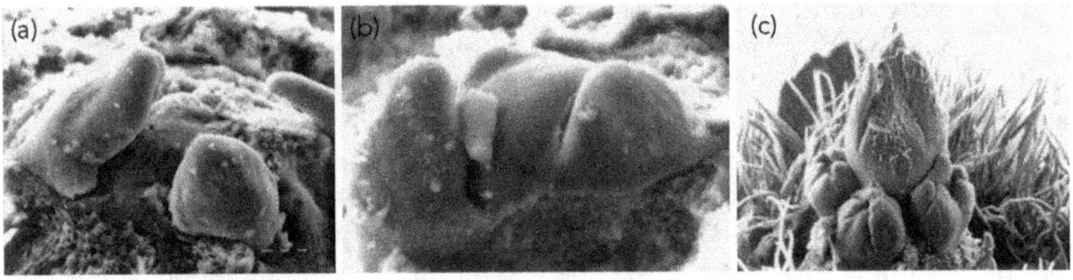

Figure 1 Bud development of 'Delicious' apple showing a flattened vegetative apex (a), doming of the apex that is among the earliest visible signs of floral development (b) and a highly differentiated flower bud (c).

There have been tremendous advancements in our understanding of genetic regulation of apple flowering over the last decade. Much of this progress has been made by applying knowledge gained from studies of the model plant *Arabidopsis* to the study of flowering in fruit trees such as apple. Flowering can be stimulated by promoting pathways where genes causing floral transition are expressed, and by expression of repressors that inhibit floral transition (Hanke et al., 2007). From an evolutionary point of view, it should not be surprising that such an important plant characteristic such as transition to flowering is controlled by the complex interaction of a number of pathways. Developmentally, the transition to flowering involves a large-scale reprogramming of gene expression within the meristem that can be described as a transcriptional cascade where floral initiator genes mediate endogenous and environmental signals and regulate flowering-integrator genes that in turn regulate floral identity genes that convert a shoot apex to an inflorescence (Mookerjee and van Nocker, 2010). Homologs of many of the flowering genes in *Arabidopsis* have been identified in apple suggesting conserved mechanisms for the regulation of flowering across species. Despite this work, the molecular mechanisms that control flowering in apple remain unclear.

Apple buds only flower once they attain a degree of complexity in terms of number of nodes within the bud. This is often referred to as the critical appendage number, and differs according to cultivar, year and rootstock (Fulford, 1966; Hirst and Ferree, 1995b; McArtney et al., 2001; McLaughlin and Greene, 1991). In general, around 18–20 nodes are required for a bud to become floral. Growth regulator applications that inhibit flowering, such as gibberellins, may function by slowing the rate of appendage development (plastochron) so that fewer buds reach a sufficient degree of complexity within the time window for flower development; other compounds such as cytokinins may increase the rate of appendage development and thereby increase flowering (Mookerjee and van Nocker, 2010). A similar mechanism appears to operate in *Arabidopsis* where hormones that shorten the plastochron also increase flowering, therefore the genetic controls may be similar.

In *Arabidopsis*, the floral identity genes AP1 and LFY promote flowering, whereas TFL1 maintains indeterminate growth and thus inhibits flower formation (Mookerjee and van Nocker, 2010). Homologs of these genes have been identified in apple and a role for their involvement in apple flowering has been proposed (Kotoda et al., 2000). The gene FLOWERING LOCUS T (FT) encodes a protein that is thought to be a major component of florigen (Lifschitz et al., 2014; Tränkner et al., 2010), and its overexpression can accelerate flowering in many plants.

Recently, studies using QTL and transcriptomic analyses have suggested that flowering control in apples is regulated by hormone-related genes, which then in turn affect flowering genes (Guitton et al., 2012; Xing et al., 2015). Several genes related to the metabolism of GA and auxin were shown to co-locate with QTLs for biennial bearing, suggesting hormone-related mechanisms for control of biennial bearing. Not only does hormone signalling mediate flower induction in apple, but it appears that sugar signalling may also be involved by regulating the expression levels of flowering genes (Xing et al., 2015). These authors proposed a hypothetical model for the regulation of flowering in apple that represents the most holistic model of flowering to date (Fig. 2). This model will likely serve as the framework for future research and understanding of the flowering process in apple.

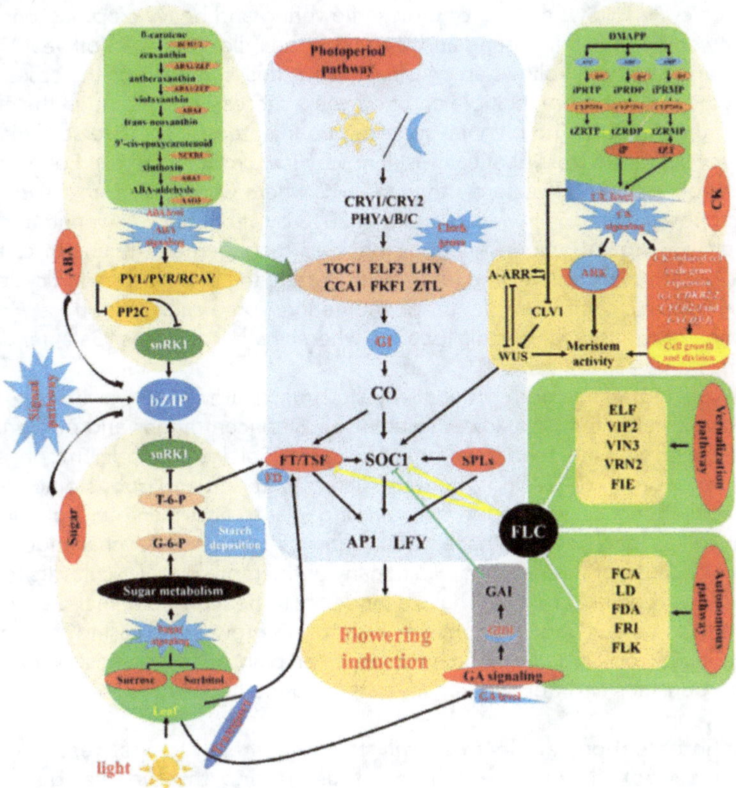

Figure 2 A proposed model of regulation of apple flowering based on carbohydrate and hormonal signalling (from Xing et al., 2015).

3 Horticultural aspects of flowering

Apple seedlings are initially juvenile (not capable of flowering), and often do not flower for seven or more years. While this is a great concern to breeders, it has little relevance to fruit growers. Trees planted in orchards are typically vegetatively propagated and consist of a scion (cultivar) grafted onto a rootstock, neither of which are usually derived directly from seed. These trees are therefore adult, not juvenile, and so are capable of flowering; however, most orchard trees do not begin producing flowers and fruit until 2–3 years after planting. *Precocity* is the length of time taken for trees to start producing fruit. The main factors regulating precocity of scions are cultivar, rootstock, tree management and environment. The degree of precocity can have a profound effect on orchard profitability, especially with high-cost orchard systems and when interest rates are high.

One of the major challenges in apple production is achieving a regular crop. *Biennial bearing* is the fluctuation in cropping from year-to-year caused by irregular flowering.

Typically, the cycle follows a trend of profuse flowering and heavy cropping one year ('on year'), followed by sparse flowering and low yields the following year ('off year'). Although biennial bearing typically follows an on-off-on trend, this is not always the case. Individual trees in an orchard can have multiple on or off years successively, although the reasons for this are not understood. The biennial cycle can be initiated by an adverse climatic event, such as a severe spring frost, that can result in very low cropping levels. For example, this occurred across many Midwestern states in 2012 where widespread frost events caused very low cropping levels. This was followed in 2013 by very heavy flowering and, in many cases, overcropping. Additionally, poor crop load management (thinning) can result in either inadequate or excessive crop loads, thus starting the biennial bearing cycle. There is a significant genetic component to this phenomenon; some cultivars such as 'Honeycrisp' exhibit a high propensity for biennial bearing whereas others such as 'Gala' tend to flower consistently from year to year.

Biennial bearing affects both productivity and fruit quality. Heavy crops in the on year typically have smaller fruit size, lower soluble solids concentrations and poor fruit colour, whereas in the off year fruit tend to be large and can exhibit low fruit calcium concentrations with a high proportion displaying disorders such as bitter pit and poor storage life. These negative consequences affect growers, packing plants and marketers of apples. This phenomenon has long been the bane of fruit growers but is a larger challenge today than ever before because of three factors. First, many of the more popular cultivars today (such as 'Honeycrisp', 'Cameo' and 'Fuji') have a tendency to be highly biennial. Secondly, these cultivars tend to have higher economic returns to the grower, increasing the opportunity cost of missing a crop, and thirdly, orchards are often planted in high-cost orchard systems (higher tree densities and trellised), again increasing the negative financial impacts of low production years.

Without intervention, apple trees will often set heavy crops, resulting in small fruit size and poor flower bud initiation, thus starting the biennial bearing cycle. The most common practice employed to reduce the severity of biennial bearing is regulation of cropping levels by chemical thinning. While crop load reduction (thinning) can be conducted manually, this is very labour-intensive and in many cases too late to achieve the most optimal outcomes. Whether thinning is carried out manually or chemically, it generally results in increased fruit size and quality, and also improves flowering for the following year (Ferree and Schmidt, 2000). Earlier thinning generally has a greater beneficial impact of increasing both fruit size and return bloom than thinning carried out later (Tromp, 2000). For this reason, many growers are chemically thinning at bloom rather than the more traditional practice of fruitlet thinning, even though earlier thinning is associated with higher risk of frost damage in some environments.

Generally speaking, strong vegetative growth is antagonistic to flowering. There are many instances of this principle in orchard management strategies. A few examples include using rootstocks, growth regulators, pruning and tree training to reduce vegetative vigour and increase flowering. For many years it has been known that rootstocks are one of the key elements of highly efficient and productive orchard systems. Rootstocks have a number of effects on tree growth and development, including vegetative vigour, precocity, carbon partitioning to the fruit and on flowering. Increased flowering on more dwarfing rootstocks is due to a higher proportion of spurs that flowered, not the production of more spurs (Hirst and Ferree, 1995a). Therefore, rootstocks influence the proportion of buds that become floral, but not the complexity at which buds make the transition from vegetative to floral (Hirst and Ferree, 1995b). Such rootstock effects on buds are likely mediated by alterations in gene expression within the scion. For example, 'Gala' trees growing on a dwarfing rootstock that promotes flowering (M.9) had higher expression of flowering-related genes than those found in 'Gala' growing on the more vigorous M.7 rootstock that flowers less (Jensen et al., 2003). It is also possible that the rootstock effect is, at least in part, indirect, and that shoot characteristics of trees growing on dwarfing rootstocks, such as shorter shoot lengths and earlier time of terminal bud set, also favour flower development.

Management practices that result in excessive vegetative vigour also tend to reduce flowering. Examples include very heavy pruning and over fertilization. Conversely, many practices that reduce vigour, such as branch bending, root pruning and scoring, often increase flower formation. Many growth regulators have been investigated for use in apple production, and a large number of these have been used routinely in commercial production. Perhaps the product with the widest application in the United States was Alar (daminozide), used extensively in the 1970s and 1980s, but now no longer approved for use on apples in the United States due to safety concerns. This material reduced shoot growth and pre-harvest drop while increasing fruit red colour, fruit firmness and flower formation for the following year's crop (Miller, 1988). This was one of a number of compounds used in the past to reduce shoot growth and increase flowering, but today prohexadione calcium is one of the few products still in commercial use. Other growth regulators can also be used to increase flowering, but without effects on reducing vegetative growth. Both naphthaleneacetic acid (NAA) and ethephon, applied during the summer, can increase flower formation, but these applications are not always effective, especially when other factors favour biennial bearing (McArtney et al., 2007). Another approach to achieving consistent flowering is to reduce flowering during on years. Applications of gibberellins made to trees during the off year have shown some potential to reduce flowering the following year, with GA_7 being more effective than GA_3 (McArtney and Li, 1998).

Adverse environmental or cultural conditions can cause reflowering of apples, where trees flower twice within a calendar year. For example, treatments such as manual defoliation, severe water stress or deblossoming can result in reflowering (Janick, 1974; Jones, 1987; Schupp and Ferree, 1987). This appears to be the result of early emergence of flower buds already initiated rather than a subsequent wave of flower formation.

Flowering tends to be more regular on stronger spurs with larger leaf area. Neilsen and Dennis (2000) found that bourse lengths greater than 10 mm tended to flower regardless of the crop load on the spur, whereas flower development in shorter bourse buds was more strongly inhibited by the number of seeds on the spur. We found similar results working with 'Gala' and 'Fuji', where flowering in bourse buds approximately 8–10 mm long was not affected by the presence of fruit on the spur (Hirst, unpublished data). In both 'Gala' and 'Honeycrisp', flowering was similar in all but the very shortest and longest bourse shoots. Leaf area of bourse shoots was important for flower formation, although spur leaves had little effect (Elsysy and Hirst, unpublished data). With two fruits on 'Honeycrisp' spurs, flowering was inhibited even at low crop loads, whereas the degree of flowering inhibition from one fruit per spur depended on the total crop load of the tree. Such applied research is helping to develop clearer profiles of the optimal fruit producing spurs in some of these highly biennial cultivars.

4 Pollination: introduction and biology

4.1 Introduction to pollination

There are a few terms relevant to pollination that are often confused. A *pollinizer* is the source of pollen, usually a nearby tree of a different apple cultivar. *Pollinators*, however, are 'agents of pollen transfer', most commonly honeybees, although other insects can also play a role. *Pollination* is the physical transfer of pollen from anthers to stigmatic surfaces. *Fertilization* is the fusion of male and female gametes to form a fertilized egg that then becomes the zygote. *Anthesis* refers to the dehiscence and release of pollen grains from the anthers, which occurs soon after flower opening.

4.2 Pollination biology

Apples are monoecious trees bearing hermaphroditic flowers. Apple flowers are typically borne terminally on spurs (shoots <5 cm in length) or longer shoots, usually in clusters of 5–6 flowers, although considerable variation can be found within a single tree. In some cultivars, such as 'Golden Delicious' and 'Gala', flowers can also be borne laterally by buds in leaf axils on the previous year's growth. Such axillary flowers are generally smaller, develop later, have lower fruit set and produce fruit of lower quality (Robbie and Atkinson, 1994; Volz et al., 1994). The gynoecium typically contains five locules each of which contains up to two seeds, although considerable variation can be found. Therefore, an apple fruit has a theoretical maximum of 10 seeds but in some cases more may be present. For example, in our work on flowering of 'Honeycrisp' apple, fruit of ~300 g in size had 0–19 seeds (Elsysy and Hirst, unpublished data).

Flowers of commercial apple cultivars typically have five pistils, each comprised of stigma, style and ovary (Fig. 3). The styles are united at the base (Pratt, 1988), so one style is not solely connected to one locule. This was borne out in a recent experiment where pistil number was manually reduced to 1–5 and flowers hand pollinated. Even with one pistil, some fruit had up to 10 seeds (Elsysy, Jahed and Hirst, unpubl. data).

Self-incompatibility (SI) systems that prevent self-fertilization, and therefore encourage outcrossing, have been widely studied (De Nettancourt, 2001; De Franceschi et al., 2012). While such mechanisms increase genetic diversity and may be evolutionarily beneficial, they may not be desirable for cultivated plants where fruit development is dependent on seed set. The stigmatic surface provides the first opportunity for the plant to discriminate the pollen landing thereon to prevent self-fertilization and encourage outcrossing. Apples exhibit gametophytic SI, where incompatibility depends on the interaction between *S*-determinant genes that are expressed in both the pistil and the pollen. Pistil *S*-locus ribonucleases (*S*-RNase) are expressed in both stigmatic and style tissue along the pollen tube growth path (Certal et al., 1999). The pollen *S* determinant is comprised of one or more pollen-expressed *S*-locus *F*-box genes called SLF (S-locus F-box) (Cheng et al., 2006; De Franceschi et al., 2012; Minamikawa et al., 2010). If there is no match in *S*-alleles between pollen and pistil, the pollen is able to inhibit stigmatic *S*-RNase and develop normally. However, if the pollen and pistil contain the same *S*-allele, ribonucleases in the pistil prevent further pollen development. Gametophytic SI not only prevents selfing, but also fertilization by closely related genotypes sharing an *S*-allele. Genes located in chromosomal positions close to the *S*-locus (i.e. in the same linkage group) may be greatly over- or under-represented in seedling populations depending on the *S*-genotype (De Franceschi et al., 2016). Up to 32 *S*-alleles have been reported in

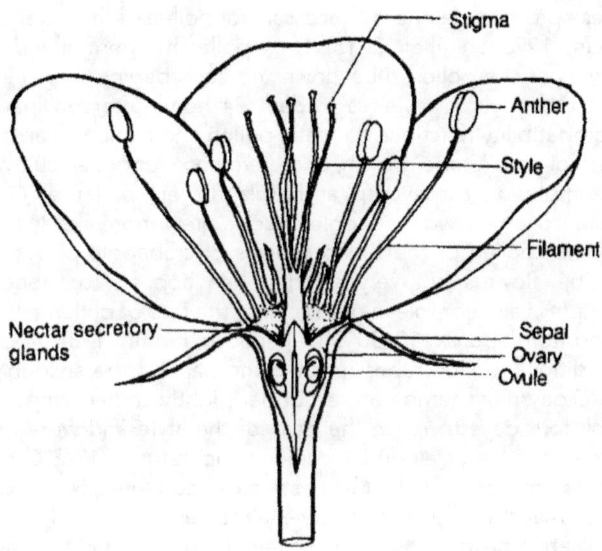

Figure 3 Diagrammatic representation of an apple flower (Pratt, 1988).

apple, but it is possible that there are some duplicates in this list, so the exact number is yet to be accurately determined. Information is available on the *S*-genotypes of many of the common apple cultivars (Broothaerts et al., 2004). Apple cultivar pairings can be classified as fully compatible (pollen and pistil differ in both *S*-loci), semi-compatible (one similar and one different *S*-locus) or incompatible (where both *S*-loci are similar) (Schneider et al., 2005). There is some evidence that fertilization and seed set is higher when flowers are pollinated with more than one source of compatible pollen, due to pollen-pollen interactions (Kron and Husband, 2006). This increases the importance of cultivar diversity in an orchard setting.

Not all flowers within a flower cluster (inflorescence) are equivalent. The apical (king) flower generally opens a few days earlier than lateral flowers. Additionally, king flowers have higher stigmatic receptivity but only for a short duration. Lateral flowers in addition to opening later are also receptive longer (~5 days); therefore, a flower cluster as a whole may be receptive for almost a week. So it appears that a type of 'two-stage' mechanism is at play, where the preferred result is for king flowers to be pollinated and fertilized, with lateral flowers playing a back-up role to ensure seed set in the event of failure of king flower fertilization (Losada and Herrero, 2013).

Pollen is haploid and contains two sperm nuclei. One of these fuses with the egg nucleus to form the zygote (eventually becoming the seed) and the other fuses with the two polar nuclei resulting in the triploid endosperm. This is the process often referred to as *double fertilization*. Occasionally, haploid pollen fuses with unreduced egg cells resulting in triploid seeds (Evans et al., 2011). While the formation of such seeds is of interest to plant biologists and apple breeders, it is not known to have any effect on fruit development.

The dehydration of apple pollen grains extends longevity but requires hydration prior to germination. Stigmatic exudates play a role in pollen recognition and hydration, and the style also supports pollen tube growth (Heslop-Harrison, 2000; Losada and Herrero, 2014). The principle reserves in mature pollen are generally carbohydrates, lipids and proteins, and these provide the energy necessary for pollen germination and early pollen tube growth (Pacini, 1996; Shi et al., 2015). Once pollen has germinated on the stigmatic surface, the growth of the pollen tube down the style begins very quickly. Significant growth of pollen tubes within styles is evident 24 hours after pollination (Jahed and Hirst, 2017a). Compatibility reactions between pollen and pistil play an important role in regulating the rate of pollen tube growth, and preference for pollen cultivars is frequently observed. For example, on 'Honeycrisp' and 'Gala' flowers, pollen of 'Golden Delicious' grew rapidly while pollen of two crabapple species grew more slowly (Jahed and Hirst, 2017a). Furthermore, pollen of these two species of crabapple performed adequately on 'Gala' and 'Fuji' flowers but very poorly when applied to 'Honeycrisp' flowers. Temperature also plays a major role in determining the rate of pollen tube growth, which are linearly related from approx. 10–30°C (Jefferies and Brain, 1984; Yoder et al., 2009). In controlled conditions, pollen tubes reached the base of the style in as little as two days under 24/7°C day/night temperatures. Under slightly cooler temperatures (18/4°C), pollen tubes took four days to reach the base of the style and were not successful in reaching the base of styles at all under cooler temperatures (13/2°C or 7/9°C) (Yoder et al., 2009). In field conditions in the Midwestern United States, 3–4 days were required for pollen tubes to reach the base of the style (Jahed and Hirst, 2017a). Caution should be used in extrapolating beyond the male and female cultivars and the environment since pollen tube growth is a complex interaction among all three of these factors (DeLong et al., 2016).

Once pollen tubes reach the base of the style they must grow through the placenta to penetrate the ovules. The pollen tube growth rate within the style is faster than within the ovary, so although the distance is short, the time taken can be as long or longer than the time taken for pollen tubes to grow down the style (Herrero and Arbeloa, 1989). Abnormalities within the ovary can limit fruit set in some cases.

Ovule receptivity is limited, and for successful fertilization, the pollen must reach the ovule while the ovule remains receptive. There is some evidence that flower quality can affect stigma receptivity and fruit set. For example, king flowers have a short but intense stigmatic receptivity, whereas the smaller lateral flowers tend to have a longer period of receptivity but a lower capacity for the amount of pollen germination on the stigma (Lesada and Herrero, 2013). Stronger flowers have a longer period of stigmatic receptivity (Williams, 1965) and king flowers have a higher potential for fruit set when both king and lateral flowers remained on the spur (Ferree et al., 2001). Therefore, it would seem that trees have a strategy for optimum performance under favourable environmental conditions (setting fruit on king flowers) with lateral flowers having longer stigmatic receptivity, being a back-up plan under less ideal pollinating conditions.

Ovule viability can be extended by cool temperatures (<10°C), but pollen tube growth is also slow during such conditions. This has led to the concept of *effective pollination period (EPP)* (Sanzol and Herrero, 2001; Williams, 1965). This is the length of time during which pollination must occur for seed set (and by extension fruit set) to be successful. It is the duration of ovule receptivity less the time taken for the pollen to germinate and grow down the style to the ovule. This assumes the stigma remains receptive, and as mentioned earlier, apples have a two-stage mechanism to extend the period of stigmatic receptivity. Lower temperatures result in slow pollen tube growth but also extend ovule receptivity, so generally increase the EPP, whereas warm temperatures increase degeneration of the stigma and reduce ovule receptivity, therefore reducing the EPP.

Because there are typically five pistils and five carpels, it may be logical to assume each pistil provides for transport of pollen tubes to each carpel to set two seeds. Apples demonstrate some level of *syncarpy*, where carpels are fused, but this may not be perfect syncarpy, and probably varies with cultivar. This was evident in some of our recent experiments (Elsysy, Jahed and Hirst, unpublished data) where we physically removed some of the pistils from 'Honeycrisp' flowers to generate fruit with a wide range of seed numbers. While there was certainly a relationship between pistil number and seed number, results were highly variable. For example, even with one pistil, fruit contained 1–11 seeds. If pistils or carpels were not fused these fruit would theoretically have contained a maximum of two seeds only.

5 Horticultural aspects of pollination

While there are a few apple cultivars that may be capable of setting fruit parthenocarpically, that is, without pollination and seed set, all common commercially important apples require cross-pollination (Matsumoto et al., 1999). However, even such parthenocarpic cultivars as 'Megumi' have higher fruit set and seed numbers when cross-pollinated. Some growth regulator treatments (especially those containing GA_4 or GA_{4+7} and 6-benzyladenine) can partially offset the need for cross-pollination and can increase the number of parthenocarpic fruit (Bukovac, 1963; McArtney et al., 2014). While this can be useful to increase fruit set where there has been a spring frost or other event to limit pollination, the growth regulator effect is unlikely to fully restore cropping (McArtney et al., 2014). In some cases, growth regulator-induced parthenocarpic fruit was slightly elongated compared with seeded fruit (Bukovac, 1963), but in the study by McArtney et al. (2014) fruit shape was unaffected.

To be successful, pollination requires several aspects to align. First, there has to be a supply of suitable pollen. This depends not only on the genotype, but also on synchrony of anthesis. Almost all apple genotypes are self-incompatible, and even cultivars previously thought to be partially self-compatible (such as 'Golden Delicious') have been shown in recent controlled experiments to also be largely self-incompatible. Triploid cultivars, such as 'Mutsu', 'Jonagold' and 'Winesap', produce very little viable pollen so that in commercial settings, they should not be considered as useful pollinizers. However, these triploid cultivars appear to be adequately pollinated by other, diploid cultivars.

In high-density, fruiting wall orchards, bees typically fly mostly within rows of trees rather than across rows (Kron et al., 2001a). Pollen can be dispersed to a considerable distance from the pollinizer source (15–18 rows), but 44–80% of pollen was deposited within three rows of the pollen source (Kron et al., 2001b). Purely from a pollination and fruit set perspective, each tree within a row should be a different genotype. However, such a planting would make orchard management very difficult, especially relating to pesticide applications. A reasonable compromise between efficacy of pollination and efficient orchard management is to plant cultivars in blocks of single cultivars no more than 4–5 rows wide (Sheffield et al., 2016). This ensures that each tree is no more than two rows away from a pollinizer. Most pollen deposited on a tree is sourced from within 20 m, again emphasizing the importance of cultivar diversity in the orchard (Kron et al., 2001a,b).

In some instances, growers wish to produce only one cultivar in their orchard, or want to grow single cultivars in blocks too large to ensure adequate cross-pollination. In such cases, pollinizer trees are often planted at a ratio of about 1:7, although this can depend on many other factors including tree size, density, training system and orchard layout (Sheffield et al., 2016). Pollinizers should be chosen that bloom profusely every year, and have bloom time and pollen compatibility with the main crop cultivar. Manchurian crab is often used as a pollinizer, but recently has been shown to be susceptible to a fungus, *Phacidiopycnis washingtonensis*, that can cause fruit rots and can lead to phytosanitary barriers (Weber, 2011, Xiao et al., 2009). Therefore, these pollinizer trees also require management, such as pruning so they do not shade neighbouring trees and pest control to ensure they do not become sources of inoculum to infect the rest of the orchard. There have been a number of comparisons of crabapple cultivars most suitable for apple orchards. These evaluations have primarily considered bloom time, bloom duration, disease resistance, compatibility and fruit set of the major local commercial cultivars grown (Kwon et al., 2015). Various university extension websites (such as http://treefruit.wsu.edu/web-article/apple-pollination/) list recommended crabapple cultivars including 'Crimson Gold', 'Indian Summer', 'Manchurian', 'Mt. Blanc', 'Mt. Evereste', 'Snodrift', 'Whitney' and 'Wickson'. Recently, there has been renewed interest in mechanical application of pollen suspensions and results have compared well, sometimes exceeding the performance of natural pollination (Hansen, 2015). This may offer a potential alternative for growers to continue sustainable apple production should bee colonies continue to decline.

Seed set is important since the number of seeds influences not only the fruit set, but aspects of fruit quality. Many factors contribute to fruit development, and across a number of experiments with many cultivars, seed number generally accounts for 20–30% of the variation of fruit size at harvest within an orchard planting. Seed number is also positively related to fruit symmetry. Fruit with low seed numbers are often misshapen to the point of being unmarketable, although there is considerable variation in the relationship between seed number and fruit shape (Brault and de Oliviera, 1995; Matsumoto et al., 2012). Such low seed number fruit also tend to have lower calcium concentrations and resultant higher levels of Ca-related fruit disorders such as senescent breakdown (Volz et al., 1996).

A number of pesticides commonly applied to orchards for pest and disease control can affect pollination. The most obvious effect is the direct impact of insecticides on bees and other pollinators, with the general recommendation being to avoid use of insecticides when flowers are open and pollinators are present. Some fungicides and their surfactants may negatively affect pollen germination and pollen tube growth directly (Kopcke et al., 2002; Yi et al., 2003). For example, the common fungicide Captan reduced pollen germination by 20%, although did not affect the number of pollen tubes reaching the basal part of the style after 20 h (Yi et al., 2003). Growth regulators such as ethylene can increase ovule senescence (Stosser and Anvari, 1982) and ethylene inhibitors such as AVG can increase fruit set presumably by extending ovule longevity and therefore EPP (Greene, 1980). Boron has long been known to play an important role in pollen germination and pollen tube growth (Stanley and Lichtenberg, 1963), and boron applications immediately after bloom can increase fruit set (Wojcik et al., 1999).

Likewise, exogenous polyamine applications can increase fruit set mainly through extending the period of ovule receptivity, although effects on increasing the rate of pollen tube growth are also possible (Costa and Bagni, 1983; Cristosto et al., 1988).

6 Conclusion

The sustainable, consistent production of apples is a two-year process that begins with adequate flower bud initiation in one year followed by fruit development in the subsequent year. The recent model proposed by Mingyu Han's research group at the Northwest Agricultural and Forestry University in China (Fig. 2) accumulates and integrates a lot of our understanding of the mechanisms controlling flower induction in apple. Their work is likely to advance research not only of the regulation of flowering, but how various control may be integrated.

Pollination and fertilization can be thought of as the first steps in fruit development. A more detailed discussion of subsequent processes, such as cell division and cell expansion, are presented further in this volume.

In recent years there has been a renewed interest in understanding the factors that contribute to pollen tube growth in order to construct models that predict pollen tube growth, fertilization and fruit set. The goal of this work is to develop tools that allow growers to regulate crop load very early in the growing season. Rather than set as much fruit as possible, and then remove the majority of that fruit to set optimum crop loads (as has been carried out traditionally), it appears to be more efficient to predict when an adequate number of fruit have set and then prevent further fruit from setting. While this approach has the potential to be more efficient and sustainable (using bloom thinning agents such as salt rather than growth regulators), it relies on accurate prediction of fertilization and fruit set. The research groups of Yoder in Virginia, Schmidt in Washington and Peck in New York are making rapid progress in developing this new approach to crop load adjustment.

7 Where to look for further information

For practical information, the extension service in many states has information online. Some examples are:

http://extension.psu.edu/plants/tree-fruit/news/2015/promoting-return-bloom-of-apple-trees
http://msue.anr.msu.edu/news/enhance_return_bloom_on_apple_varieties_by_using_summer_naa
http://www.uvm.edu/~fruit/?Page=treefruit/tf_horticulture/AppleHortBasics/cropload management.html&SM=tf_submenu.html
http://msue.anr.msu.edu/topic/apples/horticulture/pollination
http://extension.missouri.edu/p/G6001
http://treefruit.wsu.edu/web-article/apple-pollination/
http://www.uvm.edu/~fruit/treefruit/tf_horticulture/AppleHortBasics/Readings/Pollination%2Band%2BFruit%2BSet%2Bof%2BFruit%2BCrops.pdf

Other sources of information are the scientific symposia organized under the auspices of the International Society for Horticultural Science. These are usually held every four years and proceedings are published in *Acta Horticulturae*. Some symposia of interest include:

- International symposium on flowering, fruit set and alternate bearing
- International symposium on integrating canopy, rootstock and environmental physiology in orchard systems
- International apple symposium
- International symposium on plant bioregulators in fruit production

8 Acknowledgements

Portions of the work reported here were carried out by students Mokhles Elsysy, Khalil Jahed, Fatemeh Sheibani and Laura Xie. We are grateful for research support from USDA-SCRI, US Department of State, the Egyptian Government, the Washington Tree Fruit Research Commission and the Return Bloom Fund of the Indiana Horticultural Society.

9 Dedication

This chapter is dedicated to the memory of Donald James Hirst.

10 References

Blitzer, E. J., J. Gibbs, M. G. Park and B. N. Danforth. 2016. Pollination services for apple are dependent on diverse wild bee communities. *Agric. Ecosyst. Environ.* 221:1–7.

Brault, A. M. and D. de Oliveira. 1995. Seed number and an asymmetry index of 'McIntosh' apples. *HortScience* 30(1):44–6.

Broothaerts, W., I. Van Nerum and J. Keulemans. 2004. Update on and review of the incompatibility (*S*-) genotypes of apple cultivars. *HortScience* 39(5):943–7.

Buban, T. and M. Faust. 1982. Flower bud induction of apple trees: internal control and differentiation. *Hort. Rev.* 4:174–203.

Bukovac, M. J. 1963. Induction of parthenocarpic growth of apple fruits with gibberellins A_3 and A_4. *Bot. Gaz.* 124(3):191–5.

Certal, A. C., A. M. Sanchez, H. Kokko, W. Broothaerts, M. M. Oliveira and J. A. Feijo. 1999. S-RNases in apple are expressed in the pistil along the pollen tube growth path. Sex. *Plant Reprod.* 12:94–8.

Costa, G. and N. Bagni. 1983. Effects of polyamines in fruit-set of apple. *HortScience* 18(1):59–61.

Chan, B. G and J. C. Cain. 1967. The effect of seed formation on subsequent flowering in apple. *Proc. Amer. Soc. Hort. Sci.* 91:63–8.

Cheng, J., Z. Han, X. Xu and T. Li. 2006. Isolation and identification of the pollen-expressed polymorphic F-box genes linked to the S-locus in apple (Malus x domestica). Sex. *Plant Reproduct.* 19: 175–83.

Crisosto, C. H., P. B. Lombard, D. Sugar and V. S. Polito. 1988. Putrescine influences ovule senescence, fertilization time, and fruit set in 'Comice' pear. *J. Amer. Soc. Hort. Sci.* 113:708–12.

De Franceschi, P., V. Cova, S. Tartarini and L. Dondini. 2016. Characterization of a new apple S-RNase allele and its linkage with the Rvi5 gene for scab resistance. *Mol. Breed.* 36:7.

De Franceschi, P, L. Dondini and J. Sanzol. 2012. Molecular bases and evolutionary dynamics of self-incompatibility in the Pyrinae (Rosaceae). *J. Exp. Bot.* 63:4015–32.

DeLong, C. N., K. S. Yoder, L. Combs, R. E. Veilleux and G. M. Peck. 2016. Apple pollen tube growth rates are regulated by parentage and environment. *J. Amer. Soc. Hort. Sci.* 141(6):548–54.

De Nettancourt, D. 2001. *Incompatibilty and Incongruity in Wild and Cultivated Plants*. Berlin: Springer.

Evans, K. M., A. Patocchi, F. Rezzonico, F. Mathis, C. E. Durel, F. Fernandez-Fernandez, A. Boudichevskaia, F. Dunnemann, M. Stanliewicz-Kosyl, L. Gianfranceschi, M. Komjanc, M. Lateur, M. Madduri, Y. Noordijk and W. E. van der Weg. 2011. Genotyping of pedigreed apple breeding material with a genomic-covering set of SSRs: trueness-to-type of cultivars and their parentages. *Mol. Breed*. 28:535–47.

Ferree, D. C., B. L. Bishop, J. R. Schupp, D. S. Tustin and W. M. Cashmore. 2001. Influence of flower type, position in the cluster and spur characteristics on fruit set and growth of apple cultivars. *J. Hort. Sci. Biotech*. 76(1):1–8.

Ferree, D. C. and J. C. Schmidt. 2000. Chemical thinning 'Fuji' apple in the Midwest. *J. Amer. Pom. Soc*. 54(2):61–7.

Foster, T., R. Johnston and A. Selezynova. 2003. A morphological and quantitative characterization of early floral development in apple (*Malus X domestica* Borkh.). *Ann. Bot*., 92:199–206.

Fulford, R. M. 1966. The morphogenesis of apple buds. III. The inception of flowers. *Ann. Bot*. 30, 207–19.

Greene, D. 1980. Effect of silver nitrate, aminoethoxyvinylglycine and gibberellins A_{4+7} plus 6-benzylamino purine on fruit set and development of 'Delicious' apples. *J. Amer. Soc. Hort. Sci*. 105:717–20.

Guitton, B., J. J. Kelner, R. Velasco, S. E. Gardiner, D. Chagne and E. Costes. 2012. Genetic control of biennial bearing in apple. *J. Exp. Bot*. 63(1):131–49.

Hanke, M. V., H. Flachowsky, A. Peil and C. Hattasch. 2007. No flower no fruit – genetic potentials to trigger flowering in fruit trees. *Genes, Genomes and Genomics*. 1:1-20.

Hansen, M. 2015. *Precision Pollination: Mechanical Pollination Could End Use of Chemicals or Hand Labor for Thinning*. GoodFruit Grower: Yakima, WA.

Herrero, M. and A. Arbeloa. 1989. Influence of the pistil on pollen tube kinetics in peach (Prunus persica). *Amer. J. Bot*. 76(10):1441–7.

Heslop-Harrison, Y. 2000. Control gates and micro-ecology: the pollen-stigma interaction in perspective. *Ann. Bot*. 85:5–13.

Hirst, P. M. and D. C. Ferree. 1995a. Rootstock effects on shoot morphology and spur quality of 'Delicious' apple and relationships with precocity and productivity. *J. Amer. Soc. Hort. Sci*. 120(4):622–34.

Hirst, P. M. and D. C. Ferree. 1995b. Rootstock effects on the flowering of 'Delicious' apple. I. Bud development. *J. Amer. Soc. Hort. Sci*. 120(6):1010–17.

Hoover, E., N. De Silva, S. McArtney and P. Hirst. 2004. Bud development and floral morphogenesis in four apple cultivars. *J. Hort. Sci. Biotech*. 79(6):981–4. https://doi.org/10.1080/14620316.2004.11511877http://www.globalsciencebooks.info/Online/GSBOnline/images/0706/GGG_1(1)/GGG_1(1)1-20o.pdf.

Janick, J. 1974. The apples of Java. *HortScience* 9:13–15.

Jensen P. J., J. Rytter, E. A. Detwiler, J. W. Travis and T. W. McNellis. 2003. Rootstock effects on gene expression patterns in apple tree scions. *Plant Mol. Biol*. 53:493–511.

Jahed, K. and P. M. Hirst. 2017a. Pollen tube growth and fruit set in apple. *HortScience* (accepted).

Jahed, K. and P. M. Hirst. 2017b. Pollen source effects on seed number, fruit quality and return bloom of apple.*J. Amer. Pomol. Soc*. (submitted).

Jefferies, C. J. and P. Brain. 1984. A mathematical model of pollen-tube penetration in apple. *Planta* 160(1):52–8.

Jones, H. G. 1987. Repeat flowering in apple caused by water stress or defoliation. *Trees* 1:135–8.

Kopcke, D., P. Baur and J. Schonherr. 2002. Inhibition of the growth of apple pollen tubes by EDTA, surfactants and fungicides. *Ann. Appl. Biol*. 140:81–6.

Kotodo, N., H. Hayashi, M. Suzuki, M. Igarashi, Y. Hatsuyama, S. Kidou, T. Igasaki, M. Nishiguchi, K. Yano, T. Shimizu, S. Takahashi, H. Iwanami, S. Moriya and K. Abe. 2010. Molecular characterization of FLOWERING LOCUS T-like genes of apple (*Malus x domestica* Borkh.). *Plant Cell Physiol*. 51(4):561–75.

Kotoda, N., M. Wada, S. Komori, S. Kidou, K. Abe, T. Masuda and J. Soejima. 2000. Expression patterns of floral meristem identity genes LFY and AP1 during flower development in apple. *J. Amer. Soc. Hort. Sci.* 125(4):398–403.

Kron, P. and B. C. Husband. 2006. The effects of pollen diversity on plant reproduction: insights from apple. *Sex. Plant Reprod.* 19:125–31.

Kron, P., B. C. Husband and P. G. Kevan. 2001a. Across- and along-row pollen dispersal in high-density apple orchards: insights from allozyme markers. *J. Hort. Sci. Biotech.* 76:286–94.

Kron, P., B. C. Husband, P. G. Kevan and S. Belaoussoff. 2001b. Factors affecting pollen dispersal in high-density apple orchards. *HortScience* 36(6):1039–46.

Kwon, S. I., J. Yoo, J. Lee, Y. S. Moon, C. Choi, H. Y. Jung, D. H. Lee, C. K. Kim and I. K. Kang. 2015. Evaluation of crab apples for apple production in high-density apple orchards. *J. Plant Biotechnol.* 42:271–6.

Lauri, P. E., E. Terouanne and J. M. Lespinasse. 1996. Quantitative analysis of relationships between inflorescence size, bearing-axis size and fruit-set—an apple tree case study. *Ann. Bot.* 77:277–86.

Latimer, J. and B. Whipker. 2012. Selecting and using plant growth regulators on floriculture crops. Virginia Polytechnic Institute and State Univ. Ext Pub. 430–102.

Lifschitz, E., B. G. Ayre and Y. Eshed. 2014. Florigen and anti-florigen – a systemic mechanism for coordinating growth and termination in flowering plants. *Front. Plant Sci.* 4(465):1–14.

Losada, J. M. and M. Herrero. 2013. Flower strategy and stigma performance in the apple inflorescence. *Scientia Hort.* 150:283–9.

Losada, J. M. and M. Herrero. 2014. Glycoprotein composition along the pistil of *Malus x domestica* and the modulation of pollen tube growth. *BMC Plant Biol.* 14:1–14.

Mata, D. A. and J. F. Botto. 2009. Manipulation of light environment to produce high-quality poinsettia plants. *HortScience* 44(3):702–6.

Matsumoto, S. 2014. Apple pollination biology for stable and novel fruit production: search system for apple cultivar combination showing incompatibility, semicompatibility, and full-compatibility based on the *S-RNase* allele database. *Intl. J. Agronomy* 2014:9, Article ID 138271.

Matsumoto, S., S. Komori, K. Kitahara, S. Imazu and J. Soejima. 1999. S-genotypes of 15 apple cultivars and self-compatibility of 'Megumi'. *J. Jap. Soc. Hort. Sci.* 68(2):236–41.

Matsumoto, S., J. Soejima and T. Maejima. 2012. Influence of repeated pollination on seed number and fruit shape of 'Fuji' apples. *Scientia Hort.* 137:131–7.

McArtney, S., D. Greene, T. Robinson and J. Wargo. 2014. Evaluation of GA_{4+7} plus 6-benzyladenine as a frost-rescue treatment for apple. *Hort Tech.* 24(2):171–6.

McArtney, S. J., E. M. Hoover, P. M. Hirst and I. R. Brooking. 2001. Seasonal variation in the onset and duration of flower development in 'Royal Gala' apple buds. *J. Hort. Sci. Biotech.* 76:536–40.

McArtney, S. J. and S. H. Li. 1998. Selective inhibition of flowering on 'Braeburn' apple trees with gibberellins. *HortScience* 33(4):699–700.

McArtney, S., D. Unrath, J. D. Obermiller and A. Green. 2007. Naphthaleneacetic Acid, Ethephon, and Gibberellin A4 + A7 have variable effects on flesh firmness and return bloom of apple. *HortTechnology* 17(1):32–8.

Miller, S. S. 1988. Plant bioregulators in apple and pear culture. *Hort. Rev.* 10:309–401.

Minamikawa, M., H. Kakui, S. Wang, N. Kotoda, S. Kikuchi, T. Koba and H. Sassa. 2010. Apple S locus region represents a large cluster of related, polymorphic and pollen-specific F-box genes. *Plant Mol. Biol.* 74:143–54.

Mookerjee, S. and S. van Nocker. 2010. Genetics of flowering in apple. In Aldwinckle, H. S. and Malnoy, M. (Eds), *Genomics of Fruits and Vegetable Crops: Apples*. Science Publishers: Enfield, NH.

Neilsen, J. C. and F. G. Dennis. 2000. Effects of seed number, fruit removal, bourse shoot length and crop density on flowering in 'Spencer Seedless' apple. *Acta Hortic.* 527:137–46.

Pacini, E. 1996. Types and meaning of pollen carbohydrate reserves. *Sex Plant Reprod.* 9:362–6.

Robbie, F. A. and C. J. Atkinson. 1994. Wood and tree age as factors influencing the ability of apple flowers to set fruit. *J. Hort. Sci.* 69(4):609–23.

Sanzol, J. and M. Herrero. 2001. The 'effective pollination period' in fruit trees. *Scientia Hort.* 90:1–17.

Schneider, D., R. A. Stern and M. Goldway. 2005. A comparison between semi- and fully compatible apple pollinators grown under suboptimal pollination conditions. *HortScience* 40(5):1280–2.

Schupp, J. R. and D. C. Ferree. 1987. Deblossoming stimulates a second bloom in 'Golden Delicious' apple trees. *HortScience* 22:958–959.

Sheffield, C. S., H. T. Ngo and N. Azzu. 2016. *A Manual on Apple Pollination*. FAO: Rome.

Shi, X., R. Zhou, Q. Wang, T. Chang, S. Feng and G. Du. 2015. Characteristics of pollen from transgenic lines of apple carrying the exogenous *CpTI* gene. *Hort. Plant J.* 1(1):3–10.

Stanley, R. G. and E. A. Lichtenberg. 1963. The effect of various boron compounds on *in vitro* germination of pollen. *Plysiol. Plant.* 16:337–46.

Stosser, R. and S. F. Anvari. 1982. On the senescence of ovules in cherries. *Sci. Hort.* 16:29–38.

Tränkner, C., S. Lehmann, H. Hoenicka, M. V. Hanke, M. Fladung, D. Lenhardt, F. Dunemann, A. Gau, K. Schlangen, M. Malnoy and H. Flachowsky. 2010. Over-expression of an *FT*-homologous gene of apple induces early flowering in annual and perennial plants. *Planta* 232:1309–24.

Tromp, J. 2000. Flower bud formation in pome fruits as affected by fruit thinning. *Plant Growth Regul.* 31(1):27–34.

Volz, R. K., I. B. Ferguson, E. W. Hewett and D. J. Woolley. 1994. Wood age and leaf area influence fruit size and mineral composition of apple fruit. *J. Hort. Sci.* 69(2):385–95.

Volz, R. K, D. S. Tustin and I. B. Fergusson. 1996. Pollination effects on fruit mineral composition, seeds and cropping characteristics of 'Braeburn' apple trees. *Scientia Hort.* 66:169–80.

Weber, R. W. S. 2011. *Phacidiopycnis washingtonensis*, cause of a new storage rot of apples in Northern Europe. *J. Phytopath.* 159(10):682–6.

Williams, R. R. 1965. The effect of summer nitrogen applications on the quality of apple blossom. *J. Hort. Sci.* 40:31–41.

Wojcik, P., G. Cieslinski and A. Mika. 1999. Apple yield and fruit quality as influenced by boron applications. *J. Plant Nutr.* 22(9):1365–77.

Xiao, C. L., Y. K. Kim and R. J. Boal. 2009. A new canker disease of crabapple trees caused by *Phacidiopycnis washingtonensis* in Washington State. *Plant Health Progress*. https://www.plantmanagementnetwork.org/pub/php/brief/2009/crabapple/,doi:10.1094/PHP-2009-0612-01-BR.

Xing, L. B., D. Zhang, Y. M. Li, Y. W. Shen, C. P. Zhao, J. J. Ma, N. An and M. Y. Han. 2015. Transcription profiles reveal sugar and hormone signaling pathways mediating flower induction in apple (*Malus domestica* Borkh.). *Plant Cell Physiol.* 56(10):2052–68.

Yi, W., S. E. Law and H. Y. Wetzstein. 2003. Pollen tube growth in styles of apple and almond flowers after spraying with pesticides. *J. Hort. Sci. Biotech.* 78(6).

Yoder, K., R. Yuan, L. Combs, R. Byers, J. McFerson and T. Schmidt. 2009. Effects of temperature and the combination of liquid lime sulfur and fish oil on pollen germination, pollen tube growth, and fruit set in apples. *HortScience* 44(5):1277–83.

Chapter 6

Advances in understanding apple fruit development

A.N. Lakso and M. C. Goffinet, Cornell University, USA

1 Introduction

The goal of sustainable apple cultivation is the production of a regulated crop of apples that provides as high a yield as possible of desired quality while allowing the development of a good and consistent return bloom and cropping in subsequent years. This chapter will examine how apples grow and ripen (basic structure, growth habit and physiology), some of the factors that support or limit growth, why fruit abscise and how growers can manipulate fruit growth and abscission to optimize cropping.

http://dx.doi.org/10.19103/AS.2016.0017.05

Unlike an annual plant, an apple tree does not put top priority every year on production of seed to maximize the probability of species continuity. The very long juvenility period of apple seedlings argues that tree establishment has priority over fruit and seed production. In mature trees, there may be stresses or lack of resources in any year that lead to very poor production of fruit. It is appropriate to conclude that evolutionary priority in the wild was given to tree survival over many years to allow seed production in years of good weather and minimal stresses. Generally, this would lead to the emphasis in poor years on vegetative growth of shoots or roots to replenish reserves and obtain more of the limiting resource, whether sunlight, water or nutrients.

A corollary to this is that it appears that the apple fruit is programmed to abscise in the first few weeks after bloom, unless there is sufficient healthy leaf area, carbohydrates, nutrients, water and light to support their growth and retention. This is manifested in the induction of fruit abscission if fruit growth rate slows beyond a critical threshold as will be discussed later.

2 Developmental sequence of apple growth

The development of the apple fruit is a complex sequence over a growing season. It is useful to begin with a brief review of the structure of the spur where the flower develops and the fruits are borne.

The apple spur is a reproductive short shoot that develops an abbreviated inflorescence (a corymbose raceme in botanical terms) with axillary flowers (Fig. 1). The spur shoot apex is transformed into a central (king) flower. The flowering spur of apple begins with the differentiation of a terminal mixed bud, that is, a bud that includes both leaf and floret primordial (Pratt, 1988). By fall, the buds of domestic apple cultivars enter the dormant season with many bud scales enclosing several-to-many embryonic leaves and up to six flower primordia. In the north temperate zone, flowers have initiated all floral whorls by late fall but will not initiate ovules (future seeds) until spring.

Before fruiting in an apple spur, flower development within the spur first takes place. This differentiation of the apple flower and fruit takes place over two growing seasons: new bourse shoots in year-1 initiate leaf and flower primordia, with each flower initiating all floral organs (sepal, petal, stamen and ovarian carpel primordia) before winter; in spring of year-2 the flower completes ovule (future seed) and stamen differentiation (see chapter on flowering by Hirst in this volume).

After flowers open (called flowering or anthesis), fruit formation results from the continued development of ovary (and floral cup) tissue through cell division, cell enlargement and intercellular space production. It is important to understand that these histogenetic activities (the formation of different tissues from undifferentiated cells) are intricately tied to within-tree and between-fruit competition. Presence of the king affects lateral fruit retention, more so in king-dominant cultivars than in non-dominant cultivars. Fruit retention and potential for development are greatly influenced by both natural and man-made limitations as discussed later.

After pollination and ovule fertilization during the bloom period, floral tissues begin to expand differentially so that the enlarging floral organs become the mature fruit. Fruit growth and the attainment of mature fruit size and form result from the multiplication of cells, cell expansion, direction of cell division and expansion relative to the flower/fruit axis,

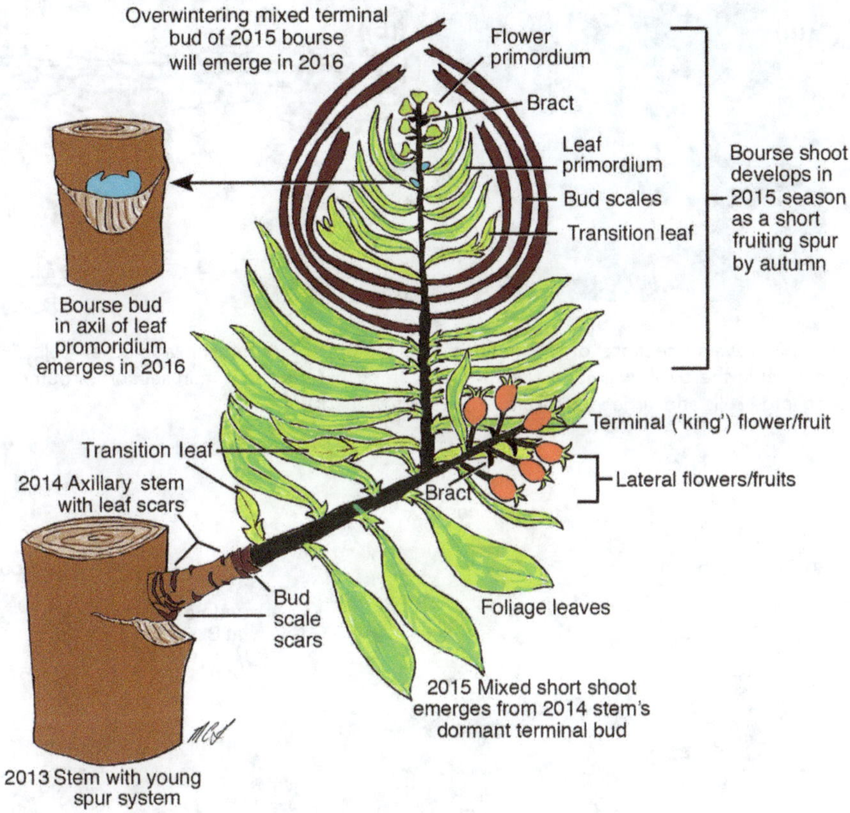

Figure 1 Apple spur origin, morphology and pattern of development, from initiation of a short shoot in the axil of a foliage leaf to winter dormancy two years later. Each year a bourse shoot emerges from the axil of a leaf just below the most basal flower of that season's flowering segment. By season's end that bourse has developed its leaves and a terminal bud containing scales and leaf and floret primordia for the next season.

as well as production of intercellular space (Fig. 2). These cellular dynamics vary among fruit tissues, resulting in the shape typical of a given cultivar (Fig. 3). Tissue structure is also greatly affected by fruit number retained as affected by orchard management practices, such as crop load adjustment, types of chemicals applied, and the amounts and application timing of chemicals.

Intense differentiation occurs in fruit tissues in the first several weeks post-bloom via cell division. Then cell division slows as cell enlargement and production of large intercellular spaces take place (Fig. 2 and 4). In later weeks, embryo growth within fertile ovules intensifies, while cell division in the fruit flesh tissue essentially ceases.

The tissue regions surrounding the five locules of the apple fruit have been described as either developing from incipient axial tissues (the receptacular theory) or from tissues common to fully or partially fused floral appendages (the appendicular theory) (cf., Pratt,

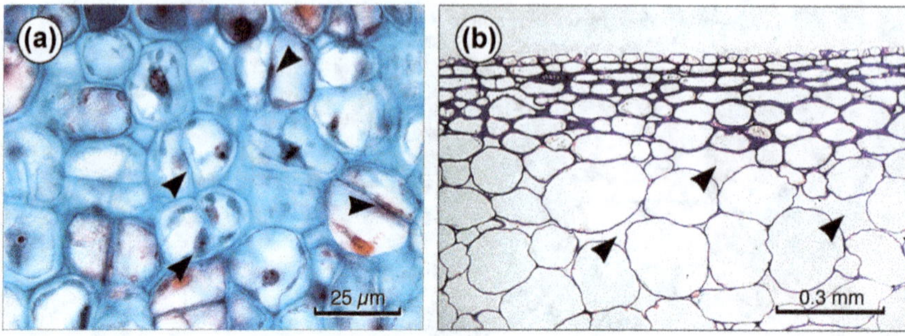

Figure 2 Transverse sections of the outer cortex region of cv. 'Empire' apple: (a) Cell divisions (arrowheads) in the ovary wall, one week post-bloom; (b) Mature fruit, thin section of outer cortex, showing large cells and large intercellular spaces (arrowheads).

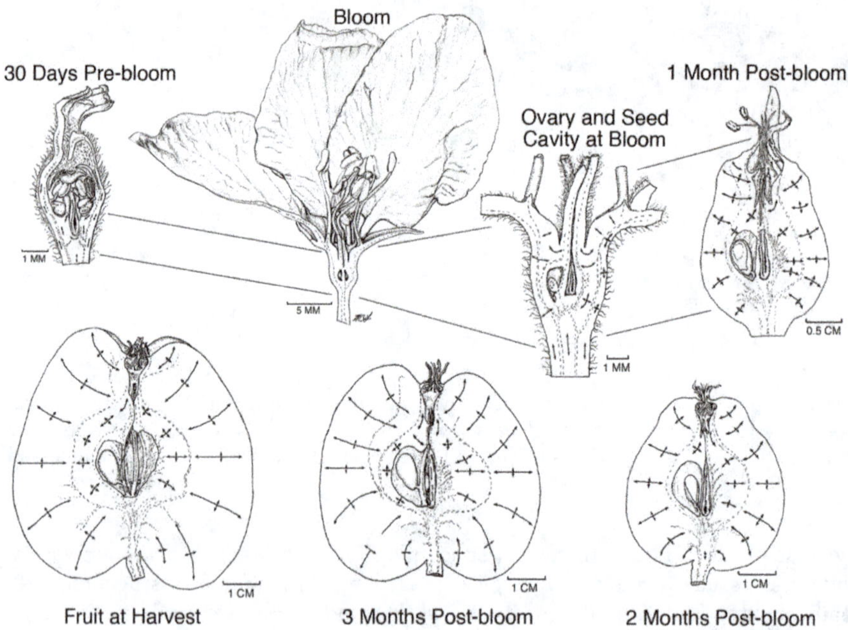

Figure 3 Seasonal growth of 'Red Delicious' fruit in relationship to major planes of cell divisions, cellular enlargement and tissue growth (arrow length and direction). As cells enlarge over the growing season, there is also a shift in the length–diameter relationship of the cells themselves in each tissue region.

1988). In this chapter we consider the apple *core* as a topographic region containing the central carpellary tissues (seeds, cartilaginous endocarp and immediately surrounding parenchyma). We label as *pith* tissue the region of parenchyma lying between the fruit's

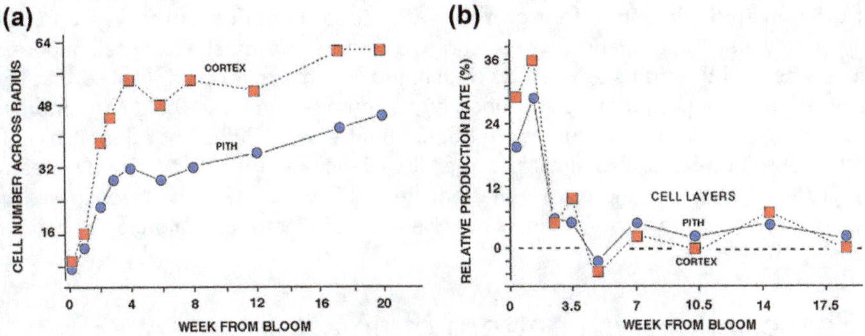

Figure 4 Cell number increase in the radial plane of the cortex and pith tissue of cv. Empire apple through fruit maturity (a), and the relative production rate of new cells (b). By four weeks post-bloom, cell divisions continue only at a very low rate.

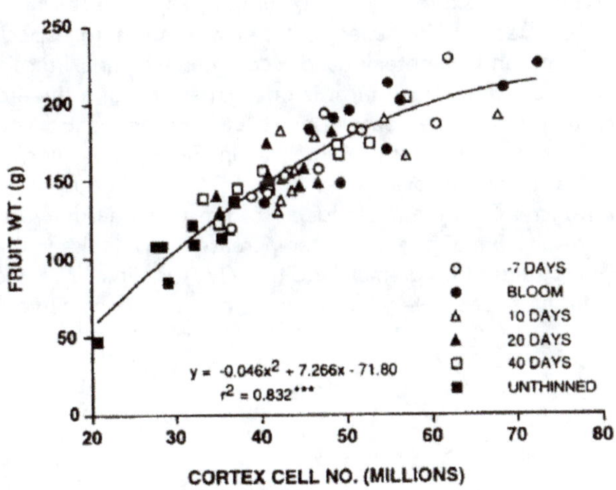

Figure 5 Relationship of final fruit weight and number of cortex cells of 'Empire' apples with variation due to thinning levels and timing of thinning (from Goffinet et al., 1995).

ten major vascular traces and this core. *Cortex* tissue is the tissue present between the ten traces and the outer epidermis.

At the end of the season the final fruit size is primarily determined by the volume of the cortex (between fruit's 10 main vascular traces and the skin). The cortex volume has three components: cell numbers, cell volume and intercellular space. There have been several studies of the relative importance of these components to variation in fruit size, with differing conclusions. Statistical differences in fruit cell number account for much of the variation in final fruit size in apple as reviewed in Westwood (1978) and Tromp et al. (2005), although very light bloom and low crop levels tend to increase both cell numbers and cell size.

If the primary limitation in fruit growth is due to competition of many young fruitlets immediately after bloom, and there are no mid- or late season stresses, cell numbers will primarily control the fruit size. A study of thinning of 'Empire' apples showed that 83% of the variation in final weight was explained by cell numbers versus 10% due to intercellular spaces and 7% due to cell volumes (Fig. 5; Goffinet et al., 1995). This relationship of fruit size to cell numbers is also similar to that for 'Delicious' fruits reported by Westwood et al. (1967). If, however, there is a limitation after cell division is complete, such as drought stress or pest injury, then cell volumes may be the main factor in final fruit size variation.

3 Seasonal growth pattern of an apple

The seasonal growth of the apple has two components. Cell division is the primary source of fruit growth in the first 7–10 days after bloom. Afterward, both cell division and expansion occur simultaneously as cell division rates slow until stopping about 4–6 weeks after bloom (except in the skin which must continue to expand greatly). About 6 weeks after bloom, fruit growth is primarily by expansion of cells and intercellular spaces.

So how should growth be expressed? It is commonly expressed as increases in diameter, weight or volume. Diameter is often used as it is easy to measure, especially if monitoring the same fruit over time, and diameter is used in commerce for final fruit size categorization. The limitation of using diameter is that it is one dimension of a three-dimensional fruit. A plot of fruit diameter over the season shows a slight lag period after bloom, then a gradually slowing increase for the rest of the season (Fig. 6). This suggests that the apple fruit gradually slows its growth over most of the season, although this is not true. For example, a 2 mm increase in diameter for a 75 mm fruit near harvest (11 grams fresh weight for 'Gala') results in a 55-times greater increase in apple tissue than would a 2 mm increase for a 5 mm fruitlet (0.2 grams). Additionally, it is difficult to accurately measure 1 mm increases or differences in very large fruit as they approach harvest.

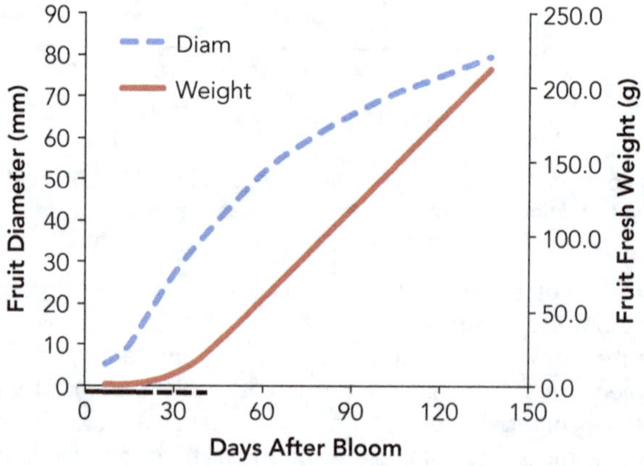

Figure 6 General seasonal curves of the fruit growth of 'Empire' apples with unlimited conditions (light crop, no water or temperature limits). Growth is expressed as diameter (dashed) or fresh weight (solid). The solid/dashed bar on the x-axis indicates the period of cell division.

Weight or volume are much preferred as expressions of growth of a three-dimensional fruit. If a fruit is allowed to grow to its genetic maximum without limitations of heavy crop competition or other stresses (water, nutrient, biotic), it grows exponentially in the first week or so after bloom (Fig. 6). Then as cell division declines and cell expansion increases, the curve of weight gradually becomes more linear. At about 4–6 weeks after bloom, the weight gain becomes quite constant and so the seasonal weight curve remains linear to harvest. This pattern of growth is called expolinear (Monteith, 2001; Lakso et al., 1995). If there are no major limitations, cell expansion adds a similar amount of weight per day until harvest, although with very heavy crops, cold weather or environmental or biotic stresses, the rate of fruit growth may decline before harvest.

Studies of seasonal growth have shown that large fruit have higher growth rates than smaller fruit as expected (e.g. 2 grams per day vs 1.2 grams per day, respectively). However, when growth per day is divided by cell numbers measured in the same fruits, the daily growth rate per cortical cell is essentially identical for fruit of either size (Lakso and Goffinet, unpublished). This emphasizes the importance of generating high cell numbers by good tree management and early crop adjustment.

Several common textbooks (Westwood, 1978; Faust, 1989; Jackson, 2003; Tromp et al., 2005) state that the fruit growth pattern of apple is a single sigmoid, with an initial exponential phase reaching maximum growth rate in mid-season, followed by a decline in growth rate. This type of curve may appear where fruit growth is limited due to mid-to-late season stress or competition as mentioned above, but it likely does not reflect the inherent growth pattern nor growth potential of the apple fruit. Interestingly, growth curves generated in cool climates tend to show a slowing of growth rate near harvest, but in warm, long-season climates, growth curves show the expolinear pattern of continued constant growth to harvest.

4 Chemical composition and seasonal changes in apples

The apple is a fruit that consists primarily of water, carbohydrates, organic acids and some secondary metabolites such as anthocyanins, aromas, flavonoids and so on. It does not produce significant amounts of lipids, oils or proteins, so it is metabolically 'cheap' to produce. That is, relatively little energy is consumed constructing 'expensive' compounds such as lipids and proteins after the initial period of cell division.

Over the season, there are continuous increases in the content of several sugars (Fig. 7), although fructose is the most common (Zhang et al., 2010). This is interesting as sorbitol, a sugar alcohol, is the only form that is transported from the leaves to the fruit, yet is not accumulated very much. Starch is formed from imported sugars but tends to peak about mid-season, and then be metabolized to low levels at maturity. Malic acid concentration, the primary organic acid named after its presence in *Malus* species, tends to develop during cell division, peak about four weeks after bloom and then gradually decline due to fruit growth. In both sugars and organic acids, the contents continue to increase until harvest, but fruit weight growth is faster, causing the concentrations per unit dry weight to decline.

Important secondary metabolites that give the apple its distinct fruit colour, texture, aroma and flavour generally develop late in the season. Reviewing these complex factors is beyond the scope of this chapter, but the reader is referred to reviews by Telias et al. (2011), Yahia (1994) and Watkins (this volume).

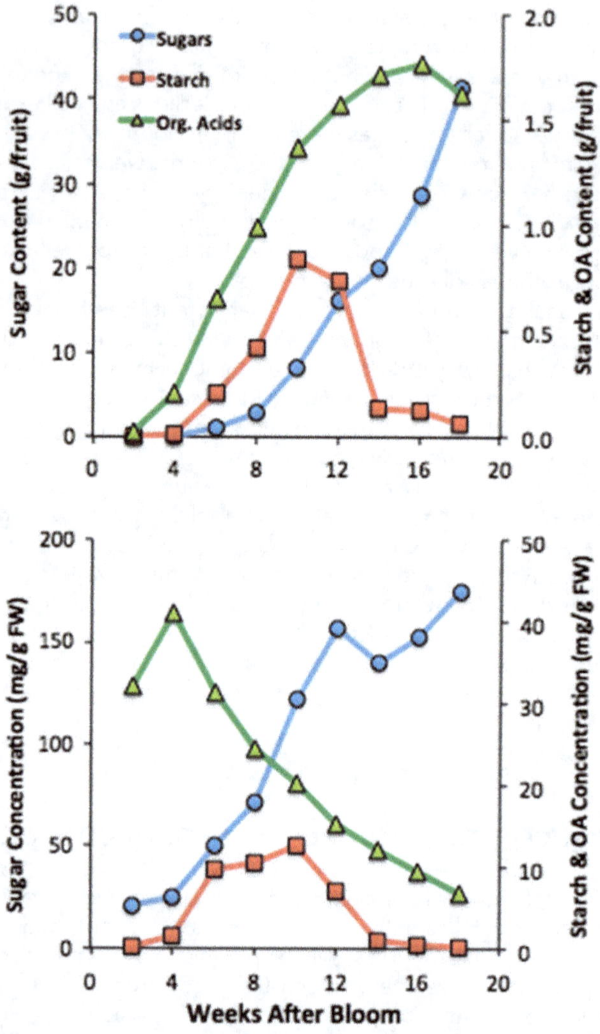

Figure 7 Seasonal patterns of key constituents of apple fruit. The primary sugars (circles) are fructose and sucrose; starch (squares) is the storage carbohydrate form and malic acid is the primary organic acid. The upper graph shows content per fruit while the lower graph shows concentration per gram fresh weight (data replotted from Zhang et al., 2010 and additional data courtesy of L. Cheng).

5 Fruit abscission and growth rate

As the apple tree produces far more flowers than it can carry as fruit to harvest, often 70–90% or more of the young apple fruitlets need to naturally fall or be thinned by hand or chemical treatment. So, why do some fruits abscise but others remain on the tree to harvest? It appears that fruit that can maintain a continuously high growth rate are retained

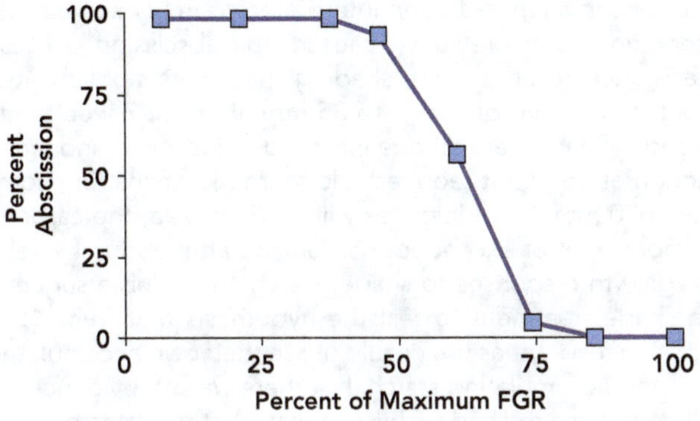

Figure 8 The relationship between percentage of 'Gala' fruit that drop versus the fruit growth rate expressed as percentage of the maximum growth rate of fruit in the population over 7 days after a chemical thinner was applied. Fruit diameters at the time of application averaged 10 mm.

on the tree; however, fruits that have slow growth or their growth slows significantly for several days will trigger the abscission process and eventually drop (Byers et al., 1991; Ward and Marini, 1999; Lakso et al., 2001; Greene and Lakso, 2013).

We have monitored the growth of thousands of fruit over many years and have found a consistent general relationship between fruit growth rate and eventual drop (Fig. 8). Although absolute values vary, the shape of the relationship is the same for many varieties as well as for fruit drop caused by different thinners, natural stresses of low light or excess competition among fruits. A decrease in growth rate that leads to fruit drop may occur for only a few days, but once it happens, the fruit is destined to drop in the next two weeks. Anything that reduces fruit growth will increase abscission, which seems to explain why so many different things (stresses, different chemicals, cloudy or hot periods, etc.) can increase fruit drop.

It is important that growth rate be expressed as percentage of the most rapidly growing fruit, not absolute growth rate. If absolute growth rate is used then cold temperatures that reduce growth rate should cause fruit drop; however, we typically find no fruit drop based on cold temperatures as cold reduces the growth of all fruit.

6 Fruit 'set'

The term 'fruit set' is often used in relation to fruit development and the lack of abscission. The term, however, is often used to mean different things for different authors and there has been no agreed-upon definition. In some cases, it refers to fruit that are simply fertilized and begin to grow shortly after bloom; however, the overwhelming majority of fertilized fruitlets abscise in the several weeks after bloom as noted above. Consequently, it has become common to refer to the fertilized fruit as the 'initial' set whereas the final number of fruit that persist to harvest as the 'final' set.

The term 'set' has a connotation that some final state has been reached, similar to the 'setting' of concrete, an irreversible change in condition. In the case of apples, at about 4 weeks after bloom the fruit number will generally not change until harvest (Byers et al., 1991). This observation has been interpreted to imply some type of physiological resistance to conditions that previously caused fruit abscission. To test this concept of physiological resistance to abscission, shading trials were applied at several stages of fruit development from 8 mm diameter to 35 mm at about 7 weeks after bloom (Lakso, Greene and Cheng, unpublished). Shade inhibited fruit growth and induced fruit drop up to 35 mm, although at 35 mm it required a longer time of shading compared with when fruit were less than 20 mm. In addition, as will be discussed, the carbohydrate supply for the final fruit numbers met or exceeded the demand after about 4 weeks after bloom.

In addition to growth responses to shade, starch and soluble sugars were analysed at intervals during shade treatment to test the hypothesis that fruits 18–20 mm or larger were more abscission resistant as they could utilize their own accumulated starch reserves. Shaded fruit stopped accumulating starch, but there was no evidence the fruit utilized its starch reserves in response to shade. This suggests that fruit starch before ripening is not available to use for growth, but is maintained for later conversion to sugar at maturity.

Although we feel that fruit set is more of an observational concept than a physiological concept, the terms 'initial set' for the population of fertilized fruit and 'final set' when no additional abscission is seen are still useful terms if used consistently. We suggest that the ambiguous Northern Hemisphere, cold climate term 'June drop' be 'dropped' for obvious reasons.

7 Role of hormones in fruit development and abscission

The development of the apple fruit depends on the quantitative supply of critical nutrients (carbohydrates, water and minerals) but is likely coordinated by hormones. The relative limitation of nutrient supply versus hormonally regulated sink demand is not clear, although likely varies with stage of development and environmental conditions. The reviews of Dr. Frtiz Bangerth (Bangerth, 2000 and 2004) summarize the research on hormones in fruit development, with many examples with apple, and are the basis of the following summary.

Fertilization of the apple flower leads to a rapid increase in hormones in the developing seeds and ovules that apparently stimulate the growth and sink activity of the receptacle that will become the cortex of the fruit. Greater numbers of seeds will generate more hormonal stimulation of growth, leading to a general positive relationship of seed number to fruit size. The effect, however, appears to be minimal above about 4–5 seeds/fruit. During the season there are variations in the concentrations of seed hormones, much higher than cortex concentrations. Immediately after bloom, the concentrations of ethylene and abscisic acid (ABA) are high, but decline before increasing again later in the season. In contrast, cytokinins increase followed by giberellins and auxins during the cell division period. After cell division is done, hormones are likely regulating cell expansion via 'expansin' genes that allow cell wall softening and expansion. In addition, hormones have been implicated in sink strength and activity, although the mechanisms (such as simply stimulating growth or more specific effects on transporters) are not clear.

Hormones certainly play a critical role in fruit abscission. Fruit produce diffusible auxin that moves from actively growing fruit to the potential abscission zone at the base of the pedicel and inhibits the formation of the abscission zone. Once the auxin transport is reduced, ethylene plays a role in the abscission process. It appears that factors such as assimilate supply, competing shoot tipping and hormone sprays affect the diffusion of auxins from the fruit positively or negatively. Considering all the many factors that can affect fruit growth and abscission, Bangerth proposes that there must be a kind of 'master transducer' that integrates all such effects via auxin diffusion. Later, current efforts to understand this via molecular biology will be addressed.

Although the regulatory control by hormones of fruit development and abscission is still only partially understood, there has been a great lack of research in the past 20 years on hormone biochemistry due to the lack of replacement of retiring researchers. Molecular biology will provide new information that will be helpful, but the lack of complementary biochemistry and physiology will slow progress.

8 Competition within the flower/fruit cluster

Clearly, all flowers and pollinated fruitlets on a tree do not complete development to fruit maturity. There is a huge number of fruit that abscise, especially in the first four weeks post-bloom. Depending on the bloom density and pollination efficacy, it is common that 70–95% of pollinated fruit ultimately abscise. As these extremely large numbers of fruit compete for limited resources shortly after bloom, there is natural abscission and reduced

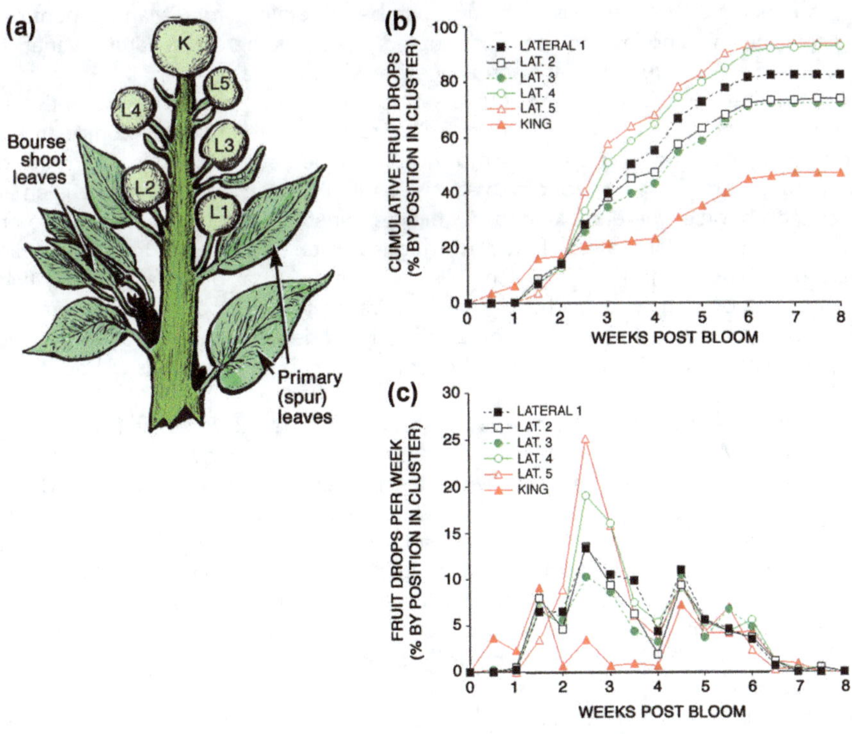

Figure 9 Flower/fruit positions within the 'Empire' apple spur (a), accumulated four-year total (1939) dropped over a two-month period at each fruit position within 430 tagged spurs via natural in-spur competition (b) and the rate of post-bloom fruit drop by cluster position (c). Kings were most likely retained until final harvest, L4 and L5 least likely, and L2 and L3 intermediate.

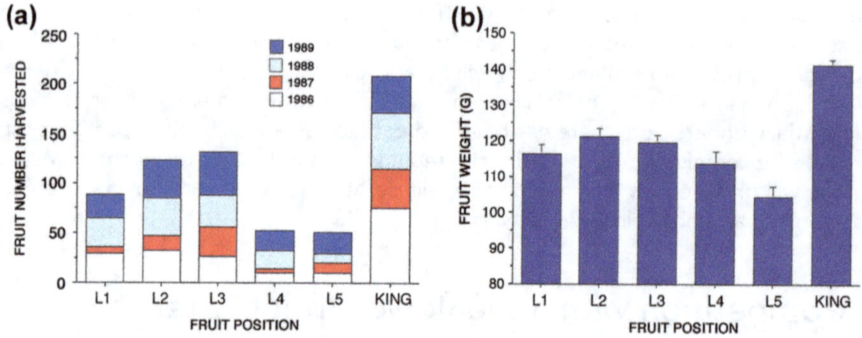

Figure 10 Fruits retained to harvest (641 fruits on 430 tagged spurs) on 'Empire' spurs on unthinned trees over a four-year period (a), and mean fruit weight (with s.e. bars) for these fruits at harvest (b) via natural in-spur competition, at each fruit position within spurs. Central king fruit were most competitive and the nearest lateral fruit (L4 and L5) least competitive. Kings were significantly largest and heaviest, L4 and L5 smallest and lightest, L2 and L3 intermediate.

size potential due to fewer cells of the final number of retained fruit. Management of crop load through thinning by growers is discussed later, but there are natural variations in competition among fruit and other organs, especially shoots.

In a detailed study of the fate of fruitlets borne on different positions within the flower cluster, Goffinet and Lakso (2016) reported a preferential retention of competing fruits. Central king position fruit was the most likely to be retained, while the most distal lateral fruits were most susceptible to abscission (Fig. 9 and 10). Thus, there is an apical dominance in force where the king fruit, which terminates the spur apex, affects retention of subordinate, axillary lateral fruits. The highest rate of drop occurs in the first few weeks post-bloom, the time in which cells are multiplying at greatest intensity (Fig. 4). Observations over many years have shown, however, that the pattern of pollinated fruit abscission over time after bloom is not consistent and depends quite strongly on weather patterns, especially light that affects carbohydrate availability to fruit, as will be discussed later.

Apical dominance within mixed buds is highly cultivar specific. For example, the cultivars 'Empire' and 'McIntosh' are weakly apical-dominant, whereas cultivars 'Red Delicious' and 'Gala' are more strongly dominant. When tagged 'Empire' spurs were thinned to only a single fruit at each of the six positions within a spur, overall per cent dropped at a given position was lower than in unthinned spurs with multiple competing fruits (Fig. 11a). Additionally, the potential for developing good fruit size increases (Fig. 11b). With no competition within the spur, the lowest laterals L1 and L2 had equivalent retention to kings. Of the retained fruit, most laterals produced fruit with weight comparable to kings. In this example, the higher than expected abscission for L3 and L4 resulted from pre-harvest abscission

'Gala' is much more king-dominant than 'Empire', as shown in a study of thinned spurs versus unthinned spurs. In unthinned spurs on unthinned trees, 80–90% of all tagged fruit dropped, although king drop was minimal. When tagged spurs were thinned to only a single fruit position, king drop remained exceedingly low, whereas lateral positions, especially those fruit nearest the king, dropped in large numbers (Fig. 12a). Final weight

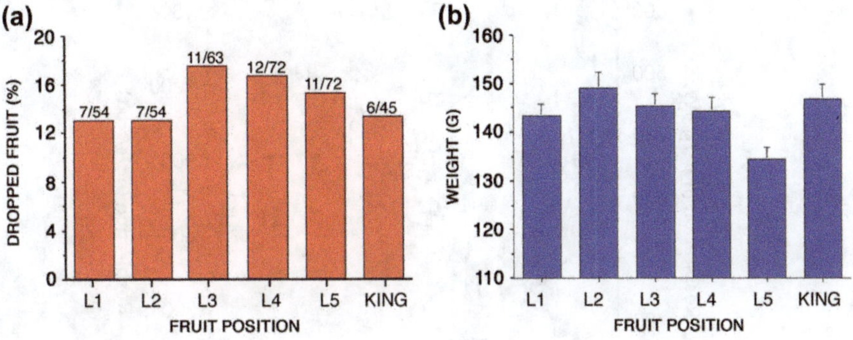

Figure 11 Percentage of fruit dropped from tagged 'Empire' spurs thinned to a single spur position (a) and mean weight (with s.e. bars) of fruit retained to harvest by position (b).

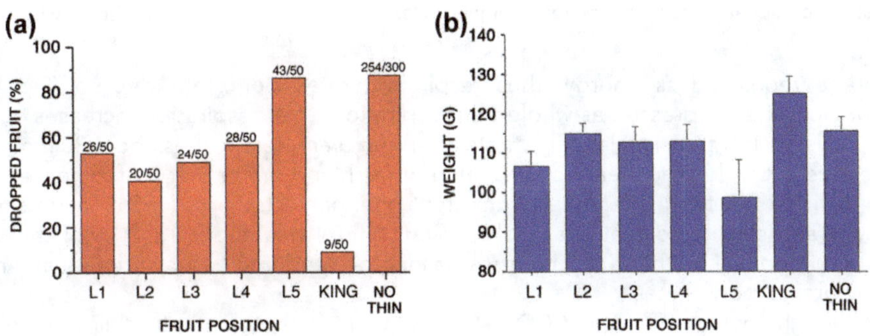

Figure 12 Percentage of fruit dropped from tagged 'Gala' spurs thinned to a single spur position versus unthinned spurs (a) and mean weight (with s.e. bars) of fruit retained to harvest by position in single-fruited versus unthinned spurs (b).

of harvested king fruit from tagged single-fruit spurs remained significantly greater than that for fruit at any other spur position (Fig. 12b).

Marini (2003) examined the response of apple growth to post-bloom chemical thinners as related to the location, king versus lateral (though not specific as to the exact lateral location) in several varieties and with several chemicals. The larger the size of the fruitlet, the greater the fruit retention, regardless of the location within the cluster.

9 Seasonal patterns of respiration and ripening

Fruit respiration that generates energy for growth and tissue maintenance is very high per gram of fruit early in the season. Cell division requires a large amount of energy. As fruit transitions to cell expansion in mid- and late season, growth and carbohydrate

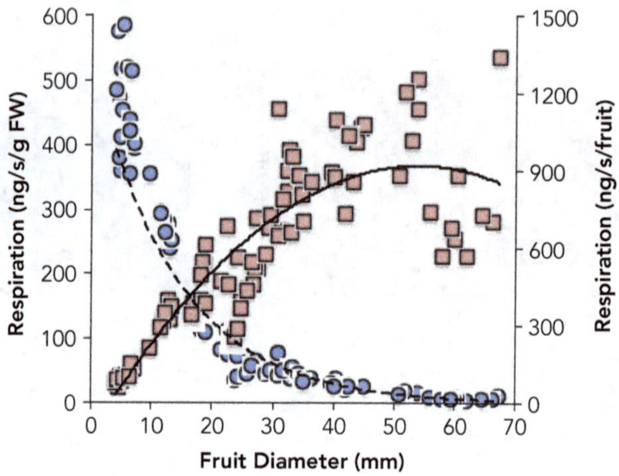

Figure 13 The developmental pattern of 'Gala' fruit dark respiration, measured at 22°C, and expressed as specific respiration per gram of fresh weight (circles) or total respiration per fruit (squares).

storage requires less energy, thus respiration rates per gram decline (Fig. 13). However, when expressed as whole fruit respiration, total respiration increases during cell division, but then is relatively stable after fruit diameter reaches about 30 mm. This is consistent with expolinear accumulation of fresh and dry matter that increases with cell division but then is relatively consistent until harvest. Jones (1981) found similar results and calculated that seasonal respiratory loss was about 12–20% of the total carbon imported by the fruit. The rest was incorporated into tissue structure and non-structural carbohydrates.

Although apple fruit have net CO_2 evolution, there is photosynthetic ability in the fruit skin. In exposed young green fruit, skin photosynthesis may completely balance dark respiration for a few hours under cool–moderate temperatures (Lakso, unpublished). This may help to partially support fruit growth of young fruits; however, placing foil shades over young fruitlets reduced fruit growth by about 20% for a day or two then recovered, but did not induce any abscission (Lakso, unpublished). Therefore, it seems that fruit photosynthesis, although clearly positive, is likely to be only a minor component over the season, but potentially critical to maintain fruit growth shortly after bloom if the fruit is strongly limited for carbohydrate import.

Apple is a climacteric fruit, as it shows an increase in respiration rate as fruit ripens. The climacteric rise is associated with increased production of the ripening hormone, ethylene, and a striking change in metabolism of the fruit. These include flesh softening, hydrolysis of starch to sugars, aroma development and changes in background colour of fruit, generally from green to yellowish (Brookfield et al., 1997; Janssen et al., 2008; see Jackson, 2003 or Tromp, 2005 for a more comprehensive review).

Ethylene production in fruit during the season is generally very low, but increases greatly, up to 1000-fold, at the end of the season as many other changes occur (see Watkins, 2002 for review). The increase is much greater than that of respiration and is an autocatalytic process as ethylene acts as a hormone.

10 Factors influencing fruit growth

The inherent growth and development pattern of apple fruit is genetically determined and is limited by a programmed change in the physiology of fruit ripening and abscission as will be discussed later. However, the amplitude of development is affected by many factors. Fruit growth requires substantial support of carbohydrates, water and nutrients and thus can be limited by deficits of these resources. Mature apple trees on dwarfing rootstocks with full crops can partition 50–75% of the total dry matter produced in season into the fruit, called the Harvest Index, in sustainably cropped orchards (Palmer, 1992; Palmer et al., 2002; Wunsche and Ferguson, 2005; Cheng and Raba, 2009). This is due in large part to the presence of the perennial existing structure and its reserves that do not need to be built from seed each year. Therefore, once established, the dwarf apple tree can invest a very large percentage of its seasonal dry matter in the fruit. In orchards with such high harvest indices, the demand for carbohydrates is very high and can be limiting for fruit growth and retention.

10.1 Carbohydrate reserves

Initial growth of shoots and flowers at budbreak in spring is supported by carbohydrate reserves in roots and branches. Gradually, as leaves are produced, there is a transition from solely depending on reserves to obtaining greater support from current photosynthesis of the leaves. Several studies have found that carbohydrate reserves reach their minimum just about bloom time then stabilize or begin to increase again, indicating that after bloom the fruit are not drawing significantly on reserves any more (Hansen, 1971; Priestley, 1960). Post-bloom manipulation of current photosynthesis by shading, photosynthesis inhibitors or CO_2 enrichment has been shown to directly affect fruit growth. This supports the contention that, after bloom or perhaps one week post-bloom, fruits are almost entirely supported by current tree photosynthesis (Byers et al., 1991; Lakso and Corelli Grappadelli, 1992; Byers, 2003; Lakso, unpublished).

10.2 Post-bloom carbohydrate support of fruit development

There are many ways to express the relative balance of crop requirements or demand to the ability or supply of the tree. Crop demand has been expressed as fruit number per tree, but the range of tree sizes due to dwarfing rootstocks necessitates finding a way to adjust for tree size. Consequently, fruit number per cm^2 of trunk cross-sectional area (TCSA), called crop load, has been adopted since the leaf area of unpruned trees or lightly pruned trees is correlated to TCSA (Westwood, 1978). Palmer (1987), however, demonstrated that with heavy or containment pruning to maintain trees within allotted spaces, the correlation of leaf area to TCSA becomes very poor.

A more physiologically based expression of crop load or 'supply:demand' is fruit per m^2 of leaf area, although even this does not directly take into account radiation energy intercepted or tree photosynthetic conversion to dry matter. Additionally, such indirect indices are given one value for the entire season, usually based on the final fruit number retained to harvest and mid-season values of leaf area or TCSA. There are obvious seasonal changes in fruit number per tree, stage of fruit development, tree canopy development and radiation interception, and general weather trends. Therefore, a single value for a dynamic season is rather impractical. For that reason a more precise model is a weather-driven

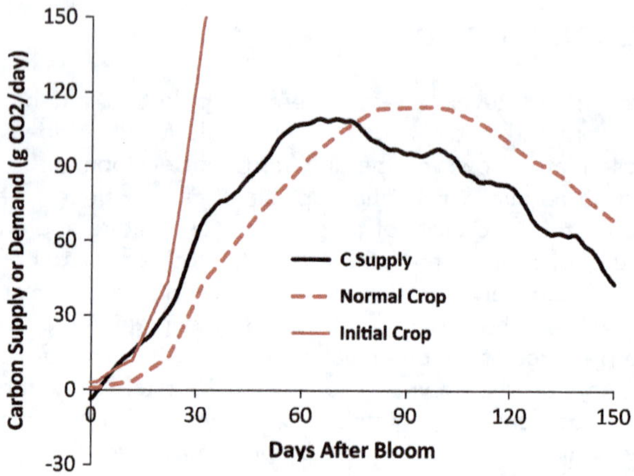

Figure 14 Simulation of seasonal patterns of carbon supply (heavy, solid line), demand of an initial crop of 1000 pollinated fruit (thin line) and demand of a normal commercial crop (300 fruit, dashed line) using long-term weather data of Geneva, New York. The MaluSim model (Lakso et al., 2001) simulated an 'Empire' orchard in Geneva, New York, to test simulations against actual performance.

dynamic simulation model of tree canopy development, radiation interception, dry matter production and organ demands (Lakso et al., 2001) to examine seasonal trends in carbon supply and demand.

Over a growing season the balance of carbohydrates supply available for crop development increases after bloom due to increasing leaf area of the tree and increasing radiation, peaking in mid-season and declining towards harvest with decreasing radiation in the autumn as estimated by a simulation model (Lakso et al., 2001) (Fig. 14). The carbohydrate demand of the crop, on the other hand, depends on the number of actively growing fruit. In this simulation, the demand of a good commercial crop of 300 'Empire' fruit is shown along with the demand of 1000 fruit that may naturally begin developing. It is clear that the demand of 1000 fruit greatly exceeds the available supply, although not until about 7–10 days after bloom (approximately 30 days after budbreak in this example with New York weather). Clearly, this is an early critical period of carbon deficits that has great impact on fruit development and fruit abscission.

Leaves on apple trees are borne on different types of shoots. Some leaves are borne on rosette-type spurs with the fruit as primary leaves developed before bloom or lateral bourse shoot leaves developed after bloom (cf., Fig. 1 and 8). Some varieties also produce fruit on long extension shoots, but most fruit is borne on short spurs. Leaves are also borne on long extension shoots or on very short lateral shoots from last year's extension shoots. Different shoot types appear to partition their photosynthates differentially to support fruit versus other competing organs such as shoots and roots.

Studies set up to determine which leaves on the trees support fruit development at different times over the season have clarified the picture. Extension shoot leaves support growth of the shoot with carbohydrates until at least 10–12 leaves have unfolded on an actively growing shoot (Quinlan and Preston, 1971; Corelli-Grappadelli et al., 1994). After that, the most basal leaves on the shoot can export carbohydrates to the fruit. However, if the shoot is shaded, as many as 20 leaves will export their carbohydrates to the shoot and

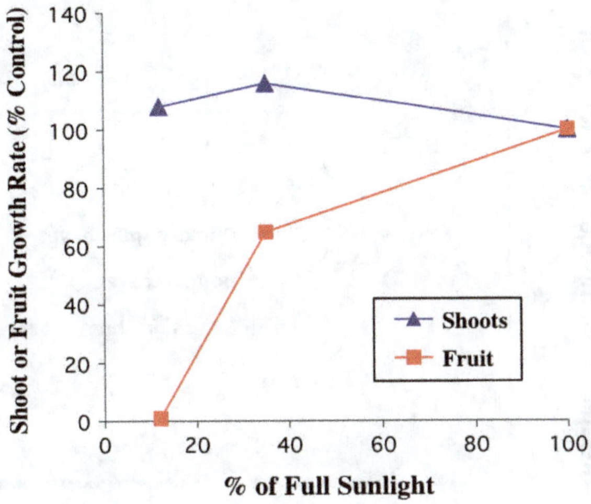

Figure 15 Effect of six days of varying light by whole-tree shading of 'Empire' on growth of active extension shoots versus fruit, beginning at fruit diameters of about 16 mm (Lakso, unpublished).

not to the fruit (Corelli-Grappadelli et al., 1994). Eventually, when extension shoots stop growing, all leaves can send their carbohydrates to fruit.

This is consistent with the practical observation that very light pruning will support fruit set and early fruit growth better than heavy heading cuts. Light pruning leaves a large number of less-vigorous shoots that stop growth quite early, and can then partition more carbohydrates to fruit development. Heavy pruning, conversely leaves fewer, but more vigorous extension shoots that compete with fruit for longer times and removes spur shoots.

In the first few critical weeks after bloom, carbohydrate support for fruit growth comes primarily from spur leaves and small 'spur-like' short lateral shoots on last year's long shoots. This was very clear when dry matter growth in fruit versus extension shoots was measured in whole trees that were shaded for six days when fruit averaged about 15–16 mm diameter (Fig. 15). Even down to about 15% full sunlight, shoots continued to grow at the same rate while fruit growth was reduced to near zero. This heavy shade to 15% of full sun completely de-fruited the trees while having no effect on shoot growth, similar to the branch shading study of Bepete and Lakso (1998).

Under limiting light and limited photosynthesis early in the season, the tree appears to give priority to extending shoots, presumably to intercept more of the limiting light. This is also consistent with other observations. Trees replenish carbohydrate reserves after bloom even though there is strong competition from many young fruitlets and shoots. Juvenility in apple seedlings (the tendency to not bloom for several years to allow the tree to become established) suggests producing a crop is of lower priority. Inhibiting active shoot growth during the post-bloom period with prohexadione-Ca leads to increased fruit retention and decreased response to chemical thinning (Greene, 2007).

The general pattern of carbohydrate support for fruit growth after bloom (Fig. 16) is that primary spur leaves (those that emerge before bloom) are initially important to support the fruit, but their importance gradually declines due to their small leaf area. Next, leaves on the lateral ('bourse') shoots in the spur develop after bloom and support the fruit unless

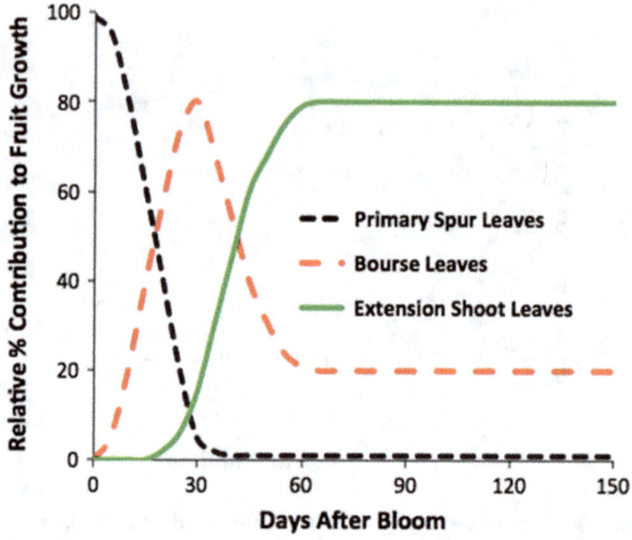

Figure 16 Generalized diagram of the changing seasonal pattern of relative support of carbohydrates for fruit growth in apple by leaves on three shoot types.

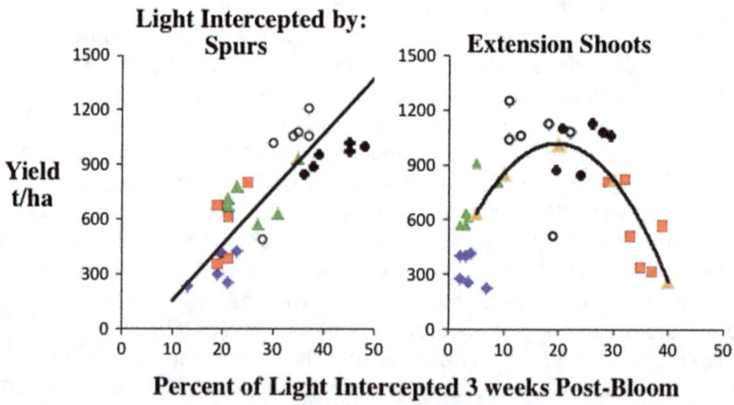

Figure 17 Relationship of orchard yield of various orchards to light interception by spurs versus by extension shoots at three weeks after bloom (re-drawn from Wunsche et al., 2000).

the bourse shoots are extremely vigorous. Finally, leaves on the extension shoots with more than about 12 leaves begin to support the fruit. Gradually, as shoots stop growing, all leaves can support fruit growth. For the last two-third of the season both bourse and shoot leaves support the fruit. Thus, it is important to expose the spur canopy of leaves to sun as much as feasible during the weeks just after bloom, as they are the primary supporters of the fruit.

This general pattern from physiological studies was confirmed more directly in trees in the field. Using a laser as an artificial sunbeam, the percentage of the sun intercepted by spur leaves versus shoot leaves was measured at a critical time three weeks post-bloom in orchards of varying yields (Wunsche et al., 1997) and values then related to the yields of those orchards (Wunsche et al., 1996 and Wunsche and Lakso, 2000). Consistent with the physiological studies, results showed that (1) at three weeks after bloom orchard yields are positively and highly related to light interception by the spur canopy, and (2) excessive light interception by extension shoots leads to lower orchard yields by reducing the light available to spurs in denser canopies (Fig. 17).

10.3 Water

Water is an essential resource for almost every process in the apple tree, providing a solvent, a transport medium for nutrients and carbohydrates, a source of turgor pressure for cell growth and a regulator of canopy temperature. Although essential, it may or may not be involved in limiting fruit development unless it is inadequate, which is discussed below.

In addition to the direct use in fruit cell turgor and growth, distribution of transpiration water to leaves and fruit delivers nutrients to the fruit-bearing shoots and the fruit directly. It is important to have good leaf area on a fruiting spur to obtain good fruit calcium uptake due to diurnal cycles of influx and efflux of xylem sap (Lang and Volz, 1998 and references therein).

Recent work from Corelli and colleagues (Morandi et al., 2011; Zibordi et al., 2014) monitored fluxes of xylem, phloem and transpiration of fruit in relation to fruit growth and environment, such as light level. Low light that reduced canopy carbohydrate production primarily reduced phloem flux into the fruit, although it was a gradual reduction indicating that the tree has some mechanism of buffering carbohydrate fluxes. This is consistent with the effects of shading on fruit abscission as 2 or 3 consecutive days of low light has a much greater thinning effect than when those days are alternated with high light days (Byers, 2003).

One aspect that has received relatively little attention is that apple fruit appear to develop xylem dysfunction later in the season (Drazeta et al., 2004) that may restrict the flow of xylem sap with its calcium and other nutrients. Loss of water permeance of the skin as the season progresses and reduced water loss rates may additionally limit late season accumulation of nutrients. Variation in skin permeance of apples at harvest can be very large even within one population although the reasons for it are not clear (Maguire et al., 1999).

Water deficits may be a critical limitation on all growth processes (see Lakso, 2003 for review). Early season water stress can inhibit fruit set (Powell, 1974) presumably by inhibiting fruit growth rates as discussed earlier. Fortunately, apple bloom occurs at an early, relatively cooler, time in the season before most summer water stress develops in humid climates. In drier climates, irrigation is generally available so water stress is less likely to be critical during bloom and early fruit growth.

Water status of leaves and fruit of an apple tree depends on not only soil water status but also evaporative demand and transpiration during the day (Lakso, 2003). Consequently, water deficits for the fruit occur more often during the summer due to higher evaporative demands regardless of the soil moisture (Powell, 1976). So water stress effects on fruit growth are more problematic in the mid- and late season typically during cell expansion.

When water deficits occur, effects depend on whether the deficit occurs during cell division or after cell division ceases. Inhibiting early cell division will lead to reduced final size as cell numbers limit size potential (Fig. 5). This combined with the continuous linear growth pattern (Fig. 6) makes deficit irrigation difficult to employ in apple compared with grape or stone fruits, which have mid-season lag periods in fruit growth when water deficits can reduce vegetative growth with little effect on fruit size.

Ebel et al. (1995) examined the interactions on seasonal fruit growth of three crop loads with full or Regulated Deficit Irrigation. They found that heavier crop loads reduced fruit growth as expected, and in addition the water stress reduced fruit growth rate. The effect of the water stress, however, was mostly at low-moderate crop loads; at higher crop loads there was less water stress effect. This was likely due to the high crop load reducing growth so much that water was not the limiting factor. Overall, the relation of fruit growth rate to mid-day stem water potential showed that fruit growth rate was not affected until potentials fell below about −1.5 to −1.6 MPa.

10.4 Temperature

Temperature can be a limiting factor for rate of apple development, especially cool-to-moderate temperatures. Warrington et al. (1999) held trees at several day/night temperatures from 13/3°C to 22/12°C for the period of 10–40 days after bloom. Expansion rates increased almost perfectly with degree-days above a base temperature of 4°C. Final fruit weights also increased with warmer early temperatures. However, final differences were less than the early growth rates suggested as the colder temperatures extended the period of cell division. After 40 days, the effect of temperature is much less (also Calderon-Zavala et al., 2005); it appears that cell division processes are much more sensitive to temperature than cell expansion and storage processes.

Although cool temperatures may often directly limit fruit growth by limiting enzyme activities, the effects of high temperature are not as clear. Most processes have temperature optima, often in the mid-20s to low-30s °C. Consequently, at temperatures over about 30°C, many different processes may be limiting. For example, high temperatures induce high evaporative demands, which may impose water stress even with adequate soil moisture. Also high temperatures will drive up demand for carbohydrates for growth and respiration of all organs while reducing the supply due to supra-optimal effects on photosynthesis, which may lead to carbohydrate limitations. Calderon-Zavala et al. (2004) found that 6 mm fruit showed increased growth up to about 31°C, but in later fruit, at 11 and 18 mm, growth rate had an optimum of 18–20°C. The temperature optimum for canopy photosynthesis was found to be about 20°C, suggesting carbohydrate shortages at high temperatures.

Temperature also has complex effects on ripening with many effects established surprisingly during the cell division period (see Tromp, 1997). Late season colour development, firmness and acid concentrations were affected by early season temperatures, although soluble solids were not. Optimal harvest windows can be predicted well from temperatures in the first 30 days after bloom (Blanpied and Silsby, 1992), leading Blanpied (1991) to conclude that the 'season of fruit ripening may be independent of growing season temperatures and light. This suggested a genetic control mechanism, which may be independent of the environmental conditions that normally prevail in apple growing regions.' Thus, it appears that once cell numbers are set, the responsiveness of fruit development by cell expansion and storage to temperature may be minimal.

10.5 Mineral nutrition

Apple fruit require adequate supplies of mineral nutrients, as do all tissues, for growth. A comprehensive study of seasonal mineral nutrient contents and concentrations of 'Gala' apple trees and fruits by Cheng and Raba (2009) provides an excellent summary of many points. Similar to carbohydrates, reserves in the tree normally provide the nutrients for growth before bloom, and then current uptake from soil takes over. As with weight growth of fruit, contents (mg or grams per fruit) show a general pattern of exponential increase during the cell division period, followed by a fairly linear or sigmoid increase until harvest, though there are some differences among the nutrients.

With a heavy crop at harvest, fruit accounted for 38% of the tree's seasonal uptake of N, whereas P and K accounted for 61% and 71% of the tree's seasonal uptake, respectively. Ca in the fruit, however, only accounted for 14% of the tree's seasonal uptake in this study (Cheng and Raba, 2009). Others have found that the seasonal uptake of Ca into fruit can similarly continue until harvest or it often appears to slow or stop in mid-season (Wilkinson, 1968; Faust, 1989; Jackson, 2003).

In terms of nutrient concentrations, which are expressed on a percentage dry weight basis, all nutrients appear to decline over the season as fruit dry weight increases faster than the nutrient contents (Fig. 18). Ca is a key nutrient in relation to fruit post-harvest storability (Faust, 1989; see chapter by Watkins in this volume), and is often a limiting factor, especially as growers push for large fruit size for profitability. Consequently, considerable applied research has been done on mid-season foliar applications of Ca to improve fruit Ca concentrations (Neilsen and Neilsen, 2003).

Flower and fruit development can be limited by nutrient deficiency during both the prior season and as carryover to the current season (Fallahi et al., 2010). As many nutrients are considered essential, by definition any of them may be limiting. In practice, most apple-growing regions have identified which nutrients may be limiting in their soils and

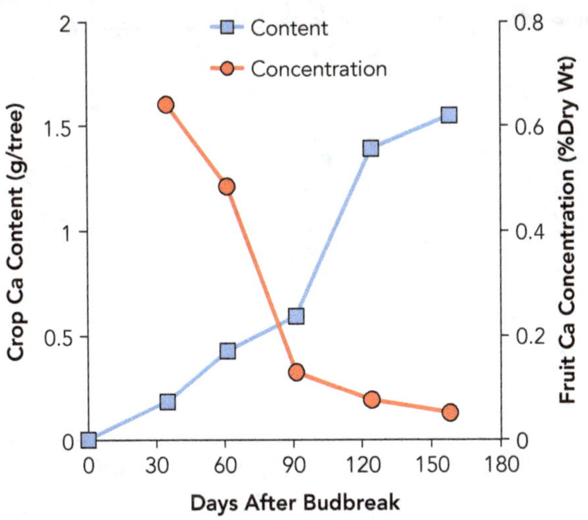

Figure 18 Seasonal pattern of total content and mean concentration (per cent dry weight) of calcium in 'Gala' apple fruit (data courtesy of L. Cheng; from study of Cheng and Raba, 2009).

fertilization regimes have been developed by decades of research. However, some more subtle effects of micronutrients that are not taken up readily from soil such as zinc and boron may still impact fruit development.

It seems that any nutrient deficiency, except possibly calcium, will likely have complex and likely indirect effects on fruit development. For example, besides direct effects on fruit cell growth or function, these may include reducing vigour leading to less leaf area per fruit and/or lower photosynthesis rates, thus reducing the carbohydrate supply to the fruit. Similarly, deficiencies may reduce root development leading to reductions of nutrient uptake, such as for K, especially during drought periods. It is difficult to determine whether a water stress affects tree physiology and demand for nutrients before the nutrient becomes limiting or vice versa. Our inadequate range of experience suggests that light, temperature or water stress most likely limits fruit growth sooner than nutrient deficiency but that nutrients may be a limit for long-term sustainability.

11 Regulation of cropping for sustainability

The apple tree generally produces more flowers and young fruitlets that it can carry sustainably to harvest. Natural drop occurs but is not sufficient or early enough to give the desired balance between yield (primarily due to fruit number) and fruit size (Fig. 19). The economically optimal crop level is the balance of enough yield to generate economic potential and the increasing economic value per fruit as a function of fruit weight allowing for good return bloom (assuming colour is adequate and fruit have no clear defects). This optimum is partly physiological and is also partly determined by the economic values based on fruit size of a given variety (see example in Byers, 2003).

Therefore, early thinning of the crop is a critical cultural practice to achieve that balance. Commercially, thinning is done as early as feasible after bloom, and sometimes during bloom, to limit fruit competition that may reduce cell division and impact potential fruit size of retained fruit. It is also important for sustainability to avoid excessive crop

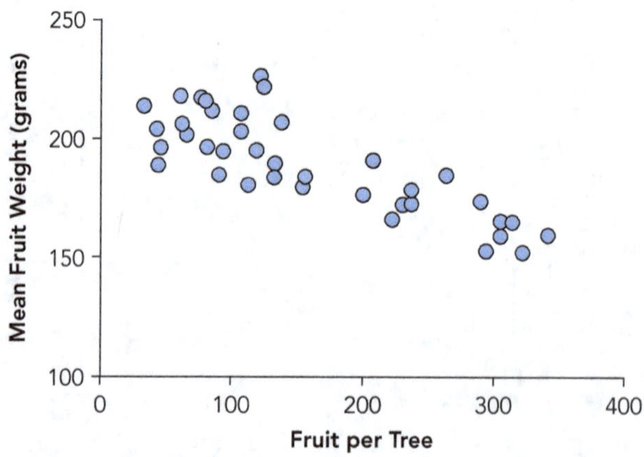

Figure 19 The relationship of final fruit weight of 'Empire' apples to the number of fruit/tree. Each point represents the mean for one tree.

competition that may inhibit flower bud development for the following year that begins generally about a month after bloom (Dennis, 2003 and Hirst in this volume).

Although hand thinning may be feasible in locations with very low labour costs, it is generally cost-prohibitive except as a late, remedial adjustment. Consequently, chemical thinners are commonly used to regulate crop level. The greatest difficulty with chemical thinning is the inconsistency of response of trees to chemical thinner application. Based on literature review, close to 50 factors may be involved in the thinning response, though there are many internal correlations among factors. Stover and Greene (2005) similarly reviewed a great many factors, especially those related to environments that affected responses to growth regulator applications.

If the basic mechanism of post-bloom fruit abscission is a critical reduction in fruit growth rate that triggers the cascade of events leading to abscission, then it is not surprising that a great number of factors or environmental conditions may be the limiting factor and affect abscission and response to chemical thinners. This is likely the reason why consistent thinning treatments, such as 10 parts per million of naphthaleneacetic acid (NAA) at the 10 mm fruit diameter stage, give inconsistent results. Greene and Costa (2013) concluded: 'Although chemical thinners were in general used since the 1940s and numerous field experiments were done varying concentration, times of application, stage of plant growth and weather conditions applied in, predictable results between years and occasions were rarely achieved.'

There have been many studies that have attempted to understand better the roles of individual factors, with experimental manipulation of variety, tree vigour, bloom density, environmental conditions or chemical used. We still do not fully understand the mechanism of action of the wide range of chemical thinners, hormones (auxins, cytokinins, ethylene, ABA) and others such as carbaryl and metamitron, likely have different mechanisms of action. Again, the most consistent effect of any thinner is the reduction of fruit growth of a portion of the fruit, leading to abscission.

Most progress in fruit thinning research has been made by varying the environment, especially solar radiation and temperature. Shading treatments for varying times and intensities and at varying stages of fruit development have been used to reduce carbohydrate supply, cause fruit abscission and accentuate the response to chemical thinners (see review by Byers, 2003; McArtney et al., 2004; Zibordi et al., 2009). In general, the effect of shade alone is relatively mild in the first week after bloom, stronger in the period of about 1–3 weeks after bloom, then fairly minimal after 3–4 weeks after bloom (Byers et al., 1991; Lakso and Corelli Grappadelli, 1992). This is consistent with estimates of carbon deficits occurring between 1 and 3 or 4 weeks after bloom as discussed earlier. Effects of low light in relation to chemical thinners was found to be minimal if occurring before thinner application, but was very strong if applied just after thinner application (Greene and Groome, 2011; Lakso, unpublished results).

Metamitron, a photosystem II photosynthesis inhibitor, has been recently approved for use in Europe as an apple thinner (McArtney and Obermiller, 2014). Terbacil, a photosynthesis inhibitor, had previously shown that carbon supply is critical to maintain fruit growth and retention though it was not registered for commercial use (Byers, 2003). It has similar effects to that of shade treatments, which we have been able to reverse with trunk injections of sorbitol in preliminary experiments (Lakso, unpublished data).

Interactions of temperature and solar radiation affect natural fruit abscission and response to thinners. Researchers and growers have observed that warmer temperatures lead to more thinning whereas cooler temperatures to less thinning (Byers, 2003). High temperatures exaggerate the effects of low radiation (Byers, 2003; Parra-Quezada et al., 2006; Yoon et al., 2011); however, sunny conditions tend to counteract warm temperatures. The integrated effects of temperature and solar radiation (warm and sunny vs cool and cloudy) have been difficult to reconcile. In addition to direct effects of temperature and radiation on tree physiology, there are other effects such as spray uptake and cuticle development, and of course effects on pollination.

12 Modelling to integrate factors in thinning

12.1 Carbohydrate supply–demand

A dynamic simulation model, MaluSim, was developed to provide a quantitative integration of the environmental factors and tree characteristics that affect the carbohydrate component of this complex system (Lakso et al., 2001). This model uses daily temperatures, solar radiation and physiological mechanisms to calculate canopy development, canopy photosynthesis, organ respiration, organ demands and distribution of dry matter. The model integrates these processes to give seasonal patterns of carbohydrate supply versus total or crop demand (Fig. 14). Simulations of a model version with constant tree parameters were compared with studies wherein the same chemical thinners were applied to an orchard at multiple intervals over many years (Robinson and Lakso, 2011). Almost every year had a unique pattern but years with particularly strong thinning reactions corresponded to years with severe carbon deficits that occurred immediately after thinner application. Deficits were due primarily to low solar radiation especially with warm temperatures (Fig. 20), consistent with shading studies just discussed. It should be noted, however, that a positive

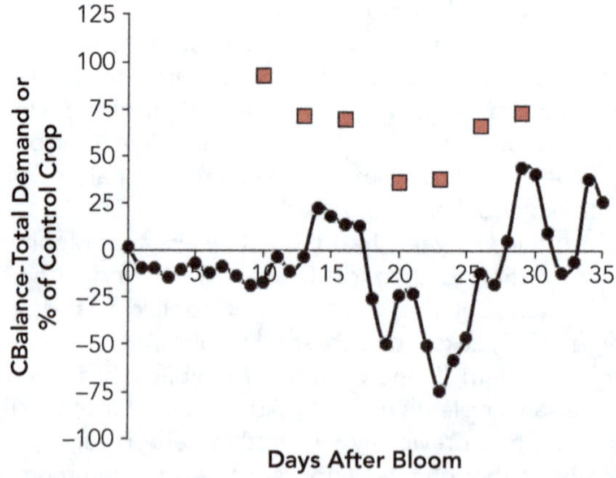

Figure 20 Example of the correlation of the patterns of daily carbon balance by simulation modelling (line) and the per cent of the control crop as a result of the same chemical thinner applications at intervals after bloom (square symbols). The simulation assumed a constant crop level equal to a crop normal for the simulated orchard and used weather data in Geneva, New York, for the year 2006. Each year gives a unique pattern.

carbon balance does not guarantee mild thinner responses as other factors that are not simulated may be more limiting to fruit retention and abscission.

Although the model with constant tree parameters only considers radiation and temperature effects on carbon balance, correlations have proven useful for researchers, extension or outreach specialists and growers. The primary value has been to alert growers to conditions of poor carbon balance that may lead to strong, and perhaps excessive, thinning. A major limitation, however, is that to be used real-time to predict thinner response, the model requires forecast weather data as it is the weather *after* treatment that is most critical. Temperature forecasts tend to be moderately accurate, but radiation forecasts are much less accurate and often are reduced to classes such as 'partly cloudy' or 'overcast'.

The model is available on the web for growers in the Northeast US with automatic regional weather data accession and thinning recommendations based on supply–demand balances (http://newa.nrcc.cornell.edu/newaTools/apple_thin; Lakso and Robinson, 2014 and 2015). Relative adjustments to the standard thinner concentrations that growers have developed based on their unique situations are recommended. These have proven useful for growers to improve the consistency of chemical thinning, and the model is being tested in many apple-growing regions.

12.2 Thinning response assay

As discussed earlier, fruit growth rate is critical to determine if fruit will be retained or abscise (Fig. 7); however, the fruit may drop from the tree a week to even three weeks after an abscission-triggering event. By the time the final drop is observed, it is too late to make any effective remedial adjustments. A simple model to assay for early prediction of thinner effect was developed based on non-destructive monitoring of the growth in diameter of a fruit sample after a thinner treatment (Greene et al., 2013). Instead of a complex curvilinear relationship, the abscission-to-fruit growth relationship was simplified: if the growth rate is >50% of the fastest growing population of fruit, it will be retained; if <50% it will abscise. Typically, it requires about one week or about 120–150 growing degree days, base temperature 4°C, to obtain a prediction that is 90+% correct. The threshold value may vary with some varieties, therefore it should be tested. This assay is rapid enough to allow follow-up treatments if needed (Greene et al., 2013).

These two modelling approaches alone and combined provide growers with improved information on the likely response to thinners and a method to assay thinner effectiveness. Along with more critical training and precision pruning methods, these models are contributing to increasingly precise crop control practices and sustainability (Robinson et al., 2016).

13 Molecular biology and apple cultivation

One of the major limiting factors in the study of fruit growth, and especially fruit abscission, has been the inability to elucidate the central mechanisms regulating development and response to environmental or cultural stimuli. From the fruit physiological studies discussed here, it appears that there is a central mechanism that integrates various support or stresses, regulates fruit growth and triggers the abscission cascade of events. A logical candidate is the regulation of cell division and growth.

Gene expression studies have begun to identify processes that may be critical to these questions, although expression of any particular genes does not necessarily control a complex process at the whole fruit or plant level. The mapping of the apple genome (Velasco et al., 2010) provides a basis for comparison and identification of genes and their functions that are needed to progress. Broad patterns of gene expression of metabolic processes over the season have been published (Janssen et al., 2008; Soglio et al., 2009). Studies of the seasonal regulation of apple fruit acid concentrations in genotypes of varying acid levels found 3066 differentially expressed genes in 648 functional classes (Bai et al., 2015). Such broad analyses show that hundreds if not thousands of genes are up- or downregulated during development which is not surprising, but also not easy to interpret cause versus effect. Other studies have focused on more specific processes. In ripening fruit, changes in expression levels of genes involved in ethylene biosynthesis, perception and cell wall degradation were identified (Li et al., 2010; Harb et al., 2012); generally confirming what was expected from the biochemistry of ripening.

Gene expression patterns relating to fruit growth and abscission are providing new insights, especially in relation to the environment or cultural practices such as chemical thinning. For fruit at the 12 mm stage, during 3 days of very low light, the first effects at 24 hours were mostly related to carbon metabolism in the fruit, consistent with carbon starvation (Zhou et al., 2008). Ethylene gene responses appeared to follow later at 72 hours after shading began. In parallel studies in the same orchard, benzyladenine thinning sprays resulted in gene expression patterns in carbon and energy metabolism similar to shading (Zhou et al., 2016).

Shade effects were further examined by Dash et al. (2012), concluding: 'shading results in coordinated changes in the expression of carbohydrate metabolism-related genes, key transcription factors related to fruit growth, and genes associated with cell production and expansion.' Conversely, reduction of crop load by prior thinning increases carbon availability to fruit, carbon metabolism and cell production-related genes leading to increased growth (Dash et al., 2013). These studies, combined with the developmental study of Malladi and Johnson (2011), physiological experiments and simulation modelling, suggest an important integrative effect of carbon and energy metabolism.

Any major change in the physiology of an organ, such as transitioning from active growth to abscission of fruit, would be expected to affect many processes. Ferrero et al. (2015) found over-expression of genes related to stress responses, hormone biosynthesis and the metabolism of auxins and ethylene. Ethylene-related genes have been found to be active in the cascade of events leading to abscission, although the timing and sequence is still unclear (Kolaric et al., 2011; Eccher et al., 2015). Botton et al. (2011) proposed a model of induction of fruit abscission that may be a helpful guide for further studies.

To have practical application, such studies require close collaboration with crop physiologists who can integrate molecular results at the organ and whole plant level. In addition to application of treatments and to interpretation of results, effective selection of trees for treatment and of the fruit for molecular analysis is critical. Interpretation is much more difficult and may lead to wrong conclusions if a fruit sample contains fruit of mixed stages of growth or abscission induction. In the post-bloom period, there is a large range of apple growth rates represented in a population of fruit. Although fewer king fruits will naturally abscise than lateral fruits, some still will slow their growth and abscise. Therefore, without individual fruit growth monitoring, it is very difficult to know in advance whether a fruit is continuing growth, has slowed or has stopped growth and induced abscission just before an experimental treatment was applied. For example, in the study of Zhou et al. (2008 and 2016), a large population of fruit were monitored for growth by measuring diameters starting several days before the shading treatment was applied. Some of the fruit had significantly slowed growth just before the treatment, and were omitted from the population to be sampled. After treatment, fruit from treated trees were sampled but only after growth monitoring identified a sample of fruit that had consistent growth rate effects. The use of fruit growth rate monitoring in advance of treatment is strongly recommended to ensure that fruit samples better represent treatments, especially in the case of critical timing studies.

A novel approach to improving the consistency of apple thinning and thus crop control is the genetic elucidation and potential for breeding or engineering self-thinning (Celton et al., 2014). Self-thinning apples may make thinning more consistent and reduce hand-thinning costs. However, obtaining a crop after a spring frost that often damages the king flower preferentially, may be riskier with self-thinning genotypes. The 5–20-fold excess of flowers over desired fruit numbers in apple trees provides a buffer in case of environmental stresses on flower survival, poor pollination or poor weather conditions post-bloom. This approach will be interesting to follow.

14 Future trends and conclusion

Important future areas for apple development research for sustainable production include the following:

- Continued testing and improvement of models that predict tree sensitivity to environmental variation and chemical thinning.
- Continued testing and improvement of fruit growth assays that provide rapid bioassays of tree response to environmental variation and chemical thinning.
- Improved understanding of the genetic control of fruit growth and abscission that will (1) provide clarification of the key processes involved, (2) allow the development of molecular-based assays to determine the physiological status of fruit during development and thinning and its potential for post-harvest storage and fruit quality, and (3) guide the development of targeted thinning chemicals. Collaboration of researchers at all levels from the molecule to the orchard will be essential.

Key fruit characteristics such as yield, fruit size, quality, resistance to disease and pests, drought resistance and so on will certainly be addressed by breeders. Gene silencing or editing will provide opportunities for more targeted variety improvement. Manipulation of traits, such as tree architecture, flower number per cluster and changes in sensitivity to thinners, well integrated with tree physiology and management, also promise to improve sustainability of fruit production.

15 Where to look for further information

For additional information on related apple physiology, see the books by Jackson (2003), Tromp et al., (2005) and D.C. Ferree and I.A. Warrington (eds), *Apples, Botany, Production and Uses*, pp. 167–94. CAB International, Wallingford, Oxon, UK.

16 References

Bai, Y., Dougherty, L., Cheng, L. and Xu, K. (2015), 'A co-expression gene network associated with developmental regulation of apple fruit acidity', *Mol. Genet. Genomics* 290, 1247–63.

Bepete, M. and Lakso, A. N. (1998), 'Differential effects of shade on early season fruit and shoot growth rates in "Empire" apple branches', *HortScience* 33, 823–5.

Blanpied, G. D. (1991), 'Environmental factors that may influence the epoch of apple ripening'. *Proc. 5th Intl. Controlled Atmos. Res. Conf.*, Wenatchee, WA, USA, 1, 83–90.

Blanpied, G. D. and Silsby, K. J. (1992), 'Predicting harvest date windows for apples', *Cornell University Cooperative Extension Information Bulletin* 221, 1–18.

Botton, A., Eccher, G., Forcato, C., Ferrarini, A., Begheldo, M., Zermiani, M., Moscatello, S., Battistelli, A., Velasco, R., Ruperti, B. and Ramina, A. (2011), 'Signaling pathways mediating the induction of apple fruitlet abscission', *Plant Physiol.* 155, 185–208.

Brookfield, P., Murphy, P., Harker, R. and MacRae, E. (1997), 'Starch degradation and starch pattern indices; interpretation and relationship to maturity', *Postharvest Biol. Technol.* 11, 23–30.

Byers, R. E. (2003), 'Flower and fruit thinning and vegetative: fruiting balance', in D. C. Ferree and I. A. Warrington (eds), *Apples: Botany, Production and Uses*, pp. 409–36. CAB International, Wallingford, Oxon, UK.

Byers, R. E., Carbaugh, D. H., Presley, C. N. and Wolf, T. K. (1991), 'The influence of low light on apple fruit abscission', *J. Hortic. Sci.* 66, 7–18.

Calderón-Zavala, G., Lakso, A. N. and Piccioni, R. M. (2004), 'Temperature effects on fruit and shoot growth in the apple (Malus domestica) early in the season', *Acta Hortic.* 636, 447–53.

Calderón-Zavala, G. (2005), 'Temperature effects on fruit and vegetative growth, carbon dioxide exchange and dry matter partitioning in the apple tree', PhD Thesis, Cornell University.

Celton, J. M., Kelner, J. J., Martinez, S., Bechti, A., Touhami, A. K., James, M. J., Durel, C. E., Laurens, F. and Costes, E. (2014), 'Fruit self-thinning: A trait to consider for genetic improvement of apple tree', *PLoS ONE* 9. 10.1371/journal.pone.0091016.

Cheng, L. and Raba, R. (2009), 'Accumulation of macro- and micronutrients and nitrogen demand-supply relationship of "Gala"/"Malling 26" apple trees grown in sand culture', *J. Am. Soc. Hortic. Sci.* 134(1), 3–13.

Corelli Grappadelli, L., Lakso, A. N. and Flore, J. A. (1994), 'Early season patterns of photosynthate partitioning in exposed and shaded apple branches', *J. Am. Soc. Hortic. Sci.* 119, 596–603.

Dash, M., Johnson, L. K. and Malladi, A. (2012), 'Severe shading reduces early fruit growth in apple by decreasing cell production and expansion', *J. Am. Soc. Hortic. Sci.* 137, 275–82.

Dash, M., Johnson, L. K. and Malladi, A. (2013), 'Reduction of fruit load affects early fruit growth in apple by enhancing carbohydrate availability, altering the expression of cell production-related genes, and increasing cell production', *J. Am. Soc. Hortic. Sci.* 138, 253–62.

Dennis, F. (2003), 'Flowering, pollination and fruit set and development', in D. C. Ferree and I. A. Warrington (eds), *Apples: Botany, Production and Uses*, pp. 153–66. CAB International, Wallingford, Oxon, UK.

Drazeta, L., Lang, A., Hall, A. J., Volz, R. K. and Jameson, P. E. (2004), 'Causes and effects of changes in xylem functionality in apple fruit', *Ann. Bot.* 93, 275–82.

Ebel, R. C., Proebsting, E. L. and Evans, R. G. (1995), 'Deficit irrigation to control vegetative growth in apple and monitoring fruit growth to schedule irrigation', *HortScience* 30, 1229–32.

Eccher, G., Begheldo, M., Boschetti, A., Ruperti, B. and Botton, A. (2015), 'Roles of ethylene production and ethylene receptor expression in regulating apple fruitlet abscission', *Plant Physiol.* 169, 125–37.

Fallahi, E., Fallahi, B., Neilsen, G. H., Neilsen, D. and Peryea, F. J. (2010), 'Effects of mineral nutrition on fruit quality and nutritional disorders in apples', *Acta Hortic.* 868, 49–60.

Faust, M. (1989), *Physiology of Temperate Zone Fruit Trees*. Wiley-Interscience, New York.

Ferrero, S., Carretero-Paulet, L., Mendes, M. A., Botton, A., Eccher, G., Masiero, S. and Colombo, L. (2015), 'Transcriptomic Signatures in Seeds of Apple (Malus domestica L. Borkh) during Fruitlet Abscission', *PLoS ONE* 10(3), e0120503.

Francesconi, A. H. D., Watkins, C. B., Lakso, A. N., Nyrop, J. P., Barnard, J. and Denning, S. S. (1996), 'Interactions of European Red Mite and crop load on maturity and quality, mineral concentrations and economic value of "Starkrimson Delicious" apples', *J. Am. Soc. Hortic. Sci.* 121, 967–72.

Goffinet, M. C., Robinson, T. L. and Lakso, A. N. (1995), 'A comparison of "Empire" apple fruit size and anatomy in unthinned and hand-thinned trees', *J. Hortic. Sci.* 70(3), 375–87.

Goffinet, M. C. and Lakso, A. N. (2016), 'Positional effects on retention potential within the spur of apple (*Malus × Domestica*) with and without competition with other fruit on the spur', *Acta Hortic.* (in press).

Greene, D. W. and Costa, G. (2013), 'Fruit thinning in pome- and stone-fruit: state of the art', *Acta Hortic.* 998, 93–102.

Greene, D. W. (2007), 'The effect of prohexadione-calcium on fruit set and chemical thinning of apple trees', *HortScience* 42, 1361–5

Greene, D. W. and Groome, P. (2011), 'Effect of thinner and time of shading on fruit set of "McIntosh" apples', *Acta Hortic.* 884, 505–10.

Greene, D. W. and Lakso, A. N. (2013), 'Prediction of initial set and the thinning response to postbloom chemical thinners on apples', *Int. J. Fruit Sci.* 13, 430–40.

Greene, D. W., Lakso, A. N., Robinson, T. L. and Schwallier, P. (2013), 'Development of a fruitlet growth model to predict thinner response on apples', *HortScience* 48(5), 584–7.

Hansen, P. (1971), '14C-studies on apple trees. VII. The early seasonal growth in leaves, flowers and shoots as dependent upon current photosynthates and existing reserves', *Physiol. Plant.* 25, 469–73.

Harb, J., Gapper, N. E., Giovannoni, J. J. and Watkins, C. B. (2012), 'Molecular analysis of softening and ethylene synthesis and signaling pathways in a non-softening apple cultivar, "Honeycrisp" and a rapidly softening cultivar, "McIntosh"', *Postharvest Biol. Technol.* 64, 94–103.

Jackson, J. E. (2003), *Biology of Apples and Pears*. Cambridge University Press, Cambridge.

Janssen, B. J., Thodey, K., Schaffer, R. J., Alba, R., Balakrishnan, L., Bishop, R., Bowen, J. H., Crowhurst, R. N., Gleave, A. P., Ledger, S., McArtney, S., Pichler, F. B., Snowden, K. C. and Ward, S. (2008), 'Global gene expression analysis of apple fruit development from the floral bud to ripe fruit', *BMC Plant Biol.* 8, 16.

Jones, H. G. (1981), 'Carbon-dioxide exchange of developing apple (Malus-Pumila-Mill) fruits', *J. Exp. Bot.* 32, 1203–10.

Kolaric, J., Plesko, I. M., Tojnko, S. and Stopar, M. (2011), 'Apple fruitlet ethylene evolution and MdACO1, MdACS5A, and MdACS5B expression after application of naphthaleneacetic acid, 6-benzyladenine, ethephon, or shading', *HortScience* 46, 1381–6.

Lakso, A. N. and Corelli Grappadelli, L. (1992), 'Implications of pruning and training practices to carbon partitioning and fruit development in apple', *Acta Hortic.* 322, 231–9.

Lakso, A. N., Corelli Grappadelli, L., Barnard, J. and Goffinet, M. C. (1995), 'An expolinear model of the growth pattern of the apple fruit', *J. Hortic. Sci.* 70(3), 389–94.

Lakso, A. N., Robinson, T. L., Goffinet, M. C. and White, M. (2001), 'Apple fruit growth responses to varying thinning methods and timing', *Acta Hortic.* 557, 407–12.

Lakso, A. N. (2003), 'Water relations of apples', in D. C. Ferree and I. A. Warrington (eds), *Apples: Botany, Production and Uses*, pp. 167–94. CAB International, Wallingford, Oxon, UK.

Lakso, A. N. and Robinson, T. L. (2014), 'Integrating physiological models in applied fruit crop research', *Acta Hortic.* 1058, 285–90.

Lakso, A. N. and Robinson, T. L. (2015), 'Decision support for apple thinning based on carbon balance modeling', *Acta Hortic.* 1068, 235–42.

Li, J. G., Zhu, H. and Yuan, R. C. (2010), 'Profiling the expression of genes related to ethylene biosynthesis, ethylene perception, and cell wall degradation during fruit abscission and fruit ripening in apple', *J. Am. Soc. Hortic. Sci.* 135, 391–401.

Lang, A. and Volz, R. K. (1998), 'Spur leaves increase calcium in young apples by promoting xylem inflow and outflow', *J. Am. Soc. Hortic. Sci.* 123, 956–60.

Maguire, K. M., Banks, N. H. and Lang, A. (1999), 'Sources of variation in water vapour permeance of apple fruit', *Postharvest Biol. Technol.* 17, 11–17.

Malladi, A. and Johnson, L. K. (2011), 'Expression profiling of cell cycle genes reveals key facilitators of cell production during carpel development, fruit set, and fruit growth in apple (Malus domestica Borkh.)', *J. Exp. Bot.* 62, 205–19.

Marini, R. P. (2003), 'Fruitlet size and position within the cluster influence apple fruitlet susceptibility to chemical thinners', *J. Hortic. Sci. Biotechnol.* 78, 813–20.

McArtney, S., White, M., Latter, I. and Campbell, J. (2004), 'Individual and combined effects of shading and thinning chemicals on abscission and dry-matter accumulation of "Royal Gala" apple fruit', *J. Hortic. Sci. Biotechnol.* 79, 441–8.

McArtney, S. J. and Obermiller, J. D. (2014), 'Use of shading and the PSII inhibitor metamitron to investigate the relationship between carbohydrate balance and chemical thinner activity in apples', *Acta Hortic.* 1042, 27–31.

Monteith, J. L. (2001), 'Fundamental equations for growth in uniform stands of vegetation', *Agric. For. Meteorol.* 104, 5–11.

Morandi, B., Zibordi, M., Losciale, P., Manfrini, L., Pierpaoli, E. and Corelli Grappadelli, L. (2011), 'Shading decreases the growth rate of young apple fruit by reducing their phloem import', *Sci. Hortic.* 127, 347–52.

Neilsen, G. H. and Neilsen, D. (2003), 'Nutritional requirements of apple', in D. C. Ferree and I. A. Warrington (eds), *Apples: Botany, Production And Uses*, pp. 267–302. CAB International, Wallingford, Oxon, UK.

Palmer, J. W. (1987), 'The measurement of leaf area in apple trees', *J. Hort. Sci.* 62, 5–10.

Palmer, J. W. (1992), 'Effects of varying crop load on photosynthesis, dry matter production and partitioning of Crispin/M.27 apple trees', *Tree Physiol.* 11, 19–33.

Palmer, J. W., Wunsche, J. W., Meland, M. and Hann, A. (2002), 'Annual dry-matter production by three apple cultivars at four within-row spacings in New Zealand', *J. Hortic. Sci. Biotechnol.* 77, 712–17.

Parra-Quezada, R., Robinson, T. L., Osborne, J. L., Lakso, A. N. and Parra-Bujanda, L. B. (2006), 'Temperature and chemical thinner affect fruit set and growth of "Empire" apple', *Acta Hortic.* 727, 327–36.

Powell, D. B. B. (1974), 'Some effects of water stress in late spring on apple trees', *J. Hortic. Sci.* 49, 257–72.

Powell, D. B. B. (1976), 'Some effects of water stress on the growth and development of apple trees', *J. Hortic. Sci.* 51, 75–90.

Pratt, C. (1988), 'Apple flower and fruit, morphology and anatomy', *Hort. Rev.* 10, 273–308.

Priestley, C. A. (1960), 'Seasonal changes in the carbohydrate resources of some 6-year-old apple trees', *Ann. Rpt. East Malling Res. Station for 1959*, 70–7.

Quinlan, J. D. and Preston, A. P. (1971), 'The influence of shoot competition on fruit retention and cropping of apple trees', *J. Hortic. Sci.* 46, 525–34

Robinson, T. L. and A. N. Lakso. 2011. 'Predicting chemical thinning response with a carbohydrate model', *Acta Hortic.* 903,743–50.

Robinson, T. L., Lakso, A. N. and Hoying, S. A. (2012), 'Advances in predicting chemical thinner response of apple using a MaluSim carbon balance model', *Acta Hortic.* 932, 223–9.

Robinson, T. L., Lakso, A. N., Greene, D. W., Reginato, G. and Rufato, A. de R. (2016), 'Managing fruit abscission in apple', *Acta Hortic.* 1119, 1–14.

Soglio, V., Costa, F., Molthoff, J. W., Weemen-Hendriks, W. M. J., Schouten, H. J. and Gianfranceschi, L. (2009), 'Transcription analysis of apple fruit development using cDNA microarrays', *Tree Gen. Genom.* 5, 685–98.

Stover, E. W. and Greene, D. W. (2005), 'Environmental effects on the performance of foliar applied plant growth regulators, a review focusing on tree fruits', *HortTechnology* 15, 214–21.

Telias, A., Bradeen, J. M., Luby, J. J., Hoover, E. E. and Allan, A. C. (2011), 'Regulation of anthocyanin accumulation in apple peel', *Hort. Rev.* 38, 357–91.

Tromp, J. (1997), 'Maturity of apple cv. Elstar as affected by temperature during a six-week period following bloom', *J. Hortic. Sci.* 72, 811–19.

Tromp, J., Webster, A. D. and Wertheim, S. J. (2005), *Fundamentals of Temperate Zone Tree Fruit Production*, Backhuys, Leiden.

Velasco, R., Zharkikh, A., Affourtit, J., Dhingra, A., Cestaro, A., Kalyanaraman, A., et al. 2010. 'The genome of the domesticated apple (Malus × domestica Borkh.)', *Nat. Genet.* 42, 833–9.

Ward, D. L. and Marini, R. P. (1999), 'Growth and development of young apple fruits following application of ethephon plus carbaryl for thinning', *HortScience*, 34, 1057–9.

Warrington, I. J., Fulton, T. A., Halligan, E. A. and de Silva, H. N. (1999), 'Apple fruit growth and maturity are affected by early season temperatures', *J. Am. Soc. Hortic. Sci.* 124(5), 468–77.

Watkins, C. B. (2002), 'Ethylene synthesis, mode of action, consequences and control', in Knee, M. (ed.), *Fruit Quality and its Biological Basis*, pp. 180–224. Sheffield Academic Press, Sheffield.

Westwood, M. N. (1978), *Temperate Zone Pomology*, W. H. Freeman, San Francisco.

Westwood, M. N., Batjer, L. P. and Billings, H. D. (1967), 'Cell size cell number and fruit density of apples as related to fruit size position in cluster and thinning method', *Proc. Am. Soc. Hortic. Sci.* 91, 51–62.

Wilkinson, B. G. (1968), 'Mineral composition of apples IX. Uptake of calcium by the fruit', *J. Sci. Food Agric.* 19, 646–7.

Wünsche, J. N., Lakso, A. N., Robinson, T. L., Lenz, F. and Denning, S. S. (1996), 'The bases of productivity in apple production systems, the role of light interception by different shoot types', *J. Am. Soc. Hortic. Sci.* 121, 886–93.

Wünsche, J. N., Lakso, A. N., Denning, S. S. and Barnard, J. (1997), 'Estimating the contribution of canopy components to direct light interception of apple trees using a laser-assisted device', *Environ. Exp. Bot.* 37, 63–72.

Wünsche, J. and Lakso, A. N. (2000), 'The relationship between leaf area and light interception by spur and extension shoot leaves to apple orchard productivity', *HortScience* 35, 1202–6.

Wunsche, J. N. and Ferguson, I. B. (2005), 'Crop load interactions in apple', *Hort. Rev.* 31, 231–90.

Yahia, E. (1994), 'Apple flavor', *Hort. Rev.* 16, 197–234.

Yoon, T. M., Robinson, T. L. and Reginato, G. H. (2011), 'Effects of temperature and light level on efficiency of chemical thinner on "Empire" apple trees', *Acta Hortic.* 903, 1085–93.

Zhang, Y., Li, P. and Cheng, L. (2010), 'Developmental changes of carbohydrates, organic acids, amino acids, and phenolic compounds in "Honeycrisp" apple flesh', *Food Chem.* 123, 1013–18.

Zhou, C., Lakso, A. N., Robinson, T. L and Gan, S. (2008), 'Isolation and characterization of genes associated with shade-induced apple abscission', *Mol. Genet. Genomics* 280, 83–92.

Zhou, C., Liu, B., Lakso, A. N., Robinson, T. L. and Gan, S. (2016), 'Identification of genes associated with BA-induced apple fruitlet abscission', *Acta Hortic.* (in press).

Zibordi, M., Domingos, S. and Corelli Grappadelli, L. (2009), 'Thinning apples via shading, an appraisal under field conditions', *J. Hortic. Sci. Biotechnol.* 84 (Special Issue), 138–44.

Zibordi, M., Manfrini, L., Losciale, P., Pierpaoli, E., Corelli Grappadelli, L. and Morandi, B. (2014), 'Weather effects on canopy gas exchanges, fruit growth and vascular flows in apple during spring and possible consequences on orchard performance', *Acta Hortic.* 1058, 343–9.

Chapter 7

Evaluating and improving rootstocks for apple cultivation

G. Fazio, USDA-ARS/Cornell University, USA

1 Introduction

The foundations of a productive and healthy orchard are the rootstocks that provide anchorage, water and nutrients essential to the above-ground portions of the trees. The utilization of composite trees has increased the efficiency of breeding productive apple trees by dividing the selection of scion traits and rootstock traits into two genetically (and functionally) different specimens, which are then brought together through grafting. The art and science of grafting scions onto rootstocks spans several millennia; it is thought that it was used initially to aid in the clonal propagation of desirable scion varieties for fruit and nut production (Janick, 2005). In these millennia, it is likely that very little attention was dedicated to the selection of a particular rootstock chosen for its properties (ease of propagation) and the properties it imparted to the scion (Tukey, 1964; Rom and Carlson, 1987; Webster, 2003; Webster and Wertheim, 2003). Clonal selection and the beginning of the science of rootstocks seems to have originated in the latter half of the last millennia, where at least for apple, certain rootstock clonal selections were identified to impart unique productivity and architectural properties (early bearing and dwarfing) onto the grafted scion variety (Monceau, 1768). It is very likely that these properties existed or were selected directly on own-rooted trees first as these trees were early bearing, inherently dwarfed and production of fruit from these curious apple plants was early and abundant compared to seedling trees (Loudon, 1822). The combination of small architecture and productivity is adapted to cultivation in fruiting gardens typical of monasteries, aristocratic and wealthy middle-class dwellings (Rivers, 1866). Someone with very practical acuity made the connection that perhaps, by grafting other apple varieties on these early bearing, dwarfed plants the scion would behave in much the same way. Specifically with

http://dx.doi.org/10.19103/AS.2016.0017.08

the 'Paradise' apple, 'Jaune de Metz' (Lindley, 1828) otherwise known as Malling 9 (and relatives) the supposition that grafting could make the scion dwarfed was correct and must have led to a revolution of some sort in apple culture because fruit fanciers and nurseries in the eighteenth and nineteenth centuries in central Europe were propagating and using these stocks to make composite dwarfed trees (Hatton, 1917). During this period of discovery there was another important element to these new rootstocks that must have been realized: the seeds from these rootstocks did not breed true, meaning that the early bearing and dwarfing properties were not wholly replicated in all seed progeny. The new rootstocks needed to be propagated clonally to maintain their properties unaltered, hence the possibility of confusion, due to uneven results if a nursery sold trees grafted on seedlings derived from these rootstocks, or had mixtures as their starting material (Hatton, 1920). Major credit needs to be given to the scientists at the East Malling Research Station in the United Kingdom, who in the midst of this confusion collected many clonal rootstocks from around Europe and painstakingly characterized each of them according to their architectural, leafing and flowering properties identified and eliminated duplicates and established what we would call today 'foundation material' of rootstocks that were named 'Malling 1-16' (Hatton, 1919). Some of these rootstocks ('Malling 9' and 'Malling 7') became very popular where 'Malling 9' (M.9) and its sport mutations became the primary rootstock that fuelled the green revolution of dwarfed apple orchards that occurred in the twentieth century in many apple production regions of the world.

In addition to clone characterization and selection, the East Malling Research Station undertook breeding activities aimed at resistance to woolly apple aphids (WAA), a major barrier for apple cultivation in the far reaches of the British Empire and other traits such as improved propagation and productivity. Following East Malling, several other countries (Russia, Germany, Poland, Sweden, Japan, Canada and the United States) hosted government- or university-supported apple rootstock breeding programmes in the latter half of the twentieth century with goals that ranged from improved disease resistance to increased propagability and productivity. All these breeding efforts utilized the initial Malling selections as the starting point for their improvement projects and crossed with parents that would complement the weaknesses of the Malling material (e.g. resistance to fire blight, cold hardiness). Narrow crosses among the Malling rootstocks resulted in two widely used rootstocks: Malling 26 (M.26) and Malling 27 (M.27) that have improved propagability, and different forms of the early bearing and dwarfing effects. Most of the dwarfing founding germplasm was interrelated and had a very narrow genetic base (Oraguzie et al., 2005; Gharghani et al., 2009).The best way to introduce new forms of disease resistance and improve on other horticultural characteristics was to make wide crosses with germplasm that exhibited the desired phenotypes (Aldwinckle et al., 1999; Momol et al., 1999; Forsline et al., 2002). The results of these wide crosses have produced a series of rootstocks that combine disease and insect resistance with productivity, and represent the second generation of rootstock technologies applied worldwide (Fischer, 1991; Wertheim, 1998; Fischer et al., 2000).

Tree fruit growers must look to alternative, economically and environmentally sustainable management schemes of production to remain competitive in the international fruit market (Robinson, 2008). They are doing this by establishing high-density plantings with much smaller trees using new cultivars. These high-density plantings may cost 10–20 times more to establish than low-density plantings, thus greatly enhancing the economic risk (Robinson et al., 2007). Potential returns of high-density plantings, however, far exceed those of low-density plantings, particularly during the first 10 years after planting, often returning the growers' initial investment much sooner than the initially less-costly, low-density plantings. The central component of high-density systems is the rootstock, the part of the tree which provides size control to allow for high-density plantings (Barritt, 2000). As part of the tree, the rootstock influences many factors in addition to tree size, particularly productivity, fruit quality, pest resistance, stress tolerance and ultimately profitability (Barden and Marini, 2001).

As our understanding of physiology of apple trees, both at the whole tree level and at the cellular level, has increased, so has the understanding of how and what scion properties are modulated by rootstocks, thus increasing the target traits that may be selected to improve whole tree performance by improving rootstock performance (Fazio and Mazzola, 2004). It makes sense then that improving rootstock performance involves two sets of very different types of traits: the inherent apple rootstock traits (rooting for propagation, lack of spines and burr knots, resistance to root pathogens, cold hardiness, etc.) that deal with the interaction between rootstocks and the environment, and scion traits that are modulated by rootstocks (tree architecture, productivity, etc.) that represent the interaction between rootstocks and scions.

2 Apple breeding methods

Breeding apple rootstocks can be a very lengthy process (Johnson et al., 2001a); there are two ways to accelerate the process: the application of marker-assisted breeding (MAB) in the pipeline and/or the intensification of later stages of field testing (Fazio et al., 2015b). The first aims to eliminate substandard germplasm (non-precocious, non-dwarfing, susceptible to diseases, etc.) from the parental and progeny pools via the development and application of robust diagnostic markers. The second is to increase the number of clonal plants tested for each elite genotype and subject them to multiple phenotyping tests and environments that represent production regions. The theoretical benefits from the application of marker technologies to breeding have been reported in publications (Bus et al., 2000; Fazio et al., 2003; Antanaviciute et al., 2012; Bassett et al., 2015). In 2011 the USDA-ARS apple rootstock breeding programme located in Geneva, New York, conducted an internal analysis of the economic impact of applying molecular markers in the breeding programme by itemizing the cost per genotype for each stage of selection. A detailed description of the rootstock breeding programme stages can be found in Fazio et al. (2015b). Briefly, stages 1 and 2 deal with parent selection, crossing, culling seedlings with disease inoculations and propagation of survivors, stages 3 and 4 deal with the establishment of plants as rootstocks in field orchards and propagation beds (see Fig. 1), stage 5 includes the evaluation of propagation beds and replicated tests on biotic and abiotic resistance, stages 6 and 7 are secondary highly replicated tests and stages 8–10 deal with pre-commercial testing with multiple varieties and multiple locations. The Geneva breeding programme elected to conduct the first round of MAB before stage 3, which involves the initial propagation of plants surviving *Phytophthora* root rot and fire blight screens. The cost of genotyping with two markers including DNA extraction and labour was about $10 per seedling. The cost to phenotype each seedling for dwarfing and precocity during stage 3 (9 years of evaluation) in 2010 dollars was $15.40/year for 9 years = $138. The cost savings by culling non-dwarfing individuals was significant, and in 2012 we were able to plant 2 orchard rows of well-replicated, high-density first-test orchard instead of the 12 previously planted.

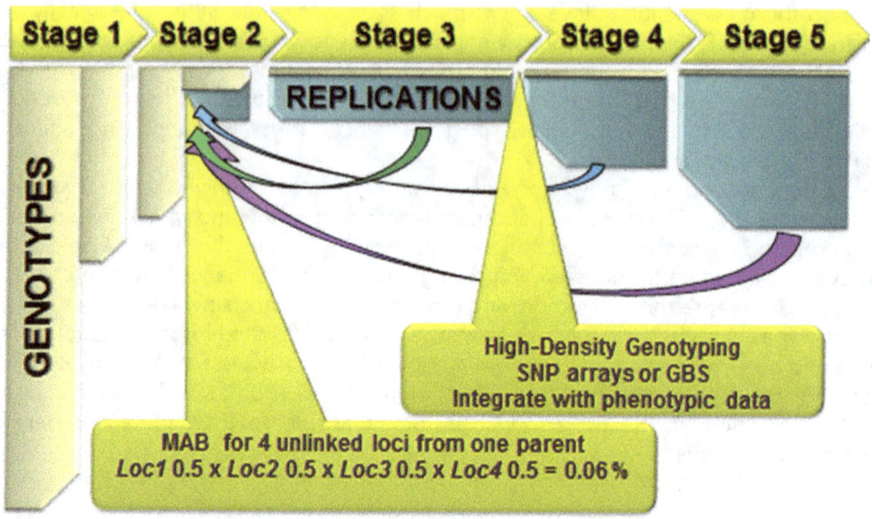

Figure 1 Integration of markers in the US national apple rootstock breeding programme has the impact of increasing highly replicated evaluation of breeding lines at earlier stages of the breeding programme allowing for time and throughput efficiency. Conventionally, stages 2–5 of the breeding programme include planting in fields as propagation sources (maintained as long as the rootstock trials last) and as rootstock trials (7–12 years). Any plant that can be culled (indicated by the arrows) at stage 2 saves the programme significant resources that would go into maintaining orchards and collecting data on useless material. In addition, maintaining the same evaluation capacity (e.g. number of plants tested in a specific stage) in stages 2–5, the programme is able to increase the number of replications and tests done in those stages, thereby increasing the confidence of selection and decreasing the time to release to industry. This cost saving allows for more in-depth genotyping applications with single-nucleotide polymorphisms (SNP) chips (Chagne et al., 2012) or genotyping by sequencing (GBS) methods (Migicovsky et al., 2016) to be applied to surviving individuals.

3 Scion traits affected by rootstocks

Until recently, the number of traits that were recognized to be modulated by apple rootstocks was pretty small: tree vigour, early bearing and water use. This list has been expanded to new architecture components such as canopy shape and bud break (sylleptic branching), and effects on fruit size and quality, on disease resistance and on nutrient availability in the scion. Perhaps the biggest breakthrough in our understanding of rootstock effects on scions is the study that monitored gene expression changes in scion tissue by different apple rootstocks (Jensen et al., 2003, 2010, 2011, 2012). At the cellular level, signals sent from the root system of different genotypes to the scion can change the expression levels of genes, which in turn change the composition of proteins and related metabolic processes and compounds in the scion. Although there are no experiments in apple that have described the opposite interaction, it is safe to assume that this dramatic change likely occurs as signalling from the scion affects the way roots behave and grow. The science behind understanding the issue of communication and affinity between scion and rootstock seems to be in its infancy and has a lot of potential as the concept of 'designer rootstocks' gets more traction in the industry.

3.1 Tree vigour

The reduction in tree vigour is perhaps the most important trait imparted by apple rootstocks to the grafted scion (Tukey, 1964). It is imparted to the scion as an early termination of overall season growth (Seleznyova et al., 2008). The benefits due to this trait in modern orchards are enormous and range from increased efficiency in picking and tree management operations, including mechanization, to the decrease of pesticide inputs, ladder accidents and other ergonomic issues (Groot, 1997a,b; Masseron and Roche, 1999; Robinson et al., 2007; James and Middleton, 2011). At the physiological level, the dwarfing trait has increased the effective light interception and partition to fruit production in the orchard and increased the production per unit area by at least 30% when compared to non-dwarfing rootstocks (Brown et al., 1985; Strong and Miller Azarenko, 1991; Atkinson et al., 1998). This means that for an industry worth $3 billion like the US industry at least $900 million are a result of the efficiency gained through dwarfing rootstocks. While the genetic components to this trait have been described to be the interaction of two main loci (Fazio et al., 2014b) and perhaps additional modifying loci (Harrison et al., 2016), it is important to mention that this is a complex trait that has fairly big interaction components and that the total effect of these components results in the overall vigour of the tree. Therefore for any scion 'S', the vigour 'V' is equal to the inherent growth dynamic genetics of the scion 'S_g', plus the dwarfing genetic components of the rootstock 'R_g', plus their interaction, plus the interaction of the whole composite tree with environmental effects 'E' such as fertility, water availability, diseases, soil type, soil pH, and soil type, or orchard management, weed competition and the like so that when scion vigour is measured, the genetic components of dwarfing rootstocks are only a part of the equation. This is exemplified by observing the effect of stunting caused by soil-borne replant disease, which has a similar effect to the dwarfing loci in apple and sometimes confuses the estimation of vigour potential of a rootstock.

Although several architectural dwarfs have been identified in domesticated and wild apple populations (Fazio et al., 2009a, 2014a), this material has not produced commercially viable rootstocks or has not been tested for similarity to the dwarfing characters offered by the alleles contained in M.9, M.8, M.13 and other Malling rootstocks belonging to the initial set selected in East Malling. The dwarfing trait has been shown to be highly heritable, modulated mainly by the combination of alleles of locus Dw1 found on chromosome 5 (Rusholme et al., 2004; Pilcher et al., 2008) and locus Dw2 found on chromosome 11 (Fazio et al., 2014b). Models that take into account some or all combinatorial allelic effects of these two loci have been able to explain upwards of 80% of the genotypic variation for dwarfing (Foster et al., 2015). The two loci interact with each other and do not necessarily seem to be additive, meaning that the lack of one dwarfing locus effect in the model negates the effect of the other. Several physiological models based on phenotypic observation have hypothesized the involvement of hormone signalling (Zhang et al., 2015; Tworkoski and Fazio, 2016), graft union anatomy (Tworkoski and Miller, 2007; Tworkoski and Fazio, 2011), hydraulic conductivity (Atkinson et al., 2003; Cohen et al., 2007), dry matter partitioning to fruit production or a combination of these (van Hooijdonk et al., 2011), while the underlying causative genes are still largely unknown.

Although current science may have a good understanding of the genetics governing this trait and may have developed robust markers for culling out undesirable material in the breeding pipeline, nothing can substitute the required time in the field to accurately measure the potential of each rootstock to dwarf the scion. Evaluation of this trait for breeding still requires a lengthy period of 7–10 years for the first observation and perhaps another 10–12 years for multi-location trials with multiple scions. As mentioned in the beginning of this section, multiple field trials are needed to evaluate rootstock interactions with different scions and environmental factors.

3.2 Inducement of early bearing in scions

Apple seedlings planted on their own roots experience a juvenile period anywhere from 4–7 years before they reach sexual maturity and bloom and fruit (Visser, 1967). Some dwarfing rootstocks have the ability to induce early bearing or reduce the juvenile period to 2 years in extreme cases (Visser and Schaap, 1967; Visser, 1973). Early bearing is a major selection criterion for improved apple rootstocks because the intensive types of cultivation of apple require a quick return on investment (early production of apples) to offset the installation and infrastructure costs to build the orchard (Cummins et al., 1995; Robinson et al., 2007, 2011). The genetic loci underlying the rootstock-induced trait 'early bearing' were first described by Fazio et al. (2014), as two loci, Eb1 and Eb2, that roughly co-located with Dw1 and Dw2, perhaps indicating that the two traits may be physiologically and genetically interconnected. Several studies have described the rootstock-induced partitioning effect of photosynthate into sexual (fruit) and vegetative portions of the tree (Seleznyova et al., 2008), comparing the effects of different rootstocks (Marini et al., 2006a; Autio et al., 2011b,c), crop loads on tree growth (Marini et al., 2012), productivity and bienniality (Marini et al., 2013). However, there is paucity in the literature about the causative elements for these rootstock effects. Breeding for this trait requires field evaluation for four years for the first observation in a replicated experimental orchard and then an additional 5 years in multi-location, multi-scion trials. Visser (1967) showed that scions with reduced juvenility also seemed to be more productive when grafted on M.9 dwarfing rootstocks, indicating the possibility of an inherent scion effect on early bearing and the need to test this scion–rootstock interaction in replicated trials.

3.3 Induction of architectural changes

Early and abundant fruit production is related to the number of flowering buds produced in the nursery phase and early establishment of the tree in the orchard (Ferree and Rhodus, 1987; Robinson et al., 1991a,b; Theron et al., 2000). This number can be influenced not only by the early bearing effects of the rootstocks discussed previously, but also by the ability of the rootstocks to produce prolific sylleptic branching (feathers on a nursery tree) in the nursery and later in the orchard. Early yield has been associated with nursery tree calliper, tree height and number of feathers. Rootstocks with wider genetic diversity than M.9 and Budagovsky 9 (B.9) have been shown to influence the production of sylleptic branches and the formation of crotch angles, that produce trees with a more open (flatter branches) structure (Fazio and Robinson, 2008a,b). This characteristic is mostly observed in rootstocks developed by the Geneva, New York, breeding programme, especially with rootstocks G.935, G.213, G.41 and G.214 (Fig. 2 and 3). The strength of these effects vary with different scions and continue through the life of the tree in the orchard, as observed

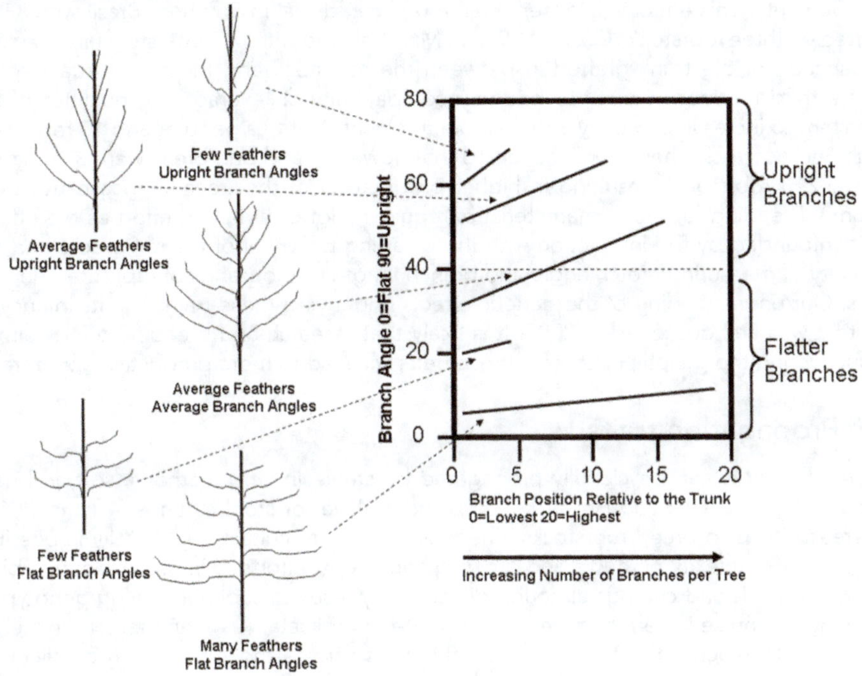

Figure 2 The number of feathers and average angle to the trunk can be captured in a simple graphic.

Figure 3 Comparison of nursery tree architecture featuring a flat branching rootstock, G.935 (right), versus an upright branching rootstock (JTE-B).

in the millions of trees planted on G.935 rootstock throughout the world. One additional characteristic that may be related to sylleptic branching is the ability of some rootstocks in the Geneva breeding programme to induce bud break and flowering in low chill environments. This effect was observed in a replicated trial in Southern Brazil with 'Gala' grafted on three rootstocks G.213, M.9 and, Marubakaido with M.9 interstem (an interstem is a section of the trunk grafted in between the rootstock and the scion usually made from a third rootstock variety to bridge incompatibility or leverage the qualities of the interstem to increase precocity of the whole tree) with 10 trees per plot and 10 replicated plots per rootstock where it was observed that flowering and bud break were 35% higher with G.213 rootstocks, resulting in higher productivity of the trees (Francescatto, pers. comm.). Breeding for these characteristics requires a lot of time and effort as the effects are confounded by the interaction with the dwarfing potential of the rootstocks and the difficulty of measuring crotch angle and branch length of thousands of replicated nursery trees. Our understanding of the genetic effects underlying these traits is in its infancy as the trait was first described in 2007. It is likely that research and breeding efforts aimed at uncovering the genetic factors for these traits will lead to more productive apple trees.

3.4 Propagation traits

Apple rootstocks can be clonally propagated by sterile *in vitro* methods, soft and hard wood cuttings (Bassuk and Howard, 1980), and by layer or stool cuttings (Adams, 2010). Whereas efforts to breed rootstocks amenable to *in vitro* culture are virtually impossible due to the complexity of media and growing conditions, efforts to improve rooting ability in layering beds and cuttings although difficult may result in superior rooting genotypes. Breeding for nursery performance can be quite complicated as many factors influence apple rootstock performance in the different nursery phases and at times may conflict with field performance. A prime example of this is the fast and easy adventitious rooting trait, highly desired in the propagation phase but correlated with the development of burr knots in the orchard – a harmful trait in certain orchard environments especially where dogwood borers and other insect borers may be present (Bergh and Leskey, 2003). Although it is possible to produce rootstocks with clean (no burr knots) shanks which also produce few or no suckers in the orchard, the selection of these orchard traits can result in difficult propagation properties with standard methods. These difficulties can be overcome with improved nursery management practices which include utilization of different propagation techniques like cuttings (Hansen, 1989; Deering, 1991) or micropropagation (Castillo et al., 2015; Geng et al., 2015), and the treatment with plant growth regulators such as prohexadione calcium in the nursery (Adams, 2010) to increase production of primary adventitious roots. The genetics of adventitious root formation have been investigated in the Geneva breeding programme revealing a complex trait with low heritability. Therefore, while it may be possible to breed for rooting traits, the importance of these traits is dwarfed by the importance of low suckering and lack of burr knots in the orchard.

Another characteristic affected by the rootstock is graft compatibility. Historically, most problems that were blamed on compatibility turned out to be virus related (Cummins and Aldwinckle, 1983; Lana et al., 1983), however, certain rootstock/scion combinations under unspecified grafting and nursery management conditions have shown a tendency for weak graft unions in very young trees (Robinson et al., 2003). Graft incompatibility can arise because of the disruption of normal healing between grafted tissues and can result in anatomical and physiological symptoms, biochemical and mechanical issues that lead to graft failure or tree death (Simons and Chu, 1983, 1985; Skene et al., 1983; Simons, 1985). Although it is likely that the method of grafting (chip budding, whip-and-tongue grafting, and machine V grafting) (Hartmann et al., 1997) has an effect on healing

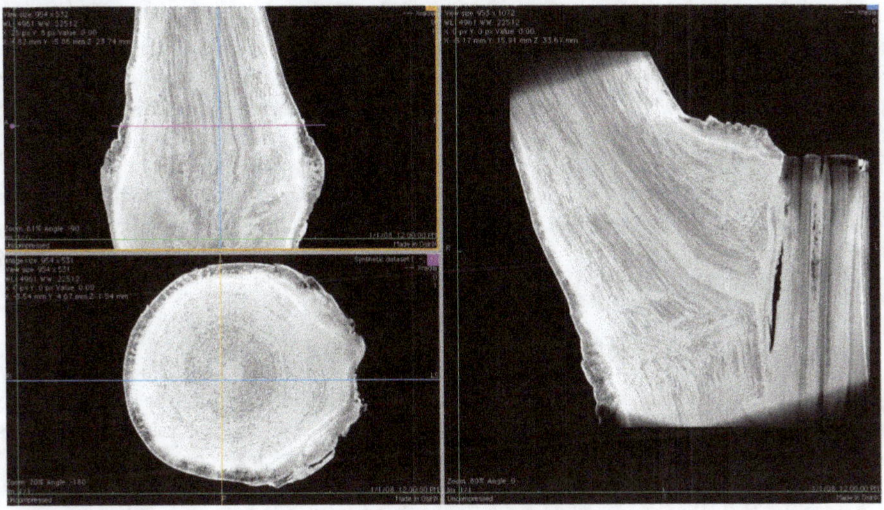

Figure 4 CAT scan of a graft union showing dead tissue where the graft was initially placed and unorganized tissue between scion and rootstock, a possible indication of weakness.

and union strength at various stages in the nursery cycle, there may be plant-growth-regulator-related and metabolic-compound-related signals that prevent the formation of a strong graft union. Efforts to understand the ability of the rootstock/scion combination to generate enough connective tissue where they meet is underway in the Geneva breeding programme through the use of X-ray tomography (CAT scans) shown in Fig. 4. Some nurseries report that large calliper stocks may not form as strong a graft union as small calliper stocks; therefore, a rootstock genotype that produces smaller calliper liners from the stool bed may be more suitable for nursery tree production. We have not been successful at evaluating finished nursery tree traits early in the breeding cycle because such traits are somewhat hard to evaluate and require high replication in order to detect rootstock effects on many different scions (Song et al. 2013).

Improvement of rootstocks sometimes means breeding interstems that can leverage dwarfing and precocity to improve the performance of a well-adapted vigorous clonal or seedling rootstock (Webster and Warrington, 1995; Luo et al., 2013). The length of the dwarfing interstem is inversely proportional to vigour induction and directly proportional to the induction of early bearing in apples (Carlson and Oh, 1975). Whereas this method of propagation is still used in some apple growing regions, most nurseries have moved away from this system because it adds extra costs (labour, materials, time, risks) to the development of a finished tree and it is not always reliable. It is possible, however, to select certain combinations of rootstocks and interstems that may leverage positive traits of root systems (disease resistance, nutrient uptake) (Ebel et al., 2000; Ma et al., 2013) and combine them with stem traits (dwarfing and early bearing) to generate even more productive trees, but no breeding programme to date has published research data on such combinations.

Apomixis, or the production of seedlings containing the same genetic information as the mother tree, has been observed in several wild apple accessions (Campbell and Dickinson,

1990). Breeding apomictic apple rootstocks has several attractive aspects in terms of nursery management as seed is more easily transported, free of viruses and provides opportunities for adjustment in production level to match demand. Although apomictic rootstock selections have been identified and evaluated, this material has not gained much traction because of inherent problems with other characteristics in this germplasm (Schmidt, 1972, 1982; Ferree, 1998) including the lack of early bearing induction so critical for modern production (Sax, 1949; Schmidt, 1970). Breeding for apomicts still continues in certain programmes in China (Sha et al., 2011; Ma et al., 2012), however, because of the difficulties related to hybridizing with apomictic species (Bisognin et al., 2008), most breeding programmes have abandoned this effort.

3.5 Drought tolerance

It is difficult to define drought tolerance without an objective reference or phenotype to measure and it is even more difficult to define in a rootstock independent of the scion-specific tolerance (Higgs and Jones, 1991; Virlet et al., 2015). The economic definition of drought tolerance (little or no loss of marketable fruit production) is different from some of the physiological definitions, which range from loss of photosynthetic activity (Massacci and Jones, 1990), to shoot and root growth under stress (Atkinson et al., 2000), to water use efficiency. Perhaps drought-tolerant rootstocks are of little value where irrigation water is available and relevant only in regions that utilize rainwater and may experience long stretches of drought. However, as climate changes and freshwater availability is threatened in traditional apple growing regions, the search for rootstocks that can thrive with less water is becoming more and more important (Ebel et al., 2001). It has long been recognized that there are differences in apple rootstock reaction to drought (Preston et al., 1972; Cummins and Aldwinckle, 1974; Ferree and Schmid, 1990), but those observations mostly dealt with spurious drought events and compared vigorous and non-vigorous types (Chandel and Chauhan, 1993; Fernandez et al., 1994). Decreased sensitivity to drought was attributed to 'Malling 9' rootstock when compared to 'Mark' (Fernandez et al., 1997) in a potted tree study. A comparison of hormonal drought response between M.9 and MM.111 rootstocks indicated that both rootstocks provided drought resistance but by mechanisms which appear to differ – M.9 produces higher levels of abscisic acid (ABA) that may regulate stomatal opening while MM.111 possesses a more extensive root system (increased soil exploration index) (Tworkoski et al., 2016). Water use efficiency, defined as the ratio of biomass produced to the rate of transpiration, and decreased sensitivity to drought (Xiang et al., 1995; Bassett et al., 2011) has been described in wild apple populations indicating the possibility of using this descriptor as a selection method. Breeding for such a complex trait may be possible only at latter stages of selection as discernment of field-meaningful data requires experiments with high replication, special equipment to control water delivery and use, and very-high-density morphological and physiological measurements. Perhaps gain can be made by selection of components of the trait such as improved root morphology, plant growth regulator signals and nutrient uptake once their effect is identified in breeding populations possessing all the other 'important' traits.

3.6 Cold tolerance

Several rootstocks seem to be tolerant to the different types of cold events that can cause injury of cambial and root tissues (Embree, 1988). Damaging cold events can be quite different in their mode of action as mid-winter events can have very different modes of action than late fall or spring cold events (Cline et al., 2012). Therefore, the methods used to evaluate sensitivity to differing cold injuring events need to address the physiological conditions specific to each event (Quamme et al., 1997; Moran et al., 2011a,b). Fluctuating temperatures in late fall, early winter and early spring are associated with hardening and de-hardening of tissues. This hardening and de-hardening process may have a strong genetic component (Forsline and Cummins, 1978), where a group of *Malus* rootstocks seem to have improved ability to be insensitive to such temperature fluctuations and remain dormant and cold-acclimated. Harvesting rootstock liners during these periods and subjecting them to increasingly low temperatures to show cambial damage is perhaps the most meaningful way to select cold hardy apple rootstocks. Observation of blackheart damage can also aid in the discernment of rootstock/scion combinations that are susceptible to mid-winter injury (Warmund and Slater, 1988; Warmund et al., 1996). Genes associated with cold response have been described for 'Gala' scions (Wisniewski et al., 2008), and similar genes may be found in apple rootstocks. However, the understanding of segregating factors that influence the different types of cold stress adaptation is virtually non-existent, making genetic or genomic-informed breeding impossible, and therefore selection relies entirely on highly replicated phenotyping.

3.7 Root morphology and architecture

Phenotypic variation in the morphology of roots has been associated with increases in yield and tolerance to abiotic stresses in several crops (Sousa et al., 2012; Chimungu et al., 2014; Lynch et al., 2014; Burton et al., 2015; Zhan et al., 2015). It makes sense that harnessing genetic and phenotypic variation in root morphology traits in apple rootstocks may improve productivity, tree size, drought tolerance, nutrient uptake, anchorage and other related whole tree functions (Eissenstat et al., 2001). Ample phenotypic variation has been characterized in wild *Malus sieversii* populations and within the Geneva apple rootstock breeding programme where genetic factors for fine root formation (highly branched fine roots) have been mapped to chromosomes 4 and 11 of the apple genome (Fazio et al., 2009b). Other traits that may be important to characterize may be the volume explored by the roots, the angle of the roots, the longevity of the roots and so on, which are all traits that are difficult to phenotype and for which robust genetic markers may be extremely useful. In Geneva, New York, the apple rootstock breeding programme measured several scion and root morphology characteristics of nursery trees of related (half-sibs) *Malus sieversii* seedlings. They found correlation between canopy volume/tree size and number of thick roots (0.38, $P < 0.001$), and a less pronounced correlation between tree size and root mass (0.25, $P < 0.001$), indicating a feedback loop between scion and root growth: the ability of the canopy to support the growth and expansion of a larger primary root system increased the vigour of young trees by their ability to produce root systems with strong primary hierarchy (Fazio et al., 2014a). Apple root systems vary in seasonal growth patterns (Eissenstat et al., 2006), which may affect their ability to forage for nutrients and water, and even colonization with beneficial mycorrhizae (Resendes et al., 2008). Root turnover rates may also play a significant role in tree nutrition and productivity as well as disease resistance or evasion as demonstrated by experiments that utilized replant-tolerant rootstocks from the Geneva breeding programme (Atucha et al., 2013; Emmett et al., 2014). Although these root traits can be targeted for MAB, the understanding of genes, gene expression patterns and physiological attributes associated with these traits in rootstocks is limited compared with our understating of scion traits; therefore, more research is needed to understand these traits before they become the subject of selection based on genetic markers.

3.8 Nutrient uptake

Another set of root-related traits deals with the genetic variation and inheritance of absorption and translocation of macro- and micronutrients by the rootstock to the scion (Tukey et al., 1962). Rootstocks have been shown to vary significantly with regard to their intrinsic ability to forage for nutrients as well as transfer them up to various sinks in the scion, including fruit, perhaps affecting organoleptic, post-harvest qualities of the fruit and disease resistance (Lockard, 1976; Westwood and Bjornstad, 1980; Om and Pathak, 1983; West and Young, 1988; Chandel and Chauhan, 1990; Rom et al., 1991; Sloan et al., 1996; Chun and Chun, 2004; Kim et al., 2004). Transgenic, cisgenic or conventional breeding approaches have been suggested to increase nutrient uptake of minerals such as zinc to improve productivity of the orchard (Swietlik et al., 2007). Most research on nutrient uptake by apple rootstocks has focused on developing the best management practices for nutrient application on a genetically restricted set of rootstocks, and it was not until a large set of genetically diverse rootstocks were observed in different soils and pH treatments that the physiological diversity of apple roots was revealed (Fazio et al., 2012). Changes in soil pH, for example, showed differences in the expected curves of absorption of metal ions such as manganese and iron, indicating that some rootstocks performed better at certain pH than others. Soil pH is one of the most important predictors of soil fertility and developing a set of rootstocks well adapted to specific pH profiles may improve orchard performance and open up more land to the cultivation of apples.

The analysis of scion nutrient concentration in leaves and fruit in several rootstock field trials in New York State have indicated the possibility that specific rootstocks may affect fruit quality of Honeycrisp apples showing that certain rootstocks are able to transfer higher calcium levels to the fruit and that the calcium-linked disorders typical of Honeycrisp are a result of scion-specific intrinsic challenges in the movement of calcium into the fruit (Fazio et al., 2015a).

Investigation of the inheritance of nutrient uptake and translocation in a full-sib population of apple rootstocks revealed quantitative trait loci (QTL) influencing scion leaf mineral concentrations of potassium (K), sodium (Na), phosphorus (P), calcium (Ca), zinc (Zn), magnesium (Mg) and molybdenum (Mo) with the most significant ones on chromosome 5 for potassium; chromosome 17 for sodium; and lower significance QTLs for calcium, copper, zinc and phosphorous (Fazio et al., 2013). Concentrations of some nutrients were highly correlated (K and P, S and P), indicating common nodes in the networked pathway that takes nutrients from the soil through the rootstocks to diverse sinks in the scion. Subsequently, leaves and fruit of Honeycrisp scion in a rootstock research plot in Summerland, Canada, featuring 31 diverse rootstocks were tested for mineral nutrient concentrations and found to have significant differences, identifying a negative correlation between phosphorous and calcium in the first season fruit (Neilsen and Havipson, 2014). The very different mechanisms (interaction with soil biota, active and passive transport, vessel composition and size, etc.) that impact absorption and transport and the fact that crop load and irrigation can also influence mineral concentrations (Neilsen et al., 2015) makes these traits difficult to improve without the aid of a robust understanding of molecular genetic factors involved. Modelling those factors to achieve a particular balance of nutrients in selected scions is therefore very complicated.

4 Disease and pest resistance

Commercial application of improved disease and insect resistance can be observed in the Geneva, New York, breeding programme. Since its inception, the programme focused on developing apple rootstocks resistant to fire blight, a North American disease caused by *Erwinia amylovora*, while maintaining the resistance to crown and root rot caused by *Phytophthora cactorum* (Aldwinckle et al., 1972; Gardner, 1977, 1980). This effort over three decades produced rootstocks that are not only resistant to fire blight and crown rot, but that are tolerant to the replant disease complex, and are also resistant to WAA (*Eriosoma lanigerum*).

4.1 Resistance to fire blight

Fire blight is a devastating disease caused by the anaerobic, gram-negative bacterium *E. amylovora,* which causes visible symptoms in blossoms, green tissues, fruit and some woody tissues of apple scions and rootstocks. Although this disease seems to have originated in the Eastern part of North America, it has now spread to most of the apple growing regions of the world. Rootstock blight on susceptible rootstocks (M.9, M.27 and M.26) can be devastating as the infection results in girdling and death of the rootstock shank eventually killing the whole tree – entire orchards and millions of trees have been destroyed because of rootstock blight. While spraying antibiotics like streptomycin can alleviate the onset of rootstock blight, genetic resistance of the rootstock is the best preventive treatment. Rootstock resistance to *E. amylovora* is found in several wild apple species and these have been utilized to breed a new series of fire blight-resistant rootstocks. There seem to be two main types of resistance in apple rootstock: a multi-genic type similar to that found in *Malus robusta* 'Robusta 5' where green tissues and flowers are not affected by the bacterium (Aldwinckle et al., 1974b; Cummins and Aldwinckle, 1974) and an ontogenic type of resistance found in Budagovsky 9 (B.9) rootstock where the green tissues are severely affected, but two-year-old and older wood seems not to react to the bacteria (Russo et al., 2008). Genetic inheritance of the 'Robusta 5' type of resistance has been described as having a strain-specific component on chromosome 3 identified as a gene belonging to the NBS-LRR class of resistance genes (Fahrentrapp et al., 2013; Broggini et al., 2014; Kost et al., 2015) and other minor QTLs on linkage groups 5, 7, 11 and 14, which do not seem to be strain-specific detected in a non-rootstock population 'Idared' × 'Robusta 5' (Wohner et al., 2014). Another locus that is non-strain specific was discovered on linkage group 7 in a rootstock population derived from a cross between 'Ottawa 3' and 'Robusta 5' (Gardiner et al., 2012). *Cis*-genic approaches with the LG03 gene proved only partially successful, suggesting a more complex pathway of resistance than just one gene recognition of the pathogen (Kost et al., 2015). The inheritance of the B.9 type of resistance is currently not known.

Screening and evaluation of resistance to fire blight is performed in two stages: the initial stage where a culture of a mix of *E. amylovora* strains are applied to actively growing shoots resulting in the elimination of any plant showing a visible necrotic lesion (Fig. 5), and

Figure 5 Side-by-side comparison of a susceptible (M.26 left) and resistant (G.41 right) rootstock inoculated with E2002a strain of *E. amylovora*. These types of inoculations are repeated several times during the process of breeding, utilizing multiple strains either independently or mixed.

Figure 6 Fire blight ooze from a blossom-inoculated tree: a sure sign that this rootstock will die killing the tree.

the second stage is where *E. amylovora* solution is sprayed during bloom onto the scion of two four-year-old finished trees in a replicated field trial (5–10 replicates) to simulate the same type of infection that would occur in a typical orchard. The second stage inoculation takes longer to appear with the initial symptoms being a small ooze point on the rootstock shank in the summer (Fig. 6), but the effects are quite dramatic in the beginning of fall when the leaves on the affected trees turn purple, indicating girdling and death of the tree (Russo et al., 2007). Combined these methods of inoculation can select for both types of resistance and have resulted in the release of several commercial rootstocks (G.65, G.11, G.16, G.30, G.202, G.41, G.935, G.214, G.814, G.213, G.969, G.890, G.222 and G.210) that show various levels of resistance to fire blight. MAB for this disease is possible but not necessary if greenhouse and field inoculation methods are available.

4.2 Replant disease complex

The specific apple replant disease complex is a syndrome observed as stunting and poor growth of young apple trees planted in soil that was previously planted with an apple or pear orchard. This complex disease causes major production losses throughout the life of the orchard. The main causative agents implicated in this syndrome are *Cylindrocarpon destructans*, *Phytophthora cactorum*, *Pythium* spp., *Rhizoctonia solani* and various pathogenic nematodes (Mazzola, 1998). The occurrence of one or more of these agents will affect the severity of the syndrome and may explain some of the site-to-site variation observed in replant land. This is one of the major problems faced by orchardists as virgin land becomes more rare, major infrastructure investments (hail nets, irrigation, etc.) become more prevalent and require a 'replant-in-place' type of renewal of the orchard and as fumigation chemistries are restricted by environmental laws (Auvil et al., 2011). The removal of the old orchard leaves a major pathogen load in the soil, which overwhelms the young root system of nursery trees. Fumigation treatments (methyl bromide, chloropicrin and nematicides) seem to be effective for less than a year as the pathogens implicated in this disease quickly recolonize the sterile soil, and fallow treatments (undesirable because they leave the land in an unproductive state) have shown mixed results, with replant symptoms sometimes appearing even after 4 years of fallow (Leinfelder and Merwin, 2006). Alternative treatments like seed meal amendments, fertilizers, compost teas and solarization have been proposed and are in various phases of research and development (Utkhede, 1999; Utkhede and Smith, 2000; Mazzola and Mullinix, 2005; Mazzola and Manici, 2012). In addition to the combination of pathogens involved in each orchard, factors like soil type, climate and other edaphic conditions seem to affect the severity of the complex, making it difficult to diagnose (Fazio et al., 2012). The effects of the disease complex are usually measured by comparing the growth of the same rootstock in sterile soil (pasteurization or chemical treatment) to a biologically active soil collected from the rhizosphere of the old orchard (Leinfelder et al., 2004; Rumberger et al., 2004; Yao et al., 2006a). A comprehensive study of multiple rootstock accessions and *Malus* species indicated that there was sufficient phenotypic diversity to enable growth in non-pasteurized soil (Isutsa and Merwin, 2000); however, the only reported commercially applicable genetic tolerance to the replant disease complex seems to be derived from progeny of 'Robusta 5' and other wild apple species (Auvil et al., 2011). Certain root genotypes have been reported to promote unique types of microbial communities, indicating a specificity or perhaps a pseudo-symbiotic effect of specific root systems that defeat the presence of pathogenic microbes (Yao et al., 2006b; Rumberger et al., 2007;

Figure 7 Seedling survivors of inoculation with *Phytophthora*. The tray was packed with live seedlings before inoculation.

St. Laurent et al., 2010). Breeding and selection for *Phytophthora* resistance performed by inoculating young seedlings (Fig. 7) (Aldwinckle et al., 1974a) and planting test orchards in replant soils might be the reason why several Geneva rootstocks have exhibited strong tolerance to apple replant disease components. New studies leveraging next-generation sequencing of *Pythium* challenged rootstock seedlings show upregulation of disease resistance-related pathways in resistant plant material indicating the possibility to select for specific resistance to *Pythium* components of replant disease (Shin et al., 2016). The placement of several apple rootstocks and breeding populations in sterile culture (micropropagation) has enabled identification of separate genetic effects of resistance to the individual replant components, as these rootstocks were inoculated with cultures of *Rhizoctonia* species and *Pythium* species independently. While this set of experiments is still ongoing (Zhu, personal communication), preliminary reports indicate segregation of QTLs affecting this trait and the possibility of developing molecular markers to select superior genotypes.

4.3 Resistance to WAA

The development of apple rootstocks resistant to WAA was one of the first breeding objectives developed in the Malling–Merton (MM) apple rootstock improvement programme as the disease pressure of these aphids made the cultivation of apples very difficult in the southern hemisphere. The donor parent of resistance to WAA was 'Northern Spy', which when crossed with several Malling selections resulted in the WAA-resistant vigorous rootstocks MM.106 (Northern Spy × M.1) and MM.111 (Northern Spy × Merton 793) (Wertheim, 1998). The 'Northern Spy' type of resistance seems to be monogenic (the Er1 locus) and has been mapped to chromosome 8 of apple. Monogenic resistance to WAA derived from 'Robusta 5' has been mapped to chromosome 17 (Er2 locus) and has been utilized extensively in the Geneva, New York, and New Zealand breeding programmes (Bus et al., 2008). Another resistance locus (Er3) from Aotea rootstock has also been mapped on chromosome 8, although it is not as effective as Er1 and Er2 (Sandanayaka et al., 2003, 2005; Sandanayaka and Backus, 2008). Phenotypic evaluation of this trait consists of rearing insects on susceptible germplasm and then transferring a specific number of insects on actively growing shoots of seedlings or replicated clones in a confined space (usually a netted greenhouse), then observing feeding and proliferation of WAA during a 2 month period after transfer (Beers et al., 2006). The monogenic nature of this type of resistance makes it amenable to utilization of *cis*-haplotype-specific markers to select parents and cull progeny that do not possess the resistance locus (Bassett et al., 2015). Other sources of WAA resistance are known in the *Malus* germplasm but very little is known about the genetic inheritance of these sources.

4.4 Tolerance of phytoplasmas and viruses

Apple viruses and phytoplasmas can cause productivity losses by interdicting basic plant functions, deforming branches and roots, and by making fruit unmarketable. To date, these pathogens are known to be spread by grafting, where infected clonal rootstocks or scions are the media for transmission (Wood, 1996; James et al., 1997; Silva et al., 2008). While the goal of apple industries throughout the world should be to work only with material that has been certified tested for viruses, phytoplasmas and other graft-transmissible agents, the eradication of these agents has been elusive due to propagation practices of some growers and homeowners that use infected sources of budwood. It is recommended that apple rootstock improvement programmes pay some attention to phenotyping apple rootstocks for susceptibility to some or all of the possible graft-transmissible viruses or phytoplasmas (Lankes and Baab, 2011). Efforts have been made in Germany and Italy to produce rootstocks resistant to the proliferation phytoplasmas (*Candidatus Phytoplasma mali*) found in certain accessions of *M. sieboldii* (Seemuller et al., 2007, 2008) and *M. sargentii* (Bisognin et al., 2008, 2009; Jarausch et al., 2008). These efforts were hampered by the polyploid and apomictic nature of the donor parents. Susceptibility to apple stem grooving virus has been observed in 'Ottawa 3' rootstocks and some of its derivatives (G.16 and G.814) which exhibited stunting or death upon being grafted with an infected scion. Some nurseries and inspection agencies have suggested maintaining certified virus-free budwood on susceptible rootstocks to reduce the possibility of introducing tainted budwood into the system. The slow decline caused by graft union necrosis among certain rootstock/scion combinations in the presence of tomato ring spot virus (ToRSV) (Tuttle and Gotlieb, 1985a,b) observed in MM.106 rootstock grafted with 'Delicious' scion is also of concern when breeding apple rootstocks. There is no conclusive data on the sensitivity of more recent apple rootstocks to ToRSV-induced graft union necrosis, but a large trial is underway in New York State to evaluate 50 genotypes for this sensitivity (Robinson, personal communication). Furthermore, there is paucity of genetic studies that describe the inheritance of susceptibility of *Malus* germplasm to viruses and phytoplasmas, making genetically informed breeding impossible. In the Geneva breeding programme, virus-sensitive parents like G.16 are being utilized for crosses, and efforts to map susceptibility loci are underway as a prerequisite to marker development to be utilized for culling susceptible seedlings before resources are wasted on growing them in larger field trials.

5 Future trends and conclusions

The varied environments where apples are grown suggest that no one rootstock will be well adapted to all environments and that coordinated, independent evaluation of new material from breeding programme be performed by local pomologists. There are some organizations in certain apple growing regions in the world that aim to independently test rootstocks in a regimented way covering multiple environments and scions (Elfving and McKibbon, 1990; Schechter et al., 1991; Usa, 1991; Kviklys, 2011). A considerable programme of tree fruit rootstock evaluation in the United States, Canada and Mexico is conducted by a group of 35+ researchers, extension specialists and industry collaborators within the CREES (cooperative research and extension services of the USDA) multi-regional project NC-140 (www.nc140.org) and in Europe through EUFRIN (www.eufrin.org). As a group the NC-140 researchers have made significant contributions to tree fruit rootstock research over the last three decades and have conducted highly coordinated impactful research for the tree fruit industry (Rom and Rom, 1991; Fernandez et al., 1995; Perry, 1996; Autio et al., 1997, 2011a,b; Barritt et al., 1997; Marini et al., 2002, 2006b). The evolution from low- to high-density plantings has raised the bar for rootstock performance and has increased the complexity of the research problems that need to be addressed in order to satisfy the needs of the industry. The tree fruit industry supply chain is complex with rootstock propagation nurseries growing and selling rootstocks to finished tree nurseries that graft the scion variety, grow finished trees and then finally sell to the growers. This complexity makes the adoption of new rootstock technologies rather viscous. This viscosity is increased by issues of local adaptability and survivability of rootstocks and also by issues regarding intellectual property. As new apple rootstocks are developed worldwide, they cannot be recommended unless there is sustained research investigating soil and climatic adaptability, root anchorage, size control, precocity, productivity, disease and pest resistance, and propagation (Autio et al., 2011c). In addition, there needs to be awareness of research into these new technologies not just by growers, but by the nurseries that will provide such material to growers. In general, field testing of rootstocks in an orchard setting requires a minimum of eight years to accurately assess the potential for improved profitability, enhancement of production efficiency and reduction of external farm inputs. It also requires testing propagation qualities in a nursery or micropropagation facility setting. With year-to-year variation in weather, this time span is necessary to obtain a true indication of rootstock performance across multiple environments typical of worldwide apple fruit production (Johnson et al., 2001a,b).

Ongoing research and results from localized evaluation experiments are reported in several settings internationally, the main ones being the recurring (every four years) ISHS symposium on 'Integrating Canopy, Rootstock and Environmental Physiology in Orchard Systems' and the symposium on 'Eucarpia Fruit Breeding and Genetics'. Apple industry associations also play a very important role in dissemination of knowledge about apple rootstocks, the premiere on which was established to promulgate the beneficial effects of adopting dwarfing apple rootstocks is the 'International Dwarf Fruit Tree Association' recently renamed 'International Fruit Tree Association' (www.ifruittree.org) because the task of conversion to dwarfing rootstocks had been largely accomplished by the constituents. Other organizations featuring coordinated international research on apple are RosBREED (www.rosbreed.org) (Iezzoni et al., 2010), FruitBreedomics (www.fruitbreedomics.com) and the Genome Database for the Rosaceae (www.rosaceae.org) are advancing the development of new knowledge about physiology, phenomics, genetics, and genomics of Rosaceaous crops and providing useful infrastructure to the development and evaluation of new apple rootstocks(Evans et al., 2012; Evans, 2013a,b; Peace et al., 2014; Chagne et al., 2015; Guan et al., 2015; Liverani et al., 2015; Mauroux et al., 2015; Fresnedo-Ramirez et al., 2016). The ultimate goal for all these organizations is to make apple growing more efficient, more environmentally friendly, more profitable for those that grow apples and more nutritious for the customers that eat apples, and the development of new apple rootstocks is an important cog in this intricate effort.

6 Where to look for further information

Further information can be found online in the fruit growing oriented magazine *New York Fruit Quarterly*. http://www.nyshs.org/fq.php and in the web pages of the Cornell Fruit Site http://www.fruit.cornell.edu/.

7 References

Adams, R. R., 2010. *Increasing The Rooting In Apple Rootstock Stoolbeds*. Cornell University, Ithaca, NY, M.S.

Aldwinckle, H. S., J. N. Cummins, R. Antoszewski, L. Harrison and C. C. Zych, 1974a. Familial differences in reaction to flood inoculation of young apple seedlings by zoospore suspension of phytophthora cactorum. *Proceedings of the XIX International Horticultural Congress IA Section VII Fruits*. Warsaw, Poland, IA.

Aldwinckle, H. S., J. N. Cummins, R. Antoszewski, L. Harrison and C. C. Zych, 1974b. Inheritance of fire blight susceptibility in some apple rootstock families. *Proceedings of the XIX International Horticultural Congress IA Section VII Fruits*, pp. 309–39 [Abstracts].

Aldwinckle, H. S., J. N. Cummins and H. I. Gustafson, 1972. Resistance to phytophthora cactorum in apple seedlings from controlled crosses. *Phytopathology* 62:743.

Aldwinckle, H. S., H. L. Gustafson, P. L. Forsline, M. T. Momol and H. Saygili, 1999. Evaluation of the core subset of the USDA apple germplasm collection for resistance to fire blight. *Acta Horticulturae* 489:269–72.

Antanaviciute, L., F. Fernandez-Fernandez, J. Jansen, E. Banchi, K. M. Evans, R. Viola, R. Velasco, J. M. Dunwell, M. Troggio and D. J. Sargent, 2012. Development of a dense SNP-based linkage map of an apple rootstock progeny using the Malus Infinium whole genome genotyping array. *Bmc Genomics* 13.

Atkinson, C. J., M. A. Else, L. Taylor and C. J. Dover, 2003. Root and stem hydraulic conductivity as determinants of growth potential in grafted trees of apple (Malus pumila Mill.). *Journal of Experimental Botany* 54:1221–9.

Atkinson, C. J., M. Policarpo, A. D. Webster and G. Kingswell, 2000. Drought tolerance of clonal Malus determined from measurements of stomatal conductance and leaf water potential. *Tree Physiology* 20:557–63.

Atkinson, C. J., M. Policarpo, A. D. Webster and A. M. Kuden, 1998. Drought tolerance of apple rootstocks: production and partitioning of dry matter. *Plant and Soil* 206:223–35.

Atucha, A., B. Emmett and T. Bauerle, 2013. Growth rate of fine root systems influences rootstock tolerance to replant disease. *Plant and Soil* 376:1–10.

Autio, W., T. L. Robinson, T. Bradshaw, J. Cline, R. M. Crassweller, C. G. Embree, E. Hoover, G. Lang, J. Masabni, M. L. Parker, R. Perry, G. L. Reighard, J. Schupp and M. Warmund, 2011a.

Performance of several dwarfing rootstocks with 'Fuji' and 'McIntosh' as scion cultivars in the 1999 NC-140 dwarf apple rootstock trials. *Acta Horticulturae* 903:319–26.

Autio, W., T. L. Robinson, J. Cline, R. M. Crassweller, C. G. Embree, E. Hoover, G. Lang, J. Masabni, M. L. Parker, R. Perry, G. L. Reighard and M. Warmund, 2011b. Performance of several semi-dwarfing rootstocks with 'Fuji' and 'McIntosh' as scion cultivars in the 1999 NC-140 semi-dwarf apple rootstock trials. *Acta Horticulturae* 903:327–34.

Autio, W., T. L. Robinson, W. Cowgill, C. Hampson, M. Kushad, G. Lang, J. Masabni, D. D. Miller, R. A. Parra-Quezada, R. Perry and C. Rom, 2011c. Performance of 'Gala' apple trees on supporter 4 and different strains of B.9, M.9, and M.26 rootstocks as part of the 2002 NC-140 apple rootstock trial. *Acta Horticulturae* 903:311–18.

Autio, W. R., J. LaMar Anderson, J. A. Barden, G. R. Brown, P. A. Domoto, D. C. Ferree, A. Gaus, R. L. Granger, R. A. Hayden, F. Morrison, C. A. Mullins, S. C. Myers, R. L. Perry, C. R. Rom, J. R. Schupp, L. D. Tukey and B. H. Barritt, 1997. Apple rootstock and scion cultivar interact to affect tree performance in the 1990 NC-140 cultivar/rootstock trial. *Compact Fruit Tree* 30:6–9.

Auvil, T. D., T. R. Schmidt, I. Hanrahan, F. Castillo, J. R. McFerson and G. Fazio, 2011. Evaluation of dwarfing rootstocks in Washington apple replant sites. *Acta Horticulturae* 903:265–71.

Barden, J. A. and R. P. Marini, 2001. Yield, fruit size, red color, and estimated crop value in the NC-40 1990 cultivar/rootstock trial in Virginia. *Journal of American Pomological Society* 55:154–8.

Barritt, B. H., 2000. Selecting an orchard system for apples. *Compact Fruit Tree* 33:89–92.

Barritt, B. H., J. A. Barden, J. Cline, R. L. Granger, M. M. Kushad, R. P. Marini, M. Parker, R. L. Perry, T. Robinson, C. R. Unrath, M. A. Dilley and F. Kappel, 1997. Performance of 'Gala' at year 5 with eight apple rootstocks in an 8-location North American NC-140 trial. *Acta Horticulturae* 451:129–35.

Bassett, C. L., D. M. Glenn, P. L. Forsline, M. E. Wisniewski and R. E. Farrell, 2011. Characterizing Water Use Efficiency and Water Deficit Responses in Apple (Malus Xdomestica Borkh. and Malus sieversii Ledeb.) M. Roem. *Hortscience* 46:1079–84.

Bassett, H., M. Malone, S. Ward, T. Foster, D. Chagne and V. Bus, 2015. Marker assisted selection in an apple rootstock breeding family. *Acta Horticulturae* 1100:25–8.

Beers, E. H., S. Cockfield and G. Fazio, 2006. Biology and management of woolly apple aphid, Eriosoma lanigerum (Hausmann), in Washington state. *Proceedings of the IOBC University of Lleida, Spain* 30:4–6.

Bergh, J. C. and T. C. Leskey, 2003. Biology, ecology and management of dogwood borer in eastern apple orchards. *Canadian Entomologist* 135:615–35.

Bisognin, C., B. Schneider, H. Salm, M. S. Grando, W. Jarausch, E. Moll and E. Seemuller, 2008. Apple proliferation resistance in apomictic rootstocks and its relationship to phytoplasma concentration and simple sequence repeat genotypes. *Phytopathology* 98:153–8.

Bisognin, C., E. Seemuller, S. Citterio, R. Velasco, M. S. Grando and W. Jarausch, 2009. Use of SSR markers to assess sexual vs apomictic origin and ploidy level of breeding progeny derived from crosses of apple proliferation-resistant Malus sieboldii and its hybrids with Malus × domestica cultivars. *Plant Breeding* 128:507–13.

Broggini, G. A. L., T. Wohner, J. Fahrentrapp, T. D. Kost, H. Flachowsky, A. Peil, M. V. Hanke, K. Richter, A. Patocchi and C. Gessler, 2014. Engineering fire blight resistance into the apple cultivar 'Gala' using the FB_MR5 CC-NBS-LRR resistance gene of Malus × robusta 5. *Plant Biotechnology Journal* 12:728–33.

Brown, C. S., E. Young and D. M. Pharr, 1985. Rootstock and scion effects on carbon partitioning in apple leaves. *Journal of the American Society for Horticultural Science* 110:701–5.

Burton, A. L., J. Johnson, J. Foerster, M. T. Hanlon, S. M. Kaeppler, J. P. Lynch and K. M. Brown, 2015. QTL mapping and phenotypic variation of root anatomical traits in maize (Zea mays L). *TAG Theoretical and Applied Genetics* 128:93–106.

Bus, V., C. Ranatunga, S. Gardiner, H. Bassett and E. Rikkerink, 2000. Marker assisted selection for pest and disease resistance in the New Zealand apple breeding program. *Acta Horitculturae* 538:541–7.

Bus, V. G. M., D. Chagne, H. C. M. Bassett, D. Bowatte, F. Calenge, J. M. Celton, C. E. Durel, M. T. Malone, A. Patocchi, A. C. Ranatunga, E. H. A. Rikkerink, D. S. Tustin, J. Zhou and S. E. Gardiner, 2008. Genome mapping of three major resistance genes to woolly apple aphid (Eriosoma lanigerum Hausm.). *Tree Genetics & Genomes* 4:223–36.

Campbell, C. S. and T. A. Dickinson, 1990. Apomixis, patterns of morphological variation, and species concepts in subfam Maloideae (Rosaceae). *Systematic Botany* 15:124–35.

Carlson, R. F. and S. D. Oh, 1975. Influence of interstem lengths of M.8 clone Malus sylvestris Mill. on growth, precocity, yield, and spacing of 2 apple cultivars. *Journal of the American Society for Horticultural Science* 100:450–2.

Castillo, A., D. Cabrera, P. Rodriguez, R. Zoppolo and T. Robinson, 2015. *In vitro* micropropagation of CG41 apple rootstock. *Acta Horticulturae* 1083:569–76.

Chagne, D., R. N. Crowhurst, M. Troggio, M. W. Davey, B. Gilmore, C. Lawley, S. Vanderzande, R. P. Hellens, S. Kumar, A. Cestaro, R. Velasco, D. Main, J. D. Rees, A. Iezzoni, T. Mockler, L. Wilhelm, E. Van de Weg, S. E. Gardiner, N. Bassil and C. Peace, 2012. Genome-Wide SNP Detection, Validation, and Development of an 8K SNP Array for Apple. *Plos ONE* 7.

Chagne, D., C. Kirk, C. Whitworth, S. Erasmuson, R. Bicknell, D. J. Sargent, K. Satish, M. Troggio and S. Kumar, 2015. Polyploid and aneuploid detection in apple using a single nucleotide polymorphism array. *Tree Genetics and Genomes* 11:94.

Chandel, J. S. and J. S. Chauhan, 1990. Effect of rootstock and water stress on foliar nutrient (Fe, Mn, Zn and Cu) composition of Starking Delicious apple. *Horticultural Journal* 3:10–14.

Chandel, J. S. and J. S. Chauhan, 1993. Effect of rootstocks and soil moisture stress on growth and vigour of apple (Malus domestica Borkh.) cv. Starking Delicious. *Punjab-Horticultural-Journal* 30:162–70.

Chimungu, J. G., K. M. Brown and J. P. Lynch, 2014. Reduced root cortical cell file number improves drought tolerance in maize. *Plant Physiology* 166:1943–55.

Chun, I. and I. J. Chun, 2004. Influence of nutrient level and rootstock on fruit quality of 'Fuji' apple trees. *Journal of the Korean Society for Horticultural Science* 45:252–5.

Cline, J. A., D. Neilsen, G. Neilsen, R. Brownlee, D. Norton and H. Quamme, 2012. Cold Hardiness of New Apple Cultivars of Commercial Importance in Canada. *Journal of the American Pomological Society* 66:174–82.

Cohen, S., A. Naor, J. Bennink, A. Grava and M. Tyree, 2007. Hydraulic resistance components of mature apple trees on rootstocks of different vigours. *Journal of Experimental Botany* 58:4213–24.

Cummins, J. N. and H. S. Aldwinckle, 1974. Breeding apple rootstocks. *HortScience* 9:367–72.

Cummins, J. N. and H. S. Aldwinckle, 1983. *Breeding Apple Rootstocks*. AVI Publishing Company Inc., Westport, CT.

Cummins, J. N., H. S. Aldwinckle and I. J. Warrington, 1995. Breeding rootstocks for tree fruit crops. *1994. New Zealand Journal of Crop and Horticultural Science* 23:395–402.

Deering, T. D., 1991. Root formation in Malus pumila 'Northern Spy' cuttings using etiolation. *Proceedings of International Plant Propagator's Society* 40:45–8.

Ebel, R. C., A. W. Caylor, J. A. Pitts and B. S. Wilkins, 2000. Mineral nutrition during establishment of Golden Delicious 'Smoothee' apples on dwarfing rootstocks and interstems. *Journal of Plant Nutrition* 23:1179–92.

Ebel, R. C., E. L. Proebsting and R. G. Evans, 2001. Apple tree and fruit responses to early termination of irrigation in a semi-arid environment. *HortScience* 36:1197–201.

Eissenstat, D. M., T. L. Bauerle, L. H. Comas, D. Neilsen, G. H. Neilsen, A. N. Lakso and D. R. Smart, 2006. Seasonal patterns of root growth in relation to shoot phenology in grape and apple. *Acta Horticulturae* 21–6.

Eissenstat, D. M., C. E. Wells and L. Wang, 2001. Root efficiency and mineral nutrition in apple. *Acta Horticulturae* 564:165–83.

Elfving, D. C. and E. D. McKibbon, 1990. Rootstock effects on spur 'Delicious' yield and pruning requirements. *Compact Fruit Tree* 23, 4–8 March 1990.

Embree, C., 1988. Apple rootstock cold hardiness evaluation. *Compact Fruit Tree* 21:99–105.

Emmett, B., E. Nelson, A. Kessler and T. Bauerle, 2014. Fine-root system development and susceptibility to pathogen colonization. *Planta* 239:325–40.

Evans, K., 2013a. Apple breeding in the pacific northwest. *Acta Horticulturae* 976:75–8.

Evans, K., 2013b. The potential impacts of genetics, genomics and breeding on organic fruit production. *Acta Horticulturae* 155–60.

Evans, K., Y. Guan, J. Luby, M. Clark, C. Schmitz, S. Brown, B. Orcheski, C. Peace, A. Lezzoni and E. van de Weg, 2012. Large-scale standardized phenotyping of apple in RosBREED. *Acta Horticulturae* 945:233–8.

Fahrentrapp, J., G. A. L. Broggini, C. Gessler, M. Kellerhals, A. Peil, M. Malnoy and K. Richter, 2013. Fine mapping of the fire blight resistance locus in Malus × robusta 5 on linkage group 3. *Acta Horticulturae* 976:499–500.

Fazio, G., H. S. Aldwinckle, G. M. Volk, C. M. Richards, W. J. Janisiewicz and P. L. Forsline, 2009a. Progress in evaluating malus sieversii for disease resistance and horticultural traits. *XII Eucarpia Symposium on Fruit Breeding and Genetics* 814:59–66.

Fazio, G., L. Cheng, M. A. Grusak and T. L. Robinson, 2015a. Apple rootstocks influence mineral nutrient concentration of leaves and fruit. *New York Fruit Quarterly* 25:11–15.

Fazio, G., S. M. Chung and J. E. Staub, 2003. Comparative analysis of response to phenotypic and marker-assisted selection for multiple lateral branching in cucumber (Cucumis sativus L.). *Theoretical and Applied Genetics* 107:875–83.

Fazio, G., P. Forsline, C. T. Chao, C. Richards and G. Volk, 2014a. Tree and Root Architecture of Malus sieversii Seedlings for Rootstock Breeding. *Acta Horitculturae* 1058:585–94.

Fazio, G., A. Kviklys, M. A. Grusak and T. L. Robinson, 2013. Phenotypic diversity and QTL mapping of absorption and translocation of nutrients by apple rootstocks. *Aspects of Applied Biology* 119:37–50.

Fazio, G., D. Kviklys, M. A. Grusak and T. L. Robinson, 2012. Soil pH, soil type and replant disease affect growth and nutrient absorption in apple rootstocks. *New York Fruit Quarterly* 20:22–8.

Fazio, G., D. Kviklys and T. Robinson, 2009b. QTL mapping of root architectural traits in apple rootstocks. *Hortscience* 44:986–7.

Fazio, G. and M. Mazzola, 2004. Target traits for the development of marker assisted selection of apple rootstocks – prospects and benefits. *Acta Horitculturae* 663:823 827.

Fazio, G. and T. Robinson, 2008a. Modification of nursery tree architecture by apple rootstocks. *Hortscience* 43:1271.

Fazio, G. and T. L. Robinson, 2008b. Modification of nursery tree architecture with apple rootstocks: a breeding perspective. *New York Fruit Quarterly* 16:13–16.

Fazio, G., T. L. Robinson and H. S. Aldwinckle, 2015b. The Geneva apple rootstock breeding program. *Plant Breeding Reviews* 39:379–424.

Fazio, G., Y. Z. Wan, D. Kviklys, L. Romero, R. Adams, D. Strickland and T. Robinson, 2014b. Dw2, a New dwarfing locus in apple rootstocks and its relationship to induction of early bearing in apple scions. *Journal of the American Society for Horticultural Science* 139:87–98.

Fernandez, R. T., R. L. Perry and D. C. Ferree, 1995. Root distribution patterns of 9 apple rootstocks in 2 contrasting soil types. *Journal of the American Society for Horticultural Science* 120:6–13.

Fernandez, R. T., R. L. Perry and J. A. Flore, 1997. Drought response of young apple trees on three rootstocks. II. Gas exchange, chlorophyll fluorescence, water relations, and leaf abscisic acid. *Journal of the American Society for Horticultural Science* 122:841–8.

Fernandez, R. T., R. L. Perry, R. M. McLean and A. R. Otero, 1994. Water stress factors and apple rootstock performance. *Compact Fruit Tree* 27:37–40.

Ferree, D. C., 1998. Performance of eight apomictic selections as apple rootstocks. *HortScience* 33:641–3.

Ferree, D. C. and W. T. Rhodus, 1987. Early performance and economic value of feathered apple trees on semi-standard rootstocks. *Journal of the American Society for Horticultural Science* 112:906–9.

Ferree, D. C. and J. C. Schmid, 1990. Performance of apple rootstock, cultivars and cultural treatments under the stress of the 1988 drought. *Research Circular Ohio Agricultural Research and Development Center* 297:25–36.

Fischer, C., 1991. Results of breeding apples for resistance to pathogens. *Nachrichtenblatt des Deutschen Pflanzenschutzdienstes* 43:147–50.

Fischer, C., M. Geibel and M. Fischer, 2000. Apple breeding in the federal centre for plant breeding research. *Acta Horitculturae* 538:225–7.

Forsline, P. L., H. S. Aldwinckle, C. Hale and R. Mitchell, 2002. Natural occurrence of fire blight in USDA apple germplasm collection after 10 years of observation. *Acta Horitculturae* 590:351–7.

Forsline, P. L. and J. N. Cummins, 1978. Screening progeny from an apple rootstock breeding program for levels of hardiness in late fall, mid-winter and early spring. *HortScience* 13:381.

Foster, T. M., J. M. Celton, D. Chagne, D. S. Tustin and S. E. Gardiner, 2015. Two quantitative trait loci, Dw1 and Dw2, are primarily responsible for rootstock-induced dwarfing in apple. *Horticulture Research* 2:11.

Fresnedo-Ramirez, J., T. J. Frett, P. J. Sandefur, A. Salgado-Rojas, J. R. Clark, K. Gasic, C. P. Peace, N. Anderson, T. P. Hartmann, D. H. Byrne, M. C. A. M. Bink, C. H. Crisosto, T. M. Gradziel and E. van de Weg, 2016. QTL mapping and breeding value estimation through pedigree-based analysis of fruit size and weight in four diverse peach breeding programs. *Tree Genetics and Genomes* 12:25.

Gardiner, S., J. Norelli, N. de Silva, G. Fazio, A. Peil, M. Malnoy, M. Horner, D. Bowatte, C. Carlisle, C. Wiedow, Y. Wan, C. Bassett, A. Baldo, J. Celton, K. Richter, H. Aldwinckle and V. Bus, 2012. Putative resistance gene markers associated with quantitative trait loci for fire blight resistance in Malus 'Robusta 5' accessions. *BMC Genetics* 13:25.

Gardner, R. G., 1977. *Breeding Apple Rootstocks Resistant to Fire Blight*. Cornell University, Ithaca, NYPhD.

Gardner, R. G., J. N. Cummins and H. S. Aldwinckle, 1980. Inheritance of fire blight resistance in Malus in relation to rootstock breeding. *Journal of the American Society for Horticultural Science* 105:912–16.

Geng, F., R. Moran, M. Day, W. Halteman and D. L. Zhang, 2015. In vitro shoot proliferation of apple rootstocks 'B 9', 'G 30', and 'G 41' grown under red and blue light. *HortScience* 50:430–3.

Gharghani, A., Z. Zamani, A. Talaie, N. C. Oraguzie, R. Fatahi, H. Hajnajari, C. Wiedow and S. E. Gardiner, 2009. Genetic identity and relationships of Iranian apple (Malus × domestica Borkh.) cultivars and landraces, wild Malus species and representative old apple cultivars based on simple sequence repeat (SSR) marker analysis. *Genetic Resources and Crop Evolution* 56:829–42.

Groot, M. J., 1997a. FPO-research on economic aspects of apple plantings. Plant spacings and rootstocks economically considered. *Fruitteelt Den Haag* 87:18–19.

Groot, M. J., 1997b. FPO-research on economic aspects of rootstock choice. M.9 not always the best choice. *Fruitteelt Den Haag* 87:18–19.

Guan, Y. Z., C. Peace, D. Rudell, S. Verma and K. Evans, 2015. QTLs detected for individual sugars and soluble solids content in apple. *Molecular Breeding* 35:135.

Hansen, O. B., 1989. Propagating apple rootstocks by semi-hardwood cuttings. *Norwegian Journal of Agricultural Sciences* 3:351–65.

Harrison, N., R. J. Harrison, N. Barber-Perez, E. Cascant-Lopez, M. Cobo-Medina, M. Lipska, R. Conde-Ruíz, P. Brain, P. J. Gregory and F. Fernández-Fernández, 2016. A new three-locus model for rootstock-induced dwarfing in apple revealed by genetic mapping of root bark percentage. *Journal of Experimental Botany* 67(6):1871–81.

Hartmann, H. T., D. E. Kester, F. T. Davies and R. L. Geneve, 1997. *Plant Propagation: Principles and Practices*. Prentice-Hall Inc., Upper Saddle River.

Hatton, R. G., 1917. Paradise apple stocks. *Journal of the Royal Horticultural Society* 42:361–99.

Hatton, R. G., 1919. Paradise apple stocks their fruit and blossom described. *Journal of the Royal Horticultural Society* 44:89–94.

Hatton, R. G., 1920. Suggestion for the right selection of apple stocks. *Journal of the Royal Horticultural Society* 45:257–68.

Higgs, K. H. and H. G. Jones, 1991. Water relations and cropping of apple cultivars on a dwarfing rootstock in response to imposed drought. *Journal of Horticultural Science* 66:367–79.

Iezzoni, A., C. Weebadde, J. Luby, C. Y. Yue, E. van de Weg, G. Fazio, D. Main, C. P. Peace, N. V. Bassil and J. McFerson, 2010. RosBREED: enabling marker-assisted breeding in Rosaceae. *International Symposium on Molecular Markers in Horticulture* 859:389–94.

Isutsa, D. K. and I. A. Merwin, 2000. Malus germplasm varies in resistance or tolerance to apple replant disease in a mixture of New York orchard soils. *HortScience* 35:262–8.

James, D., P. A. Trytten, D. J. Mackenzie, G. H. N. Towers and C. J. French, 1997. Elimination of apple stem grooving virus by chemotherapy and development of an immunocapture RT-PCR for rapid sensitive screening. *Annals of Applied Biology* 131:459–70.

James, P. and S. Middleton, 2011. The productivity and economic comparison of high-density production systems for 'Cripps Pink' and 'Cripps Red' apples in South Australia. *Acta Horticulturae* 903:611–18.

Janick, J., 2005. The origins of fruits, fruit growing, and fruit breeding. *Plant Breeding Reviews* 25:255–320.

Jarausch, W., C. Bisognin, T. Peccerella, B. Schneider and E. Seemuller, 2008. Development of resistant rootstocks for the control of apple proliferation disease. *Acta Horticulturae* 781:381–5.

Jensen, P. J., N. Halbrendt, G. Fazio, I. Makalowska, N. Altman, C. Praul, S. N. Maximova, H. K. Ngugi, R. M. Crassweller, J. W. Travis and T. W. McNellis, 2012. Rootstock-regulated gene expression patterns associated with fire blight resistance in apple. *BMC Genomics* 13.

Jensen, P. J., I. Makalowska, N. Altman, G. Fazio, C. Praul, S. N. Maximova, R. M. Crassweller, J. W. Travis and T. W. McNellis, 2010. Rootstock-regulated gene expression patterns in apple tree scions. *Tree Genetics & Genomes* 6:57–72.

Jensen, P. J., T. W. McNellis, N. Halbrendt, J. W. Travis, N. Altman, C. A. Praul, S. N. Maximova, R. M. Crassweller and I. Makalowska, 2011. Rootstock-regulated gene expression profiling in apple trees reveals genes whose expression levels are associated with fire blight resistance. *Acta Horticulturae* 903:87–93.

Jensen, P. J., J. Rytter, E. A. Detwiler, J. W. Travis and T. W. McNellis, 2003. Rootstock effects on gene expression patterns in apple tree scions. *Plant Molecular Biology* 53:493–511.

Johnson, W. C., H. S. Aldwinckle, J. N. Cummins, P. L. Forsline, H. T. Holleran, J. L. Norelli, T. L. Robinson, J. W. Palmer and J. N. Wunsche, 2001a. The new USDA-ARS/Cornell University apple rootstock breeding and evaluation program. *Acta Horticulturae* 557:35–40.

Johnson, W. C., J. N. Cummins, H. T. Holleran, S. A. Hoying, T. L. Robinson, J. W. Palmer and J. N. Wunsche, 2001b. Orchard trial performance of elite Geneva series rootstocks. *Acta Horticulturae* 557:63–7.

Kim, M., K. Ko, M. S. Kim and K. C. Ko, 2004. Relation of bitter pit development with mineral nutrients, cultivars, and rootstocks in apples (Malus domestica Borkh). *Korean Journal of Horticultural Science & Technology* 22:43–9.

Kost, T. D., C. Gessler, M. Jansch, H. Flachowsky, A. Patocchi and G. A. L. Broggini, 2015. Development of the first cisgenic apple with increased resistance to fire blight. *PLoS ONE* 10:e0143980.

Kviklys, D., 2011. Fruit rootstock research in Europe performed by EUFRIN Rootstock group. *Acta Horticulturae* 903:349–53.

Lana, A. F., J. F. Peterson, G. L. Rouselle and T. C. Vrain, 1983. Association of tobacco ringspot virus with a union incompatibility of apple. *Phytopathologische Zeitschrift* 106:141–8.

Lankes, C. and G. Baab, 2011. Screening of apple rootstocks for response to apple proliferation disease. *Acta Horticulturae* 903:379–83.

Leinfelder, M. M. and I. A. Merwin, 2006. Rootstock selection, preplant soil treatments, and tree planting positions as factors in managing apple replant disease. *Hortscience* 41:394–401.

Leinfelder, M. M., I. A. Merwin, G. Fazio and T. Robinson, 2004. Resistant rootstocks, preplant compost amendments, soil fumigation, and row repositioning for managing apple replant disease. *Hortscience* 39:841.

Lindley, J., 1828. *The Pomological Magazine; or, Figures and Descriptions of the Most Important Varieties of Fruit Cultivated in Great Britain*. London, J. Ridgway.

Liverani, A., F. Brandi, A. Leone, S. Sirri and D. Giovannini, 2015. Screening controlled pollination peach seedlings for mildew and brown rot resistance at CRA-FRF in the frame of the Fruitbreedomics EU Project. *Acta Horticulturae* 1084:187–93.

Lockard, R. G., 1976. Effect of apple rootstocks and length and type of interstock on leaf nutrient levels. *Journal of Horticultural Science* 51:289–96.

Loudon, J. C., 1822. An Encyclopædia of Gardening. Printed by A & R Spottiswoode, New-Street-Sqare. London

Luo, J., F. Wang, M. Han, X. Zhao, L. Lin, R. Wang, J. Luo, F. Wang, M. Y. Han, X. M. Zhao, L. P. Lin and R. H. Wang, 2013. Anatomical mechanism of two apple dwarf interstocks. *Journal of Northwest A & F University – Natural Science Edition* 41:124–32.

Lynch, J. P., J. G. Chimungu and K. M. Brown, 2014. Root anatomical phenes associated with water acquisition from drying soil: targets for crop improvement. *Journal of Experimental Botany* 65:6155–66.

Ma, L., C. Hou, X. Zhang, H. Li, D. Han, Y. Wang, Z. Han, L. Ma, C. W. Hou, X. Z. Zhang, H. L. Li, D. G. Han, Y. Wang and Z. H. Han, 2013. Seasonal growth and spatial distribution of apple tree roots on different rootstocks or interstems. *Journal of the American Society for Horticultural Science* 138:79–87.

Ma, R., Y. Huang, G. Sha, Y. Shi, J. Li, X. Gong, M. Li, R. Q. Ma, Y. Huang, G. L. Sha, Y. J. Shi, J. Li, X. H. Gong and M. Li, 2012. Identifying hybrids of apomictic apple rootstocks with SSR and flow cytometry. *Journal of Fruit Science* 29:461–5.

Marini, R. P., J. L. Anderson, W. R. Autio, B. H. Barritt, J. Cline, W. P. Cowgill, Jr., R. C. Crassweller, R. M. Garner, A. Gauss, R. Godin, G. M. Greene, C. Hampson, P. Hirst, M. M. Kushad, J. Masabni, E. Mielke, R. Moran, C. A. Mullins, M. Parker, R. L. Perry, J. P. Prive, G. L. Reighard, T. Robinson, C. R. Rom, T. Roper and J. R. Schupp, 2006a. Performance of 'Gala' apple trees on 18 dwarfing rootstocks: ten-year summary of the 1994 NC-140 rootstock trial. *Journal of the American Pomological Society* 60:69–83.

Marini, R. P., W. R. Autio, B. Black, J. Cline, W. R. Cowgill, Jr., R. M. Crassweller, P. A. Domoto, C. Hampson, R. Moran, R. A. Quezada, T. Robinson, D. L. Ward and D. Wolfe, 2013. Return bloom on 'Golden Delicious' apple trees as affected by previous season's crop density on three rootstocks at 11 locations. *Journal of the American Pomological Society* 67:72–9.

Marini, R. P., W. R. Autio, B. Black, J. A. Cline, W. Cowgill, R. Crassweller, P. Domoto, C. Hampson, R. Moran, R. A. Parra-Quezada, T. Robinson, M. Stasiak, D. L. Ward and D. Wolfe, 2012. Summary of the NC-140 apple physiology trial: the relationship between 'Golden Delicious' fruit weight and crop density at 12 locations as influenced by three dwarfing rootstocks. *Journal of the American Pomological Society* 66:78–90.

Marini, R. P., J. A. Barden, J. A. Cline, R. L. Perry and T. Robinson, 2002. Effect of apple rootstocks on average 'Gala' fruit weight at four locations after adjusting for crop load. *Journal of the American Society for Horticultural Science* 127:749–53.

Marini, R. P., B. H. Barritt, G. R. Brown, J. Cline, W. P. Cowgill, Jr., R. M. Crassweller, P. A. Domoto, D. C. Ferree, J. Garner, G. M. Greene, C. Hampson, P. Hirst, M. M. Kushad, J. Masabni, E. Mielke, R. Moran, C. A. Mullins, M. Parker, R. L. Perry, J. P. Prive, G. L. Reighard, T. Robinson, C. R. Rom, T. Roper, J. R. Schupp and E. Stover, 2006b. Performance of 'Gala' apple on four semi-dwarf rootstocks: a ten-year summary of the 1994 NC-140 semi-dwarf rootstock trial. *Journal of the American Pomological Society* 60:58–68.

Massacci, A. and H. G. Jones, 1990. Use of simultaneous analysis of gas-exchange and chlorophyll fluorescence quenching for analysing the effects of water stress on photosynthesis in apple leaves. *Trees: Structure and Function* 4:1–8.

Masseron, A. and L. Roche, 1999. Apple orchard management systems: technical and economic status report after a 10-year study. Part II: results and discussion. *Infos Paris* 149:40–6.

Mauroux, J. B., B. Quilot-Turion, T. Pascal, P. Lambert, M. Troggio, I. Verde, D. Micheletti and M. J. Aranzana, 2015. Building high-density peach linkage maps based on the ISPC 9K SNP chip for mapping mendelian traits and QTLs: benefits and drawbacks. *Acta Horticulturae* 1084:113–18.

Mazzola, M., 1998. Elucidation of the microbial complex having a causal role in the development of apple replant disease in Washington. *Phytopathology* 88:930–8.

Mazzola, M. and L. M. Manici, 2012. Apple replant disease: role of microbial ecology in cause and control. *Annual Review of Phytopathology* 50:45–65.

Mazzola, M. and K. Mullinix, 2005. Comparative field efficacy of management strategies containing Brassica napus seed meal or green manure for the control of apple replant disease. *Plant Disease* 89:1207–13.

Migicovsky, Z., K. M. Gardner, D. Money, J. Sawler, J. S. Bloom, P. Moffett, C. T. Chao, H. Schwaninger, G. Fazio, G.-Y. Zhong and S. Myles, 2016. Genome to phenome mapping in apple using historical data. *The Plant Genome* 9:2.

Momol, M. T., H. S. Aldwinckle, P. L. Forsline, W. F. Lamboy and H. Saygili, 1999. Fire blight resistance and horticultural evaluation of wild Malus populations from Central Asia. *Acta Horticulturae* 489:229–33.

Monceau, D.d., 1768. Traité des arbres fruitiers: contenant leur figure, leur description, leur culture. A Paris, Chez Saillant, Libraire.

Moran, R. E., Y. P. Sun, F. Geng, D. L. Zhang and G. Fazio, 2011a. Cold temperature tolerance of trunk and root tissues in one- or two-year-old apple rootstocks. *Hortscience* 46:1460–4.

Moran, R. E., D. Zhang and Y. Sun, 2011b. Cold temperature tolerance of G.16 and G.935 apple roots. *Acta Horticulturae* 903:289–93.

Neilsen, G. and C. Havipson, 2014. 'Honeycrisp' apple leaf and fruit nutrient concentration is affected by rootstock during establishment. *Journal of the American Pomological Society* 68:178–89.

Neilsen, G. H., D. Neilsen, S. H. Guak and T. Forge, 2015. The effect of deficit irrigation and crop load on leaf and fruit nutrition of fertigated 'Ambrosia'/'M.9' apple. *Hortscience* 50:1387–93.

Om, H. and R. K. Pathak, 1983. Influence of stock and scion on macro-nutrient contents of apple leaves. *Indian Journal of Plant Physiology* 26:337–43.

Oraguzie, N. C., T. Yamamoto, J. Soejima, T. Suzuki and H. N.d. Silva, 2005. DNA fingerprinting of apple (Malus spp.) rootstocks using simple sequence repeats. *Plant Breeding* 124:197–202.

Peace, C. P., J. J. Luby, M. C. A. M. Bink, A. F. Lezzoni and W. E. van de Weg, 2014. A strategy for developing representative germplasm sets for systematic QTL validation, demonstrated for apple, peach, and sweet cherry. *Tree Genetics and Genomes* 10:1679–94.

Perry, R. L., 1996. Summary of NC-140 rootstock trials for apples in Michigan. *Pennsylvania Fruit News* 76:34–5.

Pilcher, R. L. R., J. M. Celton, S. E. Gardiner and D. S. Tustin, 2008. Genetic markers linked to the dwarfing trait of apple rootstock 'Malling 9'. *Journal of the American Society for Horticultural Science* 133:100–6.

Preston, A. P., J. Ingram and C. M. Bamber, 1972. Apple rootstock studies: fifteen years' growth and cropping on twelve clones at Luddington. *Experimental Horticulture* 24:19–24.

Quamme, H. A., R. T. Brownlee, B. H. Barritt and F. Kappel, 1997. Cold hardiness evaluation of apple rootstocks. *Acta Horticulturae* 451:187–93.

Resendes, M. L., D. R. Bryla and D. M. Eissenstat, 2008. Early events in the life of apple roots: variation in root growth rate is linked to mycorrhizal and nonmycorrhizal fungal colonization. *Plant and Soil* 313:175–86.

Rivers, T., 1866. *The Miniature Fruit Garden; or, The Culture of Pyramidal and Bush Fruit Trees.* New York, Orange Judd & Company.

Robinson, T., 2008. The evolution towards more competitive apple orchard systems in the USA. *Acta Horticulturae* 772:491–500.

Robinson, T., H. Aldwinckle, G. Fazio, T. Holleran and J. Janick, 2003. The Geneva series of apple rootstocks from Cornell: performance, disease resistance, and commercialization. *Acta Horticulturae* 622:513–20.

Robinson, T. L., A. M. DeMarree and S. A. Hoying, 2007. An economic comparison of five high density apple planting systems. *Acta Horticulturae* 732:481–9.

Robinson, T. L., S. A. Hoying and G. H. Reginato, 2011. The Tall Spindle planting system: principles and performance. *Acta Horticulturae* 903:571–9.

Robinson, T. L., A. N. Lakso and S. G. Carpenter, 1991a. Canopy development, yield, and fruit quality of 'Empire' and 'Delicious' apple trees grown in four orchard production systems for ten years. *Journal of the American Society for Horticultural Science* 116:179–87.

Robinson, T. L., A. N. Lakso and Z. B. Ren, 1991b. Modifying apple tree canopies for improved production efficiency. *HortScience* 26:1005–12.

Rom, C. R. and R. C. Rom, 1991. Rootstock effects on foliar nutrient content of 'Starkspur Supreme' in the NC-140 rootstock trials. *Compact Fruit Tree* 24. Grand Rapids, Michigan, USA, 24–27 February 1991.

Rom, C. R., R. C. Rom, W. R. Autio, D. C. Elfving and R. A. Cline, 1991. Foliar nutrient content of 'Starkspur Supreme Delicious' on nine clonal apple rootstocks. *Fruit Varieties Journal* 45:252–63.

Rom, R. C. and R. F. Carlson, 1987. *Rootstocks for Fruit Crops*. Wiley, New York.

Rumberger, A., I. A. Merwin and J. E. Thies, 2007. Microbial community development in the rhizosphere of apple trees at a replant disease site. *Soil Biology & Biochemistry* 39:1645–54.

Rumberger, A., S. R. Yao, I. A. Merwin, E. B. Nelson and J. E. Thies, 2004. Rootstock genotype and orchard replant position rather than soil fumigation or compost amendment determine tree growth and rhizosphere bacterial community composition in an apple replant soil. *Plant and Soil* 264:247–60.

Rusholme, R. L., S. E. Gardiner, H. C. M. Bassett, D. S. Tustin, S. M. Ward and A. Didier, 2004. Identifying genetic markers for an apple rootstock dwarfing gene. *Acta Horticulturae* 663:405–9.

Russo, N. L., T. L. Robinson, G. Fazio and H. S. Aldwinckle, 2007. Field evaluation of 64 apple rootstocks for orchard performance and fire blight resistance. *Hortscience* 42:1517–25.

Russo, N. L., T. L. Robinson, G. I'azio and H. S. Aldwinckle, 2008. Fire blight resistance of Budagovsky 9 apple rootstock. *Plant Disease* 92:385–91.

Sandanayaka, W. R. M. and E. A. Backus, 2008. Quantitative comparison of stylet penetration behaviors of glassy-winged sharpshooter on selected hosts. *Journal of Economic Entomology* 101:1183–97.

Sandanayaka, W. R. M., V. G. M. Bus and P. Connolly, 2005. Mechanisms of woolly aphid [Eriosoma lanigerum (Hausm.)] resistance in apple. *Journal of Applied Entomology* 129:534–41.

Sandanayaka, W. R. M., V. G. M. Bus, P. Connolly and R. Newcomb, 2003. Characteristics associated with Woolly Apple Aphid Eriosoma lanigerum, resistance of three apple rootstocks. *Entomologia Experimentalis Et Applicata* 109:63–72.

Sax, K., 1949. The use of Malus species for apple rootstocks. *Proceedings. American Society for Horticultural Science* 53:219–20.

Schechter, I., D. C. Elfving and J. T. A. Proctor, 1991. Canopy development, photosynthesis, and vegetative growth as affected by apple rootstocks. *Fruit Varieties Journal* 45:229–37.

Schmidt, H., 1970. Problems in breeding for apomictic apple rootstocks. *Proceedings of the Angers Fruit Breeding Symposium*. Angers, France, 14–18 September 1970, pp. 269–83.

Schmidt, H., 1972. Reaction of 25 apomictic apple rootstock selections to inoculation with mixtures of 'latent' viruses. *Journal of Horticultural Science* 47:151–7.

Schmidt, H., 1982. Evaluation for rootstock characters in apomictic apple selections. *XXIst International Horticultural Congress*. The Hague, Netherlands.

Seemuller, E., E. Moll and B. Schneider, 2007. Malus sieboldii-based rootstocks mediate apple proliferation resistance to grafted trees. *Bulletin of Insectology* 60:301–2.

Seemuller, E., E. Moll and B. Schneider, 2008. Apple proliferation resistance of Malus sieboldii-based rootstocks in comparison to rootstocks derived from other Malus species. *European Journal of Plant Pathology* 121:109–19.

Seleznyova, A. N., D. S. Tustin and T. G. Thorp, 2008. Apple dwarfing rootstocks and interstocks affect the type of growth units produced during the annual growth cycle: precocious transition to flowering affects the composition and vigour of annual shoots. *Annals of Botany* 101:679–87.

Sha, G. L., X. H. Gong, Y. Huang, Y. C. Shao, T. Yin, Y. J. Hao and H. R. Shu, 2011. 'Qingzhen 1' and 'Qingzhen 2', two apple rootstocks with high apomitic fruit setting ability. *Acta Horticulturae* 903:159–62.

Shin, S., P. Zheng, G. Fazio, M. Mazzola, D. Main and Y. Zhu, 2016. Transcriptome changes specifically associated with apple (Malus domestica) root defense response during Pythium ultimum infection. *Physiological and Molecular Plant Pathology* 94:16–26.

Silva, F. N., O. Nickel, T. V. M. Fajardo and A. Bogo, 2008. Biological multiple indexing and RT-PCR detection of latent viruses in apple plants. *Tropical Plant Pathology* 33:157–61.

Simons, R. K., 1985. Influence of the graft union on performance of apple trees. *Compact Fruit Tree* 18:67–70.

Simons, R. K. and M. C. Chu, 1983. Graft union development: Granny Smith/EM 26 – specific growth characteristics between stock and scion. *Compact Fruit Tree* 16:73–82.

Simons, R. K. and M. C. Chu, 1985. Graft union characteristics of M.26 apple rootstock combined with 'Red Delicious' strains – morphological and anatomical development. *Scientia Horticulturae* 25:49–59.

Skene, D. S., H. R. Shepherd and B. H. Howard, 1983. Characteristic anatomy of union formation in T- and chip-budded fruit and ornamental trees. *Journal of Horticultural Science* 58:295–9.

Sloan, R. C., Jr., F. B. Matta and F. J. Killebrew, 1996. Effect of cultivar and foliar nutrients on fire blight susceptibility in apple. *Research Report Mississippi Agricultural and Forestry Experiment Station* 21:7.

Sousa, S. M. D., R. T. Clark, F. F. Mendes, A. C.d. Oliveira, M. J. V.d. Vasconcelos, S. N. Parentoni, L. V. Kochian, C. T. Guimaraes, J. V. Magalhaes, S. M. de Sousa, A. C. de Oliveira and M. J. V. de Vasconcelos, 2012. A role for root morphology and related candidate genes in P acquisition efficiency in maize. *Functional Plant Biology* 39:925–35.

St. Laurent, A., I. A. Merwin, G. Fazio, J. E. Thies and M. G. Brown, 2010. Rootstock genotype succession influences apple replant disease and root-zone microbial community composition in an orchard soil. *Plant and Soil* 337:259–72.

Strong, D. and A. Miller Azarenko, 1991. Dry matter partitioning in 'Starkspur Supreme Delicious' on nine rootstocks. *Fruit Varieties Journal* 45:238–41.

Swietlik, D., C. Vann, M. Wisniewski, T. Artlip, J. L. Norelli and L. Kochian, 2007. The effect of transporter genes on zinc stress in apple (Malus × domestica Borkh). *Acta Horticulturae* 738:345–51.

Theron, K. I., W. J. Steyn, G. Jacobs, M. Bodson and M. N. J. Verhoyen, 2000. Induction of proleptic shoot formation on pome fruit nursery trees. *Acta Horticulturae* 514:235–43.

Tukey, H. B., 1964. *Dwarfed Fruit Trees, for Orchard, Garden and Home*. Macmillan, New York.

Tukey, R. B., R. Langston and R. A. Cline, 1962. Influence of rootstock, bodystock and interstock on the nutrient content of apple foliage. *Proceedings American Society for Horticultural Science* 80:73–8.

Tuttle, M. A. and A. R. Gotlieb, 1985a. Graft union histology and distribution of tomato ringspot virus in infected McIntosh/Malling Merton 106 apple trees. *Phytopathology* 75:347–51.

Tuttle, M. A. and A. R. Gotlieb, 1985b. Histology of delicious malling Merton 106 trees affected by apple union necrosis and decline. *Phytopathology* 75:342–7.

Tworkoski, T. and G. Fazio, 2011. Physiological and morphological effects of size-controlling rootstocks on 'Fuji' apple scions. *Acta Horticulturae* 903:865–72.

Tworkoski, T. and G. Fazio, 2016. Hormone and growth interactions of scions and size-controlling rootstocks of young apple trees. *Plant Growth Regulation* 78:105–19.

Tworkoski, T., G. Fazio and D. M. Glenn, 2016. Apple rootstock resistance to drought. *Scientia Horticulturae*.

Tworkoski, T. and S. Miller, 2007. Rootstock effect on growth of apple scions with different growth habits. *Scientia Horticulturae* 111:335–43.

USA, The Ohio State University, North Carolina, 1991. Performance of 'Starkspur Supreme Delicious' apple on 9 rootstocks over 10 years in the NC-140 cooperative planting. *Fruit Varieties Journal* 45:192–9.

Utkhede, R. S., 1999. Biological treatments to increase apple tree growth in replant problem soil. *Allelopathy Journal* 6:63–8.

Utkhede, R. S. and E. M. Smith, 2000. Impact of chemical, biological and cultural treatments on the growth and yield of apple in replant-disease soil. *Australasian Plant Pathology* 29:129–36.

van Hooijdonk, B., D. Woolley, I. Warrington and S. Tustin, 2011. Rootstocks modify scion architecture, endogenous hormones, and root growth of newly grafted 'Royal Gala' apple trees. *Journal of the American Society for Horticultural Science* 136:93–102.

Virlet, N., E. Costes, S. Martinez, J. J. Kelner and J. L. Regnard, 2015. Multispectral airborne imagery in the field reveals genetic determinisms of morphological and transpiration traits of an apple tree hybrid population in response to water deficit. *Journal of Experimental Botany* 66:5453–65.

Visser, T., 1967. Juvenile period and precocity of apple and pear seedlings. *Euphytica* 16:319–20.

Visser, T., 1973. The effect of rootstocks on growth and flowering of apple seedlings. *Journal of the American Society for Horticultural Science* 98:26–8.

Visser, T. and A. A. Schaap, 1967. Preselection for juvenile period, flowering and picking time in apple seedlings. *Euphytica* 16:109–21.

Warmund, M. R., W. R. Autio, J. A. Barden, J. N. Cummins, P. A. Domoto, C. G. Embree, R. L. Granger, F. D. Morrison, J. R. Schupp and E. Young, 1996. Blackheart injury in 'Starkspur Supreme Delicious' on 15 rootstocks in the 1984 NC-140 cooperative planting. *Fruit Varieties Journal* 50:55–62.

Warmund, M. R. and V. J. Slater, 1988. Hardiness of apple and peach trees in the NC-140 rootstock trials. *Fruit Varieties Journal* 42:20–4.

Webster, A. D., 2003. Breeding and selection of apple and pear rootstocks. *Acta Horticulturae* 622:499–512.

Webster, A. D. and I. J. Warrington, 1995. Rootstock and interstock effects on deciduous fruit tree vigour, precocity, and yield productivity. *1994. New Zealand Journal of Crop and Horticultural Science* 23:373–82.

Webster, A. D. and S. J. Wertheim, 2003. Apple rootstocks. In D. C. Ferree and I. J. Warrington (Eds), *Apples: Botany, Production and Uses*, pp. 91–124. CABI Publishing, Wallingford.

Wertheim, S. J., 1998. *Rootstock Guide: Apple, Pear, Cherry, European Plum. Proefstation Voor de Fruitteelt* (Fruit Research Station), Wilhelminadorp, The Netherlands.

West, S. and E. Young, 1988. Effects of rootstock and interstock on seasonal changes in foliar nutrient (N, P, K, Ca) composition of 'Delicious' and 'Golden Delicious' apple. *Fruit Varieties Journal* 42:9–13.

Westwood, M. N. and H. O. Bjornstad, 1980. Mineral nutrient content of leaves of several apple (Malus) species. *Compact Fruit Tree* 13:67–71.

Wisniewski, M., C. Bassett, J. Norelli, D. Macarisin, T. Artlip, K. Gasic and S. Korban, 2008. Expressed sequence tag analysis of the response of apple (Malus × domestica 'Royal Gala') to low temperature and water deficit. *Physiologia Plantarum* 133:298–317.

Wohner, T., M. V. Hanke, H. Flachowsky, A. Peil, K. Richter and G. A. L. Broggini, 2014. Investigation on fire blight resistance in the cross population 'Idared' × Malus × robusta 5 with different Erwinia amylovora strains. *Acta Horticulturae* 1056:277–80.

Wood, G. A., 1996. Past and present status of virus and phytoplasma diseases in apple rootstocks in New Zealand. *New Zealand Journal of Crop and Horticultural Science* 24:133–41.

Xiang, B., M. Cheng, X. Li, Q. Jin, Y. Liu, X. Xie, B. X. Xiang, M. H. Cheng, X. L. Li, Q. Jin, Y. Q. Liu and X. L. Xie, 1995. A study of drought resistance in the rootstock germplasms of apple. *Journal of Southwest Agricultural University* 17:381–5.

Yao, S. R., I. A. Merwin, G. S. Abawi and J. E. Thies, 2006a. Soil fumigation and compost amendment alter soil microbial community composition but do not improve tree growth or yield in an apple replant site. *Soil Biology & Biochemistry* 38:587–99.

Yao, S. R., I. A. Merwin and M. G. Brown, 2006b. Root dynamics of apple rootstocks in a replanted orchard. *Hortscience* 41:1149–55.

Zhan, A., J. P. Lynch and A. Zhan, 2015. Reduced frequency of lateral root branching improves N capture from low-N soils in maize. *Journal of Experimental Botany* 66:2055–65.

Zhang, H., H. An, Y. Wang, X. Zhang, Z. Han, H. Zhang, H. S. An, Y. Wang, X. Z. Zhang and Z. H. Han, 2015. Low expression of PIN gene family members is involved in triggering the dwarfing effect in M9 interstem but not in M9 rootstock apple trees. *Acta Physiologiae Plantarum* 37:104.

Chapter 8

Advances in marker-assisted breeding of apples

K. Evans and C. Peace, Washington State University, USA

1 Introduction

A key strategy to ensure sustained and sustainable apple production is breeding. New apple scion and rootstock cultivars can be genetically improved to better suit changing growing conditions, modern production systems and consumer preferences. Apple breeding is a long-term, resource-consuming, genetically complicated venture (Fig. 1). A juvenile phase of typically five years and up to ten years can be reduced by use of a precocious rootstock, but the generation time is greater than for many other crops. A plot of hundreds to thousands of apple seedlings takes up considerable land area, even on a dwarfing rootstock. Ongoing regular maintenance of plants over large areas for multiple years adds to the expense. Understanding both the genetic relationships among germplasm individuals and the genetic influences of important traits is fundamental to successful apple breeding. Apple is a self-incompatible, highly heterozygous, ancient autotetraploid of 17 chromosomes, features that have made genetic studies difficult. Many of our well-known apple cultivars have been around for decades, if not centuries, and a reasonably high proportion of them arose as chance seedlings rather than through controlled hybridization (Morgan and Richards, 2002) and hence parentage is unknown. As cultivars have been distributed via dormant buds around the world to either growers

http://dx.doi.org/10.19103/AS.2016.0017.09

for testing opportunities in commercial production or to germplasm collections, there have been, not surprisingly, labelling and propagation mistakes, which have generated confusion when cultivars have been compared among locations (Evans et al., 2011). The inability to readily determine identity, parentage, probability of shared alleles and general relatedness among breeding germplasm individuals has proved to be a long-term challenge for apple breeders. Dissection of the genetic control of apple traits, so that numerous target attributes, such as disease- and pest-hardy trees with high productivity for fruit that are crisp, flavourful, and with a long market life can be effectively assembled with available resources, is similarly challenged. The inherent difficulties described above have relegated it, and its supporting genetic research, almost entirely to the public sector in the United States. Yet apple is one of the most economically valuable fruit crops in the world and has significant contributions to human health (Boyer and Liu, 2004; Hyson, 2011). Therefore, solving breeding challenges will have large and positive consequences.

After the spread of Mendel's principles in the early 1900s, apple geneticists, including breeders, considered whether apple traits could be dissected into discrete genetic units. In a 1959 report, apple geneticist Nils Nybom lamented that apple breeding was still mostly 'pre-Mendelian', not only because of the long generation time hampering studies but also because most traits appeared to be of polygenic inheritance. Only two traits of breeding interest were identified as exhibiting Mendelian inheritance: fruit acidity and red flesh colour, and these were not linked with each other (Nybom, 1959). Although

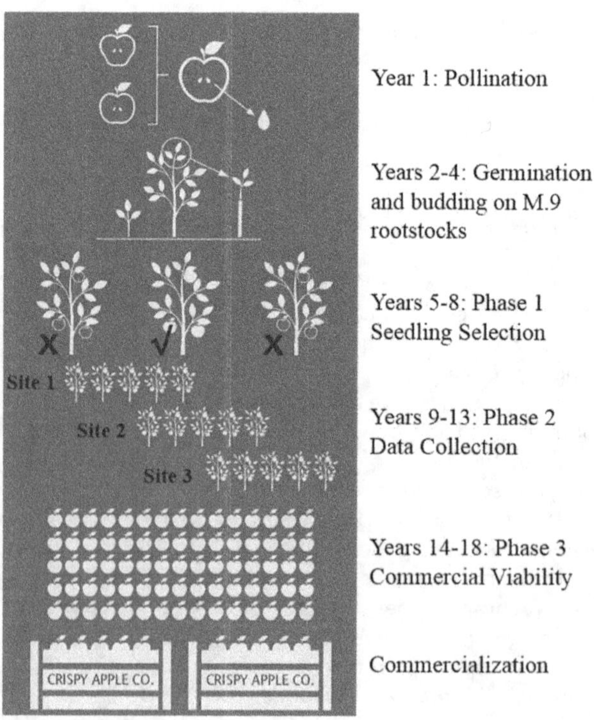

Figure 1 Schema and timeframe for the Washington State University apple breeding program.

studies in other organisms were demonstrating the diagnostic power for one Mendelian genetic factor to predict the presence of another via the principle of indirect selection, the biology of apple breeding germplasm rendered it recalcitrant. Great gains were envisaged if young seedlings could be selected for performance in traits that could only otherwise be evaluated at great cost in replicated field trials (any traits with relatively low heritability), with laborious protocols (e.g. susceptibility screening for many diseases, gas chromatography for fruit sugar profiles) or at reproductive maturity (flower and fruit traits such as cross compatibility, flowering and fruit maturity timing, fruit flavour and size, and yield). Trait performance diagnostics using genetic assays where distinct outcomes can be associated with alleles of a trait locus is known as 'marker-assisted selection' (MAS). Beyond MAS, advances in other agricultural crops, especially with the advent of DNA-based genetic markers (Paterson et al., 1991; Powell et al., 1996), demonstrated that direct assays of the allelic state of the DNA of breeding-relevant individuals could also be used to establish unique identity, ascertain parentage, reveal genetic relationships within sets of germplasm and unravel the genetic architecture of traits of interest. All of these applications of genetic markers to aid breeding planning and operations are collectively known as 'marker-assisted breeding' (MAB).

Despite its promise, apple breeding has been slow to benefit from MAB technology. Quick-fix solutions in breeding are rarely what they seem (Simmonds, 1991). Challenges have arisen in establishing required foundational resources, focusing on translation to breeding practice, and accessing timely and cost-effective DNA-based diagnostics services.

2 Advances in apple MAB

Increasingly, the challenges are being resolved such that MAB is currently a viable approach and effective contributor to apple genetic improvement. Although decades in the making, numerous independent and collaborative research projects worldwide have identified and targeted the key scientific needs and made new discoveries along the way. The direct involvement of apple breeding programmes in these endeavours and public research funding have been critical elements for success. Presently, MAB can be considered conventional for apple, because most apple breeding programmes worldwide use DNA-based information to support some breeding decisions.

3 History of apple MAB

3.1 Establishing required foundational resources

The first step towards MAB in apple required the development of fundamental genomics resources, which accumulated from more than two decades of lively and collaborative research. Research from individual programmes and institutions increasingly amalgamated into international collaborations such as the European Apple Genome Mapping Project (King et al., 1991), DARE (Lespinasse et al., 2000), HiDRAS (Gianfranceschi and Soglio, 2004), RosBREED (Iezzoni et al., 2010), and FruitBreedomics (Laurens et al., 2010). The painstaking establishment of genomics resources for apple took longer than initially anticipated and effectively diverted attention from the application of genomics to cultivar development. The early promise of having and using trait-predictive DNA tests was not fulfilled by the end of the twentieth century. Research funds and new students and trainees in apple 'breeding' did not work directly on cultivar development but rather were drawn to the attractive new field of 'genomics' and 'biotechnology'. The funding and attention to obtain the genomics resources described below were underestimated yet largely essential to establishing a sufficient foundation from which to develop MAB tools and practices.

Early attention was focused on transferring molecular marker technologies to apple from other crops. Inheritance patterns of apple isozymes were the first to be described (Chevreau and Laurens, 1987), and were quickly used to establish an early linkage map of apple (Weeden and Lamb, 1987) and even establish linkage to a trait of interest, self-incompatibility (Manganaris and Alston, 1987). High heterozygosity in this outcrossing crop helped mapping efforts; however, the multiple ancestrally duplicated regions in the apple genome hindered progress by indicating two homoelogous positions for many markers. Randomly distributed, neutral DNA marker types of RAPDs (Weeden et al., 1992), then AFLPs (Maliepaard et al., 1998), SSRs (Hokanson et al., 1998) and eventually SNPs (Chagné et al., 2012; Khan et al., 2012) detected sufficient genetic polymorphism to enable development of linkage maps with high degrees of genome coverage (Hemmat et al., 1994; Maliepaard et al., 1998; Lin et al., 1998; Liebhard et al., 2003; Kenis and Keulemanns, 2005; Silfverberg-Dilworth et al., 2006). Genetic linkage maps were subsequently used for discovering, localizing and characterizing Mendelian trait loci (MTLs) and quantitative trait loci (QTLs) underlying traits of interest (e.g. Conner et al., 1998; King et al., 2001, and various others reviewed by Korban and Tartarini 2009). Most of the early successes were identification of MTLs for resistances to diseases and pests (Evans and James, 2003; Gessler et al., 2006). Fruit quality traits in apple are typically controlled by multiple QTLs; one clear exception was the locus influencing malic acid content, the major source of acidity in apple fruit, identified in the 1950s (Nybom, 1959) and mapped as part of the large collaborative European Apple Genome Mapping project in the 1990s (Maliepaard et al., 1998).

Resistances to diseases and pests were early targets, not only because they were in the most part MTLs but also because they were relatively easy to phenotype. Standardized protocols with clear evaluation scales and illustrative photographs such as those developed as part of the large European projects (EAGMAP, DARE and HiDRAS) meant that replicated phenotypic data could be compared among collaborators and large datasets amassed (King et al., 1998). Of course, breeders are extremely interested in fruit quality traits; however, in apple they are difficult to phenotype accurately as fruit quality is heavily influenced by uncontrolled variation in climatic conditions during production and in production management systems, as well as in the maturity of fruit at harvest and when evaluated (Kromdijk et al. 2014). Collecting data from trees with a standardized crop load, at standardized maturity at harvest as well as at fixed post-harvest time-points, is also crucial. Although some instrumental measurements are available for apple quality, none have yet surpassed the human palate despite its susceptibility to subjectivity and fatigue (Evans et al., 2010b). Standardized phenotyping protocols for apple used in the U.S. RosBREED project (Evans et al., 2012a; https://www.rosbreed.org/sites/default/files/files/RosBREED_2010-Phenotyping_protocol_Malus.pdf) were based upon those developed in the European projects.

Into the first decade of the new millennium, research efforts extended to obtaining large-scale DNA sequence resources and constructing physical genetic maps anchored to their genetic counterparts (Newcomb et al., 2006; Gasic et al., 2009) that ultimately provided whole-genome maps of key apple cultivars and of apple in general (Velasco et al., 2010). Such high-resolution genomic resources in turn have enabled the development of genome scans in the form of SNP arrays (Chagné et al., 2012; Bianco et al., 2014, 2016) that can illuminate further the genomic composition of apple. Databases have become vital to expanded research capacity for researchers to collate and make accessible the vast datasets, led in breadth and depth by the Genome Database for Rosaceae (Jung et al., 2004, 2008, 2014).

3.2 Focusing on translation to breeding practice

With genetic markers and genomics resources accessible, apple geneticists had the opportunity to develop and deliver knowledge and tools for the benefit of practical cultivar development. The earliest and subsequently most prevalent outcomes for practical apple breeding have been in two major areas: knowledge about germplasm identity/relatedness and knowledge about trait genetic architecture. For example, RAPD markers were used to understand relatedness among apple cultivars and between these and *Malus* wild species (Dunemann, 1993) and to ascertain unique identity for a new cultivar release (Tancred et al., 1994). AFLP markers were useful for discriminating among apple cultivars and especially detecting a distinct gene pool for Portuguese cultivars (Goulão et al., 2001). MTL and QTL analyses since the late 1990s have been effective for determining the number and genomic locations of loci for traits of breeding interest as well as the relative effects and genetic action of their alleles (Korban and Tartarini, 2009), valuable knowledge for guiding breeding decisions in achieving genetic improvement.

The first trait-preditive genetic markers to be applied in breeding were those based on MTLs for disease resistance (Ru et al., 2015). MTLs are loci with functional allelic variation within investigated germplasm that explains the bulk of the phenotypic variation. The first reported application of marker-assisted seedling selection (MASS, Peace and Norelli, 2009) was for resistance to apple scab, targeting the *Rvi6* (formerly *Vf*) locus (Bus et al., 2000). Genotyping for *Rvi6* apple scab resistance was considered more reliable than phenotyping and had the further advantage of detecting homozygous resistant plants (Tartarini et al., 1999).

However, MASS based on trait loci accounting for less phenotypic variation than a typical MTL, that is, QTLs, did not smoothly flow from the availability of genetic maps and abundance of reported QTLs. The subsequent steps required to translate those resources and knowledge to information readily useable in routine selection decisions were underestimated. Although both were established and new professionals were now well trained in generating genomics resources, they were rarely trained in apple breeding, rarely trained in conversion of genomics resources and knowledge to practical tools, and not trained in conducting MASS because little had yet occurred. Therefore, coordinated efforts were undertaken to direct genomics resources and knowledge to practical breeding application (Peace and Norelli, 2009), with the large-scale, multi-institution projects RosBREED, from 2009 to 2014 (Iezzoni et al., 2010), FruitBreedomics, from 2010 to 2014 (Laurens et al., 2010), and 'RosBREED 2', from 2014 to 2019 (www.rosbreed.org), established to explicitly 'bridge the chasm' between genomics and breeding application.

Because QTLs were discovered and characterized in single families, with at most four alleles segregating and the genetic backgrounds limited to those created by combining the two parents, performance predictions were relevant for those same families and perhaps those closely related, but not for breeding germplasm in general. Pedigree-Based

Analysis (PBA) statistics and software were developed to address this bottleneck, enabling QTL analyses to be conducted on multiple, variably sized, related families, so that the QTLs characterized are representative of their effects in those families (van de Weg et al., 2004; Bink et al., 2008; Peace and Norelli, 2009). Such germplasm typifies apple breeding programmes; PBA is therefore suited to identifying and simultaneously validating QTLs in breeding germplasm, streamlining their translation to practical application. However, typical routine phenotypic evaluation of breeding germplasm does not extend to individuals obviously deficient for a critical trait threshold (such as having small fruit size or poor taste), yet QTL analysis needs such data to identify contrasting alleles, so that such alleles can be selected against in MAS. Furthermore, QTL characterization leading to robust assays typically requires more detailed phenotypic evaluation over more trees, more years and more samples (such as number of fruit) than routinely performed in breeding. To solve such issues, apple germplasm resources used for PBA (Peace et al., 2014; Bink et al., 2014) have been treated as special genetic stocks, requiring extra funding to maintain and evaluate. To further ensure that QTL validation in breeding germplasm is not a bottleneck, Peace et al. (2014) proposed and used a method that considers the 'average allelic representation' of important breeding parents as the major criterion in determining the composition of germplasm sets to undergo QTL analysis for multiple traits of interest. The U.S. Apple Reference Germplasm Set established under these considerations consisted of 467 apple individuals with 154 cultivars, elite selections, their ancestors and 313 seedlings in 21 families of 3–28 seedlings each (Peace et al., 2014). Once QTLs controlling specific traits have been discovered and characterized, focused attention is required to convert them into trait performance-predictive assays. Reported trait-predictive DNA tests for apple are described in the Evaluation tools Section 4.2.

In practice, routine MAB requires consideration of the logistics of ongoing breeding operations. Ru et al. (2015) reported that integrating DNA testing into traditional breeding operations was the most commonly perceived challenge among a small group of U.S. Rosaceae crop breeders. MASS especially requires DNA testing operations that fit within the timing and infrastructure of each specific breeding programme. Edge-Garza et al. (2014) described a DNA extraction method adapted to apple that has streamlined tissue-sampling procedure, easing a critical logistical issue for breeding programmes. With this method, a single technician can obtain tissue and extract DNA from 3200 seedlings per week. Gaining experience with routine MAB helps identify opportunities to further streamline operations, both for the breeding programmes according to their own idiosyncrasies and for the service providers of DNA-based diagnostics. The case study below (Section 6) describes several examples.

3.3 Accessing DNA-based diagnostics services

Once assays are available for identity and relatedness and/or trait predictions, breeding programmes benefit from access to service providers who are able to provide readily interpretable DNA-based information that is timely and cost-effective (Ru et al., 2015). These service providers might be for-profit commercial companies, not-for-profit services of institutions such as universities who mostly seek to cover costs and often via research collaborations with the breeding programme, or the laboratories of breeding programmes themselves. Accessing the former two entities divests the breeding programme of involvement in technical details that can distract from practical cultivar development, while a major advantage of the latter is complete tailoring to the breeding programme's specific needs. Even when outsourcing, upfront costs of initially investing in MAB prior to benefits returning can be an obstacle for some breeding programmes (Ru et al., 2015).

High costs associated with DNA testing, or at least the perception of such, as well as lack of access to services have historically limited widespread application where relevant DNA tests were available (Yue et al., 2012). More than a decade ago, Frey et al. (2004) described an efficient laboratory procedure for quickly and cheaply genotyping thousands of apple seedlings for apple scab and powdery mildew resistance, proving that the genotyping step is not the limiting factor in practical deployment of MASS. Commercial costs of DNA extraction and screening one to several markers is currently available for several dollars or less and prices continue to fall, rendering DNA testing a sensible alternative to phenotypic selection for many breeding programmes and situations. To identify general and specific situations in which MASS is expected to be cost-efficient for any given breeding programme, Edge-Garza et al. (2015) examined changes over many fixed and variable factors and developed a spreadsheet-based tool. General conclusions were that MASS can often be very cost-efficient, even when just a single DNA test is used and especially where a large proportion of seedlings are culled at an early stage using DNA information (Edge-Garza et al., 2015).

Besides cost-efficiency, the desired outcome is DNA information that is relevant to the breeding programme's germplasm, readily interpretable by personnel of the particular breeding programme, and obtained in a timeframe that enables ready incorporation into ongoing breeding operations. MASS has particularly short windows (often only weeks) during which tissue sampling, DNA extraction, genotyping and data interpretation must be completed for thousands of plants (Peace and Norelli, 2009; Edge-Garza et al., 2010). Service providers that are technically well equipped and attuned to the needs of apple breeding and willing to tailor their services to the idiosyncrasies of individual clients are the most desirable. To date, most DNA testing has been conducted by research laboratories closely associated with public breeding programmes, thus effectively subsidized, but a transition to sustainable commercial services is underway. In Europe, the FruitBreedomics project fuelled the establishment of 'MAB Services', three commercial service centres situated in Spain, France and Italy and attached to the expertise of particular research institutions (http://www.fruitbreedomics.com/index.php/the-results/fruitbreedomics-mab-services). In the United States, the RosBREED project developed three temporary 'DNA Testing Portals', one of which supports apple breeding programmes in particular (https://www.rosbreed.org/breeding); however, commercial service providers will be relied upon from late 2017. In New Zealand, Slipstream Automation, a spin-off company of Plant & Food Research Ltd, provides tailored MAB support, including for Plant & Food Research's apple breeding programme and other clients.

4 Current tools in MAB for apple

Genetic marker tools currently used directly for apple breeding applications can be considered to fall within two major categories: characterization and evaluation. Characterization is for identity/relatedness, that is, to ascertain uniqueness, determine identity (confirm trueness-to type), verify parentage and pedigree, reveal ancestry and calculate relatedness. Evaluation is for genetic potential, that is, revealing the alleles carried by individuals at MTLs or QTLs in order to predict performance for traits of interest

– such as for disease resistance/susceptibility, *S*-locus incompatibility genotypes, levels of fruit quality components, as well as genomic predictions based on genome-wide markers covering numerous small-effect alleles. Not considered here are marker tools for dissecting trait genetics, such as in genetic linkage mapping, fine-mapping and candidate gene approaches, because those are for research upstream of breeding application and have been dealt with exhaustively in past reviews (e.g. Peace and Norelli, 2009; Korban and Tartarini, 2009) as well as in Chapter 2 (Mapping and exploiting the apple genome by Dhingra) of this book. Similarly, not considered here are DNA-based tools for understanding genomic structure, the physiological basis of traits, or epigenetics, such as those for DNA sequencing, cytogenetics, gene expression and DNA methylation patterns.

4.1 Characterization

Characterization using genetic markers is currently performed effectively with SSRs that can be assayed on a range of genotyping platforms. Although the International Union for the Protection of New Varieties of Plants, the official convention for ensuring distinctness, uniformity and stability of new plant cultivars, still relies entirely on phenotypic descriptors (e.g. UPOV 2015 for apple rootstocks), DNA marker information is well suited. Uniqueness of new cultivars can be readily established with just a handful of multi-allelic SSRs. Evans et al. (2010a) described the use of a set of seven SSRs primarily to establish the uniqueness, as well as ascertain the parentage, of a new cultivar released commercially. A major effort to determine trueness-to-type of hundreds of apple cultivars and selections used a genome-spanning set of 80 SSRs (Evans et al., 2011) conveniently multiplexed in 21 assays (Patocchi et al., 2009b). The same study successfully used the set of SSRs to ascertain and deduce parentage within the large germplasm set. A subset of just 16 of the SSRs was used to reveal relatedness among almost 2500 accessions of 14 European germplasm collections (Urrestarazu et al., 2016). Because the SSRs were well chosen to be informative, the cumulative probability of identity was extremely low (1.3×10^{-22}, the probability of two individuals with the same genotypic profile across all 16 markers not being identical), enabling detection of hundreds of synonymous individuals within and across collections. Furthermore, previously unknown parentages for more than 30 old cultivars could be deduced (Urrestarazu et al., 2016). Similarly, Salvi et al. (2014) used 27 SSRs to determine identity and parentage of more than 100 apple cultivars. SSR marker information on apple breeding germplasm can be obtained with various marker platforms, helping ensure that any breeding programme has ready access to necessary services. LI-COR and ABI DNA sequencers/analysers (e.g. Patocchi et al., 2009b; Salvi et al., 2014; Urrestarazu et al., 2016) are particularly useful for streamlined multiplexing of SSRs, while polyacrylamide gel electrophoresis (e.g. Evans et al., 2010a) and even low-technical-demand agarose gel electrophoresis are suitable for this marker type. Although DNA extraction is commonly conducted using Qiagen DNeasy kits (e.g. Patocchi et al., 2009b; Urrestarazu et al., 2016), extraction methods resulting in low quantity and/or quality DNA are also readily useable because SSRs are based on a simple PCR assay.

4.2 Evaluation

Marker tools for evaluation currently in use are mostly based on locus-specific PCR assays. Trait-predictive DNA tests available at present for marker-assisted parent selection (MAPS) and MASS primarily use the simple PCR marker classes SSR and

SCAR (Table 1). Such marker types have simple DNA extraction needs, as described above, and enable versatile single-locus screening for cost-efficiency and data-handling efficiency. As often conducted for characterization (e.g. Patocchi et al., 2009b), SSR and SCAR markers for evaluation purposes can be combined into multi-locus assays via multiplexing, for example, several markers are combined into a single PCR reaction for apple fruit texture and several other traits (D. Edge-Garza, pers. comm.). Although some SNP-based marker assays for apple trait loci have been reported (Table 1) and are increasingly in development (Chagné et al., 2016b), they can have limited versatility as locus-specific assays. The bi-allelic nature of SNPs is often associated with a lack of transferability across breeding families when used as single markers to tag a trait locus (S. Gardiner pers. comm.). Unless they lie on the causal mutation of a significant phenotypic difference, the breeding utility of SNPs requires consolidation into multi-marker, multi-allelic haplotypes.

Multi-marker SNP-based assays are now available for characterization and evaluation applications at three levels of genome-spanning resolution. The International RosBREED SNP Consortium apple 8K SNP array v1 (Chagné et al., 2012) using the Illumina platform, contains 7867 SNPs covering the apple genome (more specifically, the apple whole-genome sequence v1 of Velasco et al., 2010), of which more than 5500 readily reveal polymorphism among apple breeding germplasm (Chagné et al., 2012), with half of those (2272) genetically mapped in a rootstock family (Antanaviciute et al., 2012) and a quarter (1091) genetically mapped in three half-sib dessert fruit families (Clark et al., 2014). The 8K array SNPs were arranged into focal points: several SNPs located in clusters, with well-spread clusters across the genome. The Illumina 20K SNP array (Bianco et al., 2014) improved upon the 8K array by including more than 3000 of the most robust SNPs and using a further 17 000 SNPs to refine the focal points by tightening their internal intervals, filling genomic gaps with additional clusters, and saturating each cluster with more SNPs. More than 15 000 of the 20K array SNPs have been genetically mapped (Bianco et al., 2014), and the SNP clusters at each focal point have been converted into 1112 multi-SNP multi-allelic loci (Voorrips et al., in press). An even higher-density SNP array, the Axiom® Apple480K SNP genotyping array, is also publicly available, and contains 275 000 SNPs that were considered 'very robust' (Bianco et al., 2016). These medium- to high-density SNP arrays have enabled investigation of the feasibility of genome-wide selection (also known as genomic selection) as a routine breeding practice for apple (Kumar et al., 2012, 2015; Muranty et al., 2015). Genome-wide selection is further enabled with the development and use of low-density SNP panels and genotype imputation (Muranty et al., 2015).

Note that despite the vaunted advantages of SNPs, SSRs are still in common and effective use for characterization purposes. However, pilot studies using SNP-based KASP™ assays appear promising for apple (Baumgartner et al., 2016), and an overall transition to multi-locus SNP array-based assays for both characterization and evaluation is expected over the next few years (see Section 7 Future Trends and Conclusion).

5 Impacts of MAB on apple breeding

Planting an orchard of a new apple cultivar is a large financial risk to a commercial grower and can be disastrous if the wrong cultivar is used. Similarly, establishing replicated

Table 1 Locus-specific, trait performance-predictive assays ('DNA tests') available for apple

Trait	Locus/loci	Marker type(s)	MTL or QTL	Reference(s)
Biotic resistance				
Scab	Various *Rvi*	SCAR, SSR, SNP	MTLs	Cheng et al., 1998; Tartarini et al., 1999; Patocchi et al., 2009a; Bus et al., 2010; Soriano et al., 2014; Cova et al., 2015; Jänsch et al., 2015; Baumgartner et al., 2016
Powdery mildew	Various *Pl*	SCAR, SNP	MTLs	Evans and James 2003; James et al., 2004; Bus et al., 2010; Jänsch et al., 2015; Baumgartner et al., 2016
Fire blight	Various	SCAR, SSR, SNP	QTLs	Khan et al., 2007; Peil et al., 2007; Le Roux et al., 2010; Gardiner et al., 2012; Jänsch et al., 2015
Aphids	Various *Sd*, *Er*, *Dp-fl*	SCAR, SSR, SNP	MTLs	Cevik and King, 2002; Bus et al., 2008, 2010; Pagliarani et al., 2016; Baumgartner et al., 2016
Alternaria blotch	*Alt*	SSR	MTL	Li et al., 2011
Azoxystrobin fungicide resistance	*Azs*	SSR	MTL	Tobutt et al., 2009
Phenology, tree architecture and productivity				
Self-incompatibility	*SI*	SCAR	MTL	Fernández-Fernández et al., 2008; Larsen et al., 2016
Seedling chlorophyll level	*Vir*, *MdPHYLLO*	SSR, SNP, CAPS	MTL	Fernández-Fernández et al., 2014; Orcheski et al., 2015
Tree habit	*Co*	SCAR, SSR	MTL	Kim et al., 2003; Morimoto and Banno 2015
Dwarfing	*Dw1*	SSR	QTL	Celton et al., 2009; Foster et al., 2015
Fruit quality				
Skin overcolour	*Rf/ MdMYB1*	CAPS, SNP	MTL	Takos et al., 2006; Zhu et al., 2011; Zhang et al., 2014; Yuan et al., 2014; Chagné et al., 2016a
Flesh colour	*Rt/Rni/ MdMYB10*	SSR, SCAR, SNP	MTL	Chagné et al., 2007; Fernández-Fernández et al., 2008; Zhang et al., 2014
Acidity	*Ma*	CAPS, SSR	QTL	Bai et al., 2012; Xu et al., 2012
Firmness	*Md-PG1*	SSR, SNP	QTL	Longhi et al., 2013; Baumgartner et al., 2016
Ethylene production and softening	*ACS1*, *ACO1*, *ACS3a*	SCAR, SNP, CAPS	QTLs	Harada et al., 2000; Zhu and Barritt 2008; Edge-Garza et al., 2010; Baumgartner et al., 2016; Dougherty et al., 2016
Bitter pit susceptibility	*Bp1*	SSR	QTL	Buti et al., 2015

breeding trials of elite selections requires that the exact individuals to be evaluated are included. Mistakes can occur in sourcing of budwood used for grafting, labelling or tree handling during the propagation process. DNA testing can be used to ensure that planted material is true-to-type, eliminating the need to wait until fruiting. Such identity verification should be a routine part of the release and propagation to commercial plant numbers; it has been relatively easy for breeders and propagators to identify a stage for testing without it being onerous or expensive (see Section 6 Case Study). Establishing the 'fingerprint' of a new cultivar is also useful as evidence in Intellectual Property disputes.

Although originally viewed to be less important than MASS, MAPS is proving to be just as important, if not more important. Previously apple breeders have had to rely almost solely on their knowledge of the phenotype of different varieties or selections when choosing which to combine in their crossing programme. With the limitations of the long juvenile phase, making test crosses to determine homozygosity, for example, can consume several years. Very little information is available on combining ability or breeding values for key traits in apple (Hampson and Sholberg, 2008; Kouassi et al., 2009). Use of markers to fully describe potential parents genotypically adds considerably to the information the breeder can use to design crosses. Breeders can often be restricted by budget or simply the logistics of applying markers for seedling selection; however, marker screening of a parent pool is straight forward and relatively inexpensive. In addition, the markers used at this stage can and should include even those that explain only a small part of the trait variation and can be used almost as soon as the locus is identified without the need for development of a high-throughput DNA test, as long as the predictions are robust across target germplasm. Multi-allelic SSRs and multi-SNP haplotypes, used in QTL characterization studies, are effective for this application. Careful interpretation of the parental genotypes leads to more efficient cross design and often reduces the need for MASS. For example, if the breeder was targeting to have at least one favourable allele from the *Md-ACS1* gene for the ethylene pathway enzyme of 1-aminocyclopropane-1-carboxylic acid synthase (Zhu and Barritt, 2008), a cross using 'Fuji' (homozygous for the favourable allele) as one parent should result in all seedlings inheriting a favourable allele.

While MAPS results in an improved cross design so that seedling populations with greater potential for success might be expected, use of MASS enables breeders to cull the seedlings with the least potential for success at the earliest possible stage from those populations (Ru et al., 2015). Ideally, markers that explain a reasonably high proportion of the genetic variance are the most effective for MASS. In apple, as in most crops, markers linked to trait loci with major-effect alleles for disease resistance were targeted first for development and their application is essential in the development of new cultivars with durable, pyramided resistances (Baumgartner et al., 2015).

In early discussions of the benefits of MAB for apple (Arús et al., 1994), one major outcome was thought to be a reduction in the time required to develop a new cultivar. Currently, there has been relatively little impact of MAB on the typical apple breeding timeline; seedlings still need to grow sufficient woody structure to support a crop for evaluation at each round of selection. However, there are ways in which use of markers might nevertheless result in faster development of new cultivars in a single generation. First, as routine MASS within a breeding programme results in a considerable cost saving (Edge-Garza et al., 2015), it also enables an increase in the initial number of seeds sown – as breeding success is directly associated with the number of seedlings evaluated, this resource reallocation should improve the overall time taken to develop a new cultivar. Second, reducing the breeding timeline by abbreviating the juvenility period with environmental control is feasible when seedling numbers have been reduced with genetic marker screening (van Nocker and Gardiner, 2014).

MAB can be particularly effective to speed cultivar development when multiple generations are required for introgressing attributes from related wild species into domestic apple, especially pest and disease resistance. The multiple generations required to produce an apple with good fruit quality attributes as well as the desired introgressed attribute (Janick et al., 2006) become more achievable when breeders can efficiently select in young seedlings for multiple traits and eliminate undesirable wild donor genome segments. DNA tests have already been used to successfully select seedlings with more durable pyramided resistances (Bus et al., 2008; Kellerhals et al., 2011; Baumgartner et al., 2015), and when combined with new breeding technologies such as FasTrack and early-flowering breeding (Flachowsky et al., 2012; van Nocker and Gardiner, 2014), they will enable faster introgression of multiple resistances to diseases such as fire blight, scab and powdery mildew (Flachowsky et al., 2011) as well as resistance to storage moulds (Norelli et al., 2014).

MASS, combined with MAPS, increases breeding efficiency, reduces costs and increases the quality of the seedling orchards; however, its routine implementation requires alteration to breeding programme operational logistics. Breeders have always needed to be extremely accurate with labelling and record-keeping, but DNA tests have revealed many cases of incorrect parentage. Although these can be frustrating when concerning a released cultivar (and can lead to multiple other breeders mistakenly using an incorrect parent), parentage mistakes can be costly when using MASS, as in some cases none of the seedlings will have the desired allelic combinations. Local sources of parental germplasm should be tested for trueness-to-type rather than breeders relying on published information regarding genotype, as there are numerous discrepancies between cultivars with the same name around the world (Evans et al., 2011). Use of automation in the molecular genetics lab has reduced errors in obtaining data when screening thousands of seedlings; however, breeders still need to consider optimizing seedling sampling for their specific conditions (see Section 6 Case Study).

6 Case study: MAB in the Washington State University apple breeding program

Adoption of MAB in the Washington State University (WSU) apple breeding program (WABP) has taken several years and continues to evolve as technology advances. The WABP, based in at the WSU Tree Fruit Research and Extension Center, Wenatchee, Washington, USA, works in collaboration with the WSU Tree Fruit Genotyping Laboratory (WTFGL), based at WSU Pullman, WA, for development and application of MAB. DNA testing services provided by the WTFGL for screening elite material and seedlings are organized in a standardized 8-by-12-well plate format. Two control wells per plate, for positive controls (such as true-to-type 'mother' trees or parents) or negative controls (left blank), provide quality assurance and orient sampling and culling operations (Fig. 2). The WTFGL has established various methods such as these for efficiently managing the high-throughput DNA testing process (Edge-Garza et al., 2012).

Figure 2 (a) Three seedling trays labelled for two DNA collection plates; (b) custom modular trays; (c) collection tubes containing silica beads with sampled leaf tissue.

6.1 Characterization

Verifying the identity of elite selections entering replicated trials and new cultivar releases is now a routine operation in the WABP. The policy for new cultivar release from WSU is to distribute virus-tested budwood to nurseries for propagation of mother trees for further commercial mass propagation of grafted trees. Leaf samples are routinely collected from all mother trees that are established from buds released from the Clean Plant Center Northwest (Prosser, WA) and are fingerprinted by the WTFGL with a battery of DNA tests to ensure that no mix-ups have occurred and provide a level of confidence in the mass-propagated germplasm that was previously impossible to obtain.

Parentage of elite WABP selections is routinely determined to verify or update breeding records. This process is often a by-product of evaluating selections for their functional genotypes at trait loci of interest. Such parentage-checking occasionally reveals incorrect paternity (Peace, 2015), which has influenced considerations of breeding value of parents. In situations where open pollination was purposely used at the time of crossing, DNA testing has often been used to deduce the paternal parent – such as for the WABP's first release, 'WA 2' (Evans et al., 2010a). Routine MASS in the WABP also readily detects seedlings with unintended parentage. Proportions of such seedlings below about 5% has not been cause for concern; however, when about 15% was detected as the average degree of unintended parentage several years ago, and up to 70% for some families, crossing operations were adjusted to reduce the opportunity for error.

6.2 Evaluation

MAPS was first applied in 2008 following validation of the genotypic effects on WABP germplasm of the *Md-ACS1* and *Md-ACO1* genes for the ethylene pathway enzymes of 1-aminocyclopropane-1-carboxylic acid synthase and oxidase (Zhu and Barritt, 2008). Crosses were designed to avoid the worst combination of the alleles. As new DNA tests were developed, all were tested routinely on the WABP parent set and allelic combinations considered when designing crosses.

Deploying DNA tests for MAPS is relatively straightforward, because the number of samples is limited and the parent trees are typically established in the orchard and well labelled. However, initial deployment of MASS, which began in 2010, proved logistically challenging (Evans et al., 2012b) as there was a scarcity of information regarding how to manage sample collection, format the resulting genotypic data and cull seedlings – while ensuring accuracy at every stage. Although breeders strive to ensure every data point in a breeding programme is accurate, as culling decisions after MASS are terminal, reducing error and improving data confidence were the main considerations when devising a deployment strategy for MASS in the WABP.

According to cost-efficiency analysis specific to the WABP (Edge-Garza et al., 2010) using a calculator developed for tree fruit breeding (Edge-Garza et al., 2015), the most cost-efficient phase of the programme to deploy MASS was in the nursery in the eight weeks between phenotypically culling seedlings with powdery mildew symptoms and budding selected seedlings onto rootstocks (Rowland et al. 2010). The silica bead method of DNA extraction was adapted for the WABP particularly because of the ease of tissue sampling within the breeding programme (Edge-Garza et al., 2014). Leaf samples were collected into deep-well 8-strip microtubes (Fig. 2), each tube containing dried silica beads, in a 96-tube microrack ('plate') and the corresponding seedlings were labelled with both cross number and plate position (e.g. A1); a written description of samples in each plate ('plate map') was completed for further assurance. Filled sample plates were transported to the WTFGL where DNA was extracted (Edge-Garza et al., 2014) and used for genetic tests. Although the nursery was the most cost-efficient stage (Edge-Garza et al., 2015), opportunities for mistakes in sampling, labelling and ultimately matching DNA test results to specific seedlings were considerable. Confidence in data-matching individual seedlings was so greatly compromised that alternative stages of MASS deployment were explored.

MASS on greenhouse-grown seedlings after powdery mildew screening and prior to shipment to the nursery was identified as the second most cost-efficient stage for MASS deployment (Edge-Garza et al., 2010, 2015). Greenhouse seedlings were more accessible and better labelled, simplifying tissue collection and culling, which reduced opportunities for mistakes (thus increasing data confidence) and was only slightly less cost-efficient than nursery screening.

A seedling planting system in a 96 (8-by-12) configuration was developed to match the 96-well collection plates. Apple seedlings grow a long tap root that needs room to extend or the health of resulting plants can be compromised. Custom, modular seedling trays were constructed from pre-owned, commercially available, flat propagation trays fitted with a custom-made wire grid attached to the flats to form a supportive frame (Fig. 2) and provide a 96-plant format. Dimensions of the pre-owned trays meant that three trays were needed for two 96-well plates (Fig. 2). Polyvinyl chloride (PVC) tubing was used to make the individual seedling 'cells' which are supported at the base by the flat and at the top by the wire grid. PVC tubes are filled with compost and tamped down with a wooden

	1	2	3	4	5	6	7	8	9	10	11	12
A	1409-.A01 CULL	1409-.A02 CULL	1409-.A03 KEEP	1409-.A04 CULL	1409-.A05 KEEP	1409-.A06 CULL	1409-.A07 CULL	1409-.A08 KEEP	1409-.A09 CULL	1409-.A10 CULL	1409-.A11 KEEP	1409-.A12 KEEP
B	1409-.B01 KEEP	1409-.B02 CULL	1409-.B03 KEEP	1409-.B04 CULL	1409-.B05 CULL	1409-.B06 CULL	1409-.B07 CULL	1409-.B08 CULL	1409-.B09 KEEP	1409-.B10 KEEP	1409-.B11 KEEP	1409-.B12 KEEP
C	1409-.C01 CULL	1409-.C02 KEEP	1409-.C03 KEEP	1409-.C04 CULL	1409-.C05 CULL	1409-.C06 CULL	1409-.C07 KEEP	1409-.C08 CULL	1409-.C09 CULL	1409-.C10 KEEP	1409-.C11 CULL	1409-.C12 CULL
D	1409-.D01 KEEP	1409-.D02 CULL	1409-.D03 CULL	1409-.D04 CULL	1409-.D05 CULL	1409-.D06 KEEP	1409-.D07 CULL	1409-.D08 CULL	1409-.D09 KEEP	1409-.D10 KEEP	1409-.D11 KEEP	1409-.D12 CULL
E	1409-.E01 KEEP	1409-.E02 CULL	1409-.E03 CULL	1409-.E04 KEEP			1409-.E07 KEEP	1409-.E08 CULL	1409-.E09 CULL	1409-.E10 CULL	1409-.E11 CULL	1409-.E12 KEEP
F	1409-.F01 KEEP	1409-.F02 KEEP	1409-.F03 KEEP	1409-.F04 CULL	1409-.F05 CULL	1409-.F06 CULL	1409-.F07 CULL	1409-.F08 KEEP	1409-.F09 CULL	1409-.F10 CULL	1409-.F11 KEEP	1409-.F12 KEEP
G	1409-.G01 CULL	1409-.G02 KEEP	1409-.G03 CULL	1409-.G04 CULL	1409-.G05 KEEP	1409-.G06 CULL	1409-.G07 CULL	1409-.G08 CULL	1409-.G09 CULL	1409-.G10 KEEP	1409-.G11 KEEP	1409-.G12 KEEP
H	1409-.H01 CULL	1409-.H02 CULL	1409-.H03 CULL	1409-.H04 CULL	1409-.H05 KEEP	1409-.H06 CULL	1409-.H07 CULL	1409-.H08 KEEP	1409-.H09 KEEP	1409-.H10 KEEP	1409-.H11 KEEP	1409-.H12 KEEP

Figure 3 Output of DNA testing results for 94 seedlings, with two blanks as negative controls, provided in an 8-by-12 plate map format (Microsoft Excel worksheet).

dowel of slightly smaller diameter which is also used to push out the potting medium and seedling when ready (i.e. for culling or planting in nursery).

The flexibility of being able to remove individual seedlings in their PVC tubes provides multiple benefits: ease of removing seedlings (due to non-germination, showing damping-off symptoms, expressing recessive lethal alleles etc.), consolidating trays to capacity and sorting seedlings within each family prior to DNA collection to standardize by growth rate; efficient culling; and ease of consolidating seedlings after culling prior to shipping to the nursery. Complicated labelling (e.g. bar coding) is unnecessary as opposite corners of the 96-plant system are labelled with the cross number and plate number, orienting the sampling and culling of seedlings to the 96-well plate. Tissue sample collection begins as soon as enough seedlings are at the desired growth stage and is much less labour-intensive than in the nursery as seedlings are closer together and more ergonomically positioned.

Data is generated from the WTFGL in the form of a spreadsheet that enables the breeder to make a priori culling decisions based on allele combinations. The spreadsheet outputs a plate map displaying the Keep/Cull decision for each seedling (Fig. 3) that is printed and readily used to determine which seedlings in the 96-plant system need to be removed.

Currently in the WABP, DNA tests for MASS focus on fruit quality in alignment with breeding objectives. In addition to the *Md-ACS1* and *Md-ACO1* genes, the trait loci *Ma* and *LG8A* (Verma, 2014) for acidity and texture, and *Md-PG1* (Longhi et al., 2013) for firmness and storability are in regular use for screening. Additional genetic tests for trait loci such as *LG1Fru* (Guan et al., 2015; Sandefur, 2016) for fructose content will be used in the 2017 season and beyond. Typically, as with *Md-ACS1* and *Md-ACO1*, selection focuses on culling individual seedlings with the worst combination of alleles, rather than selecting those deemed to have the best alleles. However, the focus on culling or keeping is influenced by robust information on the heritability of the trait and the predictiveness of the DNA test (Ru et al., 2016). In total, more than 45 000 WABP seedlings have been screened for fruit quality traits using MASS since 2010. Of those, approximately 44% were retained for propagation and further testing.

7 Future trends and conclusion

Several imminent tools and technological advances are poised to strongly influence how apple MAB is conducted in the next few years. Fundamentally, improvements to the publicly available apple whole-genome sequence expected in early 2017 and led by researchers at Fondazione Edmund Mach (San Michele all'Adige, Italy) will ensure that existing and future genome-wide tools for breeding truly consider the entire genome and that laboratory assays for trait-influencing alleles always rely on correct genomic positioning. Decreasing prices, greater streamlining of sample handling and improved data management capabilities will render genotypic information on breeding germplasm readily accessible to any breeding programme.

The current set of trait-predictive DNA tests used by any given breeding programme (such as in the case study above) is a mere handful, and the outcome of each test can usually be considered as an independent selection criterion. However, a search for reported QTLs and MTLs on the Genome Database for Rosaceae (Jung et al., 2014) in mid-2016 returns more than 1000 hits. With 17 chromosomes totalling approximately 1200–1400 cM in genetic length (Silfverberg-Dilworth et al., 2006; Velasco et al., 2010; Antanaviciute et al., 2012; Ben Sadok et al., 2015), each trait locus discovered must be closely linked with numerous others. The trajectory of increasingly known linkage relationships among an increasing number of reported trait loci can only be set to continue in the future. As more trait-predictive DNA tests targeting loci of interest are developed, accounting for the network of linkage relationships will become unavoidable in making breeding decisions.

For predicting performance based on individually significant-effect trait loci, multi-marker arrays that simultaneously genotype dozens or more loci are not expected to completely replace marker types in current common use such as SSRs, SCARs and single-locus SNPs, at least not in the near future. However, as trait-predictive assays of various marker types as well as newly discovered QTLs are being increasingly converted to SNP-based assays, the opportunity arises to take advantage of the relative simplicity of SNP marker scoring (Baumgartner et al., 2016) and combine SNP assays for many trait loci into single multi-SNP assays. The main challenges to date for widespread application of multi-SNP assays, besides the availability of trait-predictive markers, have been DNA extraction and cost. SNP assays require relatively high-quality DNA, which hampers the ability of breeding programmes to collect tissue quickly for large sample numbers; however, resolution of this technical issue is imminent (M. Cook, Slipstream Automation, New Zealand, and C. Saski of Clemson University Genomics Institute, pers. comm.). A multi-marker assay means that all markers must be paid for, even if the breeder expects genotypes at only one or a few loci to indicate genetic inferiority for half or more of the seedlings. However, the costs of multi-SNP assays are beginning to fall into the range, and soon below, costs of genotyping several SSRs/SCARs. Therefore, in the next year or two, breeders are expected to have access to cost-effective multi-SNP assays for trait performance predictions; such assays will provide information at more loci than are needed for MASS and therefore bioinformatics solutions will be needed to relieve breeding programmes of data overload. For evaluation of prospective or current parents and elite selections or commercially released cultivars, the issues of tissue sampling and cost are almost non-existent, because of the much higher ratio of benefit to cost per plant than for seedlings.

Genome-wide selection, or genomic selection, appears feasible for widespread application in apple breeding (Kumar et al., 2012, 2015; Muranty et al., 2015), specifically for genetic improvement of those traits for which more than three significant QTLs are segregating in a programme's germplasm (V. Whitaker, University of Florida, and R. Bernardo, University of Minnesota, pers. comm.). Where locus-specific DNA tests predict whether seedlings in a family under selection are likely to perform beyond a critical trait threshold, such assays are expected to be used widely to efficiently reduce family sizes, together with any simple and early phenotypic assays, prior to conducting genome scans and subsequent genomic selection. Researchers at Plant & Food Research Ltd, New Zealand, have already successfully performed this two-step approach (S. Gardiner, D. Chagné and S. Kumar, pers. comm.).

Genome scans currently available for apple should continue to perform their roles effectively over the next several years. Although whole-genome DNA sequencing is becoming affordable for key parents and cultivars, such information will remain in research use for the next few years only as it would be purpose overkill and data overload for breeding. The 8K and 20K apple arrays (Chagné et al., 2012; Bianco et al., 2014) efficiently reveal recombination patterns in derived breeding germplasm, and will continue to enable breeders to understand patterns of inheritance resulting from their crossing and selection efforts – a key breeding need. Targeting and tracing inherited genomic segments, directly in cultivar development as well as in pre-breeding, is about to become commonplace. The jury is still out on whether the current medium-density arrays or to-be-developed lower-resolution SNP arrays (such as the 384-SNP array; Muranty et al., 2015) that provide sufficient data for specific purposes and germplasm will become the norm; the primary factors will be cost and data handling ease. For breeding application opportunities at higher genome resolution, the 480K array (Bianco et al., 2016) is at the equivalent genomic resolution of the Illumina OmniExpress human SNP array (Illumina Inc., San Diego, CA) that is in common use for a range of applications in human research and direct-to-consumer testing. While the 480K apple array has enormous potential value for revealing new parent–offspring relationships, and, upstream to breeding, discovering new alleles by genome-wide association studies (Bianco et al., 2016), further fascinating applications beckon in evaluating breeding germplasm, inspired by advances in human genetics.

More accessible and cost-effective tools will expand the spectrum of breeding operations that use increasingly sophisticated DNA information within and across breeding programmes. With DNA-based information available in real time for wild accessions and pre-breeding families, hundreds to thousands of potential parents, seedlings and elite selections, and new cultivar releases, opportunities to improve breeding effectiveness will be devised by creative breeders. For example, it is now considered possible to shave years off the typical breeding timeline, if the breeder's confidence in the genetic potential of selected seedlings is sufficiently increased by the use of numerous reliable DNA tests. Bassett et al. (2015) reported that Plant & Food Research's rootstock breeding programme is revolutionized with the use of DNA tests for MTLs for dwarfing and pest and disease resistance that enables bypassing of costly and time-consuming single-tree evaluation (the equivalent of skipping Phase 1 of Fig. 1). Another scenario recently reported from the same institute (Kumar, 2016) has such a large proportion of seedlings from a population being culled from MAB followed by genomic selection that the breeder moves the few retained seedlings directly into replicated trials, also skipping single-tree trials. Genome-wide knowledge on prospective parents, including full evaluation of their genotypes at all significant trait loci, will not only streamline MASS planning and operations, but is also expected to reveal enticing new cross combinations.

In view of the rapid increase in the availability of genomic data, the lack of high-quality phenotypic data is quickly becoming the limiting factor in the development of robust predictive DNA tests. Breeder collaborations with pre- and post-harvest physiologists, pathologists and biological systems engineers are resulting in the development and application of high-resolution and even remote sensing technology or phenomics, to provide the big datasets necessary to dissect some of the phenotypic evaluation into constituent parts (Desnoues et al., 2016; Allard et al., 2016; Sankaran et al., 2016), increasing the likelihood of identifying associated DNA markers. As this technology advances, it will become more important for breeders to expand these collaborations to ensure accurate interpretation of the data as well as to incorporate bioinformatics expertise into their programmes.

Advances described above will be facilitated by improvements in managing increasingly complex breeding information for effective decision-making. In recognition of this need, the Genome Database for Rosaceae (GDR), originally established to serve as a publicly accessible repository for genomics data in the Rosaceae family (Jung et al., 2004), has expanded to include genetics (Jung et al., 2008) and breeding data and tools (Jung et al., 2014). A breeding database with associated analytical tools and programme management modules was developed to facilitate the integration of breeding with molecular genetics and genomics information (Evans et al., 2013), and further much-needed breeding information management tools continue to roll out (Jung et al., 2016). While access to such DNA information will add complexity to breeding decisions, effective use of such information will mean greater breeding impact.

The future of MAB for apple is 'DNA-informed breeding' – routine use of DNA information across the genome and across the spectrum of breeding operations. While MAB implies the use of genetic markers for individually significant trait loci, that is, MTLs and QTLs, and is often considered synonymous with the specific applications of MAS, that is, MAPS and MASS, the broader term of DNA-informed breeding encompasses emerging breeding applications such as genome-wide selection and hopefully wrests back the role of actual breeding to the breeder. In this context, DNA information is just additional knowledge available for breeders' consideration, and how it is used will be up to each breeder's creativity, available germplasm and other physical resources. In any case, an increasing proportion of apple breeding programmes are routinely using DNA information in an increasing number of ways, concomitantly enhancing the efficiency, accuracy and pace of new apple cultivar development worldwide for sustainable production.

8 Where to look for further information

The following references are good general introductions to apple breeding, the fundamental principles of which change little over time: Brown, 1975; Janick et al., 1996; Brown, 2012. A general introduction to the genomics of key apple traits is Korban and Tartarini (2009). The Rosaceae International Genomics Initiative research community meets every two years at the International Rosaceae Genomics Conference; information about this group and other Rosaceae community conferences can be found on the community webpage of the GDR (www.rosaceae.org). The European Association for Research on Plant Breeding (Eucarpia) Fruit section meets every four years and publishes proceedings as Acta Horticulturae (www.eucarpia.org). The authors highly recommend readers access the GDR (www.rosaceae.org; Jung et al., 2014); it is a community research hub beyond its genomics capacity, housing a searchable publication database and links to community projects such as RosBREED (www.rosbreed.org) and FruitBreedomics (www.fruitbreedomics.com). Key participants in the development and application of MAB in apple include the New Zealand Institute for Plant & Food Research Ltd (Palmerston North and Havelock North, New Zealand), WSU (Pullman and Wenatchee, USA), Institute de Recherche en Horticulture et Semences, INRA (Angers, France), Fondazione Edmund Mach (San Michele all'Adige, Italy) and Agroscope (Wädenswil, Switzerland).

9 Acknowledgements

The authors thank Drs Sue Gardiner and Jim Luby for their thorough and thoughtful review of this chapter. This work was partially funded by the Washington Tree Fruit Research Commission, USDA's National Institute of Food and Agriculture–Specialty Crop Research Initiative projects, 'RosBREED: Enabling Marker-Assisted Breeding in Rosaceae' (2009-51181-05808) and 'RosBREED: Combining Disease Resistance and Horticultural Quality in New Rosaceous Cultivars' (2014-51181-22378), and USDA Hatch funds provided to the Department of Horticulture, WSU.

10 References

Allard, A., Laurens, F. and Costes, F. (2016). Character integration, breeding goals compatibility and selection indexes using genome wide breeding values: a study case. Program/Abstracts S606, *8th International Rosaceae Genomics Conference*, 21–24 June, Angers, France.

Antanaviciute, L., Fernández-Fernández, F., Jansen, J., Banchi, E., Evans, K. M., Viola, R., Velasco, R., Dunwell, J. M., Troggio, M. and Sargent, D. J. (2012). Development of a dense SNP-based linkage map of an apple rootstock progeny using the *Malus* Infinium whole genome genotyping array. *BMC Genomics* 13:203.

Arús, P., Messeguer, R., Viruel, M., Tobutt, K., Dirlewanger, E., Santi, F., Quarta, R. and Ritter, E. (1994). The European Prunus mapping project. In *Progress in Temperate Fruit Breeding*, 305–8. Kluwer Academic Press, the Netherlands.

Bai, Y., Dougherty, L., Li, M., Fazio, G., Cheng, L. and Xu, K. (2012). A natural mutation-led truncation in one of the two aluminum-activated malate transporter-like genes at the *Ma* locus is associated with low fruit acidity in apple. *Molecular Genetics & Genomes* 287:663–78.

Bassett, H., Malone, M., Ward, S., Foster, T., Chagné, D. and Bus, V. (2015). Marker assisted selection in an apple rootsock breeding family. *Acta Horticulturae* 1100:25–8.

Baumgartner, I. O., Kellerhals, M., Costa, F., Dondini, L., Pagliarani, G., Gregori, R., Tartarini, S., Leumann, L., Laurens, F. and Patocchi, A. (2016). Development of SNP-based assays for disease resistance and fruit quality traits in apple (*Malus* × *domestica* Borkh.) and vaildation in breeding pilot studies. *Tree Genetics & Genomes* 12:35.

Baumgartner, I. O., Patocchi, A., Frey, J., Peil, A. and Kellerhals, M. (2015). Breeding elite lines of apple carrying pyramided homozygous resistance genes against apple scab and resistance against powdery mildew and fire blight. *Plant Molecular Biology Reporter* 33: 573–1583.

Ben Sadok, I., Tiecher, A., Galvez-Lopez, D., Lahaye, M., Lasserre-Zuber, P., Bruneau, M., Hanteville, S., Robic, R., Cournol, R. and Laurens, F. (2015). Apple fruit texture QTLs: year and cold storage effects on sensory and instrumental traits. *Tree Genetics & Genomes* 11:119.

Bianco, L., Cestaro, A., Linsmith, G., Muranty, H., Denancé, C., Théron, A., Poncet, C., Micheletti, D., Kerschbamer, E., Di Pierro, E. A., Larger, S., Pindo, M., van de Weg, E., Davassi, A., Laurens, F., Velasco, R., Durel, C.-E. and Troggio, M. (2016). Development and validation of the Axiom®Apple 480K SNP genotyping array. *Plant Journal* 86:62–74.

Bianco, L., Cestaro, A., Sargent, D. J., Banchi, E., Derdak, S., Di Guardo, M., Salvi, S., Jansen, J., Viola, R., Gut, I., Laurens, F., Chagné, D., Velasco, R., van de Weg, E. and Troggio, M. (2014). Development and validation of a 20K single nucleotide polymorphism (SNP) whole genome genotyping array for apple (*Malus* × *domestica* Borkh.). *PLoS ONE* 9:e110377.

Bink, M. C. A. M., Boer, M. P., ter Braak, C. J. F., Jansen, J., Voorrips, R. E. and van de Weg, W. E. (2008). Bayesian analysis of complex traits in pedigreed plant populations. *Euphytica* 161:85–96.

Bink, M. C. A. M., Jansen, J., Madduri, M., Voorrips, R. E., Durel, C. -E., Kouassi, A. B., Laurens, F., Mathis, F., Gessler, C., Gobbin, D., Rezzonico, F., Patocchi, A., Kellerhals, M., Boudichevskaia, A., Dunemann, F., Peil, A., Nowicka, A., Lata, B., Stankiewicz-Kosyl, M., Jeziorek, K., Pitera, E., Soska, A., Tomala, K., Evans, K.M., Fernández, F., Guerra, W., Korbin, M., Keller, S., Lewandowski, M., Plocharski, W., Rutkowski, K., Zurawicz, E., Costa, F., Sansavini, S., Tartarini, S., Komjanc, M., Mott, D., Antofie, A., Lateur, M., Rondia, A., Gianfranceschi, L. and van de Weg, W. E. (2014). Bayesian QTL analyses using pedigreed families of an outcrossing species, with application to fruit firmness in apple. *Theoretical & Applied Genetics* 127:1073–90.

Boyer, J. and Liu, R. H. (2004). Apple phytochemicals and their health benefits. *Nutrition Journal* 3:5.

Brown, A. G. (1975). Apples. In Janick J.and Moore J. N. (eds), *Advances in Fruit Breeding*. Purdue University Press, West Lafayette, pp.3–37.

Brown, S. (2012). Apple. In Badenes M. L. and Byrne D. H. (eds), *Fruit Breeding*. New York, USA: Handbook of Plant Breeding: Springer US, pp.329–67.

Bus, V. G.M., Bassett, H. C.M., Bowatte, D., Chagné, D., Ranatunga, C. A., Ulluwishewa, D., Wiedow, C. and Gardiner, S. E. (2010). Genome mapping of an apple scab, a powdery mildew and a woolly apple aphid resistance gene from open-pollinated Mildew Immune Selection. *Tree Genetics & Genomes* 6:477–87.

Bus, V. G.M., Chagné, D., Bassett, H. C. M., Bowatte, D., Calenge, F., Celton, J. -M., Durel, C. -E., Malone, M. T., Patocchi, A., Ranatunga, A. C., Rikkerink, E. H. A., Tustin, D. S., Zhou, J. and Gardiner, S. E. (2008). Genome mapping of three major resistance genes to woolly apple aphid (*Eriosoma lanigerum* Hausm.). *Tree Genetics & Genomes* 4:223–36.

Bus, V., Gardiner, S., Bassett, H., Ranarunga, C. and Rikkerink, E. (2000). Marker assisted selection for pest and disease resistance in the New Zealand apple breeding programme. *Acta Horticulturae* 538:541–7.

Buti, M., Poles, L., Caset, D., Magnago, P., Fernández-Fernández, F., Colgan, R. J., Velasco, R. and Sargent, D. J. (2015). Identification and validation of a QTL influencing bitter pit symptoms in apple (*Malus* × *domestica*). *Molecular Breeding* 35:258.

Celton, J. -M., Tustin, D. S., Chagné, D. and Gardiner, S. E. (2009). Construction of a dense genetic linkage map for apple rootstocks using SSRs developed from *Malus* ESTs and *Pyrus* genomic sequences. *Tree Genetics & Genomes* 5:93–107.

Cevik, V. and King, G. J. (2002). High-resolution genetic analysis of the Sd-1 aphid resistance locus in *Malus* spp. *Theoretical & Applied Genetics* 105:346–54.

Chagné, D., Carlisle, C. M., Blond, C., Volz, R., Whitworth, C. J., Oraguzie, N. C., Crowhurst, R. N., Allan, A. C., Espley, R. V., Hellens, R. P. and Gardiner, S. E. (2007). Mapping a candidate gene (*MdMYB10*) for red flesh and foliage colour in apple. *BMC Genomics* 8:212.

Chagné, D., Crowhurst, R. N., Troggio, M., Davey, M. W., Gilmore, B., Lawley, C., Vanderzande, S., Hellens, R. P., Kumar, S., Cestaro, A., Velasco, R., Main, D., Rees, J., Iezzoni, A., Mockler, T.,

Wilhelm, L., van de Weg, E., Gardiner, S. E., Bassil, N. and Peace, C. (2012). Genome-wide SNP detection, validation, and development of an 8K SNP array for apple. *PLoS ONE* 7:e31745.

Chagné, D., Kirk, C., How, N., Whitworth, C., Fontic, C., Reig, G., Sawyer, G., Rouse, S., Poles, L., Gardiner, S. E., Kumar, S., Espley, R., Volz, R. K., Troggio, M. and Iglesias, I. (2016a). A functional genetic marker for apple red skin coloration across different environments. *Tree Genetics & Genomes* 12:67.

Chagné, D., Kirk, C., Vanderzande, S., Edge-Garza, D., Roniotis, J., Patocchi, A., Costa, F., Volz, R., Gardiner, S., Norelli, J. L., Kumar, S., Troggio, M., Poles, L., Sargent, D., Padmarasu, S., Velasco, R., Peace, C. P. and Bassil, N. (2016b). Marker validation for MAS in apple. Program/Abstracts S603, *8th International Rosaceae Genomics Conference*, 21–24 June, Angers, France.

Cheng, F. S., Weeden, N. F., Brown, S. K., Aldwinckle, H. S., Gardiner, S. E. and Bus, V. G. (1998). Development of a DNA marker for *Vm*, a gene conferring resistance to apple scab. *Genome* 41:208–14.

Chevreau, E. and Laurens, F. (1987). The pattern of inheritance in apple (*Malus* x *domestica* Borkh.): further results from leaf isozyme analysis. *Theoretical & Applied Genetics* 75:90–5.

Clark, M. D., Schmitz, C. A., Rosyara, U. R., Luby, J. J. and Bradeen, J. M. (2014). A consensus 'Honeycrisp' apple (*Malus* × *domestica*) genetic linkage map from three full-sib progeny populations. *Tree Genetics & Genomes* 10:627–39.

Conner, P. J., Brown, S. K. and Weeden, N. F. (1998). Molecular-marker analysis of quantitative traits for growth and development in juvenile apple trees. *Theoretical & Applied Genetics* 96:1027–35.

Cova, V., Bandara, N. L., Liang, W., Tartarini, S., Patocchi, A., Troggio, M., Velasco, R. and Komjanc, M. (2015). Fine mapping of the *Rvi5* (*Vm*) apple scab resistance locus in the 'Murray' apple genotype. *Molecular Breeding* 35:200.

Desnoues, E., Baldazzi, V., Genard, M., Lambert, P., Confolent, C., Heurtevin, L. and Quilot-Turion, B. (2016). From gene to phenotype: genetic control of modeling of sugar metabolism during peach fruit development. Program/Abstracts S403, *8th International Rosaceae Genomics Conference*, 21–24 June, Angers, France.

Dougherty, L., Zhu, Y. and Xu, K. (2016). Assessing the allelotypic effect of two aminocyclopropane carboxylic acid synthase-encoding genes *MdACS1* and *MdACS3a* on fruit ethylene production and softening in *Malus*. *Horticulture Research* 3:16024.

Dunemann, F. (1993). Molecular classification of *Malus* with RAPD markers. In Schmidt H. and Kellerhals M. (eds), *Progress In Temperate Fruit Breeding*. Proceedings of the Eucarpia Fruit Breeding Section Meeting. Institute for Ornamental Plant Breeding, Federal Centre for Breeding Research on Cultivated Plants, Ahrensburg, Germany.

Edge-Garza, D. A., Luby, J. and Peace, C. (2015). Decision support for cost-efficient and logistically feasible marker-assisted seedling selection in fruit breeding. *Molecular Breeding* 35:223.

Edge-Garza, D. A., Rowland, T. V. Jr, Haendiges, S. and Peace, C. (2014). A high-throughput and cost-efficient DNA extraction protocol for the tree fruit crops of apple, sweet cherry, and peach relying on silica beads during tissue sampling. *Molecular Breeding* 34:2225–8.

Edge-Garza, D., Rowland, T. Jr., Konishi, B., Brutcher, L., Evans, K. and Peace, C. (2012). Quality assurance improves resource savings in routine marker-assisted seedling selection for the Washington apple breeding program. 18 January 2016. http://rgc6.org/Book-of-abstracts. (abstr.), p. 89.

Edge-Garza, D., Zhu, Y. and Peace, C. (2010). Enabling marker-assisted seedling selection in the Washington Apple Breeding Program. Acta Horticulturae 859:369–73.

Evans, K. M., Barritt, B. H., Konishi, B. S., Dilley, M. A., Brutcher, L. J. and Peace, C. P. (2010a). 'WA 2' apple. *HortScience* 45:668–9.

Evans, K. M., Brutcher, L. J., Konishi, B. S. and Barritt, B. H. (2010b). Correlation of sensory analysis with physical textural data from a computerized penetrometer in the Washington State University apple breeding program. *HortTechnology* 20:1026–9.

Evans, K., Guan, Y., Luby, J., Clark, M., Schmitz, C., Brown, S., Orcheski, B., Peace, C., van de Weg, E. and Iezzoni, A. (2012a). Large-scale standardized phenotyping of apple in RosBREED. *Acta Horticulturae* 945:233–8.

Evans, K. M. and James, C. M. (2003). Identification of SCAR markers linked to *Pl-w* mildew resistance in apple. *Theoretical & Applied Genetics* 106:1178–83.

Evans, K., Jung, S., Lee, T., Brutcher, L., Cho, I., Peace, C. and Main, D. (2013). Addition of a breeding database in the Genome Database for Rosaceae. Database 2013:bat078.

Evans, K., Konishi, B., Brutcher, L., Edge-Garza, D., Rowland, T. and Peace, C. (2012b). The logistical challenges of marker-assisted seedling selection in an apple breeding program. Accessed 18 January 2016. http://rgc6.org/Book-of-abstracts. (abstr.), p. 92.

Evans, K. M., Patocchi, A., Rezzonico, F., Mathis, F., Durel, C., Fernández-Fernández, F., Boudichevskaia, A., Dunemann, F., Stankiewicz-Kosyl, M., Gianfranceschi, L., Komjanc, M., Lateur, M., Madduri, M., Noordijk, Y. and van de Weg, W. E (2011). Genotyping of pedigreed apple breeding material with a genome-covering set of SSRs: trueness-to-type of cultivars and their parentages. *Molecular Breeding* 28:535–47.

Fernández-Fernández, F., Evans, K. M., Clarke, J. B., Govan, C. L., James, C. M., Marič, S. and Tobutt, K. R. (2008). Development of an STS map of an interspecific progeny of *Malus*. *Tree Genetics & Genomes* 4:469–79.

Fernández- Fernández, F. F., Padmarasu, S., Sŭrbanovski, N., Evans, K. M., Tobutt, K. R. and Sargent, D. J. (2014). Characterisation of the virescent locus controlling a recessive phenotype in apple rootstocks (*Malus pumila* Mill.) *Molecular Breeding* 33:373–83.

Flachowsky, H., Le Roux, P., Peil, A., Patocchi, A., Richter, K. and Hanke, M. (2011). Application of a high-speed breeding technology to apple (Malus × domestica) based on transgenic early flowering plants and marker-assisted selection. *New Phytologist* 192:364–77.

Flachowsky, H., Szankowski, I., Waidmann, S., Peil, A. Tränkner, C. and Hanke, M. (2012). The *MdTFL1* gene of apple (*Malus* × *domestica* Borkh.) reduces vegetative growth and generation time. *Tree Physiology* 32:1288–301.

Foster, T. M., Celton J.-M., Chagné D., Tustin, D. S. and Gardiner S. E. (2015). Two quantitative trait loci, *Dw1* and *Dw2*, are primarily responsible for rootstock-induced dwarfing in apple. *Horticulture Research* 2:15001.

Frey, J. E., Frey, B., Sauer, C. and Kellerhals, M. (2004). Efficient low-cost DNA extraction and multiplex fluorescent PCR method for marker-assisted selection in breeding. *Plant Breeding* 123:554–7.

Gardiner, S. E., Norelli, J. L., de Silva, N., Fazio, G., Peil, A., Malnoy, M., Horner, M., Bowatte, D., Carlisle, C., Wiedow, C., Wan, Y. Z., Bassett, C. L., Baldo, A. M., Celton, J. -M., Richter, K., Aldwinckle, H. S. and Bus, V. G. M. (2012). Putative resistance gene markers associated with quantitative trait loci for fire blight resistance in *Malus* 'Robusta 5' accessions. *BMC Genetics* 13:25.

Gasic, K., Gonzalez, D. O., Thimmapuram, J., Liu, L., Malnoy, M., Gong, G., Han, Y., Vodkin, L. O., Aldwinckle, H. S., Carroll, N. J., Orvis, K. S., Goldsbrough, P., Clifton, S., Pape, D., Fulton, L., Martin, J., Theising, B., Wisniewski, M. E., Fazio, G., Feltus, F. A. and Korban, S. S. (2009). Comparative analysis and functional annotation of a large Expressed Sequence Tag collection of apple. *Plant Genome* 2:23–38.

Gessler, C., Patocchi, A., Sansavini, S., Tartarini, S. and Gianfranceschi, L. (2006). *Venturia inaequalis* resistance in apple. *Critical Reviews in Plant Sciences* 25:473–503.

Gianfranceschi, L. and Soglio, V. (2004). The European Project HiDRAS: innovative multidisciplinary approaches to breeding high quality disease resistant apples. *Acta Horticulturae* 663:327–30.

Goulão, L., Cabrita, L., Oliviera, C. M. and Leitão, J. M. (2001). Comparing RAPD and AFLP$_{TM}$ analysis in discrimination and estimation of genetic similarities among apple (*Malus domestica* Borkh.) cultivars. *Euphytica* 119:259–70.

Guan, Y., Peace, C., Rudell, D., Verma, S. and Evans, K. (2015). QTLs detected for individual sugars and soluble solids content in apple. *Molecular Breeding* 35:1–13.

Hampson, C. R. and Sholberg, P. L. (2008). Estimating combining ability for fire blight resistance in apple progenies. *Acta Horticulturae* 793:337–44.

Harada, T., Sunako, T., Wakasa, Y., Soejima, J., Satoh, T. and Niizeki, M. (2000). An allele of the 1-aminocyclopropane-1-carboxylate synthase gene (*Md-ACS1*) accounts for the low level of ethylene production in climacteric fruits of some apple cultivars. *Theoretical & Applied Genetics* 101:742–6.

Hemmat, M., Weeden, N. F., Manganaris, A. G. and Lawson, D. M. (1994). Molecular marker linkage map for apple. *Journal of Heredity* 85:4–11.

Hokanson, S. C., Szewc-McFadden, A. K., Lamboy, W. F. and McFerson, J. R. (1998). Microsatellite (SSR) markers reveal genetic identities, genetic diversity and relationships in a *Malus x domestica* Borkh. core subset collection. *Theoretical & Applied Genetics* 97:671–83.

Hyson, D. A. (2011). A comprehensive review of apples and apple components and their relationship to human health. *Advances in Nutrition* 2:408–20.

Iezzoni, A., Peace, C., Weebadde, C., Luby, J., Yue, C. Y., Bassil, N., Bink, M., Brown, S., Byrne, D., Clark, J., Crisosto, C., Davis, T., Evans, K., Fazio, G., Finn, C., Gallardo, K., Gasic, K., Gradziel, T., Hancock, J., Jussaume, R., Luby, J., Main, D., McCracken, V., Oraguzie, N., Reighard, G., Stone, A., Taylor, M., van de Weg, E., Wang, D. and Xu, K. (2010). RosBREED: Enabling marker-assisted breeding in Rosaceae. *Acta Horticulturae* 859:389–94.

James C. M., Clarke J. B. and Evans, K. M. (2004). Identification of molecular markers linked to the mildew resistance gene *Pl-d* in apple. *Theoretical & Applied Genetics* 110:175–81.

Janick, J., Cummins, J. N., Brown, S. K. and Hemmat, M. (1996). Apples. In Janick, J. and Moore, J. N. (eds), *Fruit Breeding, Volume I: Tree and Tropical Fruits*. John Wiley and Sons, Inc, New York.

Janick, J., Goffreda, J. C. and Korban, S. S. (2006). 'Co-op 39' (CrimsonCrispTM) Apple. *HortScience* 41(2):465–6.

Jänsch, M., Broggini, G. A. L., Weger, J., Bus, V. G. M., Gardiner, S. E., Bassett, H. and Patocchi, A. (2015). Identification of SNPs linked to eight apple disease resistance loci. *Molecular Breeding* 35:45.

Jung, S., Ficklin, S. P., Lee, T., Cheng, C. H., Blenda, A., Zheng, P., Yu, J., Bombarely, A., Cho, I., Ru, S., Evans, K., Peace, C., Abbott, A. G., Mueller, L. A., Olmstead, M. A. and Main, D. (2014). The Genome Database for Rosaceae (GDR): year 10 update. *Nucleic Acids Research* 42(1):D1237–44.

Jung, S., Jesudurai, C., Staton, M., Du, Z., Ficklin, S., Cho, I., Abbott, A., Tomkins, J. and Main, D. (2004). GDR (Genome Database for Rosaceae): integrated web resources for Rosaceae genomics and genetics research. *BMC Bioinformatics* 5:130.

Jung, S., Lee, T., Cheng, C. H.,Gasic K., Campbell T., Evans K., Peace C. and Main D. (2016). Marker validation for MAS in apple. Program/Abstracts GDR Workshop, *8th International Rosaceae Genomics Conference*, 21–24 June, Angers, France.

Jung, S., Staton, M., Lee, T., Blenda, A., Svancara, R., Abbott, A. and Main, D. (2008). GDR (Genome Database for Rosaceae): integrated web-database for Rosaceae genomics and genetics data. *Nucleic Acids Research* 36:D1034–D1040.

Kellerhals, M., Franck, L., Baumgartner, I. O., Patocchi, A. and Frey, J. E. (2011). Breeding for fire blight resistance in apple. *Acta Horticulturae* 896:385–9.

Kenis, K. and Keulemans, J. (2005). Genetic linkage maps of two apple cultivars (*Malus × domestica* Borkh.) based on AFLP and microsatellite markers. *Molecular Breeding* 15:205–19.

Khan, M. A., Durel, C.-E., Duffy, B., Drouet, D., Kellerhals, M., Gessler, C. and Patocchi, A. (2007). Development of molecular markers linked to the 'Fiesta' linkage group 7 major QTL for fire blight resistance and their application for marker-assisted selection. *Genome* 50:568–77.

Khan, M. A., Han, Y., Zhao, Y. F. and Korban, S. S. (2012). A high-throughput apple SNP genotyping platform using the GoldenGate$_{TM}$ assay. *Gene* 494:196–201.

Kim, M. Y., Song, K. J., Hwang, J. H., Shin, Y. U. and Lee, H. J. (2003). Development of RAPD and SCAR markers linked to the *Co* gene conferring columnar growth habit in apple (*Malus pumila* Mill). *Journal of Horticultural Science & Biotechnology* 78:512–17.

King, G. J., Alston, F. H., Battle, I., Chevreau, E., Gessler, C., Janse, J., Lindhout, P., Manganaris, A. G., Sansavini, S., Schmidt, H. and Tobutt, K. R. (1991). The 'European Apple Genome Mapping Project'-developing a strategy for mapping genes coding for agronomic characters in tree species. *Euphyica* 56:89–94.

King, G. J., Alston, F. H., Brown, L. M., Chevreau, E., Evans, K. M., Dunemann, F., Janse, J., Laurens, F., Lynn, J. R., Maliepaard, C., Manganaris, A. G., Roche, P. A., Schmidt, H., Tartarini,

S., Verhaegh, J. and Vrielink, R. (1998) Multiple field and glasshouse assessments increase the reliability of linkage mapping of the Vf source of scab resistance in apple. *Theoretical & Applied Genetics* 96:699–708.

King, G. J., Lynn, J. R., Dover, C. J., Evans, K. M. and Seymour, G. B. (2001). Resolution of quantitative trait loci for mechanical measures accounting for genetic variation in fruit texture of apple (*Malus pumila* Mill.). *Theoretical & Applied Genetics* 102:1227–35.

Korban, S. S. and Tartarini, S. (2009). Apple structural genomics. In Folta, K. M. and Gardiner, S. E. (eds), *Genetics and Genomics of Rosaceae, Plant Genetics and Genomics: Crops and Models* 6. Springer, New York.

Kouassi, A. B., Durel, C.-E., Costa, F., Tartarini, S., van de Weg, E., Evans, K., Fernández- Fernández, F., Govan, C., Boudichevskaja, A., Dunemann, F., Antofie, A., Lateur, M., Stankiewicz-Kosyl, M., Soska, A., Tomala, K., Lewandowski, M., Rutkovski, K., Zurawicz, E., Guerra, W. and Laurens, F. (2009). Estimation of genetic parameters and prediction of breeding values for apple fruit quality traits using pedigreed plant material in Europe. *Tree Genetics & Genomes* 5:659–72.

Kromdijk, J., Bertin, N., Heuvelink, E., Molenaar, J., de Visser, P. H. B, Marcelis, L. F. M. and Struik, P. C. (2014). Crop management impacts the efficiency of quantitative trait loci (QTL) detection and use: case study of fruit load × QTL interactions. *Journal of Experimental Botany* 65:11–22.

Kumar, S. (2016). Prediction of apple fruit phenotypes using genome-wide markers: progress and challenges. Program/Abstracts S608, *8th International Rosaceae Genomics Conference*, 21–24 June, Angers, France.

Kumar, S., Chagné, D., Bink, M. C. A. M., Volz, R. K., Whitworth, C. and Carlisle, C. (2012). Genomic selection for fruit quality traits in apple (*Malus* × *domestica* Borkh.). *PLoS ONE* 7:e36674.

Kumar, S., Molloy, C., Muñoz, P., Daetwyler, H., Chagné D. and Volz, R. (2015). Genome-enabled estimates of additive and non-additive genetic variances and prediction of apple phenotypes across environments. *G3 Genes Genomes Genetics* 5:2711–8.

Larsen, B., Ørgaard, M., Toldam-Andersen, T. B. and Pedersen, C. (2016). A high-throughput method for genotyping S-RNase alleles in apple. *Molecular Breeding* 36:24.

Laurens, F., Durel, C.-E., Patocchi, A., Peil, A., Salvi, S., Tartarini, S., Velasco, R. and van de Weg, E. (2010). Review on apple genetics and breeding programs and presentation of a new initiative of a news European initiative to increase fruit breeding efficiency. *Journal of Fruit Science* 27:102–7.

Le Roux, P.-M., Khan, M. A., Broggini, B., Duffy, C., Gessler, C. and Patocchi, A. (2010). Mapping of quantitative trait loci for fire blight resistance in the apple cultivars 'Florina' and 'Nova Easygro'. *Genome* 53:710–22.

Lespinasse, Y., Durel, C.-E., Parisi, L., Laurens, F., Chevalier, M. and Pinet, C. (2000). A European project: D. A. R. E. Durable Apple Resistance in Europe. Durable resistance of apple to scab and powdery mildew: one step towards an environmental friendly orchard. *Acta Horticulturae* 538:197–200.

Li, Y., Zhang, L., Zhang, Z., Cong, P. and Cheng, Z. (2011). A simple sequence repeat marker linked to the susceptibility of apple to Alternaria blotch caused by *Alternaria alternata* apple pathotype. *Journal of the American Society for Horticultural Science* 136:109–15.

Liebhard, R., Koller, B., Gianfranceschi, L. and Gessler, C. (2003). Creating a saturated reference map for the apple (Malus x domestica Borkh.) genome. *Theoretical & Applied Genetics* 106:1497–508.

Lin, M. J., Shin, Y. U., Yae, B. W., Hwang, J. H., Song, K. J. and Suh, E. J. (1998). Linkage mapping of interspecific population between *Malus domestica* cv. Fuji and *M. baccata* based on RAPD markers. *Journal Korean Society for Horticultural Science* 39:316–22.

Longhi, S., Hamblin, M. T., Trainotti, L., Peace, C. P., Velasco, R. and Costa, F. (2013). A candidate gene based approach validates Md-PG1 as the main responsible for a QTL impacting fruit texture in apple (*Malus* x *domestica* Borkh). *BMC Plant Biology* 13:37.

Maliepaard, C., Alston, F. H., Van Arkel, G., Brown, L. M., Chevreau, E., Dunemann, F., Evans, K. M., Gardiner, S. E., Guilforrd, P., van Heusden, A. W., Janse, J., Laurens, F., Lynn, J. R., Manganaris, A. G., Den Nijs, A. P. M., Periam, N., Rikkerink, E., Roche, P., Ryder, C., Sansavini, S., Schmidt, H., Tartarini, S., Verhaegh, J., Vrielink-van Ginkel, M. and King, G. J. (1998). Aligning male and female linkage maps of apple (*Malus pumila* Mill.) using multi-allelic markers. *Theoretical & Applied Genetics* 97:60–73.

Manganaris, A. G. and Alston, F. H. (1987). Inheritance and linkage relationships of glutamate oxaloacetate transaminase isoenzymes in apple. 1. The gene *GOT-1*, a marker for the *S* incompatibility locus. *Theoretical & Applied Genetics* 74:154–61.

Morgan, J. and Richards, A. (2002). *The New Book of Apples*. Ebury Press, London.

Morimoto, T. and Banno, K. (2015). Genetic and physical mapping of *Co*, a gene controlling the columnar trait of apple. *Tree Genetics & Genomes* 11:807.

Muranty, H., Troggio, M., Sadok, I. B., Al Rifaï, M., Auwerkerken, A., Banchi, E., Velasco, R., Stevanato, P., van De Weg, W. E., Di Guardo, M. and Kumar, S. (2015). Accuracy and responses of genomic selection on key traits in apple breeding. *Horticulture Research* 2:15060.

Newcomb, R. D, Crowhurst, R. N., Gleave, A. P., Rikkerink, E. H. A., Allan, A. C., Beuning, L. L., Bowen, J. H., Gera, E., Jamieson, K. R., Janssen, B. J., Laing, W. A., McCartney, S., Nain, B., Ross, G. S., Snowden, K. C., Souleyre, E. J. F., Walton, E. F. and Yauk, Y. K. (2006). Analyses of expressed sequence tags from apple. *Plant Physiology* 141:147–66.

Norelli, J. L., Wisniewski, M. and Droby, S. (2014). Identification of a QTL for postharvest disease resistance to *Penicillium expansum* in *Malus sieversii*. *Acta Horticulturae* 1053:199–203.

Nybom, N. (1959). On the inheritance of acidity in cultivated apples. *Hereditas* 45:332–50.

Orcheski, B., Parker, R. and Brown, S. (2015). Pale green lethal disorder in apple (*Malus*) is caused by a mutation in the *PHYLLO* gene which is essential for phylloquinone (vitamin K1) biosynthesis. *Tree Genetics & Genomes* 11:131.

Pagliarani, G., Dapena, E., Miñarro, M., Denancé, C., Lespinasse, Y., Rat-Morris, E., Troggio, M., Durel, C.-E. and Tartarini, S. (2016). Fine mapping of the rosy apple aphid resistance locus *Dp-fl* on Linkage Group 8 of the apple cultivar 'Florina'. *Tree Genetics & Genomes* 12:56.

Paterson, A. H., Tanksley, S. D. and Sorrels, M. E. (1991). DNA markers in plant improvement. *Advances in Agronomy* 46:39–90

Patocchi, A., Frei, A., Frey, J. and Kellerhals, M. (2009a). Towards improvement of marker assisted selection of apple scab resistant cultivars: Venturia inaequalis virulence surveys and standardization of molecular markers alleles associated with resistance genes. *Molecular Breeding* 24:337–47.

Patocchi, A., Fernández-Fernández, F., Evans, K., Gobbin, D., Rezzonico, F., Boudichevskaia, A., Dunemann, F., Stankiewicz-Kosyl, M., Mathis, F., Durel, C.-E., Gianfranceschi, L., Costa, F., Toller, C., Cova, V., Mott, D., Komjanc, M., Barbaro, E., Kodde, L., Rikkerink, E., Gessler, C. and van de Weg, W. E. (2009b). Development and test of 21 multiplex PCRs composed of SSRs spanning 80% of the apple genome. *Tree Genetics & Genomes* 5:211–23.

Peace, C. (2015). Parentage & pedigree – a Technology Interfacing article. Community Breeders' Page, RosBREED Quarterly Newsletter 5(2):8. www.rosbreed.org/articles/community-breeders/parentage-pedigree

Peace, C. P., Luby, J. J., van de Weg, W. E., Bink, M. C. A. M. and Iezzoni, A. F. (2014). A strategy for developing representative germplasm sets for systematic QTL validation, demonstrated for apple, peach, and sweet cherry. *Tree Genetics & Genomes* 10:1679–94.

Peace, C. P. and Norelli, J. L. (2009). Genomics approaches to crop improvement in the Rosaceae. In Folta and Gardiner (Eds), Genetics and Genomics of Rosaceae. Springer, NY.

Peil, A., Garcia-Libreros, T., Richter, K., Trognitz, F. C., Trognitz, B., Hanke, M.-V. and Flachowsky, H. (2007). Strong evidence for a fire blight resistance gene of *Malus robusta* located on linkage group 3 detected by rapid genome scanning. *Plant Breeding* 126:470–6.

Powell, W., Morgante, M., Andre, C., Hanafey, M., Vogel, J., Tingey, S. and Rafalski, A. (1996). Comparison of RFLP, RAPD, AFLP, and SSR (microsatellite) markers for germplasm improvement. *Molecular Breeding* 2:225–38.

Rowland, T., Edge-Garza, D., Evans, K. and Peace, C. (2010). Routine marker-assisted selection in the Washington apple breeding program provides resource savings. Abstracts of the 5th International Rosaceae Genomics Conference, Stellenbosch.

Ru, S., Hardner, C., Carter, P. A., Evans, K., Main, D. and Peace, C. (2016). Modelling of genetic gain for single traits from marker-assisted seedling selection in clonally propagated crops. *Horticulture Research* 3:16015.

Ru, S., Main, D., Evans, K. M. and Peace, C. P. (2015). Current applications, challenges, and perspectives of marker-assisted seedling selection in Rosaceae tree fruit breeding. *Tree Genetics & Genomes* 11:8.

Salvi, S., Micheletti, D., Magnago, P., Fontanari, M., Viola, R., Pindo, M. and Velasco, R. (2014). One-step reconstruction of multi-generation pedigree networks in apple (*Malus × domestica* Borkh.) and the parentage of Golden Delicious. *Molecular Breeding* 34:511–24.

Sandefur, P. (2016). Enhancing efficiency in tree-fruit breeding by developing trait-predictive DNA tests. PhD diss., Washington State University.

Sankaran, S., Slaughter, D. C., Carter, A. H., Kalcsits, L. A., Okamuro, J. K., Kirchhoff, H. and Poland, J. (2016). Integrating science and engineering to address needs in high-throughput phenotyping. 2016 American Society of Agricultural and Biological Engineers (ASABE) Annual International Meeting, 17–20 July, 2016, Orlando, FL.

Silfverberg-Dilworth, E., Matasci, C. L., van de Weg, W. E., van Kaauwen, M. P. W., Walser, M., Kodde, L. P., Soglio, V., Gianfranceschi, L., Durel, C.-E., Costa, F., Yamamoto, T., Koller, B., Gessler, C. and Patocchi, A. (2006). Microsatellite markers spanning the apple (Malus x domestica Borkh.) genome. *Tree Genetics & Genomes* 2:202–24.

Simmonds, N. W. (1991). Bandwagons I have known. Tropical Agriculture Association Newsletter. December 1991:7–11.

Soriano, J. M., Madduri, M., Schaart, J. G., van der Burgh, A., van Kaauwen, M. P. W., Tomic, L., Groenwold, R., Velasco, R., van de Weg, E. and Schouten, H. J. (2014). Fine mapping of the gene *Rvi18* (*V25*) for broad-spectrum resistance to apple scab, and development of a linked SSR marker suitable for marker-assisted breeding. *Molecular Breeding* 34:2021–32.

Takos, A. M., Jaffé, F. W., Jacob, S. R., Bogs, J., Robinson, S. P. and Walker, A. R. (2006). Light-induced expression of a *MYB* gene regulates anthocyanin biosynthesis in red apples. *Plant Physiology* 142:1216–32.

Tancred, S. J., Zeppa, A. G. and Graham, G. C. (1994). The use of the PCR-RAPD technique in improving plant variety rights description of a new Queensland apple (*Malus domestica*) cultivar. *Australian Journal of Experimental Agriculture* 34:665–7.

Tartarini, S., Gianfranceschi, L., Sansavini, S. and Gessler, C. (1999). Development of reliable PCR markers for the selection of the *Vf* gene conferring scab resistance in apple. *Plant Breeding* 118:183–6.

Tobutt, K., Clarke, J., Boškovic´, G., De laRosa, R. and Boškovic´, R. (2009). A gene for susceptibility to the fungicide azoxystrobin in apple and a tightly linked microsatellite marker. *Plant Breeding* 128:312–16.

Urrestarazu, J., Denancé, C., Ravon, E., Guyader, A., Guisnel, R., Feugey, L., Poncet, C., Lateur, M., Houben, P., Ordidge, M., Fernández-Fernández, F., Evans, K. M., Paprstein, F., Sedlak, J., Nybom, H., Garkava-Gustavsson, L., Miranda, C., Gassmann, J., Kellerhals, M., Suprun, I., Pikunova, A. V., Krasova, N. G., Torutaeva, E., Dondini, L., Tartarini, S., Laurens, F. and Durel, C.-E. (2016). Analysis of the genetic diversity and structure across a wide range of germplasm reveals prominent gene flow in apple at the European level. *BMC Plant Biology* 16:130.

van de Weg, W. E., Voorrips, R. E., Finkers, R., Kodde, L. P., Jansen, J. and Bink, M. C. A. M. (2004). Pedigree genotyping: A new pedigree based-approach of QTL identification and allele mining. *Acta Horticulturae* 663:45–50.

van Nocker, S. and Gardiner, S. E. (2014). Breeding better cultivars, faster: applications of new technologies for the deployment of superior horticultural tree crops. *Horticulture Research* 1:14022.

Velasco, R., Zharkikh, A., Affourtit, J., Dhingra, A., Cestaro, A., Kalyanaraman, A., Fontana, P., Bhatnagar, S. K., Troggio, M., Pruss, D., Salvi, S., Pindo, M., Baldi, P., Castelletti, S., Cavaiuolo, M., Coppola, G., Costa, F., Cova, V., Dal Ri, A., Goremykin, V., Komjanc, M., Longhi, S., Magnago, P.,

Malacarne, G., Malnoy, M., Micheletti, D., Moretto, M., Perazzolli, M., Si-Ammour, A., Vezzulli, S., Zini, E., Eldredge, G., Fitzgerald, L. M., Gutin, N., Lanchbury, J., Macalma, T., Mitchell, J. T., Reid, J., Wardell, B., Kodira, C., Chen, Z., Desany, B., Niazi, F., Palmer, M., Koepke, T., Jiwan, D., Schaeffer, S., Krishnan, V., Wu, C., Chu, V. T., King, S. T., Vick, J., Tao, Q., Mraz, A., Stormo, A., Stormo, K., Bogden, R., Ederle, D., Stella, A., Vecchietti, A., Kater, M. M., Masiero, S., Lasserre, P., Lespinasse, Y., Allan, A. C., Bus, V., Chagné, D., Crowhurst, R. N., Gleave, A. P., Lavezzo, E., Fawcett, J. A., Proost, S., Rouzé, P., Sterck, L., Toppo, S., Lazzari, B., Hellens, R. P., Durel, C. -E., Gutin, A., Bumgarner, R. E., Gardiner, S. E., Skolnick, M., Egholm, M., Van de Peer, Y., Salamini, F. and Viola, R. (2010). The genome of the domesticated apple (*Malus × domestica* Borkh.). *Nature Genetics* 42:833–9.

Verma, S. (2014). A Predictive Genetic Knowledge For Apple. PhD diss., Washington State University.

Voorrips, R. E., Bink, M. C. A. M., Kruisselbrink, J. W., Koehorst-van Putten, H. J. J. and van de Weg, W. E. (2016). PediHaplotyper: software for consistent assignment of marker haplotypes in pedigrees. *Molecular Breeding* 36:119.

Weeden, N. F. and R. C. Lamb (1987). Genetics and linkage analysis of 19 isozyme loci in apple. *Journal of the American Society for Horticultural Science* 112:865–72.

Weeden, N. F., Timmerman, G. M., Hemmat, M., Kneen, B. E. and Lodhi, M. A. (1992). Inheritance and reliability of RAPD markers. In Hoisington D. and McNab A (eds), *Proc Symp Appl RAPD Technol Plant Breed*. Crop Sciences Society of America, Minneapolis, pp. 12–17.

Xu, K., Wang, A. and Brown, B. (2011). Genetic characterization of the *Ma* locus with pH and titratable acidity in apple. *Molecular Breeding* 30:899.

Yuan, K., Wang, C., Wang, J., Xin, L., Zhou, G., Li, L. and Shen, G. (2014). Analysis of the *MdMYB1* gene sequence and development of new molecular markers related to apple skin color and fruit-bearing traits. *Molecular Genetics and Genomics* 289:1257–65.

Yue, C., Gallardo, R. K., McCracken, V., Luby, J., McFerson, J., Liu, L. and Iezzoni, A. (2012). Technical and socio-economic challenges to setting and implementing priorities in North American rosaceous fruit breeding programs. *HortScience* 47:1320–7.

Zhang, X. J., Wang, L. X., Chen, X. X., Liu, Y. L., Meng, R., Wang, Y. J. and Zhao, Z. Y. (2014). A and *MdMYB1* allele-specific markers controlling apple (*Malus* x *domestica* Borkh.) skin color and suitability for marker-assisted selection. *Genetics and Molecular Research* 13:9103–14.

Zhu, Y. and Barritt, B. H. (2008). Md-ACS1 and Md-ACO1 genotyping of apple (*Malus × domestica* Borkh.) breeding parents and suitability for marker-assisted selection. *Tree Genetics & Genomes* 4:555–62.

Zhu, Y., Evans, K. M. and Peace, C. P. (2011). Utility testing of an apple skin color MdMYB1 marker in two progenies. *Molecular Breeding* 27:525–32.

Part 2

Cultivation techniques

Chapter 9

Innovations in apple tree cultivation to manage crop load and ripening

Stefano Musacchi, Washington State University, USA; and
Duane Greene, University of Massachusetts, USA

1 Introduction

Modern pomology, introduced in Europe in the 1950s–1960s, continues to innovate and develop new techniques for the management of orchard fruit trees. Initially, innovation was influenced by increased production costs and the breeding of new varieties. This push created specialized orchards where trees were planted at high density, and fruited quickly, but had a shorter life span. Now, the sustainability of these specialized orchards is one of the major topics of research. Many investigators are interested in how to make the orchard more efficient and how to reduce the amount of sprays necessary for maintenance through the development of new training systems and cultivars resistant to disease. Training systems for apple are chosen in relation to the cultivar and rootstock vigour, and to the ability to crop fruits with a high percentage of blush or over-colour, especially for bi-coloured cultivars.

http://dx.doi.org/10.19103/AS.2016.0017.11

In the 1950s, apple orchards generally had a few hundred (200–400) free-standing and full-canopied trees per hectare. Higher density plantings, with the goal of developing a small spindle canopy or a fruiting wall, became widespread in Italy by the use of the palmette system developed by Baldassari, an innovative grower in Ferrara in the mid 1950s (Sansavini, 1993). In the late 1970s, this evolved from a regular shape to a free-style tree, where feathers (side branches on the nursery tree) were used to build branch scaffolds (Mitov and Zocca, 1966). The main consequence of this trend was the standardization of planar systems with a medium (400–1,100-trees/ha) planting density. The next step took place in the 1980s, when new cultivars that were more compact were introduced, leading to the widespread adoption of the spindle-type system. For this system, planting density increased to 2,000–2,500 trees per hectare, with a reduction in distance between trees in the row to less than 1–1.5 m and 3.5–4 m between rows. Depending on the vigour, the height of the tree ranged between 2.5 and 3.5 m.

The spindle became the most widespread training system for apple in Europe (Sansavini and Corelli, 1998). This system is easy to manage, both during the training period and afterwards for routine pruning (Lespinasse and Delort, 1986; Sansavini and Corelli, 1990). This system reduces costs during the early stage of orchard establishment, because there is no need to tie the tree to wires, a time-consuming and labour-intensive task. It also reduces the time when the tree is unproductive and brings it into early production of large volumes of fruits. Additionally, the spindle allows orchard management from the ground and increases the efficiency of working machinery and sprayers (Sansavini, 1997). Finally, the spindle system increases harvest efficiency, because around 70% of the fruit can be harvested from ground level without the use of platforms or ladders.

From the 1990s into current times, approaches have diversified in to High-Density Plantings (HDP). The current standard has a planting density range between 2,000 and 4,000 trees per hectare, depending on the type of cultivar/rootstock combinations and the soil fertility. High-density orchards in Europe generally use M.9 rootstock, while the Medium-Density Planning (MDP) are grafted with M.26 and, in the case of spur varieties, M.M.106. In the United States, new high-density orchards are being planted with rootstocks from the Cornell University (Geneva) breeding programme – G.41, G.935, G.11 – because of their combined tolerance to fire blight, woolly aphid and replant disease or disorder.

Also during this time period, Ultra High Density Planting (UHDP) orchards with density beyond 5,000 and even 8,000–12,000 trees per hectare with the super spindle system were also planted.

Planting density in HDP can change depending on elevation, soil fertility, training type and cultivar characteristics (many new cultivars have red skin, or are bi-colour) (Sansavini and Corelli Grappadelli, 1997; Musacchi et al., 2007).

Nurseries also play an important role in HDP development as growers can purchase a tree with the desired characteristics best suited to their orchard design. Alongside the traditionally chip-budded trees from a two-year cycle, there are two-year knip plants, or trees obtained with the June bud (1-year-old) and bench-budded plants from a one-year cycle. These different types of plants are suitable for spindle and angle canopy training systems. Furthermore, the innovative 'bi-axis' system is a pre-formed, split-branch tree created in the nursery that obviates the need for heading back in the orchard (topping) and the year delay in shoot-system formation (Musacchi, 2007).

A necessary prerequisite for a profitable orchard includes proper site selection, appropriate cultivars, and rootstock and planting systems that are compatible for the location. However, this is not a guarantee of success. Pome (apple and pear) fruits are considered high-value crops that are very expensive to plant, maintain and bring into production. Other management decisions are required to achieve successful and profitable orchards, and as the planting density increases, orchard management becomes more important. Infrastructure is required to handle nutrition through fertigation applied in small amounts on a daily basis. This technique reduces and optimizes the use of fertilizer and increases the sustainability of the orchard. Managers must prune to minimize the growth habit problems specific to different cultivars. Frequent use of plant growth regulators (PGRs) may be necessary to thin fruit, improve fruit appearance or modify growth, techniques that are largely adopted in modern orchards.

Under most growing conditions, the use of Plant Growth Regulators (PGRs) is required to achieve the highest potential quality and productivity from an orchard. The range and extent of use of PGRs by the tree fruit industry is unmatched by any other industry. They are used to control a wide range of responses important to apple producers including nursery tree production, thinning, preharvest drop control, enhanced flowering and return bloom, alteration of fruit growth, enhancement of fruit red colour, improvement in fruit cosmetic appeal, regulation of ripening, and appropriate vegetative growth control.

The primary reason for using PGRs is to increase the value and quality of the crop rather than increase production. Apple orchards are expensive to establish and are equally expensive to manage and harvest a crop. PGRs are expensive to use; however, the enhancement in the value of the crop far exceeds the high cost that is incurred to use them.

The use of PGRs generally requires governmental approval and they can only be used following approval and registration by one or more government jurisdictions. Since countries, states and provinces differ in their view of their importance and safety of specific compounds, PGRs are not universally used or available throughout the world. Further, tree structure, cultivars grown and differing climates may require extensive use of specific PGRs in one area, but they may not be needed, or deemed to be unimportant in another growing environment.

This chapter aims to provide an overview of the training systems and management techniques for enhancement of economic value and environmental sustainability of apple production.

2 Training system

2.1 Planar canopy

It is possible to group all the Italian systems developed after 1960 under the general name of 'Palmette'. Other systems developed include the United States' 'Penn State thin-wall' (Tukey, 1978), New Zealand's 'Lincoln canopy system' and 'Ebro Trellis system' (Dunn and Stolp, 1981; Tustin et al., 1989), and France's 'Solen system' (Lespinasse, 1989).

Palmette

Palmette was one of the first training systems used for apple in Italy. Trees from the nursery are planted on vigorous rootstocks, and scaffold branches that are more or less symmetrically located on the tree are brought in along the row and tied to supporting

Figure 1 (a) Cv. Hi Early. Regular palmette of apple over 30 years old. (b) Modern planar canopy with a tri-axis tree structure.

wires, while branches growing into the row space are removed. The aim is to form a continuous fruiting wall as soon as possible after planting. However, it is important to avoid the crossing of scaffolds and to limit wall thickness that might cause shading. Planting density varies from 400 to 1,100 trees per hectare. This system requires intensive work during the first three years to develop a continuous fruiting wall. Support frames such as posts and wires are necessary to correct the position and direction of the main scaffolds (Sansavini and Corelli, 1990).

The palmette-style apple tree was used for a short time in Europe following the adoption of dwarfing rootstocks like M.9. It is still possible to find examples of palmette in areas such as the Rio Negro in Argentina (Fig. 1a). The most recent type of palmette adopted is referred to as free palmette, and it uses feathers produced in the nursery. The main training approach is to do a minimum amount of pruning in the early years and allow the tree to mature structurally, avoiding unwanted excess vigour induced by heavy pruning. The free palmette system is still a valid training system, especially in heavy clay soil conditions that require vigorous rootstocks. The orchard density ranges from 800 to 1,100 trees per hectare. Limitations for this training system include wide planting distances, more pruning and harvesting costs (operations that account for almost 60% of production costs) as trees can exceed 4 m in height with a natural tendency to carry a disproportionate amount of the crop at the top of the tree. This system would be suitable for mechanized thinning, pruning and harvest.

Bi-axis system

A new approach for reducing vigour in close plantings is represented by the longitudinal Y system called 'bi-axis'. This system requires a trellis structure with posts and 4 wires equally spaced at a distance of 70 cm between wires. Bi-axis trees can be created either in the nursery or by heading back the tree in the field (Musacchi, 2008; Musacchi et al., 2009); however, one more year is necessary to develop the canopy structure if the bi-axis is created in the field. The nursery technique to produce bi-axis trees utilizes a double chip-budding or a bench-graft and has been patented as 'Bibaum®'. Trees are planted with the double axis oriented in the direction of the row which results in a flat and narrow canopy with a depth of 70–80 cm. Dividing the vigour between two axes presents the undeniable advantage of better control of the vegetative growth. This system doubles the number of shoots and reduces their length to half in comparison to spindle trees planted at the same distance (Dorigoni et al., 2011). This vigour control results in a reduction in pruning time. The bi-axis system combines the advantages of the palmette system, such as mechanization of thinning, pruning and harvest, and avoids many of the flaws in the palmette system. In addition, the reduced tree spacing and tree height allow optimization of the distribution of fruits in the canopy. Fruit distribution at the top of the tree is similar to that at the bottom of the tree (Musacchi et al., 2014), resulting in more fruits being exposed to the light, which assures good red colour development in bi-coloured varieties. This is an advantage over the spindle system where a large proportion of the crop (70%) is in the bottom part of the tree, resulting in a dramatic reduction of the red colour due to shading (Musacchi et al., 2014) (Fig. 2a,b,c).

At planting, the two axes will bend in opposite directions and with a crotch angle of 30–40 degrees to make the basal angle of insertion wider. When vegetative growth starts, the two axes will be oriented vertically to build a 'double axis' tree. This technique doubles the number of productive axes per hectare. Differences between the cross-sectional areas of the trunks can be reduced with manipulation during the initial orientation; the weaker axis can be positioned more vertically than the vigorous axis. Regardless, for apple, yield and fruit size are not affected by the reduction of the trunk cross-sectional area of one of the two axes.

At the end of the second year, the tree height will be 2–2.5 m. Pruning is simple and is required to eliminate the shoots that are too vigorous and that are growing in the internal part of the Y. A good rule is to eliminate all the shoots that have a diameter 30% greater than that of the trunk where they are inserted. It is strongly recommended to leave a stub to avoid producing "blind wood," especially in the type IV cultivars such as 'Granny Smith'. Ideally, there will be 20–25 small branches on each axis. It is not recommended to top the tree axes. If the trunk presents blind areas in the second year, it is possible to girdle the trunk or notch over the dormant buds to induce bud-break (see section about scoring and girdling). In the third year, the tree should completely fill the space. The pruning rules are the same as the second year and it is possible to start branch renewal starting from the third year. It is recommended to renew 20% of the structure annually to minimize branches ageing. For a bi-axis system, proposed planting distances are 3.3 × 1.2 m with 2,525 trees per hectare. A modified version of planar canopy is represented by a tri-axis structure of the tree (Fig. 1b).

2.2 Vertical canopy

Spindle

The spindle is the most prevalent training system for orchards of a medium-high density (2,000–3,000 trees/ha) because it allows for intensification of planting density without excessive labour costs. Even the initial structure to support trees is simple and requires just posts and wires. Training a tree as a spindle requires a feathered nursery tree to build the first scaffold of branches. This, in association with the use of dwarfing rootstocks (M.9 and related strains), allows for significant production in the second year after planting (7–8 t/ha) (Fig. 3). However, if the quality of the tree is poor, it is possible to cut the tree back to 0.80–1 m from the ground and stimulate new structural shoot growth (Sansavini and Corelli, 1997).

Figure 2 (a) Cv. Fuji. Bi-axis grafted on M9 (third leaf). (b) Cv. Gala. Bi-axis trees grafted on M9 T337 (third leaf). (c) Cv. Rosy Glow trained as bi-axis in South Tyrol, Italy.

Figure 3 Cv. Rosy glow trained as spindle (second leaf).

The spindle and its derived forms are characterized by a central axis and lateral productive branches, usually 5–7 positioned in a circle, which do not have a precise pattern but do not overlap. This arrangement allows for good light interception and photosynthetic activity. The tree should assume a cone shape, so branches at the top of the tree must be shorter than those at the bottom. Also, the top branches can be removed and replaced with younger, less vigorous, lateral branches. However, lateral branches with an acute angle to the trunk must be removed to leave a stub, or bent through tying or applying weight.

Pruning interventions are limited and focus on the elimination of excessively numerous or vigorous branches that could compromise the tree's vegetative balance. The branches may have an angle up to 40° with respect to vertical axis or may be bent downward depending on the growth habit of the cultivar. The first scaffold branches should be 45 cm from the ground; however, this height may be 40 cm if the branches are inclined, or 70–80 cm if the branches are bent. It is essential that the branches retain a good open crotch angle, and if necessary, they must be immediately opened further with the help of small weights or wire ties. The central axis is left free to grow and the branches must be periodically renewed with appropriate cuts (Sansavini and Corelli Grappadelli, 1997; Musacchi and Farina, 1998).

Slender spindle

There are two versions of this training system: one from Northern Europe (Wertheim, 1968) and one more commonly found in Southern Europe called 'French Vertical Axis System' (Lespinasse, 1980; Lauri and Lespinasse, 2000). The Northern version was originally designed to increase the planting density, enhance early bearing, and reduce tree height (Robinson et al., 2006). This training system was also used as a multi-row system (Wagenmakers et al., 1991). The Southern version is particularly suited to forming tall fruiting walls (3–3.5 m). It has a single vertical trunk, maintained at the desired height by regularly heading back, and fruiting branches regularly spaced along the trunk. Renewal pruning occurs every three years to remove pendulous branches and to maintain fruiting on young spurs (Lauri, 2009). The fundamental basis of this training system is pruning young trees according to their fruiting type instead of imposing an artificial tree shape (Lauri and Lespinasse, 2000). For cultivars with basitonic growth (Red Chief, Oregon spur), it is important to promote the development of the leader and avoid upright branches in the lower part of the tree. The contrary is applied in acrotonic (Granny Smith, Rome Beauty) cultivars where lateral shoots in the upper part of the tree are pinched back to promote the growth of lower branches (Lespinasse and Delort, 1986). For medium-vigour cultivars like 'Gala' and 'Rosy Glow', as a slender spindle, the planting distance can range from 3.3 × 0.8 m to 3.3–3.5 m × 0.9–1 m. In the case of vigorous cultivars such as 'Fuji' that react to plant density, 3,788 to 3,367 trees/ha would be appropriate (Table 1).

Solaxe

The Solaxe system is the combination of two training systems: the 'Solen' (Lespinasse, 1989) and the 'Axe vertical' (Lauri and Lespinasse, 2000). This training system is based on the progressive formation of the tree structure, and is developed to establish an equilibrium between vegetative and fruiting activity (Fig. 4). This situation can be obtained with the use of permanent fruiting branches. For this form, both the central axis and the fruiting branches grow freely, whereas competing vegetative shoots are removed. As the branch ages, secondary fruiting formations like brindles and fruiting spurs will develop.

The Solaxe technique represents movement away from hard pruning operations that direct development of tree structure, towards management that is compatible with the natural growth behaviour of the tree. Less pruning allowed natural growth to occur, and created greater equilibrium within the tree. A better understanding of tree fruiting habits and the function of different fruiting structures led to the development of this new, more efficient training system (Lauri et al., 1997).

Table 1 Planting distance for the main training system

Training system	Planting distance	Cultivar	Rootstocks of reference
Palmette	4-4.5 × 2-2.5 m	All standard varieties	MM 106, M4
Bi-axis	3.3 × 1.2 m	All standard varieties	M9
Spindle	3.5 × 1.2 m	All standard varieties	M9
Slender spindle	3.3 × 0.8-0.9 m	All standard varieties	M9
Tall spindle	3.3 × 0.9 m	All standard varieties	M9 or Geneva 11
Solaxe-centrifugal	4.0 × 1.1-1.5 m	Vigorous varieties – type IV	M9 or more vigorous rootstocks
Super spindle	2.5-2.75 × 0.3-0.35 m	Medium vigour varieties or spur type	M9 or dwarfing genotypes like P22
V-system	3.5 × 0.5-0.7m	All standard varieties	M9
Y-system	4-4.5 × 0.7-1m	All standard varieties	M9 or some Geneva 11, Geneva 41

Figure 4 Cv. Fuji. Trained as Solaxe.

A final variation on the Solaxe training method is the 'conduite centrifuge'. It is a freer training system where trees have a central axis. On the main axis, several fruiting branches are developed and neither the top nor the lateral branches are cut back. Trees trained with this method appear 'wild' and the management of fruiting density and production uses the 'extinction' technique (Larrive and Lespinasse, 2002). The planting distances for the Solaxe-centrifugal system rise to 4 m between rows and 1.1–1.5 m in the row as a function of the vigour (Table 1).

Tall spindle

The tall spindle production system is the US version of the spindle developed by Robinson and colleagues (2006). The authors consider the tall spindle a mix of ideas and techniques from several training systems previously described, such as the slender spindle, the vertical axis, the Solaxe system and the super spindle system (Robinson, 2011). This system utilizes the European concept of feathered trees from the nursery with at least 10–15 feathers per tree at a planting density of 2,500–3,000 trees/ha. Pruning at planting is minimal and feathers longer than 25 cm must be bent to below horizontal to control the vigour and enhance early bearing. Lateral branches can be bent in the same fashion in summer to limit growth activity. The main difference when compared to the Vertical Axis and Slender Spindle (Robinson et al., 2006) is that branches are not considered permanent and it is possible to remove them when they become too large. All the thick branches in the top zone of the tree must be removed to optimize light distribution in the canopy. A stub can be left after branch removal to produce horizontal renewal shoots. The mature height of a tall spindle tree is 3 m and requires a structure with 3–4 wires to support the crop weight.

Super spindle

This training system is used in Ultra High Density Plantings (UHDP) (10,390–13,333 trees/ha) and is an adaptation of a pear training system introduced by De Longo-Zanini in 1939 in Italy. Utilizing this system the authors produced 80 t of the pear cultivar 'Passe Crassane'. More recently the Super Spindle system can be found in Germany and is practised mainly on apple and pear, with current experimentation on cherries (Curzel, 1994; Musacchi and Lugli, 2014). It was derived from the slender spindle when the distances between the trees in the row became less than 50 cm (Werth, 1989). Trees are planted at 30–35 cm in the row with 2.5 to 2.75 m between rows. With an increase in tree density, the tree shape changes from conical to cylindrical (Werth, 1989; Nüberlin, 1993). The super spindle has a columnar, unique, central axis with short lateral fruiting formations (Sansavini and Corelli, 1997). A dwarfing rootstock, similar to the weaker strains of M.9, is necessary to manage the vigour. This type of training system requires less labour, fruits earlier, has a greater yield, and increases light interception (Mantinger, 1998). The large number of trees per hectare, however, inflates initial establishment costs. To reduce the planting costs, it is possible to graft the trees directly in the field. There are not enough data to prove the economic advantages of switching to this training system for apple, especially in flat areas like the Po valley in Italy or other areas where the cost of land is less compared to Northern areas of cultivation like the South Tyrol (Sansavini and Corelli, 1997).

2.3 Angled canopy

V-system

The V-system is utilized in many areas of the world because it maximizes light interception and yield by creating two walls of production (Robinson and Lakso, 1989; Robinson and Lakso, 1991). Higher production is possible because of an increase in planting density to over 5,000 trees/ha, which is obtained by reducing the planting distances. The main drawbacks of these systems are the high costs of the structures and wires to support the trees and the laborious pruning when compared to the single row system. The traditional V-system in Europe is obtained by alternately tilting the trees in a row at a 10–15° angle from vertical so that the entire row is structured on two inclined planes. The trees are developed with small, permanent branches starting at the base and with branch size decreasing with increasing height. The vegetative activity is controlled by the strong competition for water and nutrients occurring in the root systems (Fig. 5).

Y-system

The Y system was successfully tested in many countries. It was developed in Australia as the 'Tatura trellis' system and requires a very expensive structure with more than 10 wires (5 per side). To obtain a double fruiting wall, a single row of trees is cut back 40–60 cm

Figure 5 Cv. Aurora. V-system with regular distribution of branches in the tree canopy.

from the ground (Sansavini and Corelli, 1990) to grow two new axes. The Y-system is technically more difficult and takes longer to form compared to the V-system, but the V-system requires approximately double the number of trees. Excessive watershoot growth can be induced in the Y system if the angle between the two axes becomes too wide (40–45°), resulting in increased pruning costs.

3 Pruning techniques related to the cultivar habit

3.1 Apple habit

It is impossible to describe pruning techniques and orchard planting systems without referring to the four cropping models originally presented by Lespinasse (1980). In this section, most popular apple cultivars from each group are denoted and each model is accompanied by an appropriate pruning regime (Lespinasse, 1980; Sansavini and Corelli, 1990). Recognition of where flowers and fruits develop on different types of bearing wood is important because their location will determine how the different trees should be pruned and trained.

Type I: 'Red Delicious' spur types

Cultivars that belong to this group have a strong tendency to develop branches at the bottom of the tree (basitony). The dominance of the trunk (leader) is not particularly strong and varies among cultivars. The majority of the fruiting spurs are located on branches that are at least two years old. The fruits remain close to the main branches, so the fruit set area is close to the main structure of the tree. The natural tendency of these cultivars is to produce short nodes with many flower buds. The leaves are highly photosynthetically efficient and present a strong tendency to produce a crop in a biennial way. Accurate thinning and bud removal is important with an emphasis on removing the flower buds located on the bottom part of the branch or that exhibit a small size (Sansavini and Corelli, 1990).

Type II: cv 'Red Delicious', 'Reine des Reinettes', 'Stayman', and 'Gloster'

These cultivars are easy to manage in the free-growing form. The main branches are inserted to the trunk and have strong wide angles. These trees normally show greater dominance of the central leader than the spur types. However, basitonic tendencies remain strong particularly on vigorous rootstocks. The majority of the fruiting spurs are located on branches that are two to four years old. Crowned brindles occur more frequently than in spur types and require the bending of the shoots and branches to encourage flowering and fruit set (Sansavini and Corelli, 1990).

Type III: cv 'Golden Delicious', 'Gala', 'Braeburn', and 'Cripps Pink'

Cultivars that belong to this group are well suited to be trained as 'vertical axis' or spindle system with fruiting branches directly on the trunk with a wide angle (60–90°). Fruiting spurs are principally situated on young wood of one to three years. There are numerous coronate twigs (short shoots). The fruiting zone moves rapidly away from the centre of the tree and causes the bending of branches. Some of these cultivars, when over-cropped can become biennial (Sansavini and Corelli, 1990).

Type IV: cv 'Rome Beauty', 'Granny Smith', 'Cosmic Crisp', and 'Cortland'

These cultivars rarely develop lateral shoots in the lower portion of the trunk and branching is located on the upper third of the trunk (leader), which gives the tree a cylindrical appearance (acrotonic tendency). Most of the fruiting spurs are located on young wood of one or two years. The fruiting zone moves towards the outside of the tree, especially in 'Granny Smith' (Sansavini and Corelli, 1990).

3.2 Pruning

Pruning can manipulate canopy porosity, crop load and quality, improve and maintain the shape of the tree, influence flowering, and modify the ratio between roots and canopy. Pruning techniques change over time in relation with the training system and the cultivar. They also evolve with the age of the tree and drive the natural bearing habits to maximize the productive stage. Pruning can affect growth, flowering, fruit maturity and mineral uptake (Ferree and Scupp, 2003).

The time of year when pruning takes place can strongly affect the tree response. A severe summer pruning can dramatically decrease the fruit size because of a reduction in leaf surface area. However, summer pruning can increase the over-colour in red varieties, reduce the vigour and enhance flower bud formation (Faust, 1989). A severe winter pruning can reduce the number of vegetative points of growth and induce an excess of shoot growth, which has a negative effect on flower bud formation.

Pruning regimes for HDP and UHDP orchard designs incorporate a number of innovations to upgrade tree governance including green pruning, bending, and heading back. These techniques combined with the traditional winter pruning help to maintain a more efficient tree. Several of the most important cultivars for these training systems are type III or IV according to the Lespinasse model. For these cultivars and systems, two types of pruning techniques are applied: bending and click.

Bending and long pruning

Historically, bending of apple shoots to open the crotch angle during the summer and winter is used to control the vigour of the cultivar and to modify the development of the vegetative and reproductive gradient (Grisvard, 1957; Giulivo, 1990). Lespinasse (1980) demonstrated that bending can increase the number of flower buds and the carbohydrate content, while reducing the size of the branches. When the branches are bent from 30° to 90° from the trunk, flower bud formation increases 64%. If the bending increases to 120°, the number of flower buds is 2.2 times higher when compared to the 30° bending.

Shoot bending during summer can reduce the vegetative activity and modify the speed of shoot growth with a positive effect on the plastochrone, or the time between successive leaf initiation events. When the plastochrone increases to 5–7 days, flower bud formation is initiated (Baldini, 1990). The stress induced by the bending can stimulate ethylene production in the internal air space within the shoot (Robitaille and Leopold, 1974; Ferree and Schupp, 2003). Also when the branches are horizontal, auxin production and movement changes and there is a reduction of apical dominance and an increase in shoot growth. Many of the training systems described in this chapter utilize bending as a way to manipulate and maintain the shape of the tree.

Bending is the basis of the pruning technique developed by Jean Marie Lespinasse and the Mafcot ('Maitrise de la fructification, concepts et techniques') research group. It is defined as 'taille longue' (long pruning) and is being tested in many areas around the world. For this method of pruning, branches and central axes are bent over at a horizontal angle without cutting them back. The fruit-bearing units are reduced by the application of 'extinction' technique (flower buds removal) to lower their density and optimize light exposure (Diemoz, 2005). Lespinasse and Delort (1986) state the essential principles to control fruiting with long pruning are: development of branches high on the main trunk, all the feathers below 1 m are removed to prevent branches touching the ground under fruit weight, and branches are bent to increase flower bud formation in the middle-upper part of the tree. To control tree vigour in the upper part of the tree, bending is necessary to keep an appropriate balance between vegetative growth and development of fruit units on the outer portion of the tree (Diemoz, 2005).

The Mafcot research group developed an instrument called 'l'Equili-fruit®' that determines the number of fruiting units that should be retained on a branch. This gauge-like instrument measures the diameter of each branch to determine the optimum number of flower buds to maintain for a correct productive bearing. This manual procedure is performed during the first two years after planting and takes approximately 150 hours per hectare each year; once the tree is formed, pruning time is reduced dramatically

Figure 6 (a) Bending at planting in a 'knip' tree. (b) Bending in a spindle tree.

(Diemoz et al., 2002a,b). Bending and long pruning can also be utilized in a spindle tree to reduce vigour and increase flower bud formation (Fig. 6a and b).

Click pruning

Lespinasse developed the Solaxe system specifically to minimize the pruning problems of type IV varieties which are characterized by terminal bearing shoots (Lauri and Lespinasse, 2000). These cultivars typically present branches with blind wood in the basal zone. For other training systems, such as spindle, on a type III or IV cultivar, it is necessary to apply the bending or click pruning technique to minimize the negative effect of the blind wood.

Click pruning was originally developed in Northern Europe to increase light penetration (Dallabetta, 2014) and is based on dormant cutting of 1-year-old wood in both the apical and basal zones (Fig. 7). When planting, feathers are bent to an angle of 40–45° to the vertical. The apical heading of the stem on 1-year-old wood ensures strong regrowth and promotes vigour in the shoots. Making a basal cut into 1-year-old wood prevents the limb from ageing and consequent exhaustion. Once apical dominance is removed, lateral buds on the branches swell instead of remaining dormant. Mohammadi et al. (2013) reported an increase of flower bud formation when using the click technique.

Click pruning requires that about 20 to 25% of the branches should be renovated every year to maintain a high fruit quality. For this reason, mature trees must have 25–30 small branches to prevent yield reduction. Ventura and Sansavini (2005) found that limb renewal can minimize alternate bearing, whereas Lauri and Lespinasse (2000) reported that renewal pruning lead to an imbalance between vegetative and reproductive activity and resulted in small fruit. Other reported positive effects of click pruning (based on branch renewal) are increased photosynthesis, shoot growth, and increased yields and

Figure 7 'Click' pruning in a spindle tree.

fruit weight in apple (Tustin et al., 1988; Warrington et al., 1996; Ming Xia et al., 2011; Dallabetta 2014).

Scoring and girdling

Scoring is based on severing the phloem while girdling involves bark removal. Both techniques cause a temporary interruption of movement in the phloem, altering the carbohydrates and hormones balance (Ferree and Scupp, 2003; McArtney and Obermiller, 2015). This technique is usually performed with a double-bladed clipper or a small saw and the wounds heal in a couple of weeks. More severe girdling, such as that resulting from a scoring knife used on grapes, takes longer to heal because a wider band of bark is removed. These cuts can promote bud-break and enhance vegetative growth on those parts of the trunk that would otherwise not occur. These techniques can be applied in spring (green tip stage) during the first year or, even better, in the second year after planting, when the root system is established and the reaction to the cut will induce a large number of shoots and a strong regrowth.

Dutch cut

For training systems with vertical axes that have short fruiting shoots extending from the trunk, lack of shoot renewal or regrowth of a vigorous upright bud following an angled cut, can occur. This can be remedied using a so-called 'Dutch cut' or cutting horizontally, instead of at an angle (Fig. 8). This often results in the growth of a bud from the lower side of the horizontal cut that develops at an appropriately wide angle from the trunk. This pruning technique results in the formation of two new shoots that are less vigorous than the initial shoot and that are usually at a wider trunk insertion angle. This type of pruning is performed in winter.

Figure 8 Dutch cut utilized to minimize blind wood.

4 Mechanical pruning and 'Mur Fruitier'

The increasing costs of production and labour shortage have encouraged apple growers to introduce mechanization of pruning and harvest for the HDP orchard (Robinson et al., 2013a). Many kinds of equipments are available to facilitate pruning, including the hedging and topping saw-bar. Some of these technologies are extensively used in Europe where orchard mechanization started over 50 years ago (Sazo, 2010). Training systems characterized by a planar canopy or continuous rows are the most suitable for this kind of pruning.

Mechanized pruning of apple and pear were tested for the first time by Sansavini in 1974 in the Po valley, Italy. Results from this study identified both advantages and disadvantages of this method. When done in summer, there are advantages of less regrowth, an increase of flower bud formation, shoots are still herbaceous and easy to cut, and there is less infection from fungi (Sansavini, 1974). However, the risk of fire blight (bacterial) infection increases if there are favourable weather conditions, so this technique is not recommended in orchards where fire blight is present and it is strongly recommended to apply copper or other products immediately after pruning. Severe leaf removal from mechanized pruning can result in fruit size reduction (Sansavini, 1974).

Mechanical pruning is performed on a cutting plane (horizontal, vertical and oblique) and generally involves the use of a sickle-bar or saw-bar. Other equipment can open windows (slot pruning) in the canopy working by at different cutting depths (Sansavini, 1974). Cain (1971) observed that mechanical pruning on 'McIntosh' apples with a slotting pruner compared with a vertical saw-bar induced better light penetration in the canopy, increased the number of spurs (Fig. 9) and improved fruit colour and yields. Mechanical pruning requires only 3–4 h/ha, but usually some follow-up hand pruning is necessary for fine-tuning, removal of vertical shoots, branches that are competing with the trunk, and branches and shoots that are growing along the row. Hand pruning may require an additional 40 h/ha, especially in the early stages of orchard establishment (Masseron, 2002).

In France, the first 'Mur Fruitier' orchard, which was developed in 1988, employed the physiological principles of Lorette (developed in 1914 and published in 1925 in English),

Figure 9 Effect of mechanical pruning on flower bud formation.

which recommended summer pruning only. The concept combines mechanical pruning with training systems. The Mur Fruitier technique is based on the concept that a tree has only two groups of organs: the main trunk structure and many small lateral short shoots. The tree structure can be obtained starting from orchards with a vertical axis, a bi-axis or a palmette tree. The height of the tree is 2.5 to 3.5 m with a maximum width of the canopy only 80 cm; long permanent branches are not required. Time of pruning influences the effect on flower bud formation; if the cut is done too early, there may be an excess of growth, if too late, there is no direct effect on flower bud formation. For this reason, mechanical pruning is performed between the end of May and beginning of June when shoots have 12 fully developed leaves (Masseron, 2002).

5 Crop load determination and effect on fruit quality

Crop load in apple can have several impacts on fruit quality. When there is a low crop load, excess vegetative growth may divert calcium away from the developing fruit, resulting in an increase in calcium-related disorders, especially bitter pit (Faust and Shear, 1968, Ferguson and Watkins, 1989; Rosemberger et al., 2004). The incidence of calcium-related disorders is often reduced by properly adjusting crop load to an acceptable level, and this will modify the tree vigour. Crop load influences photosynthesis (Palmer et al., 1997) and dry matter production and allocation (Avery, 1975; Heim et al., 1979; Palmer, 1992). Finally, an excessive crop load negatively affects fruit quality by reducing fruit size, inhibiting colour development, and lowering sugars and acidity (Embree et al., 2007; Hampson and Kemp, 2003).

Control of the crop load can therefore improve the capacity of the tree to grow and produce high quality fruit. This is achieved in part by initially counting buds on 1.5–2% of the tree before pruning. Ideally, flower buds on spurs and brindilla (one-year-old shoots with terminal flower bud) are counted with the goal of producing one fruit per spur and two per brindilla. The number of flowers on a brindilla can depend on its length. After counting, bud differentiation is checked by viewing the flower buds under a microscope or dissecting scope. Usually this practice is done in fall and should be repeated later for verification. After pruning, buds should be re-counted to affirm that the desired number were retained.

To determine the number of buds to prune, production goals can be estimated with the following calculation:

Production Target = number of fruits per tree × number of trees per hectare × average fruit weight

It is a common practice to add a 30–40% more of buds just to be sure that, if the winter is too severe, the production level is not affected. This formula is applied to avoid over cropping in heavy blooming years, low production in light blooming years and help regulate biennial bearing (Crassweller et al., 2005; Embree and Nichols, 2005; Robinson and Watkins, 2003; Robinson et al., 2013b; Robinson et al., 2014). An excessive number of fruits can induce alternate bearing in a cultivar, especially in combination with lack of, poor or late thinning (Pellerin et al., 2011, 2012).

The concept of crop load management for the Mur Fruitier orchard is based on the number of fruits per m^2 of canopy surface. The ideal number of fruits per m^2 is calculated to be 25. The productive surface in a Mur Fruitier orchard is around 13,000–17,000 m^2/ha. To achieve a crop load of 60–70 t/ha of fruits with an average weight of 180 g, it is necessary to have 335,000–390,000 fruits/ha (Masseron, 2002).

6 Tree production in the nursery

Advances and innovations introduced by the nursery industry in recent years mean it is now possible to purchase a given type of tree suitable for any desired orchard design. All trees planted in modern apple orchards are formed by the union of two bionts. The rootstock and the cultivar are produced separately; the two can be combined in numerous ways to create a tree with characteristics suitable for specific training systems.

Nurseries offer traditional chip-budded trees in a two-year cycle, two-year knip plants obtained with the June bud (1-year-old), and bench-budded plants in a 1-year cycle. All of these types of trees are generally suitable for the spindle training system. The innovative 'bi-axis' system is planted with nursery-formed, split-branch trees created with the 'Bibaum®' technique of double budding or bench grafting. These trees require orchard topping and a year delay in shoot-system formation (Lugli and Musacchi, 1993; Musacchi, 2008).

The goal for all tree types is to create feathers that speed the early formation of branches. This makes it possible to rapidly bring an orchard into production, depending on the variety (Sansavini and Zocca, 1965; Mitov and Zocca, 1966). Only in cases where the tree produced by the nursery is completely devoid of feathers the orchardist need to head the stem to 50–60 cm to produce a new leader and lateral shoots (Fig. 10).

Figure 10 Apple trees in the nursery.

6.1 Two-year nursery production cycle

The two-year production cycle of trees in the nursery begins in late February to March with the planting of rootstocks. The first year is dedicated to growing rootstocks to the prime size for budding. The second year is utilized to grow the canopy of the budded scion (Lugli and Musacchi, 1993).

The vegetative buds of the desired variety must be collected from a mother field of differentiated and matured trees that (in most cases) are certified to be virus free. The shoots from which they are taken are defoliated by hand or by machine. In the case of the T-budding, a short section of the petiole must be saved to make the coupling operations easier (Hartmann and Kester, 1990). It is better to use buds located in the middle part of the shoot because basal buds do not swell as well as and the apical buds are still immature.

Budding time runs from late July to mid-September, depending on the cambial activity of the rootstocks and the weather conditions. The cambium must be active enough to generate a good callus, which will heal the wound and rebuild tissues between the rootstock and the scion variety. Budding on the rootstock is done 10–15 cm above the soil level. Budding at 5 cm or lower may increase the possibility of scion rooting after the tree is planted. Placing the bud is performed by T-budding, an old technique, or with the more innovative practice of chip-budding. Chip-budding was developed in the 1970s and it increases the time frame when budding is possible, allows for better contact between the two bionts, promotes a faster merging and more solid union, creates better tree uniformity, leads to a greater number of feathers because of accelerated growth, and has a union that is less sensitive to cold and fungi spores (Howard et al., 1974; Ligonniere, 1981). Chip-budding also expands the time that trees can be propagated because 'slipping' of the bark is not necessary. Ten to twenty days after budding, the wound is healed and the bud remains dormant until the following spring. The rootstock's canopy above the budding is mechanically removed by a wood-chipping machine during winter. The stem just above the bud is cut at an oblique angle to minimize the effect of wood necrosis in the graft union to make the union stronger (Lugli and Musacchi, 1993).

In the spring of the second year, the shoot from the swelling bud grows very fast, and in one month, it can be 50–60 cm tall. During this vegetative stage, branching agent products based on cytokines and gibberellins can be applied in the nursery to induce feather formation, depending on where the first scaffold of branches is desired. Treatments to promote feathers can be repeated 4–6 times as the stem grows to construct a pre-formed tree with 5–6 good feathers and many short shoots in the upper zone of the tree. All the shoots growing 50 cm or less from the soil surface must be removed when they are still herbaceous because they will not produce branches. In the fall of the second year, trees are ready to be lifted and stored in a cold room until planted in the orchard.

6.2 Two-year cycle with side graft

This type of tree production differs from that described above because the variety is placed with a side graft in the second year starting in February or later, depending on the climate zone. This method is useful to recover rootstocks where budding was unsuccessful in the previous year.

The side graft method utilizes a bud-stick with one or two buds. One end of the bud-stick is shaped like a triangle and it interlocks with the rootstock cut to receive it. The wood necessary for this kind of graft must be collected during winter and cold-stored to stay dormant (Lugli and Musacchi, 1993).

6.3 One-year short cycles

It is possible to produce a tree in one year with two different techniques: chip-budding in the spring or bench grafting during the winter.

Chip-budding in spring

This technique is practised mainly in regions characterized by a long growing season. Rootstocks are planted in late February – early March. The chip-budding is performed in April or May utilizing material collected in winter and cold-stored. When the wound is healed from budding, it is possible to remove the rootstock's canopy and force bud swell. The budding operations are identical to those described for the second year of the dormant chip-budding (Lugli and Musacchi, 1993). Sometimes for this technique, the percentage of bud take is low.

Bench grafts

Bench grafting is an important technique to produce a tree in one year. It is performed during the dormant season and cultivar wood is grafted on a bare rooted rootstock that may or may not be dormant. Scion and rootstock must be of equal diameter, and middle and upper diameter classes are preferred (9–11 mm and over 11 mm). A deep V shape is machine cut and the two pieces of wood are grafted together. This graft can be performed by hand using the whip-and-tongue technique (Lugli and Musacchi, 1993). After grafting, the wound is covered with paper tape and soaked in melted wax to cover all the wounded surfaces. The grafted combination must be stored at 15–20°C for 10–15 days to allow the wounds to heal. The grafted tree should be stored at 4–6°C and transferred to the orchard when the risk of a hard freeze is low (Fig. 11a,b and c).

6.4 'Knip tree'

The knip tree is the best choice for a spindle training system and allows production in the same year of planting if all the management techniques are done correctly. By the second year in the field, it is possible to obtain a significant crop of 10–15 t/ha.

Knip is a Dutch word, which means cut, and this explains why the basal zone of the tree is older when compared to the upper zone. This technique started when unsold trees were planted and headed back in the nursery and allowed to grow a new leader (Ravaglia and Corelli Grappadelli, 1990). Italian nurseries developed a two-year cycle to produce knip trees; one-year-old trees are headed back to 50–70 cm, depending on where the first scaffold of branch needs to (or should) be located. A single shoot is selected and treated with growth regulator to induce feather formation. These trees usually develop feathers with a wider crotch angle and early bearing (Fig. 12).

Figure 11 (a) Bench graft using the whip-and-tongue technique. (b) Bench grafts sealed with wax. (c) Bench grafts just planted in the nursery.

Figure 12 'Knip' tree. The upper part of the tree is 1-year-old, while the lower part is 2 years old.

6.5 Bi-axis tree from the nursery

Bi-axis trees are created in the nursery with the 'Bibaum®' technique. This is where two buds of the same variety are grafted at the same level on the rootstock and the two axes are developed from the inserted buds. Another possible method to achieve a bi-axis is to bench-graft a bare rooted rootstock with two buds. In this case, the insertion point of the two axes will be at different heights and may make the resulting tree unbalanced especially if the management in the nursery is not performed correctly and the axes have different diameters (Musacchi, 2008) (Fig. 13).

6.6 Potted plants

Pre-potted trees are widely used for many evergreen species like olive and citrus but are new in the apple market. The main techniques to produce potted trees is dormant mini-chip-budding on rootstocks from *in vitro* micropropagation, or bench grafting if the rootstocks are obtained by layering or stool-bedding. There are advantages for the nursery that can be summarised as: produce a tree in one year, more convenient shipping, easier to commercialize, and recovery in inventories because trees can be reused the following year. Benefits to growers include reduced transplant stress leading to reduced

Figure 13 Bi-axis (Bibaum®) technique produces tree with two axes in the nursery and has been patented by an Italian nursery.

initial tree loss, and trees come into production earlier (Ravaglia and Corelli Grappadelli, 1990; Lugli and Musacchi 1993) (Fig. 14).

6.7 Nursery application of plant growth regulators (PGRs)

Not all varieties respond the same way to the techniques above, and early branches are often too vigorous, too erect, and at an angle that is close to insertion. Alternative methods to obtaining the desired branches include the localized application of PGRs typically with a mix of gibberellin (GA_{4+7}) and benzyl adenine (BA). Promalin® (1.8% GA_{4+7} and 1.8% BA) can induce feather production in varieties like 'Red Chief' and 'Golden Delicious' (Forshey, 1982; Musacchi and Costa, 1992). Products based on GA_{4+7} plus BA inhibit flowering but may produce longer feathers. BA alone does not inhibit flowering, so this would be useful in trees where immediate production is desired (Elfving and Visser, 2007; Sazo and Robinson, 2011). Some nurseries with well-formed trees (like knip trees or those produced in a 2-year cycle) apply a product like ethephon in the fall that releases ethylene and increases flower bud formation.

Figure 14 Tree produced in container.

7 Control of vegetative growth using plant growth regulators (PGRs)

Management of the growth of apple trees is an ongoing activity. The goal is to maintain a balance between vegetative growth and cropping but management of this is only partially under the control of the orchardist through pruning and tree training. Frost, or freeze damage during the winter may kill flowers, thus reducing the crop. Under these circumstances terminal growth is frequently excessive and requires supplemental control. The use of PGRs is the most effective, convenient and cost-effective tool available and these can be used occasionally or yearly depending on the growth control required. There are two primary growth retardants in general use today: prohexadione-calcium (Pro-Ca) and ethephon (Ethrel®).

7.1 Growth control using prohexadione-calcium

Pro-Ca is a gibberellin biosynthesis inhibitor. Because it blocks GA synthesis late in the biosynthetic pathway (between GA_{20} and GA_1) (Evans et al., 1999), it has relatively few other effects that cannot be attributed directly to inhibition of GA synthesis. The primary reason to apply Pro-Ca is to reduce terminal growth by reducing internode length. This reduces crowding between trees due to shoot competition, increases light penetration into the tree, improving fruit quality, and reduces the amount of pruning required.

There are three Pro-Ca formulations available: Apogee® and Kudos® contain 27.5% a.i., whereas, Regalis® is a 10% a.i. formulation. Pro-Ca at 27.5% may precipitate out of solution if 'hard' water is used so to remedy this situation an equivalent weight of ammonium sulphate to Apogee® or Kudos® must be added to the spray tank. Alternatively, a commercial water conditioner may be used following label directions. Regalis® does not require water adjustment since it is formulated with the proper amount of ammonium sulphate. A non-ionic or organosilicate surfactant is recommended to use with all formulations. Pro-Ca should not be applied with calcium sprays since they will reduce its effectiveness.

Apple shoot growth is most rapid early in the season, starting near bloom. To achieve the most growth control the initial Pro-Ca application should be made as soon as there is sufficient leaf area to absorb Pro-Ca, as it requires 10–14 days after application for growth control to become effective. Measureable and commercially significant growth retardation can be achieved with an initial application of 60 $mg \cdot L^{-1}$ better growth control is possible with rates of 120–240 $mg \cdot L^{-1}$. Generally, a second application is required 2–4 weeks after the first; further follow-up applications are frequently needed. As the season progresses, acceptable growth control can be achieved with lower rates. The amount applied and the number of applications required for season-long growth control depends on several factors including the cultivar, its vigour, the rootstock, soil, the weather, moisture during the growing season and the geographical location. Usually, more applications are required in orchards in warm climates where the growing season is longer. Typically, 2 to 3 applications of Pro-Ca would be used for season-long control in the major fruit-growing areas.

Pro-Ca often results in increased fruit set by reducing the intensity or extent of fruitlet drop. This is the same physiological period where chemical thinners cause fruit abscission (see later section). Pro-Ca inhibits drop of some fruits that would normally be encouraged to drop by post-bloom chemical thinners. Therefore, when Pro-Ca is used in blocks where thinners are or may be applied, it is recommended that more aggressive thinning should be done to achieve an appropriate final fruit set.

Pro-Ca provides excellent control of shoot fire blight (*Erwinia amylovora*) even though it does not have any antibacterial activity against the fire blight bacteria. It is believed to act by thickening the cell walls in the shoot, thus making them less permeable and susceptible to infection. The initial application should be made early, at petal fall of the king flower. Since effective control depends on growth retardation, the spray will only confer resistance to fire blight when growth retardation has been initiated, 10–14 days after application. Rates of 6 to 12 oz/100 gal (125–250 $mg \cdot L^{-1}$) are generally used. Pro-Ca does not afford any direct protection from blossom blight. Pro-Ca is a very good alternative to the use of antibiotics for control of shoot fire blight bacteria. Since Pro-Ca works by altering the cell structure of the inhibited shoot, the bacteria cannot build up resistance to overcome the Pro-Ca action. Antibiotics can be reserved for blossom blight and control of shoot blight following a traumatic event in the orchard such as hail.

7.2 Control of vegetative growth with ethephon

Ethephon is a very effective growth retardant especially when applied from bloom to a week after bloom. It is also a good fruit thinner when applied over the same period of time (see later section). Consequently, when ethephon is used to control growth it may also cause significant crop reduction and therefore should not be used routinely for growth control on bearing trees due to its thinning potential. Ethephon may be useful in orchards that have lost a crop due to frost. Application of 500 $mg{\cdot}L^{-1}$ when shoots are 10–15 cm will retard growth and help the trees to re-establish the delicate balance between vegetative growth and cropping.

Ethephon may be used to control growth of non-bearing trees, but this use is now far less common because of the dominant use of precocious dwarfing rootstocks in new plantings. On nonbearing trees, rates of 300–500 $mg{\cdot}L^{-1}$ are commonly applied when the terminal growth reaches 7–12 cm. Growth control is more difficult on trees propagated on vigorous rootstocks. Ethephon will enhance flower bud formation on these more vigorous trees.

7.3 Promotion of flowering on bearing trees

Apple trees tend to be biennial, producing a heavy crop of small fruits one year followed by a year when production is low or when there is no crop. Cultivars differ in their biennial bearing tendencies. NAA and ethephon are typically used to thin fruits (see later section), but they also have the capability to increase flower bud formation, independent of their ability to thin. If these PGRs are frequently applied at low rates, and not during the thinning 'window of opportunity' (bloom to 25 mm), they have the ability to increase flower bud formation without causing thinning. There is a general consensus, based on experience that either 5 $mg{\cdot}L^{-1}$ NAA or 100 $mg{\cdot}L^{-1}$ ethephon may be effective when fruits reach 25–30 mm in diameter. The spray should be repeated three more times at 7–10 day intervals. The effectiveness of these sprays differs from year to year and there is no guarantee that they will actually increase flowering. This is however the best current option available to influence the following season's flowering on bearing trees once a crop has been set. This practice should not be used on certain cultivars, for example, 'McIntosh', 'Macoun' or 'Honeycrisp', because there is the possibility of advancing ripening if ethephon is used during this time period.

8 The physiological basis of chemical thinning: current situation and new perspectives

8.1 Chemical thinning

Chemical thinning has been practised in orchards since it was first recognized in the 1930s to help alleviate biennial bearing. Initial efforts were made to learn how to most effectively use the thinners available. Many empirical thinning studies were done, but variability and lack of consistency of the thinning responses persisted. More recently, progress has been made and more emphasis has been placed on attempting to determine how individual thinners work, gaining a greater understanding of how environmental conditions influence fruit set and thinner response and recognizing the changing susceptibility of fruits to thinners as they go through different physiological stages of development. As the lack of available labour for hand-thinning orchards has become an ever-increasing problem facing the long-term sustainability of the apple industry, the interest in chemical thinners has continued to grow.

A number of studies have been done to determine the mode of action of specific thinners. While there is no consensus on a specific mode of action, it appears that they may act by either influencing the amount carbohydrate available to the developing fruits or altering the auxin production or movement in the fruits. Bangerth (2004) reported that young fruits that persist to harvest have a continuous supply of auxin moving from the developing fruits, down the pedicels and through the abscission zone. Abscising fruit has a reduced supply of auxin moving from the fruit through the abscission layer, and this triggers the production of enzymes that results in the destruction of the cells in the abscission zone. When the carbon supply available to young developing fruitlets drops to a critical low level, abscission is irreversibly initiated. When the carbohydrate available to the young developing fruit is reduced either by shading (Byers et al., 1990a) or by applying photosynthetic inhibitors (Byers et al., 1990b), the fruit could be induced to abscise. Lakso et al. (2006) has shown that conditions that result in low availability of carbohydrates during peak carbohydrate demand periods result in fruit abscission. It may be that the carbohydrate deficit may serve as the precipitating event which then signals the reduction and movement of auxins to the abscission zone.

Variability of the weather has been recognized as a major contributing factor in the lack of thinning consistency, even in the same block using the same thinning and the same rates in different years. Many researchers have made contributions to help quantify and clarify the weather influence on thinning. Williams and Edgerton (1981) identified 15 factors that could influence thinning response, but temperature and solar radiation before, during and in the 3–4 days following thinner application have emerged as the factors that most often influence response.

Apple may be chemically thinned over a wide period of time starting as early as bloom and extending until the fruit reaches about the 25 mm fruit growth stage. As a fruit grows, it undergoes physiological changes that of necessity require different thinning strategies using different thinning chemicals. During bloom, the objective of applied thinners is to damage the stigma or interfere with pollen germination, pollen tube growth or fertilization. Once the king flower is successfully pollinated, the mode of action of most commonly used blossom thinners is to inflict some type of lethal damage to the ovary. Blossom thinners are routinely used in arid growing regions; if they are applied in humid growing climates, these compounds may cause some phytotoxicity. In addition to potential damage, growers appear to be somewhat reluctant to thin before the pollination period ends and when the danger of frost has subsided so that they can assess the initial set. Thinning decisions in humid regions should be reevaluated since thinning at bloom provides the best opportunity to increase fruit size and ensure return bloom (Batjer and Hoffman, 1951); the perceived danger of thinning at bloom may be more a perception than actually documented.

Petal fall, immediately after bloom, offers another important thinning opportunity. The bloom period has passed and the danger of frost has passed. Meaningful thinning can occur, but over thinning rarely happens if thinners are applied at this time (Greene, 2002). Physiologically, cell division is occurring and seed development is starting; however, visible changes are small, subtle and difficult to measure.

Fruits are most vulnerable to applied thinners during the 8–14 mm growth stage when typically there are 3–5 fruitlets on each spur and they are in the log growth phase. Storage carbohydrate has been nearly depleted and the primary sources of carbohydrate are now supplied by the leaves associated directly with the spurs. Import of photosynthate from other portions of the tree does not generally occur until later when shoot growth reaches about 10 inches (25 cm). Chemical thinners applied at this time create a stress which may amplify the differences between photosynthate required for all fruits to persist and the ability of the spurs to supply the required carbohydrate for continued growth of all fruit. When a carbohydrate deficiency occurs, an internal signal is given, resulting in the abscission of the weakest fruit.

As fruits grow larger (>15 mm), they become progressively more difficult to thin (Greene, 2002; Schwallier, 1996). At this stage of development, ethylene production declines, there is less intra-fruit competition due to earlier fruit abscission and fruits are starting to accumulate starch. Carbohydrate deficits are less likely unless unusual weather conditions occur, which would then allow thinners to cause some fruit abscission. As fruits approach 17–20 mm diameter, they regain more sensitivity to ethylene and ethephon, or 1-amino-cyclopropane-1-carboxylic acid (ACC) application appears to be more effective. Once fruit has grown larger than about 25 mm, further reduction in crop load is generally very difficult.

8.2 Thinning chemicals

Thinning chemicals may be roughly placed in two categories: caustic thinners and hormone thinners. Caustic thinners are generally applied during bloom with the objective of damaging the flower, preventing the pollen germination or inhibiting pollen tub growth and thus preventing fertilization of the ovule. The king flower is generally the largest flower in the cluster and it is the first flower to open so the goal is to assure that the king flowers set fruit while inhibiting set of the remaining lateral flowers. Attempts are made to time the blossom-thinning application to allow pollination and fertilization of the king and then apply a blossom thinner to inhibit set of the remaining lateral flowers in the cluster. Sometimes the window of opportunity to apply blossom thinner is narrow and can be measured in hours.

A concerted effort was made to find a replacement for Elgetol (DNOC, sodium 4, 6-dinitro-ortho-cresylate) when its registration was cancelled in 1990, but few replacement thinners are in general use or even remain registered. Lime sulphur is a very commonly used blossom thinner, frequently combined with fish oil or other oils to enhance activity. Ammonium thiosulfate (ATS) is a foliar nitrogen fertilizer. Although not specifically registered for use as a thinner on apples, if ATS is applied to trees at bloom at a rate of 2–4 gal/acre (18.8–37.5 L·ha^{-1}) it has been shown to be a viable thinner.

Blossom thinners are routinely used in dry, low-humidity climates. If blossom thinners are applied before wet or high-humid periods it is not uncommon to have a significant amount of foliage damage. Spur leaves are very important to support fruit cell division/early growth. The observation has been made that application of blossom thinners in wet conditions may lead to thinning, but may not manifest in increased fruit size at harvest due to foliage damage. Blossom thinning removes the fruit early, thus providing the best opportunity to assure return bloom. It therefore remains a viable option to use, especially on cultivars that are noted for biennial bearing.

The majority of thinning is done using hormone-type thinners. Unlike caustic thinners that act by causing some type of damage, hormonal thinners act by creating some form of stress and interacting with physiological events in the plant, which are frequently modified by the weather. Fruit physiological changes occur as the fruit continues to grow. Therefore, selection of thinner(s) may change over time as the fruit grows and undergoes these developmental changes. The mode of action of specific thinners is not definitively known, although many investigators have identified physiological events that may be altered by thinners such as reduction in photosynthesis, increase in respiration, limitation of photosynthate transport or enhanced production of ethylene (Dennis, 2002). Disruption of one or a combination of these could result in fruit abscission. As a fruit increases in diameter above 15 mm, it becomes progressively more difficult for thinners to cause fruit abscission. In general, when fruit diameter exceeds 25 mm, it no longer responds to hormonal thinners (Greene, 2002).

The timing of application and the circumstances under which individual thinners are used have previously been reviewed: Byers (2003), Greene (2002), Schwallier (1996), Wertheim (2000) and Williams and Edgerton (1981).

Naphthaleneacetic acid (NAA)

NAA is the oldest and perhaps the most potent of the thinners now in general use. It may be used at concentrations from 3 to 20 mg·L^{-1}, but the rates used generally are between 5 and 12 mg·L^{-1}. It may be applied between bloom and when the fruit grows to about 15 mm in size. Frequently leaf epinasy is noted following application. When higher rates are applied on fruits (>15 mm), especially during very warm weather, reduced fruit size or no increase in fruit size may be observed even though crop load is reduced. It is frequently chosen for use on cultivars known to be reluctant to thin. Because NAA has the reputation for being a strong thinner, growers use it with caution. It is commonly combined at lower rates with other thinners, especially carbaryl.

Naphthaleneacetamide (NAAm, NAD)

NAD was tested and evaluated at the same time that NAA was evaluated. NAA became the thinner of choice because it was more potent. NAD is generally used at a rate of 40 to 50 mg·L^{-1} and the time of application is generally at bloom or petal fall. It can cause pygmy fruit formation when applied late and on 'Delicious'. It can cause leaf epinasty, but it is far less severe than that caused by NAA. NAD is frequently used in combination with carbaryl and then followed by another thinner at the 8–14 mm stage.

Carbaryl

Carbaryl is perhaps the most versatile thinner in general use. It is an insecticide and a mild thinner that can be used over a wide range of developmental stages from petal fall until the fruit reaches 18–20 mm. It is a popular thinner since it rarely over thins. Unfortunately, it is very toxic to bees, so the earliest time it should be used is after bloom, when bee hives have been removed from the orchard. Unlike most thinners, the thinning response is not concentration dependent.

Benzyladenine (BA)

BA is the newest thinner to be registered for use on apples. Effective thinning concentrations are 50 to 150 mg·L^{-1}, although the most commonly used rates are between 75 and 100 mg·L^{-1}. BA, when used alone, is a mild thinner, but when combined with carbaryl, it is considered a potent thinner that may be similar in strength to NAA. BA can also increase fruit size beyond the size that can be attributed to a reduction in crop load (Wismer et al., 1995).

Ethephon

Ethephon is generally not considered a mainstream thinner in most growing regions. Flower and fruit susceptibility to ethephon varies depending on the stage of development. Ethephon is an effective blossom thinner (Jones et al., 1990), but it loses some of its effectiveness as the fruit develops. Fruits redevelop thinning sensitivity between 16 and 22 mm diameter (Marini, 1996). Since it is the only thinner that can thin large-sized fruits somewhat consistently, it is sometimes referred to as a 'last chance thinner'. The primary use of ethephon now appears to be in orchards where growers have failed to achieve acceptable thinning during the normal thinning time.

8.3 Precision thinning

Thinning is the most important management activity a grower is required to do. Insufficient thinning will lead to the production of small fruits that have poor quality and thus lower value. With most cultivars, insufficient thinning will lead to poor return bloom that will reduce or eliminate crop load potential the following year. Excessive thinning will result in reduced yield of poor-quality fruits. Establishment and growing costs have escalated to a point where large fluctuations in income are no longer economically sustainable.

In the past, crop load was adjusted in many growing areas primarily during the period from bloom until fruitlets reach about 15 mm diameter. Chemical thinners were used to accomplish this. However, chemical thinning has a long history of being erratic and inconsistent. Much of this is attributed to varying weather conditions prior to and following thinner application. Attempts have been made over the past several years to incorporate information gained from research coupled with the development of computer models to make the thinning process more predictable and to result in consistent and acceptable final crop loads. This process is now termed Precision Crop Load Management, which involves three activities: 1. precision dormant pruning adjustment, and the use of computer models to 2. predict thinning prior to thinner application and 3. early prediction of effectiveness of thinning sprays.

Precision dormant pruning

Precision thinning starts by growers determining the goal number of fruits to harvest from a tree or unit. This is a deviation from earlier orchard practices, where crop loads per tree were generally not set. Setting fruit crop goals is quite easy on new HDP, where tree numbers and total desired yield per acre are used. It is much more difficult on larger trees where branch subunits are relied upon.

Thinning at pruning time offers several advantages. It reduces early competition among flowers. Weaker flowers may be pruned off, thus allowing resources and new growth on a tree to be diverted to the large flower buds that have a greater potential to produce large fruits.

Once the target number of fruits per tree is established, the number of flower buds on a tree needs to be determined. This may be challenging at first, but with a little practice, it can be done somewhat accurately. First, select five uniform and representative trees in a block and count all flower buds. Next, calculate the blossom cluster averages for the counted trees. Finally, multiply the predetermined desired number of fruits on each tree by 1.5, which will give, for most cultivars, the number of flower buds that should remain after pruning.

Assuming an orchard is planted at a 12 × 3 ft (3.7 m × 0.92 m) spacing, it would contain 1,210 trees. If the desired yield is 59,300 kg·ha^{-1}, then 100 fruits per tree would be the desired final crop load. If you multiply 100 fruits by 1.5 the target number of flowers to remain should be 150. Trees would be pruned to achieve that number. Initially, pruning followed by counting would be required to fine-tune the process.

Apple carbon balance model

Precision thinning is a process to methodically adjust crop load to a predetermined and calculated level. This method utilizes two recently developed computer models, the apple carbon balance model and the fruit growth model. These two models work together to allow methodical reduction in crop load.

The apple tree carbon balance model was developed by Alan Lakso at Cornell University using a mathematical model to estimate apple tree photosynthesis, respiration and growth of fruits, leaves, roots and woody structures (Lakso et al., 2006, 2007). The model uses daily maximum and minimum temperatures and solar radiation to calculate the daily production of carbohydrate and allocates the available carbohydrate to the organs in the tree including the developing fruits. Temperature and light are primarily responsible for the carbon balance in a tree. In low-light and high-temperature conditions there would be a carbon deficit and thinning would be easy. Conversely, high light and lower temperatures favour carbohydrate accumulation and thinning would be difficult. The model quantifies the carbon balance using local weather station inputs. The data can then be used as a

Figure 15 Application of Promalin® to elongate the shape of the fruit in 'Delicious'.

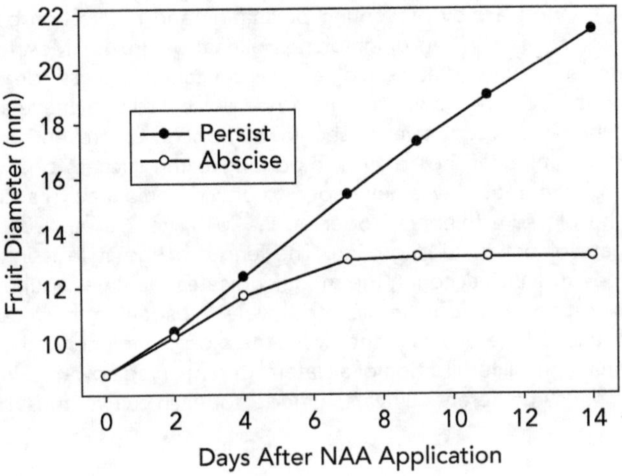

Figure 16 Growth of a fruit that will persist (solid circles) and a fruit that will abscise (open circles) following an NAA thinning spray during the 7–10 mm fruit size stage. The Fruit Growth Model predicts that a fruit will abscise if the growth rate of a fruit slows to 50% or less of the growth rate of the fastest growth fruit over the period from 4 to 7 days after thinner application. At 7 days after thinner application, the growth rates of fruit destined to drop are significantly different at $P = 0.01$ percent level (**) from the fruit that will persist.

guide to determine if thinning at a given time is appropriate and, if so, provide guidance about the strength of the thinner needed.

Apple fruit growth model

This model is based on the observation that fruits that abscise following thinner application, slow growth well in advance of the time that they actually abscise (Greene et al., 2013). Generally, 3–4 days is required for a thinner to affect fruit growth. In this model, spurs are tagged prior to thinner application. At 3–4 days after thinner application, all fruits in the spurs are measured. Fruits must be at least 6 mm in diameter for initial measurement. A second measurement is made 4–5 days after the initial measurement (Fig. 16) and the growth rate of each fruit is calculated. Fruits with a growth rate of less than 50% of the fastest-growing fruits on a tree are predicted to abscise. This model requires establishing a desired end crop load. The model predicts the final crop load if no further thinning is done. If the predicted crop load is insufficiently reduced, then a follow-up thinner spray is applied. Fruits will then be measured again to determine if the crop load has been lowered to the target level. Implementation and details for grower use are outlined by Robinson et al. (2013b) and Schwallier and Irish-Brown (2015).

Pollen tube growth model

Blossom thinning is done routinely in areas that are typically arid. It has the advantage over post-bloom thinning since it can have the greatest effect on fruit size and return bloom

(Greene, 2002). Thinners act by preventing pollination and fertilization by damaging the anthers, stigma and style of flowers or inhibiting pollen tube growth (McArtney et al., 2006). Blossom thinners can be very erratic since timing is critical for success. Ideally, there should be a sufficient number of king flowers open and be pollinated to achieve a full crop. Once this is accomplished a blossom thinner should be applied to prevent pollination of the remaining flowers. Timing of the spray is important to the success of blossom thinners. The pollen tube growth model was developed to provide guidance to accurately time the application of the blossom thinners (Yoder et al., 2013; Peck et al., 2015). Temperature influences the rate of pollen tube growth. The length of the style is initially measured, and taking temperature into account, the model indicates the time required for the pollen tube to grow through the style to fertilize the ovule. As soon as the model predicts the pollen tube has reached the ovules in the ovary, the blossom thinner should be applied to discourage pollination and fertilization of all lateral or remaining flowers. Since pollen tubes grow at variable rates in different cultivars, a model for each cultivar must be established.

9 Improving fruit appearance

Appearance of the fruit often plays a dominant role in purchasing decisions by customers and greatly impacts the price that the grower receives from the packing house. Good cultural techniques and management of the tree are an essential prerequisite but frequently PGRs may be required to improve fruit appearance even when the tree management is excellent.

9.1 Fruit russet control

Russet is a physiological disorder that occurs during the early stages of fruit development. Rapidly dividing epidermal cells increase in size faster than the cuticle can expand. The cuticle cracks, exposing some of the underlying epidermal cells. As a result, cambium is formed, producing corky periderm russeted areas (Faust and Sherer, 1972). Water often plays a critical and central role in russet formation, as it tends to be produced under conditions of high humidity or frequent rain or dew. These conditions lead to an apple developing a thin cuticle that is prone to cracking under high turgor conditions (Faust and Sherer, 1972). Some varieties are also genetically pre-disposed towards russet.

Foliar sprays of GA_{4+7} can either eliminate fruit russeting or reduce its severity (Taylor, 1975). When applied near petal fall at rates of 25 to 250 $mg{\cdot}L^{-1}$, GAs are known to inhibit flower bud formation even when applied at moderate rates (Tromp, 1982). By making multiple sprays (generally 4) of GA_{4+7} at a rate of 10 $mg{\cdot}L^{-1}$, fruit russeting can be reduced with minimal effect on return bloom (Miller, 1988). High rates may improve russet control but the increased control may not be justified because of the possible adverse effect on return bloom (Eccher and Boffelli, 1981).

Scarf skin is a disorder of apples that is characterized as fruits having grey flecking and/or a milky appearance on the surface. Presence on the fruit leads to a reduction in grade and therefore lower returns to the grower. Treatment of susceptible cultivars with multiple applications of GA_{4+7} (as described for russet) will reduce the incidence of scarf skin, but it may not reduce the incidence to commercially acceptable levels (Ferree et al., 1984).

9.2 Fruit shape

Fruit shape can be a significant component for customer recognition of variety when choosing an apple to purchase. Cultivars that are blocky and have prominent calyx lobes are perceived by some customers as being favourable. Apple elongation and calyx lobe development are promoted by low temperatures during early fruit growth (Shaw, 1914), typical in the growing conditions in Washington state. The Washington apple industry has been able to market shape, especially for 'Delicious' as a quality parameter. Cool nights cause increased production of endogenous GAs. Abbott Laboratories developed a proprietary product, Promalin®, to elongate the shape of 'Delicious'. It contains equal quantities of GA_{4+7} and 6-benzyladenine (BA) and was registered for use in 1979. The application of this product near bloom allows growers in other regions that do not have the timely cool temperatures at night to produce the characteristic elongated 'Delicious' with prominent calyx lobes. Unrath (1974) demonstrated that when Promalin® was applied at 25 $mg \cdot L^{-1}$ at bloom time, fruit length/diameter ratio and weight increased. Using GA rates above 50·$mg \cdot L^{-1}$ could result in a reduction in return bloom. GA_4 and GA_7 increase fruit elongation but GA_7 is known to reduce flower bud formation more than GA_4 (Tromp, 1982).

10 Preharvest application of plant growth regulators

The apple is a climacteric fruit; the onset of the climacteric is generally considered to signal a definitive stage and marks the development of maturation and ripening (Watkins, 2003). In a climacteric fruit, ethylene advances the timing of the initiation of the climacteric and advances ripening. Ethylene is an endogenous hormone that plays a dominant role in the ripening process in apples and the initiation of preharvest drop. Stressful conditions (drought, high temperature) can trigger abscission prior to the fruit reaching an acceptable level of maturity. Additionally, if orchardists do not have sufficient labour to harvest fruits in a timely manner, drop control measures would be required. Since drop is stimulated by ripening or induced stresses and these represent major economic losses, strategies are needed to be employed to control this abscission to avoid losses.

10.1 Delay preharvest drop

Many commercially important apple cultivars suffer from the malady of fruits dropping before they have reached the maturity that would indicate optimum commercial harvest date (Greene et al., 2014). With drop-prone cultivars such as 'McIntosh', 'Macoun' and 'Honeycrisp', it is often normal to have 20% of the fruit drop prior to harvest and in some years, it is not uncommon to have up to 50% drop. There are three commercially available drop control compounds: naphthaleneacetic acid (NAA), aminoethoryvinylgrycine (AVG, ReTain®) and 1-methycyclopropene (1-MCP, Harvista™).

Aminoethoxyvinylglycine (AVG)

AVG is a naturally occurring PGR that was discovered in the early 1970s. Its primary mode of action is to inhibit ethylene biosynthesis (Boller et al., 1979). Bangerth (1978) was the

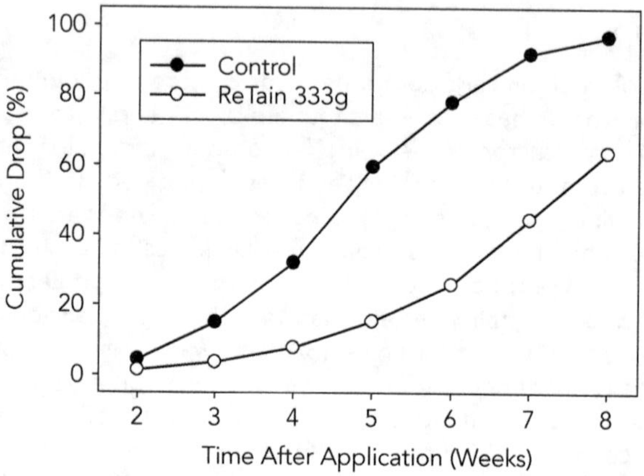

Figure 17 Preharvest drop of 'McIntosh' apples sprayed with one package of ReTain® (333g/acre, 825g/ha) 3–3.5 weeks before anticipated harvest. Each data point represents the average cumulative drop over 7 years (2009–15).

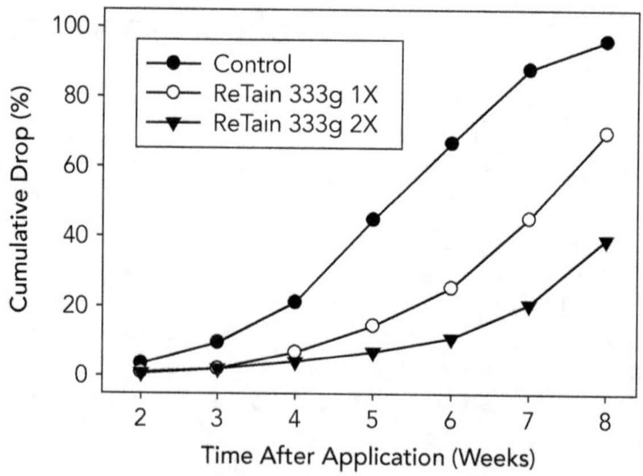

Figure 18 Preharvest drop of 'McIntosh' apples following one application of ReTain® at 333g/acre (825g/ha) 3–3.5 weeks before anticipated harvest and two applications of ReTain® at 3–3.5 weeks before anticipated harvest and again two weeks later. Each data point represents the average cumulative drop over 3 years (2013–15).

first to demonstrate that it could inhibit preharvest drop of apples. Development of this compound was not continued because of the high cost of production. In response to the withdrawal of registration of daminozide in 1989, an important drop control PGR, development of AVG was resumed. Abbott Laboratories registered a proprietary formulation of AVG in 1997, ReTain® to control preharvest drop and delay ripening of apples.

The extent of and the strategy for preharvest drop control is influenced by cultivar, the time of application, the objective of the application and the amount applied.

For most cultivars where drop control is a significant problem, the initial application is made 3–4 weeks before the anticipated start of normal harvest. Following application, it generally requires 10–14 days for drop control to become effective (Greene and Schupp, 2004; Schupp and Greene, 2004; Greene, 2006). If an application is made later, there is a chance that early drop may occur before AVG becomes effective. Application earlier may result in early drop control but loss of drop control later during the harvest season.

For most drop-prone cultivars, an initial application of 1 package (333 g of 15% ReTain® formulation per acre) is used. This rate will frequently control drop and restrict drop loss to about 20% for 5 and 6 weeks following application (Fig. 17). It has long been known that higher rates of AVG will produce greater drop control. In 2015, the ReTain® label was updated to allow application of two 333 g packages of ReTain® in a season. This additional AVG further extended the effective period of drop control (Fig. 18). A lower (half) rate is usually used on low ethylene-producing cultivars such as 'Gala' and 'Honeycrisp'. If full rates are used on these cultivars, ripening may be delayed too long (3–4 weeks). The maximum amount of ReTain® that can be applied to an orchard is 2 packages/acre/year. This amount may be applied at one time or it can be applied in two or more instalments. The timing and amounts applied at each time depend on the objectives of the grower. The earlier in the season ReTain® is applied, the greater the delay in ripening. Maximum delay in ripening would occur if the higher rates of ReTain® were applied 4 weeks before harvest. Frequently growers use a split application strategy to have less reduction in ripening and use later application(s) to extend the period of drop control.

A 100% organosilicone surfactant will enhance foliar uptake and improve performance, but not all surfactants are equally effective. Currently, Silwet L-77 and Sylguard 309 are recommended at the rate of 0.1% v/v. Other organosilicate surfactants may be effective, but they would require testing. The organosilicone surfactant also provides some rainfastness if the spray has had time to dry (Greene, 2006).

1-methylcyclopropene (1-MCP, Harvista™)

1-MCP is an inhibitor of ethylene action that acts by binding irreversibly to ethylene receptors in the treated tissue. If binding sites are occupied by 1-MCP, a plant is unable to respond to ethylene even if ethylene is present (Fan et al., 1999). It is widely used for application to harvested fruit in sealed rooms. Because 1-MCP is a gas, application to trees in the orchard provides logistical challenges. Initial studies used a sprayable formulation that was applied immediately after mixing with oil and water, thus minimizing loss of efficacy due to off gassing. Application of Harvista™ is now done by a proprietary in-line injection system that is done under the direction of its development company AgroFresh. AgroFresh is currently refining the application process with the development of a sprayer specifically designed to apply Harvista™.

Harvista™, like ReTain®, may also delay red colour development, starch degradation, flesh firmness loss, degradation of starch and delay watercore development. Following application, drop control may be achieved within 3–4 days. Since it can be applied closer to normal harvest, fruits have a longer time to mature normally on the tree and to develop red colour compared with ReTain®, which must be applied earlier. Loss of drop control with Harvista™ is associated with ripening fruit starting to generate new ethylene-binding sites, which can then respond to ethylene being produced by the fruit. A follow-up application of Harvista™ may saturate newly generated binding sites, thus significantly extending the period of drop control.

Naphthaleneacetic acid (NAA)

It has been known since the 1930s that auxins can inhibit preharvest drop of apples. Over the past 75 years several auxins have been registered for preharvest drop control, but NAA remains the only auxin registered for use for this purpose.

The standard recommendation for drop control is to make the initial application before the first sound fruit starts to drop. NAA at 10 $mg{\cdot}L^{-1}$ is frequently used and provides drop control lasting 7–9 days. A second supplemental application can be made to extend the period of drop control to about 14 days. NAA may advance ripening on apples when used to control drop, the amount of stimulation depends on the cultivar used, the weather following application, time interval between application and harvest, and the amount of NAA applied. Higher rates of NAA, warm weather (mid 80°F, 35°C) and a delay in the harvest of treated fruits are likely to stimulate ripening and lead to reduced post-harvest storage life. Drop control following NAA application can be seen within 2–3 days of application, unless significant drop was already occurring at the time of application.

NAA is sometimes used in conjunction with ReTain®. Since ReTain® requires 10–14 days for drop control to be initiated, NAA may be included in the application for near-term drop control. A minimum of a half rate of ReTain® is required to counteract the tendency for NAA to advance ripening. Robinson et al. (2010) and Yuan and Carbaugh (2007) reported that addition of NAA in the second of a ReTain® split application enhanced the drop control achieved by a ReTain® application alone. An alternative approach to drop control by NAA was reported by Marini et al. (1993). Application of several low rates of NAA (5 $mg{\cdot}L^{-1}$) starting 4 weeks before anticipated harvest resulted in an excellent drop control of 'Delicious' apples that was equal to or better than a higher rate made at the traditional time closer to harvest. This approach has been useful on 'Delicious', but it has not been successful on more ethylene-sensitive cultivars such as 'McIntosh'.

11 Future trends in research

Periodically changes happen in the design and management of orchards. These shifts are generally due to research results (genetics and breeding) or to specific needs of the industry to make the production more sustainable. One example was the adoption of the dwarfing rootstock M9, which allowed for a reduction of canopy size, leading to training systems maximizing the increased light interception and the possibility to mechanize some operations. The next big revolution for training systems will be driven by the necessity to utilize robots or specialized equipment to perform harvest in an automated way, which may be due to the lack of labour and specialized workers. Other aspects that will drive new solutions will be the breeding of new genotypes resistant to disease (especially fire blight, scab and powdery mildew), with better internal quality and greater productivity.

12 Where to look for further information

A good source of additional information is the *Acta Horticulturae* published by the International Society for Horticultural Science (ISHS) – http://www.ishs.org/acta-horticulturae.

Two working groups are important to complete the information enclosed in this chapter: orchard and plantation systems (65, 114, 160, 243, 322, 349, 451, 513, 557, 732, 903, 1058) and plant growth and bioregulators in fruit production (34, 80 120, 134, 137, 179, 239, 394, 463, 653, 727, 774, 884, 1042). An extended publication on apple training systems and PGRs is also available in the book entitled *Apples*, edited by Ferree and Warrington and published in 2003 by CABI publishing (ISBN 0 851995926).

13 References

Avery, D. J. 1975. Effects of fruits on photosynthetic efficiency. In H. C. Pereira (ed.) *Climate and the Orchard*. CAB, Farnham Royal, UK, pp. 110–12.

Bangerth, F. 1978. The effect of a substituted amino acid on ethylene biosynthesis, respiration, ripening and preharvest drop of apple fruit. *J. Amer. Soc. Hort Sci.* 103: 401–4.

Bangerth, F. 2004. Internal regulation of fruit growth and abscission. *Acta Hort.* 636: 235–48.

Baldini, E. 1990. *Arboricoltura Generale*. Bologna, Italy: Clueb.

Batjer, L. P. and Hoffman, M. B. 1951. Fruit thinning with chemical sprays. U. S. Dept. Agr. Circ. 867. Washington, DC.

Boller, R. C., Herner, R. C. and Kende, H. 1979. Assay for the enzymatic formation of an ethylene precursor 1-aminocyclopropene-1-carboxylic acid. *Planta* 145: 293–303.

Byers, R. E., Barden, J. A. and Carbaugh, D. H. 1990a. Thinning of spur 'Delicious' apples by shade, terbacil, carbaryl, and ethephon. *J. Amer. Soc. Hort. Sci.* 115: 9–13.

Byers, R. E., Barden, J. A., Young, R. E. and Carbaugh, D. H. 1990b. Apple thinning by photosynthetic inhibition. *J. Amer. Soc. Hort. Sci.* 115: 14–19.

Byers, R. E. 2003. Flower and fruit thinning and vegetative: fruiting balance. In D. C. Ferree and I. J. Warrington (eds), *Apples Botany, Production and Uses*. Cambridge, MA: CABI Publishing, pp. 409–36.

Cain, J. C. 1971. Effects of mechanical pruning of apple hedgerows with a slotting saw on light penetration and fruiting. *J. Amer. Soc. Hort. Sci.* 96: 664–7.

Crassweller, R., McNew, R., Azarenko, A., Barrit, B., Belding, R., Berkett, L., Brown, S., Clemens, J., Cline, J., Cowgill, W., Ferree, D., Garcia, E., Greene, D., Greene, G., Hampson, C., Merwin, I., Miller, D., Miller, S., Moran, R., Obermiller, J., Rosenberger, D., Rom, T., Roper, T., Schupp, J., and Stover, E. 2005. Performance of apple cultivars in the 1995 NE-183 regional project planting: I. Growth and yield characteristics. *J. Amer. Pom. Soc.* 59: 18–27.

Curzel, G. 1994. Impianti ultrafitti a 'superspindle': tecnica o teoria? Rivista di frutticoltura e ortofloricoltura. n. 12.

Dallabetta, N. 2014. Effect of training systems and pruning methods on fruit quality in apple. Ph.D. Alma Mater Studiorum – University of Bologna, Italy.

Dennis, F. G. 2002. Mechanisms of action of apple thinning chemicals. *HortScience* 37: 471–4.

Diemoz, M., Vittone, G. and Pantezzi, T. 2002a. Potatura del melo: possibili evoluzioni. Terra Trentina: 21–23.

Diemoz, M., Vittone, G., Peano, M. and Pantezzi, T. 2002b. La potatura lunga del melo con diradamento delle lamburde. *L'informatore agrario*. 25: 43–4.

Diemoz, M. 2005. Allevamento del melo: Il solaxe e la 'conduite Centrifuge'. Institut agricole regional.

Dorigoni, A., Lezzer, P., Dallabetta, N., Serra, S. and Musacchi, S. 2011. Bi-axis: An alternative to slender spindle for apple orchards. *Acta Hort.* 903: 581–8.

Dunn, J. T. and Stolp, M. 1981. Apple production for mechanization harvesting on the Lincoln Canopy System. New Zealand Agriculture Engineering Institute. Extension Bulletin N. 10.

Eccher, T. and Boffelli, G. 1981. Effects of application of GA_{4+7} on russeting, fruit set, and shape of 'Golden Delicious' apples. *Scientia Hort.* 14: 307–14.

Elfving, D. C. and Visser, D. B. 2007. Improving the efficacy of cytokinin applications for stimulation of lateral branch development in young sweet cherry trees in the orchard. *HortScience* 42: 251–6.

Embree, C. G. and Nichols, D. S. 2005. Bio-regulators improve fruit size and color and reduce crop load and annual bearing of Honeycrisp apples. *Can. J. Plant Sci.* 85: 453–5.

Embree, C. G., Myra, M. T., Nichols, D. S. and Wright, A. H. 2007. Effect of blossom density and crop load on growth, fruit quality, and return bloom in 'Honeycrisp' apple. *HortScience* 42 (7): 1622–5.

Evans, J. R., Evans, R. R. Rogusci, C. L. and Rademacher, W. 1999. Mode of action, metabolism and uptake of BAS 125W prohexadione-calcium. *HortScience* 34: 1200–1.

Fan, X., Blakenship, S. M. and Mattheis, J. P. 1999. 1-methylcyclopropene inhibits apple ripening. *J. Amer. Soc. Hort. Sci.* 124: 690–5.

Faust, M. and Shear, C. B. 1968. Corking disorders of apples: a physiological and biochemical review. *Bot. Rev.* 34: 441–69.

Faust, M. and Sherer, C. B. 1972. Russeting of apples, an interpretive review. *HortScience* 7: 233–5.

Faust, M. 1989. *Physiology of Temperate Zone Fruit Trees*. Hoboken, MI: Wiley.

Ferguson, I. B. and Watkins, C. B. 1989. Bitter pit in apple fruit. *Hort. Rev.* 11: 289–355.

Ferree, D. C., Ellis, M. A. and Bishop, B. L. 1984. Scarf skin on 'Rome Beauty': time of origin and influence of fungicides and GA_{4+7}. *J. Amer. Soc. Hort. Sci.* 109: 422–7.

Ferree, D. C. and Schupp, J. R. 2003 Pruning and training physiology. In D. C. Ferree and I. J. Warrington (eds), *Apples, Botany, Production and Uses*. Cambridge, MA: CABI Publishing, pp. 319–44.

Forshey, C. G. 1982. Branching responses of young apple trees to applications of 6-benzyladenino purine and gibberellins A_{4+7}. *J. Amer. Soc. Hort. Sci.* 107: 128–32.

Giulivo, C. 1990. Principi fisiologici della potatura. In La potatura degli alberi da frutto negli anni '90 Verona 27/4/1990: 9–38.

Greene, D. W. 2002. Chemicals, timing, and environmental factors involved in thinner efficacy on apple. *HortScience* 37: 9–13.

Greene, D. W. and Schupp, J. R. 2004. Effect of aminoethoxyvinylglycine (AVG) on preharvest drop, fruit quality and maturation of 'McInosh' apples. II. Effect of timing and concentration relationships and spray volume. *HortScience* 39: 1036–41.

Greene, D. W. 2006. An update of preharvest drop control of apples with aminoethoxyvinylglycine (ReTain). *Acta Hort.* 727: 311–19.

Greene, D. W., Lakso, A. N., Robinson, T. L. and Schawallier, P. 2013. Development of the fruit growth model to predict thinner responses to apple. *HortScience* 48: 584–7.

Greene, D. W., Krupa, J. and Autio, W. 2014. Factors influencing preharvest drop of apples. *Acta Hort.* 1042: 231–5.

Grisvard, P. 1957. *La taille des arbres fruitiers*. Montpellier, Paris: La Maison Rustique.

Hampson, C. R. and Kemp, H. 2003. Characteristics of important commercial apple cultivars. In D. C. Ferree and I. J. Warrington (eds), *Apples, Botany, Production and Uses*. Cambridge, MA: CABI Publishing, pp. 61–90.

Hartmann, H. T. and Kester, D. E. 1990. Propagazione delle piante. Edagricole, Bologna.

Heim, G., Landsberg, J. J., Watson, R. L. and Brain, P. 1979. Ecophysiology of apple trees; dry matter production and partitioning in young 'Golden Delicious' apple trees. *J. Appl. Ecol.* 16: 179–94.

Howard, B. H., Skene, D. S. and Coles, J. S. 1974. The effects of different grafting methods upon the development of one-year-old nursery apple-trees. *J. Hort. Sci.* 49: 187–295.

Jones, K. M., Koen, T. B., Oakford, M. J. and Bound, S. A. 1990. Thinning 'Red Fuji' apples using ethephon at two timings. *J. Hort Sci.* 65: 381–384.

Lakso, A. N., Robinson, T. L. and Greene, D. W. 2006. Integration of environment, physiology, and fruit abscission via carbon balance modeling-implications for understanding growth regulator response. *Acta Hort.* 727: 321–6.

Lakso, A. N., Robinson, T. L. and Greene, D. W. 2007. Using and apple tree carbohydrate model to understand thinning responses to weather and chemical thinners. *N. Y. Fruit Q.* 15 (3): 17–20.

Larrive, G., and Lespinasse, J. M. 2002. Conduite du pommier – Experimentation sur la technique de l'extinction dans le bassin Grand-Sud-Ouest. *L'Arboricolture Fruitiere* 558: 35–9.

Lauri, P. E., Terouanne, E. and Lespinasse, J. M. 1997. Relationship between the early development of apple fruiting branches and the regularity of bearing – an approach to the strategies of various cultivars. *J. Hort Sci.* 72 (4): 519–30.

Lauri, P. E. and Lespinasse, J. M. 2000. The vertical axis and solaxe systems in France. *Acta Hort.* 513: 287–96.

Lauri, P. E. 2009. Developing a new paradigm for apple training. *The Compact Fruit Tree.* 42 (2): 17–9.

Lespinasse, J. M. 1980. La conduit du Pommier II. L'axe vertical la renovation de verges. INVUFLEC (Ed.), Paris.

Lespinasse, J. M. and Delort, F. 1986. Apple tree management in vertical axis: appraisal after ten years of experiments. *Acta Hort.* 160: 120–55.

Lespinasse, J. M. 1989. A new fruit training system: the Solen. *Acta Hort.* 243: 117–20.

Ligonniere, G. 1981. Le chip-budding. *Le Fruit Belge* 396: 235–7.

Lorette, L. 1925. *Lorette System of Pruning.* London: Martin Hopkinson & company Ltd.

Lugli, S. and Musacchi, S. 1993. Tecniche e cicli produttivi delle pomacee in vivaio. *Frutticoltura* 1: 25–33.

Mantinger, H. 1998. Sistemi di impianto ed evoluzione tecnica della melicoltura in Alto Adige. Frutticoltura ad alta densità. Edagricole.

Marini, R. P., Byers, R. E. and Sowers, D. L. 1993. Repeat applications of NAA control preharvest drop of 'Delicious' apples. *J. Hort. Sci.* 68: 247–53.

Marini, R. P. 1996. Chemically thinning spur 'Delicious' apples with carbaryl, NAA, and ethephon at various stages of fruit development. *HortTechnology* 6: 241–6.

Masseron, A. 2002. Pommier: le mur fruitier. Ed. Centre technique interprofessionnel des fruits et legumes, Paris.

McArtney, S. J., Palmer, J., Davies, S. and Seymour, S. 2006. Effects of lime sulfur and fish oil on pollen tube growth, leaf photosynthesis and fruit set in apple. *HortScience* 41: 357–60.

McArtney, S. J. and Obermiller, J. D. 2015. Effect of notching, 6-benzyladenine and 6-benzyladenine plus gibberelllinA4+7 on bud break and shoot development from paradormant buds on the leader of young apple trees. *HortTechnology* 25: 233–7.

Miller, S. S. 1988. Plant bioregulators in apple and pear culture. *Hort. Rev.* 10: 309–401.

Ming Xia, L., GuiJun, G. and GangShuan, B. 2011. Effect of renewal pruning on apple photosynthetic ability and fruit quality in final full productive stage. *J. Northwest A & F Uni.* 39 (1): 179–85.

Mitov, P. and Zocca, A. 1966. La diversa attitudine ad emettere rami anticipate nelle cultivar di melo. *Frutticoltura* 1: 71–9.

Mohammadi, A., Mahmoudi, M. J. and Rezaee, R. 2013. Vegetative and reproductive responses of some apple cultivars (Malus domestica Borkh.) to heading back pruning. *Int. J. AgriScience* 3 (8): 628–35.

Musacchi, S. and Costa, G. 1992. Controllo della morfogenesi dei rami anticipati nel melo in vivaio. *Frutticoltura* 3: 103–6.

Musacchi, S. and Farina, M. 1998. Forme di allevamento per impianti di melo ad alta densità. *Informatore Agrario* 22: 65–9.

Musacchi, S. 2007. Training systems and soil management for Southern European pear orchards. *Acta Hort.* 772: 447–57.

Musacchi, S., Ancarani, V. and Sansavini, S. 2007. Primi risultati dei nuovi portinnesti clonali del melo della serie Geneva® (CG) innestati con Golden Delicious. *Frutticoltura* 10: 44–9.

Musacchi, S. 2008. Bibaum®: a new training system for pear orchard. *Acta Hort.* 800: 763–8.

Musacchi, S., Lanzoni, A., Ancarani, V. and Serra, S. 2009. Primi risultati del confronto fra fusetto e sistema a doppio asse (Bibaum) nel pero. Rivista di frutticoltura e di ortofrutticoltura. 9: 28–35.

Musacchi, S., Bucci, D., Ancarani, V., Gagliardi, F. and Serra, S. 2014. Investigation of 'Modì®' Habitus in Relation to Training Systems. *Acta Hort.* 1058: 121–8.

Musacchi, S. and Lugli, S. 2014. High density planting for sweet cherry orchards. *Acta Hort.* 1020: 489–95.

Nüberlin, F. 1993. The super spindle apple orchard system. *Compact Fruit Tree.* 26: 17.

Palmer, J. W. 1992. Effects of varying crop load on photosynthesis, dry matter production and partitioning of Crispin/M.27 apple trees. *Tree Physiol.* 11: 19–33.

Palmer, J. W., Giuliani, R. and Adams, H. M. 1997. Effect of crop load on fruiting and leaf photosynthesis of 'Braebrun'/M26 apple trees. *Tree Physio.* 17: 741–6.

Peck, G. M., Combs, L. D., DeLong, C. and Yoder, K. 2015. Precision apple flower thinning using organically approved chemicals. *Int. Symp. Innov. Integr. Org. Hort.* 1137: 47–52.

Pellerin, B. P., Buszard, D., Iron, D., Embree, C. G., Marini, R. P., Nichols, D. S., Neilsen, G. H. and Neilsen, D. 2011. A theory of blossom thinning to consider maximum annual flower bud numbers on biennial apple trees. *HortScience* 46 (1): 40–2.

Pellerin, B. P., Buszard, D., Georgallas, A. and Nowakowski, R. J. 2012. A novel framework to consider endogenous hormonal control of apple tree flowering. *HortScience* 47 (5): 589–92.

Ravaglia, G. F., Corelli Grappadelli, L. 1990 Le piante in vaso nel vivaismo frutticolo. *Frutticoltura* 5: 65–70.

Robinson, T. L. and Lakso, A. N. 1989. Light interception, yield and fruit quality of 'Empire' and 'Delicious' apple trees grown in four orchard system. *Acta Hort.* 243: 175–84.

Robinson, T. 2011. Advances in apple culture worldwide. *Rev. Bras. Frutic.* 33 (spe1): 37–47

Robinson, T. L. and Lakso, A. N. 1991. Bases of yield and production efficiency in apple orchard systems. *J. Amer. Soc. Hort. Sci.* 116: 188–94.

Robinson, T. L. and Watkins, C. B. 2003. Cropload of 'Honeycrisp' affects not only fruit size but many quality attributes. *New York Fruit Q.* 11 (3): 7–10.

Robinson, T. L., Hoying, S. A. and Reginato, G. H. 2006. The tall spindle apple production system. *N. Y. Q.* 14 (2): 21–8.

Robinson, T. L., Hoying, S., Ungerman, K. and Kviklys, D. 2010. ReTain combined with NAA controls preharvest drop of 'McIntosh' apples better than either chemical alone. *N. Y. Fruit Q.* 18 (3): 9–13.

Robinson, T. 2011. Advances in apple culture worldwide. *Rev. Bras. Frutic.* 33 (spe1): 37–47.

Robinson, T. L., Hoying, S., Sazo M., DeMarree, A. and Dominguez, L. 2013a. A vision for apple orchard systems of the future. *N. Y. Fruit Q.* 21 (3): 11–16.

Robinson, T., Lakso, A., Greene, D. and Hoying, S. 2013b. Precision crop load management. *N. Y. Fruit Q.* 21 (2): 3–9.

Robinson, T., Hoying, S., Sazo, M. M. and Rufati, A. 2014. Precision crop load management. Part 2. *N.Y. Fruit Q.* 22 (1): 9–13.

Robitaille, H. A. and Leopold, A. C. 1974. Ethylene and the regulation of apple stem growth under stress. *Physiologia Plantarum.* 32: 301–4.

Rosenberger, D. A., Schupp, J. R., Hoying, S. A., Cheng, L. and Watkins, C. B. 2004. Controlling bitter pit in "Honeycrisp" apples. *HortTech.* 14 (3): 342–9.

Sansavini, S. and Zocca, A. 1965. La diversa attitudina ad emettere rami anticipatinelle cultivar di pero. *Frutticoltura* 3: 233–41.

Sansavini, S. 1974. Primi risultati di una ricerca triennale sulla potatura meccanica del melo, del pero e del pesco. *L'Italia Agricola* 111 (11): 62–85.

Sansavini, S. and Corelli, G. L. 1990 La potatura e le forme di allevamento del melo. In La potatura degli alberi da frutto negli anni '90 Verona 27/4/1990.

Sansavini, S. 1993. Modelli d'impianto, portinnesti e forme d'allevamento del pero. In La coltura del pero per la produzione integrate.

Sansavini, S. 1997. Nuovi impianti e qualità delle mele. Rivista di frutticoltura e ortofloricoltura. n.1.

Sansavini, S. and Corelli Grappadelli, L. 1997. Modelli d'impianto, portinnesti e forme d'allevamento per la melicoltura di pianura. In Melicoltura di pianura: quale futuro Verona 12 aprile 1997: 127–72.

Sansavini, S. and Corelli, G. L. 1998. La potatura e forme di allevamento del melo. Frutticoltura ad alta densità. Edagricole.

Sazo, M. M. and Robinson, T. 2011. The use of plant growth regulators for branching of nursey tree in NY state. *N. Y. Fruit Q.* 19 (2): 5–9.

Schwallier, P. G. 1996. *Apple Thinning Guide*. Sparta, MI: Great Lakes Publishing Co.

Schwallier, P. G. and Irish-Brown, A. 2015. Predicting apple fruit set model. *N. Y. Fruit Q.* 23 (1): 15–18.

Schupp, J. R. and Greene, D. W. 2004. Affect of aminoethoxyvinylglycine, (AVG) on pre-harvest drop, fruit quality and maturation of 'McIntosh' apples. I. Concentration and timing of dilute applications of AVG. *HortScience* 39: 1030–5.

Shaw, J. K. 1914. A study of variations in apples. *Mass. Agric. Expt. Sta. Bull.* 149: 29–36.

Taylor, B. K. 1975. Reduction of apple skin russeting by gibberellinA4+7. *J. Hort Sci.* 50: 169–72.

Tromp, J. 1982. Flower-bud formation in apple as affected by various gibberellins. *J. Hort. Sci.* 57: 277–82.

Tukey, L. D. 1978. The thin-wall trellis hedgerow system. *Acta Hort.* 65: 261–6.

Tustin, D. S., Hirst, P. M. and Warrington, I. J. 1988. Influence of orientation and position of fruiting laterals on canopy light penetration, yield and fruit quality of 'Granny Smith' apple. *J. Amer. Soc. Hort. Sci.* 113 (5): 693–9.

Tustin, S., Hirst, P., Warrington, I. and Stanley, J. 1989. Light distribution and fruit quality trough multi-layered trellis apple canopies. *Acta Hort.* 243: 209–12.

Unrath, C. R. 1974. The commercial implication of gibberellin A4 A7 plus benzyladenine for improving shape and yield of 'Delicious' apples. *J. Amer. Soc. Hort. Sci.* 99: 381–4.

Ventura, M. and Sansavini, S. 2005. Tipologia di rami a frutto e qualità delle mele cvs Braeburn, Golden Delicious e Jonagold. *Frutticoltura* 11: 52–61.

Wagenmakers, P. S., Nijsse, F. and de Gendt, C. M. E. 1991. Planting systems and light climate. Annual Report, Wilhelminadorp. 29–40.

Warrington, I. J., Stanley, C. J., Tustin, D. S., Hirst, P. M. and Cashmore, W. M. 1996. Light transmission, yield distribution, and fruit quality in six tree canopy forms of 'Granny Smith' apple. *J. Tree Fruit Product.* 1 (1): 27–54.

Watkins, C. B. 2003. Principles and practices of postharvest handling and stress. 2003. In D. C. Ferree and I. J. Warrington (eds), *Apples, Botany, Production and Uses*. Cambridge, MA: CABI Publishing.

Werth, K. 1989. Die superspindel. Hole investition mit hohen Gewinnchancen und holem Risiko. *Obstbau Weinbau* 10: 273–6.

Wertheim, S. J. 1968. The training of the slender spindle. No 7. Proefstation Fruiteelt, Wilhelmindadorp, NL.

Wertheim, S. J. 2000. Developments in chemical thinning of apple and pear. *Plant Growth Regulat.* 31: 85–100.

Williams, M. W. and Edgerton, L. J. 1981. Fruit thinning of apples and pears with chemicals. U. S. Dept. Agr. Bul. 289, Washington, DC.

Wismer, P. T., Proctor, J. T. A. and Elfving, D. C. 1995. Benzyladenine affects cell division and cell size during apple fruit thinning. *J. Am. Soc. Hort. Sci.* 120: 802–7.

Yoder, K. S., Peck, G. M., Combs, L. D. and Byers, R. E. 2013. Using the pollen tube growth model to improve apple bloom thinning for organic production. *Acta Hort.* 1001: 207–14.

Yuan, R. and Carbaugh, D. H. 2007. Effect of NAA, AVG, and 1-MCP on ethylene biosynthesis, preharvest drop, fruit maturity, and quality of 'Golden Supreme' and 'Golden Delicious' apples. *HortScience* 42: 101–5.

Chapter 10

Advances in soil and nutrient management in apple cultivation

G. H. Neilsen, D. Neilsen and T. Forge, Summerland Research and Development Centre Agriculture and Agri-Food Canada; and K. Hannam, Natural Resources Canada

1 Introduction

Soil and nutrient management of apple orchards has a long history, as indicated by several detailed earlier reviews (Boynton and Oberly, 1966; Neilsen and Neilsen, 2003). Many standard textbooks concerned with the physiology and commercial production of temperate tree fruits, including apple, have chapters highlighting key principles to achieve optimum management of soil, nutrient and water resources (Childers et al., 1995; Tromp et al., 2005). Cumulative historical research has led to the widespread development and use of midsummer nutrient standards for apple leaves (Shear and Faust, 1980) and detailed examination of factors affecting mineral concentration of fruit (Perring and Pearson, 1984) with a particular emphasis on mineral nutrients such as Ca that influence fruit quality (Faust, 1989). Information has been compiled to examine different management strategies for apple, including foliar nutrient applications (Swietlik and Faust, 1984) or fertigated applications of nutrients with irrigation water (Neilsen and Neilsen, 2008). The evolution of apple orchards to high-density planting systems that increase yield and fruit nutrient demand per unit of land has stimulated research to investigate methods to improve tree and fruit performance on infertile, coarse-textured soils (Neilsen and Neilsen, 2006a). Similarly, the widespread adoption of herbicides to control weed competition renewed

http://dx.doi.org/10.19103/AS.2016.0017.12

interest in orchard floor management options that might mitigate detrimental effects associated with reduced organic inputs to orchard soil (Merwin, 2003). Soil and nutrient management can thus be seen to be constantly evolving in response to new challenges.

It is the purpose of this chapter to first explore emerging issues that are likely to influence future approaches to soil and nutrient management in apple orchards. These include climate change and variability, which has been cited as one of the major threats to sustaining horticultural production in the coming century (Warrington, 2008). There is also worldwide recognition that long-term sustainable field production of agricultural crops depends on maintaining soil quality and hence on developing methods to minimize soil degradation (FAO and ITPS, 2015). Society is also increasingly concerned with human activities, including agricultural production practices that might result in diminishing the quality of water resources that are increasingly viewed as finite and vital to future human health (Pimentel, 2011). Future decreases in availability and increases in costs of conventional fertilizers may pose a direct threat to the maintenance of the nutritional health of apple orchards (Vance, 2001).

The chapter will conclude with an assessment of new orchard soil and nutrient management strategies, with a potential to mitigate the emerging environmental, social and resource limitation challenges identified above. Improved nutrient-use efficiency by precision nutrient management techniques including fertigation, targeted foliar nutrition, use of new fertilizer formulations and better sensing of nutrient limitations are of particular interest. Similar improvements in precision water management, discussed in greater detail in Chapter 13, will also affect future soil and nutrient management. Thus, implications of improved irrigation scheduling and conservative irrigation techniques, including partial and deficit irrigation and increased use of mulches, will be assessed relative to implications regarding soil and nutrient management. In addition to incremental development of techniques to improve the precision of nutrient and water management, the implications of major changes in production systems will be discussed. This will include the adoption of organic and integrated production systems that place a central emphasis upon maintaining long-term soil quality and organic matter content. Greater adoption of improved genetic material resulting from improved molecular and genetic understanding of mineral nutrition has the potential to alter nutrient management strategies. Finally, the consequences of changing apple production areas in response to changing climate will be discussed relative to soil and nutrient management.

2 Emerging issues affecting soil and nutrient management

2.1 Climate change and variability

We are living in an era of massive, rapid and global climate change caused by continuing increases in atmospheric concentrations of the major greenhouse gases (GHGs) carbon dioxide (CO_2), methane (CH_4) and nitrous oxide (N_2O), which are higher now than at any time in the past 800 years (Fig. 1). As a consequence, the most up-to-date technical report from the Intergovernmental Panel on Climate Change (IPCC) indicates a high probability that the recent trend of increased surface air temperatures, including both maximum and minimum temperatures over land, will continue (Stocker et al., 2013). At a global

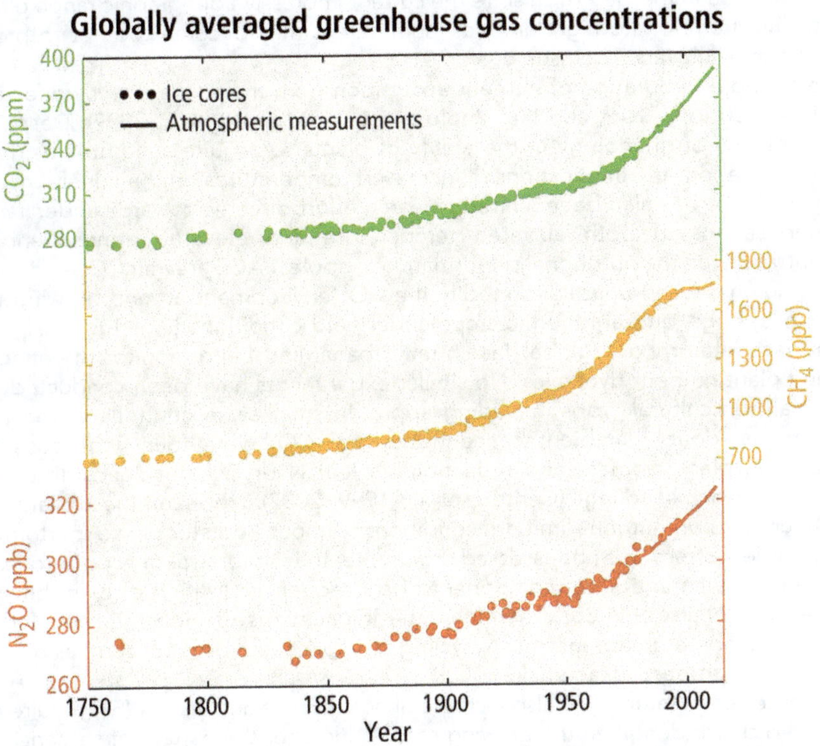

Figure 1 Observed changes in atmospheric greenhouse gas concentrations. Atmospheric concentrations of carbon dioxide (CO_2, green), methane (CH_4, orange), and nitrous oxide (N_2O, red). Data from ice cores (symbols) and direct atmospheric measurements (lines) are overlaid (Figure 1.3 in IPCC, 2014: Climate Change 2014: Synthesis Report).

level, increased air temperatures contribute to elevated levels of atmospheric moisture, because the amount of water contained in air increases with temperature, but the net effect on precipitation patterns is difficult to predict; reduced precipitation is projected for some regions while increased precipitation is projected for others. Regional shifts in the proportion of annual snow and rainfall are also expected. As well, greater climate variability, for example, changes in the intensity and frequency of extremes in temperature and precipitation, is anticipated.

Current and anticipated patterns of increased air temperatures have had, and will continue to have, major effects on the growth cycle of temperate tree fruits, including apple. Abnormally high winter temperatures can disrupt the normal cycle of dormancy and increase the possibility of lack of chilling (Campoy et al., 2011). Advances in the seasonal timing of apple growth and development (phenology) are associated with earlier flowering (Fujisawa and Kobayashi, 2010) and harvest date (Perry et al., 1987). Even on irrigated sites, greater water stress has been associated with detrimental effects on fruit quality, including increased incidence of sunburn and reduced fruit size

(Caprio and Quamme, 1999). Despite the current impressive geographic range of apple production, from low to high latitudes, some areas may become too hot to produce current apple cultivars (Thomson et al., 2014).

The timing and duration of nutrient acquisition from the soil is directly affected by the timing of apple root and leaf deployment (Nord and Lynch, 2009). Despite the important role of phenology in the ability of plants to acquire soil nutrients, little is understood about the implications of increased temperatures on the mineral nutrition of apple. There has also been limited research effort directed towards understanding the implications of both elevated temperature and elevated atmospheric CO_2 concentration, on the nutritional performance of apple. FACE (free air CO_2-enrichment) technology has been utilized to modify the CO_2 environment around growing trees, grasslands, bogs, and agronomic crops under field conditions (Liberloo et al., 2007; https://facedata.ornl.gov/global_face.html#UrbanaMinus) and monitor the effects on soil and plant nutrient dynamics. No similar experiments have been conducted using perennial horticultural crops, including apple. In small-scale controlled greenhouse experiments, double ambient concentrations of CO_2 were found to cause large increases in plant biomass and reductions in leaf N dry weight concentrations for cherry and peach seedlings (Centrito et al., 1999; 2002). However, these plants were grown under non-nutrient-limiting conditions without rootstocks, so performance under nutrient stress was not assessed. For apple, the withdrawal of N prior to leaf fall is identified as an important component of subsequent utilization of N in the spring (Neilsen and Neilsen, 2003). Therefore, the implications of high CO_2 coupled with increased temperature for internal N cycling of apple could be a concern, as observed for many forest trees (Lukac et al., 2010).

Increased temperatures will also increase annual evapotranspiration (ET) requirements of apple orchards, lengthen the growing season and alter the pattern of seasonal water requirements (Neilsen et al., 2006a). Although climate change effects on precipitation are less predictable on a regional scale, it is likely that the industry, as a whole, will increase the use of irrigation to supplement water availability and improve orchard resilience to water stress. This trend is likely to stimulate adoption of associated technologies, including fertigation (Bar-Yosef, 1999), whereby nutrients are supplied via irrigation (Neilsen et al., 1999).

There is growing recognition that soil carbon sequestration may be a key tool for mitigating the effects of climate change (Palm et al., 2014; Sperrow, 2014). It has been stated that augmentation of soil organic matter is a top priority for enhancing soil structure and fertility, but also to counter the atmospheric greenhouse effect via carbon sequestration (Rosenweig and Hillel, 2000). Perennial apple orchards may have the potential to act as C sinks by maintaining long-term C storage in the soil and short- to medium-term storage in the wood, but more work is needed (Zanotelli et al., 2014; Deurer et al., 2012). Effective use of cover crops to increase C sequestration has been advocated (Poeplau and Don, 2015), but few relevant studies have been conducted on orchards. Thus, the implications of many management strategies for orchard C sequestration are little known. This is even more pertinent for the emissions from orchards of other potent GHGs, such as N_2O (Swarts et al., 2016). Similarly, the importance of orchard management practices that reduce the production of other major GHGs such as NO_2 will become increasingly recognized. This will require assessment of the effects that irrigation method, timing and duration, and orchard floor C management (amendments, mulches and cover crops) can have on N_2O and CH_4 evolution (Fentabil et al., 2016).

2.2 Soil degradation

There is an increasing recognition that soils provide a range of ecosystem services that are relevant far beyond primary agriculture. Recent reports by the United Nations, developed in response to the designation of 2015 as the International Year of Soil by the Global Soil Partnership, have demonstrated the central importance of sustainable soil management for poverty reduction, human and environmental health, and economic prosperity. Thus, soil degradation is increasingly being recognized as a societal problem, rather than as just an agricultural or environmental problem.

One of the major challenges in producing high-value horticultural crops, including apples, has always been the need to overcome unfavourable soil conditions. Historically, tools developed to meet the nutritional requirements of apples have facilitated the identification of soil limitations to nutrient acquisition and have led to the development of methods to overcome these limitations (Neilsen and Neilsen, 2003). Apple trees are known to be more sensitive to excess salinity than many crops (Maas, 1987) and are also prone to growth problems at both low (<5.5, Hoyt and Neilsen, 1985) and high (>8.0, Neilsen et al., 1993) pH. Furthermore, retarded apple growth has long been associated with toxicities resulting from lead arsenate residues in orchards (Benson, 1976). Thus, the negative implications of degraded soils were previously recognized via evolutions in apple orchard soil management. Research is ongoing to identify the consequences of current apple production systems on long-term soil quality (Goh et al., 2001).

Poorly practiced intensive agriculture poses a major threat to soil quality and agricultural productivity (Gomiero, et al., 2011). Soil degradation was identified as a major limitation to the sustainable use of soils (FAO and ITPS, 2015). For example, conversion of productive agricultural soils to other competing economic activities, including urban and industrial development, and uncontrolled erosion result in the physical disappearance of productive soils. Adverse soil chemical changes, such as soil acidification, can result from atmospheric acid deposition and/or excessive application of acidifying fertilizers. Soils can also become contaminated by trace metals, such as arsenic, cadmium and nickel, which occur naturally in some soils but can also result from atmospheric pollution or contamination from pesticides or fertilizers. Similarly, soil salinization can occur naturally via inadequate landscape drainage but can also be caused or compounded by poor irrigation and fertigation practices. Furthermore, soil compaction, caused by excessive machine or animal traffic, can destroy the physical structure of the soil, reducing aeration and increasing the risk of waterlogging. As well, declining soil organic matter content, associated with over-cultivation or prolonged herbicide use, reduces the water- and nutrient-holding capacity of the soil and decreases soil biological activity and diversity. Such degradation in soil biological activity has negative feedback effects on important soil properties and functions such as aggregate formation, which is crucial to soil porosity and water-holding capacity, decomposition and nutrient cycling, and the suppression of soil-borne pests and pathogens. The concept of soil health recognizes the linkage between soil biology and chemical and physical aspects of soil quality (Doran and Zeis, 2000). Accordingly, research on orchard soil management practices is increasingly being directed at more holistic improvement of overall soil health rather than at correcting deficiencies in specific soil chemical or physical properties (Merwin, 2003). Enhanced organic matter inputs, via cover crops, organic amendments and mulches, are central to the improvement of soil health, linking soil health management with C sequestration as described above.

It is also recognized that 'replant disease' complex, a soil disorder, which can result in growth inhibition and poor establishment of apples planted in previous orchard soils, may reflect degradation of soil biological diversity (Mazzola and Manici, 2012). Consequently, improved strategies for management of soil-borne pathogens and plant-parasitic nematodes implicated in replant disease complex are being sought (Mazzola and Manici, 2012). These strategies may include practices such as compost amendments to enhance overall soil health and thereby induce general soil disease suppression (van Schoor et al., 2009; Forge et al., 2016a,b) as well as practices that more specifically target the causal organisms (Mazzola and Manici, 2012).

2.3 Degradation of water resources

Water availability is another factor which can reduce yield of agricultural crops from optimum values (Mueller et al., 2012). As rapid climate change continues, water availability is becoming a significant concern for many growing regions. There is also a growing recognition that both population growth and economic development will place even greater demands on food production systems and the water required to maintain these systems (Fereres et al., 2011). This has renewed focus on precision irrigation methods that improve water-use efficiency, thereby maximizing the crop yield per unit of applied water (Behboudian and Mills, 1997; Monaghan et al., 2013). The water savings achieved using microirrigation (e.g. drip, or mini-sprinkler) technologies, relative to more traditional overhead sprinkler irrigation systems, are particularly relevant for high-value horticultural production, including apple orchards, in semi-arid regions with limited water supplies (Neilsen et al., 1995c). Deficit and partial root zone drying approaches (Leib et al., 2006), which reduce the quantity of water applications while maintaining or improving the quality of fruit, have also been applied in apple production. Improved on-farm water-use efficiency can also be facilitated by monitoring field soil water variability, using inexpensive soil moisture sensors and *in situ* sensor arrays (Greenwood et al., 2010), and by using effective plant-based indicators of moisture stress (Jones, 2004), as discussed in further detail in chapter 13. The technical challenges associated with wireless sensor networks for agriculture, including irrigation management, were recently reviewed by Ojaha et al. (2015).

In addition to concern over the quantity of water available to agriculture, declining water quality, as a consequence of inadequate retention of nutrients within the root zone on agricultural sites, has also become a concern. The problem has become particularly pertinent for nitrogen; numerous studies have reported elevated NO_3-N concentrations in receiving bodies, including freshwater lakes (Neilsen et al., 1982) and groundwater (Spalding and Exner, 1993). Irrigated and heavily fertilized horticulture crops have been frequently implicated (Nightingale, 1972; Pionke et al., 1990). The problem is particularly acute for orchard crops, including apple, when high volumes of water pass through the root zone. Such a situation can commonly occur on coarse-textured soils, which are easily over-irrigated in an attempt to overcome water stress. Soil solution monitoring in apple orchards has demonstrated that soil NO_3-N concentrations are particularly sensitive to the quantity of applied water (Neilsen et al., 1998). Leaching of N can be reduced by N retention within the root zone, but this requires precise additions of water on coarse-textured soils and possible modification of the soil via organic amendments, to improve its water-holding capacity. A nine-year lysimeter study with apple trees confirmed the importance of irrigation quantity to the drainage loss of N, but also indicated the important role that cover crops can play in reducing N leaching losses (Stevenson and Neilsen, 1990); in this study, 40% to 45% of applied fertilizer N reappeared in drainage waters under grass but 88% to 115% of applied N appeared in drainage waters under bare soils. Thus, it may be possible and, even desirable, to reduce N application rates on irrigated sites where the orchard floor vegetation has been reduced (e.g. using herbicide) without compromising crop yield.

In practice, over-fertilization with N occurs in many apple orchards, especially when N applications exceed 150 kg N/ha, because such high application rates exceed tree demand (Weinbaum et al., 1992). A recent study indicated that better control of the quantity and duration of fertigation is key to controlling orchard N losses to groundwater (Baram et al., 2016). Applying irrigation at rates designed to meet ET demands would be one way of reducing excess N leaching (Neilsen et al., 2009b). Weinbaum et al. (1992) identified a number of other factors that result in over-fertilization with N on perennial crops; these include ineffective soil N testing, which fails to account for the N supplied by recycled organic materials and residual soil organic matter. There has only been limited research on soil and plant indicators of excessive N application. For example, limited use has been made of end of the growing season soil N testing, although high available soil N at this time could indicate a need to reduce future N applications. Similarly, leaf N concentrations associated with decreased harvest fruit colour and firmness have not been defined for apple cultivars (Neilsen et al. 2009b). N fertilization rates also often fail to take into account increases in annual N demand (and other nutrients) caused by high crop loads.

Although most research on degradation of water quality has been concerned with retention of N in horticultural production systems, excess P in receiving waters has been recognized as a major pollutant because it stimulates excessive algal growth (eutrophication) at relatively low concentrations (Miller et al., 1982). Despite recognition of the critical role that P availability plays in improving apple tree establishment, especially on soils suffering from replant disorder (Neilsen and Yorston, 1991), and the ability of fertigation to increase P mobility in the soil (Neilsen et al., 2008b), few studies have been conducted on factors affecting P leaching from apple orchards soils.

2.4 Changes in fertilizer availability and management

The future availability and cost of non-renewable commercial fertilizers is also a growing concern. This issue is particularly acute for P, whose supply depends upon the availability of fossil rock phosphate (Cordell et al., 2009). Using 2007–8 extraction rates and current known world supplies, the International Plant Nutrition Institute has estimated that quantities of rock phosphate will last 93 years and quantities of potash will last 235 years (Fixen and Johnston, 2011). These estimates suggest that the supply of P and K fertilizer will be adequate for the foreseeable future; it is anticipated that higher cost, lower quality reserves could be used to extend supply. The cost of N fertilizer is directly correlated with the fluctuating cost of natural gas, which is used in its manufacture. Thus, the apparently inevitable prospect of higher fertilizer prices will provide economic incentives to improve fertilizer-use efficiency through improved horticultural practices, including optimization of the rate, timing and placement of fertilizers. Higher fertilizer costs will also provide incentives to develop new fertilizer formulations that will improve nutrient-use efficiency and minimize environmental impacts. These efforts parallel those required to avoid unnecessary nutrient losses, which can degrade water resources, as discussed previously.

Renewed consideration will be given to alternative fertilization strategies to use effective plant growth-promoting organisms and organic fertilizers to replace or supplement traditional inorganic fertilization.

3 Precision nutrient management: fertigation and targeted foliar application

Improving the precision of nutrient use in orchards is important to avoid unnecessary nutrient applications and inefficiency in their use. The potential for achieving more precise nutrient applications will be discussed for fertigation, targeted foliar applications, new fertilizer sources and future improvements in the sensing of nutrient limitations.

3.1 Fertigation

Fertigation involves the simultaneous application of nutrients with irrigation water; this technology has been developed for a range of horticultural crops (Bar Yosef, 1999; Haynes, 1985; Neilsen et al., 1999). Fertigation provides the potential to more closely synchronize nutrient supply with plant nutrient demand. By fertigating with N, which easily leaches from orchard soils and is frequently applied at the highest rates of all nutrients, acceptable apple performance can be achieved at lower rates of fertilizer application. In fertigated 'Delicious' apple orchards, the quantity of N removed in harvested fruit did not increase when application rates exceeded 150 kg N/ha because yield did not increase at higher fertigation rates (Klein et al., 1989). Similarly, a series of studies, conducted over six growing seasons that monitored the quantity of N in harvested fruit and senescent leaves from a high-density apple orchard on dwarfing Malling 9 (M.9) rootstock indicated only modest annual N requirements (9 to 40 kg N/ha) (Table 1; Neilsen et al., 2009b). The study was undertaken using five different apple cultivars, all on M.9 rootstock, which were fertigated daily at two N levels (25 and 125 kg N/ha/yr) beginning 4 weeks post-bloom; irrigation was scheduled daily in response to the previous day's ET (Neilsen et al., 2009b). The high N application rate was associated with fruit quality decline, including

Table 1 Estimates of annual nitrogen requirements for high-density apple[z] as indicated by nitrogen removal in senescent leaves and harvested fruit

	N Content	
Cultivar/Rootstock/Age	g/tree	kg/ha
Golden Delicious/M.9 end of yr 1[y]	2.8	9.2
Elstar/M.9 end of yr 4	10.2	33.1
Gala/M.9 end of yr 3	10.5	34.4
Gala/M.9 end of yr 6	12.2	39.8

[z]Plantings at 1 m × 3 m spacing.
[y]No fruit, leaves only.
Adapted from Neilsen and Neilsen (2002).

Table 2 Harvest fruit firmness and red colour as affected by N rate for Ambrosia, Cameo, Fuji, Gala and Silken apple cultivars on M.9 rootstock during the first 6 fruiting seasons

	Fruit firmness (N)					
Year	Yr 1	Yr 2	Yr 3	Yr 4	Yr 5	Yr 6
N rate[z]						
25 kg N/ha	85.7	82.4	83.8[y]	80.9	89.9	84.7
125 kg N/ha	84.7	80.9	80.6	78.7	87.3	81.4
Significance	NS	*	*	***	***	***
	Red colour (% solid red)x					
N rate						
25 kg N/ha	85	80	89	85	85	83[y]
125 kg N/ha	80	80	75	80	80	77
Significance	NS	NS	*	*	NS	*

[z]Fertigated daily for 4 weeks, post-bloom as calcium nitrate (15.5N-0P-0K).
[y]Interaction with cultivar, not all cultivars behave the same.
[x]Not measured for green skin cultivar, Silken.
*, *** Means significantly different at $P \leq 0.05$, $P \leq 0.001$, respectively, or not significantly different (NS).
Adapted from Neilsen et al. (2009b).

decreased fruit firmness and colour (Table 2). It was determined that optimum fruit quality and adequate N availability could be achieved through the cumulative annual application of 40 kg N/ha, applied daily for a minimum of 4 weeks during the growing season.

The timing of fertigation can also be manipulated so that it more closely matches the periods of greatest tree N demand. The importance of recycled N reserves for early season N requirements of apple has long been recognized (Oland, 1959; Titus and Kang, 1982). Minimal uptake of soil applied N occurs in apple until rapid shoot growth; thus, the common practice of applying fertilizer N during the dormant season is inefficient (Millard and Neilsen, 1989). Delaying N applications until post-bloom, when apple tree N demands are high due to simultaneous shoot, root and fruit growth, can be more fully exploited in fertigated apple orchards, where N is delivered directly to the roots because it is applied in the irrigation water (Neilsen and Neilsen, 2002). This principle was illustrated using an experiment involving sequential harvests of newly planted 'Golden Delicious' apples that had been fertigated with isotopically labelled ^{15}N calcium nitrate fertilizer at different times of the year (Neilsen et al., 2001). Variation in the ^{15}N composition of different plant parts was used to determine the timing of incorporation of N fertilizer into new tissue. Root uptake of N did not occur until after shoot growth had commenced (Fig. 2). Applications of N fertilizer early in the growing season were less efficient than applications later in the growing season. Similar studies involving sequential partitioning of fruiting apple trees fertilized with isotopically labelled N have indicated that much of tree growth prior to petal fall is dependent on remobilization of stored (unlabelled N), whereas fertigated N absorbed by roots primarily occurs later and is largely allocated to developing shoots

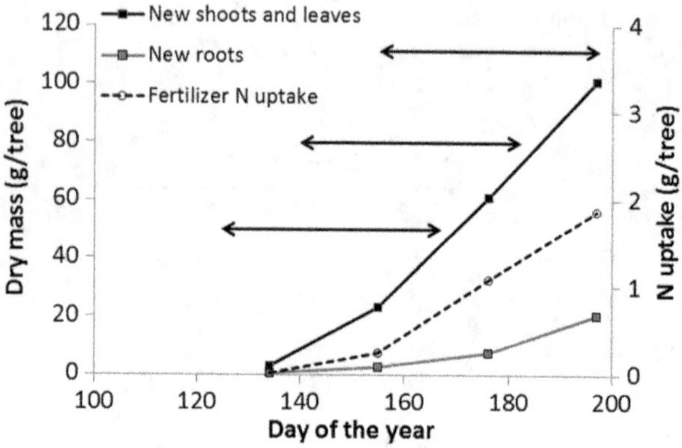

Figure 2 Relationship of daily N application timing, N1, N2, N3 (2–8, 5–11, 8–14 weeks after planting, respectively) to growth of Golden Delicious/M.9 apple (Neilsen and Neilsen, 2002).

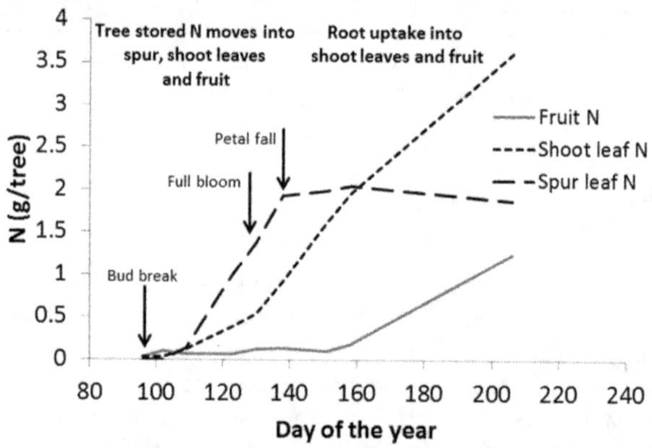

Figure 3 Uptake of fertigated ^{15}N-fertilizer into sequentially harvested two-year-old Fuji/M.9 relative to tree phenology. Adapted from Guak et al. (2003) and Neilsen et al. (2006a).

and, after the end of cell division, to developing fruits (Fig. 3). The optimum timing for applications of N via fertigation (in the 4 to 6 weeks period post-petal fall) can thus be defined based on tree phenology, offering improved precision compared to single-time broadcast N applications. Furthermore, phenological-based fertilizer applications can be used to overcome the effects of climate variability on the timing of the annual growth cycle of the apple tree.

It is not only major plant nutrients, such as N, that can be effectively applied via fertigation but also the trace nutrient boron (B), for which deficiency symptoms are well known for apple (Neilsen and Neilsen, 2003). For example, in the second year of a long-term N fertigation experiment on four apple cultivars, blossom-blast B-deficiency symptoms coincided with leaf B concentrations near and below the 20 mg B/kg deficiency threshold (Fig. 4). B fertigation was equally effective as foliar sprays at preventing

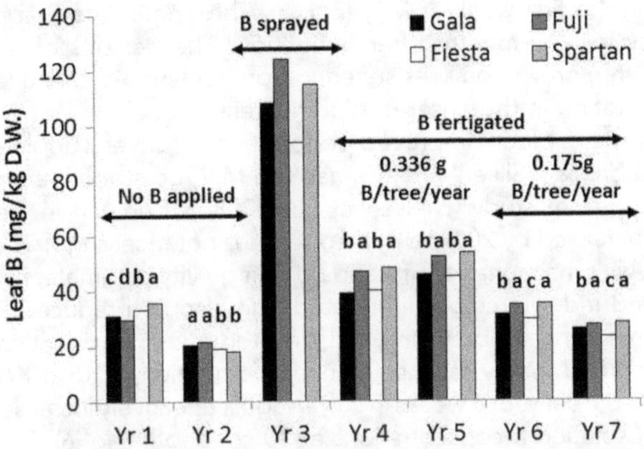

Figure 4 Leaf boron (B) concentration of four apple cultivars in response to B application method over the first seven growing seasons. Adapted from Neilsen et al. (2004c).

B-deficiency symptoms, and increased leaf B concentrations above deficiency values, indicating that elevated B was not a consequence of surface contamination of the leaves (which can occur after foliar sprays). B adequacy was achieved via modest cumulative annual fertigation amounts ranging from 0.175 g B/tree (approx. 0.3 kg/ha) to 0.336 g B/tree (approx. 0.6 kg/ha); B was applied daily for 4 to 8 weeks from late May to late July. Such modest unit area applications would have been difficult to apply to the soil using traditional broadcast fertilizer application methods.

Fertigation can also be effective at improving apple tree uptake of nutrients such as P, which is considered to have reduced availability in the soil, due to precipitation and chemical exchange processes which decrease its solubility. This is illustrated by changes in extractable soil P measured at 10–20 cm depth beneath a drip emitter in an apple orchard where a single P application of 17.5 g P as ammonium phosphate (10-34-0) per tree was

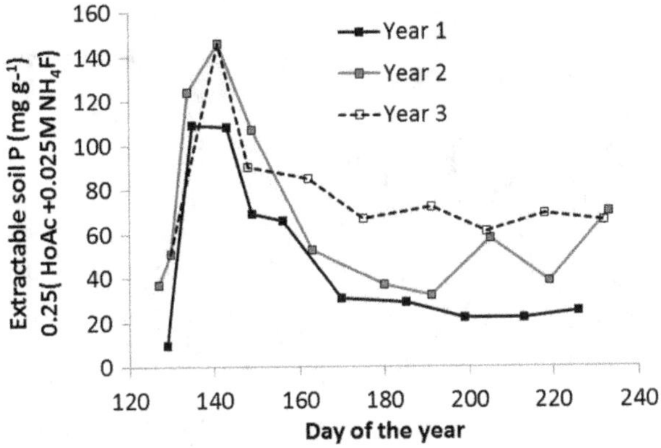

Figure 5 Exchangeable soil phosphorus (P) at 10–20cm depth below drip emitters after fertigation of 17.5 g P as ammonium polyphosphate (10-34-0) annually in the first week of May (Neilsen et al., 1997).

made annually in the first week of May (Fig. 5). The single-time P application increased soil P availability for 2–3 months after fertigation. The benefits of fertigated P include improved apple tree growth on sites suffering from replant disorder (Neilsen and Yorston, 1991) and acceleration in the onset of bloom (Neilsen et al., 1990). A single application of 20 g of P per tree at bloom improved performance of several apple cultivars, including 'Fuji', 'Gala', 'Ambrosia', 'Silken' and 'Cameo' on M.9 rootstock over the first five fruiting seasons of a long-term study (Neilsen et al., 2008b). Cumulative yield of trees which received P was increased by 20% compared to trees not receiving P. These yield increases were associated with maintenance of standard fruit quality parameters at harvest (e.g. size, soluble solids and red skin colour). P-treated fruit also had reduced incidence of water core, improved resistance to browning and elevated fruit antioxidant content.

In a coarse-textured loamy sand soil, annual fertigation of 15 g K/year as potassium chloride (KCl, 0-0-60) daily for 6 weeks in July–August effectively increased soil K availability by elevating soil solution K concentrations at 30-cm depth (Fig. 6). Potassium fertigation maintained leaf K concentrations above deficiency at this site, increasing yield, fruit size and red colour of 'Gala', 'Fuji', 'Fiesta' and 'Spartan' cultivars relative to trees not receiving K (Neilsen et al., 2004b). Further research comparing the effectiveness of various K sources, which included soluble forms of potassium chloride, potassium/magnesium sulphate, potassium sulphate or potassium thiosulphate, found that all were equally effective at maintaining adequate K nutrition (Neilsen and Neilsen, 2006a). The fertigation of large amounts of K in apple orchards requires vigilance due to the potential to develop high fruit K/Ca ratios that are associated with reduced apple storage quality in certain apple cultivars such as 'Spartan' (Wolk et al., 1998). Increasing relative fruit Ca concentrations is not readily achieved via fertigation.

It is important to avoid degradation of soil quality in the soil volume receiving fertigated nutrients. This includes avoiding the development of excessive soil salinity from over-fertilization and avoiding soil pH decline or associated depletion of soil bases including K (Neilsen et al., 2004b). Fertigation of ammonium-containing fertilizers is an acidifying process because of the transformation of ammonium to nitrate (Parchomchuk et al., 1993). It may be necessary to avoid ammonium fertigation on soils with limited resistance to acidification and susceptibility to development of K deficiency (Neilsen et al., 1995b).

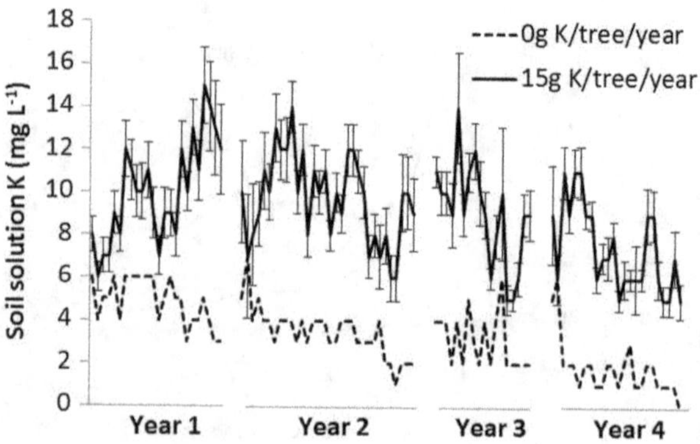

Figure 6 Soil solution potassium (K) concentration at 30 cm depth directly beneath the drip emitter as affected by daily fertigation on a sandy loam orchard soil (Neilsen et al., 2004b).

3.2 Targeted foliar application

Improved precision of nutrient applications for apple can also be achieved by foliar nutrient applications at unit area application rates much lower than those applied via broadcast or even fertigation applications to the soil. This has also been the preferred method for elevating and maintaining apple leaf concentrations above deficiency levels for essential micronutrients such as zinc (Zn), manganese (Mn), iron (Fe) and copper (Cu), which are all required by apple trees at low quantities (Swietlik and Faust, 1984). Foliar sprays have also been the preferred method for increasing the Ca concentrations of apple fruit; this is a necessary precaution for avoiding Ca-related disorders of the fruit, such as bitter pit (Neilsen and Neilsen, 2003). Pre-harvest applications of soluble Ca salts, such as food-grade calcium chloride, can be applied directly onto the fruit (Peryea et al., 2007). This method can be effective in overcoming localized Ca deficiencies that develop within the fruit due to competition with leaves (Vang-Petersen, 1980) and declining xylem supply of Ca to the apple late in the growing season (Drazeta et al., 2004). Soil Ca applications have been largely ineffective at curing such problems.

Much less effort has been directed towards the environmental implications of using foliar nutrition as a mechanism of avoiding indiscriminate or excessive soil fertilization. Applications of excess N in orchards have been implicated in groundwater pollution. A 2-yr study on mature, fruiting, 6-yr-old 'Gala' apple trees compared the effects of foliar sprays or soil applications of 50 g N/tree/year biweekly from mid-May to mid-August (Dong et al., 2005); foliar N applications produced much lower potentially leachable soil nitrate but maintained tree N status, yield and fruit quantity. Late season foliar urea applications are also efficiently taken up and remobilized by the tree in the subsequent spring, thereby improving the performance of apple trees that have low N status late in the growing season (Cheng et al., 2002). Thus, targeted foliar N applications could improve the precision of orchard N management, especially when applied late in the season, when soil N applications should be minimized to prevent nitrate leaching.

4 Precision nutrient management: improved fertilizer forms and sensing of nutrient limitations

4.1 Improved fertilizer forms

New fertilizer formulations are under development to improve nutrient-use efficiency and minimize environmental impacts (Chien et al., 2009). Compared with other nutrients, N-use efficiency is especially low, ranging from 25% to 50% for many perennial crops, including apple (Weinbaum et al., 1992). Thus, enhanced efficiency N fertilizers, including slow- or controlled-release forms of urea, or formulations containing denitrification, nitrification and/or urease inhibitors may have the potential to reduce nitrate leaching or denitrification losses (Chien et al., 2009). Similarly, new forms of P fertilizer have been proposed that reduce P fixation; these include the use of water-soluble polymers (Avail), which reduce the rate at which P is converted from water-soluble to -insoluble forms (Gordon and Tindall, 2006). The use of engineered nanoparticles, as effective nutrient sources or materials that enhance fertilizer effectiveness or stimulate plant growth, is in its infancy and has recently been reviewed (Liu and Lal, 2015). However, rigorous research must first be conducted to assess the effects of nanoparticles on non-target organisms and to better understand agro-ecological ramifications of these materials (Bindraban et al., 2015). In general, there have been limited field assessments of any of these new fertilizer formulations in apple orchards.

Plant growth-promoting organisms have also been identified as having the potential to improve plant growth (Owen et al., 2015). A range of bacterial and fungal bio-inoculants with possible beneficial properties have been isolated; these include organisms that produce hormones to stimulate growth; that improve plant nutrition by increasing root area; that fix N or solubilize P; or that protect against plant disease by sequestering harmful substances. Preliminary testing of these organisms, including vesicular-arbuscular mycorrhizal fungi, is underway in horticultural production systems (Rouphael et al., 2015). Symbiotic arbuscular mycorrhizal fungi (AMF) and some groups of rhizosphere-specific bacteria and fungi have the potential to enhance plant uptake of nutrients, particularly P, and have been developed as commercial bio-inoculants (Owen et al., 2015). AMF have been studied more extensively than other groups of bio-inoculants. Enhanced acquisition of P, Cu and Zn appears to be the main benefit of AMF to most plant species, but increased resistance to root pathogens can also be a significant benefit (Forge et al., 2001; Ridgeway et al., 2008).

A large number of controlled inoculation studies using pasteurized or sterilized soil have demonstrated benefits of AMF inoculation of apple (reviewed in Forge et al., 2001). Data on the performance of AMF inoculants in field soils and under field conditions are more limited. Ridgeway et al. (2008) studied the effects of adding supplemental AMF spore inocula on apple seedling growth in soil from an apple replant site and from nearby pasture. AMF colonization of apple roots was lower in the non-inoculated replant soil than in the non-inoculated pasture soil, but inoculation had a positive effect on AMF colonization and apple growth in both soils. Plenchette et al. (1981) and Forge et al. (2001) transplanted apple plants that had been inoculated under greenhouse conditions to field sites; the latter study planted inoculated and non-inoculated plants into both fumigated and non-fumigated plots. Both studies demonstrated enhanced nutrient uptake and growth of AMF-inoculated plants. In the study of Forge et al. (2001), these benefits were only statistically significant for plants transplanted to fumigated plots. Indigenous AMF in non-fumigated soil resulted in non-inoculated plants being colonized to levels comparable to the inoculated plants, indicating that better understanding of inoculum potential in replant soils is needed before the practical benefits of AMF inoculants can be realized. Few data are available, however, to indicate whether tree fruit production practices, organic or not, result in AMF populations inadequate for healthy apple tree growth. Miller et al. (1985) studied AMF colonization of apple in 18 rootstock plantings across the United States and concluded that there was adequate inoculum potential at all sites, but infection levels varied inversely with soil concentrations of P and Zn.

Non-AMF bio-inoculants that have potential utility include species and strains of rhizosphere-colonizing bacteria and fungi (Owen et al., 2015). Like AMF, many of these potential bio-inoculants appear to have multiple benefits including enhanced resistance to pathogens, tolerance to abiotic stress and P mobilization in the rhizosphere (Owen et al., 2015). Few published studies have documented the utility of these prospective bio-inoculants for improving replant success of apple. Several commercial formulations of the fungus *Trichoderma* were tested under greenhouse conditions for their effects on replant disease symptoms (Kandula et al., 2010). The commercial formulations improved growth relative to untreated control soil but not as well as fumigation.

Research by Mazzola and Manici (2012) suggests that 'replant disease', commonly observed when replanting apple in old orchard soils, may reflect an unhealthy soil microbiome, which might be mitigated by creating a healthy balance of soil microorganisms. To date, however, the response of tree growth to applications of plant growth-promoting microorganisms has been less consistent than conventional fertilizer applications.

Increased use of organic sources to supply plant nutrients would expand fertilization options. Use of certified organic amendments, including composts, mulches and cover crops as green manures, are currently central to organic horticultural production (Neilsen et al., 2009a). It has been shown that acceptable growth and yield performance can be maintained in perennial apple orchards (Neilsen et al., 2014) and annual vegetable crop production (Hernandez et al., 2016), even when relying solely on organic nutrient sources, rather than conventional inorganic fertilizers. Predicting the quantity of nutrients available from different organic sources can be more problematic than from inorganic sources. For example, the availability of mineral N from organic N sources can vary considerably among materials (Gaskell and Smith, 2007). In part, this is because the stability and residence time of organic matter within soil profiles remain poorly understood (Bingham and Cotrufo, 2016). There are similar uncertainties in the availability of soil organic P (Bunemann, 2015), despite the fact that organic waste products represent a potentially significant source of plant-available P (Brod et al., 2015). Although research is required to better define nutrient availability from organic amendments, it has been suggested that organic fertilization can be sustainable if sufficient nutrients can be returned to the soil to match crop nutrient uptake and other losses (Condron et al., 2000). Alternatively, replacing conventional chemical fertilization with some proportion of organic sources could be used to conserve inorganic fertilizer sources while improving soil biological function and quality (Tao et al., 2015; Hatfield and Walthall, 2015).

4.2 Improved sensing of nutrient limitations

Traditionally, determination of the nutritional health of apple orchards has relied on chemical analyses of sampled plant tissues at key stages during the annual growth cycle. Apple leaf and fruit samples have most commonly been assessed using regionally developed nutrient standards that require determination of the mineral content of mid-terminal leaf samples collected mid-season from new-year extension shoots of representative vigour. Such determinations, while useful for monitoring the long-term nutritional health of apple orchards, are frequently made too late in the growing season to allow adjustment of the current year's fertilization cycles. Seasonal calibration curves have sometimes been developed locally to adjust standard mid-season values to other times of the year.

The use of handheld visible and/or infrared radiation sensors to estimate chlorophyll concentration within leaves is a non-destructive method of monitoring that can be made throughout the growing season (Heege and Thiessen, 2013). This method can be coupled with fertigation to adjust fertilizer additions during the season. A study using non-destructive SPAD-501 meter readings to determine N and chlorophyll status of apple trees is illustrative of the potential and limitations of non-destructive sensors for nutrient determination (Neilsen et al., 1995a). In this study, there was a good relationship between leaf N and chlorophyll content early but not later in the growing season and the relationship differed by apple cultivar (Table 3). This technique showed promise as a means of identifying a need for N application, but not the quantity required. As well, high within-tree variability indicated that a systematic sampling procedure, similar to standardized

Table 3 Coefficient of determination (r^2) for leaf N concentration and leaf chlorophyll content (SPAD reading) for four apple cultivars during a single growing season (n = 30)

	Date				
Cultivar	27 May	17 June	08 July	29 July	20 August
Fuji	0.66**	0.60**	0.44**	0.10	0.01
Spartan	0.64**	0.68**	0.70**	0.23**	0.12
Fiesta	0.59**	0.62**	0.89**	0.26**	0.10
Gala	0.52**	0.73**	0.44**	0.10	0.10

** Significant at $P \leq 0.01$, otherwise not significant.
Adapted from Neilsen et al. (1995a).

procedures developed for leaf sampling for chemical analyses would be required to achieve consistent results. It is likely that any new sensing technology developed for rapid nutrient analysis would have to undergo similar evaluation and standardization of methods.

5 Precision water management

Retention of soluble nutrients within the root zone of irrigated orchards will be improved by advances in precision irrigation, as discussed in detail in chapter 13. The intention of Section 5 will be to highlight the potential effects of some of these irrigation innovations on altering apple orchard soil and nutrient management strategies.

5.1 Improved irrigation scheduling

In a drip-irrigated 'Gala' apple orchard on M.9 rootstock, scheduling of irrigation in order to apply daily water requirements to meet previous day's ET, as measured by an atmometer linked to an irrigation system, was compared to applying daily irrigation at a fixed rate to exceed maximum transpiration (Neilsen and Neilsen, 2002). Drainage of water at 1.5 m depth (below the root zone), as measured by passive capillary wick samplers (PCAPs, Boll et al., 1992), was much higher during cooler months (May, June and September) from unscheduled irrigation when fixed rate irrigation quantities exceeded tree water requirements (Fig. 7a). This orchard also received fertigated N applications for 6 weeks post-bloom with the consequence that N drainage losses were much higher in June when irrigation was unscheduled and fertigation was underway (Fig. 7b). Furthermore, the annual pattern of N loss from the root zone was measured by PCAPs when irrigation was scheduled annually according to evaporative demand and N was fertigated post-bloom for 12-weeks in a ' Braeburn' apple orchard on M.26 rootstock (Neilsen et al., 2008a). Data from the 2004 growing season was representative of the three-year study with comparisons possible between high and low N rates over two microirrigation methods (drip and microsprinkler). Largest N leaching losses occurred when N application rates were high and when high water applications were made during periods of high evaporative demand (Fig. 8). High N leaching loss was also possible at these times from drip irrigation systems

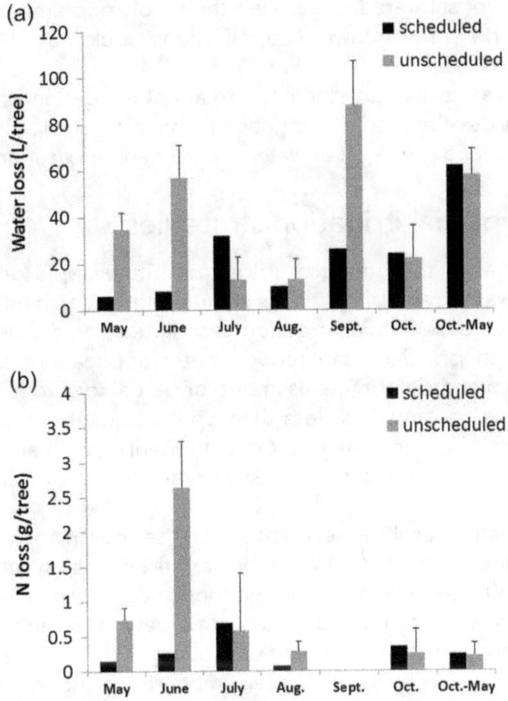

Figure 7 Water (a) and nitrogen (b) loss beneath the root zone as measured by passive capillary wick samplers at 0.75 m depth for two-year-old Gala/M.9 apple trees during the growing season in response to irrigation either applied to meet evaporative demand (scheduled) or applied at a fixed daily rate (unscheduled). Adapted from Neilsen and Neilsen (2002).

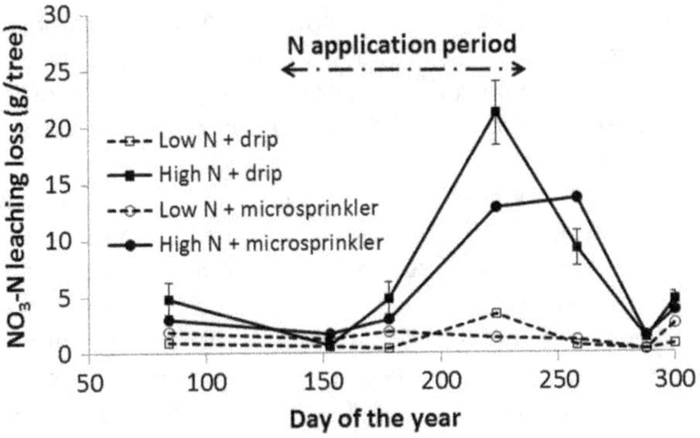

Figure 8 Nitrogen leaching losses as measured by passive capillary wick samplers at 0.75 m depth for low (2.0 mM) and high (12.0 mM) solution N concentration through drip and microsprinkler systems in a Braeburn/M.26 orchard. Adapted from Neilsen et al. (2008a).

within a small volume of saturated soil around the in-row drip emitters (data not shown). A shorter fertigation period with lower N applications would have been the most useful practice for reducing N leaching losses (Fig. 8).

Collectively, these results indicate the need to adapt fertigation strategies so that they can be used with new developments in irrigation technology, such as irrigation scheduling, to improve the control and precision of water and nutrient applications.

5.2 Deficit and partial irrigation strategies

In addition to changes in the timing of irrigation, future apple growers may have to cope with reduced water availability; this is stimulating research into the water savings that are possible via conservative irrigation techniques, as discussed in chapter 13. Conservative irrigation practices can reduce water applications by irrigating only a portion of the whole orchard floor (i.e. as in drip or partial root zone drying), or by using deficit irrigation strategies, in which less than optimal quantities of water are applied in an attempt to maintain or improve fruit quality (Behboudian and Mills, 1997). It will be important to assess the implications of such practices on orchard nutrition and soil management.

The use of drip irrigation results in substantial water savings in apple orchards compared to sprinkler irrigation, which wets the whole orchard floor. In a 3-year comparative study, drip irrigation used 60% to 70% of the water consumed using sprinkler irrigation, even though both systems were controlled by an atmometer that adjusted daily irrigation quantities to replace the water lost to ET the previous day (Neilsen et al., 2016). Restricting the volume of wetted soil requires a greater emphasis on monitoring and maintaining optimum soil quality in the irrigated root zone. This problem will be particularly acute for coarse-textured sandy soils, where the wetted volume is often smaller than in fine-texture soils, and which have a limited nutrient buffering capacity. The problems of soil acidification and the development of K deficiency three to five years after commencement of fertigation have been previously discussed in Section 3.1.

Apple crop load reduction can be utilized to maintain fruit size when water applications are restricted due to reduced water availability (Naor et al., 2008). It can therefore be viewed as an extreme form of deficit irrigation and provide useful insight into the nutritional consequences of suboptimal water applications. The consequences for leaf and fruit nutrition were measured over a 3-year period in an 'Ambrosia' apple orchard on M.9 rootstock that was subjected to factorial combinations of three levels of crop load and several season-long irrigation treatments, ranging from 50% ET replacement (applied to one or both sides of the tree row) to a well-watered control (100% ET), to increasing water deficits each year (50%, 25%, 18% ET) (Neilsen et al., 2015); all trees received a recommended NPKB-fertigation programme. Leaf N, Ca, Mg, P and B concentrations indicated that crop load had greater effects on tree nutrition than any deficit irrigation strategy. Only leaf K concentrations were near deficiency by the end of the study; this occurred across all treatments, implying that K-fertigation rates became inadequate as the trees aged. Fruit Ca concentrations, which should be maintained above 4.0 mg Ca/100 g FW for maximum apple storage quality, were affected by both irrigation and crop load treatments (Table 4). Lower rates of irrigation reduced fruit size more than they reduced the accumulation of Ca in the fruit, given that maximum fruit Ca concentrations were observed in the smallest fruit (which were produced in plots receiving the lowest water application). Concentrations of fruit Ca were lowest in the low crop load treatments, suggesting that deficit irrigation applied to sensitive cultivars

Table 4 Fruit size and fruit Ca concentration at harvest for Ambrosia apple as affected by deficit irrigation and crop load treatments

	Fruit size (g)			Fruit Ca (mg/100g FW)				
					Yr 2			
Irrigation (I)[z]	Yr 1	Yr 2	Yr 3	Yr 1	Low	Medium	High	Yr 3
I1	250a	216a	187a	3.5	3.4	3.8	4.0	4.5c
I2	230b	193a	159b	3.9	3.6	4.6	5.2	5.6b
I3	223b	211a	152b	3.8	3.4	4.0	4.6	5.4b
I4	228b	157b	98c	4.0	4.3	6.1	7.0	7.7a
Significance	****	***	****	NS				****
Crop load (CL)[y]								
Low	264a	244a	233a	3.2c				4.3c
Medium	246b	199b	159b	3.7bc				5.8b
High	182c	139c	96c	4.6a				7.3a
	****	****	****					****
CL × I interaction	NS	NS	NS	NS		*		NS
Interaction SE						0.05		

[z]Irrigation I1 (100% evapotranspiration (ET) both sides daily), I2 (50% ET both sides daily), I3 (50% ET one side daily), I4 (every 2 days, both sides at 50% ET (Yr 1), 25% (Yr 2) and 18% ET (Yr 3)).
[y]Low crop load targeted at 2.5, 3 and 3.75 fruit cm^2 TCSA, Yr 1–3 ; medium crop load at 4.5, 6 and 7.5 fruit cm^2 TCSA, Yr 1–3 and high crop load at 9, 12 and 15 fruit/cm^2 TCSA, Yr 1–3.
NS, *, ***, **** Non-significant or means significantly different at $P \leq 0.05$, ≤ 0.001 or $P \leq 0.0001$, respectively.
Adapted from Neilsen et al. (2015, 2016).

could increase the susceptibility of apples to Ca-related physiological disorders when the crop load is light.

In general, deficit irrigation strategies that reduce tree vigour and yield are expected to decrease orchard nutrient demand, but underlying nutritional limitations at a site would still have to be considered.

5.3 Mulching

In general, incorporation of mulches in orchard floor management offers potential for a more precise and effective use of water and a potential for conserving water use in irrigated orchards. In-row mulching frequently improves the growth and yield of apple trees, as discussed subsequently under organic and integrated systems. In part, this may demonstrate that mulch can improve water-use efficiency in orchards by reducing evaporation from the soil surface and alter surface run-off characteristics (Adhikari et al., 2016). In a three-year study conducted in three different orchards, application of a spray-on-newsprint/straw mulch resulted in consistently higher gravimetric soil moisture contents at the 0–15 cm depth relative to herbicide-treated check treatments. (Samples were collected within 25 cm of drip irrigation emitters (Cline et al., 2011).) The ability of many mulches to maintain elevated soil moisture contents creates the potential for water savings on irrigated sites. Straw mulch lowered irrigation water use by approximately 10% in a mature cherry orchard when averaged over microsprinkler and double-lateral drip irrigation systems in Oregon (Yin et al., 2012). Similar water savings were apparent in a 24-chamber lysimeter study when water was applied to maintain soil moisture content in chambers with and without surface paper mulch planted (or not) to 'Gala' apple trees on M.9 rootstock. Season-long water requirements indicated a similar pattern in both years of the study, with less water required in mulched lysimeters (Fig. 9). In more humid regions that are prone to high-intensity precipitation, mulches can increase the surface infiltration capacity, significantly reducing surface run-off and loss of nutrients or fungicides (Atucha et al., 2011). The observed benefits of mulches may be unique to organic mulches. Neilsen et al. (2003b) observed that woven plastic fabric mulch resulted in lower water-holding capacity (at -500kPa), infiltration rate and wet aggregate stability than several organic mulch combinations. In addition to providing barriers to evaporation, radiation and heat flux, organic mulches fuel biological activity that can have positive feedback effects on soil structure (aggregation) and water-holding capacity.

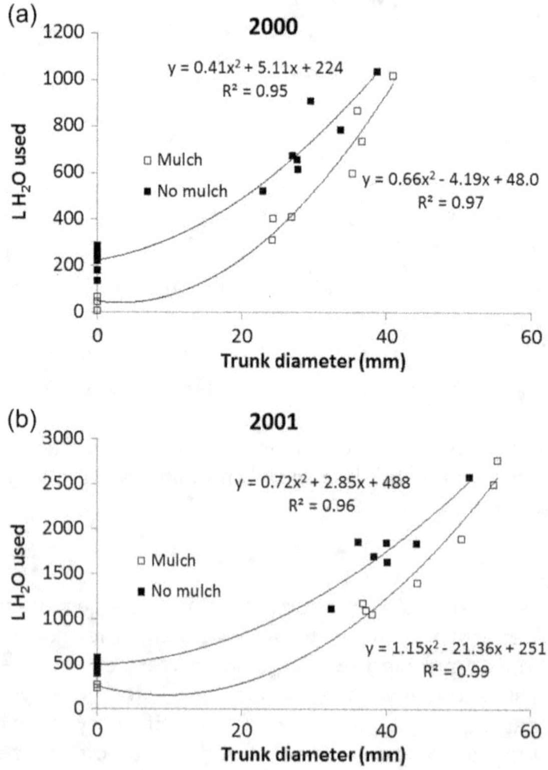

Figure 9 The effect of mulch on water consumption (a) from 15 June to 23 August 2000 and (b) from 1 May to 30 September 2001 by Gala/Ottawa3 apple trees growing in lysimeter compartments with a 2.25 m^2 surface area when water was applied to maintain soil moisture content at field capacity.

6 Altered production systems and altered production areas

6.1 Altered production systems

Organic and integrated systems

Organic apple production has expanded worldwide (Granatstein et al., 2015) as a result of increased consumer demand and the potential for greater environmental sustainability, due to reduced inputs of pesticides and fertilizers (Reganold et al., 2001). With the goal of maintaining soil health by increasing soil organic matter content (Neilsen et al., 2009a), organic soil and nutrient management practices are similar to that of integrated fruit production, which emphasizes balanced nutrient cycles and maintenance of soil fertility while allowing use of some agricultural chemicals (Avilla and Riedl, 2003). The nature of integrated and organic production system methods and their general adoption are discussed in greater detail in Chapters 25 and 26, respectively. This discussion will focus on potential benefits of their increased use on soil quality, plant nutrition and apple orchard performance.

Benefits associated with increased soil applications of organic matter in apple production include the potential to improve soil fertility by direct addition of multiple essential nutrients; improve soil structure thereby increasing soil aeration; improve nutrient and water retention capacity; and intensify biological and biochemical activity in the soil, thereby buffering the soil against soil-borne pests and adverse conditions.

Organic matter contains a wider range of plant nutrients than inorganic fertilizer (Table 5). Therefore, in contrast to chemical fertilizer, addition of organic materials involves co-application of multiple essential plant nutrients to the soil. There can be considerable variation in nutrient concentrations among organic materials (Table 5), for example, biosolids and vegetable wastes tend to have higher concentrations of P, while biosolids tend to have much higher concentrations of Zn and Cu. However, nutrient concentrations in organic soil amendments are frequently low compared to chemical fertilizers. For example, N values in organic amendments frequently range from 1 to 4% compared to 15.5 to 46% for chemical fertilizers. Furthermore, the short-term availability of nutrients contained in organic material can be less than total nutrient contents, which are determined after digestion of organic materials with acid. For example, concentrations of available (i.e. mineral) N can vary widely from − 2% (N immobilized) per year to 54% per year for a range of organic amendments, as determined under field conditions (Table 6; Gale et al., 2006). Carbon/nitrogen ratios are, in general, useful indicators of N availability, with low C/N ratios associated with high annual N supplying potential. Thus, if Agassiz compost (Table 5) were used as a substitute for an annual chemical fertilizer application of 55 kg N/ha (120 kg urea (45-0-0)/ha) and assuming a typical moisture content of 50% for the compost with the potential to mineralize 30% of total N annually, about 12 tonnes/ha (fresh weight) of compost would be required. Such applications of compost would result in total (not necessarily readily available) annual per hectare inputs of approximately 50 kg P, 60 kg K, 134 kg Ca, 20 kg Mg, 0.9 kg Zn and 0.15 kg Cu. The sustained addition of organic amendments in apple orchards therefore has the long-term potential to reduce annual fertilization requirements by augmenting soil nutrition supplying capacity. However, the timing of nutrient availability is also an important issue. This is illustrated for N, as it is

Table 5 Total nutrient concentration of selected organic sources for potential use in apple orchards as surface mulches or incorporated amendments

Organic source	Description	Total nutrient concentration[z]										
		C	N	C/N	P	K	Ca	Mg	Zn	Mn	Fe	Cu
		% dw							mg/kg dw			
Envirowaste	Poultry/food waste compost	37.2	1.8	20.7	0.5	0.75	5.0	0.43	128	227	5750	92
Agassiz[x]	Vegetable (80% vol) plus poultry manure waste (20%)	39.3	2.5	15.7	2.3	2.7	6.1	0.9	417	514	5220	66
OgoGrow	Composted biosolids plus wood waste	37.9	1.9	19.9	1.8	0.6	1.2	0.6	234	nm[y]	nm	nm
Riverside	Forestry (small branches, twigs, needles) wood waste ground to 10mm	35.8	0.8	44.8	0.1	0.3	1.2	0.2	62	785	6780	393
GVRD[x]	Dewatered biosolids City of Vancouver	nm	3.3	–	1.1	0.5	2.4	0.2	758	235	12737	977
Dairy solids	Composted	50.2	1.4	35.1	0.6	0.9	1.3	0.3	166	nm	2296	45

[z]Determined for N after Kjeldahl digestion and for others by emission spectroscopy after nitric/perchloric digestion.
[y]Not measured.
[x]Possibly immature composting process.

Table 6 Variation in N concentration, C/N ratio and potentially available nitrogen (PAN) in first year for various organic amendments as determined in the field[z]

Organic material	N%	C/N ratio	Estimated PAN[y]	
			Mean	Range
Broiler litter	3.84	9.5	42	25–54
Broiler litter compost	4.05	8.8	38	28–40
Dairy solids	1.62	27.0	9	5–17
Dairy solids compost	1.99	19.8	6	−2 to 16
Rabbit manure	3.00	11.5	27	NA[x]
Rabbit manure compost	1.80	10.0	22	NA
Yard trimmings	2.00	13.3	19	12–28
Yard trimmings compost	1.72	16.5	5	−10 to 19
Pelleted fish	9.4	4.5	77	NA
Canola meal	5.7	8	60	NA
Feather meal	13.7	4	99	NA

[z]Adapted from Gale et al. (2006). Appearing in Neilsen et al. (2014).
[y]PAN estimated from field studies of N uptake by corn (*Zea mays* L.) after incorporation of the amendment into the soil, with reference to standard curves of fertilizer recovery efficiency. Ranges are given for materials that were assessed at two sites (north-western Oregon and south-western Washington) in each of two years. Other materials were only tested in two locations in one year. Canola and feather meals were only assessed once.
[x]NA, not available.

better to maintain high N in the 4 to 6 weeks immediately post-bloom when rapid fruit and shoot growth occurs simultaneously with the formation of next year's fruit buds. Excessive late season N applications may reduce fruit quality resulting in softer, more poorly coloured fruit (Neilsen and Neilsen, 2003). The predictability of the timing of N supply from organic sources relative to apple tree N demands requires further research and may be more difficult to achieve than for readily available chemical fertilizers.

It is desirable to maintain a porous soil structure in apple orchards to allow ready penetration of roots, oxygen and water through the soil profile. The importance of soil organic matter in encouraging the creation of soil micro-aggregates that are resistant to decomposition has long been recognized (Dawson and Smith, 2007). Particulate soil C (particles larger than 53 μm) is considered to be the important 'active' fraction of organic matter, supporting microbial growth that in turn promotes aggregation and improved texture and C-sequestration potential (Neilsen et al., 2014). Particulate C values doubled beneath bark mulches after six years (Neilsen et al., 2014; Fig. 10). In contrast, particulate C values were very low within the herbicide strip of nearby conventionally managed orchard on the same silt loam soil series, averaging 3 g C/kg of soil, much lower than any of the organic soil management treatments in Fig. 10.

Organic amendments have a large capacity to retain nutrients (Table 7), with high cation exchange capacities (CEC) that can be 10-fold greater per unit area weight than typical

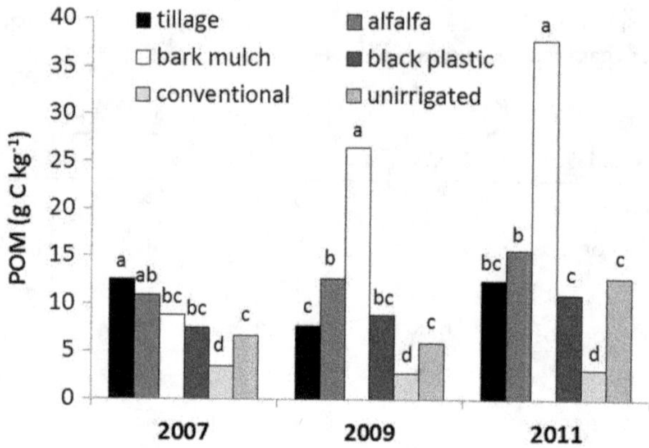

Figure 10 Particulate organic matter (POM > 53microns) –C as affected by organically acceptable soil management treatments 2, 4 and 6 years after treatment initiation. Columns within years significantly different at $P \leq 0.05$ (*) if labelled with different letters. Adapted from Neilsen et al. (2014).

Table 7 Cation exchange capacity (CEC) and exchangeable Ca, Mg and K as measured on a range of composts relative to a typical apple orchard soil

	CEC[z]	Ca[z]	Mg[z]	K[z]
Compost/soil	cmol (+)/kg soil dry weight			
C1 Vermicomposted dairy solids	175	16.6	15.4	21.6
C2 Peat	91	8.4	1.6	0.3
C3 Composted pulp waste	98	12.5	2.0	1.4
C4 Municipal biosolids compost	90	9.0	8.7	8.6
C5 Agricultural compost (80% vol vegetation, 20% vol poultry manure)	107	17.6	7.9	54.1
Osoyoos sandy loam	4.8	2.7	0.5	0.2

[z]Total exchange capacity (CEC) and exchangeable Ca, Mg and K measured after extraction in ammonium acetate at pH 7.0 (Hendershot et al., 1993)

coarse-textured apple orchard soils. Similarly, high moisture holding capacity is indicated by moisture contents (by weight) that frequently exceed 100% (Fig. 11). However, there can be a wide range in both the total exchangeable nutrient content of organic soil amendments, as illustrated for K in composts (Table 7). Furthermore, measurable improvements in, for example, the CEC of a loamy sand orchard soil would require as much as 20% organic material, by volume, to be added to the soil (Fig. 12).

A major effect of increasing soil organic matter is to increase the number and diversity of soil organisms, with the potential to stimulate beneficial biochemical reactions or suppress populations of soil-borne pests or pathogens. For example, in the previously cited experiment comparing several organically acceptable soil management strategies, microbial biomass and number of non-parasitic fungivorous nematodes were elevated when soil management involved surface application of bark mulch (Table 8; Forge et al., 2015).

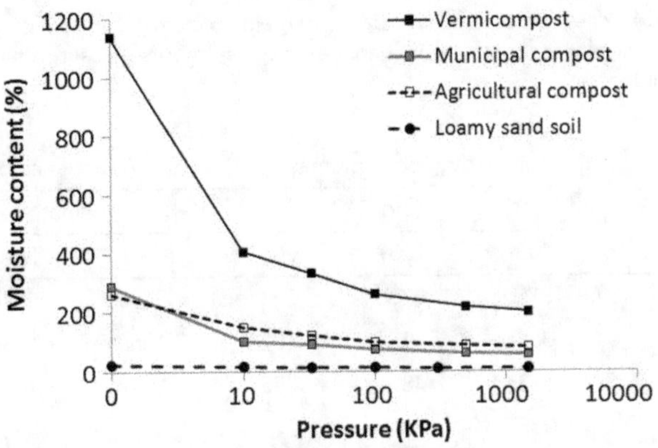

Figure 11 Moisture content over a range of imposed pressures for selected composts relative to a loamy sand soil.

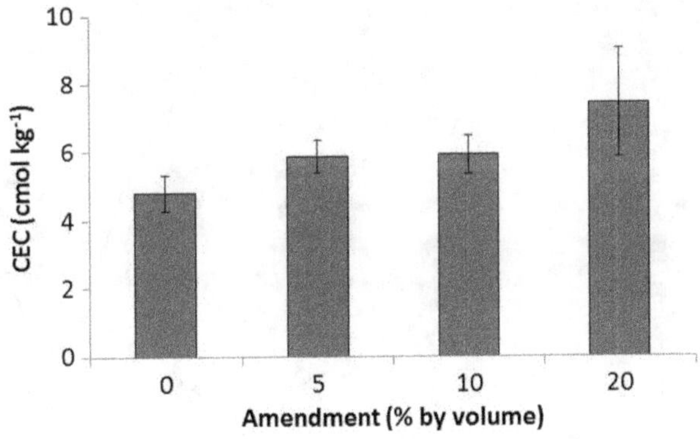

Figure 12 Cation exchange capacity (CEC) of a loamy sand orchard soil as affected by laboratory incubation with municipal compost.

Parasitic *Pratylenchus penetrans* nematode population densities were similarly suppressed after maintenance of an alfalfa hay mulch in a high-density 'Braeburn' apple orchard (Fig. 13; Forge et al., 2013)

Despite the many benefits to orchard soil quality associated with augmenting organic matter in orchards, successful apple orchards have been established on soils with a wide range of organic matter contents, and it has not been possible to directly link organic matter levels to orchard performance. This lack of relationship may be due to the fact that the active fraction of organic matter, which supports most biological activity and is the source of most nutrient mineralization, is usually small (<25%) relative to total soil organic matter contents. A number of randomized, replicated experiments have been published over the past 20 years in apple orchards from a worldwide range of production areas that allow some assessment of the effects of organic additions, which enhance active fraction organic matter, on apple tree performance relative to conventional orchard fertilizers and orchard floor management practices (Table 9). Many of these trials were established

Table 8 Microbial biomass carbon and abundance of major trophic groups of non-parasitic nematodes as affected by organically acceptable soil management treatments measured two, four and six years after experiment initiation

	Microbial biomass (mg C/g)[z]			Abundance of trophic groups (nematodes/100 ml soil)[y]		
	2007	2009	2011	Bact	Fung	OmPred
Annual compost/tillage	316bc	228b	397abc	979ab	562b	21b
Alfalfa mulch	361ab	492a	502ab	1366a	693b	69a
Bark mulch	273cd	481a	795a	1042ab	1020a	33b
Black plastic	235d	183b	273bc	676b	556b	14b
Significance	****	****	**	*	*	***

[z]As determined after chloroform fumigation and extraction with 0.25 M K_2SO_4 (Wu et al., 1990).
[y]Main effects over years. Bact=bacterivorous nematodes; Fung=fungivorous and root hair feeding nematodes; Ompred= omnivorous + predacious nematodes. *, **, ***, **** Means significantly different at P ≤ 0.05, P ≤ 0.01, P ≤ 0.001, P ≤ 0.0001, respectively.
Adapted from Forge et al. (2013).

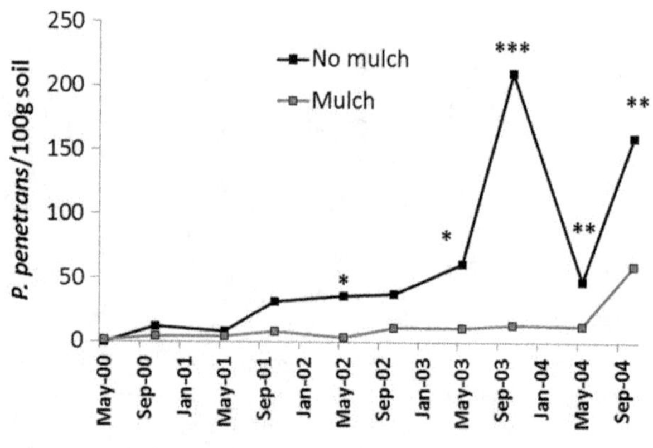

Figure 13 Effect of alfalfa hay mulch on populations of *Pratylenchus penetrans* (root lesion nematode) in Cleman very fine sandy loam orchard soil at 15–30 cm depth (Forge et al., 2013). *, **, *** Means significantly different at P ≤ 0.05, P ≤ 0.01, P ≤ 0.001, respectively.

on sites with challenging soil conditions, including limited fertility, susceptibility to water stress or the existence of a replant disorder. Most of these studies involved high-density apple orchards planted on dwarfing rootstocks, which have greater sensitivity to environmental stresses than trees on more vigorous rootstocks. Organic additions have been made in different ways, often within the tree row, as amendments (A) incorporated within the soil profile or as maintained surface mulches (M). Measurement variables typically included tree growth (increased trunk cross-sectional area (TCSA) or yield (Y)). Few assessments were made of treatment effects on fruit quality, despite the importance of fruit size and colour to grower returns. However, using TCSA and Y as indicators of treatments' success, it is apparent that use of mulches has generally been more effective than incorporation of soil amendments. In about two-thirds of the trials (18 of 27 apple orchard experiments), improved TCSA or Y were reported after mulching whereas use of amendments was successful only about one-third of the time (5 of 18 experiments) (Table 9). These statistics demonstrate that superior performance relative to conventional soil management practices is not always achieved. It is, however, encouraging that similar growth could result via use of organic amendments rather than soil fumigation in some studies. That being said, increased adoption of fully organic or integrated production systems will not eliminate the need to monitor other nutrient and soil limitations that may decrease apple orchard performance. For example, if sites have fertile soils or effective fertigation regimes, organic additions may not further improve growth and yield of apple orchards (Neilsen et al., 2004a). Similarly, organic applications could be ineffective when an overriding limitation, such as severe replant disease or potassium deficiency, is unaffected by the treatment.

Increased use of superior genetic material

It has long been recognized that the nutritional requirements of different apple cultivars and rootstocks vary. Traditional apple breeding programmes have emphasized the need to improve disease and pathogen resistance without compromising the acquisition and utilization of basic resources, such as light, nutrients and water, which are critical for producing apples with sufficiently good texture, size and colour, and storage capability to satisfy consumer-driven market demands (Neilsen et al., 2009a). As indicated in Section 2.1, changing climatic conditions will alter the geographic range over which apples grow, allowing production in soils not previously used for apple orchards. This situation has stimulated calls to adapt breeding programmes to produce new crops with improved mineral nutrition (Pilbeam, 2015). It has been proposed to take advantage of advances in molecular and genetic understanding by using an analysis of quantitative trait loci (QTL) to identify nutrient-efficient crops (Hirel et al., 2007). The genetic complexity of identifying QTL that control the acquisition of individual nutrients, which depend on nutrient uptake, assimilation, translocation and recycling, and which depend on the availability of other nutrients (especially N), will be formidable.

Rootstocks have been widely used in apple production to control tree vigour and increase precocity and unit area yield. Recently, the breeding of rootstocks for improved genetic potential to absorb and translocate nutrients to grafted scions is under investigation in New York (Fazio et al., 2013). The complexity of this task is illustrated by a recent three-year study of the 'Honeycrisp' cultivar on 31 dwarfing rootstocks in southern British Columbia, Canada (Neilsen and Hampson, 2014). It was noteworthy that nearly all the rootstocks achieved adequate N levels throughout the year, suggesting that N fertigation strategies were adequate for all rootstocks. However, none of the rootstocks tested were able to demonstrate consistent success in achieving adequate scion leaf levels for all nutrients and would have required fertilizer additions (Table 10). There were also differences in the nutritional performance of selected rootstocks on 'Honeycrisp' in this study relative to the performance of the same rootstocks on 'Golden Delicious' in the previously referenced New York study (Fazio et al., 2013; Neilsen and Hampson, 2014). The possibility of scion x rootstock interactions complicates the fact that environment x genetic interactions, resulting from the range of physical, chemical and biological conditions in which roots operate, already make it difficult to isolate the effects of rootstock on apple tree nutrition. Nevertheless, future improvements in genetic tools may offer new ways of mitigating nutrient stresses resulting from soil limitations.

Table 9 Randomized, replicated trials assessing the comparative effects of incorporated organic amendments (A) or surface applied organic mulches (M) on improving apple yield (Y) and canopy size (trunk cross-sectional area (TCSA)

Location/duration	Cultivar/spacing	Comparisons	Comment/outcome
		Incorporated amendment (A)	
Braun et al. (2010) Nova Scotia Canada, 6 yrs	Honeycrisp M.4 1.8 m × 4.2 m	Composted hog manure (17% by volume). Planting row trench vs fumigation and untreated control.	Replant site. Compost equivalent to fumigation. Y, TCSA exceeds control.
Merwin (2013) New York, USA, 10yrs	Empire M.26, M.7, G.30, G.210	Compost amendment vs fumigation and untreated control.	Replant site. Y similar among treatments. Rootstock has most effect on Y, TCSA.
Neilsen et al. (2004a) British Columbia (BC), Canada, 1 yr	Braeburn M.26 1.25 m × 3 m	Biosolids (90t/ha) rotovated to 30 cm depth vs untreated control.	Replant site. Shoot growth increased by biosolids.
Reganold et al. (2001) Wash., USA, 5 yrs	Golden Delicious M.9 1.4 m × 3.2 m	Compost applied at 919 kg/ha/yr for 2 yrs. vs conventional fertilization.	Y, TCSA similar, sometimes smaller fruit for compost.
		Incorporated amendment (A) and surface mulch (M)	
Forge et al. (2013) Wash, USA, 6 yrs	Braeburn M.9 1.2 m × 2.6 m	Tree row (1.5 m) strip. Alfalfa mulch 0.1 m depth (M). Compost dairy solids 45t/ha tree row at planting (A) vs unammended.	Replant site. Only M increased Y, TCSA but performance still suboptimal.
Neilsen et al., 2003a BC Canada, 6 yrs	Spartan M.9 1.25 m × 3.5 m	Tree row (2.0 m) strip. Shredded paper, alfalfa mulches (M) maintained annually. Biosolids amendment (A) 45t/ha every third year vs untreated check.	Paper mulch: largest TCSA. M, A greater Y than check.
Neilsen et al. (2004a) BC, Canada, 2 yrs	Gala M.9 1 m × 3 m	Tree row (2.0 m) strip. Wood chip mulch (M, 4cm). Biosolids amendment (A) 45t/ha incorporated to 0.2 m depth vs untreated control	Neither M or A increase Y or TCSA
Neilsen et al. (2007) BC, Canada, 6 yr	Golden Delicious M.9 1.9 m × 3.0 m	Tree row (2.0 m) strip. Paper mulch (M) vs biosolids amendment (A) 40mT/ha (Yr 1) + 20 mT/ha (Yr 4).	M increased Y and TCSA. Site had marginal K nutrition.

Rom et al. (2010) Arkansas, USA, 2 yrs	Enterprise M.26 2.0 m × 4.0 m	Tree row (2.0 m) strip. Paper, wood chip, monthly mow and blow mulches (M) vs urban green compost (A) (0.1 m depth) vs inorganic and organic fertilizer and unfertilized check	Compost (A), Woodchips (M) increased TCSA and Y relative to other treatments.
Van Schoor et al. (2009) South Africa, 2 yrs	Golden Delicious seedling (n = 2), Fuji M.793 (n = 1). Spacing not specified	Surface compost (A) (straw and manure, 15 kg/tree) wheat straw mulch (M) (0.1 m) vs fumigation and untreated control.	Three replant sites. Compost (A) and mulch (M) equivalent shoot growth to fumigated, improved relative to control.
		Surface Mulch (M)	
Cline et al. (2011) BC, Canada, 5 yrs	Gala, Granny Smith, Ambrosia, Honeycrisp M.9 0.6–1.0 m × 3–3.5 m	Spray on fibre mulch (M) vs herbicide strip tree row treatments.	Four field sites. TCSA improved by M. Y increase measured at single site.
Granatstein and Mullinix (2008) Washington, USA, 4yrs	Red Delicious M.26 1.0 m × 4.3 m	Woodchip (0.1 m), shredded paper (0.1 m), alfalfa mulch (M) vs bare control.	Alfalfa M: highest TCSA and Y. Other mulches: TCSA exceeds control.
Hogue et al. (2010) BC, Canada, 6 yrs	Gala M.9 1 m × 3 m	Spray on fibre mulch (M) relative to herbicide strip tree row (2 m) treatments.	Site had low K availability. M had higher TCSA and Y compared to herbicide treatment.
Merwin (2013) New York, USA, 18 yrs	Empire M.9/MM.111	Composted hardwood bark mulch (M) maintained at 0.1 m depth vs herbicide strip treatments.	Long-term TCSA best with M. Y equivalent to early season herbicide treatment.
Merwin et al. (1995), 4 yrs	Liberty, Empire Mark Freedom M.9/MM.111	Wood chip/hay mulch 0.1 m depth renewed annually vs herbicide strip tree row (1.8 m) treatments.	Two sites. No differences in TCSA and Y.

Table 10 Number of rootstocks (31 total), annually and cumulatively with low nutrient concentrations according to recommended regional commercial values

		Low rootstocks (n)			Cumulative	
Tissue/Nutrient	Recommended Value[z]	Year 1	Year 2	Year 3	(n)	(%)
Leaf Zn	>12 mg/kg	29	29	24	82	88
Leaf P	>2.0 g/kg	21	15	29	65	70
Leaf Mg	>2.6 g/kg	13	31	12	56	60
Fruit Ca	>4 mg/100gFW[y]	nm[x]	nm	16	16	52
Leaf Mn	>25 mg/kg	0	18	18	36	39
Leaf K	>13 g/kg	0	1	29	30	32
Leaf Ca	>10 g/kg	11	15	0	26	28
Leaf Fe	>45 mg/kg	1	17	7	25	27
Leaf N	>19 g/kg	1	2	0	3	3
Leaf B	>20 mg/kg	0	0	0	0	0
Leaf Cu	>4 mg/kg	0	0	0	0	0

[z]British Columbia Ministry of Agriculture and Lands.
[y]Fruit nutrient concentration on a fresh weight (FW) basis, leaf concentration on a dry weight basis.
[x]Not measured (nm).
Adapted from Neilsen and Hampson (2014).

Rootstock genotype can have significant influences on growth in replant soil and some newly released apple rootstocks appear to be more resistant or tolerant of the replant disease complex than traditional widely used rootstocks such as M.9 (Auvil et al., 2011; Robinson et al., 2012). Growth and productivity of these rootstocks have been compared in fumigated and non-fumigated soil at multiple sites across North America (Auvil et al., 2011; Robinson et al., 2012), but more research is needed to clarify their responses to different components of the replant disease complex that may vary among sites. The use of such rootstocks could reduce the need for pre-plant fumigation, which can have non-target negative effects on soil health that may in turn affect soil structure and nutrient uptake (Forge et al., 2016a).

6.2 Altered production areas

The cumulative effect of a warming climate is likely to alter the spatial boundaries of production areas of many temperate fruit crops, including apple. In California, for example, warmer winter temperature may prevent many fruit and nut crops from undergoing sufficient chilling to overcome seasonal dormancy (Luedling et al., 2009), making many 'warmer' sites less favourable for production. In contrast, winter freeze events have traditionally limited the poleward expansion of many fruit crops. In southern British Columbia, Canada (a traditional northern apple-growing region), the extent of apple production may migrate in response to a decrease in the occurrence of lethal minimum winter temperatures and a shift in the timing of freeze events to earlier in the winter (Quamme et al., 2010). Minimal work on this subject has been conducted for apple, which has a wide range of cultivars and rootstocks that vary in chilling and dormancy requirements. It will be possible to alter apple production areas for climatic reasons but this may require modification of soil and nutrient management strategies to overcome possible soil limitations. It will be important to consider soil physical, chemical and biological limitations prior to orchard establishment due to the extra effort and cost associated with their amelioration (Barden and Neilsen, 2003).

7 Conclusions

In conclusion, humans have a long history of managing orchard soils by manipulating water and nutrient availability to improve apple yield and quality. However, there are looming challenges, such as rapid anthropogenic climate change, soil and water quality degradation, and diminishing availability of chemical fertilizers, which require the use and development of management practices that protect or improve soil quality and maximize resource-use efficiency. Precision nutrient strategies, e.g., fertigation, targeted foliar nutrient applications and non-destructive sensing technologies, can be used to accurately diagnose and correct nutrient deficiencies. As well, careful manipulation of fertilizer formulations, to reduce fertilizer losses, and protection or enhancement of the soil biotic community to improve fertilizer uptake, can also be effective at improving fertilizer-use efficiency and soil health. Similarly, careful control of irrigation rate, method and timing, and applications of surface mulches can demonstrably improve water-use efficiency, reduce leaching and protect water quality and availability. Altered production systems, such as organic or integrated horticultural systems, which focus on protecting soil health by avoiding the use of chemical pesticides and fertilizers and maintaining soil organic matter pools, have gained widespread public support and have been adopted by producers around the world. In order to further adapt to the growing challenges imposed by climate change, the regions best suited to apple production may shift in the coming years; new cultivars and rootstocks may also prove key to maintaining the productivity of apple cultivation into the future.

8 Where to look for further information

For those wishing to broaden their understanding of soil and nutrient management of apple orchards, much information can be obtained from the earlier CABI book edited by Ferree and Warrington (2003) containing 24 chapters spanning the botany, production and utilization of apple. Several chapters are particularly pertinent discussing soil and site considerations for apple orchards, nutritional and water requirements, and orchard floor management strategies in addition to the requirements for integrated and organic apple production. An excellent introductory textbook describing the nature and properties of soils has recently been published by Weil and Brady (2016). Recently the United Nation Food and Agriculture Organization (FAO) has published voluntary sustainable soil management guidelines accessible at http://www.fao.org/documents/card/en/c/0549ec19-2d49-4cfb-9b96-bfbbc7cc40bc/.

A classic publication summarizing much historical research on fruit tree nutrition edited by Childers (1966), including a chapter on apple (Boynton and Oberley, 1966), provides a perspective on the evolution of apple tree nutritional strategies. The Horticultural Reviews series, edited since 1979 by Jules Janick, contains state-of-the-art reviews of pertinent research on horticultural crops with information relevant to soil and nutrient management of apple. Through the years several pertinent articles have been compiled including the development of plant tissue nutrient standards (Shear and Faust, 1980), root distribution and effectiveness (Atkinson, 1980), orchard floor vegetation management (Hogue and Neilsen, 1987), foliar nutrition (Swietlik and Faust, 1984), zinc nutrition (Swietlik, 1999), organic production (Raviv, 2009) and fruit Ca deficiencies (Tonetto de Freitas and Mitcham, 2012). Similarly reviews specific to various critical soil and nutrition topics such as fertigation (Neilsen and Neilsen, 2008) are periodically published in other venues.

Many countries have national horticultural societies which organize annual meetings to update horticultural research including orchard nutrient and water management (in the United States see the American Society for Horticultural Science (ASHS), http://www.ashs.org). The International Society of Horticultural Science (ISHS) (http://www.ishs.org) serves a similar international function with general horticultural meetings every fourth year. Within this umbrella organization are numerous working groups with F20 (Mineral Nutrition of Fruit Crops) organizing periodic symposium pertinent to the subject of this chapter. National soil science societies including the Soil Science Society of America (http://www.soils.org) and the Canadian Soil Science Society (http://www.csss.ca) in addition to the International Union of Soil Sciences (http://www.iuss.org) serve similar functions to continuously highlight new findings in soil science research pertinent to all crops.

In North America over the past 40 years, regional rootstock trials have been coordinated by the NC-140 Regional Research Project (http://www.nc140.org) providing a mechanism of international collaborative trials concerning new rootstocks. Recently, plans are underway to include a root to fruit nutrition component of these multi-site trials.

Strong research traditions in fruit tree nutrient and soil management, including apple currently exist in the Pacific Northwest of North America, including the Universities of Washington, Oregon and Idaho and Agriculture Canada's Research and Development Centre in Summerland, British Columbia. Cornell University and the University of California at Davis also have strong programmes. In Europe, the Italian universities of Bologna and Bolzano have made significant contributions to the field.

9 References

Adhikari, R., Bristow, K. L., Casey, P. S., Freischmidt, G., Hornbuckle, J. W. and Adhikari, B. (2016). Preformed and sprayable polymeric mulch film to improve agricultural water use efficiency. *Agricultural Water Management* 169: 1–13.

Atkinson, D. (1980). The distribution and effectiveness of the roots of tree crops. *Horticultural Reviews* 2: 424-90.

Atucha, A., Merwin, I. A., Purohit, C. K. and Brown, M. G. (2011). Nitrogen dynamics and nutrient budgets in four orchard groundcover management systems. *HortScience* 46: 1184–93.

Auvil, T. D., Schmidt, T. R., Hanrahan, I., Castillo, F., McFerson, J. R. and Fazio, G. (2011). Evaluation of dwarfing rootstocks in Washington apple replant sites. *Acta Horticulturae* 903, 265–71.

Avilla, J. and Riedl, H. (2003). Integrated fruit production for apples-principles and guidelines. In D. Ferree and I. J. Warrington (Eds) *Apples, Botany, Production and Uses*, CABI publishing, Wallingford, Oxon, UK, pp. 539–49.

Baram, S., Couvreur, V., Harter, T., Read, M., Brown, P. H., Hopmans, J. W. and Smart, D. R. (2016). Assessment of orchard N losses to groundwater with a vadose zone monitoring network. *Agricultural Water Management* 172: 83–95.

Bar-Yosef, B. (1999). Advances in fertigation. *Advances in Agronomy* 65: 1–76.

Barden, J. A. and Neilsen, G. H. (2003). Selecting the orchard site, site preparation and orchard planning and establishment. In D. Ferree and I. J. Warrington (Eds) *Apples, Botany, Production and Uses,* CABI publishing, Wallingford, Oxon, UK, pp. 237–65.

Behboudian, M. H. and Mills, T. M. (1997). Deficit irrigation in deciduous orchards. *Horticultural Reviews* 21: 105–31.

Benson, N. R. (1976). Retardation of apple tree growth by soil arsenic residues. *Journal of the American Society for Horticultural Science* 101: 251–3.

Bindraban, P. S., Dimkpa, C., Nagarajan, L., Roy. A. and Rabbinge, R. (2015). Revisiting fertilisers and fertilisation strategies for nutrient uptake by plants. *Biology and Fertility of Soils* 51: 897–911.

Bingham, A. H. and Cotrufo, M. F. (2016). Organic nitrogen storage in mineral soil: implications for policy and management. *Science of the Total Environment* 551–2: 116–26.

Boll, J., Steenhuis, T. S. and Selker, J. S. (1992). Fiberglass wicks for sampling water and solutions in the vadose zone. *Soil Science Society of America Journal* 56: 701–7.

Boynton, D. and Oberly, G. H. (1966). Apple nutrition. In N. F. Childers (Ed.), *Nutrition of Fruit Crops., Horticultural Publications,* New Brunswick, NJ, pp. 1–50.

Braun, P. G., Fuller, K. D., McRae, K. and Fillmore, S. A. E. (2010). Response of 'Honeycrisp[R]' apple trees to combinations of pre-plant fumigation, deep ripping, and hog manure compost incorporation in a soil with replant disease. *HortScience* 45: 1702–7.

Brod, E., Ogaard, A. F., Hansen, E., Wragg, D., Haraldsen, T. K. and Krogstad, T. (2015). Waste products as alternative phosphorus fertilisers part I: inorganic P species affect fertilisation effects depending on soil pH. *Nutrient Cycling in Agroecosystems* 103: 167–85.

Bunemann, E. K. (2015). Assessment of gross and net mineralization rates of soil organic phosphorus-a review. *Soil Biology and Biochemistry* 89: 82–98.

Campoy, J. A., Ruiz. D. and Egea, J. (2011). Dormancy in temperate fruit trees in global warming context: A review. *Scientia Horticulturae* 130: 357–72.

Caprio, J. M. and Quamme, H. A. (1999). Weather conditions associated with apple production in the Okanagan Valley of British Columbia. *Canadian Journal of Plant Science* 79: 129–37.

Centritto, M., Lee, H. S. J. and Jarvis, P. G. (1999). Interactive effects of elevated [CO_2] and drought on cherry (*Prunus avium*) sedlings. *New Phytologist* 141: 129–40.

Centritto, M., Lucas, M. E. and Jarvis, P. G. (2002). Gas exchange, biomass, whole-plant water-use efficiency and water uptake of peach (*Prunus persica*) seedlings in response to elevated carbon dioxide concentration and water availability. *Tree Physiology* 22: 699–706.

Cheng, L., Dong, S. and Fuchigami, L. H. (2002). Urea uptake and nitrogen remobilization by apple leaves in relation to tree nitrogen status in autumn. *Journal of Horticultural Science and Biotechnology* 77: 13–18.

Chien, S. H., Prochnow, L. I. and Cantarella, H. (2009). Recent developments of fertilizer production and use to improve nutrient efficiency and minimize environmental impacts. *Advances in Agronomy* 102: 267–322.

Childers, N. F. (1966). Temperate to Tropical Fruit Nutrition, Horticultural Publications, Rutgers, NJ, USA.

Childers, N. F., Morris, J. R. and Sibbet, G. S. (1995). *Modern Fruit Science*, Horticultural Publications, Gainesville, FL, USA.

Cline, J., Neilsen, G., Hogue, E., Kuchta, S. and Neilsen, D. (2011). Spray-on-mulch technology for intensively grown irrigated apple orchards: influence on tree establishment, early yields, and soil physical properties. *HortTechnology* 21: 398–411.

Condron, L. M., Cameron, K. C., Di, H. J., Clough, T. J., Forbes, E. A., Mclaren, R. G. and Silva, R. G. (2000). A comparison of soil and environmental quality under organic and conventional farming systems in New Zealand. *New Zealand Journal of Agricultural Research* 43: 443–66.

Cordell, D., Drangert, J. O. and White, S. (2009). The story of phosphorus: global food security and food for thought. *Global Environmental Change* 19: 292–305.

Dawson, J. J. C. and Smith, P. (2007). Carbon loss from soil and its consequence for land-use management. *Science of the Total Environment* 382: 165–90.

Deurer, M., Muller, K., Kim, I., Huh, K. Y., Young, I., Jun, G. I. and Clothier, B. E. (2012). Can minor compaction increase soil carbon sequestration? A case study in a soil under a wheel-track in an orchard. *Geoderma* 183/184: 74–9.

Dong, S., Neilsen, D., Neilsen, G. H. and Fuchigami, L. H. (2005). Foliar N application reduces soil NO_3-N leaching loss in apple orchards. *Plant and Soil* 268: 357–66.

Doran, J. W. and Zeis, M. R. (2000). Soil health and sustainability: managing the biotic component of soil quality. *Applied Soil Ecology* 15: 3–11.

Drazeta, L., Lang, A., Hall, A. J., Volz, R. K. and Jameson, P. E. (2004). Causes and effects of changes in xylem functionality in apple fruit. *Annals of Botany* 93: 275–82.

FAO and ITPS. (2015). Status of the World's Soil Resources (SWSR)-Main Report. Food and Agriculture Organization (FA0) of the United Nations and Intergovernmental Technical Panel on Soils (ITPS), Rome, Italy.

Faust, M. (1989). *Physiology of Temperate Zone Fruit Trees*, John Wiley and Sons, New York.

Fazio, G., Kviklys, D., Grusak, M. A. and Robinson, T. (2013). Phenotypic diversity and QTL mapping of adsorption and translocation of nutrients by apple rootstock. *Aspects of Applied Biology* 119: 37–50.

Fentabil, M. M., Nichol, C. F., Jones, M. D., Neilsen, G. H., Neilsen, D. and Hannam, K. D. (2016). Effect of drip irrigation frequency, nitrogen rate and mulching on nitrous oxide emissions in semi-arid climate: An assessment across two years in an apple orchard. *Agriculture, Ecosystems and Environment* 235: 242-52..

Fereres, E., Orgaz, O. and Gonzalez-Dugo, V. (2011). Reflections on food security under water scarcity. *Journal of Experimental Botany* 62: 4079–86.

Ferree, D. C. and Warrington, I. J. (2003) Apples Botany, Production and Uses, CABI Publishing, CAB International, Wallingford, Oxon, UK.

Fixen, P. E. and Johnston, A. (2011). World fertilizer reserves: a view to the future. *Journal of Science of Food and Agriculture* 92: 1001–5.

Forge, T., Muehlchen, A., Hackenberg, C., Neilsen. G. and Vrain, T. (2001). Effects of preplant inoculation of apple (*Malus domestica* Borkh.) with arbuscular mycorrhizal fungi on population growth of the root-lesion nematode, *Pratylenchus pentrans Plant Soil* 236: 185–96.

Forge, T., Neilsen, G., Neilsen, D., Hogue, E. and Faubion, D. (2013). Composted dairy manure and alfalfa mulch affect soil ecology and early production of 'Braeburn' apple on M.9 rootstock. *HortScience* 48: 645–51.

Forge, T., Neilsen, G., Neilsen, D., O'Gorman, D. O., Hogue, E. and Angers, D. (2015). Organic orchard soil management practices affect soil biology and organic matter. *Acta Horticulturae* 1076: 77–83.

Forge, T. A., Neilsen, G. H. and Neilsen, D. (2016a). Organically acceptable practices to improve replant success of temperate tree-fruit crops. *Scientia Horticulturae* 200: 205–41.

Forge, T., Kenney, E., Hashimoto, N., Neilsen, D. and Zebarth, B. (2016b). Compost and poultry manure as preplant soil amendments for perennial fruit crops: comparative effects on root lesion nematodes, soil quality and risk of nitrate leaching. *Agriculture, Ecosystems and Environment* 223, 48–58.

Fujisawa, M. and Kobayashi, K. (2010). Apple (*Malus pumilia var. domestica)* phenology is advancing due to rising air temperature in northern Japan. *Global Change Biology* 16: 2651–60.

Gale, E. S., Sullivan, D. M., Cogger, C. G., Bary, A. I., Hemphill, D. D. and Myhre, E. A. (2006). Estimating plant available nitrogen release from manures, composts and specialty products. *Journal of Environmental Quality* 35: 2321–32.

Gaskell, M. and Smith, R. (2007). Nitrogen sources for organic vegetable crops. *HortTechnology* 17: 431–41.

Goh, K. M., Pearson, D. R. and Daly, M. J. (2001). Effects of apple tree production systems on some important soil physical, chemical and biological quality parameters. *Biological Agriculture and Horticulture* 18: 269–92.

Gomiero, T., Pimentel, D. and Paoletti, M. G. (2011). Is there a need for a more sustainable agriculture? *Critical Reviews in Plant Science* 30: 6–23.

Gordon, B. and Tindall, T. (2006). Fluid P performance improved with polymers. *Fluid Journal* 14: 12–13.

Grantatstein, D. and Mullinix, K. (2008). Mulching options for Northwest organic and conventional growers. *HortScience* 43: 45–50.

Granatstein, D., Kirkby, E., Ostenson, H. and Willer, H. (2015). Global situation for organic tree fruits. Scientia. *Horticulturae*, http://dx.doi.org/10.1016/j.scenta2015.12.008.

Greenwood, D. J., Zhang, K., Hilton, H. W. and Thompson, A. J. (2010). Opportunities for improving irrigation efficiency with quantitative models, soil water sensors and wireless technology. *Journal of Agricultural Science* 148: 1–16.

Guak, S., Neilsen, D., Millard, P., Wendler, R. and Neilsen, G. H. (2003). Determining the role of N remobilization for growth of apple *(Malus domestica Borkh.)* trees by measuring xylem-sap N flux. *Journal of Experimental Botany* 54: 2121–31.

Haynes, R. J. (1985). Principles of fertiser use for trickle irrigated crops. *Fertilizer Research* 6: 235–55.

Hatfield, J. L. and Walthall, C. L. (2015). Soil biological fertility: foundation for the next revolution in agriculture? *Communications in Soil Science and Plant Analysis* 46: 753–62.

Heege, H. J. and Thiessen, E. (2013). chapter 6 Sensing of crop properties. In H. J. Heege (Ed.), *Precision in Crop Farming: Site Specific Concepts and Sensing Methods: Applications and Results*, Springer Science + Business Media, Dordrecht, Germany, pp. 103–41.

Hendershot, W. H., Lalande, H. and Duquette, M. (1993). Ion exchange and exchangeable cations. In M. R. Carter (Ed.),Soil Sampling and Methods of Analysis Lewis Publishers, Boca Raton, Florida 33431, pp. 167–76.

Hernandez, T., Chocano, C., Moreno, J.-L. and Garcia, C. (2016). Use of compost as an alternative to conventional inorganic fertilizer in intensive lettuce (*Lactuca sativa L.)* crops-effects on soil and plant. *Soil & Tillage Research* 160: 14–22.

Hirel, B., Le Gouis, J., Ney, B. and Gallais, A. (2007). The challenge of improving nitrogen use efficiency in crop plants: towards a more central role for genetic variability and quantitative genetics within integrated approaches. *Journal of Experimental Botany* 58: 2369–87.

Hogue, E. J., Cline, J. A., Neilsen, G. and Neilsen, D. (2010). Growth and yield responses to mulches and cover crops under low potassium conditions in drip-irrigated apple orchards on coarse soil. *HortScience* 45: 1866–71.

Hogue, E. J. and Neilsen, G. H. (1987). Orchard floor vegetation management. *Horticultural Reviews* 9: 377-430.

Hoyt, P. B. and Neilsen, G. H. (1985). Effects of soil pH and associated cations on growth of apple trees planted in old orchard soils. *Plant and Soil* 86: 395–401.

IPCC (2014): Climate Change 2014: Synthesis Report. Contribution of Working Groups I, II and III to the Fifth Assessment Report of the Intergovernmental Panel on Climate Change [Core Writing Team, R. K. Pachauri and L. A. Meyer (eds)]. IPCC, Geneva, Switzerland, 151p.

Jones, H. G. (2004). Irrigation scheduling: advantages and pitfalls of plant-based methods. *Journal of Experimental Botany* 55: 2427–36.

Kandula, D. R. W., Jones, E. E., Horner, I. J. and Stewart, A. (2010). The effect of *Trichoderma* bio-inoculants on specific apple replant disease (SARD) symptoms in apple rootstocks in New Zealand. *Australasian Plant Pathology* 39: 312–18.

Klein, I., Levin, I., Bar-Yosef, B., Assaf, R. and Berkovitz, A. (1989). Drip nitrogen fertigation of 'Starking Delicious' apple trees. *Plant and Soil* 119: 305–14.

Leib, B., Caspari, H., Redulla, C., Andrews, P. and Jabro, J. (2006). Partial rootzone drying and deficit irrigation of 'Fuji'apples in a semi-arid climate. *Irrigation Science* 24: 85–99.

Liberloo, M. I., Tulva, I., Raim, O., Kull, O. and Cuelemans, R. (2007). Photosynthetic stimulation under long-term CO_2 enrichment and fertilization is sustained across a closed *Populus* canopy profile (EUROFACE). *New Phytologist* 173: 537–49.

Liu, R. and Lal, R. (2015). Potentials of engineered nanoparticles as fertilizers for increasing agronomic productions. *Science of the Total Environment* 514: 131–9.

Luedeling, E., Zhang, M., Luedeling, V. and Givertz, E. H. (2009). Sensitivity of winter chill models for fruit and nut trees to climatic changes expected in California's central valley. *Agriculture, Ecosystems and Environment* 133: 23–31.

Lukac, M., Calfapietra, C., Lagomarsino, A. and Lareto, F. (2010). Global climate change and tree nutrition: effects of elevated CO_2 and temperature. *Tree Physiology* 30: 1209–20.

Maas, E. V. (1987). Salt tolerance of plants. In B. R. Christie (Ed.), *CRC Handbook of Plant Science in Agriculture* Vol. II, CRC Press, Boca Raton, Florida, USA, pp. 57–75.

Mazzola, M. and Manici, L. M. (2012). Apple replant disease: role of microbial ecology in cause and control. *Annual Review of Phytopathology* 50: 45–65.

Merwin, I. A. (2003). Orchard floor management systems. In D. Ferree and I. J. Warrington (Eds), *Apples, Botany, Production and Uses*, CABI publishing, Wallingford, Oxon, UK, pp. 303–18.

Merwin, I. A. (2013). Soil management affects orchard soil health and tree productivity. *Washington State Horticultural Association Annual Meeting* 2013: 5.

Merwin, I. A., Rosenberger, D. A., Engle, C. A., Rist, D. L. and Fargione, M. (1995). Comparing mulches, herbicides, and cultivation as orchard ground cover management systems. *HortTechnology* 5: 151–8.

Millard, P. and Neilsen, G. H. (1989). The influence of nitrogen supply on the uptake and remobilization of stored N for the seasonal growth of apple trees. *Annals of Botany* 63: 301–9.

Miller, M. H., Robinson, J. B., Coote, D. R., Spires, A. C. and Draper, D. W. (1982). Agriculture and water quality in the Canadian Great Lakes basin: III. Phosphorus. *Journal of Environmental Quality* 11: 487–93.

Monaghan, J. M., Daccache, A., Vickers, L. H., Hess, T. M., Weatherhead, E. K., Grove, I. G. and Knox, J. W. (2013). More 'crop per drop': constraints and opportunities for precision irrigation in European agriculture. *Journal of Science of Food and Agriculture* 93: 977–80.

Miller, D. D., Domoto, P. A. and Walker, C. (1985). Mycorrhizal fungi at eighteen apple rootstock plantings in the United States. *New Phytologist* 100: 379–91.

Mueller, N. D., Gerber, J. S., Johnston, M., Ray, D. K., Ramankutty, N. and Foley, J. A. (2012). Closing yield gaps through nutrient and water management. *Nature* 490.7419: 254–7.

Naor, A., Naschitz, S., Peres, M. and Gal, Y. (2008). Response of apple fruit size to tree water status and crop load. *Tree Physiology* 28: 1255–61.

Neilsen, D. and Neilsen, G. H. (2002). Efficient use of nitrogen and water in high-density apple orchards. *HortTechnology* 12: 19–25.

Neilsen, D. and Neilsen, G. (2008). Fertigation of deciduous fruit trees: apple and sweet cherry. In P. Imus and M. R. Price (Eds), *Fertigation: Optimizing the Utilization of Water and Nutrition*, International Potash Institute, Horgen, Switz, pp. 76–88.

Neilsen, D., Hogue, E., Hoyt, P. B. and Drought, B. G. (1993). Oxidation of elemental sulphur and acidulation of calcareous orchard soils in southern British Columbia. *Canadian Journal of Soil Science* 73: 103–114.

Neilsen, D., Hogue, E. J., Neilsen, G. H. and Parchomchuk, P. (1995a). Using SPAD-502 values to assess the nitrogen status of apple trees. *HortScience* 30: 508–12.

Neilsen, D., Hoyt, P. B., Parchomchuk, P., Neilsen, G. H. and Hogue, E. J. (1995b). Measurement of the sensitivity of orchard soils to acidification. *Canadian Journal of Soil Science* 75: 391–5.

Neilsen, D., Millard, P., Herbert, L., Neilsen, G. H., Hogue, E. J., Parchomchuk, P. and Zebarth, B. (2001). Remobilization and uptake of N by newly planted apple trees in response to irrigation method and timing of N application. *Tree Physiology* 21: 513–21.

Neilsen, D., Neilsen, G. H., Gregory, D., Forge, T. and Zebarth, B. (2008a). Drainage losses of water, N and P from micro-irrigation systems in a young, high density apple planting. *Acta Horticulturae* 792: 483–90.

Neilsen, D., Neilsen, G., Guak, S. and Forge, T. (2016). Consequences of deficit irrigation and crop load reduction on plant water relations, yield, and quality of 'Ambrosia' apple. *HortScience* 51: 98–106.

Neilsen, D., Neilsen, G. H., Herbert, L., Millard, P. and Guak, S. (2006a). Allocation of dry matter and N to fruit and shoots in dwarf apple in response to sink size and N availability. *Acta Horticulturae* 721: 33–40.

Neilsen, D., Parchomchuk, P., Neilsen. G. H. and Hogue, E. J. (1998). Using soil solution monitoring to determine the effects of irrigation management and fertigation on nitrogen availability in high-density apple orchards. *Journal of the American Society of Horticultural Science* 123: 706–13.

Neilsen, D., Smith, C. A. S., Frank. G., Koch, W., Alila, Y., Merritt, W. S., Taylor, W. G., Barton, M., Hall, J. W. and Cohen, S. J. (2006b). Potential impacts of climate change on water availability for crops in the Okanagan basin, British Columbia. *Canadian Journal of Soil Science* 86: 921–36.

Neilsen, G., Forge, T., Angers, D., Neilsen, D. and Hogue, E. (2014). Suitable orchard floor management strategies in organic apple orchards that augment soil organic matter and maintain tree performance. *Plant and Soil* 378: 325–35.

Neilsen, G. and Hampson, C. (2014). 'Honeycrisp apple leaf and fruit nutrient concentration is affected by rootstock during establishment. *Journal of the American Pomological Society* 68: 178–89.

Neilsen, G. H. and Neilsen, D. (2003). Nutritional requirements for apple. In D. Ferree and I. J. Warrington (Eds), *Apples, Botany, Production and Uses*, CABI publishing, Wallingford, Oxon, UK, pp. 267–302.

Neilsen, G. H. and Neilsen, D. (2006a). Response of high density apple orchards on coarse textured soil to form of potassium applied by fertigation. *Canadian Journal of Soil Science* 86: 749–55.

Neilsen, G. and Neilsen, D. (2006b). Strategies for nutrient and water management of high density apple orchards on coarse-textured soils. *Annals of Warsaw Agricultural University-SGGW, Horticulture and Landscape Architecture* 27: 181–92.

Neilsen, G. H. and Yorston, J. (1991). Soil disinfection and monoammonium phosphate fertilization increase precocity of apples on replant problem soils. *Journal of American Society for Horticultural Science* 116: 651–4.

Neilsen, G. H., Culley, J. L. B. and Cameron, D. R. (1982). Agriculture and water quality in the Canadian Great Lakes basin: IV. Nitrogen. *Journal of Environmental Quality* 11: 493–7.

Neilsen, G. H., Hogue, E. J., Forge, T., and Neilsen, D. (2003a). Mulches and biosolids affect vigor, yield and leaf nutrition of fertigated high density apple. *HortScience* 38: 41–5.

Neilsen, G. H., Hogue, E. J., Forge, T. and Neilsen, D. (2003b). Surface application of mulches and biosolids affect orchard soil properties after 7 years. *Canadian Journal of Soil Science* 83: 131–7.

Neilsen, G. H., Hogue, E. J., Forge, T., Neilsen, D. and Kuchta, S. (2007). Nutritional implications of biosolids and paper mulch applications in high density apple orchards. *Canadian Journal of Plant Science* 87: 551–8.

Neilsen, G. H., Hogue, E. J., Neilsen, D. and Forge, T. (2004a). Use of organic applications to increase productivity of high density apple orchards. *Acta Horticulturae* 638: 347–56.

Neilsen, G. H., Hogue, E. J. and Parchomchuk, P. (1990). Flowering of apple trees in the second year is increased by first-year P fertilization. *HortScience* 25: 1247–50.

Neilsen, G. H., Lowery, D. T., Forge, T. A. and Neilsen, D. (2009a). Organic fruit production in British Columbia. *Canadian Journal of Plant Science* 89: 677–92.

Neilsen, G. H., Neilsen, D., Guak, S. and Forge, T. (2015). The effect of deficit irrigation and crop load on leaf and fruit nutrition of fertigated 'Ambrosia'/'M.9' apple. *HortScience* 50: 1387–93.

Neilsen, G. H., Neilsen, D. and Herbert, L. (2009b). Nitrogen fertigation concentration and timing of application affect nitrogen nutrition, yield, firmness and color of apples grown at high density. *HortScience* 44: 1425–31.

Neilsen, G. H., Neilsen, D., Herbert, L. C. and Hogue, E. J. (2004b). Response of apple to fertigation of N and K under conditions susceptible to the development of K deficiency. *Journal of the American Society for Horticultural Science* 129: 26–31.

Neilsen, G. H., Neilsen, D., Hogue, E. J. and Herbert, L. C. (2004c). Zinc and boron nutrition management in fertigated high density apple orchards. *Canadian Journal of Plant Science* 84: 823–8.

Neilsen, G. H., Neilsen, D. and Peryea, F. (1999). Response of soil and irrigated fruit trees to fertigation or broadcast application of nitrogen, phosphorus, and potassium. *HortTechnology* 9: 393–401.

Neilsen, G. H., Neilsen, D., Toivonen, P. and Herbert, L. (2008b). Annual bloom-time phosphorus fertigation affects soil phosphorus, apple tree phosphorus nutrition, yield, and fruit quality. *HortScience* 43: 885–90.

Neilsen, G. H., Parchomchuk, P. and Neilsen, D. (1997). Distribution of soil P and K as affected by NP-fertigation in high density apple orchards. *Acta Horticulturae* 448: 439–47.

Neilsen, G. H., Parchomchuk, P., Neilsen. D., Berard, R. and Hogue, E. J. (1995c). Leaf nutrition and soil nutrients are affected by irrigation frequency and method for NP-fertigated 'Gala' apple. *Journal of the American Society for Horticultural Science* 120: 971–6.

Nightingale, H. I. (1972). Nitrates in soil and groundwater beneath irrigated and fertilized crops. *Soil Science* 114: 300–11.

Nord, E. A. and Lynch, J. P. (2009). Pant phenology: a critical controller of soil resource acquisition. *Journal of Experimental Botany* 60: 1927–37.

Ojha, T., Misra, S. and Raghuwanshi, N. S. (2015). Wireless sensor networks for agriculture: The state-of-the-art in practice and future challenges. *Computers and Electronics in Agriculture* 118: 66–84.

Oland, K. (1959). Nitrogenous reserves of apple trees. *Physiologica Plantarum* 12: 594–648.

Owen, D., Williams, A. P., Griffith, G. W. and Withers, P. J. A. (2015). Use of commercial bio-inoculants to increase agricultural production through improved phosphorus acquisition. *Applied Soil Ecology* 86: 41–54.

Palm, C., Blanco-Canqui, H., DeClerck, F., Gatere, L. and Grace, P. (2014). Conservation agriculture and ecosystem services: an overview. *Agriculture, Ecosystems and Environment* 187: 87–105.

Parchomchuk, P., Neilsen, G. H. and Hogue, E. J. (1993). Effects of drip fertigation of NH_4-N and P on soil pH and cation leaching. *Canadian Journal of Soil Science* 73: 157–64.

Perry, K. B., Blakenship, S. M. and Unrath, C. R. (1987). Predicting harvest date of 'Delicious' apples using heat unit accumulations. *Agricultural and Forest Meteorology* 39: 81–8.

Perring, M. A. and Pearson, K. (1984). The distribution of calcium, magnesium and phosphorus in Cox's Orange Pippin and Spartan apple fruit. *Journal of the Science of Food and Agriculture* 35: 897–9.

Peryea, F. J., Neilsen, G. H. and Faubion, D. (2007). Start-timing for calcium chloride spray programs influences fruit calcium and bitter pit in 'Braeburn' and 'Honeycrisp' apples. *Journal of Plant Nutrition* 30: 1213–27.

Pilbeam, D. J. (2015). Breeding crops for improved mineral nutrition under climate change conditions. *Journal of Experimental Botany* 66: 3511–21.

Pimentel, D. (2011). Food for thought: a review of the role of energy in current and evolving agriculture. *Critical Reviews in Plant Sciences* 30: 35–44.

Pionke, H. B., Sharma, M. L. and Hirschberg, K.-J. (1990). Impact of irrigated horticulture on nitrate concentration in groundwater. *Agriculture, Ecosystems and Environment* 32: 119–32.

Plenchette, C., Furlan, V. and Fortin, J. A. (1981). Growth stimulation of apple trees in unsterilized soil under field conditions with VA mycorrhiza inoculation. *Canadian Journal of Botany* 59: 2003–8.

Poeplau, P. and Don, A. (2015). Carbon sequestration in agricultural soils via cultivation of cover crops-a meta-analysis. *Agriculture, Ecosystems and Environment* 200: 33–41.

Quamme, H., Cannon, A. J., Neilsen, D., Caprio, J. M. and Taylor, W. G. (2010). The potential impact of climate change on the occurrence of winter freeze events in six fruit crops grown in the Okanagan valley. *Canadian Journal of Plant Science* 90: 85–93.

Raviv, R. (2009). Sustainability of organic horticulture. *Horticultural Reviews* 36:257-87.

Ridgeway, H. J., Kandula, J. and Stewart, A. (2008). Arbuscular mycorrhiza improve apple rootstock growth in soil conducive to specific apple replant disease. *New Zealand Plant Protection* 61: 48–53.

Robinson, T., Autio, W., Clements, J., Cowgill, W., Embree, C., Gonzalez, V., Hoying, S., Kushad, M., Parker, M., Parra, R. and Schupp, J. (2012). Rootstock tolerance to apple replant disease for improved sustainability of apple production. *Acta Horticulturae* 940, 521–8.

Reganold, J. P., Glover, J. D., Andrews, P. K. and Hinman, H. R. (2001). Sustainability of three apple production systems. *Nature* 410:19 April: 926–30.

Rom, C. R., Garcia, M. E., McAfee, J., Friedrich, H., Choi, H. S., Johnson, D. T., Popp, J. and Savin, M. (2010). The effects of groundcover management and nutrient source during organic orchard establishment. *Acta Horticuturae* 873: 105–13.

Rosenzweig, C. and Hillel, D. (2000). Soils and global climate challenges and opportunities. *Soil Science* 165: 47–56.

Rouphael, Y., Franken, P., Schneider, C., Schwarz, D., Giovanetti, M., Aglonucci, M., De Pascale, S., Bonini, P. and Colla, G. (2015). Arbuscular mycorrhizal fungi act as biostimulants in horticultural crops. *Scientia Horticuturae* 196: 91–108.

Shear, C. B. and Faust, M. (1980) Nutritional ranges of deciduous tree fruits and nuts. *Horticultural Reviews* 2: 142-63.

Spalding, R. F. and Exner, M. E. (1993). Occurrence of nitrate in groundwater- a review. *Journal of Environmental Quality* 22: 392–402.

Sperrow, M. (2014). An enhanced method for using the IPCC approach to estimate soil organic storage potential on U.S. agricultural soils. *Agriculture, Ecosystems and Environment* 193: 96–107.

Stevenson, D. S. and Neilsen, G. H. (1990). Nitrogen additions and losses to drainage in orchard-type irrigated lysimeters. *Canadian Journal of Soil Science* 70: 11–19.

Stocker, T. F., Qin, D., Plattner, G.-K. V., Alexander, L. V., Allen, S. K., Bindoff, N. L., Breon, F. M., Church, J. A., Cubasch, U., Emori, S., Forster, P., Friedlingstein, P., Gillett, N., Gregory, J. M., Hartmann, D. L., Jansen, E., Kirtman, B., Knutti, R., Krishna Kumar, K., Lemke, P., Marotzke, J., Mason-Delmotte, V., Meehl, G. A., Mokhov, I. I., Piao, S., Ramaswamy, V., Randall, D., Rhein, M., Rojas. M., Sabine, C., Shindell, D., Talley, L. D., Vaughan, D. and S. P. Xie. (2013). Technical Summary. In. Climate Change 2013: The Physical Science Basis. Contribution of Working Group 1 to the Fifth Assessment Report of the Intergovernmental Panel on Climate Change [Stocker, T. F., Qin, D., Plattner, G. -K., Tignor, M., Allen, S. K., Boschung, J., Nauels, A., Xia, Y., Bex, V. and P. M. Midgley (eds)]. Cambridge University Press, Cambridge, United Kingdom and New York, NY, USA.

Swarts, N., Montagu, K., Oliver, G., Southam-Rogers, L., Hardie, M., Corkrey, R., Rogers, G. and Close, D. (2016). Benchmarking nitrous oxide emissions in deciduous tree cropping systems. *Soil Research* 54: 500–11.

Swietlik, D. (1999). Zinc nutrition of horticultural crops. *Horticultural Reviews* 23: 109-78.

Swietlik, D. and Faust, M. (1984). Foliar nutrition of fruit crops. *Horicultural Reviews* 6: 287–355.

Tonnetto de Freitas, S. and Mitcham, E. J. (2012). Factors involved in fruit calcium disorders. *Horticultural Reviews* 40:107-46.

Tao, R., Liang, Y., Wakelin, S. A. and Chu, G. (2015). Supplementing chemical fertilizer with an organic component increases soil biological function and quality. *Applied Soil Ecolology* 96: 42–51.

Thomson, T., McCaskill, M., Goodwin, I., Kearney, G. and Lolicato, S. (2014). Potential impacts of rising global temperatures on Australia's pome fruit industry and adaptation strategies. *New Zealand Journal of Crop and Horticultural Science* 42: 21–30.

Titus, J. S. and Kang, S.-M. (1982). Nitrogen metabolism. *Horticultural Reviews* 4: 204–46.

Tromp, J., Webster, A. D. and Wertheim, S. J.(2005). (Eds), *Fundamentals of Temperate Zone Fruit Production*, Backhuys Publishers, Leiden, The Netherlands.

Vance, C. P. (2001). Symbiotic nitrogen fixation and phosphorus acquisition plant nutrition in a world of declining renewable resources. *Plant Physiology* 127: 390–7.

Vang-Petersen, O. (1980). Calcium nutrition of apple trees: a review. *Scientia Horticulturae* 12: 1–9.

Van Schoor, L., Denman, S. and Cook, N. C. (2009). Characterisation of apple replant disease under South African conditions and potential biological management strategies. *Scientia Horticulturae* 119: 153–62.

Warrington, I. J. (2008). Climate change and horticulture. *Chronica Horticulturae* 48: 3–4.

Weil, R. R. and Brady, N. C. (2016) The Nature and Properties of Soils (15th Edition), Pearson Education Limited Publishers, Harlow, Essex, UK.

Weinbaum, S. A., Johnson, R. S. and DeJong, T. M. (1992). Causes and consequences of overfertilization in orchards. *HortTechnolgy* 2: 112–21.

Wolk, W. D., Lau, O. L., Neilsen, G. H. and Drought, B. G. (1998). Factors and time of sample collection for correlating storage potential of 'McIntosh', 'Spartan', and 'Golden Delicious' apples. *Journal of American Society for Horticultural Science* 123: 104–9.

Wu, J., Joergensen, R. G., Pommerening, B., Chaussod, R. and Brookes, P. M. (1990). Measurement of soil microbial biomass C by fumigation-extraction an automated procedure. *Soil Biology and Biochemistry* 22: 1167–9.

Yin, X., Long, L. E., Huang, X.-L., Jaja, N., Bai, J., Seavert, C. F. and le Roux, J. (2012). Transitional effects of double-lateral drip irrigation and straw mulch on irrigation water consumption, mineral nutrition, yield, and storability of sweet cherry. *HortTechnolology* 22: 484–92.

Zanotelli, D., Montagnani, L., Manca, G., Scandellari, F. and Tagliavini, M. (2014). Net ecosystem carbon balance of an apple orchard. *European Journal of Agronomy* 63: 97–104.

Chapter 11

Mechanization and automation for apple production

Q. Zhang, M. Karkee and L. R. Khot, Washington State University, USA

1 Introduction

1.1 Use of manual labour

Washington State produces about 60% of the US apple crop, which is sold mostly for fresh market consumption. Like elsewhere in the world, in order to sustain fresh market apple production, the US apple industry requires a large, seasonal, suitably skilled workforce to perform labour-intensive field work including pruning, thinning and harvesting. Orchard labour is one of the most significant and variable contributors to the cost of apple production. For example, labour represents 46% of 'Gala' apple variable production costs on a full production year (Gallardo et al., 2010). In addition to cost, the availability of orchard-skilled labour is uncertain at best. Over half of the workers employed in agriculture in the United States are unauthorized immigrants, primarily from Mexico (Calvin and Martin, 2010; Hertz and Zahniser, 2013). As shown by a recent Pew Research Center study, net migration from Mexico to the United States has been negative over the past five years (Gonzalez-Barrera, 2015). This migration trend is similar in other parts of the world as well, reinforcing the uncertainty of farm labour availability and the need to minimize dependence on labour for the long-term sustainability of the global apple industry.

http://dx.doi.org/10.19103/AS.2016.0017.14

1.2 The move towards mechanization and automation

One potential solution to the farm labour challenge is the adoption of mechanization and automation/robotic technologies. Agricultural mechanization has made revolutionary changes in field crop production technology, making it possible to achieve high yields using minimal farm labour. The tree fruit industry is increasingly interested in seeking labour-saving technologies such as mechanized pruning and harvesting. Federal and state government agencies, stakeholder commodity groups and machinery manufacturers have invested substantial resources into the research and development of multiple levels of mechanization solutions. These range from simple mechanical aids to fully autonomous field robots for training, pruning, thinning, spraying and harvesting of apples and other tree fruit crops.

Europe leads the world in orchard mechanization, with commercially available products such as orchard platforms and mechanical pruning and thinning machines already in use by tree fruit growers. The US tree fruit industry, which was slower than Europe in moving towards simple, narrow, accessible and productive (SNAP) fruiting wall architectures (a type of training system that creates very narrow two-dimensional canopies in which the majority of branches and fruits are visible to machines), is now actively seeking mechanization solutions to meet labour and other regulatory and fresh market consumer-driven challenges. For example, fruiting wall architectures could provide a machine-accessible orchard system where a mechanical or robotic solution for fresh market apple harvesting that is capable of picking more than 95% of the fruit with less than 5% of harvest-induced cullage could be possible. Furthermore, our discussions with a number of growers and researchers around the world have indicated that a target harvest speed of about 60 apples per minute resulting in a labour reduction of more than 80% would be essential to be economically competitive with current harvest solutions.

2 Levels of mechanization

In this chapter, agricultural mechanization is defined as using an intermediate device to convert the energy delivered from a power source into an action or activity of a machine, which then replaces human labour performing designated work. The aim is to increase worker productivity through use of some of the simplest and most commonly used mechanical devices in modern crop production including tractors, harvesters, sprayers, trucks, airplanes, and other off-road vehicles.

One key aspect of increasing the efficiency of field operations is the attempt to make these operations more uniform. However, in contrast to field crops, tree fruit crops such as apples are inherently four-dimensional in nature; in addition to their three-dimensional physical architecture, there is also variation among individual trees. This makes it very difficult, if not impossible, to introduce uniform, mechanized field operations without damaging the plant or diminishing the quality of the produce because of the random and hard-to-control impact of the machines on the plants and produce. This additional complexity of the apple crop contributes greatly to the current heavy dependence on human labour in fresh market apple production. Therefore, in this section, we briefly review three important levels of mechanization, and their potential to help in addressing the labour challenge in apple production.

The first and most basic level of mechanization is the introduction of labour-assist tools and technologies. A labour-assist technology is designed to provide field workers with a means of performing production tasks more efficiently, thereby improving their productivity. In subsequent sections, we discuss labour-assist tools for training and pruning, thinning and harvesting, and show how these can improve labour productivity and worker safety in various field operations. However, we also show that there are limitations on the potential of such tools to achieve substantial labour savings. For example, in Section 3.1 we discuss reports on use of harvest-assist platforms that have a wide variation in ability to improve labour productivity depending on the uniformity of the fruit distribution, homogeneity of worker skill level, and reliability of the machine among other factors (Duraj et al., 2010; Lewis and Hanrahan, 2014). Labour-assist tools and technologies have the potential to help address labour-related issues in apple production; however, they may not be sufficient to keep labour use to a level that is sustainable in the long run.

The second level of mechanization involves improving certain apple production operations using human-operated machines to carry out various tasks such as chemical application and crop harvesting. We show how the mechanization of operations reduces dependence on human labour, but consider how this level of mechanization faces problems in ensuring operations are sufficiently targeted and selective.

The third level of mechanization involves automation and robotics taking over the role of human workers in performing production operations. This level of mechanization has the greatest potential to reduce dependence on manual labour and, if successfully implemented, may also allow appropriately targeted and selective operations. However, it is the most technologically challenging level of mechanization, and more investment and research are required if it is to reach its potential.

3 Training and pruning

In apple production, training and pruning are crucial for establishing the basic structure of trees, creating adequate light penetration channels for better fruit quality, and making the trees easier to maintain and harvest. Training is performed by tying tree branches to trellis wires or other structures. Pruning is typically completed by physically cutting and removing parts of a tree in accordance with guidelines, and this requires skilled workers. Pruning generally consists of dormant pruning, summer/fall pruning and/or meristem manipulation (see Chapter xx Musacchi and Greene).

In this section, we review existing labour-assist platforms and powered hand tools, and then consider the use of non-selective mechanical pruning technology. Finally, we consider the potential for future development of automated, robotic, selective pruning technology.

3.1 Labour-assist platforms

The use of a ladder is common in performing many orchard operations, including tree pruning and training. It takes time for workers to climb up and down the ladders to perform their productive work, and additional time moving ladders from one working position to another. Furthermore, the improper use of a ladder can lead to serious injuries or even death of orchard workers. Therefore, the use of a labour-assist device to replace ladders, such as orchard platform machines, would enable orchard workers to increase

Figure 1 Orchard workers performing tree training tasks on a self-propelled and self-steered orchard platform in a Washington orchard trained in a fruiting wall system (photo courtesy of Karen Lewis).

their productivity and improve safety. Orchard platforms are designed to position two or more workers close to the work zone, enabling them to perform various tasks more effectively and safely. The platforms can be either tractor-pulled or self-propelled.

Tractor-pulled platforms are the most basic kind of labour-assist platform, and can satisfactorily offer the basic functionality of positioning workers close to the working zone without using a ladder. However, they require a tractor driver, not located on the platform, to keep the platform moving in between tree rows. Because this driver is separated from his colleagues, he may find it difficult to move the platform at the optimum speed to support the workers on the platform in performing their work most efficiently. The addition of a driver, who is not directly involved in training or pruning, also lowers the overall productivity of the team of workers.

Self-propelled platforms position the driver close to orchard workers and make it easier to synchronize platform speed with workers' pace. Further mechanization can take the form of auto-steer technology. Some platform machines have adopted either self- or auto-steer technologies to eliminate the dedicated driver on the platform, thus further improving overall labour efficiency.

One commonly used self-steered system on orchard platforms uses a steering mechanism to pull the platform towards one tree row at a preset slow rate. When a contact sensor installed on the side of the platform touches the trees, a steering controller will create a correction signal to steer the platform away from the row to keep the platform on a proper course to perform the orchard work. Alternatively, an auto-steered system can use a global positioning system (GPS) receiver or a laser scanning system to detect the platform position with respect to tree rows. In such systems, a desired path to perform various tasks is planned which the platform can follow automatically, regardless of whether the path includes straight line segments, curves, or missing specific rows. Figure 1 shows one such self-steered orchard platform being used in a Washington orchard supporting a group of four workers performing a tree training task.

One issue observed while using mobile platform technology is the inconsistency in productivity among individual workers on the same platform. In almost all field operations, it is unavoidable that some workers work faster than others. To ensure everyone is able to complete their task, the moving speed of a platform is generally controlled to allow the slowest worker sufficient time to finish his or her task, which draws the productivity of the entire group down. One solution to help improve the productivity of a group working on the same platform is to create individually adjustable working zones to allow the faster workers to cover a larger zone, so that all workers on the platform can complete the job at the same time.

Another major obstacle in adopting platforms in orchard operations is the inconsistency in tree architecture and row spacing from one commercial orchard to the next. Trials conducted in commercial orchards at different locations of the United States revealed that the use of mobile platforms could increase labour efficiency by 25–105% (depending on the nature of the task) in machine-accessible fruiting wall orchards compared to using ladders (Anonymous, 2016). However, such an advantage was not obvious when the platforms were used in orchards with traditional tree architectures. Mechanization solutions, therefore, for tree fruit production need to be developed in parallel with machine-accessible orchard systems.

Although this section has focused on labour-assist platforms for training and pruning operations, they may also be used to improve efficiency and productivity of other orchard operations, such as thinning and harvesting. The issues involved in choosing a platform, and potential problems such as varying working speeds of labourers and the accessibility of the orchard system, will be similar with these other orchard operations as discussed above for training and pruning.

3.2 Powered hand tools

Commonly, tree pruning is completed using a hand pruner, long-handled lopping shears, or a pruning saw. Hand pruning is a highly labour-intensive operation, and it also requires workers to have adequate knowledge and skill to do the job well. The use of some kind of powered pruning equipment, such as pneumatic-, hydraulic-machine are commercially available in various parts or electric-powered pruners, could certainly reduce the required force to cut large-sized branches, which is very important to reduce worker fatigue and improve their productivity during long hours of pruning work. Some unconfirmed field studies by growers suggested that one could achieve about 50% labour-saving in orchard pruning by simply switching from using hand tools to using air-powered pruners. Such labour savings could be further improved if the use of power tools is combined with orchard platforms which could position the workers close to the working zone without using ladders.

3.3 Non-selective mechanical training and pruning

As training and pruning in high-density fruiting wall orchards often requires the removal of substantial amounts of wood, the use of powered hand tools may not be sufficient to reach the desired productivity. Dormant pruning constitutes 20% or more of total pre-harvest production costs (Gallardo et al., 2010; Hansen, 2011) for some varieties of apples. Mechanical pruning, particularly for dormant pruning, could offer the required increase in productivity; however, pruning a tree well requires a high level of experience and skill, as well as ample strength, making it a very challenging task to satisfactorily meet pruning quality requirements using existing mechanical pruning technologies. In this section we review the available mechanical pruning technologies, describe how they may be combined with manual pruning, and then review some remaining challenges.

Researchers have been working on developing machines for mass or non-selective pruning of fruit trees for several decades. Early research focused on pruning systems in which a cutting tool runs above tree rows while maintaining a certain distance from the centre of the tree canopies. With this approach, everything above a certain height and beyond a certain distance from the trunk and main branches was removed. For example, when such a machine was used for topping and side hedging of citrus trees, a 50% reduction in pruning cost, in comparison with hand pruning, was achieved (Moore, 1958). Another study on using a mass pruning machine for blueberries achieved a good yield with mass topping of one half of the bushes every two years immediately after the harvest (Davies and Crocker, 1994).

Different types of mass pruning machines are commercially available in various parts of the world. Commonly, these machines use rotating discs or sickle bars as pruning mechanisms. Some machines available on the market use overlapping cutting discs mounted on a shaft that hedge tree canopies at a certain height/width (Fig. 2a). Another group of machines use a similar concept to mount sickle-bars on the shaft (Fig. 2b). Machines with both of these pruning mechanisms can include reconfigurable sections, which make the machines more flexible for pruning trees with complex canopy architectures as different sections of the pruning shaft (or bar) can be oriented into desired directions to prune and hedge canopies. Commercially available machines provide options for self-propelled or tractor-powered operation. Pruning mechanisms are generally hydraulically powered.

Although mass pruning methods and machines have been researched widely and commercialized to some extent, widespread adoption of the machines is still lacking. One of the challenges in adopting these machines for mechanized pruning is the complexity of tree canopies, particularly for those in classical orchard architectures. Reproducibility of results from mass pruning is challenging as it involves numerous variables due to complex orientation, position and size of branches on the trees. As apple orchards are evolving towards simpler, narrower, more accessible and productive (SNAP) systems (also called fruiting wall systems), there is renewed opportunity for mechanized pruning during dormancy as well as growing seasons.

Even though these machines have shown promising results in research and commercial applications, further studies are necessary to establish the proven benefit. More

Figure 2 (left) A disc-type pruning machine being evaluated at a WSU research farm (photo courtesy of Matthew Whiting); (right) summer pruning of apple trees with a sickle-bar machine (Hill, 2015).

importantly, these machines lack selectivity in pruning tree branches, which is essential in maintaining plant health and structure while ensuring a high level of fruit quality and yield. Therefore, as also suggested by various studies (Sevilla, 1979; Carbonneau, 1979; Martí and González, 2010), mass pruning machines may be most useful when used in conjunction with human or (in the future) robotic pruners that can fine-tune the pruning following a mass pruning operation. So, for example, growers can use mechanical pruners to do a box pruning to cut branches on the tops and sides of trees, then use skilled pruning workers to perform a fine pruning to ensure the required pruning quality. While achieving the desired level of pruning quality, such an approach could also substantially reduce pruning labour as the mass pruning can take care of the bulk of the pruning job. Another potential approach for utilizing mass pruning would be to alternate mass pruning and manual or robotic pruning over years.

The mechanical pruning systems discussed in this section achieve a certain level of performance when integrated with manual pruning to fine-tune and clean up the operation after mechanical pruning (Sevilla, 1979; Carbonneau, 1979; Martí and González, 2010). However, mechanical (non-selective) pruning is susceptible to over-pruning, which can stimulate excessive growth of shoots and increased pest pressure, leading to reduced quality of fruit (Moore and Gough, 2007). The technique may also result in inadequate pruning, which will keep older unproductive shoots in the canopies, leading to smaller fruit size and lower yield. For example, Jensen et al. (1980) found that machine pruning could be inconsistent with different operators. Velazquez and Gonzalez (2010) showed that mechanical pruning in citrus orchards reduced production and worsened the vegetative development of the trees.

3.4 Automation and robotics for pruning

Given the problems with non-selective, mechanical pruning, an automated or robotic machine for selective pruning which can identify the branches or parts on the tree that need to be pruned is essential to reduce manual labour while maintaining or improving the yield and quality of fruit crops.

In selective pruning operations, unproductive and diseased branches and shoots are removed selectively from trees. Automated, selective pruning has been studied for various crops including vineyards with a few prototype machines being demonstrated in commercial fields or in the early stages of commercialization. Selective pruning relies on understanding often subjective knowledge and skill that experienced workers, growers and horticulturists have and then creating objective rules from such knowledge to make pruning decisions for automated operations. A robotic or automated pruning system will also have to incorporate a sensing or machine vision system to register the 3D structure of trees. Such machines must include capabilities for identification of pruning branches on apple trees using pruning rules or strategies, an actuating robotic manipulator and an end-effector to complete a pruning operation.

Robotic systems have been studied for selective pruning for a number of decades. For example, Sevilla (1985) used modelling and simulation to study a robotic manipulator for grapevine pruning. The manipulator was tested in the laboratory environment, and it was concluded that robotic pruning of grapevines was technically feasible. Ochs and Gunkel (1993) also worked on a machine vision system for a grapevine pruner. Similarly, Kondo et al. (1993, 1994) developed a manipulator and vision system for a multipurpose vineyard robot. However, all of these robotic systems focused on grapevines, which have a relatively more uniform and organized canopy architecture than apple trees.

Researchers are also investigating robotic pruning techniques for apples. For example, algorithms and techniques have been developed at Washington State University Center for Precision and Automated Agricultural Systems to convert expert knowledge in pruning into simple rules that could be implemented by a machine. A 3D vision-based system was then proposed to create structures of 'tall spindle' apple trees and to identify and locate pruning branches using the specified pruning rules (Karkee et al., 2014; Karkee and Adhikari, 2015).

Researchers at Purdue University have been investigating various methods for 3D reconstruction of grapevines and fruit trees (Lehnert, 2015). In collaboration with a California-based private company, their team, including vision robotics scientists and engineers, has demonstrated a prototype machine for pruning grapevines. However, the development of a commercially adoptable robotic machine for selective pruning of tall spindle apple trees requires more focused studies in various aspects of this technology, including human–machine collaboration and an expert system for pruning rule creation appropriate for various modern apple orchards.

4 Thinning

Under most conditions, trees will set too much fruit to have high-quality produce with the desired yield, and require removal of a proportion of either the bloom or young fruit, in order to maintain the optimal crop load to achieve the size, colour and other quality measures desired for fresh market consumption. Thinning bloom or fruitlets helps ensure good profitability for the current season as well as good return bloom and crop yield for the following season. However, there is a difficult balancing act when it comes to thinning as both over-thinning and under-thinning will be detrimental for fruit quality and yield, and therefore for profitability (see Chapters xxx Hirst, Lakso and Goffinet, and Musacchi and Greene).

Thinning can be done using chemical thinners or by physical means during and shortly after the bloom period. It can also be done later in the season to remove some small fruits or those growing too close together. To optimize the operation, generally both flower and green fruit thinning are practised in commercial orchards. These canopy and crop-load management operations have primarily been performed manually by workers plucking the blossom clusters or fruitlets off using their fingers. Such manual thinning allows workers to accurately remove flowers or fruitlets for the best fruit distribution, and therefore achieve optimum fruit size and fruit quality. However, it requires a large seasonal labour force and is a tedious process, often mind-numbing for workers after long hours of labour.

To improve efficiency and reduce dependence on a large labour force, both hand-held thinning devices and fully mechanized thinning are promising. In this section we review existing labour-assist tools for manual thinning, non-selective chemical and mechanical thinning of both blossoms and fruits, and finally recent research and development in the area of automated and robotic thinning.

4.1 Labour-assist tools for manual thinning

A typical hand-held mechanical thinning device is a portable string-style thinner, which may be electrical-, air-, hydraulic- or gas powered. Some of the designs include a telescopic

Figure 3 Evaluation of a gasoline-powered hand-held mechanical thinner with telescopic capability in a Washington sweet cherry orchard.

handle allowing the user to reach the upper tree canopies from the ground (Fig. 3). Such features provide operators with the capability of having full control of the thinning process to effectively remove flowers or fruitlets from any tree training style including vase, trellis or central leader, and even the naturally grown tree systems. More importantly, a hand-held thinner also allows the operator to perform semi-selective thinning to remove flowers or fruitlets according to the crop-load management goal on each individual branch (Wang et al. 2013).

4.2 Non-selective chemical and mechanical thinning

4.2.1 Blossom thinning

Chemical thinning relies on the application of plant growth regulators, such as synthetic versions of growth hormones in the auxin and cytokinin families, to tree canopies (Extension, 2016). Application of chemicals for thinning is similar to applying any other crop protection chemicals in orchards using sprayers (e.g. an air-blast sprayer; see Section 5.1). However, the deposition and coverage of thinning chemicals is more sensitive than that of other chemicals, as the effect of not being able to apply the right amount at the right time could be more damaging in terms of crop quality and yield.

Mechanical blossom thinning machines, particularly string thinners, have been around for several decades, but have increased in application in recent years. Studies have shown that mechanical thinning of apples can reduce labour input while maintaining or increasing fruit size (Schupp et al., 2008; Baugher et al., 2010). A study based on 6374 acres of peaches and nectarines in the province of Ontario, Canada, indicated that the use of mechanical blossom thinners could help to reduce thinning costs by 50% and realize savings of approximately \$1.6 million (Slingerland, 2009).

String thinning machines developed by either commercial companies or research programmes (e.g. Darwin String Thinner, NP Seymour LTD, UK; Bonner String thinner, University of Bonn) have been used over the last decade by farmers as well as researchers around the world with promising results (Veal et al., 2008; Damerow and Blanke, 2009;

Figure 4 A mechanical string thinner with three individual heads was evaluated in Washington orchards for its effectiveness in bloom removal, fruit quality and yield (photo courtesy of Karen Lewis).

Miller et al., 2010). This type of machine generally consists of one or more shafts with spindles that can be rotated at a certain speed to achieve a desired level of blossom thinning (Fig. 4). Spindles are wrapped with thinning strings attached in various configurations. Optimization of string material, string shape and size, string density, and speed of spindle rotation has been an important research area over the last decade. Similar to pruning machines (see Section 3.2), thinning machines can be designed to include flexible segments (Damerow and Blanke, 2009) that can be positioned and oriented dynamically to match different types of tree canopies. As alternatives to string thinners, drum shakers (Miller et al., 2011) and trunk shakers (Rosa et al., 2008) have also been evaluated for blossom thinning in peaches and other crops, with potential to apply in apples.

There are a few limitations of these thinning machines. First, the machines are suitable for thinning fruit trees trained in a few specific architectures such as bi-axis, Y-trellis, and Upright Fruiting Offshoots (UFO) fruiting wall systems. In addition, these machines are applicable for only non-selective thinning, and they can cause damage to branches and leaves due to excessive power of the rotating spindle or other types of impact on tree canopies. Some of these issues could be avoided if a robotic thinning solution could be developed, and the potential for this will be discussed in Section 4.3.

4.2.2 Young fruit thinning

Chemical thinning of young fruit is similar to chemical blossom thinning in terms of the machines and methods used; however, the chemicals used for green fruit thinning are often different. Mechanical thinning of young fruit has also been studied (e.g. Schupp et al., 2008; Miller et al., 2011); however, commercial adoption of mechanical green fruit thinning is limited, partially due to a lack of consistency in the results and to damage to remaining fruit and tree branches. Some growers also practice chemical or mechanical thinning followed by hand thinning. If machine or chemical thinning has been applied to the blossom, hand thinning of young fruit is particularly important to ensure the desired load and spatial distribution of fruit is achieved.

4.3 Automation and robotics for selective/targeted thinning

One limitation of chemical thinning is the high level of variability and uncertainty of the results caused by various known and unknown factors such as environmental and weather parameters, cultivar, canopy density, canopy type, and current density of blossom or young fruit. Mechanical thinning faces similar issues, because it is difficult to target a desired location in a tree canopy for thinning with a mechanical tool such as a Darwin thinner. The possibility of an automated or robotic machine for blossom thinning is an area of recent research and development (Emery et al., 2010; Lyons et al., 2015). A robotic machine will have the potential to achieve a precise level of blossom thinning, young fruit thinning or both using its capability to detect target areas for thinning and to manoeuvre a robotic end-effector to remove the desired level of blossom or fruit from the target locations. Robots could also be developed to work at the individual blossom or young fruit level so that highly precise thinning results could be achieved. Automated or robotic operations have the potential to improve the precision of the operation and reduce the uncertainty of the thinning outcome. In addition, automated operation will avoid operator fatigue and monotony, and will potentially increase the speed of operation.

Researchers have reported efforts in developing sensing, machine vision, manipulator, and end-effector technologies for robotic thinning operations (Nielsen et al., 2012; Wouters et al., 2013; 2015; Lyons et al., 2015) and a selective automated thinning technique was recently patented (Lyons and Heinemann, 2016). A string thinner with a number of adjustable thinning heads (similar to one in Fig. 4) can be used as an automated machine for targeted thinning operations. For individual blossom/fruit or small cluster level operations, a robotic end-effector with higher precision may be essential.

The overall development of automated or robotic thinning is still in its infancy. More comprehensive research and development in all aspects of robotic thinning is essential before this technology can become commercially adoptable.

5 Pest and disease control

To effectively control pests and diseases that adversely affect yield and fruit quality, growers need to spray orchards using recommended crop protectants in a timely manner. Spraying is performed almost entirely using machines and therefore requires a minimal amount of human labour input. Crop spraying is, however, inherently hazardous for operators driving a spraying machine in the confined orchard environment. In addition, poor chemical applications in adverse micro-meteorological conditions can reach beyond the intended target and hamper the surrounding orchard ecosystem. Automated and precision spray application technology can help to address the above issues and to increase orchard labour health and safety. In this section, we review existing application technologies, and then discuss the pertinent efforts to develop variable rate spraying technology for use in tree fruit production.

5.1 Application technologies

Application technologies are critical for effective pest and disease management in apple production. The axial fan air-blast sprayer is one of the most commonly used types of equipment for this purpose (Fig. 5). Such sprayers encompass a spray tank pulled by a tractor and have a power-take-off-driven axial fan that generates an air-assist to push spray droplets from nozzles to the intended canopies. Many variants of air-assisted sprayers have been developed and utilized commercially, including conventional air-blast, tower, canon and tunnel sprayers and, more recently, electrostatic sprayers.

Typically, a sprayer system includes a chemical tank, spray nozzles, air-assist fan(s), DC-powered mechanical pump, agitator and flow return valves as the key components. The air-assist rate depends on the fan size, blade angle and speed. Axial fan sprayers are robust, easy to operate, maintain, and modify, and offer ease of orchard entrance; however, if not configured correctly, they result in poor coverage in large canopies. For such sprayers, air flow adjustments are tedious and applications often cause excessive drift in small canopy crops.

Tower sprayers address some of the issues of the axial fan sprayers as they offer improved coverage in the upper canopy by reducing the distance between the nozzles and the target (Fox et al., 2008). Often, a tower sprayer consists of one or more axial fans vertically arranged in the tower attachment. It blows the spray material horizontally towards the canopy to have a uniform coverage on the target. However, tower sprayers have limited access to non-trellised systems or tree architectures with excessive inter-row branches. The variants designed for tall canopies need careful manoeuvring on uneven, hilly topography and require more powerful tractors. Also, changing and optimizing the sprayer application volume to match the canopy characteristics throughout the season is complicated and not intuitive with such systems.

Tunnel-shaped sprayers, sometimes called recycling sprayers, use shields on either side of the canopy row to collect the over-spray application (Planas et al., 2002). This configuration reduces the amount of spray material used and has less chemical drift than other sprayer designs (Jamar, 2010). However, as spray material is recycled from sprayed canopies, it can result in disease spreading between the trees. In addition, the design of tunnel sprayers for crops such as apples with tall canopies is challenging.

Figure 5 A typical axial fan air-blast sprayer used in apple production.

Electrostatic sprayers use electrical charge to create positively charged droplets, which are then attracted to the negatively charged canopies (Fig. 6). They are thus ideal to reduce off-target deposition (Kirk et al., 2001) and are suited for greenhouse, field crop and orchard spraying applications. Electrostatic spraying is a new technology and is being offered as a uniform coverage solution to reduce material and fuel use, time spent spraying and hence to reduce the cost of spray applications. However, there are many concerns about the maintenance and the performance of such sprayers in different environments, and further validation of the technology is required before large-scale adoption.

Traditionally, spray applications in apple crop production were based on the application rate per unit area covered by the ground vehicle. Row and tree spacing was used to estimate the agrochemical application rates per unit area. This method accounted for neither canopy height/width nor variability in canopy growth stages (e.g. leafing, half and fully developed canopies). With economic and environmental implications becoming a concern among the growers and stakeholder community, efforts are now being made to optimize spray coverage to complement the canopy attributes. Optimization of conventional sprayers to have canopy-specific applications can be implemented by (1) changing each nozzle tip for required output rates, (2) configuring nozzle orientations, (3) adjusting axial fan speed to adjust the air-assist and (4) switching off pertinent nozzle combinations. Nonetheless, such optimization may not be adequate to produce optimal coverage in canopies, and may result in drifting of sprayed material. Variable rate technologies that quantify the canopy attributes and vary the application rate and strength of air-assist accordingly are thus critically needed and are discussed in the next section.

5.2 Variable rate spraying

Although tree fruit production has transitioned from traditional tree architecture to more efficient high-density architectures, development and adoption of new sprayer technologies has been somewhat slow. In recent years, advancements in non-contact canopy sensing, geo-referenced remote variability mapping, and variable rate droplet-generation technologies, along with embedded system advances have enabled the

Figure 6 An electrostatic sprayer used in orchard spray applications (photo courtesy of Matthew Whiting).

industry to move towards the use of variable rate sprayer technologies (Gonzalez-de-Soto et al., 2016). Several prototypes have been developed and tested for variable rate applications in a range of horticultural crops. Some examples include variable rate sprayers for use in nurseries, orchards and vineyards (Escolà et al., 2013; Chen et al., 2013; Khot et al., 2014).

Efforts have been made to either retrofit existing sprayers or develop a new prototype. Configuration and operation of a typical variable rate sprayer is ideally governed by the way canopy variability is mapped. Canopy height, volume, width, area and sometimes density can be estimated using ground-based non-contact sensors. Ultrasonic and light detection and ranging (LiDAR) sensors that work on time-of-flight have been commonly used to estimate some of the above-mentioned canopy attributes. Ultrasonic sensors provide a point measurement of distance to canopy which can then be used to determine presence or absence of canopy (i.e. gaps in canopy) as well as to estimate canopy volume. Arrays of ultrasonic sensors, ideally one per nozzle or for a section of nozzles, are needed to determine canopy height and volume due to the point sampling nature of the sensors. By contrast, a LiDAR sensor can scan a wider field of view (e.g. up to 270°) with angular resolution as high as 0.25° that can map an entire canopy cross-section for accurate estimation of height, volume and density among other attributes. Accuracy, range, resolution, outdoor ruggedness and cost are some of the factors that govern canopy mapping sensor selection. Irrespective of sensor type, canopy data then can be provided to an embedded control system integrated with predefined decision rules for actuation of variable rate delivery of liquid and air-assist. Canopy data can also be geo-referenced using a GPS receiver, which can also provide equipment travel speed for adjusting the rate of applications.

Delivering the variable rate chemical onto target canopy surfaces can be achieved by (1) adjusting application rate at individual or sections of nozzles and (2) adjusting air assistance according to released spray volume (droplets). Pulse width modulation (PWM)-based solenoid valve – enabled spray nozzles can be used to accomplish the prior task. Such nozzles can be either switched on/off or pulsed at variable rates depending on the canopy characteristics. The latter can be achieved by an actuator-enabled louvre that diverts air from the axial fan air-blast sprayer. Alternate designs include localized air-assist for a section of nozzles.

Various implementations of the approaches defined above are available in the literature. For example, Chen et al. (2013) modified a vineyard sprayer to be a variable rate sprayer for nursery and orchards. It used a LiDAR sensor to map the canopy with 0.5° angular resolution in the 180° field of view. Sensor data was converted to determine canopy occurrence, height, width, volume and density such that PWM-based solenoid value – controlled nozzles could vary the rate of application. Similarly, Khot et al. (2014) retrofitted an existing axial fan air-blast orchard sprayer with PWM solenoid value – controlled nozzles for variable spray (liquid) rate and actuator controlled air-diverting louvres for air-assist control (Fig. 7). A 0.25° angular resolution LiDAR to scan citrus canopy occurrence, height and volume in 270° field of view was used. Pérez-Ruiz et al. (2015) retrofitted an axial fan air-blast sprayer for use in olives and tree fruit orchards. Ultrasonic sensors spaced every 0.5 m detected canopy characteristics up to 2 m horizontally in four vertical sections and PWM-enabled nozzles varied the application rates accordingly.

It should be noted that, for any of the above systems, it is vital to have robust electronics which can understand input from canopy sensors and perform real-time decision-making, analogues to rate controllers, on liquid and air-assist actuation. Such electronics also

Figure 7 Retrofitted axial fan air-blast sprayer with variable rate nozzles and actuator-driven air-assist louvre for effective agrochemical applications in citrus (photo courtesy of Reza Ehsani).

need to be appropriately configurable with robust 'decision rules' for changes in crop characteristics.

Researchers at Michigan State University (http://www.canopydelivery.msu.edu/) in collaboration with Washington State University have been working towards an alternative technology called the 'Solid State Canopy Delivery System', where micro-spray nozzles placed in high-density apple canopies can effectively deliver nutrients and other chemicals in accordance with site topography conditions throughout the production season. In addition, semi- or autonomous aerial sprayer technology for precision chemical applications is on the horizon (Giles and Billing, 2015). Both of the above emerging technologies are expected to reduce chemical usage and associated spray drift.

6 Technologically-assisted apple harvesting

Fresh market produce is still hand harvested, and harvesting is the most intensive and expensive labour component in apple production. A large, seasonal, semi-skilled workforce is required to harvest apples for the fresh market. Studies show that when apple pickers are in the orchard, only 30% of their time is spent picking fruit, while the rest of it is spent positioning and climbing ladders, and unloading collected fruit (Calvin and Martin, 2010; Hertz and Zahniser, 2013). Dependence on the large, seasonal, semi-skilled workforce for harvesting also creates one of the most critical labour-related risks in apple production: that of not having enough of the right people at the right time to perform the time-sensitive harvest.

There have been a lot of efforts directed into developing mechanical harvest solutions for fresh apple production around the world, which have also recently received venture capitalist attention. In general, the development of mechanical harvest technologies has achieved much success, but still faces a few major challenges to meet the high standard of accuracy, speed, robustness and cost to satisfy the long-term sustainability of the fresh market apple industry.

In this section, we review harvest-assist machines, mechanized bin management and bulk harvesting techniques. The important topic of robotic apple harvesting will then be reviewed in detail in the next section.

6.1 Harvest-assist machines

One practical, currently available solution to reduce the dependence on seasonal labour is the use of harvest-assist machines for more efficient harvest and increased consistency in fruit handling. Several harvest-assist machines are commercially available to apple growers and many offer in-field fruit-conveying and bin-filling functions to help increase picker efficiency while controlling the handling-induced fruit damage at a low level. One common design of this type of machine includes the integration of a bin-filling mechanism onto a mobile orchard platform. Such a machine can support a crew of several human pickers, both on the platform and on the ground, placing their hand-picked apples on some type of fruit conveyor that carries the fruit to a bin filler to complete the loading process at or near the locations where the fruit is picked (Fig. 8). By removing the need to use ladders and to carry fruit between the trees and the bins, these machines not only increase pickers' time efficiency but also reduce their fatigue, helping to substantially improve the productivity of the crew.

Several different fruit-conveying and bin-filling mechanisms can be seen on apple harvest-assist machines. One example includes mechanical systems consisting of flexible conveyors and a bin filler allowing four to eight pickers to work as a team. While designs may vary in terms of arrangement of conveying belts and bin-filling mechanisms used, this type of system uses a series of mechanical conveyors to deliver apples from the pickers to the bins. Fingered conveying belts are commonly used on mechanical conveyors. An advantage of using fingered conveying belts is that once an apple is placed on the

Figure 8 A crew of human pickers harvesting apple using a harvest-assist machine.

conveyor, it normally does not touch another apple until it reaches the bin, which should help reduce handling-induced apple bruising during the harvest process.

Another type of system is the vacuum system, which conveys fruits from the picker to the bin filler using a vacuum tube. One unique feature of vacuum-conveying systems is the use of a fruit deceleration mechanism to slow down the delivered apple, allowing it to be gently loaded into the bin. The effectiveness and robustness of the fruit deceleration mechanism is critical in keeping the handling-induced apple-bruising rate to a very low level.

Sorting apples into two or three quality grades during harvest is another form of harvest-assist mechanization (Lehnert, 2013). A computer vision-based automated sorting apparatus is integrated into an apple harvest-assist machine with multiple mechanical fruit conveyors and two to three bin-filling mechanisms. Such platforms can support a number of people picking apples, and can sort harvested apples according to their colour, size, shape and weight, and then fill into corresponding bins. This technology will enable growers to separate out defective fruit in the orchard, resulting in less postharvest disease/pest problems and lower transport, postharvest storage and packing costs, which helps growers improve their overall harvest efficiency and product safety.

6.2 Bin management

Typically, bins capable of holding 400 kg of fruit are used in the Pacific Northwest (PNW) region of the United States for apple storage and transportation. In Washington State alone, there are more than 7 million bins to be handled in orchards during harvest, so good bin management is essential in ensuring a smooth operation. It is a common practice in the PNW that empty bins are distributed before harvest between tree rows at positions assumed to be most convenient for pickers to unload their picked apples. As harvest progresses, the picking spots will gradually move away from bins being filled, which will result in a longer time for pickers to walk back and forth between picking spots and the bin, resulting in a substantial loss of productive time. In-orchard labour-assist bin management using tractor-mounted forklifts could help to reduce the non-productive time of pickers by re-positioning empty or half-full bins close to picking spots.

Adding bin management capability onto harvest-assist mobile platforms is an obvious way to improve overall productivity, and several such combinations are already commercially available in the US and European markets. While the specific designs vary, the basic principle of operation is fairly similar: the platform picks up one or more bins as it moves between the harvested tree rows, carries those bins allowing pickers to load their picked fruit, and releases filled bins to the ground behind the platform. Completely autonomous bin handling such as the one currently being investigated at Washington State University is another potential solution (Ye et al., 2012).

6.3 Bulk harvesting

Fruit can be detached from the trees in bulk with various techniques including trunk or branch shaking, and canopy vibration. Detached fruit can then be captured before it hits the ground. This technique of bulk or mass harvesting is used in harvesting various types of tree fruit crops, primarily those destined for the processing market. Below, we provide an overview of the bulk harvesting techniques investigated in the past for harvesting apples, and then discuss a new approach being investigated for bulk harvesting of fresh market apples.

6.3.1 Overview

Several different mass harvesting systems have been developed and used commercially. Canopy shaking deploys fingers or rods to apply different patterns of vibratory motion to canopies to detach fruit. Alternatively, fruit can be pushed from the canopy using special techniques such as a rod-press mechanism (Peterson et al., 1985). Fruit falling from trees can be captured and collected into containers with inverted umbrella or inclined catching surfaces. Different variants of these machines have been developed over the years, including a machine with a catching surface with a padded rod conveyor covered by decelerating strips. Later examples of these machines have included an integrated impacting device or 'rapid deployment actuator' (RDA), which has been found to be more effective for shaking the trees (Peterson et al., 1999). With RDA, trees are impacted with a high level of force causing fruit to potentially fly some distance, which makes fruit capturing and collection more challenging. One way to improve fruit collection with mass harvesting systems has been the use of a mirrored or a two-sided harvesting system with catching surfaces extending to both sides of a tree row (Fig. 9; Peterson and Wolford, 2003).

Fruit can also be detached by mechanical trunk shaking which minimizes the damage to trees and fruit that RDA can potentially cause. A tree will respond differently to different excitation frequencies during shaking, and fruit removal normally occurs when the detachment force exceeds the pedicel-fruit tensile force (Markwardt et al., 1964). Several studies have investigated mechanical shakers (Whitney, 1977; Parameswarakumar and Gupta, 1991; Polat et al., 2006) with the goal of optimizing shaking frequency, amplitude and shaking duration for maximum fruit-removal efficiency. Recent studies have found that effective shaking frequency could vary between apple cultivars. It was found that 15–20 Hz could be optimal for detaching 'Jazz' apples with potential to minimize fruit damage (He et al., 2016).

Although mechanical harvesting techniques have been commercially successful in harvesting fruit destined for the processing market, the major limitation of such techniques

Figure 9 A mirrored fruit-harvesting machine developed and evaluated by Peterson and Wolford (2003) harvesting apples.

for fresh market apple harvesting is the unacceptable level of bruising and other types of fruit damage (e.g. cuts and punctures) induced by the harvesting and/or collection mechanisms. Fruit catching is one of the major sources of such damage; it can be minimized by the use of shock-absorbing (or foam) materials to cover catching surfaces (Ortiz et al., 2011). Localized or targeted detachment and capturing has the potential to further decrease harvest-induced fruit damage to a level acceptable for fruit for the fresh market. This will be discussed in Section 6.3.2.

6.3.2 Targeted shake-and-catch harvesting

Study and evaluation of innovative methods for fruit detachment and catching mechanisms are essential for the development of commercially adoptable mass harvesting systems for fresh market apples so that the desired level of fruit-removal efficiency and fruit quality can be achieved. Targeted bulk harvesting systems, also called targeted shake-and-catch systems, address some of the issues of fruit removal and quality, by harvesting a number of fruits from a localized, target branch of a tree trained in modern fruiting wall architectures. Fruit would be collected very close to where they are detached using specially designed catching mechanisms, and with incorporation of fruit buffering and separation techniques, the frequency and extent of fruit-to-fruit and fruit-to-catching surface contact are minimized. Reduction in such events will help keep fruit damage to an acceptable level for fresh market consumption.

In a current study at Washington State University, various designs of shaking and catching mechanisms for a targeted shake-and-catch harvesting system are being tested (Fig. 10). Investigation includes (1) the understanding and optimization of energy transfer to apples located in various parts of tree canopies trained in modern fruiting wall architectures, (2) evaluation of novel shaking and catching mechanisms for reducing fruit damage and (3) integration of catching surface with tree canopies.

Fruit drop tests were used to identify potential sources of bruising and to identify catching surface materials that may reduce fruit bruising. A dynamic test was also conducted to investigate the energy transmission at different locations in the limbs with different levels

Figure 10 A research prototype developed and evaluated by WSU researchers for targeted shake-and-catch harvesting of apples in tree canopies trained in formal fruiting wall system.

of shaking frequency and amplitude. The results show that fruit-removal efficiency and quality depended on the cultivars and tree architectures and some cultivars (e.g. 'Envy') could generally be detached when the acceleration was higher than 5 g within the first 5 s of shaking (unpublished results). The majority of unremoved fruit remained on thin and long branches that were relatively distant from the shaking location.

A fruit-catching mechanism using air suspension as a cushion has also been investigated to minimize fruit bruising during mechanical harvesting. It is expected that the fruit catching with air suspension can significantly reduce the percentage of damaged apples as it decelerates and isolates fruit before they hit the catching surface. The technique is being evaluated with apple cultivars with different bruising susceptibilities.

The ultimate success of targeted shake-and-catch harvesting relies on improvement and adoption of canopy architecture that creates sections (one branch or more) of tree canopy in which localized shaking and catching is possible. The team at WSU is exploring potential improvement in apple canopy structures that would allow catching surfaces to be inserted into the canopy at different levels without damaging fruit, branches or the catching surface itself. Other considerations for improving the performance and therefore adoptability of a targeted shake-and-catch harvesting system could include reducing the number of layers in a formal apple tree architecture, and inclusion of a mirrored machine close to the harvesting zone from both sides of the tree rows.

7 Robotic apple harvesting

The targeted shake-and-catch systems discussed in the previous section have the potential to make bulk harvesting of fresh market apples possible in the future for certain varieties grown in certain tree architectures. However, it is anticipated that no universal harvesting solution is likely with a bulk harvesting method. For some apple varieties, particularly the high value varieties such as 'Honeycrisp', robotic harvesting could be a viable solution to reduce labour demand and increase long-term sustainability. Automated or robotic machines designed for individual apple picking consist of a set of sensors to detect and locate fruit, a robotic manipulator to access fruit in the canopies, and a robotic end-effector to detach the fruit. Even though no robotic system has been commercialized for apple harvesting, prototypes have been developed with varying levels of success in terms of harvesting efficiency (per cent of fruit harvested) and cycle time (average time taken to harvest one fruit).

7.1 Brief historical perspective

In 1984, French scientists developed the first prototype of a self-propelled robot for fruit harvesting (the MAGALI, D'Esnon et al., 1987). The MAGALI prototype used three-degrees-of-freedom (3 DOF) manipulation with an end-effector shaped as a suction cup. The vision system (colour camera) coordinates coincided with the robot coordinates. More than 50% of 'Golden Delicious' apples were detected and harvested, 75% of harvested fruit still had the stem attached. The average harvest time was 4 s per apple (Li et al., 2011; D'Esnon et al., 1987).

A prototype robot that had the capability to perform various human and machine functions was developed within the Agribot project in Spain (Ceres et al. 1998). The system included an operator driving a robotic harvester and a laser range-finder to detect the fruit. A computer then estimated the precise position of the fruit, computed the picking sequences and controlled the motion of the picking arm and gripper-cutter. The picking arm had a 3 DOF design with one vertical and two horizontal rotation axes, plus the rotation on the end-effector with the gripper-cutter. The system took 2 s to pick one fruit, with the gripper-cutter performance identified as the bottle-neck.

Kataoka et al. (1999) developed an end-effector for robotic apple harvesting which was used to study and analyse the dynamics of fruit separation. The harvester hand detached the apple by making a rotation around the abscission layer; the probability of success for the harvester was more than 90%.

The Autonomous Fruit Picking Machine (AFPM) consisted of an industry robot manipulator (Panasonic VR006L), a 2D horizontal stabilization unit, a 7th external vertical axis to enlarge the operation range, a safety scanning device, a central control unit, a touch panel PC with Human Machine Interface, a canopy and all around curtain to even out light conditions, and finally, a fruit-gripper design consisting of a silicone funnel with a camera mounted inside (Baeten et al., 2007). The flexible gripper guaranteed a firm grip without damaging the fruit and served as the mouth of a vacuum cleaner. Eighty percent of the apples with diameters ranging from 6 to 11 cm were harvested and the overall cycle time was 8–10 s per apple with 70% of the apple stems attached. Features that needed improvement included the communication between the vision-PC and the central control unit, the image processing, and the picking cycle period (to reduce to 5 s or less).

Setiawan et al. (2004) and Bulanon and Kataoka (2010) have also developed robotic harvesting systems in the past decade; both use an end-effector on a controllable arm to remove the apple from the tree. Bulanon and Kataoka (2010) designed a fruit detection system and an end-effector for harvesting 'Fuji' apples. The end-effector simulated the rotating wrist action commonly used in hand harvesting. They reported a success rate of 89% for picking fruit; however, the processing time required for their system to locate fruit is the major drawback when considering a commercially applicable harvesting solution.

These systems indicate that the functionality of robotic machines for selective fruit picking is promising. There are still a few technical challenges such as limited robustness, sluggish response and insufficient harvest rate, in addition to the high cost that need to be addressed before the technology can be commercially adopted.

7.2 Fruit detection and localization

Locating the fruit on the tree is the first step for robotic harvesting. Machine vision is the most commonly used approach because it is capable of detecting fruit based on colour, shape, texture and other parameters (Bulanon et al., 2010; Tabb et al., 2006; Hannan and Burks, 2004). Other sensing technologies, such as laser scanners, are also successfully used to determine the relative position of the apple from the robot.

While it is reasonable to state that the technologies for locating fruit on a tree have been well developed and validated at the conceptual level, the literature also reveals various technical obstacles (Gongal et al., 2015; Bulanon et al., 2010; Tabb et al., 2006; Hannan and Burks, 2004). For example, it was difficult to detect apples covered by other objects using a machine vision approach and it took a long processing time to locate an apple in a complicated canopy using laser scanning approach. It is safe to say that the development phase has now evolved to application-related studies. Current research focuses include, but are not limited to, finding solutions for (1) how to improve the reliability of detecting fruit under changing natural light, (2) how to improve detection of partially or fully occluded fruit in different tree canopy architectures and (3) what will be the optimal canopy structure that facilitates mechanical harvest without sacrificing fruit yield and quality.

7.3 Fruit detachment

Detachment of fruit is the central, and also the most challenging, task in robotic (or pick-and-place) harvesting of apples for the fresh market. Many fruit detaching studies have been reported (Gongal et al., 2015); however, only limited success has been achieved due to the complexity of the motions involved in fruit detachment and/or the apple's sensitivity to bruising.

Almost all reported robotic harvesting studies use some type of end-effector modified from industrial robots. Many successful studies into developing robotic end-effectors for picking different fruits, such as strawberry and kiwi fruit (Scarfe et al., 2009), have been conducted in the past two decades. A few commercial prototypes using different approaches have been introduced and tested by technology innovation firms. Major obstacles preventing this advanced robotic technology from being adopted by tree fruit growers worldwide include the low productivity and the high overall costs. Based on published data (Section 7.1), a cycle time of 3–5 s or slower for picking a fruit from easy-to-reach canopies has been achieved using conventional robotic approaches. A much faster cycle time of up to 1 s per fruit was reported as achievable using non-conventional approaches claimed by private technology innovators (based on their public presentations). Different fruit-grabbing methods, such as a suction cup or gripper-and-cutter, could noticeably affect the productivity under different tree canopies.

The path planning method to find optimal trajectories for end-effectors to reach and pick fruit at different locations presents a significant remaining challenge for improving harvest productivity. Such a technical challenge may have to be addressed using an integrated horticultural-engineering approach to make the canopy more accessible and to design effective and efficient robotic end-effectors for tree fruit harvesting. An alternative or complementary approach is to improve productivity through the use of a multi-arm picking robot; however, increasing the number of arms would obviously increase the complexity and cost of the system (including operation, maintenance and malfunction costs in addition to initial purchase cost). Still to be determined are what will be the acceptable number of hands for productivity and what overall cost of harvesting solutions will be economical for apple growers in specific geographic regions.

7.4 Fruit placement

Placing a fruit involves dropping an individual fruit directly into a container or into a conveying system that moves the fruit to a container. It is expected that dropping a fruit into a container or a conveying system will be addressed sufficiently when fruit-picking technology is studied and integrated into a robotic system. In the system where fruit is not directly dropped into a container, conveying or transportation and bin/container filling mechanisms are also needed. Among all core technologies, fruit conveying and bin filling is the most matured technology, with some designs being commercially adopted and applied in the various mechanical or mechanical-assist harvest machines discussed in Section 6.1. Examples of such technologies include mechanical conveyors, vacuum-transporting tubes and mechanical/hydraulic fruit decelerators. Even though these machines have been commercially adopted on a smaller scale, it is noted that reducing the extent and frequency of impacts on fruit during transportation and bin/container filling remains a challenge to improve overall fruit quality.

8 Future trends and conclusion

Despite extensive efforts in developing robotic machines for various orchard operations including harvesting, pruning and thinning, commercial success has been limited. Major obstacles include the complexity and uncertainty of the farming environment, the speed, robustness and reliability of the technology and equipment, economies of scale and the high cost of both purchasing and operation, plus the high skill requirements for operators and technical services. End-effector technology to detach fruit from limbs without damaging the harvested and neighbouring fruit is one critical area. Fruit identification and localization is another area requiring significant improvement. Improving overall cycle time for fruit detection, detachment and placing in a bin or container would be another area of innovation. The human–machine–tree interaction and improvement in horticultural systems are crucial too. Research teams are continuing to work on developing new and improved canopy architectures that would increasingly simplify robotic operations in orchards by presenting fruit, branches and other parts of tree canopy, in the most accessible location and orientation possible. In addition, the optimal solution can be achieved through exploitation of the capability of robotic systems and human intelligence in a collaborative environment.

Scientists at Washington State University are focusing on human machine collaboration, understanding of the human picking processes, and designing robotic manipulators and end-effectors for effective robotic picking for development of a commercially adoptable robotic harvester. A machine vision system with hierarchical fruit detection combined with

Figure 11 A prototype robotic apple harvester developed by WSU. The machine was evaluated and 'Jazz' and 'Envy' apple trees trained in modern fruiting wall architecture.

a harvesting method with human–machine collaboration has achieved close to 100% fruit detection in a fruiting wall apple orchard (Silwal et al., 2016). The first prototype of the robotic harvesting system developed by the team achieved the picking cycle time of 5–7 s with a relatively economic robotic manipulator (Fig. 11). The investigators are working on improving the prototype to achieve a 2–4 s cycle time while keeping the design cost-effective.

Innovation in tree architecture/physiology, breeding and genetics that would improve crop characteristics for mechanical harvesting is a key requirement. Measures that are critically important for consideration within the biological programmes include (1) loosening of the abscission layer when fruit are ready for harvest so that only a minimal amount of force is needed to detach fruit, (2) increasing the uniformity of fruit distribution, maturity and quality within a tree canopy so that fruit can be harvested at single pass with minimal fruit-to-fruit contact and (3) minimizing the potential of fruit cullage through innovative training and blossom/green fruit thinning techniques. A systems approach, including various areas of expertise such as horticulture, engineering and socio-economic analysis, will be essential in developing integrated solutions to meet these challenges and provide robotic approaches that can achieve optimal results. Economic analysis to determine the break-even point for manual and mechanical harvesting is also key.

A major obstacle to harvest mechanization is the fact that many farmers send their produce to packers and processors (as opposed to packing in the field), who are usually equipped to deal with manually harvested produce or machine-picked produce but not both, making it hard for one farmer to mechanize in isolation. Packers and processors can encourage or discourage mechanization by adjusting their standards. Mechanical harvesting does not eliminate the need for hand labour to harvest small fields, to pick what the machines miss, or to meet special customer requests. If a machine is less selective and gentle than hand labour, additional labour may be required to sort out undesirable products or debris.

In recent years, there have been increased levels of interest and effort from local, state and federal funding organizations in the United States (e.g. Washington Tree Fruit Research Commission, Washington State Specialty Crop Block Grant, National Robotics Initiative), Europe (e.g. European Research Council) and in other parts of the world, in developing labour-saving robotic technologies for specialty crops including apples. This has generated opportunities over the last decade for researchers to increase their efforts in this area. These extensive research efforts in automation and robotics for agriculture (e.g. a number of research projects funded by National Robotics Initiative of US government and European Research Council) show a potential for developing robotic solutions for pruning, thinning, harvesting and other aspects of apple production in the near future. However, the exact time frame cannot be easily predicted as there are multiple factors that will play a role in shaping the speed and direction of progress, including the cost and availability of labour and economy of scale.

Large-scale adoption of new technologies can change the structure of the apple industry which is currently concentrated with the largest growers supplying the vast majority of produce to the fresh market. Depending on the type of machinery available, mechanization may accelerate the trend towards fewer and larger producers. Larger growers are the most likely to invest in expensive technological solutions as the fixed investment can be spread over more acreage. An alternative to buying specialized machinery could be to use a service module, where a third party buys the machinery and provides technology-driven services to multiple growers. Some growers may worry that if they do not purchase their own machines, field operations may not occur at the optimal time. In other cases, technological solutions may be available in a range of configurations appropriate for farms of different sizes. Once growers invest in specialized machinery for a particular crop, they may reduce the number of commodities produced on their farms.

Mechanization and automation are labour-saving technologies aiming to enhance apple production by reducing dependency on seasonal farm labour. Based on the amount of human labour required in using different equipment, those labour-saving technologies could be categorized into operator-assist, mechanical and robotic systems. Research and development of labour-saving solutions exist at both public research institutes and private technology innovation firms, and this will help to meet the demands of different growers in different regions.

9 Where to look for further information

Researchers at Washington State University Center for Precision and Automated Agricultural Systems (CPAAS) have been working on developing several labour-saving technologies from a hand-held mechanical blossom thinner, harvest labour management system, mechanical and robotic apple harvesting systems to an autonomous bin management robot for fresh market apple production (www.cpaas.wsu.edu).

10 References

Anonymous. 2016. Mobile Orchard Platforms. Available at http://extension.psu.edu/plants/tree-fruit/resources/specialty-crop-innovations/mobile-platforms. Accessed 29 February 2016.

Baugher, T. A., J. Schupp, K. Ellis, J. Remcheck, E. Winzeler, R. Duncan, S. Johnson, K. Lewis, G. Reighard, G. Henderson, M. Norton, A. Dhaddey and P. Heinemann. 2010. String blossom thinner designed for variable tree forms increases crop load management efficiency in trials in four United States peach-growing regions. *HortTechnology*, 20:409–14.

Calvin, L., and P. Martin. 2010. The U.S. Produce Industry and Labor: Facing the Future in a Global Economy. Economic Research Report No. (ERR-106), USDA-ERS, Washington, DC.

Carbonneau, A. 1979. La taille mechanique de la vigne: Des experiences austrialieness. *Progress Agricole et Viticole* (abstract in English, paper in French).

Ceres, R., J. L. Pons, A. R. Jiménez, A. R. Martín and L. Calderón. 1998. Design and implementation of an aided fruit-harvesting robot (Agribot). *Industrial Robot*, 25(5):337–46.

Chen, Y., H. Zhu and H. E. Ozkan. 2013. Development of a variable-rate sprayer with laser scanning sensor to synchronize spray outputs to tree structures. *Transactions of the ASABE*, 55(3):773–81.

Damerow, L., and M. M. Blanke. 2009. A novel device for precise and selective thinning in fruit crops to improve fruit quality. *Acta Horticulturae*, 824:275–80.

Davies, F. S, and T. E. Crocker. 1994. Pruning Blueberries in Florida. Florida Cooperative Extension Fact Sheet HS-77.

D'Esnon, A. G., G. Rabatel, R. Pellenc, A. Journeau and M. J. Aldon. 1987. Magali: A self-propelled robot to pick apples. ASAE Paper No. 87-1037, St. Josept, MI, ASAE.

Duraj, V., J. A. Miles, D. J. Tejeda, E. J. Mitcham, W. V. Biasi, L. Asín, R. B. Elkins, J. M. Meyers and J. Abreu. 2010. Comparison of platform versus ladders for harvest in northern California pear orchard. In *XI International Pear Symposium* 909, pp. 241–9.

Emery, K. G., D. M. Faubion, C. S. Walsh and Y. Tao. 2010. Development of 3-D range imaging system to scan peach branches for selective robotic blossom thinning. ASABE Paper No. 1009202, St. Josept, MI, ASABE.

Escolà, A., J. R. Rosell-Polo, S. Planas, E. Gil, J. Pomar, F. Camp, J. Llorens and F. Solanelles. 2013. Variable rate sprayer. Part 1 – Orchard prototype: Design, implementation and validation. *Computers and Electronics in Agriculture*, 95:122–35.

Extension. 2016. Apple Fruit Thinning. Available at: http://articles.extension.org/pages/69070/apple-fruit-thinning. Accessed 29 March 2016.

Fox, R. D., R. C. Derksen, H. Zhu, R. D. Brazee and S. A. Svensson. 2008. A history of air-blast sprayer development and future prospects. *Transactions of the ASABE*, 51(2):405–10.

Gallardo, R. K., M. Taylor and H. Hinman. 2010. 2009 Cost Estimates of Establishing and Producing Gala Apples in Washington. Washington State University Extension Factsheet FS005E, Pullman, WA.

Giles, D. K., and R. C. Billing. 2015. Deployment and performance of a UAV for crop spraying. *Chemical Engineering Transactions*, 44: 307–12.

Gongal, A., S. Amatya, M. Karkee, Q. Zhang and K. Lewis. 2015. Sensors and systems for fruit detection and localization: A review. *Computers and Electronics in Agriculture*, 116:8–19.

Gonzalez-Barrera, A. 2015. More Mexicans Leaving Than Coming to the U.S. Pew Research Center. Available at: http://www.pewhispanic.org/files/2015/11/2015-11-19_mexican-immigration_FINAL.pdf. Accessed 2 March 2016.

Gonzalez-de-Soto, M., L. Emmi, M. Perez-Ruiz, J. Aguera and P. Gonzalez-de-Santos. 2016. Autonomous systems for precise spraying- evaluation of a robotized patch sprayer. *Biosystems Engineering*, doi:10.1016/j.biosystemseng.2015.12.018.

Hansen, M. 2011. Fruiting Walls Suit Machinery. *Good Fruit Grower*, 62(2):24–5.

He, L., H. Fu, D. Sun, M. Karkee and Q. Zhang. 2016. A shake and catch harvesting system for 'Jazz' apple trees trained in vertical fruiting wall. 2016 ASABE Annual International Meeting Paper, St. Joseph, MI, ASABE.

Hertz, T., and S. Zahniser. 2013. Is there a farm labor shortage? *American Journal of Agricultural Economics*, 95(2):476–81.

Hill, C. 2016. Mechanical Pruning of Pink Lady and Kanzi Apple Varieties. Available at: http://horticulture.wsu.edu/wp-content/uploads/2012/04/Chelsea-Hill-Poster.pdf?9982c6. Accessed 26 March 2016.

Ingels, C., P. M. Geisel, C. L. Unruh and P. M. Lawson. 2016. Fruit Trees: Thinning Young Fruit. University of California ANR publications. Available at: http://homeorchard.ucdavis.edu/8047.pdf. Accessed 29 March 2016.

Jamar, L., O. Mostade, B. Huyghebaert, O. Pigeon and M. Lateur. 2010. Comparative performance of recycling tunnel and conventional sprayers using standard and drift-mitigating nozzles in dwarf apple orchards. *Crop Protection*, 29(6):561–6.

Jensen, F. L. July 1980. Effects of Mechanical Pruning on Grapes. *California Agriculture*. Available at: https://ucanr.edu/repositoryfiles/ca3407p33-72258.pdf.

Karkee, M., and B. Adhikari. 2015. A method for three-dimensional reconstruction of apple trees for automated pruning. *Transactions of the ASABE*, 58(3):565–74.

Karkee, M., B. Adhikari, S. Amatya and Q. Zhang. 2014. Identification of pruning branches in tall spindle apple trees for automated pruning. *Computers and Electronics in Agriculture*, 103:127–35.

Khot, L. R., R. Ehsani, J. M. Maja, J. M. Campoy, C. Wellington and A. Al-Jumaili. 2014. Evaluation of deposition and coverage of an air-assisted sprayer, and two airblast sprayers in citrus orchards. *Transactions of the ASABE*, 57(4):1007–13.

Kirk, I. W., W. C. Hoffmann and J. B. Carlton. 2001. Aerial electrostatic spray system performance. *Transactions of the ASABE*, 44(5):1089–92.

Kondo, N., Y. Shibano, K. Mohri and M. Monta. 1993. Basic studies on robot to work in vineyard 1: Manipulator and harvesting hand. *Journal of Japanese Society of Agriculture and Machinery*, 55(6): 85–94

Kondo, N., Y. Shibano, K. Mohri and M. Monta. 1994. Basic studies on robot to work in vineyard 2: Discriminating, position detecting and harvesting experiments using a visual sensor. *Journal of Japanese Society of Agriculture and Machinery*, 56(1): 45–53.

Lehnert, C., T. Perez and C. McCool. 2015. Optimisation-based Design of a Manipulator for Harvesting Capsicum. IEEE International Conference on Robotics and Automation (ICRA) Workshop on Robotics in Agriculture. Seattle, WA.

Lehnert, R. 2013. In-orchard sorting. 18 November 2013 Issue, Good Fruit Grower. Available at http://www.goodfruit.com/in-orchard-sorting/. Accessed 10 March 2016.

Lewis, K., and I. Hanrahan. 2014. Field Testing of Two Harvest Aids in Washington Apple Orchards. A presentation. Available at http://www.crec.ifas.ufl.edu/harvest/pdfs/presentations/Lewis_ISHS_Session_7.pdf. Accessed 20 March 2016.

Li, S., S. Yao, Y. Zhang, L. Meng and Z. Lu. 2011. Grasp planning analysis and strategy modeling for underactuated multi-fingered robot hand using in fruits grasping. *African Journal of Agricultural Research*, 6(9):2086–98.

Lyons, D., and P. Heinemann. 2016. Selective Automated Blossom Thinning. U.S. Patent 20,160,057,940.

Lyons, D. J., P. H. Heinemann, J. R. Schupp, T. A. Baugher and J. Liu. 2015. Development of a selective automated blossom thinning system for peaches. *Transactions of the ASABE*, 58(6):1447–57.

Markwardt, E. D., R. W. Guest, J. C. Cain and R. L. Labelle. 1964. Mechanical cherry harvesting. *Transactions of the ASAE*, 7(1):70–4, 82.

Martí, B. V., and E. F. González. 2010. The influence of mechanical pruning in cost reduction, production of fruit, and biomass waste in citrus orchards. *Applied Engineering in Agriculture*, 26(4):531–40.

Miller, S., and T. Tworkoski. 2010. Blossom thinning in apple and peach with an essential oil. *HortScience*, 45:1218–25.

Miller, S. S., J. R. Schupp and T. A. Baugher. 2011. Performance of mechanical thinners for bloom or green fruit thinning in peaches. *HortScience*, 46(1):43–51.

Moore P. W. 1958. Mechanical pruning for citrus. *California Agriculture*, 11:7–13.

Moore-Gough, C., and R. E. Gough. 2007. Pruning Fruit Trees in Montana. Available at: http://msuextension.org/publications/YardandGarden/MT199215AG.pdf. Accessed 2 November 2011.

Nielsen, M., D. C. Slaughter and C. Gliever. 2012. Vision-based 3D peach tree reconstruction for automated blossom thinning. *Industrial Informatics, IEEE Transactions*, 8(1):188–96.

Ochs, E. S., and W. W. Gunkel. 1993. Robotic Grape Pruner Field Performance Simulation. ASABE Paper No. 933528. St. Joseph, MI, ASAE.

Ortiz, C., J. Blasco, S. Balasch and A. Torregrosa. 2011. Shock absorbing surfaces for collecting fruit during the mechanical harvesting of citrus. *Biosystems Engineering*, 110(1):2–9.

Parameswarakumar, M., and C. P. Gupta. 1991. Design parameters for vibratory mango harvesting system. *Transactions of the ASAE*, 34(1):14–20.

Pérez-Ruiz, M., P. Gonzalez-de-Santos, A. Ribeiro, C. Fernandez-Quintanilla, A. Peruzzi, M. Vieri, S. Tomic and J. Agüera. 2015. Highlights and preliminary results for autonomous crop protection. *Computers and Electronics in Agriculture*, 110:150–61.

Peterson, D. L., B. S. Bennedsen, W. C. Anger and S. D. Wolford. 1999. A systems approach to robotic bulk harvesting of apples. *Transactions of the ASAE*, 42(4):871–6.

Peterson, D. L., S. S. Miller and T. S. Kornecki. 1985. Over-the-row harvester for apples. *Transactions of the ASAE*, 28(5):1393–7.

Planas, S., F. Solanelles and A. Fillat. 2002. Assessment of recycling tunnel sprayers in mediterranean vineyards and apple orchards. *Biosystems Engineering*, 82(1):45–52.

Peterson, D. L., and S. D. Wolford. 2003. Fresh–Market Quality Tree Fruit Harvester Part II: Apples. *Applied Engineering in Agriculture*, 19(5):545–8.

Polat, R., I. Gezer, M. Guner, E. Dursun, D. Erdogan and H. C. Bilim. 2006. Mechanical harvesting of pistachio nuts. *Journal of Food Engineering*, 79(4):1131–5.

Rosa, U. A., K. G. Cheetancheri, C. J. Gliever, S. H. Lee, J. Thompson and D. C. Slaughter. 2008. An electro-mechanical limb shaker for fruit thinning. *Computer and Electronics in Agriculture*, 61:213–21.

Scarfe, A. J., R. C. Flemmer, H. H. Bakker and C. L. Flemmer. 2009. Development of an Autonomous Kiwifruit Picking Robot. 4th International Conference on Autonomous Robots and Agents. Wellington, New Zealand.

Schupp, J. R., T. A. Baugher, S. S. Miller, R. M. Harsh and K. M. Lesser. 2008. Mechanical thinning of peach and apple trees reduces labor input and increases fruit size. *HortTechnology*, 18:660–70.

Setiawan A. I., T. Furukawa and A. Preston. 2004. A low-cost gripper for an apple picking robot. *IEEE International Conference on Robotics and Automation*, pp. 4448–53.

Sevilla, F. 1983. Etude de la faisabilte d une mechanization de la vigne en France. *Progress Agricole et Viticole* (abstract in English, paper in French).

Sevilla, F. 1985. A Robot to Prune the Grapevine. *AgriMation*, 1:190–9, St. Joseph, MI, ASAE.

Slingerland, K. 2009. Mechanical blossom thinning - dream or reality. Available at: http://www.uoguelph.ca/plant/treefruit/documents/. Accessed 9 March 2016.

Veal, D., L. Damerow and M. M. Blanke. 2008. Selective mechanical thinning to regulate fruit set, improve quality and overcome alternate bearing in fruit crops. In IX International Symposium on Integrating Canopy, Rootstock and Environmental Physiology in Orchard Systems 903, pp. 775–81.

Velazquez, M. B., and E. F. Gonzalez. 2010. The influence of mechanical pruning in cost reduction, production of fruit, and biomass waste in citrus orchards. *American Society of Agricultural and Biological Engineers*, 26(4):531–40.

Wang, M., H. Wang, Q. Zhang, K. M. Lewis and P. A. Scharf. 2013. A hand-held mechanical blossom thinning device for fruit trees. *Transactions of the ASABE*, 29(2):155–60.

Whitney, J. D. 1977. Design and performance of an air shaker for citrus fruit removal. *Transactions of the ASAE*, 20(1):52–6.

Wouters, N., B. De Ketelaere, J. De Baerdemaeker and W. Saeys. 2013. Hyperspectral waveband selection for automatic detection of floral pear buds. *Precision Agriculture*, 14(1):86–98.

Wouters, N., B. De Ketelaere, T. Deckers, J. De Baerdemaeker and W. Saeys. 2015. Multispectral detection of floral buds for automated thinning of pear. *Computers and Electronics in Agriculture*, 113:93–103.

Ye, Y., L. He, J. Zhou, Q. Zhang, M. Wang and K. Lewis. 2012. A remotely controlled bin-dog for transferring bins in orchards environment. ASABE Paper No. 131596530, St. Joseph, MI, ASABE.

Chapter 12

Sustainable approaches to control postharvest diseases of apples

W. J. Janisiewicz and W. M. Jurick II, USDA-ARS, USA

1 Introduction

An increase in apple production and a growing consumer market starting at the beginning of the last century created the necessity for storing apples for extended periods of time. At that point, there was little information on apple ripening, storage or the diseases that developed in storage. Heavy losses from physiological disorders and decays caused by fungi were often experienced. Although direct assessments were not made, the losses can be estimated to be between 20 and 50% (Kader, 1992). Losses as high as 15% and 22.6% to various decays on apples in storage were still reported from Ireland and France, respectively, in the late 1960s (Bondoux, 1967; Swinburne, 1970) and even more recently, 5% losses were reported due to apple blue mould (caused by *Penicillium expansum*) at the New York terminal market (Cappelini et al., 1987). Total postharvest losses are difficult to determine and they are often underestimated because, to be complete, they should encompass losses occurring during storage, handling, transportation, retail and at the consumer level (Moline, 1984). Nevertheless, the need to reduce these losses has been great.

http://dx.doi.org/10.19103/AS.2016.0017.15

Control of postharvest diseases and disorders of apples went through several stages during the past century. Initially, studies were conducted on the effects of harvest time, respiration after harvest, and storage at low temperature and high humidity on apple quality and decay development (Culpepper and Caldwell, 1927; Eustace, 1908; Morse, 1908). Maintaining apple quality under different storage conditions, and the effects of these conditions on physiological disorders such as Jonathan spot and soft scald, apple scald or internal breakdown (Plagge and Gerhardt, 1930; Plagge and Maney, 1924, 1925) dominated early research. During this period, postharvest decays of apples from several countries were described in various monographs, which were summarized by Kidd and Beaumont (1924) and Rose (1924). This was followed by a comprehensive monograph on postharvest diseases of apples, pear and quinces (Rose et al., 1933). At that time, the control recommendations for the major postharvest decays of apples; blue mould, grey mould (caused by *Botrytis cinerea*), brown rot (caused by *Monilinia* spp.) and bull's-eye rot (caused by *Neofabraea malicorticis, N. perennans* and *N. alba*), emphasized careful handling to prevent apple bruising and prompt cold storage after harvest (Wright and Smyth, 1954). Recommendations for brown rot and bull's-eye rot also included orchard sprays with sulphur and Bordeaux mixtures, respectively. The first chemical reported to successfully control decays on harvested apples was sodium chloro-2-phenylphenate, applied as a disinfecting wash (English, 1948; English et al., 1948). It was recommended for postharvest use on apples in the second edition of the monograph by Rose et al. (1951). From then onward, fungicides have been routinely recommended for drenching fruit before storage, and several biocides, mainly chlorine and acetic acid, for use on packing lines and in the sterilization of contaminated bins to reduce conidia concentration of pathogens which resulted in significant reductions in losses (Eckert and Kolbezen, 1964; Eckert and Ogawa, 1988; Hardenburg and Spalding, 1972; Harvey, 1978; Pierson, 1966; Rosenberger, 2001, 2009; Sholberg, 1998b; Smith, 1962; Sutton et al., 2014; Watkins et al., 2014). Later, drenches with hot water, calcium or biocontrol agents were also recommended for suppression of blue mould decay (Hansen et al., 2010; Sutton et al., 2014; Watkins et al., 2014). The control measures against newly emerging diseases important in the western United States, for example Sphaeropsis rot and Phacidiopycnis rot, include pre- and postharvest application of fungicides as well as the removal of cankers and twigs with dieback symptoms (Kim and Xiao, 2008; Xiao and Kim, 2008).

Extensive reliance on fungicides to control postharvest fruit decays during the past several decades has exposed the limitations of this approach. This has been demonstrated by the development of resistance to fungicides in major pathogens and the increasing difficulties in developing new fungicides (Baraldi et al., 2003; Bertrand and Saulie-Carter, 1978; Kim and Xiao, 2010; Li and Xiao, 2008; Rosenberger and Meyer, 1981; Spot and Cervantes, 1986; Xiao and Kim, 2010; Xiao et al., 2014). The extensive use of fungicides also raises concerns about their effect on the environment by leaving residues in the air, soil and water, and the effect on non-target microorganisms (Andrews and Kenerley, 1978; Lamichhane et al., 2016; Vorstermans and Creemers, 2011). The potential negative effects on human health, especially for fungicide applicators and consumers ingesting fungicide residues with fruits, have resulted in social pressure demanding a reduction in the use of fungicides (Bertolini, 2008; Harris, 2013; Janisiewicz, 2013; Kader, 1992; Kennedy et al., 1998; Kuchler et al., 1996; Vorstermans and Creemers, 2011; Xiao and Boal, 2009). Thus, the search for alternative approaches that are more sustainable in controlling postharvest diseases of fruits has intensified during the past three decades (Bertolini and Pratella, 2008; Kennedy et al., 1998; Palou et al., 2016). The main focus has been on exploring the usefulness of various natural products, food additives and substances generally regarded as safe (GRAS), as well as physical treatments (Janisiewicz and Conway, 2010; Palou et al., 2016), biological control (Janisiewicz and Korsten, 2002; Janisiewicz, 2013) and natural fruit resistance either by induction in harvested fruit or by exploration in breeding programmes (Janisiewicz and Conway, 2010; Janisiewicz et al., 2008, 2016; Johnson et al., 1997; Jurick et al., 2011; Stevens et al., 2005). Most of these approaches are inadequate by themselves; however, when used in combination they may rival fungicide treatments. In this chapter we will describe various alternative approaches to synthetic fungicides for controlling postharvest diseases of apples that are more sustainable and safer to human health and the environment.

2 Natural plant-derived products

2.1 Plant volatile compounds

Several plant volatile compounds have been evaluated for their effect against major postharvest pathogens of apples. The most effective include acetaldehyde, hexanal and *trans*-2-hexenal.

2.1.1 Acetaldehyde

Acetaldehyde (Aa) occurs naturally in fruits and vegetable (Nurtsen, 1970). Aa vapour was shown to effectively control blue mould on 'Red Delicious' and 'Golden Delicious' apples (Stadelbacher and Prasad, 1974). Exposure of *P. expansum* conidia to Aa vapour at a concentration of 0.5% for 180 min at 21°C prevented conidia germination on potato dextrose agar (PDA) medium and reduced decay on puncture inoculated apples by 94%. No phytotoxicity was observed. Shorter exposures at higher concentrations were equally effective without any phytotoxicity. There was a direct correlation between the concentration and the exposure time. However, the effectiveness of this compound was only demonstrated at 21°C and RH 65% and not under cold storage conditions. Although this is a naturally occurring compound and it can be produced by some yeasts, the potential for using it to control fruit decays has diminished, as it has been classified as a Group 1 carcinogen (Secretan et al., 2009).

2.1.2 Hexanal and *trans*-2-hexanal

Hexanal is a precursor of the esters that provide a characteristic apple aroma (Pillard, 1986). Fumigation with 100 ppm hexanal completely inhibited *P. expansum* and *B. cinerea* growth on PDA medium during a 48 h exposure interval; however, growth resumed after additional incubation in ambient air (Song et al., 1998). Concentrations of 450 ppm were required for complete prevention of growth after removal from the treatment and subsequent incubation in the air for 120 h. In a test on wounded 'Golden Delicious' apples treated with 8–12 µmol/L hexanal, almost all fruit developed lesions that were about half the size of the control treatment at 22°C (Fan et al., 2006). At 4°C with 5–7 µmol/L concentrations, the incidence of blue mould was reduced by 33% after 6 weeks of storage. To make this treatment commercially feasible, Sholberg and Randall (2007) combined the preharvest application of a fungicide, cyprodinil, with hexanal fumigation at low rates (2–3 mg/L) for 24 h immediately after harvest. This treatment reduced infections on wounded apples from 10% to 1%.

Fumigation treatment of apples with *trans*-2-hexenal vapours at 12.5 μL/L for 24 h and incubation for 7 days at 20°C reduced the incidence of apple blue mould between 50% and 98%, depending on cultivar, without any negative effect on fruit quality (Neri et al., 2006b). This was the most effective compound out of several plant volatiles tested (Neri et al., 2006a). This compound also inhibited conidial germination of *Neofabraea alba*, the causal agent of lenticel rot of apples *in vitro*, but had little effect on controlling decay (Neri et al., 2009).

2.1.3 Allyl isothiocyanates

Volatile allyl isothiocyanates (AITC) are products of the enzymatic hydrolysis of glucosinolates that occur in many plants, especially in crucifers (e.g. sinigrin). They were shown to be inhibitory to germination and mycelial growth of *P. expansum* (Delaquis and Sholberg, 1997; Mari et al., 1993) and reduced the incidence of blue mould on pears by as much as 90% after 24 h exposure to an AITC-enriched atmosphere (Mari et al., 2002). No phytotoxic effects were observed. Delaying treatment for 24–48 h after inoculation with *P. expansum* also significantly reduced decay. Pre-storage treatment with AITC has been suggested as a possible application. It is very likely that this treatment will also be effective on apples.

2.2 Alkylresorcinols

Alkylresorcinols (AR) are a group of phenolic lipids that naturally occur in plants, including fruit (mango) and grains, and exhibit antimicrobial activity that may be involved in resistance to various diseases (Droby et al., 1987; Hassan et al., 2007; Landberg et al., 2008; Reiss, 1989). Rye bran is rich in AR and has been used to extract pure AR for testing its effectiveness against various fungi including apple pathogens *P. expansum* and *N. perennas* (causing bull's-eye rot) (Dey et al., 2013; Tahir et al., 2014). AR applied as an emulsion spray reduced spore germination and mycelial growth of both fungi by 50% on PDA medium, and blue mould between 17% and 52%, and bull's-eye rot between 31% and 78%, depending on cultivar of apples stored at 2°C for 12 weeks. Although this compound looks promising because no negative effect on apple skin appearance and organoleptic quality was apparent, more work is needed especially on effective concentrations on different apple cultivars in long-term storage.

2.3 Essential oils

Essential oils (EO) are obtained from plant materials mainly by steam distillation in commercial production; however, they can be also obtained by fermentation, pressing, and cold or hot enfleurage (Van de Braak and Leijten, 1999). There are a few thousand known EO of which approximately 300 are used predominantly as fragrances, flavours and pharmaceuticals. The antimicrobial properties of some EO have been known for a long time and their potential for application in foods has been widely investigated (see review by Burt, 2004; Deans and Ritchie, 1987). The phenols eugenol, thymol and carvacrol, and terpenoids menthol and eucalyptol have been most widely studied for their application in foods (Serrano et al., 2008). Results from studies on the use of EO for postharvest control of decays on fruits and vegetables have been mixed. Although tests *in vitro* often indicated good antifungal activity, tests on fruit are less impressive. Postharvest application of eugenol, formulated with ethoxylate or Tween 80, to several apple cultivars was ineffective in controlling lenticel infection caused by *Phlyctema vagabunda*, blue mould (*P. expansum*), grey mould (*B. cinerea*) and brown rot (*Monilinia fructigena*), and when heated to 50°C, these treatments induced phytotoxicity and cuticle damage occurred. However, eugenol formulated with soy lecithin reduced the diseases without significant cuticle damage (Amiri et al., 2008). EO from several Mediterranean plants inhibited grey mould and blue mould development on apples, with emulsions of savoury and thyme oils at 1% being the most effective and having greater efficacy on 'Granny Smith' and 'Red Chief' than on 'Golden Delicious' (Lopez-Reyes, 2010). In general, the efficacy of the EO declined with length of storage and was dependent on cultivar. Thus, for the treatment to be effective during extended storage, it would have to be applied repeatedly. Garlic extract (40% and 50%) applied to apples reduced blue mould and grey mould more in a curative application than in a protective application, and in combination with 1% clove oil, further increased the efficacy of the protective application on the three apple cultivars tested (Daniel et al., 2015).

It appears that the potential usefulness of EO for controlling postharvest diseases of apples is restricted to applications combined with other alternative treatments, often with limited activity, that can result in an additive or synergistic effect that provides commercially adequate control (Janisiewicz and Conway, 2010; Mari et al., 2003; Spadaro and Gullino, 2014; Usall et al., 2015). The balance between the effective concentration of EO and phytotoxicity is a significant limitation to this approach. Before EO can be used commercially for controlling postharvest diseases, more research must be conducted on determining the relationships between an effective EO concentration (used individually or

Figure 1 Portable automated bin drencher developed for testing various decay control treatments applied to apples and pears before storage. A small reservoir tank allows for testing small volumes on an entire bin. This is particularly important with biocontrol agents, where limited amounts are available because they are often grown in small fermenters under laboratory conditions (Janisiewicz and Peterson, 2005).

in mixtures), exposure time, formulation and phytotoxicity. Also, the effect of EO on the organoleptic qualities of different cultivars of apples must be determined.

3 GRAS substances and sanitizers

3.1 Bicarbonate salt

Sodium bicarbonate (SBC, baking soda) was recommended for controlling postharvest decays of citrus fruits almost a century ago (Barger, 1928); however, with the advent of synthetic fungicides, its use has diminished. A renewed interest in this compound has occurred recently through demonstrations of its effectiveness on apples in combination with newly developed alternative methods, in particular biological control, hot water and hot air, and calcium chloride treatments, and also with other treatments on citrus, peaches and cherries (see review by Janisiewicz and Conway, 2010). On 'Golden Delicious' apples, treatment with 0.3% or 1.0% SBC significantly reduced blue mould, but not bitter rot, after 4 months of storage at 0°C followed by 14 days at 20°C (Conway et al., 2004). In a subsequent study, 2% SBC greatly improved control of blue mould when combined with two antagonists (see Section 8), *Cryptococcus laurentii* and *Metschnikowia pulcherrima*, applied either individually or in combination (Conway et al., 2005). In commercial trials, drenching bins Fig. 1 containing artificially wounded 'Golden Delicious' apples treated with 2% SBC reduced incidence of blue mould decay from 28 to 18% and from 72 to 35% after 5.5 months storage at ~1°C in CA (1.5% O_2, 2.0% CO_2) in two consecutive years, respectively (Janisiewicz et al., 2008b). In a drop inoculation test on wounded 'Golden Delicious' apples, 2% SBC reduced the incidence of grey mould and blue mould by 70% and 29%, respectively; in combination with the antagonist *Candida oleophila*, no grey mould developed and blue mould incidence was only ~2% (Droby et al., 2003). The reduction of postharvest decays on apples and other fruits under various conditions after treatment with SBC testifies to its usefulness, especially in combination with other alternative treatments, and as a cheap GRAS substance, it should be given serious consideration for wide commercial applications.

3.2 Calcium

Treatment of apples with calcium chloride ($CaCl_2$) after harvest reduces both physiological disorders such as bitter pit, internal breakdown and fruit softening (Bangerth et al., 1972; Conway et al., 2002; Mason et al., 1975; Reid and Padfield, 1975; Scott and Wills, 1979) and decays such as blue mould, grey mould and bitter rot (caused by *Colletotrichum gloeosporioides*, telomorph *Glomerella cingulata*) (Conway et al., 1991). The beneficial effect of calcium is attributed to increasing resistance in the fruit by making cell walls less accessible to fruit-softening enzymes or cell-wall-degrading enzymes produced by the pathogen rather than any direct effect it may have on the pathogens (Conway et al., 1988; Grant et al., 1973; Knee, 1978). Pressure infiltration of 'Golden Delicious' apples at 103 kPa with 2% $CaCl_2$ was more effective than dipping or vacuum infiltration (Conway, 1982), and after storing apples for 6 months at 0°C, inoculation with the pathogens and subsequent incubation at 20°C for up to two weeks, the incidence of bitter rot, grey mould and blue mould were reduced by 70%, 59% and 37%, respectively (Conway et al., 1991). Similar $CaCl_2$ treatments of 'Gala' apples stored at 1°C for 6 months reduced blue mould by 38%, and in combination with an antagonist, *Pseudomonas syringae*, by ~80% (Conway et al., 1999). Calcium treatment combined with many other antagonists and heat treatments increased control of postharvest decays on various fruits (see review by Janisiewicz and Conway, 2010); dipping apples in $CaCl_2$ has been practised in some commercial packinghouses. However, to obtain the best results in reducing postharvest decays, concentrations of calcium in the tissue should be increased to a higher level, which can be achieved only by pressure infiltration (Conway, 1982). The uptake of calcium is affected by differences in cultivars, fruit maturity, growing season and even the orchard, creating the potential for peel injury if too much calcium is infiltrated, which complicates the use of this treatment (Conway et al., 2002).

3.3 Chitosan

Chitosan is derived mostly from the chitin of outer shell of crustaceans, mainly shrimp, crab and krill, by partial N-deacetylation (Sanford, 1989). Its structure is very similar to cellulose and, like cellulose, is not water soluble, which limits its application. Chitosan is approved as a food additive in the United States and has been used, for example, in food preservation, processing or deacidification of fruit juices (Zhang et al., 2011). It has wide antimicrobial activity and has been shown to inhibit the development of various fungal structures and postharvest decays of fruits, including those caused by *B. cinerea* and *P. expansum* on apples (El-Ghaouth et al., 2000a,b; Rabea et al., 2003; Yang et al., 2012; Yu et al., 2012; Zhang et al., 2011). It has also been shown to induce resistant responses against postharvest pathogens in apples and other fruits (Bautista-Banos et al., 2006; de Capdeville et al., 2002). A 2% aqueous suspension of chitosan reduced the lesion sizes of blue mould decay by as much as 85% on 'Red Delicious' when applied in a drop application test to intact fruits which were later wounded and inoculated with the pathogen. Glycol chitosan, a water-soluble form of chitosan, when used alone as the bioactive coating at 0.5%, reduced the incidence of blue mould and grey mould decay of 'Red Delicious' apples by 30%, and in combination with a biocontrol agent *Candida saitoana*, by 80% and 84%, respectively (El-Ghaouth et al., 2000a,b). The use of glycol chitosan as a bioactive coating is an attractive approach, because it has higher wettability, which allows for a uniform spread on fruit upon drying resulting in adequate adhesion, cohesion and durability to function properly. This may be especially beneficial when applying in combination with other treatments to increase the efficacy of control. It also eliminates the need for combining chitosan with surfactants, such as Tween 80 (Choi et al., 2002). Growing the biocontrol agent *Rhodotorula mucilaginosa* (a yeast) in nutrient yeast dextrose broth medium amended with 0.5% chitosan increased biocontrol activity of the yeast against postharvest pathogens of strawberries by stimulating its chitinase activity that degrades the pathogens' mycelial cell walls (Zhang et al., 2014). It will be worthwhile to determine if this approach would also work with antagonists in an apple system; for this to be commercially valuable, the antagonists will also have to be amenable to large-scale fermentation production systems.

3.4 Other chemicals

The potential of other substances such as silicon, selenium, potassium sorbate, nisin or ethanol has been explored for controlling pathogens causing postharvest diseases of apple. However, either low efficacy limits their usefulness or there is a lack of appropriate follow-up tests on fruit to determine their potential effectiveness after the original *in vitro* research was published (Fadda et al., 2015; see review by Janisiewicz and Conway, 2010; Wu et al., 2014, 2016).

3.5 Acetic acid vapour and peracetic acid

Sholberg and Gauce (1995) were the first to demonstrate control of blue mould on wounded 'Golden Delicious' apples by fumigation with acetic acid (AA). Sholberg (1998b) also tested the effectiveness of several other short-chain organic acids against postharvest decays; however, AA was the most effective against *P. expansum* and other postharvest pathogens. Concentrations of 2.7 mg AA/L were sufficient to completely control the disease under laboratory conditions. High relative humidity in storage increased the effectiveness and consistency of the control at 5 and 20°C. In a large-scale test with naturally infected fruit that was fumigated with AA and then stored in air for 3 months at 1°C, a 50% reduction in decay was observed compared to the control (Sholberg et al., 2001). Unfortunately, AA treatment may cause lenticel burning that increases in severity with the number of fumigation cycles, and varies with the quantity of fruit, container type and apple cultivar. Thus, the fumigation with AA can be safely used only for sanitization of picking bins and large storage rooms (Sholberg, 2004, 2009).

Peracetic acid, which is derived from the treatment of AA with hydrogen peroxide, is also considered to be a good sanitizer, and has been successfully used on 'Braeburn' apples to reduce artificial contamination with *E. coli* O157:H7. Reductions in the bacterial population increased as the exposure time increased from 5 to 15 min at which point a 5-log reduction was achieved (Wisniewsky et al., 2000). No visible changes in the appearance or texture of the apples were observed after the treatment, however apples were rinsed with water to remove peracetic acid residues. At a concentration of 80 ppm, peracetic acid reduced *P. expansum* conidia by no more than 2-log units on the surface of various apple cultivars artificially inoculated with the fungus (Salomão et al., 2008). Such a reduction may not be adequate for effective decay control, especially when higher concentrations of pathogen spores occur in the handling system.

3.6 Chlorine and ozone

Chlorinated and ozonated water have been used as postharvest treatments for fruits and vegetables because of their biocidal activity; their usefulness in controlling fruit decays is limited mainly to dump tanks and flume water (Horovitz and Cantalejo, 2014; Salomão et al., 2008; Smilanick, 2003; Suslow, 1997). Both can inhibit spore germination of the major fruit decaying fungi; however, they are strongly effected by organic matter and are not effective in controlling decays originating from wound infection (Spotts and Cervantes, 1992; Spotts and Peters, 1980). Both are more useful as sanitizers against bacterial and viral contamination of produce (Suslow, 1997, 2004). Research from the 1940s indicates the usefulness of 1–3 ppm ozone gas as a sanitizer to lower spore counts on the walls of storage rooms and packaging more so than for direct control of apple decays where a concentration of 3.25 ppm may cause injury to fruit (see review by Horovitz and Cantalejo, 2014). There are some reports on the usefulness of ozone gas to control fruit diseases, for example it was reported that the shelf-life of apples was extended for several weeks by exposure to 5–6 mg/m^3 ozone gas for four hours every day, but no follow-up research has been conducted after this initial report (Bazarova, 1982). Ozone gas is toxic to humans and special protection is needed for workers as the time-weighted average exposure for an 8 h work day is only 0.1 ppm according to the federal exposure limit (Smilanick, 2003). A postharvest dip of apples infected with flyspeck and sooty blotch in chlorine (800 ppm for 7 min) followed by 30 sec brushing and a rinse, resulted in an increase of Extra Fancy-grade apples from 25% to 55% and 100% for 'Jonathan' and 'Golden Delicious', respectively (Batzer et al., 2002). In the same study, treatment with a mixture of hydrogen peroxide and peracetic acid (80 ppm for 7 min) was less effective. As much of apple handling in packinghouses is in water, chlorine is often added to kill spores of fungal pathogens causing fruit decays and also to protect against contamination with food-borne pathogens (Kelley, 2004; Suslow, 2004). As pH and organic content have significant effects on the antimicrobial activity of chlorine special equipment must be used to maintain the desirable concentration of active chlorine.

4 Heat treatment

Heat treatments of apples before storage have been developed for controlling pathogens and insects, enhancing fruit quality by slowing fruit ripening and senescence, and reducing storage disorders (Klein and Lurie, 1990; Lurie, 2001; Lurie et al., 1998; Porritt and Lidster, 1978; Smith and Lay-Yee, 2000). In controlling postharvest decays, the main effect is directly on the pathogens, although there may be some defence reactions induced in apples (Conway et al., 1999; Lurie, 1998; Maxin et al., 2012a,b). Heat may be applied to apples as hot dry air, a hot water dip, or a short hot water rinse and brushing (Burchill, 1964; Fallik, 2004; Fallik et al., 2001; Klein and Lurie, 1990; Porritt and Lidster, 1978). Heat treatments alone are only partially effective in reducing fruit decays on apples kept for long periods of time in cold storage; however, in combination with other treatments, for example calcium infiltration, SBC and biological control, can provide a commercially acceptable level of control (Conway et al., 1999, 2005; Leverentz et al., 2000). Results from heat treatments with respect to the efficacy of decay control, sensitivity to heat injury or the effect on fruit quality are often cultivar dependent, which puts an additional burden on the commercial development of this treatment.

4.1 Hot air

The most commonly reported hot air treatment of preclimacteric apples is 38°C for four days. This temperature and duration of exposure effectively controlled blue mould from natural infection (Porritt and Lidster, 1978) and artificial inoculation with *P. expansum* or *B. cinerea* (Conway et al., 1999; Fallik et al., 1995; Klein et al., 1997). This treatment also reduced fruit softening and respiration on 'Anna' and 'Granny Smith' as well as enhanced the colour and overall quality of 'Golden Delicious' apples without causing heat injury after several months storage at 0 or 1°C (Fallik et al., 2001; Klein and Lurie, 1990; Porritt and Lidster, 1978; Sams et al., 1993). Hot air treatment changed the structure of the epicuticular wax of 'Golden Delicious' apples, removing the typical wide network of deep cracks, resulting in lower penetration of $CaCl_2$ during dip or pressure infiltration (Roy et al., 1994; Lurie et al., 1996). This should be taken into consideration when using an integrated approach that includes hot air prior to treatment with $CaCl_2$ (Sams et al., 1993; Conway et al., 1999). When wounded 'Red Fuji' and 'Gala' apples were subjected to hot air treatment (38°C, 4d) and subsequent inoculation with *P. expansum*, *B. cinerea* or *C. acutatum*, wound healing occurred, resulting in significant decay reduction on 'Red Fuji' but not on 'Gala' apples (Shao et al., 2010).

4.2 Hot water dips, vapour and rinsing with brushing

The earliest experimentation with heat treatment for controlling postharvest decays of apples was conducted on 'Cox's Orange Pippin' infected with *Gloeosporium* spp. (Burchill, 1964; Edney and Burchill, 1967). Gloeosporium rot from natural infection was reduced to negligible levels by dipping apples in 45°C water for 6 minutes and storing at 2.8°C for 5 months (Burchill, 1964). Higher temperatures caused fruit damage, and temperatures lower than 40°C were ineffective. In subsequent studies, the effectiveness of hot water and steam vapour were compared on 'Cox's Orange Pippin' artificially inoculated with the fungus. Again, 45°C was the most effective temperature in controlling decay without causing fruit injury, and the hot water dip treatment appeared to be slightly more effective than the steam vapour treatment. Delaying the heat treatment for 3–9 days after inoculation increased the effectiveness of the treatments. This delay allowed spores to germinate, making the fungus more susceptible to the heat treatment. Treatment of organically grown 'Topaz' apples with hot water at 53°C for 2 min reduced *Gloeosporium* spp. decay from 56.5% to 2.9% and from 94.4% to 15.7% after 4 and 6 months of storage at 1°C, respectively (Trierweiler et al., 2003). Sensory evaluation of heated and non-heated fruit showed no differences except on the appearance of the apple because of browning of the peel, indicating that further refinement of the treatment is necessary for apples designated for fresh market. Hot water dips at 50°C for 3 min were very effective in reducing decays on 'Elstar' apples inoculated with conidia of *Neonectria galligena* and *P. expansum*, and to a lesser extent on apples inoculated with *B. cinerea* and stored at 2°C for several weeks (Maxin et al., 2012a). Heat treatment, in addition to a direct effect on pathogen viability, also reduced decay on 'Elstar' apples inoculated with *P. expansum* after heat treatment, indicating the activation of a defence response in the fruit. Skin damage on this cultivar begins to occur at 51°C. On 'Ingrid Marie' and 'Pinova' apples, this treatment was also effective against decays caused by natural infection of *N. alba*, *N. perennans*, *M. fuctigena*, *C. acutatum*, *Phacidiopycnis washingtonensis*, *Cladosporium* spp. and *N. galligena* after 3 months of storage at 2°C followed by 18 days at 18°C.

Hot water rinsing has been shown to be effective in protecting other fruits and vegetables from postharvest decays (Fallik, 2004). Fallik et al. (2001) compared hot air (38°C, 4 d) with hot water rinsing and brushing (55°C, 15 sec) and found both treatments to be equally effective in controlling blue mould on naturally infected or artificially inoculated with *P. expansum* 'Golden Delicious' apples stored for 4 months at 1°C followed by 10 days at 20°C. However, hot water rinsing and brushing at 60°C for 20 sec increased decays caused by *P. expansum*, *Mucor* spp., *Phoma exigua* and *P. washingtonensis* on 'Ingrid Marie' and 'Pinova' apples, which was attributed to increased sensitivity of the fruit due to heat damage (Maxin et al., 2012b).

Despite problems with heat damage and the need to establish optimal conditions for individual cultivars, short hot water dipping is an attractive treatment as part of an integrated system because it could be relatively easily adopted for application on packing lines for treatment of fruit before, during and after storage.

5 Controlled atmosphere

The research on controlled atmosphere (CA) storage of apples with reduced O_2 and elevated CO_2 dates back to 1920 with the pioneering work by F. Kidd and C. West in England, followed in the 1930s by R. Smock in the United States (see Olsen, 1986; updated 2010). A fungistatic effect of CA on postharvest pathogens of apple such as *P. expansum* or *B. cinerea* was first observed *in vitro* (Littlefield et al., 1966; Wells and Uota, 1970), and then demonstrated on strawberries where oxygen levels of 0.25% were effective in reducing decay caused *by B. cinerea* but not by *Rhizopus stolonifer* (Couey et al., 1966). As it became apparent that apples stored in CA senesce less and have higher quality than those stored in air, parameters for atmosphere composition in CA

Figure 2 Blue mould development on 'Golden Delicious' apples wound-inoculated with *Penicillium expansum* alone (control treatment) (a) or in combination with a mixture of two antagonists and sodium bicarbonate (2% w/v) (b), and stored in air (trays on the left) or in CA (trays on the right) storage for 4 months at 1°C. Note slightly reduced and lighter in colour decay symptoms on control treatment apples stored in CA in comparison with apples stored in the air. Complete decay control was achieved on apples protected with the antagonist mixture and sodium bicarbonate and stored in CA, while apples with the same treatments but stored in air still exhibited some decay (Conway et al., 2007; Janisiewicz, 2013).

storage were developed for individual cultivars (Drake, 1999; Olsen, 1986). Anecdotal evidence indicated that apples stored in CA had less decay than those stored in air, which agrees with general observations that apples exhibiting a lesser degree of senescence are less susceptible to disease. Later it was shown that CA storage suppresses postharvest diseases by both extending the period of host resistance and by the direct fungistatic effect on the pathogens (Sommer, 1985). Sitton and Patterson (1992) showed conclusively that storing 'McIntosh', 'Golden Delicious' and 'Delicious' apples in CA with CO_2 at 2.8% reduced the development of decay caused by *B. cinerea*, *P. expansum* and *Pezicula malicorticis* after 61 days at 0°C and maintained high fruit quality. Low O_2 had limited fungistatic activity and at levels above 3% appears to favour decay development. Storing organic 'Topaz' apples in CA (1% O_2, 3% CO_2) at 1°C reduced *Gloeosporium* decay from 56.5% to 24.1% and from 94.4% to 41.4% after 4 and 6 months of storage, respectively (Trierweiler, 2003).

Levels of decay control on apples stored in CA have not been adequate to rely on this approach alone for controlling decays in commercial storage. Instead, CA can be used as an additional tool that could be easily combined with other alternative treatments for combating postharvest decays (Conway et al., 2007; Tahir et al., 2009; Fig. 2). To achieve optimal decay control by CA treatment without compromising fruit quality, different parameters such as the cultivar, physiological age of the fruit, susceptibility to physiological disorders such as superficial scald, soft scald, brown core, internal browning, and susceptibility to low O_2 and high CO_2 injuries, as well as prevailing pathogens in the region and length of storage should be considered.

6 Irradiation with UV-C

The application of UV-C irradiation, in addition to having induced resistance activity at low irradiation levels of 7.5 kJ m^{-2} (see section on induced resistance), has been shown to have direct germicidal activity against postharvest pathogens such as *B. cinerea*, *C. gloeosporioides* and *Cryptosporiopsis perennans* (now *Neofabraea perennans*, causing bull's-eye rot) (Bartnicki et al., 2010; Marquenie et al., 2002; Stevens et al., 1996, 2005). The germicidal effect was shown to be greatly enhanced against *B. cinerea* and *C. acutatum* by including a dark period of four hours immediately after irradiation (Janisiewicz et al., 2015, 2016b). It is presumed that this prevents activation of the light-induced repair mechanism for repairing DNA damaged during exposure of the fungus to UV-C. Irradiation of apples with UV-C may serve two purposes, a reduction of food-borne bacterial pathogens and control of fruit decays caused by fungi. Exposing the surface of 'Red Delicious' apples artificially inoculated with *Escherichia coli* O157:H7 to a UV-C dose of 24 mW cm^{-2} reduced populations of this bacterium by 3.3-log units (Yuan et al., 2004). Under commercial conditions, a UV-C dose of 0.0069 kJ m^{-2} reduced bull's-eye rot on artificially inoculated 'Fuji' and 'Gala' apples by 56% and 69%, respectively, and on naturally infected apples by 54% and 85%, respectively (Bartnicki et al., 2011). In the same study, UV-C treatment was as effective as a hot water (50°C) spray for 12 sec. UV-C treatment of apples on sorting lines has been used occasionally in packinghouses before experimentally proven efficacy (Janisiewicz, personal observations); however, with the new information, its use will only increase due to its relatively high efficacy, adaptability to current practices and presumed compatibility with other alternative treatments (Wilson et al., 1997).

7 Natural sources of resistance

7.1 Induction of host defence mechanisms

Plants have evolved complex coordinated mechanisms to defend themselves against fungal, bacterial and insect pathogens. Innovative studies in plant defence have been conducted in the model plant *Arabidopsis thaliana* due to its short life cycle, genetic tractability, and the broad molecular toolkit developed. Over the past 10 years, fundamental scientific breakthroughs in plant innate immunity, induced resistance and preformed defences have been translated from *Arabidopsis* to fruit and vegetable crops with the end goal of achieving sustainable control. The bulk of these investigations have focused on physical and chemical aspects to prime host plant defences and induce resistance. Here we discuss apple fruit defences induced by UV-C irradiation, heat treatment, wound responses, antioxidant signalling and systemic acquired resistance (SAR) inducers (Terry and Joyce, 2004).

Earlier mentioned physical treatments of apple fruit may also induce host defences and have largely entailed the use of low hermetic doses of UV-C and heat treatments (i.e. hot air or water). Hot air treatment at 38°C for 4 days was shown to be cultivar dependent on its impact on wound healing and defence protein activity (Shao et al., 2010). Hot water treatment of apples at 45°C revealed levels of blue mould control from 30 to 100% and was accompanied by the induction of genes involved in heat shock that may also mediate resistance (Spadoni et al., 2015). Low hermetic doses of UV-C reduced disease incidence and severity of bitter rot and *Alternaria* rot on 'Golden Delicious' apples by 70% and 40%, respectively (Stevens et al., 1996, 2005; Wilson et al., 1994). Positioning apples with the stem end towards the irradiation source induced resistance to *C. gloeosporioides* more than those positioned on their side, or with two-side rotation, suggesting the presence and concentration of signal transduction receptors located at the stem end of the fruit (Stevens et al., 2005). While these tactics independently provided varying levels of control they have not been adopted by industry; however, they could be combined with other alternatives to achieve higher levels of control. Both heat and UV-C treatments not only induced resistance in the host but, as mentioned previously, also have biocidal effects on the pathogen. It is possible that the resistance genes induced by heat and UV-C can be activated in commercial apple varieties like 'Gala', 'Fuji' and 'Honeycrisp' using molecular approaches. These inducible defence pathways could be amplified in the absence of the stimulus (i.e. UV-C and heat) to create higher levels of resistance to decay in existing commercial apple varieties.

Treatment of apple fruit with chemical inducers that activate plant defence pathways had been studied mainly using salicylic acid (SA), jasmonate (JA), β-amino butyric acid and ethylene. A study by Quaglia et al. (2011) showed that apple fruit treated with methyl jasmonate, β-amino butyric acid or acibenzolar-*S*-methyl (SA analogue) induced the accumulation of pathogenesis-related (PR) proteins in apple, but was found to be ineffective against *P. expansum* infection. However, the antimicrobial activity of these compounds and their direct effects on fungal growth were demonstrated. It is possible, therefore, that the timing or intensities at which the host response was elicited may be altered to provide a better means of control in the future.

Molecular genetic studies using pathogen-inoculated fruit corroborates findings from physiological investigations using SAR inducers and shows that apple defence mechanisms are complex and multi-factorial. For example, Yin et al. (2013) showed that resistance in apple fruit to *Diplocarpon mali* involves the production of reactive oxygen species (ROS), SA/JA signalling and PR protein expression, which are hallmarks of plant defence. It is clear that more fundamental research in priming host defence responses using chemical elicitors is needed in apple. In-depth investigations combining the use of genomic, metabolomic and proteomic approaches may uncover primary pathways, master regulators or potent chemical inducers of resistance that can lead to defence reactions resulting in higher levels of resistance to postharvest decay.

Apples are hand harvested and stored for extended periods of time, during which fruit may succumb to bruises, fingernail scratches, stem punctures and stem-end pulls that are the main route of infection for major pathogens such as *Penicillum* spp. and *B. cinerea*. Numerous investigations concerning the role of wound responses and physical barriers to prevent decay in apple have been conducted as a result. Lakshininarayana et al. (1987) demonstrated wounds healing in two different apple cultivars, at a pre-climacteric stage of maturity, which resulted in higher levels of phenolic compounds, tannins, lignins and callose. Ultimately, healed wounds were more resistant to *B. cinerea* and *P. expansum* than fresh wounds due to the formation of physical barriers that prevented pathogen establishment. More recent studies by Torres et al. (2003) have shown the important role of hydrogen peroxide (H_2O_2) synthesis and antioxidant enzyme production in resistance to blue mould of immature 'Golden Delicious' apples. Their results indicate that harvest date and fruit maturity play pivotal roles in modulating resistance. Additional studies by Su et al. (2011) and Vilanova et al. (2014) showed that wound-induced activation of H_2O_2 accompanied by antioxidant enzyme activity and lignification decreased with host maturity indicating that host metabolism in apple fruit is an important component of resistance to decay and may be a future target for manipulation in the host to mitigate decay.

7.2 Genetic sources of resistance and their application for postharvest decay control

Resistance to disease and insects has been routinely considered an important component of apple breeding programmes worldwide. However, the focus of these efforts has been towards field production with little attention towards resistance to postharvest decay (Janisiewicz et al., 2008). The genus *Malus* is composed of 27 wild species of which *M. sieversii* Lebed. is considered the progenitor of the domesticated apple (Juniper et al., 1999). Central Asia is the centre of origin for apples, and wild *Malus sieversii* forests are still present in Kazakhstan (Dzhangaleiv, 2003). In the 1990s, the USDA conducted several collecting trips for wild apple germplasm resulting in the Kazak collection in Geneva, New York (Forsline et al., 2003). This wild apple collection includes a wide variety of horticultural traits, represents a wide array of genetic diversity and has been the focus of identifying sources of resistance to physiological disorders (scald, drought), field diseases (fire blight, scab) and postharvest decay (blue mould, bitter rot).

The Kazak wild apple germplasm collection has been evaluated for resistance to two of the most economically important postharvest decay pathogens in the United States, *P. expansum* and *C. acutatum* (Janisiewicz et al., 2008a; Jurick II et al., 2011). These pioneering studies revealed previously undocumented levels of resistance to blue mould and bitter rot, and in some cases, immunity to both diseases. These studies uncovered that resistance to both pathogens is rare and that the distribution of phenotypes with regard to decay tolerance is skewed towards susceptibility. They also showed that accessions resistant against one pathogen (i.e. *P. expansum*) can be susceptible to the other (i.e. *C. acutatum*) and vice versa. Results from these studies have reinforced the utility of the wild apple germplasm collection for the identification of resistance to blue mould and bitter rot, which are currently lacking in commercial apple cultivars (Spotts et al., 1999). Phenotypic data from these studies are being used by breeders to find quantitative trait loci (QTL) associated with resistance to blue mould and bitter rot and to develop DNA-based markers (Norelli, 2013). Such tools can assist breeders in screening crosses to incorporate resistance into new varieties against postharvest decay pathogens to help abate decay (Iwata et al., 2016).

Time course studies of wound responses in wild apple accessions with varying levels of resistance have shown that resistance is positively correlated with the speed of the wound response that prevents the pathogen from infecting the host and causing decay (Janisiewicz et al., 2016a). No decay developed, even from wounds inoculated with pathogen concentration of 10^5 conidia/mL, on most resistant/immune accessions. Histochemical analysis of the wound area revealed that high levels of ROS were detected immediately after wounding in both resistant/immune and highly susceptible accessions, and that callose and lignin/suberin play minor roles in the defence response. These studies indicate that durable resistance in physiologically mature wild apples is governed by several mechanisms. This presents new challenges in explaining the mechanisms of resistance as some of these results, especially with regard to ROS, are contrary to the earlier reports in the literature for cultivated apples. Since it appears that a constitutive resistance/immunity exists in some accessions, profiles of phenolic compounds from these apples may be the first logical step to explore and are currently underway by Janisiewicz and colleagues.

Conducting omics-based studies including comparative genomics, transcriptome sequencing, metabolomics and proteomics to discover molecular regulators, biochemical pathways and DNA-based markers mediating resistance in wild apple fruit to postharvest decay is essential to explain the basis of the observed resistance. They can be followed by comparative/functional genomics to explain the genetic mechanisms and their regulators which can then be transferred to existing commercial cultivars to make them resistant to decay. A two-pronged approach for creating new apple cultivars using molecular markers for disease resistance and genome-editing technologies to modify existing cultivars would result in less decay and increased fruit quality. DNA-based markers for resistance, utilized in marker-assisted breeding, could be used by breeders to make new apple cultivars with superior disease resistance. Genome-editing technologies involving CRISPR/Cas9 (Hsu et al., 2014; Ledford, 2015; van der Oost et al., 2009) may be used to manipulate molecular regulators and edit promoters of apple fruit defence genes to enhance decay resistance in popular apple cultivars. With forthcoming fundamental developments in determining mechanisms and regulators of host resistance, application of these breakthroughs will give rise to new possibilities for controlling postharvest decays, and are limited only by the scope of scientific creativity, imagination and resources.

8 Biological control

Biological control of postharvest diseases of temperate, subtropical and tropical fruits has been the most widely studied alternative to synthetic fungicides during the past three decades. Research conducted in more than two dozen countries resulted in several commercial products originally developed mostly for apples such as Bio-Save® (a.i. *Pseudomonas syringae*), Boni Protect (a.i. *Aureobasidium pullulans*), Candifruit (a.i. *Candida sake*), Nexy (a.i. *Candida oleophila*), Pantovital (a.i. *Pantoea agglomerans*), and Yield Plus (a.i. *Cryptococcus albidus*) and for citrus fruits, Aspire (a.i. *Candida oleophila*) (Droby et al., 1998; EFSA, 2012; Janisiewicz and Jeffers, 1997; Jijakli et al., 1993; Leibinger et al., 1997; Nunes et al., 2002; Usall et al., 2001). As more information is accumulated on the efficacy of some of these products the registration has expanded to other fruits and vegetables.

Bio-Save® is the only biocontrol product currently being used for control of postharvest decays on apples and pears in the United States, and is a good example for a case study (Janisiewicz and Jeffers, 1997; Janisiewicz and Marchi, 1992). Aspire™, based on *Candida oleophila*, is the other registered product; however, it was taken off the market three years after its introduction in 1995. Bio-Save® use has been increasing since its large-scale introduction in 1995, and the original registration for postharvest use on apples, pears and citrus fruit has been expanded to cherries, potato and sweet potato, with registrations for additional commodities pending (http://jetharvest.com/; Janisiewicz, personal information). This product has been very effective in controlling blue mould of apples (Janisiewicz and Jeffers, 1997). In three out of four large-scale trials, on wounded and artificially inoculated with *P. expansum* conidia 'Golden Delicious' and 'Delicious' apples, it reduced the incidence of blue mould decay to below 4% compared with controls where decay incidence ranged from 40% to 92% (Janisiewicz and Jeffers, 1997). The reduction did not differ from standard 0.05% thiabendazole (TBZ) treatment. In the fourth test (conducted as a first in the series), the decay incidence on Bio-Save® treated apples was 29% and 96% on control. In all four trials, control by an unformulated *P. syringae* antagonist suspension was no different from TBZ treatment. Selection of this antagonist is a result of a massive

Figure 3 An example of the enhancement of blue mould biocontrol on 'Golden Delicious' apples with an antagonist mixture. Apples were wound-inoculated with *Penicillium expansum* and treated with either suspensions of individual yeast antagonists (ST1-A24 or ST4-E14, upper boxes) or a mixture of these antagonists (ST1-A24 x ST4-E14). The antagonists were mixed in equal proportions and the total biomass of the mixture was equal to the biomass in treatment with an individual antagonist suspension.

screening process of microorganisms isolated from apple and pear fruit that lead to finding several very effective bacterial and yeast antagonists (Janisiewicz et al., 1987; Janisiewicz and Roitman, 1988). Currently there are more than two dozen programmes worldwide working on the development of potential antagonists for biological control of postharvest decays on various fruits primarily using bacteria and yeasts isolated from the fruit surfaces (Janisiewicz and Korsten, 2002; Mari et al., 2014). Selecting a screening method in a search for the antagonists may predetermine the mechanism of biocontrol by which an antagonist controls the pathogen. For example, *in vitro* dual culture plate method, which selects an antagonist based on its ability to inhibit growth of pathogen mycelium on agar media, selects for antagonists producing inhibitory compounds such as antibiotics, siderophores (compounds sequestering iron) or lytic enzymes (e.g. chitinase) (Calvente et al., 1999; Chen et al., 2016; Di Francesco et al., 2016; Janisiewicz and Roitman, 1988). A membrane method may be used to select antagonists colonizing different fungal structures including appressoria, which may be important for fungi such as *Colletotrichum* spp. or *Monilinia* spp. that have a latent infection stage on the fruit (Janisiewicz et al., 2011). A direct screening on fruit, although much more laborious, results in the selection of antagonists with potentially different mechanisms of biocontrol, which, in addition to those previously mentioned, may include competition for limited nutrients and space, direct parasitism, induced resistance or interference with pathogenicity factors (Di Francesco et al., 2016; Manso and Nunes, 2011; Zhang et al., 2016). Thus, it is important to use more than one method to increase the pool of antagonists with different mechanisms of biocontrol. Some antagonists may operate by more than one mechanism. Antagonists producing antibiotics are often very effective but may be difficult to commercialize because of potential problems with registration. However, they may lead to development of effective fungicides; for example, *Pseudomonas cepacia* (now *Burkholderia cepacia*) has been shown to be a very effective antagonist against blue mould and grey mould on apples and pears, grey mould on strawberries, brown rot on peaches and nectarines and blue mould on citrus in postharvest applications (Janisiewicz and Roitman, 1988; Smilanick et al., 1993). It produces an antibiotic compound identified as pyrrolnitrin (Roitman et al., 1990), which alone was also very effective in controlling blue mould and grey mould of apples, grey mould of strawberries and raspberries, blue mould on lemons, and *B. cinerea* on rose flowers (Janisiewicz et al., 1991; Takeda et al., 1990; Goulart et al., 1992; Hammer et al., 1993; Janisiewicz and Roitman, 1990). Several analogues of this compound have been produced and evaluated for disease control activity, and one of them, fludioxonil, is currently one of the most widely used postharvest fungicides against fruit decays worldwide.

For biocontrol agents to succeed in controlling decays originating from wound infections they must be good wound colonizers, the main court of entry for major apple pathogens. This may have additional benefits by protecting wounds against potential food-borne pathogens such as *E. coli* O157:H7 which can grow exponentially in apple wounds (Janisiewicz et al., 1999a). *P. syringae* (used in Bio-Save®) prevented *E. coli* from increasing its populations in apple wounds (Janisiewicz et al., 1999b). Biocontrol effectiveness can be increased by using mixtures of mutually compatible antagonists (Janisiewicz and Bors, 1995; Fig. 3), adding nutrients that are more beneficial to antagonists than to the pathogen they control (Janisiewicz et al., 1992), or growing antagonists in media that promote their survival and mechanism of biocontrol (Abadias et al., 2001; Zhang et al., 2014; Zeng et al., 2015).

Many different yeasts isolated from various habitats, in addition to those mentioned previously, have been evaluated for biocontrol activity against postharvest decays of apple. Unfortunately, after initial reports, few have been developed for commercial use. There may be several reasons for this, including poor performance of a biocontrol agent in more stringent tests, poor prospects for mass production and/or formulation of the biocontrol agent, concerns about human safety, lack of long-term commitment by the investigators and the reluctance of granting agencies to fund such long-term projects. Judging from the products on the market, the development of biological control against postharvest decays of fruit takes approximately ten years from inception to commercial production (Janisiewicz, 1998).

Large-scale commercial application of the biological control against postharvest disease of apples revealed its limitations. Although in many cases biocontrol can be as effective as fungicide treatments, it has inherent limitations such as lack of eradicative activity, more rapid decline in performance on ripe fruit and sensitivity to postharvest practices including high temperatures in drying tunnels, incompatibility with antioxidants (DPA, ethoxyquin) or chlorine residues on fruit handled in chlorinated water before storage (e.g. pre-sizing). Thus, biological control can be applied as the stand-alone treatment only when precautions are taken considering these limitations; otherwise it should be used as the main treatment in integrated control with other alternatives to synthetic fungicides described in previous sections (Di Francesco and Mari, 2014; Janisiewicz, 2013; Janisiewicz and Conway, 2010; Mari et al., 2014).

9 Integrated control

There is increasing agreement among investigators working with control of postharvest decays of fruits that replacing synthetic fungicides with a single alternative treatment may provide adequate control but only under some specific circumstances; in most situations it will be inadequate because of the limitations of the alternative treatments. Much of the recent research has been focused on developing integrated control that either encompasses biological control as the pillar in combination with other compatible treatment(s), or is just a simple combination of alternative treatments (see reviews by Di Francesco and Mari, 2014; Janisiewicz and Conway, 2010; Palou et al., 2016; Romanazzi et al., 2016; Usall et al., 2015). This approach is well represented by the hurdles concept

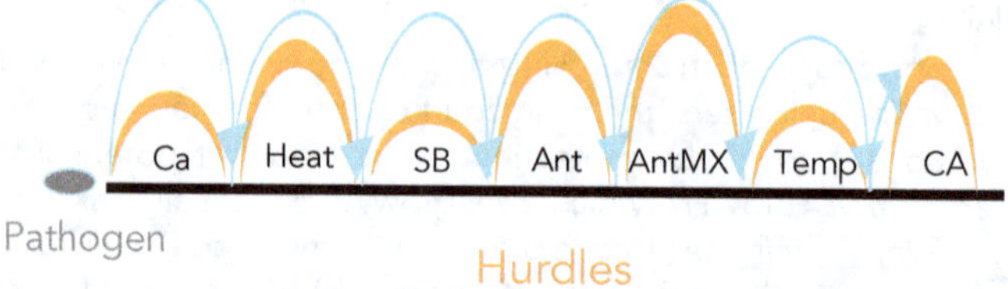

Ca = calcium, Heat = 38C for 4 d, SB = sodium bicarbonate, Ant= antagonist, AntMX = antagonist mixture; Temp = low storage temperature, CA = controlled atmosphere storage

Figure 4 Hurdle concept for controlling postharvest diseases on pome fruits. With each additional treatment (hurdle) reducing fruit decay by a certain amount, an additive or even synergistic effect can be achieved resulting in complete control (Janisiewicz, 2008).

developed for apples (Janisiewicz, 2008, 2013; Janisiewicz et al., 2008b; Fig. 4), which follows the original idea developed for food preservation (Leistner, 2000). Here, each additional treatment reduces the incidence/severity of the decay by a certain percentage, which may be additive or synergistic, and eventually results in the pathogen not being able to overcome the final hurdle, resulting in control of the fruit decay. Figure 4 represents apple treatments described in earlier sections, but the combination of treatments can vary as long as the treatments are mutually compatible and do not cause fruit damage (Di Francesco and Mari, 2014; Errampalli and Brubacher, 2006; Janisiewicz and Conway, 2010; Palou et al., 2016).

10 Future trends and conclusions

The need for sustainable and safe protection methods against postharvest decays of apples, and a rapidly growing organic market require the development of alternative approaches to synthetic fungicides. Advances made during the past two decades indicate that we are on the verge of developing new approaches that require integration of several alternative methods for apple decay control under a broad range of conditions. The extent of implementation of these methods will depend on the market forces as consumer demand for apples free of synthetic pesticides is growing rapidly and the supply of organic apples is often inadequate. As some of the alternatives, such as biological control, have been used in commercial packinghouses to satisfy the most urgent needs, transition to an integrated approach with a wider spectrum of activity and less limitations is expected to occur in the near future. The discovery of resistance to postharvest apple decays among the wild apple germplasm, for the first time, gives hope for the ultimate, genetic solution to achieve long-lasting durable resistance. This approach, however, will take a considerable amount of time and will require the long-term commitment of researchers and supporting institutions.

11 Where to look for further information

Much of the general information about controlling postharvest diseases of apples can be found on the websites of university extension services from major apple growing regions in the United States such as Washington State (http://postharvest.tfrec.wsu.edu), New York (http://www.scaffolds.entomology.cornell.edu), Pennsylvania (http://extension.psu.edu/plants/tree-fruit/diseases/apple-diseases) and Oregon (http://extension.oregonstate.edu/yamhill/search/postharvest%252Bapple/0/4). The second edition of Compendium of Apple and Pear Diseases and Pests, published by The American Phytopathological Society, includes postharvest diseases and their control. Most of the above sources provide information on control of postharvest apple decays that is focused on the use of fungicides. Agricultural Handbook No. 376, *Market Diseases of Apples, Pears and Quinces*, published by U.S. Government Printing Office in 1971, describes major postharvest diseases, except those described more recently, and also includes information on other decay control approaches including sanitation and cultural practices. The Research Institute of Organic Agriculture in Switzerland (FiBL) website (http://www.fibl.org/en/homepage.htm) provides in-depth information on alternatives to synthetic fungicides, and various educational institutions such as the University of Hohenheim in Stuttgart (https://www.uni-hohenheim.de/eur-organic), University of Göttingen in collaboration with the University of Kassel-Witzenhausen in Göttingen (https://www.uni-goettingen.de/en/100751.html) and EUR-Organic consortium (http://www.eur-organic.eu/) offer individual courses and graduate degrees in organic agriculture. The latter also includes the websites of various educational and research organizations focused on sustainable agricultural practices.

12 References

Abadias, M., Teixidó, N., Usall, J., Benabarre, A. and Viñas, I. (2001), 'Viability, efficacy, and storage stability of freeze-dried biocontrol agent *Candida sake* using different protective and rehydration media', *J. Food Prot.*, 64(6), 856–61.

Amiri, A., Dugas, R., Pichot, A. L. and Bompeix, G. (2008), '*In vitro* and *in vitro* activity of eugenol oil *(Eugenia caryophylata)* against four important postharvest apple pathogens', *Int. J. Food Microbiol.*, 126(1), 13–19.

Andrews, J. H. and Kenerley, C. M. (1978), 'The effect of pesticide program on non-target epiphytic microbial populations of apple leaves', *Can. J. Microbiol.*, 24(9), 1058–72.

Bangerth, F., Dilley, D. R. and Dewey, D. H. (1972), 'Effect of postharvest calcium treatments on internal breakdown and respiration of apple fruits', *Amer Soc Hort Sci J.*, 97, 679–82.

Baraldi, E., Mari, M., Chierici, E., Pondrelli, M., Bertolini, P. and Pratella, G. C. (2003), 'Studies on thiabendazole resistance of *Penicillium expansum* of pears: pathogenic fitness and genetic characterization', *Plant Pathol.*, 52(3), 362–70.

Barger, W. R. (1928), 'Sodium bicarbonate as citrus fruit disinfectant', *Calif. Citrogr.*, 13, 164–74.

Bartnicki, V. A., Valdebenito-Sanhueza, R. M., do Amarante, C. V. T., de Castro, L. A. S., Rizzati, M. R., and de Souza, J. A. V. (2010), 'Água aquecida e radiação UV-C no controle pós-colheita de Cryptosporiopsis perennans em maçãs', *Pesquisa Agropecuária Brasileira*, 45(2), 124–31.

Bartnicki, V. A., Valdebenito-Sanhueza, R. M., Do Amarante, C. V. T. and Steffens, C. A. (2011), 'Tratamentos hidrotérmico e com radiação UV-C no controle pós-colheita da podridão olho-de-boi em uma linha comercial de seleção de maçãs', *Rev. Bras. Frut.*, (33), 737–45.

Batzer, J. C., Gleason, M. L., Weldon, B., Dixon, P. M. and Nutter Jr., F. W. (2002), 'Evaluation of postharvest removal of sooty blotch and flyspeck on apples using sodium hypochlorite, hydrogen peroxide with peroxyacetic acid, and soap', *Plant Dis.*, 86(12), 1325–32.

Bautista-Baños, S., Hernandez-Lauzardo, A. N., Velazquez-Del Valle, M. G., Hernández-López, M., Barka, E. A., Bosquez-Molina, E. and Wilson, C. L. (2006), 'Chitosan as a potential natural compound to control pre and postharvest diseases of horticultural commodities', *Crop Protec.*, 25(2), 108–18.

Bazarova, V. I. (1982), 'Use of ozone in storage of apples', *Food Sci. Tech. Abstracts.*, 4(11), J1653.

Bertolini, P. and Pratella, G. C. (2008), 'Novel approaches for the control of postharvest diseases and disorders', *COST Action 924, Proc. International Congress.* 3–5 May 2007, Lo Scarabeo, Bologna, Italy.

Bertrand, P. F. and Saulie-Carter, J. L. (1978), 'The occurrence of benomyl-tolerant strains of *Penicillium expansum* and *Botrytis cinerea* in the Mid-Columbia region of Oregon and Washington', *Plant Dis. Reporter.*, 62, 302–5.

Bondoux, P. (1967), 'Les maladies cryptogamiques des poires et des pommes au cours de l'entreposage', *Ann. Inst. Nat. Rech. Agron., Paris, Sér. C (Annls Epiphyt.)*, 18, 509–50.

Burchill, R. T. (1964), 'Hot water as a possible post-harvest control of Gloeosporium rots of stored apples', *Plant Pathol.*, 13(3), 106–7.

Burt, S. (2004), 'Essential oils: their antibacterial properties and potential applications in foods – a review', *Int. J. Food Microbiol.*, 94(3), 223–53.

Calvente, V., Benuzzi, D. and de Tosetti, M. I. S. (1999), 'Antagonistic action of siderophores from *Rhodotorula glutinis* upon the postharvest pathogen *Penicillum expansum*', *Int. Biodeter. Biodegr.*, 43(4), 167–72.

Cappelini, R. A., Ceponis, M. J. and Lightner, G. W. (1987), 'Disorders in apple and pear shipments to the New York market, 1972–1984', *Plant Dis.*, 71, 852–6.

Chen, X., Zhang, Y., Fu, X., Li, Y. and Wang, Q. (2016), 'Isolation and characterization of *Bacillus amyloliquefaciens* PG12 for the biological control of apple ring rot', *Postharvest Biol. Technol.*, 115, 113–21.

Choi, W. Y., Park, H. J., Ahn, D. J., Lee, J. and Lee, C. Y. (2002), 'Wettability of chitosan coating solution on Fuji apple skin', *J. Food Sci.*, 67(7), 2668–72.

Conway, W. S. (1982), 'Effect of postharvest calcium treatment on decay of Delicious apples', *Plant Dis.*, 66, 402–3.

Conway, W. S., Gross, K. C., Boyer, C. D. and Sams, C. E. (1988), 'Inhibition of *Penicillium expansum* polygalacturonase activity by increased apple cell wall calcium', *Phytopathology* 78, 1052–5.

Conway, W. S., Janisiewicz, W. J., Klain, J. D. and Sams, C. E. (1999), 'Strategy for combining heat treatment, calcium infiltration and biological control to reduce postharvest decay of 'Gala' apples', *HortSci.*, 34(4), 700–4.

Conway, W. S., Janisiewicz, W. J., Leverentz, B., Saftner, R. A. and Camp, M. J. (2007), 'Control of blue mold of apple by combining controlled atmosphere, antagonist mixture, and sodium bicarbonate', *Postharvest Biol. Technol.*, 45(3), 326–32.

Conway, W. S., Leverentz, B., Janisiewicz, W. J., Blodgett, A. B., Saftner, R. A. and Camp, M. J. (2004), 'Integrating heat treatment, biocontrol and sodium bicarbonate to reduce postharvest decay of apple caused by *Colletotrichum acutatum* and *Penicillium expansum*', *Postharvest Biol. Technol.*, 34(1), 11–20.

Conway, W. S., Leverentz, B., Janisiewicz, W. J., Saftner, R. A. and Camp, M. J. (2005), 'Improving biocontrol using antagonist mixtures with heat and/or sodium bicarbonate to control postharvest decay of apple fruit', *Postharvest Biol. Technol.*, 36(3), 235–44.

Conway, W. S., Sams, C. E., Abbott, J. A. and Bruton, B. D. (1991), 'Postharvest calcium treatment of apple fruit to provide broad-spectrum protection against postharvest pathogens', *Plant Dis.*, 75(6), 620–2.

Conway, W. S., Sams, C. E. and Hickey, K. D. (2002), 'Pre- and postharvest calcium treatment of apple fruit and its effect on quality', *Acta Hort.*, 594, 413–19.

Couey, H. M., Follstad, M. N. and Uota, M. (1966), 'Low-oxygen atmospheres for control of postharvest decay of fresh strawberries', *Phytopathology* 56(12), 1339.

Culpepper, C. W. and Caldwell, J. S. (1927), 'The relation of atmospheric humidity to the deterioration of evaporated apples in storage', *J. Agric. Res.*, 35(10), 889–906.

Daniel, C. K., Lennox, C. and Vries, F. A. (2015), '*In vivo* application of garlic extract in combination with clove oil to prevent postharvest decay caused by *Botrytis cinerea*, *Penicillium expansum* and *Neofabraea alba* on apples', *Postharvest Biol. Technol.*, 99, 88–92.

de Capdeville, G., Wilson, C. L., Beer, S. V. and Aist, J. R. (2002), 'Alternative disease control agents induce resistance to blue mold in harvested 'Red Delicious' apple fruit', *Phytopathology*, 92(8), 900–8.

Deans, S. G. and Ritchie, G. (1987), 'Antibacterial properties of plant essential oils', *Int. J. Food Microbiol.*, 5(2), 165–80.

Delaquis, P. J. and Sholberg, P. L. (1997), 'Antimicrobial activity of gaseous allyl isothiocyanate', *J. Food Protec.*, 60(8), 943–7.

Dey, E. S., Ahmadi-Afzadi, M., Nybom, H. and Tahir, I. (2013), 'Alkylresorcinols isolated from rye bran by supercritical fluid of carbon dioxide and suspended in a food-grade emulsion show activity against *Penicillium expansum* on apples', *Arch. Phytopathol. Plant Protec.*, 46(1), 105–19.

Di Francesco, A. and Mari, M. (2014), 'Use of biocontrol agents in combination with physical and chemical treatments: efficacy assessment', *Stewart Postharvest Rev.*, 10(1), 1–4.

Di Francesco, A., Martini, C. and Mari, M. (2016), 'Biological control of postharvest diseases by microbial antagonists: how many mechanism of action?', *Eur. J. Plant Pathol.*, 1–7. DOI: 10.1007/s10658-016-0867-0.

Drake, S. (1999), 'Short-Term Controlled Atmosphere Storage of Apples', WSU-TFREC/Postharvest information network. http://postharvest.tfrec.wsu.edu/pages/PC99B (Accessed 29 October 2010).

Droby, S., Cohen, L., Daus, A., Weiss, B., Horev, B., Chalutz, E., Katz, H., Jeren-Tzur, M. and Shachnai, A. (1998), 'Commercial testing of Aspire: a yeast preparation for the biological control of postharvest decay of citrus', *Biol. Control*, 12(2), 97–101.

Droby, S., Prusky, D., Jacoby, B. and Goldman, A. (1987), 'Induction of antifungal resorcinols in flesh of unripe mango fruits and its relation to latent infection by *Alternaria alternata*', *Physiol. Molec. Plant Pathol.*, 30(2), 285–92.

Droby, S., Wisniewski, M., El Ghaouth, A. and Wilson, C. (2003), 'Influence of food additives on the control of postharvest rots of apple and peach and efficacy of the yeast-based biocontrol product Aspire', *Postharvest Biol. Technol.*, 27(2), 127–35.

Dzhangaliev, A. D. (2003), 'The wild apple tree of Kazahkstan', *Hort. Rev.*, 29, 263–303.

Eckert, J. W. and Kolbezen, M. J. (1964), '2-Aminobutane salts for control of post-harvest decay of citrus, apple, pear, peach, and banana fruits', *Phytopathology*, 54(8), 978.

Eckert, J. W. and Ogawa, J. M. (1988), 'The chemical control of postharvest diseases: deciduous fruits, berries, vegetables and root/tuber crops', *Ann. Rev. Phytopathol.*, 26(1), 433–69.

Edney, K. L. and Burchill, R. T. (1967), 'The use of heat to control the rotting of 'Cox Orange Pippin' apples by *Gloeosporium* spp.', *Ann. Appl. Biol.*, 59(3), 389–400.

El Ghaouth, A., Smilanick, J. L., Brown, G. E., Wisniewski, M. and Wilson, C. L. (2000a), 'Application of *Candida saitoana* and glycolchitosan for the control of postharvest diseases of apple and citrus fruit under semi-commercial conditions', *Plant Dis.*, 84, 243–8.

El Ghaouth, A., Smilanick, J. L. and Wilson, C. L. (2000b), 'Enhancement of the performance of *Candida saitoana* by the addition of glycolchitosan for the control of postharvest decay of apple and citrus fruit', *Postharvest Biol. Technol.*, 19(1), 103–10.

English, H. (1948), 'Disinfectant washes for control of decay in apples and pears', *Phytopathology*, 38(11), 914.

English, H., Wright, T. R. and Smith, E. (1948), 'Chemicals cut storage losses', *Better Fruit*, 43 (2), 9–11.

Errampalli, D. and Brubacher, N. R. (2006), 'Biological and integrated control of postharvest blue mold (*Penicillium expansum*) of apples by *Pseudomonas syringae* and cyprodinil', *Biol. Control*, 36(1), 49–56.

European Food Safety Authority (2012), 'Conclusion on the peer review of the pesticide risk assessment of the active substance *Candida oleophila* strain O', *EFSA J.*, 10(11), 2944 [27 p.] doi:10.2903/j.efsa.2012.2944. Available www.efsa.europa.eu/efsajournal.

Eustace, H. J. (1908), 'Investigation on some fruit diseases', *N.Y. Bulletin*, 297, New York, Agr. Exp. Sta. Geneva.

Fadda, A., Barberis, A., D'Aquino, S., Palma, A., Angioni, A., Lai, F. and Schirra, M. (2015), 'Residue levels and performance of potassium sorbate and thiabendazole and their co-application against blue mold of apples when applied as water dip treatments at 20 or 53°C', *Postharvest Biol. Technol.*, 106, 33–43.

Fallik, E. (2004), 'Prestorage hot water treatments (immersion, rinsing and brushing)', *Postharvest Biol. Technol.*, 32(2), 125–34.

Fallik, E., Grinberg, S., Gambourg, M., Klein, J. D. and Lurie, S. (1995), 'Prestorage heat treatment reduces pathogenicity of *Penicillium expansum* in apple fruit', *Plant Pathol.*, 45(1), 92–7.

Fallik, E., Tuvia-Alkalai, S., Feng, X. and Lurie, S. (2001), 'Ripening characterisation and decay development of stored apples after a short pre-storage hot water rinsing and brushing', *Inn. Food Sci. Emer. Technol.*, 2(2), 127–32.

Fan, L., Song, J., Beaudry, R. M. and Hildebrand, P. D. (2006), 'Effect of hexanal vapor on spore viability of *Penicillium expansum*, lesion development on whole apples and fruit volatile biosynthesis', *J. Food Sci.*, 71(3), M105–M109.

Forsline, P. L., Aldwinkle, H. S., Dickson, E. E., Luby, J. J. and Hokanson, S. C. (2003), 'Collection, maintenance, characterization, and utilization of wild apples of Central Asia', *Hort. Rev.*, 29, 1–61.

Goulart, B. L., Hammer, P. E., Evensen, K. B., Janisiewicz, W. and Takeda, F. (1992), 'Pyrrolnitrin, captan + benomyl, and high CO_2 enhance raspberry shelf life at 0 or 18C', *J. Amer. Soc. Hort. Sci.*, 117(2), 265–70.

Grant, G. T., Morris, E. R., Rees, D. A., Smith, P. J. and Thom, D. (1973), 'Biological interactions between polysaccharides and divalent cations: the egg-box model', *FEBS Letters*, 32(1), 195–8.

Hammer, P. E., Evensen, K. B. and Janisiewicz, W. J. (1993), 'Postharvest control of *Botrytis cinerea* on cut rose flowers with pyrrolnitrin', *Plant Dis.*, 77(3), 283–6.

Hansen, J., Xiao, Ch-L and Kupferman, G. (2010), 'Bin sanitation: an effective way to reduce codling moth and fungal decay spores', WSU- Postharvest Information Network, http://postharvest.tfrec.wsu.edu/REP2010B.pdf.

Hardenburg, R. E. and Spalding, D. H. (1972), 'Postharvest benomyl and thiabendazole treatments, alone and with scald inhibitors, to control blue and gray mold in wounded apples', *Amer. Soc. Hort. Sci. J.*, 97, 154–8.

Harris, D. (2013), 'Organics at the crossroads: trends in organic retailing in the United States', In, D. Granatstein, P. K., Andrews, S. D., Bishop and W. Janisiewicz (Eds), *Proceedings of the International Organic Fruit Research Symposium 'Organic Fruit 2012: From Research to Practice'*, held on 18–21 June in Leavenworth, WA, *Acta Horticulturae* 1001, 377–82.

Harvey, J. M. (1978), 'Reduction of losses in fresh market fruits and vegetables', *Ann. Rev. Phytopathol.*, 16(1), 321–41.

Hassan, M. K., Dann, E. K., Irving, D. E. and Coates, L. M. (2007), 'Concentration of constitutive alk(en)ylresorcinol in peel of commercial mango varieties and resistance to postharvest anthracnose', *Physiol. Mol. Plant Pathol.*, 71(4), 158–65.

Horovitz, S. and Cantalejo, M. J. (2014), 'Application of ozone for postharvest treatments of fruits and vegetables', *Crit. Rev. Food Sci. Nutr.*, 54, 312–39.

Hsu, P. D., Lander, E. S. and Zhang, F. (2014), 'Development and application of CRISPR-Cas9 for genome engineering', *Cell*, 157, 1262–78.

Iwata, H., Minamikawa, M. F., Kajiya-Kenegae, H., Ishimori, M. and Hayashi, T. (2016), 'Genomics-assisted breeding in fruit trees', *Breeding Sci.*, 66(1), 100–15.

Janisiewicz, W. J. (1987), 'Postharvest biological control of blue mold on apples', *Phytopathology*, 77(3), 481–5.

Janisiewicz, W. J. (1998), 'Biological control of postharvest diseases of pome fruits: from inception to commercial product', In *COST 914 – COST 915 Joint Workshop – Non Conventional Methods for the Control of Postharvest Disease and Microbial Spoilage*, Publications of the European Communities, Luxembourg, pp. 71–6.

Janisiewicz, W. J. (2008), 'Commercial applications and future prospects for the use of biological control after harvest', In P. Bertolini (Ed.), *Proceedings of the COST 924 International Congress on 'Novel Approaches for the Control of Postharvest Diseases and Disorders'*, Bologna, Italy, 3–5 May 2007, pp. 9–18.

Janisiewicz, W. J. (2013), 'Biological Control of Postharvest Diseases: Hurdles, Successes and Prospects', In D. Granatstein, P. K., Andrews, S. D., Bishop and W. Janisiewicz (Eds), *Proceedings of the International Organic Fruit Research Symposium 'Organic Fruit 2012: From Research to Practice'*, held on 18–21 June in Leavenworth, WA, *Acta Horticulturae*, 1001, 273–84.

Janisiewicz, W. J. and Conway, W. S. (2010), 'Combining biological control with physical and chemical treatments to control fruit decays after harvest', *Stewart Postharvest Rev.*, 9(1), 1–16.

Janisiewicz, W. J., Conway, W. S., Brown, M. W., Sapers, G. M., Fratamico, P. and Buchanan, R. L. (1999a), 'Fate of *Escherichia coli* O157:H7 on fresh-cut apple tissue and its potential for transmission by fruit flies', *Appl. Environ. Microbiol.*, 65(1), 1–5.

Janisiewicz, W. J., Conway, W. S. and Leverentz, B. (1999b), 'Biological control of apple decays of apple can prevent growth of *Escherichia coli* O157:H7 in apple wounds', *J. Food Protec.*, 62(12), 1372–5.

Janisiewicz, W. J. and Jeffers, S. N. (1997), 'Efficacy of commercial formulations of two biofungicides for control of blue mold and gray mold of apples in cold storage', *Crop Protec.*, 16(7), 629–33.

Janisiewicz, W. J. and Korsten, L. (2002), 'Biological control of postharvest diseases of fruits', *Ann. Rev. Phytopathol.*, 40(1), 411–41.

Janisiewicz, W. J. and Marchi, A. (1992), 'Control of storage rots on various pear cultivars with saprophytic strain of *Pseudomonas syringae*', *Plant Dis.*, 76(6), 555–60.

Janisiewicz, W. J., Nichols, B., Bauchan, G., Chao, C. T. and Jurick II, W. M. (2016a), 'Wound responses of wild apples suggest multiple resistance mechanism against blue mold decay', *Postharvest Biol. Technol., 117*, 132–40.

Janisiewicz, W. J., Peterson, D. L., Yoder, K. S. and Miller, S. S. (2005), 'Experimental bin drenching system for testing biocontrol agents for control of postharvest decay of apples', *Plant Dis.*, 89(5), 487–90.

Janisiewicz, W. J., Pimenta, R. S. and Jurick II, W. M. (2011), 'A novel method for selecting antagonists against postharvest fruit decays originating from latent infections', *Biol. Control*, 59(3), 384–9.

Janisiewicz, W. J. and Roitman, J. (1988), 'Biological control of blue mold and gray mold on apple and pear with *Pseudomonas cepacia*', *Phytopathology*, 78(12), 1697–700.

Janisiewicz, W. J. and Roitman, J. (1990), 'Biological control of postharvest roots in fruits using *Pseudomonas cepacia* and pyrrolnitrin produced therefrom', United States Patent No: 4,975,277.

Janisiewicz, W. J., Saftner, R. A., Conway, W. S. and Forsline, P. L. (2008a), 'Preliminary evaluation of apple germplasm from Kazakhstan for resistance to postharvest blue mold in fruit caused by *Penicillium expansum*', *HortSci.*, 43, 420–6.

Janisiewicz, W. J., Saftner, R. A., Conway, W. S. and Yoder, K. S. (2008b), 'Control of blue mold decay of apple during commercial controlled atmosphere storage with yeast antagonists and sodium bicarbonate', *Postharvest Biol. Technol.*, 49(3), 374–8.

Janisiewicz, W. J., Takeda, F., Glenn, D. M., Camp, M. J. and Jurick II, W. M. (2016b), 'Dark period following UV-C treatment enhances killing of *Botrytis cinerea* conidia and controls gray mold of strawberries', *Phytopathology*, 106, 386–94.

Janisiewicz, W. J., Takeda, F., Jurick II, W., Nichols, B., Wolford, S. and Glenn, D. M. (2015), 'A novel approach to control gray mold, anthracnose, and powdery mildew on strawberry using low-dose UV-C irradiation', *Phytopathology*, 105(Suppl.), S4.64 (Abstract).

Janisiewicz, W. J., Usall, J. and Bors, R. (1992a), 'Nutritional enhancement of biocontrol of blue–mold on apples', *Phytopathology*, 82 (11), 1364–70.

Janisiewicz, W., Yourman, L., Roitman, J. and Mahoney, N. (1991), 'Postharvest control of blue mold and gray mold of apples and pears by dip treatment with pyrrolnitrin, a metabolite of *Pseudomonas cepacia*', *Plant Dis.*, 75(5), 490–4.

Jijakli, M. H., Lepoivre, P., Tossut, P. and Thonart, P. (1993), 'Biological control of *Botrytis cinerea* and *Penicillium* spp. on postharvest apples by two antagonistic yeasts', *Meded. Fac. Landbouwwet. Univ. Gent.*, 58, 1349–58.

Johnson, G. I., Highley, E. and Joyce, D. C. (1998), 'Disease Resistance in fruit', *Proceeding of an International Workshop*, held at Chiang Mai, Thailand, 18–21 May 1997, *ACIAR Proceedings* No 80, p. 233.

Juniper, B. E., Watkins, R. and Harris, S. H. (1999), 'The origin of the apple', *Acta Hortic.*, 484, 27–33.

Jurick II, W. M., Janisiewicz, W. J., Saftner, R. A., Vico, I., Gaskins, V. L., Park, E., Forsline, P. L., Fazio, G. and Conway, W. S. (2011), 'Identification of wild apple germplasm (*Malus* spp.) accessions with resistance to the postharvest decay pathogens *Penicillium expansum* and *Colletotrichum acutatum*', *Plant Breed.*, 130(4), 481–6.

Kader, A. A. (Ed.) (1992), *Postharvest Technology of Horticultural Crops*. Publication 3311, University of California, Div. Agric. Natur. Rec., p. 535.

Kennedy, I. R., Skerritti, J. H., Johnson, G. I. and Highley, E. (1998), 'Seeking Agricultural Produce Free of Pesticide Residues', *Proceedings of an International Workshop*, held in Yogyakarta, Indonesia, 17–19 February 1998, *ACIAR Proceedings* No 85, p. 406.

Kidd, M. N. and Beaumont, A. (1924), 'Apple rot fungi in storage', *Trans. Brit. Mycol. Soc.*, 10(1), 98–120.

Kim, Y. K. and Xiao, C. L. (2008), 'Distribution and incidence of Sphaeropsis rot in apple in Washington State', *Plant Dis.*, 92(6), 940–6.

Kim, Y. K. and Xiao, C. L. (2010), 'Resistance to pyraclostrobin and boscalid in populations of *Botrytis cinerea* from stored apples in Washington State', *Plant Dis.*, 94(5), 604–12.

Klein, J. D., Conway, W. S., Whitaker, B. D. and Sams, C. E. (1997), '*Botrytis cinerea* decay in apples is inhibited by postharvest heat and calcium treatments', *J. Am. Soc. Hort. Sci.*, 122(1), 91–4.

Klein, J. D. and Lurie, S. (1990), 'Prestorage heat treatment as a means of improving poststorage quality of apples', *J. Am. Soc. Hort. Sci.*, 115(2), 265–9.

Knee, M. (1978), 'Properties of polygalacturonate and cell cohesion in apple fruit cortical tissue', *Phytochemistry* 17(8), 1257–60.

Kuchler, F., Chandran, R. and Ralston, K. (1996), 'The linkage between pesticide use and pesticide residues', *Am. J. Alt. Agric.*, 11, 161–7.

Lakshininarayana, S., Sommer, N. F., Polito, V. and Fortlage, R. J. (1987), 'Development of resistance to infection of *Botrytis cinerea* and *Penicillium expansum* in wounds of mature apple fruit', *Phytopathology* 77, 1674–8.

Lamichhane, J. R., Dachbrodt-Saaydeh, S., Kudsk, P. and Messéan, A. (2016), 'Toward a reduced reliance on conventional pesticides in European agriculture', *Plant. Dis.*, 100 (1), 10–24.

Landberg, R., Kamal-Eldin, A., Salmenkallio-Marttila, M., Rouau, X. and Åman, P. (2008), 'Localization of alkylresorcinols in wheat, rye and barley kernels', *J. Cereal Sci.*, 48(2), 401–6.

Ledford, H. (2015), 'CRISPR, the disruptor', *Nature* 522 (7554), 20–4.

Leibinger, W., Breuker, B., Hahn, M. and Mendgen, K. (1997), 'Control of postharvest pathogens and colonization of the apple surface by antagonistic microorganisms in the field', *Phytopathology*, 87(11), 1103–10.

Leistner, L. (2000), 'Basic aspects of food preservation by hurdle technology', *Int. J. Food Microbiol.*, 55(1), 181–6.

Leverentz, B., Janisiewicz, W. J., Conway, W. S., Saftner, R. A., Fuchs, Y., Sams, C. E. and Camp, M. J. (2000), 'Combining yeasts or a bacterial biocontrol agent and heat treatment to reduce postharvest decay of 'Gala' apples', *Postharvest Biol. Technol.*, 21(1), 87–94.

Li, H. X. and Xiao, C. L. (2008), 'Characterization of fludioxonil-resistant and pyrimethanil-resistant phenotypes of *Penicillium expansum* from apple', *Phytopathology* 98(4), 427–35.

Littlefield, N. A., Wankier, B. N., Salunkhe, D. K. and McGill, J. N. (1966), 'Fungistatic effects of controlled atmospheres', *Appl. Microbiol.*, 14(4), 579–81.

Lopez-Reyes, J. G., Spadaro, D., Gullino, M. L. and Garibaldi, A. (2010), 'Efficacy of plant essential oils on postharvest control of rot caused by fungi on four cultivars of apples in vivo', *Flavour Fragr. J.*, 25(3), 171–7.

Lurie, S. (1998), 'Postharvest heat treatments', *Postharvest Biol. Technol.*, 14(3), 257–69.

Lurie, S. (2001), 'Physical treatments as replacements for postharvest chemical treatments', *IV International Conference on Postharvest Science*, 553, 533–6.

Lurie, S., Fallik, E. and Klein, J. D. (1996), 'The effect of heat treatment on apple epicuticular wax and calcium uptake', *Postharvest Biol. Technol.*, 8(4), 271–7.

Lurie, S., Fallik, E., Klein, J. D., Kozar, F. and Kovacs, K. (1998), 'Postharvest heat treatment of apples to control San Jose scale (*Quadraspidiotus perniciosus* Comstock) and blue mold (*Penicillium expansum* Link) and maintain fruit firmness', *J. Amer. Soc. Hort. Sci.* 123(1), 110–14.

Manso, T. and Nunes, C. (2011), '*Metschnikowia andauensis* as a new biocontrol agent of fruit postharvest diseases', *Postharvest Biol. Technol.*, 61(1), 64–71.

Mari, M., Bertolini, P. and Pratella, G. C. (2003), 'Non-conventional methods for the control of post-harvest pear diseases', *J. Appl. Microbiol.*, 94(5), 761–6.

Mari, M., Di Francesco, A. and Bertolini, P. (2014), 'Control of fruit postharvest diseases: old issues and innovative approaches', *Stewart Postharvest Rev.*, 10(1), 1–4.

Mari, M., Iori, R., Leoni, O. and Marchi, A. (1993), '*In vitro* activity of glucosinolate-derived isothiocyanates against postharvest fruit pathogens', *Ann. Appl. Biol.*, 123(1), 155–64.

Mari, M., Leoni, O., Iori, R. and Cembali, T. (2002), 'Antifungal vapour-phase activity of allyl-isothiocyanate against *Penicillium expansum* on pears', *Plant Pathol.*, 51(2), 231–6.

Marquenie, D., Lammertyn, J., Geeraerd, A. H., Soontjens, C., Van Impe, J. F., Nicolaı, B. M. and Michiels, C. W. (2002), 'Inactivation of conidia of *Botrytis cinerea* and *Monilinia fructigena* using UV-C and heat treatment', *Int. J. of Food Microbiol.*, 74(1), 27–35.

Mason, J. L., Jasmin, J. J. and Granger, R. L. (1975), 'Softening of 'McIntosh' apples reduced by a post-harvest dip in calcium chloride solution plus thickeners', *HortSci.*, 10, 524–5.

Maxin, P, Weber, R. W. S., Pedersen, H. L. and Williams, M. (2012a), 'Hot-water dipping of apples to control *Penicillium expansum*, *Neonectria galligena* and *Botrytis cinerea*: effects of temperature on spore germination and fruit rots', *Europ. J. Hort. Sci.*, 77, 1–9.

Maxin, P, Weber, R. W. S., Pedersen, H. L. and Williams, M. (2012b), 'Control of a wide range of storage rots in naturally infected apples by hot-water dipping and rinsing', *Postharvest Biol. Technol.*, 70, 25–31.

Moline, H. E. (1984), 'Postharvest Pathology of Fruits and Vegetables: Postharvest losses in perishable crops', *UC Bulletin 1914*, Agric. Exp. St. Univ. Cal. Berkeley, CA, p. 80.

Morse, F. W. (1908), *The Respiration of Apples and its Relation to their Keeping*, vol. 135. New Hampshire Agricultural Experiment Station, New Hampshire College of Agriculture and the Mechanic Arts.

Neri, F., Mari, M., Brigati, S. and Bertolini, P. (2009), 'Control of *Neofabraea alba* by plant volatile compounds and hot water', *Postharvest Biol. Technol.*, 51(3), 425–30.

Neri, F., Mari, M. and Brigati, S. (2006a), 'Control of *Penicillium expansum* by volatile compounds', *Plant Pathol.*, 55(1), 100–5.

Neri, F., Mari, M., Menniti, A. M., Brigati, S. and Bertolini, P. (2006b), 'Control of *Penicillium expansum* in pears and apples by *trans*-2-hexenal vapors', *Postharvest. Biol. Technol.*, 41(1), 101–8.

Norelli, J. L. (January 2013), 'Genetic analysis of *Malus sieversii* PI 613981 for resistance to postharvest apple fruit decay caused by *Penicillium expansum* (blue mold)', In *Plant and Animal Genome XXI Conference* (Abstract PO418).

Nunes, C., Usall, J., Teixidó, N., Fons, E. and Viñas, I. (2002), 'Post-harvest biological control by *Pantoea agglomerans* (CPA-2) on Golden Delicious apples', *J. Appl. Microbiol.*, 92(2), 247–55.

Nursten, H. E. (1970), 'Volatile compounds: The aroma of fruits', In A. C. Hulme (Ed.), *The Biochemistry of Fruit and their Products*, Academic Press, London and New York, pp. 239–68.

Olsen, K. (1986), 'Views on CA Storage of Apples', WSU-TFREC/Postharvest information network, http://postharvest.tfrec.wsu.edu/pages/N4I2C (Accessed 29 October 2010).

Palou, L., Ali, A., Fallik, E. and Ramanazzi, G. (2016), 'GRAS, plant and animal-derived compounds as alternatives to conventional fungicides for the control of postharvest diseases of fresh horticultural produce', *Postharvest Biol. Technol.* (in press).

Pierson, C. F. (1966), 'Fungicides for the control of blue-mold rot of apples', *Plant Dis. Reporter*, 50(12), 913–15.

Pierson, C. F., Ceponis, M. J. and McColloch, L. P. (1971), 'Market diseases of apples, pears, and quinces', *Agricultural Research Service, United States Department of Agriculture*, Agriculture Handbook No 376.

Pillard, N. M. M. (1986), 'Evolution of the capacity of aldehyde production by crushed apple tissue., during an extended storage of fruits', In G. Charalambous (Ed.), *The Shelf Life of Foods and Beverages. Proceedings of 4th International Flavor Conference*, Elsevier, Amsterdam, pp. 368–78.

Plagge, H. H. and Gerhardt, F. (1930), *Acidity Changes Associated with the Keeping Quality of Apples under Various Storage Conditions* (No. 131), Agricultural Experiment Station, Iowa State College of Agriculture and Mechanic Arts.

Plagge, H. H. and Maney, T. J. (1924), 'Apple storage investigations', I. Jonathan-spot and soft-scald. II. Apple-scald and internal breakdown (No. 222). Agricultural Experiment Station, Iowa State College of Agriculture and Mechanic Arts.

Plagge, H. H. and Maney, T. J. (1925), *Cold Storage Investigation with Wealthy Apples-Fifth Progress Report* (No. 230), Agricultural Experiment Station, Iowa State College of Agriculture and Mechanic Arts.

Porritt, S. W. and Lidster, P. D. (1978), 'The effect of pre-storage heating on ripening and senescence of apples during cold storage', *J. Amer. Soc. Hort. Sci.*, 103, 584–7.

Quaglia, M., Ederli, L., Pasqualini, S. and Zazzerini, A. (2011), 'Biological control agents and chemical inducers of resistance for postharvest control of *Penicillium expansum* Link on apple fruit', *Postharvest Biol. Technol.*, 59(3), 307–15.

Rabea, E. I., Badway, M. E. T., Stevens, C. V., Smagghe, G. and Steurbaut, W. (2003), 'Chitosan as antimicrobial agent: applications and mode of action', *Biomacromolecules*, 4(6), 1457–65.

Reid, M. S. and Padfield, C. A. S. (1975), 'Control of bitter pit in apples with lecithin and calcium', *New Zealand J. Agric. Res.*, 18(4), 383–5.

Reiss, J. (1989), 'Influence of alkylresorcinols from rye and related compounds on the growth of food-borne molds', *Cereal Chem.*, 66(6), 491–3.

Roitman, J. N., Mahoney, N. E. and Janisiewicz, W. J. (1990), 'Production and composition of phenylpyrrole metabolites produced by *Pseudomonas cepacia*', *Appl. Microbiol. Biotechnol.*, 34(3), 381–6.

Romanazzi, G., Smilanick, J. L., Feliziani, E. and Droby, S. (2016), 'Integrated management of postharvest gray mold on fruit crops', *Postharvest Biol. Technol.*, 113, 69–76.

Rose, D. H. (1924), 'Diseases of apples on the market', *U.S. Dep. Agri. Dep. Bul.*, 1253, p. 24.

Rose, H. D., Brooks, C., Fisher, D. F. and Bratley, C. O. (1933), 'Market Diseases of Apples, Pears and Quinces', *Miscellaneous Publication No. 168*, U.S. Department of Agriculture, Agricultural Research Service, Washington, DC.

Rosenberger, D. A. (2001), 'Postharvest decay control without fungicides', In Storage Workshop 2001: *Apple Handling and Storage*, NRAES Bull. 153, Northeast Regional Agricultural Engineering Service, Ithaca, NY, pp. 21–3.

Rosenberger, D. A. (2009), 'Fungicides, biocides, and sanitizers for managing postharvest pathogens in apples', *New York Fruit Quarterly*, 17, 3–6.

Rosenberger, D. A. and Meyer, F. W. (1981), 'Postharvest fungicides for apples: development of resistance to benomyl, vinclozolin, and iprodione', *Plant Dis.*, 65, 1010–13.

Roy, S., Conway, W. S., Watada, A. F., Sams, C. E., Erbe, E. F. and Wergin, P. W. (1994), 'Heat treatment affects epicuticular wax structure and postharvest calcium uptake in 'Golden Delicious' apples', *HortSci.*, 29(9), 1056–8.

Salomão, B. C. M., Aragão, G. M. F., Churey, J. J. and Worobo, R. W. (2008), 'Efficacy of sanitizing treatments against *Penicillium expansum* inoculated on six varieties of apples', *J. Food Protec.*, 71(3), 643–7

Sams, C. F., Conway, W. S., Abbott, J. A., Lewis, R. J. and Ben-Shalom, N. (1993), 'Firmness and decay of apples following postharvest pressure infiltration of calcium and heat treatment', *J. Amer. Soc. Hort. Sci.*, 118(5), 623–7.

Sandford, P. (1989), 'Chitosan: commercial uses and potential applications', In G. Skjak-Braek, T. Anthosen, and P. Standford (Eds), *Chitin and Chitosan. Sources, Chemistry, Biochemistry. Physical Properties and Applications*. Elsevier Applied Science, London and New York, pp. 51–69.

Scott, K. J. and Wills, R. B. H. (1979), 'Effects of vacuum and pressure infiltration of calcium chloride and storage temperature on the incidence of bitter rot and low temperature breakdown of apples', *Crop Pasture Sci.*, 30(5), 917–28.

Secretan, B., Straif, K., Baan, R., Grosse, Y., El Ghissassi, F., Bouvard, V. and Cogliano, V. (2009), 'A review of human carcinogens – Part E: tobacco, areca nut, alcohol, coal smoke, and salted fish', *The Lancet Oncology*, 10(11), 1033–4.

Serrano, M., Martinez-Romero, D., Guillen, F., Velverde, J. M., Zapata, S. C., Castillo, S. and Valero, D. (2008), 'The addition of essential oils to MAP as a tool to maintain the overall quality of fruits', *Trends Food Sci. Technol.*, 19(9), 464–71.

Shao, X., Tu, K., Tu, S., Su, J. and Zhao, Y. (2010), 'Effect of heat treatment on wound healing in 'Gala' and 'Red Fuji' apple fruit', *J. Ag. Food Chem.*, 58(7), 4303–9.

Sholberg, P. L. and Gaunce, A. P. (1995), 'Fumigation of fruits with acetic acid to prevent postharvest decay', *HortSci.*, 30(6), 1271–5.

Sholberg, P. L. (1998a), Postharvest strategies that reduce risk of pome fruit decay. In *1998 Washington Tree Fruit Postharvest Conference Proceedings*, pp. 50–4. http://postharvest.tfrec.wsu.edu/PC2004F.pdf.

Sholberg, P. L. (1998b), 'Fumigation of fruit with short-chain organic acids to reduce the potential of postharvest decay', *Plant Dis.*, 82(6), 686–93.

Sholberg, P. L. (2004), 'Bin and storage room sanitation', WSU – Postharvest Information Network, http://postharvest.tfrec.wsu.edu/PC2004F.pdf.

Sholberg, P. L. (2009), 'Control of postharvest decay by fumigation with acetic acid or plant volatile compounds', *Fresh Produce*, 3 (SI1), 80–6.

Sholberg, P. L., Cliff, M. and Moyls, A. L. (2001), 'Fumigation with acetic vapor to control decay of stored apples', *Fruits*, 56(5), 355–66.

Sholberg, P. L. and Randall, P. (2007), 'Fumigation of stored pome fruit with hexanal reduces blue mold and gray mold decay', *HortSci.*, 42(3), 611–16.

Sitton, J. W. and Patterson, M. E. (1992), 'Effect of high-carbon dioxide and low-oxygen controlled atmospheres on postharvest decays of apples', *Plant Dis.*, 76(10), 992–5.

Smilanick, J. L. (2003), 'Use of ozone in storage and packing facility', *WSU – TFRES Postharvest Information Network, 2003 Proceedings*, http://postharvest.tfrec.wsu.edu/PC2003H.pdf.

Smilanick, J. L., Denis-Arrue, R., Bosch, J. R., Gonzalez, A. R., Henson, D. and Janisiewicz, W. J. (1993), 'Control of postharvest brown rot of nectarines and peaches by Pseudomonas species', *Crop Protec.*, 12(7), 513–20.

Smith, K. J. and Lay-Yee, M. (2000), 'Response of 'Royal Gala' apples to hot water treatment for insect control', *Postharvest Biol. Technol.*, 19(2), 111–22.

Smith, W. L. (1962), 'Chemical treatments to reduce postharvest spoilage of fruits and vegetables', *Botan. Rev.*, 28(3), 411–45.

Sommer, N. F. (1985), 'Role of controlled environment in suppression of postharvest diseases', *Can. J. Plant Pathol.*, 7(3), 331–9.

Song, J., Leepipattwit, R., Deng, W. and Beaudry, R. M. (1998), 'Hexanal vapor acts as residueless antifungal agent that enhances aroma biosynthesis in apple fruit', *Acta Hortic.*, 464, 219–24.

Spadaro, D. and Gullino, M. L. (2014), 'Use of essential oils to control postharvest rots on pome and stone fruit', In D. Prusky and M. L. Gullino (Eds), *Post-harvest Pathology. Plant Pathology in the 21st Century, Contributions to the 10th International Congress, ICPP 2013*. Switzerland, Springer, pp. 101–10.

Spadoni, A., Guidarelli, M., Phillips, J., Mari, M. and Wisnewski, M. (2015), 'Transcriptional profiling of apple fruit in response to heat treatment: involvement of a defense response during *Penicillium expansum* infection', *Postharvest Biol. Technol.*, 101, 37–48.

Spotts, R. and Cervantes, L. A. (1986), 'Population, pathogenicity and benomyl resistance of *Botrytis* spp., *Penicillum* spp. and *Mucor piriformis* in packinghouses', *Plant Dis.*, 70, 106–8.

Spotts, R. A. and Cervantes, L. A. (1992), 'Effect of ozonated water on postharvest pathogens of pear in laboratory and packinghouse tests', *Plant Dis.*, 76, 256–9.

Spotts, R. A., Cervantes, L. A., Mielke, E. A. (1999), 'Variability in postharvest decay among apple cultivars', *Plant Dis.*, 83, 1051–4.

Spotts, R. A. and Peters, B. B. (1980), 'Chlorine and chlorine dioxide for control of 'd'Anjou' pear decay', *Plant Dis.*, 64, 1095–7.

Stadelbacher, G. J. and Prasad, K. (1974), 'Postharvest decay control of apple by acetaldehyde vapor', *J. Amer. Soc. Hort. Sci.*, 99, 364–8.

Stevens, C., Khan, V. A., Wilson, C. L., Lu, J. Y., Chalutz, E. and Droby, S. (2005), 'The effect of fruit orientation of postharvest commodities following low dose ultraviolet light-C treatment on host induced resistance to decay', *Crop Protec.*, 24, 756–9.

Stevens, C., Wilson, C. L., Lu, J. Y., Khan, V. A., Chalutz, E., Kabwe, M. K., Haung, Z., Adeyeye, O., Pusey, L., Wisnewski, W. E. and West. M. (1996), 'Plant hormesis induced by ultraviolet light-C for controlling postharvest diseases of tree fruits', *Crop Prot.*, 15, 129–34.

Su, J., Tu, K., Cheng, L., Tu, S., Wang, M., Xu, H. and Zhan, G. (2011), 'Wound-induced H_2O_2 and resistance to *Botrytis cinerea* decline with ripening of apple fruit', *Postharvest Biol. Technol.*, 62, 64–70.

Suslow, T. S. (1997), 'Postharvest chlorination', Publication 8003. University of California. http://danrcs.ucdavis.edu.

Suslow, T. S. (2004), 'Ozone applications for postharvest disinfection of edible horticultural crops', Publication 8133 University of California, http://anrcatalog.ucdavis.edu/pdf/8003.pdf.

Sutton, T. B., Alswinckle, H. S., Agnello, A. M. and Walgenbach, J. F. (Eds) (2014), *Compendium of Apple and Pear Diseases and Pests*. American Phytopathological Society.

Swinburne, T. R. (1970), 'Fungal rotting of Apples. I. A survey of the extent and cause of current fruit losses in Northern Ireland', *Record of Agricultural Research, Ministry of Agriculture, Northern Ireland*, 18(1), 15–19.

Tahir, I. I., Ahmadi-Afzadi, M., Nybom, H. and Dey, E. S. (2014), 'Rye bran alkylresorcinols inhibit growth of *Penicillum expansum* and *Neofabrea perennans* in vitro and in vivo on different apple cultivars', *Eur. J. Hort. Sci.*, 79(4), 218–25.

Tahir, I. I., Johansson, E. and Olsson, M. E. (2009), 'Improvement of apple quality and storability by a combination of heat treatment and controlled atmosphere storage', *HortSci.*, 44(6), 1648–54.

Takeda, F., Janisiewicz, W. J., Roitman, J., Mahoney, N. and Abeles, F. B. (1990), 'Pyrrolnitrin delays postharvest fruit rot in strawberries', *HortSci.*, 25(3), 320–2.

Terry, L. A. and Joyce, D. C. (2004), 'Elicitors of induced disease resistance in postharvest horticultural crops: a brief review', *Postharvest Biol. Technol.*, 32(1), 1–13.

Torres, R., Valentines, M. C., Usall, J., Viñas, I. and Larrigaudiere, C. (2003), 'Possible involvement of hydrogen peroxide in the development of resistance mechanism in 'Golden Delicious' apple fruit', *Postharvest Biol. Technol.*, 27, 235–42.

Trierweiler, B., Schirmer, H. and Tauscher, B. (2003), 'Hot water treatment to control Gloeosporium disease on apples during long-term storage', *J. Appl. Bot.*, 77(5/6), 156–9.

Usall, J., Casals, C., Sisquella, M., Palou, L. and De Cal, A. (2015), 'Alternative technologies to control postharvest diseases of stone fruits', *Stewart Postharvest Rev.*, 11(4), 1–6.

Usall, J., Teixido, N., Torres, R., de Eribe, X. O. and Viñas, I. (2001), 'Pilot tests of *Candida sake* (CPA-1) applications to control postharvest blue mold on apple fruit', *Postharvest Biol. Technol.*, 21(2), 147–56.

Van de Braak, S. A. A. J. and Leijten, G. C. J. J. (1999), *Essential Oils and Oleoresins: A Survey in the Netherlands and other Major Markets in the European Union*. CBI, Centre for the Promotion of Imports from developing Countries, Rotterdam, p. 116.

van der Oost, J., Jore, M. M., Westra, E. R., Lundgren, M. and Brouns, S. J. J. (2009), 'CRISPR-based adaptive and heritable immunity in prokaryotes', *Trends Biochemical Sci.*, 34(8), 401–7.

Vilanova, L., Viñas, I., Torres, R., Usall. J., Buron-Moles, G. and Teixido, N. (2014), 'Increasing maturity reduces wound response lignifications processes against *Penicillium expansum* (pathogen) and *Penicillium digitatum* (non-host pathogen) infection in apples', *Postharvest Biol. Technol.*, 88, 54–60.

Vorstermans, B. and Creemers, P. (2011), 'Current status of development of biological control products for postharvest use in Europe', *Acta Hort. (ISHS)*, 905, 313–18.

Watkins, C. B., Kupferman, E. and Rosenberger, D. A. (2014), 'Apple', In K. C. Gross, C. Y. Wang and M. Saltveit (Eds), *The Commercial Storage of Fruits, Vegetables, and Florist and Nursery Stocks*, Agriculture Handbook Number 66, http://www.ba.ars.usda.gov/hb66/contents.html (Accessed 29 January 2016).

Wells, J. M. and Uota, M. (1970), 'Germination and growth of five fungi in low-oxygen and high-carbon dioxide atmospheres', *Phytopathology*, 60(1), 50–3.

Wilson, C. L., El Ghaouth, A., Chalutz, E., Droby, S., Stevens, C., Lu, J. Y., Khan, V. and Arul, J. (1994), 'Potential of induced resistance to control postharvest diseases of fruits and vegetables', *Plant Dis.*, 78(9), 837–44.

Wilson, C. L., Upchurch, B., El Ghaouth, A., Stevens, C., Khan, V., Droby, S. and Chalutz, E. (1997), 'Using an on-line UV-C apparatus to treat harvested fruit for controlling postharvest decay', *HortTechnol.*, 7(3), 278–82.

Wisniewsky, M. A., Glatz, B. A., Gleason, M. L. and Reitmeier, C. L. (2000), 'Reduction of *Escherichia coli* O157:H7 counts on whole fresh apples by treatment with sanitizers', *J. Food Protec.*, 63(6), 703–8.

Wright, T. R. and Smyth, E. (1954), 'Relation of bruising and other factors to blue mold decay of Delicious apples', *U.S. Dep. Agric. Cir.*, 935, p. 15.

Wu, Z., Yin, X., Bañuelos, G. S., Lin, Z-Q., Zhu, Z., Liu, Y., Yuan, L. and Li, M. (2016), 'Effect of selenium on control of postharvest gray mold of tomato fruit and possible mechanism involved', *Frontiers Microbiol.*, 6, 1441.

Wu, Z. L., Yin, X. B., Lin, Z. Q., Bañuelos, G. S., Yuan, L. X., Liu, Y. and Li, M. (2014), 'Inhibitory effect of selenium against *Penicillium expansum* and its possible mechanisms of action', *Current Microbiol.*, 69(2), 192–201.

Xiao, C. L. and Boal, R. J. (2009), 'Residual activity of fludioxonil and pyrimethanil against *Penicillium expansum* on apple fruit', *Plant Dis.*, 93(10), 1003–8.

Xiao, C. L. and Kim, Y. K. (2008), 'Postharvest fruit rot in apples caused by *Botrytis cinerea, Phacidiopycnis washingtonensis* and *Sphaeropsis pyriputrescens*', *Plant Health Progress*, 1–13.

Xiao, C. L. and Kim, Y. K. (2010), 'Control of postharvest diseases in apples with reduced-risk fungicides', *Stewart Postharvest Review*, 6(1), 1–6.

Xiao, C. L., Kim, Y. K. and Boal, R. J. (2014), 'First report of occurrence of pyrimethanil resistance in *Penicillium expansum* from stored apples in Washington State', *Plant Dis.*, 98(7), 924–8.

Yang, L. Y., Zhang, J. L., Bassett, C. L. and Meng, X. H. (2012), 'Difference between chitosan and oligochitosan in growth of *Monilinia fructicola* and control of brown rot in peach fruit', *LWT-Food Sci. Technol.*, 46(1), 254–9.

Yaun, B. R., Sumner, S. S., Eifert, J. D. and Marcy, J. E. (2004), 'Inhibition of pathogens on fresh produce by ultraviolet energy', *Int. J. Food Microbiol.*, 90(1), 1–8.

Yin, L., Zou, Y., Li, M., Ke, X., Li, C., Liang, D. and Ma, F. (2013), 'Resistance of *Malus* plants to *Diplocarpon mali* infection is associated with the antioxidant system and defense signaling pathways', *Physiol. Mol. Plant Pathol.*, 84, 146–52.

Yu, T., Yu, C., Chen, F., Sheng, K., Zhou, T., Zunun, M., Abudu, O., Yang, S. and Zheng, X. (2012), 'Integrated control of blue mold in pear fruit by combined application of chitosan, a biocontrol yeast and calcium chloride', *Postharvest Biol. Technol.*, 69, 49–53.

Zeng, L., Yu, C., Fu, D., Lu, H., Zhu, R., Lu, L., Zheng, X. and Yu, T. (2015), 'Improvement in the effectiveness of *Cryptococcus laurentii* to control postharvest blue mold of pear by its culture in β-glucan amended nutrient broth', *Postharvest Biol. Technol.*, 104 (1), 26–32.

Zhang, H., Ge, L., Chen, K., Zhao, L. and Zhang, X. (2014), 'Enhanced biocontrol activity of *Rhodotorula mucilaginosa* cultured in media containing chitosan against postharvest diseases in strawberries: possible mechanisms underlying the effect', *J. Agric. Food Chem.*, 62, 4214–24.

Zhang, H., Li, R. and Liu, W. (2011), 'Effects of chitin and its derivative chitosan on postharvest decay of fruits: a review', *Int. J. Molec. Sci.*, 12(2), 917–34.

Zhang, Q., Young, D., Zhang, Y., Shi, X., Li, B. and Li, G. (2016), '*Streptomyces rochei* A-1 induces resistance and defense-related responses against *Botryospheria dothidea* in apple fruit during storage', *Postharvest Biol. Technol.*, 115, 30–7.

Chapter 13

Advances in postharvest handling and storage of apples

Christopher B. Watkins, Cornell University, USA

1 Introduction

The average consumer may recognize different apple cultivars and associated colour, flavour and taste differences at the market, but likely is unaware of the differences in ripening physiology that impact availability of the fruit that he/she is looking at. The term 'apple' suggests the existence of a single fruit type with uniformity of genetics and subsequently of maturation and ripening that does not exist because the apple represents a highly variable range of fruit types arising from wild *Malus* species. These fruits were originally characterized by small size, high acidity, and low sugar concentrations and therefore not highly desirable for humans. In contrast, modern fruit cultivars typically have large fruit size, less acidity, high sugar concentrations, and a high percentage coverage and intensity of red skin colour. These factors are critical for marketability as they relate directly to appearance, eating quality and flavour, all of which are desired by consumers. Consumers buy with their eyes, but thereafter, an integration of texture, acid/sugar

http://dx.doi.org/10.19103/AS.2016.0017.16

balance and aroma volatiles that results in the eating quality and flavour expected for any given apple cultivar becomes critical. Most commercial cultivars have originated from a relatively few founder cultivars such as 'Rosemary Russet', 'Jonathan', 'Golden Delicious', 'Red Delicious', 'Cox's Orange Pippin' and 'McIntosh' (Khan et al., 2014; Noiton and Alspach, 1996). Cultivars also have arisen from chance seedlings or bud sports from a single cultivar or developed from breeding programmes. Newly developed cultivars such as 'Jazz™', 'Pink Lady®' and 'Honeycrisp' have become increasingly popular while older ones are losing market appeal.

In general, early season apples have short storage potential, while late season apples have long storage potential, a reflection of faster rates of maturation and ripening in early than late harvested fruit. Each cultivar has specific ripening patterns, and these can range from the low chilling requiring 'Anna', which softens rapidly and is marketed directly or stored for only a few days, to 'Fuji' and 'Idared', which can be stored for long periods even with less than optimal refrigeration. Maturation and ripening rates can further be affected by growing region. For example, 'Cox's Orange Pippin' is an early season apple with short storage potential when grown in New Zealand, but a mid- to late season apple with long-term storage potential when grown in England.

The success of the modern orchard is based on high yields of the fruit of uniform size, colour and internal quality characteristics such as firmness, sugar concentrations and acidity that is desired for each specific cultivar. Harvest management and crop storage under optimal conditions is a critical factor in maintaining sustainability of the orchard operation; deteriorating fruit quality after harvest represents loss of investments in the orchard level as well as contributing to food waste. Reduced fruit quality can occur as a result of loss of desirable attributes due to over-ripening, physical injuries such as bruising, and development of pathological (see Chapter 15) and physiological disorders. Therefore, postharvest handling of apples, as for most horticultural products, is largely concerned with the maintenance of product quality after harvest, rather than its improvement. Meeting market criteria for any cultivar represents an integration of the genetics controlling ripening, growing climate and orchard management practices, maturity of the fruit at harvest, and the marketing plan, which can range from days to 12 months or more. Although factors that can contribute to fruit acceptability, such as starch conversion to sugars and development of aroma volatiles, can occur after harvest, the ideal apple in the market is typically close to that at harvest, especially firm fruits with crisp texture characteristics.

The apple industries' success in maintaining fruit quality after harvest can vary depending on investments in the orchard and storage facilities, grower decisions and skills, worker training, and maintenance of the cold chain including proper storage conditions during transport and at the market. Although important at all scales of production, attention to these factors becomes critical in those industries relying on a highly competitive, long-term storage market.

Regardless of industry size, however, overall postharvest objectives from growers to consumers are to:

1 Reduce metabolic rates that result in undesirable changes in colour, composition, texture, flavour and nutritional status.
2 Reduce water loss that can result in loss of saleable weight, reduction in appearance factors such as shrivelling, softening and loss of crispness.
3 Minimize bruising, friction damage and other mechanical injuries.

4 Reduce spoilage caused by pathogen infection, especially of damaged or wounded tissues.
5 Prevent physiological disorders such as superficial scald, low-temperature disorders and bitter pit.

An additional objective has become of increasing importance, that is, the reduction of chemical usage, avoidance of contamination and maintenance of food safety. Greater concerns exist about application of postharvest treatments such as diphenylamine (DPA), fungicides and edible coatings (waxes), and contamination, especially of organic apples. In addition, a major change in rules governing harvest, storage management and transport of horticultural produce is being implemented as a result of the modernization of food safety standards in the United States under the auspices of the Food and Drug Administration Food Safety Modernization Act. For example, the rule on Sanitary Transportation of Human and Animal Food aims to protect foods from farm to table by preventing practices during transportation that create food safety risks. These include cleanliness of vehicles and transportation equipment; measures taken during transportation to ensure food safety, such as adequate temperature controls and preventing contamination of ready-to-eat food from touching raw food; protection of food from contamination by non-food items in the same load or previous load and protection of food from cross-contact.

This chapter outlines the recent advances in understanding the various factors associated with fruit maturation and quality assessment, and postharvest technologies. Since a review of these factors by Watkins (2003), substantial progress has been made in knowledge of 'omics of fruit ripening, the understanding of apple quality and its measurement, preharvest management of fruit ripening, development of additional harvest indices and technologies for maintaining fruit quality after harvest.

2 Fruit quality assessment

Appearance is the main sensory characteristic that consumers take into account when evaluating fruit quality and making their purchase decisions (Jaeger et al., 2016; Nicolai et al., 2007). Once purchased, textural properties are considered by consumers as the primary drivers of apple quality, followed by flavour characteristics, typically soluble solids concentration (SSC), titratable acidity (TA) (Harker et al., 2008; Jaeger et al., 1998) and internal defects such as bruising and physiological disorder (Jaeger et al., 2016). Postharvest evaluations usually focus on texture (firmness), SSC and TA. Formal sensory analyses for evaluation of flavour are also highly desirable from a postharvest perspective, but are limited by a number of factors (Corollaro et al., 2014): expense and limitations due to small numbers of assessments because humans are used as 'instruments'; they cannot be used in real time, an important factor as high variability among fruits requires large sample sizes.

Texture analyses of mechanical and acoustic components of apple cultivars have been related to firmness, crispness and juiciness attributes judged by expert panellists (Costa et al., 2011). Corollaro et al. (2014) expanded these approaches to develop a sensory-instrumental analysis based on objective measurements of flesh appearance, texture, and taste based on chemical and colorimeter data. Application of such methods remains limited, however, until they can be simplified and made easier to use in practice.

Apple flavour is a complex combination of many individual sugars, acids and aroma volatiles (Dixon and Hewett, 2000) and is a somewhat subjective phenotype that integrates multiple sensory systems, ultimately manifesting itself at brain level (Kumar et al., 2015b). Sensory sweetness is estimated by SSC (Harker et al., 2002). Esters contribute more than 80% of volatile compounds and produce fruity aromas. The major volatiles affecting fruity and sweet sensory attributes are acetate, butyl acetate, 2-methylbutyl acetate and hexyl acetate, while those contributing to sour, bitter and grassy attributes are ethyl acetate, ethyl butanoate, 1-butanol, ethanol and α-farnesene (Karlsen et al., 1999). Instrumental measurements of volatiles by an electronic nose, a biomimetic instrument that can perceive aroma in a similar way to the human olfactory system (Nicolai et al., 2014), are not yet commercially available. However, research with electronic nose devices includes evaluation of the effects of cultivar, maturity and ripening stages at harvest and during shelf life periods, and postharvest treatments (Baietto and Wilson, 2015). Interestingly, as discussed in Section 6.3, not all volatiles are necessarily desirable as some, especially esters, can be perceived as markers of 'over-ripeness'.

Fruit quality is obviously a key factor for both plant breeders and postharvest scientists as the goal is to maintain the quality at harvest for as long as required to meet market demands. While both types of scientists should work closely together, evaluation of breeding populations is difficult because large numbers of fruits necessary for comprehensive evaluation of quality during storage are typically not available until advanced selections have been made.

3 Harvest maturity and harvest indices

Fruit maturity at the time of harvest is a major factor influencing storability of any apple cultivar. Fruits that are harvested earlier and therefore less mature within the harvest

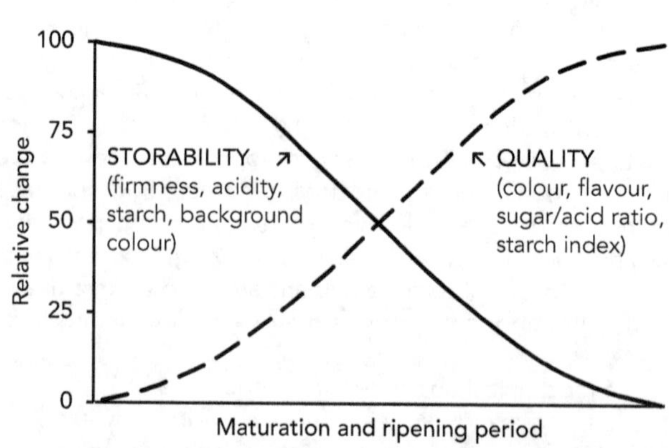

Figure 1 The relationship between the storability of a ripening apple and the quality characteristics desired by the consumer. As a fruit ripens, the quality attributes such as colour and flavour increase, but the potential storage period declines.

window tend to have longer storage potential, but have lower quality characteristics, than fruits harvested later when they are more mature. Therefore, storage potential is inevitably a compromise with quality (Fig. 1). An early harvested fruit will maintain firmness more easily over extended storage periods. It will often have lower flavour quality as it will have been harvested at the early stages of maturity before, or close to, the onset of autocatalytic ethylene production. Such fruits can also be more susceptible to disorders such as superficial scald and bitter pit. In contrast, a fruit with full flavour quality at harvest, for example in a pick-your-own operation or at a local farmers market, will have limited storability. Faster ripening is most likely associated with high ethylene production associated with the respiratory climacteric. These fruits can soften rapidly and lose green background colour, as well as being susceptible to disorders such as senescent breakdown.

An apple is not marketable if it does not meet consumer expectations, which can vary according to the market. The impact of supermarket requirements on standards such as uniform size and colour, and freedom from blemish, has sometimes superseded eating quality, although requirements for minimum standards of firmness and SSC by wholesale buyers is increasing. Red colour for most cultivars is a major factor in marketability, and for some cultivars, even if fruits are ripe and ready for harvest, they cannot be sold without meeting colour standards (% coverage), for example, those regulated by the USDA to meet Fancy or Extra Fancy grades.

Red colour aside, many methods have been developed and investigated to determine fruit maturity (Watkins, 2003). Most of these tests are not necessarily related to maturity *sensu stricto* but rather to quality, for example, being affected by exposure to sunlight in the case of firmness and SSC. Therefore, the term 'harvest indices' is preferred (Watkins, 2003). Ethylene production or internal ethylene concentration (IEC) is a major physiological marker and therefore it is an important determinant of harvest timing. However, its measurement requires expensive gas chromatography and laboratory facilities limiting its use to larger-scale storage operations. As a result, the starch pattern index (SPI), flesh firmness, SSC, background or ground colour and to a lesser extent TA are more commonly used. No single harvest index should be used to decide harvest timing, as each has limitations. Formal maturity programmes in which the samples are collected before and during harvest periods exist in many growing regions, and are increasingly common for large-scale storage facilities. Data are used by field personnel and storage operators to develop best guidelines for optimal harvest times for immediate sale, or desired storage periods.

The SPI, however, is currently popular as a guide to harvest and many starch charts are available to rate the starch iodine staining patterns from numerical ranges of 1–6 and 1–10 (Blanpied and Silsby, 1992; Reid et al., 1982; Travers et al., 2002). The popularity of the SPI lies in its ease of use, despite significant flaws including high variation in interpretation of staining patterns (Peirs et al., 2002); failure of the fruit to accumulate full starch complements, markedly different starch concentrations at similar SPIs; and incomplete staining of starch because amylose (AM), but not amylopectin, is stained by iodine (Doerflinger et al., 2015a).

Flesh firmness is typically measured using pressure testers (penetrometers). Handheld penetrometers are often used by field personnel, but are subject to significant operator variation (Harker et al., 1996; LehmanSalada, 1996). Drill-mounted penetrometers can be more accurate than handheld testers, but increasingly automated pressure testers (Chauvin et al., 2010; DeLong et al., 2000) are being used. SSC is measured using refractometers, and increasingly with digital readings.

There is continued interest in non-destructive evaluation of apple ripening and quality traits, such as colour, firmness, dry matter and SSC (Betemps et al., 2012; Lu, 2004; Merzlyak et al., 2003), that could be used as harvest indices. Reviews on fruit quality evaluation using visible (vis) and near-infrared (NIR) technology including extensive coverage of apples are available (Nicolai et al., 2007; Wang et al., 2015).

The loss of chlorophyll can be a useful harvest index for apples as degreening is a feature of ripening (Knee, 1972; Solovchenko et al., 2005; Zude-Sasse et al., 2002), and colour cards have been used to assess background or ground colour for many years. However, this can be difficult for many red cultivars and high colouring strains. Recently, a small handheld delta absorbance (DA) meter (Sintéleia, Bologna, Italy) has been commercialized, which measures the absorption differences (I_{AD}) of wavelengths at 670 and 720 nm and functions essentially as electronic colour chart without interference by anthocyanins and other pigments. From the interactance (I) spectra, fruit absorbance (A) is calculated using the Beer–Lambert's law ($A=\log_{10} I^{-1}$). The I_{AD} is calculated as $I_{AD}=A_{670}-A_{720}$ where A_{670} and A_{720} are the A values measured at 670 and 720 nm, respectively (Ziosi et al., 2008). Recent research has detected good correlations between I_{AD} values and other harvest indices in various apple cultivars such as 'Starking', 'Granny Smith' and 'Pink Lady' (Farneti et al., 2015b; Nyasordzi et al., 2013), 'Gala' (Costamagna et al., 2013; Toivonen and Hampson, 2014), 'Braeburn' and 'Cripps Pink' (Zanella et al., 2013) and 'Honeycrisp' (DeLong et al., 2014). However, although good relationships between I_{AD} values and other harvest indices existed for 'Aurora Golden Gala™', 'Fuji' and 'Royal Gala', these became weaker for any indices other than chlorophyll a content if 1-MCP or controlled atmosphere (CA) storage was applied after harvest (Toivonen and Hampson, 2014). Furthermore, the plant growth regulators (PGRs), Retain and Harvista, that modulate ripening of the fruit (described in the next section) weakened the relationships between I_{AD} values and other harvest indices, especially IEC (Doerflinger et al., 2016). Therefore, in an industry where storage operators are receiving fruits from untreated and Retain- and Harvista-treated blocks, the usefulness of I_{AD} values as a harvest index to predict storability of fruit lots may be decreased.

Research by McGlone et al. (2002) demonstrated that NIR spectra predicted SSC in the fruit after storage rather than at harvest, and that dry matter concentration (DMC) determined by NIR could be a better predictor of SSC; total carbohydrate concentrations, rather than insoluble or soluble forms were likely assessed. DMC predictions of apples at harvest were strongly correlated with SSC after storage (McGlone et al., 2003), and subsequently DMC was proposed as a quality metric for apples (Palmer et al., 2010). The research of Palmer et al. (2010) and others who have measured DMC to investigate orchard factors on fruit quality (Palmer et al., 2013; Saei et al., 2011; Serra et al., 2016) employed destructive measurements of DMC and SSC. Increasing evaluation of DMC and other quality factors, including pigments such as chlorophyll is likely with the availability of handheld devices such as Nirvana (Integrated Spectronics, Australia) (Kumar et al., 2015a) and the F-750 Produce Quality meter (Felix Instruments, USA).

4 Plant growth regulators

PGRs are used by growers to manage harvest by preventing preharvest drop, extending the harvest period to allow greater fruit size to develop, to manage labour requirements for harvest, and/or delay maturation and ripening processes in order to obtain more uniform postharvest responses. Three PGRs are used commercially, sometimes in combination, but availability of each is restricted regionally.

1 Naphthaleneacetic acid (NAA) is a synthetic auxin that prevents abscission and acts as a potent 'stop–drop' treatment. For cultivars such as 'McIntosh' it has often been used as an emergency application in seasons where fruit drop starts prior to commercially acceptable fruit colour, and in specific cases where ethylene is applied by preharvest spray to induce colour development in regions where colour development is delayed by climate (Stover et al., 2003). Although abscission is delayed, however, ethylene production by the fruit can be stimulated resulting in faster fruit softening (Yuan and Carbaugh, 2007). As a result, NAA-treated fruits should not be stored for long periods.
2 Aminoethoxyvinylglycine (AVG), marketed as ReTain, is an inhibitor of 1-aminocyclopropane-1-carboxylic acid (ACC) synthase activity and thereby delays ethylene production. The recommended timing of AVG sprays on trees was originally 4 weeks before anticipated harvest, but label changes now allow intervals as close as a week before harvest to be used. AVG delays fruit drop and fruit ripening, and can have postharvest benefits such as maintenance of firmness (Greene and Schupp, 2004; Schupp and Greene, 2004). AVG is used commonly in North America, but is not registered for use in Asia and Europe.
3 1-Methylcyclopropene (1-MCP) is an inhibitor of ethylene perception that is widely used to inhibit fruit ripening after harvest (see Section 6). A preharvest formulation of 1-MCP has been developed (Harvista™) to allow applications in the field. Harvista™ delays fruit drop and fruit ripening, and has postharvest benefits by maintaining quality factors such as firmness that are more pronounced than AVG (McArtney et al., 2008; Varanasi et al., 2013; Watkins et al., 2010). Product development is continuing and commercial registration, while currently restricted to the United States, is likely to become available worldwide.

Recent emphasis has been on combinations of these PGRs. Yuan and Carbaugh (2007) found that NAA in combination with AVG or preharvest 1-MCP treatments was more effective than NAA, AVG or 1-MCP alone, and prevented the side effect of NAA-induced softening. Greater expression of genes involved in ethylene biosynthesis and cell wall degradation in NAA-treated fruit is reduced by AVG (Li and Yuan, 2008). Expression of ethylene biosynthetic and receptor genes are inhibited by preharvest 1-MCP (Varanasi et al., 2013).

5 Harvest, handling and grading operations

Commercial harvest is usually carried out by teams of pickers. This operation represents the first step where significant loss of fruit quality can occur due to handling damage. It has been estimated that 40% of bruising can occur in the field (Funt et al., 1999) and training of pickers in proper harvesting techniques such as twisting of the fruit to avoid stem removal, gently placing the fruit into harvesting bags and then into field bins to avoid impact injuries is essential. Some cultivars such as Gala and Honeycrisp require multiple picks to ensure that only fruit with red colouration and at correct maturity, required to meet grade standards, are harvested. Repeated harvests of fruits are more expensive than where the tree are stripped of all fruits at one harvest, but are essential to ensure maximum market returns for these cultivars can be achieved. Other factors that affect susceptibility of the fruit to bruising damage include orchard management systems and harvest date (Eckhoff et al., 2009; Opara, 2007). In modern systems, bin identification is established in the field to identify orchard block, cultivar, date of harvest and often picking team. Initially, these processes were put into place for quality management such as obtaining data on pack out percentages within orchard blocks, but are becoming increasingly important because of food safety requirements for traceability.

In a typical storage operation, bins of fruits are transported from the orchard and moved immediately into refrigerated rooms to be cooled as quickly as possible. Depending on the scale of operation, and especially where lots of fruits from different orchards are stored, the first step, however, is quality control. Samples of fruits from bins are inspected to assess quality factors such as size and colour, incidences of damage, bruising, bitter pit, russet and other defects. (Information about any quality issues can be immediately reported back to the harvest teams to ensure that these losses are reduced from next harvest.) Harvest indices such as flesh firmness, SSC, SPI and where equipment is available, IEC are also measured. Presence of watercore in susceptible cultivars is assessed as severe watercore can negatively affect storability. Cold storage, especially with CA storage and postharvest treatments (see following sections), is expensive, and so the condition of fruits upon receipt at the storage should be a key factor on deciding whether fruits are sold immediately, stored or sent to processing. The percentage of fruits that can be packed to meet 'Extra Fancy' and 'Fancy' colour requirements, for example, in the United States to meet USDA or Washington State apple standards (Anon., 1974), as well as optimum size requirements, greatly affects financial returns to the grower.

Cultivars that are susceptible to superficial scald development are treated with DPA by drenching, thermofogging or aerosol technologies. DPA is often co-applied with fungicides, and sometimes with calcium when drenching is used. 1-MCP is applied as a gas to sealed storage rooms either once when the room is filled or during loading in the case of cultivars such as McIntosh where the recommended interval between harvest and treatment can be as little as three days to obtain maximum responses of fruits to 1-MCP.

Grading and packing of apples after harvest varies greatly in scale depending on the size of the industry involved. In Farmers' Markets, for example, there is often less focus on uniformity of the fruit, and the fruits are displayed as field harvested in a range of container types from baskets to cardboard or wooden crates, with little sorting other than to exclude bruised, infected or damaged fruits. Most retail operations such as in supermarkets are more demanding. In some packing sheds, fruits of uniform size are graded visually and packed by hand into cartons before they are shipped to the market. However, the vast majority of fruits in western apple storages are graded using commercial packing lines. The capacity of these lines in large-scale operations, as found in apple-growing regions such as Washington State, is typically 50–60 bins/h but can be as much as 100 bins/h. Grading lines can be much smaller in a single grower operation with a limited number of storage rooms.

Grading may take place soon after harvest where fruits are graded by size and colour (pre-sizing) and then stored until needed for marketing. The advantage of this method is that the storage operator knows the volume of fruits in each market category, and that only high-quality fruits are stored. A normal grading and packaging operation has several standard processes that are similar internationally. Bins of fruits are removed from air or CA storage and dumped into the grading line, usually using water, so that fruits float out of the bins into water flumes and are carried by water flow to the conveyer belt while minimizing damage. The water is typically chlorinated in North America, but is typically unchlorinated in many European countries owing to concerns about carcinogens produced by the interaction of chlorine with organic matter. The first step in the process is elimination of leaves and then the apples receive visual inspection to remove any decayed or damaged fruits. The apples are then washed, waxed and then dried, and then hand sorted again to detect any decay or other culls, and then electronically sorted by size and colour. Fruits are then packed into cartons, labelled, stacked onto pallets by hand or by robotic equipment, and then cooled prior to transport to markets.

Internal quality assessment of fruits to determine factors such as SSC and presence of physiological disorders can also take place during the grading process if appropriate equipment is available. Advances in grading technology continue to be made, with emphasis on non-destructive assessments of quality. Optical techniques include vis/NIR spectroscopy, time- and space-resolved spectroscopy, machine vision and multispectral imaging, while mechanical techniques include impact analysis, vibration analysis and ultrasound (Nicolai et al., 2014). Applications of NIR spectroscopy include SSC, firmness, pH, acidity, dry matter, alcohol-insoluble solids, stiffness, sugars, background colour, SPI, bruises, sensory attributes and physiological disorder such as brown heart (Nicolai et al., 2007). SSC is the most consistent attribute measured, and commercial grading lines equipped with NIR sensors are widely available. Machine vision systems are also used widely for detection of colour, size, shape and surface defects (Zou and Zhao, 2015). Other techniques that may be useful for apples include X-ray radiography and tomography, magnetic resonance imaging, mass spectroscopy, gas sensors and electronic noses, but largely limited thus far to the laboratory (Nicolai et al., 2014).

6 Postharvest storage technologies

Common storage, where low temperatures are maintained by using cellars or in caves without refrigeration, is still used in some parts of the world. In China, for example, 'Fuji' apples are routinely stored until March under these conditions, but such long air storage periods are rarely used in developed countries because of higher quality standards expected by consumers. Postharvest innovations from early utilization of refrigerated storage to the commercial development of CA technology are the mainstay of apple industries around the world, and year-round storage capacity can be obtained with appropriate cultivars. Refrigeration, and to a much lesser extent, CA technology, is also used for transport of apples. The benefits of low storage temperature and CA storage on product quality are in part mediated through reducing respiration rates and in part reducing the production and action of ethylene. The recent commercialization of technologies based on 1-MCP, which prevents ethylene binding, has provided an additional tool for maintaining fruit quality, and is widely used internationally by apple industries. However, regardless of the technology used – refrigeration, CA storage and/or 1-MCP – the greatest limitation to storage of apples remains the inherent physiology that affects the ripening rates of any specific cultivar.

6.1 Refrigeration

In general, the lower the storage temperature, the longer the storage life of a given apple cultivar as long as the freezing temperature is not reached. The recommended storage temperature for most cultivars is 0°C, but can be as high as 4°C depending on sensitivity to chilling injuries (Watkins, 2003). Optimal temperatures can be different for fruits under air than CA storage conditions, for example, for 'McIntosh' 0°C and 2–3°C in air and CA, respectively.

Refrigeration technology has developed with greater attention to uniform maintenance of temperatures by adequate refrigeration capacity, uniform air distribution, minimal temperature differences between the evaporator coil and the air temperature, and precise temperature control systems (Thompson, 2016). Relative humidity control is usually achieved by minimizing temperature variation in the room and by operating the evaporator coil, with a high surface area, at a temperature close to the set point temperature of the room. Improved refrigeration technologies are likely reducing energy usage and thereby increasing sustainability. As discussed below, the use of CA and 1-MCP has provided additional opportunities for maintaining quality with less energy usage.

Cooling of apples is usually passive, where fruits are placed into refrigerated rooms with normal airflow. Although this method is inefficient because air flows *around* rather than *through* bins of apples, loss of condition is typically minimal because of the relatively slow ripening rate of apples. However, undesirable softening, especially of cultivars that soften rapidly, can occur if rooms are filled quickly and beyond the refrigeration capacity of the room. New rooms are usually equipped with greater refrigeration capacity than that required for only maintaining low temperatures. When capacity is limited, cooling can be improved by stacking bins appropriately to maximize airflow through the bins, or by using modified room cooling whereby empty rooms are used to cool fruits overnight before being moved to the final storage (Bartsch and Blanpied, 1990). Forced air cooling is sometimes used to cool apples packed in cartons (Watkins, 2003).

6.2 Controlled atmosphere (CA) storage

CA storage of apples, usually refers to lower O_2 and higher CO_2 partial pressures (commonly expressed as %) in the atmosphere around the fruit. The ability to accurately control storage atmospheres has improved greatly due to storage room quality (better sealed rooms with reduced gas exchange/leakage), and technologies for control of O_2 and CO_2 in the rooms. Over the past 50 years, CA technology has progressed from controlled ventilation (6–18% O_2; 3–15% CO_2) to conventional CA (2–5% O_2; 2–5% CO_2) to low O_2 (LO; 1.5–2% O_2; 1–3% CO_2) to ultralow O_2 (ULO; 0.8–1.2% O_2; 0.5–2% CO_2) storage (Hoehn et al., 2009). Rapid CA, where O_2 in the storage atmosphere is reduced to 3–5% within a few days of room closing is now common. These developments have led to better maintenance of fruit quality over longer storage periods. Additional CA technologies include initial low O_2 stress (ILOS) where O_2 levels are maintained as low as 0.25–0.7% for short periods after harvest, followed by LO or ULO storage (Wang and Dilley, 2000; Zanella, 2003). A driving force behind these stress treatment has been a desire to find non-chemical methods to control superficial scald, especially given the recent reduction of allowable DPA residues of apples in Europe.

The most recent advance in CA technology is dynamic CA (DCA) storage, which is based on the principle that lowering partial pressures of O_2 to the lowest tolerated by the fruit,

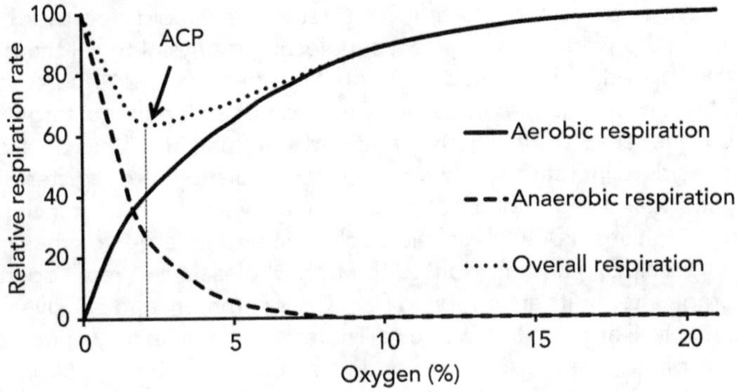

Figure 2 Aerobic, anaerobic and overall respiration rate of a generalized plant product in response to oxygen in the storage environment. The anaerobic compensation point (ACP) represents the lowest oxygen concentration achievable without injurious fermentation and represents the concentration at which the maximum storage potential can be obtained. Modified from Prange et al. (2003).

without causing excessive anaerobic respiration. Maintaining low O_2, typically <0.8% and CO_2 is <1.5% can result in further maintenance of quality over that achieved even by ULO, and further reduce risk of superficial scald development. Fruit responses to low O_2 can now be 'sensed' in real time as it reaches the anaerobic compensation point (ACP) below which fermentation occurs (Prange et al., 2003) (Fig. 2), so it is possible to alter the O_2 partial pressures in response to changes in fruit physiology that affect the ACP in storage over time, hence the term 'dynamic'. Available technologies are DCA-chlorophyll fluorescence (DCA-CF; HarvestWatch™) (Prange et al., 2013), ethanol (dynamic control system (DCS) (Schouten et al., 1997) and respiratory quotient (Bessemans et al., 2016; Gasser et al., 2008). Regardless of the sensor system used, the O_2 concentration is typically increased about 0.2% above the critical threshold to avoid damage to the fruit. As a result, firmness and colour retention is greater for DCA-stored than ULO or standard CA-stored fruits (DeLong et al., 2004b; Gabioud Rebeaud et al., 2015; Veltman et al., 2003; Zanella and Rossi, 2015).

6.3 1-Methylcyclopropene (1-MCP)

1-MCP is a competitive inhibitor of ethylene perception that binds irreversibly to ethylene-binding sites in plant cells, thereby preventing ethylene binding and the eliciting of subsequent signal transduction and translation. Use of 1-MCP as an ethylene inhibitor was patented 20 years ago by Sisler and Blankenship (1996). The commercialization of 1-MCP as the SmartFresh℠ Quality System by AgroFresh Inc. led to rapid adoption of 1-MCP-based technologies for many horticultural industries and by 2016, regulatory approval for use of 1-MCP had been obtained in almost 50 countries. 1-MCP is applied to harvested fruits in sealed rooms as a gas, released from powder or tablets. Current options for application of SmartFresh™ include SmartFresh ProTabs, SmartFresh SmartTabs and SmartFresh InBox (AgroFresh, 2016a). Competing products are now available: FYSIUM® from PACE International and TruPick from Decco/ESSENTIV.

1-MCP has several characteristics that have resulted in rapid approval by regulatory authorities around the world: it is a gaseous molecule at physiological temperatures and therefore easily applied, it is applied for short time periods (≤24 hours), it is active at low concentrations (≤1 μL L^{-1}), it has a unique mode of action that is targeted given the absence of ethylene receptors in invertebrates and mammals and it has an excellent safety profile with no residues in treated produce. 1-MCP formulations are registered for use on a wide variety of fruits and vegetables but apple is in many ways the 'poster child' for 1-MCP as a technology (Watkins, 2008). First, the apple is stored in large volumes in air and CA storage, and then shipped and distributed through wholesale and retail operations to the consumer. Storage after fruits are removed from CA is common, and abusive temperatures can occur during retail and shelf life periods. Therefore, the inhibitory effects of 1-MCP on ethylene and respiration rates, softening and loss of acidity, which result in maintenance of quality from shipping to when the consumer eats the fruit, have resulted in demand for treated products. Second, quality characteristics of the apple that are desired by consumers, such as crisp fracturable texture and sugar/acid balance, are similar to those found in the fruit at harvest. 1-MCP also has a major benefit of controlling superficial scald.

The availability of 1-MCP-based technology has resulted in a vast literature where many aspects related to its usage have been explored (Watkins, 2015). Examples of these are shown for apples in Table 1, and many of these findings have been incorporated

Table 1 Summary of factors that have been shown to affect responses of apple fruit to 1-methylcyclopropene (1-MCP) with selected references

Factor	Selected references
Cultivar	(Bai et al., 2005; Watkins, 2008)
Maturity	(Cocci et al., 2014; Doerflinger et al., 2015b; Jung et al., 2006; Mir et al., 2001; Rebeaud and Gasser, 2015; Watkins et al., 2000)
Storage type (DCA, ULO, CA compared with air)	(Brackmann et al., 2012; Rebeaud and Gasser, 2015; Watkins and Nock, 2005; Watkins et al., 2000; Zanella and Rossi, 2015)
Storage temperature	(Fan and Mattheis, 2001; Kittemann et al., 2015a; Mir et al., 2001; Toivonen and Lu, 2005)
Time between harvest and 1-MCP treatment	(DeEll et al., 2008; Kittemann et al., 2008; Rebeaud and Gasser, 2015; Talamini do Amarante et al., 2010; Tatsuki et al., 2007; Watkins and Nock, 2005)
Delayed CA storage after 1-MCP treatment	(DeEll and Ehsani-Moghaddam, 2012; Watkins and Nock, 2012)
Repeated 1-MCP applications	(DeEll and Ehsani-Moghaddam, 2013; DeEll et al., 2016; Lu et al., 2013; Nock and Watkins, 2013b; Nock and Watkins, 2013a)
DPA metabolism and degradation	(Arquiza et al., 2005; Rudell et al., 2006)
Non-target materials such as wooden bins	(Ambaw et al., 2011; Ambaw et al., 2013; Rodriguez-Perez et al., 2009; Vallejo and Beaudry, 2006)

into guidelines aimed to maximize effectiveness of SmartFresh[SM] for each apple industry (AgroFresh, 2016b). These guidelines, for example, provide definition of the maximum time between harvest and treatment, which can range from 3 days for 'McIntosh' to 10 days for 'Fuji' to modifications for delaying CA establishment after 1-MCP treatment. The SmartFresh[TM] label was modified in the United States in 2009 and Canada in 2011 to allow multiple 1-MCP treatments, thereby permitting application of 1-MCP to the fruit more than once while storage rooms are being loaded, or during storage. 1-MCP application to the fruit before rooms are completely filled can improve maintenance of quality and more effectively reduce incidence of superficial scald and senescent disorders (Lu et al., 2013; Nock and Watkins, 2013), and reduce the need for rapid CA (Watkins and Nock, 2012).

One concern about 1-MCP use has always been flavour, as aroma volatiles are so closely linked with ethylene production (Schaffer et al., 2007; Yang et al., 2016). 1-MCP, as with CA, especially low ULO, is known to inhibit production of aroma volatiles (Bai et al., 2005; Fan and Mattheis, 1999; Ferenczi et al., 2006), although reduced aroma volatile production as a result of DCA storage was less than for 1-MCP treatment (Raffo et al., 2009). However, several studies have shown that consumers prefer 1-MCP treated fruits to untreated fruits (Marin et al., 2009; Pre-Aymard et al., 2005; Tomic et al., 2016; Wunsche and Heyn, 2015). These responses by consumers reinforce the key role of texture in consumer acceptability. While production of volatiles can be reduced by 1-MCP, some of them, especially esters are associated with over-ripening of the fruit and therefore, they are not necessarily desirable.

An interesting feature of 1-MCP is its effects on physiological disorders. Not only has understanding these effects been important in managing susceptibility of cultivars to disorders, but 1-MCP has provided a powerful research tool to understand the role of ethylene in their development (see Section 9).

7 Comparing and assessing storage technologies

7.1 Comparing DCA and 1-MCP as technologies

1-MCP and DCA are new technologies that have had significant impacts on apple industries. Interestingly, while both technologies are available worldwide, the use in Europe, especially in Italy, is much greater than elsewhere. In North America, for example, few commercial storages are using DCA technology, and these are mainly restricted to Washington organic storages. In part, the divergence may relate to differences in availability of the two technologies. 1-MCP was registered for use on apples in North and South Americas as early as 2001, but not in the Europe Union until 2005. By contrast, DCA-CF was available in 2002, and in Europe where 1-MCP was not available, coincided with significant building of new high-quality storage facilities suitable for DCA application. Thus DCA represented the logical step in CA technology for maintenance of fruit quality at a time when 1-MCP was not available. Both technologies have advantages and disadvantages as listed below. Important factors for selecting either of them include the availability of the technology relative to scale of a particular industry, how the technologies are integrated into existing infrastructure, availability of financial investment required for high-quality storages suitable for DCA and demand for the fruit without any postharvest chemical treatment. Some storage operators are using both technologies.

1-MCP

Advantages

- Applied as a gas, active at very low concentrations ($<1\ \mu L\ L^{-1}$) and application period less than 24 hours.
- No investment in computerized CA storage technology required.
- High-quality storage rooms are not required.
- No risk of low oxygen injury associated with operator error, or fruits of mixed quality.
- 1-MCP application can occur as required at harvest or during storage.
- Treatment can be adapted to any room size, and partial or full rooms can be treated.
- Rooms can be opened and closed, without loss of fruit quality.
- Incorporation into packaging materials possible.
- Maintains quality of treated products in air.
- Controls some physiological disorders such as superficial scald.
- Maintains quality, especially texture and acidity, during entire marketing chain after storage.

Disadvantages

- Not listed for use on organic fruit.
- Cost of 1-MCP for application.
- Can increase risk of physiological disorder development, especially CO_2-related injuries, resulting in modification of handling protocols, or avoidance of highly susceptible cultivars such as 'Braeburn'.

DCA

Advantages

- Chemical free and therefore suitable for organic fruit.
- Easily installed within existing CA storage facilities, if they are of high-quality construction.
- Can provide control of physiological disorders such as superficial scald.

Disadvantages

- Requires high-quality storage rooms to enable economic maintenance of low O_2 concentrations.
- Higher initial investment, especially in sensors and computerized CA storage technology.
- Requires selection of representative uniform fruit.
- Requires training of storage operator to interpret changes in product responses to O_2 concentrations.
- Costly for container-based operations.
- Greater potential for loss of quality after removal of products from CA storage.

7.2 Storage technologies and sustainability

Temperature control contributes to the costs, carbon footprint and total energy use of products, and a compromise between maintaining storage conditions that result in maximum storage life and energy costs of operation exists (East et al., 2013). Under CA conditions, cycling of refrigeration capacity, for example turning off evaporator fans, resulted in 30 to 60% energy savings without loss of apple quality (East et al., 2013; Yost, 1984). Inhibiting ethylene production by treatment of fruits with 1-MCP alone or in combination with ULO and/or DCA technology may also allow warmer temperatures to be used without loss of quality, and with associated energy savings (Kittemann et al., 2015b; Mazzurana et al., 2016; McCormick et al., 2010). 1-MCP treatment can increase storage periods of fruits kept under non-refrigerated conditions (Mir et al., 2001, 2013), providing potential to maintain quality under suboptimal storage conditions. Warmer storage temperatures may not always be desirable, however, as responses may be cultivar specific, for example 1-MCP treated 'Empire' apples are more susceptible to browning at 2°C than at 0.5°C (Watkins, 2008).

8 Postharvest treatments

Depending on the particular growing region and marketing plan, apples may be treated immediately after harvest, or treated after removal from storage during packaging operations. 1-MCP is a postharvest treatment but is discussed above as a technology because of its effects on fruit ripening and storability. Other common treatments at the time of harvest are the application of fungicides to control decay (see chapter 15), and the antioxidant, DPA, to control the physiological disorder, superficial scald, on susceptible cultivars, but also secondarily, other disorders such as carbon dioxide injury (see Section 9). Edible coatings (waxes) are applied routinely to fruits at the time of packing. Other treatments such as heat treatments are applied to meet narrow requirements, for example non-chemical control of decay.

8.1 Diphenylamine (DPA)

DPA was shown to be a potent inhibitor of superficial scald development (Smock, 1961), and was quickly registered for commercial use in the United States and around the world due to its low cost, easy application and low toxicity. DPA and the technology of DPA application have been extensively reviewed (Little and Holmes, 2000). Standard application practice has been DPA dips or drenches, typically 1–2 g L^{-1}, as permitted by label, with maximum residue levels (MRL) of between 5 and 10 mg kg^{-1}, depending on the country. Food safety concerns associated with recirculating drenches lead to the development of alternative application methods, including non-circulating DPA applications, where only the desired amount of DPA is applied to fruits in each bin, and increasingly, DPA applications by thermofogging (PACE, 2016) and aerosol technologies (DECCO, 2016).

The future of DPA is uncertain, however, as although DPA is considered safe, the European Commission withdrew authorization for DPA use in 2009, and as of March 2014, established the MRL for apples (and pears) at 0.01 mg kg^{-1} based on the precautionary principle related to gaps in data related to DPA by-products. Research in Europe has shown that DPA contamination of fruit storage infrastructures can result in cross-contamination of apples during storage, even in rooms kept free of DPA-treated fruits for 3 years (Robatscher et al., 2012). Cross-contamination risk is a major concern for countries such as South Africa and the United States that export the fruit to Europe.

DPA prevents superficial scald by inhibiting the oxidation of α-farnesene and the subsequent cell damage by free radical activity (Lurie and Watkins, 2012). Low O_2, ILOS and DCA probably inhibit scald by reducing α-farnesene production as well as its oxidation, while 1-MCP inhibition of ethylene production results in less availability of α-farnesene. 1-MCP, DCA and other low O_2 strategies provide alternatives to DPA, and have been used commercially with great success in Italy and elsewhere (Prange et al., 2015).

DPA has other effects on fruit physiology and susceptibility physiological disorders, for example, inhibiting ripening processes such as degreening of the fruit (Lurie et al., 1989; Tomic et al., 2016). Because DPA has been used for so long, storage operators are not always aware of its prophylactic benefits on other disorders. One example was the unrecognized benefits of DPA in inhibiting CO_2 injury; when recommendations were made to stop using DPA on the 'Empire' apple, which has low susceptibility to scald, massive fruit losses occurred as a result of external CO_2 injury (Burmeister and Dilley, 1995). Development of internal CO_2 injuries can also be inhibited by DPA (Contreras et al., 2014; Mattheis and Rudell, 2008). 1-MCP treatment has also been shown to further increase susceptibility of the fruit to CO_2 injury (DeEll et al., 2003; Fawbush et al., 2008; Nock and Watkins, 2013). Susceptibility to injury decreases with pre-storage in air before the fruits are subjected to CAs (Watkins et al., 1997), but 1-MCP maintains this sensitivity (Fawbush et al., 2008). If DPA is not used, additional handling steps have to be put in place to minimize risk of injury, and in fact, use of DPA on CO_2 injury-susceptible cultivars, has increased as a result of increased 1-MCP applications.

8.2 Edible coatings

Edible coatings (waxes) are thin layers of edible material applied to the surfaces of fresh fruits and vegetables in addition to, or as a replacement for, natural protective waxy coatings. They can extend the shelf life of apples by reducing moisture loss and modifying the internal atmospheres, as well as improving their appearance (Dhall, 2013; Falguera et al., 2011; Serrano et al., 2015). Commonly used waxes in the United States are made from natural ingredients certified by the Food and Drug Administration as safe to eat and they come from sources such as carnauba (extracted from leaves of the Brazilian palm, *Copernicia prunifera*), candelilla wax (derived from a reed like desert plant of the genus *Euphorbia*), and food-grade shellac (excreted by female lac beetles, *Kerria lacca*). The potential of edible coatings for apples as carriers or delivery systems for antimicrobials, texture enhancers and nutraceuticals appears to be largely unrealized.

Although waxes might be useful to maintain quality of air-stored fruit, they are normally applied on the packing line only to those fruit graded for appropriate size and colour to enhance cosmetic appeal. Waxing is not traditionally used in some countries, for example China, but wax coatings on apples are widely accepted in most markets. Opposition to their use does exist, however. Waxes have been the subject of ill-informed scaremongering, for example, claims that they are carcinogens and/or trap pesticides within the fruit. Other problems with waxes are due to international differences in registration of other components in the wax. For example, morpholine is used as an emulsifier in waxes, and fruits with wax containing morpholine shipped from the United States to Europe have been rejected, not because of health concerns but because no application was ever made to approve this additive in the EU, and therefore it should not be present on food sold in Europe.

Synthetic waxes are not allowable on organic fruits, but shellac and carnauba are permissible. Vegans will not eat apples coated with shellac because it is excreted by female lac beetles; in contrast, according to Kosher rules, shellac can be used even though it is derived from an insect, because it is excreted (Regenstein et al., 2003).

8.3 Heat treatments

Heat treatments, applied as hot water, hot air or vapour heat, can inhibit physiological disorders such as superficial scald, reduce decay and kill insects (Fallik et al., 2001; Lurie, 1998). Heat treatment is not subject to regulations that affect its adoption, although when used for quarantine purposes, there are strict parameters established to guarantee insect mortality. Commercial application of heat treatments has been challenging for several reasons including difficulties associated with treating bulk loads of fruit, the narrow temperature x time margin between the desired product response and heat damage and the lower decay control efficacy relative to fungicides. In addition, variability of product responses can be caused by maturity stage, growing conditions and handling between harvest and treatment. Application of heat treatments is generally restricted to specific situations where alternative options for market access are limited. Hot water treatments are used to a limited extent to reduce decay of organic apples in Denmark and Germany (Maxin et al., 2014) and heat treatment is part of commercial processing of fresh-cut apple slices (FreshAppeal, 2016).

Intermittent warming is a form of heat treatment that has also received extensive study as a means of alleviating chilling injury of horticultural products, including superficial scald of apples (Lurie and Watkins, 2012). However, in a commercial environment such treatments are logistically challenging for any horticultural product (Biswas et al., 2016), primarily due to the inability to warm and cool large volumes of fruits in uniform patterns, and the expense of refrigeration for bulk cooling.

9 Physiological disorders

A large number of physiological disorders associated with senescence, mineral imbalances, low O_2 and/or high CO_2 concentrations, and low storage temperatures have been described for apple (Smock, 1977; Snowdon, 2008; Wilkinson and Fidler, 1973). Susceptibility to disorders is a function of cultivar, harvest date, storage conditions and storage length. Several of these have gained increased attention because of their occurrence on new cultivars such as 'Honeycrisp', because of the effects of 1-MCP, and potential unavailability of DPA. The availability of 1-MCP has also provided an experimental tool to further understand the interaction of ethylene with disorder development.

Major disorders of apples include:

1 *Watercore* is characterized by a translucent watersoaked appearance, which is nitially associated with the vascular bundles of the coreline, but which can extend from the

core to the skin surface (Marlow and Loescher, 1984). Susceptible cultivars include 'Fuji' and 'Red Delicious'. Features that are consistently associated with watercore are flooding of the intercellular spaces, low levels of reducing sugars and high sorbitol concentrations. Watercored tissue has a lower intercellular air space volume, reduced permeance to gas diffusion and increased internal CO_2 concentrations, and accumulation of ethanol and acetaldehyde can occur in apples with watercore even under ambient O_2 conditions (Argenta et al., 2002). Watercore symptoms can be expressed as radial or block types (Harker et al., 1999). The temporal development of watercore, the morphology and severity of symptoms can differ among growing regions, apparently due to environment factors rather than maturity. 1-MCP can delay loss of watercore during storage, and can increase risk of internal browning disorder development, especially under CA storage conditions. Protocols to manage the effects of 1-MCP on watercore have included using air instead of CA storage, delaying imposition of low O_2 conditions and delaying 1-MCP application (Mattheis, 2008).

2 *Senescent breakdown*, which as its name implies is associated with senescent fruit, typically harvested late and/or stored for too long. The disorder is less commonly observed with modern handling and storage protocols such as rapid cooling to optimal storage temperatures after harvest. However, senescent breakdown may still be found in fruits with low calcium and high nitrogen, especially when fruits are large. Inhibition of ethylene by 1-MCP, and therefore of ripening, can reduce disorder incidence (DeLong et al., 2004a; Lee et al., 2016; Moran and McManus, 2005).
3 *Bitter pit* is the physiological breakdown of cells under the skin and/or throughout the fruit, often at the calyx end of the fruit either before harvest or during storage (Ferguson and Watkins, 1989). It can cause serious losses in apple cultivars such as 'Cox's Orange Pippin', 'Delicious', 'Idared', 'Cortland' and 'Honeycrisp'. In addition to genetic susceptibility (Volz et al., 2006), climate and orchard management practices play important roles (Ferguson and Watkins, 1992). Calcium concentrations, especially in relation to other minerals, in the fruit tissues are important, with low calcium concentrations typically resulting in higher susceptibility of the fruit to bitter pit. No direct effects of DPA or 1-MCP on susceptibility to bitter pit have been described, and reported effects of 1-MCP are inconsistent (Gago et al., 2015; Pesis et al., 2009). However, 1-MCP may induce a blotch-like disorder in fruits that are susceptible to bitter pit development.
4 *Superficial scald* is a disorder thought to result from chilling injury (Watkins et al., 1995). The disorder occurs after long-term cold storage and usually manifests as brown or black patches on the fruit skin. Scald is an expression of cell death within the surface layers of cells in localized regions as a result of free radical activity associated with oxidation of α-farnesene (reviewed by Lurie and Watkins, 2012). Both 1-MCP and DPA can inhibit scald development, but by different mechanisms. α-Farnesene production and accumulation in the fruit skin is closely associated with autocatalytic ethylene production during ripening, and therefore inhibition of ethylene production by 1-MCP results in the absence of the substrate for oxidation and subsequent free radical injury to cells (Fan and Mattheis, 2000; Watkins et al., 2000). By contrast, DPA acts by inhibiting oxidation of α-farnesene rather than affecting its availability (Arquiza et al., 2005).
5 *External and internal CO_2 injuries* are typically associated with CA storage of apples, although they can also occur in air storage where poor ventilation results in excessive

CO_2 concentrations, for example, cartons or ship holds (Snowdon, 2008). Susceptible cultivars include 'Braeburn', 'Fuji', 'McIntosh', 'Cortland', 'Empire' and 'Honeycrisp'. Susceptibility of the fruit to CO_2 injury is at its highest during early CA storage, and the risk of injury can be reduced by maintaining low levels of CO_2 in storage atmospheres during this early period of storage. Concerns about susceptibility of the fruit to injury have formed the basis of specific recommendations for some cultivars, for example, keeping CO_2 concentrations low at 2% for 4 to 6 weeks before increasing to 5% for the reminder of the storage period for 'McIntosh' (Blanpied, 1977), delaying CA application for 2 weeks and maintaining CO_2 concentrations lower than 1% for 'Braeburn' (Elgar et al., 1998), or conditioning at warmer temperatures before storage prior to CA application in the case of 'Honeycrisp' (Contreras et al., 2014; Lum et al., 2016).

DPA is a powerful inhibitor of CO_2 injuries (Burmeister and Dilley, 1995; Contreras et al., 2014; Fawbush et al., 2008; Lee et al., 2012b; Watkins et al., 1997). 1-MCP enhances risk of CO_2 injuries (DeEll et al., 2003; Fawbush et al., 2008; Hatoum et al., 2014b; Watkins and Nock, 2004) and has led to greater use of DPA by some storage operators to reduce risk. The mechanism of 1-MCP action on CO_2 injury is not known, but by inhibiting ethylene production, it appears to prevent the reduction of injury susceptibility associated with delays prior to CA establishment (Fawbush et al., 2008).

6 *Soft scald and soggy breakdown*. The symptoms of soft scald (syn. deep scald, ribbon scald) are the development of sharply defined brown lesions on the skin of the apple, which can extend into the flesh. Susceptible cultivars include 'Fuji', 'Jonathan', 'McIntosh', 'Delicious', 'Golden Delicious' and 'Honeycrisp'. Soggy breakdown is an internal disorder that can occur alone or with soft scald, and in worst cases a complete ring of soft, brown, spongy tissue is present. Both disorders are forms of low-temperature injury, being worse in fruits stored below 2–3°C. These disorders have received attention due to increasing production volumes of 'Honeycrisp', and a conditioning treatment of 10°C for 7 days before storage at 3°C (Watkins et al., 2004) has been implemented commercially. DPA and 1-MCP also decrease but do not always control the disorders (Watkins et al., 2004; Wills and Scott, 1982). Greater maturity of fruits at harvest is a major disposing factor implicated in the occurrence of soft scald (Fan and Mattheis, 2000; Tong et al., 2003; Wargo and Watkins, 2004). Wet conditions and cool temperatures prior to harvest may increase susceptibility of the fruit to the disorder (Lachapelle et al., 2013; Moran et al., 2009).

7 *Firm flesh browning* is a serious disorder of 'Empire' apples (Watkins and Liu, 2010). Although the current storage temperature recommendation for this cultivar is 2°C to reduce risk of chilling injury development without excessive softening, 1-MCP treatment can increase flesh browning incidence of the fruit at these warmer storage temperatures (Watkins, 2008), especially in later harvested fruits (Doerflinger et al., 2015b). Development of flesh browning has become a serious limiting factor for the long-term CA storage of the cultivar (DeEll et al., 2007; Fawbush et al., 2009). It has been hypothesized that flesh browning may result from inhibition of ethylene production by the fruit that results in abnormal metabolism; this inhibition may occur either at low chilling-injury-inducing temperatures (0.5°C) with and without 1-MCP, or at warmer temperatures (3–4°C) when the fruit has been treated with 1-MCP (Jung and Watkins, 2011). Analysis of the development of flesh browning in 'Empire' apples has revealed that the disorder originates in the stem end of the fruit (Lee et al., 2012a). Interestingly, a new disorder of 'Gala' apples, known as stem end browning, has a similar pattern of development (Lee et al., 2013, 2016).

Table 2 Summary of 'omic (genomic, proteomic and metabolomic) studies of major physiological disorders of apple fruit with selected references

Disorder	Selected references
Bitter pit	(Krawitzky et al., 2016)
Superficial scald	(Busatto et al., 2014; Farneti et al., 2015a; Leisso et al., 2013; Mditshwa et al., 2016; Rudell et al., 2011; Rudell et al., 2009) Du et al., 2016
External and internal carbon dioxide injury	(Gapper et al., 2013; Hatoum et al., 2014a; Hatoum et al., 2016; Mellidou et al., 2014; Vandendriessche et al., 2013)
Soft scald	(Leisso et al., 2015)
Firm flesh browning	(Lee et al., 2012c)

8 *Diffuse skin browning* (DSB) is a new disorder that to date has been limited to 'Golden Delicious' clones treated with 1-MCP (Larrigaudiere et al., 2010). Although it occurs in the early harvested fruit, and is similar in visual symptoms to superficial scald, it is not inhibited by DPA. DSB is associated with sensitivity of clones to russeting. DSB is controlled by employing progressive cooling strategies (Larrigaudiere et al., 2010).

Recent advances in 'omic technologies have enabled extensive studies on genomics and metabolomics and to a lesser extent, proteomics, of several physiological disorders. The objectives of these studies includes gaining knowledge of the aetiology of disorder development, improving prediction and diagnosis of storage disorders to inform treatment and storage management decisions, and developing biomarker-based diagnostic protocols. These studies are significantly advancing knowledge, especially given the paucity of information about disorders other than superficial scald. Even for that disorder, while the α-farnesene hypothesis for scald development is well accepted, extensive metabolomic differentiation between untreated fruit and those treated with DPA or 1-MCP well before visual symptoms of damage has been shown (Rudell et al., 2009). A detailed review of 'omic studies of physiological disorders is outside of the scope of this chapter, but examples are listed in Table 2. In general, both primary and secondary metabolism are affected, including energy-related (linked to the tricarboxylic acid cycle and the electron transport chain), oxidative stress and lipid-related genes (related to membrane alterations and fatty acid oxidation). Future prospects for prediction and diagnostic tools for physiological disorders are good resulting in selection and management of fruit lots that will store without unexpected losses associated with variations in orchard management and seasonal conditions.

10 Summary

Apples are widely grown around the world and under diverse management systems. The modern orchard, however, is one that is based on intensive plantings of uniform dwarf or semi-dwarf trees as discussed in other chapters of this book. The drive towards these intensive plantings results from reduced pruning and labour costs, more uniform maturity and quality, and efficient harvest. The apples produced by these systems represent a much greater investment than is the case for traditional orchards and therefore optimal harvest, handling and storage management procedures are essential for sustainability of the grower. These changes are occurring at a time of consumer demand for high-quality apples with excellent sensory attributes with minimal chemical usage, but at the same time for defect-free fruits that meet the highest food safety standards. Management practices that ensure good agricultural practices, including product traceability during harvest, handling, storage and transport are being reinforced by food safety and/or organic legislation.

To meet these demands, plant breeders will continue to develop new and exciting cultivars for consumers, with attention to desirable ripening characteristics, such as lower ethylene production and slower softening rates; storage characteristics, including resistance to physiological and pathological disorder; and handling ability, for example skin characteristics that are necessary for transport needs such as low susceptibility to bruising. These demands will also be met not only by a focus on standard harvest indices such as colour (red and background) SPIs, firmness and SSC is critical, but increasingly on application of quality tools such as non-destructive measurement of dry matter to predict quality of the fruit after storage.

Storage technologies have advanced greatly with improvements in refrigeration technology and CA storage management such as rapid CA technology as a means of maintaining quality and reducing waste. In the last decade or so, two new technologies, 1-MCP and DCA, have been integrated into many apple storage operations. The commercialization of 1-MCP has led to improved fruit quality maintenance throughout the entire marketing chain and especially for the consumer. DCA, on the other hand, has resulted from advances in technology, which now allow measurement of physiological responses of the fruit when stressed and therefore safe storage of the fruit at oxygen concentrations previously assumed to be too risky to apply under commercial conditions. Together with improved grading equipment, including internal quality analysers, the ability to provide the consumer with high-quality apples continues to improve.

11 Where to look for further information

For additional information on postharvest handling, see books by B.H. Wills and J.B. Golding 2016. Postharvest. An Introduction to the Physiology and Handling of Fruit and Vegetable. 6th edition, CABI, Boston. ISBN: 9781786391483. Good resources that cover apples and associated storage technologies are found in USDA Handbook 66, The Commercial Storage of Fruits, Vegetables, and Florist and Nursery Stocks, revised 2016. K.C. Gross, C.Y. Wang and M. Saltveit (eds). https://www.ars.usda.gov/ARSUserFiles/oc/np/.../CommercialStorage.pdf

12 References

AgroFresh. (2016a). http://www.agrofresh.com/web/portal/smartfresh-technology.

AgroFresh. (2016b). SmartFresh™ Quality System, Apple Use Recommendations. *AgroFresh Inc, PA.*

Ambaw, A., Beaudry, R., Bulens, I., Delele, M. A., Ho, Q. T., Schenk, A., Nicolai, B. M. and Verboven, P. (2011). Modeling the diffusion-adsorption kinetics of 1-methylcyclopropene (1-MCP) in apple fruit and non-target materials in storage rooms. *Journal of Food Engineering* 102, 257–65.

Ambaw, A., Verboven, P., Defraeye, T., Tijskens, E., Schenk, A., Opara, U. L. and Nicolai, B. M. (2013). Effect of box materials on the distribution of 1-MCP gas during cold storage: A CFD study. *Journal of Food Engineering* 119, 150–8.

Anon (1974). chapter 16-403 WAC: Standards for apples marketed within the State of Washington. http://apps.leg.wa.gov/WAC/default.aspx?cite=16-403&full=true.

Argenta, L., Fan, X. T. and Mattheis, J. (2002). Impact of watercore on gas permeance and incidence of internal disorders in 'Fuji' apples. *Postharvest Biology and Technology* 24, 113–22.

Arquiza, J., Hay, A. G., Nock, J. F. and Watkins, C. B. (2005). 1-Methylcyclopropene interactions with diphenylamine on diphenylamine degradation, α-farnesene and conjugated trienol concentrations, and polyphenol oxidase and peroxidase activities in apple fruit. *Journal of Agricultural and Food Chemistry* 53, 7565–70.

Bai, J. H., Baldwin, E. A., Goodner, K. L., Mattheis, J. P. and Brecht, J. K. (2005). Response of four apple cultivars to 1-methylcyclopropene treatment and controlled atmosphere storage. *HortScience* 40, 1534–8.

Baietto, M. and Wilson, A. D. (2015). Electronic-nose applications for fruit identification, ripeness and quality grading. *Sensors* 15, 899–931.

Bartsch, J. A. and Blanpied, G. D. (1990). Refrigeration and Controlled Atmosphere Storage for Horticultural Crops. *Bulletin 22, Northeast Regional Agricultural Engineering Service (NRAES)*, 45pp.

Bessemans, N., Verboven, P., Verlinden, B. E. and Nicolai, B. M. (2016). A novel type of dynamic controlled atmosphere storage based on the respiratory quotient (RQ-DCA). *Postharvest Biology and Technology* 115, 91–102.

Betemps, D. L., Carlos Fachinello, J., Galarca, S. P., Portela, N. M., Remorini, D., Massai, R. and Agati, G. (2012). Non-destructive evaluation of ripening and quality traits in apples using a multiparametric fluorescence sensor. *Journal of the Science of Food and Agriculture* 92, 1855–64.

Biswas, P., East, A. R., Hewett, E. W. and Heyes, J. A. (2016). Intermittent warming in alleviating chilling injury-a potential technique with commercial constraint. *Food Bioprocess Technology* 9, 1–15.

Blanpied, G. D. (1977). Requirements and recommendations for eastern and midwestern apples. *In 'Controlled atmospheres for the storage and transport of perishable agricultural commodities' [see FSTA (1979) 11 8J1407]*. 225–30.

Blanpied, G. D. and Silsby, K. J. (1992). Predicting harvest date windows for apples *Cornell Coop. Ext. Bul.* 221, 12.

Brackmann, A., de Gasperin, A. R., Both, V., Pavanello, E. P., Weber Schorr, M. R. and Anese, R.d.O. (2012). Application of 1-methylcyclopropene, initial low oxygen stress and storage in ultralow oxygen on Fuji apples quality. *Ciencia Rural* 42, 1386–91.

Burmeister, D. M. and Dilley, D. R. (1995). A scald-like controlled-atmosphere storage disorder of Empire apples - a chilling injury-induced by CO_2. *Postharvest Biology and Technology* 6, 1–7.

Busatto, N., Farneti, B., Tadiello, A., Vrhovsek, U., Cappellin, L., Biasioli, F., Velasco, R., Costa, G. and Costa, F. (2014). Target metabolite and gene transcription profiling during the development of superficial scald in apple (*Malus x domestica* Borkh). *BMC Plant Biology* 14.

Chauvin, M. A., Ross, C. F., Pitts, M., Kupferman, E. and Swanson, B. (2010). Relationship between instrumental and sensory determination of apple and pear texture. *Journal of Food Quality* 33, 181–98.

Cocci, E., Sacchetti, G., Rocculi, P. and Dalla Rosa, M. (2014). Response of Pink Lady(R) apples to post-harvest application of 1-methylcyclopropene as a function of applied dose, maturity at harvest, storage time and controlled atmosphere storage. *Journal of the Science of Food and Agriculture* 94, 2691–8.

Contreras, C., Alsmairat, N. and Beaudry, R. (2014). Prestorage conditioning and diphenylamine improve resistance to controlled-atmosphere-related injury in 'Honeycrisp' apples. *HortScience* 49, 76–81.

Corollaro, M. L., Aprea, E., Endrizzi, I., Betta, E., Dematte, M. L., Charles, M., Bergamaschi, M., Costa, F., Biasioli, F., Grappadelli, L. C. and Gasperi, F. (2014). A combined sensory-instrumental tool for apple quality evaluation. *Postharvest Biology and Technology* 96, 135–44.

Costa, F., Cappellin, L., Longhi, S., Guerra, W., Magnago, P., Porro, D., Soukoulis, C., Salvi, S., Velasco, R., Biasioli, F. and Gasperi, F. (2011). Assessment of apple (*Malus x domestica* Borkh.) fruit texture by a combined acoustic-mechanical profiling strategy. *Postharvest Biology and Technology* 61, 21–8.

Costamagna, F., Giordani, L. and Costa, G. (2013). Use of I_{AD} index to define harvest time and characterize ripening variability at harvest in 'Gala' apple fruit. *Acta Horticulturae* 998, 117–23.

DECCO (2016). http://deccous.com/products.

DeEll, J. and Ehsani-Moghaddam, B. (2013). Effects of rapid consecutive postharvest 1-methylcyclopropene treatments on fruit quality and storage disorders in apples. *HortScience* 48, 227–32.

DeEll, J. R., Ayres, J. T. and Murr, D. P. (2007). 1-Methylcyclopropene influences 'Empire' and 'Delicious' apple quality during long-term commercial storage. *HortTechnology* 17, 46–51.

DeEll, J. R., Ayres, J. T. and Murr, D. P. (2008). 1-Methylcyclopropene concentration and timing of postharvest application alters the ripening of 'McIntosh' apples during storage. *HortTechnology* 18, 624–30.

DeEll, J. R. and Ehsani-Moghaddam, B. (2012). Delayed controlled atmosphere storage affects storage disorders of 'Empire' apples. *Postharvest Biology and Technology* 67, 167–71.

DeEll, J. R., Lum, G. B. and Ehsani-Moghaddam, B. (2016). Effects of multiple 1-methylcyclopropene treatments on apple fruit quality and disorders in controlled atmosphere storage. *Postharvest Biology and Technology* 111, 93–8.

DeEll, J. R., Murr, D. P., Willey, L. and Porteous, M. D. (2003). 1-Methylcyclopropene (1-MCP) increases CO_2 injury in apples *Acta Horticulturae* 600, 277–80.

DeLong, J., Prange, R., Harrison, P., Nichols, D. and Wright, H. (2014). Determination of optimal harvest boundaries for Honeycrisp™ fruit using a new chlorophyll meter. *Canadian Journal of Plant Science* 94, 361–9.

DeLong, J. M., Prange, R. K. and Harrison, P. A. (2004a). The influence of 1-methylcyclopropene on 'Cortland' and 'McIntosh' apple quality following long-term storage. *HortScience* 39, 1062–5.

DeLong, J. M., Prange, R. K., Harrison, P. A. and McRae, K. B. (2000). Comparison of a new apple firmness penetrometer with three standard instruments. *Postharvest Biology and Technology* 19, 201–9.

DeLong, J. M., Prange, R. K., Leyte, J. C. and Harrison, P. A. (2004b). A new technology that determines low-oxygen thresholds in controlled-atmosphere-stored apples. *HortTechnology* 14, 262–6.

Dhall, R. K. (2013). Advances in edible coatings for fresh fruits and vegetables: a review. *Crit. Rev. Food Science Nutrition* 53, 435–50.

Dixon, J. and Hewett, E. W. (2000). Factors affecting apple aroma/flavour volatile concentration: a review. *New Zealand Journal of Crop and Horticultural Science* 28, 155–73.

Doerflinger, F. C., Miller, W. B., Nock, J. F. and Watkins, C. B. (2015a). Relationships between starch pattern indices and starch concentrations in four apple cultivars. *Postharvest Biology and Technology* 110, 86–95.

Doerflinger, F. C., Nock, J. F., Al Shoffe, Y., Shao, X. and Watkins, C. B. (2016). Non-destructive maturity assessment of 'Empire' apples treated with preharvest inhibitors of ethylene production with a delta absorbance (DA) meter. *Acta Horticulturae* 1119:227–33. DOI 10.17660/ActaHortic.2016.1119.32.

Doerflinger, F. C., Rickard, B. J., Nock, J. F. and Watkins, C. B. (2015b). An economic analysis of harvest timing to manage the physiological storage disorder firm flesh browning in 'Empire' apples. *Postharvest Biology and Technology* 107, 1–8.

East, A. R., Smale, N. J. and Trujillo, F. J. (2013). Potential for energy cost savings by utilising alternative temperature control strategies for controlled atmosphere stored apples. *International Journal of Refrigeration* 36, 1109–17.

Eckhoff, L., Kopcke, D. and Dierend, W. (2009). Influence of harvest date and storage method on bruises sensitivity of apples. *Erwerbs-Obstbau* 51, 55–65.

Elgar, H. J., Burmeister, D. M. and Watkins, C. B. (1998). Storage and handling effects on a CO_2 related internal browning disorder of 'Braeburn' apples. *HortScience* 33, 719–22.

Falguera, V., Pablo Quintero, J., Jimenez, A., Aldemar Munoz, J. and Ibarz, A. (2011). Edible films and coatings: Structures, active functions and trends in their use. *Trends Food ScienceTechnology* 22, 292–303.

Fallik, E., Tuvia-Alkalai, S., Feng, X. and Lurie, S. (2001). Ripening characterisation and decay development of stored apples after a short pre-storage hot water rinsing and brushing. *Innovative Food Science & Emerging Technologies* 2, 127–32.

Fan, X. and Mattheis, J. P. (2000). Reduction of ethylene-induced physiological disorders of carrots and iceberg lettuce by 1-methylcyclopropene. *HortScience* 35, 1312–14.

Fan, X. T. and Mattheis, J. P. (1999). Impact of 1-methylcyclopropene and methyl jasmonate on apple volatile production. *Journal of Agricultural and Food Chemistry* 47, 2847–53.

Fan, X. T. and Mattheis, J. P. (2001). 1-methylcyclopropene and storage temperature influence responses of 'Gala' apple fruit to gamma irradiation. *Postharvest Biology and Technology* 23, 143–51.

Farneti, B., Busatto, N., Khomenko, I., Cappellin, L., Gutierrez, S., Spinelli, F., Velasco, R., Biasioli, F., Costa, G. and Costa, F. (2015a). Untargeted metabolomics investigation of volatile compounds involved in the development of apple superficial scald by PTR-ToF-MS. *Metabolomics* 11, 341–9.

Farneti, B., Gutierrez, M. S., Novak, B., Busatto, N., Ravaglia, D., Spinelli, F. and Costa, G. (2015b). Use of the index of absorbance difference (I_{AD}) as a tool for tailoring post-harvest 1-MCP application to control apple superficial scald. *Scientia Horticulturae* 190, 110–16.

Fawbush, F., Nock, J. F. and Watkins, C. B. (2008). External carbon dioxide injury and 1-methylcyclopropene. *Postharvest Biology and Technology* 48, 92–8.

Fawbush, F., Nock, J. F. and Watkins, C. B. (2009). Antioxidant contents and activity of 1-methylcyclopropene (1-MCP)-treated 'Empire' apples in air and controlled atmosphere storage. *Postharvest Biology and Technology* 52, 30–7.

Ferenczi, A., Song, J., Tian, M., Vlachonasios, K., Dilley, D. and Beaudry, R. (2006). Volatile ester suppression and recovery following 1-methylcyclopropene application to apple fruit. *Journal of the American Society for Horticultural Science* 131, 691–701.

Ferguson, I. B. and Watkins, C. B. (1989). Bitter pit in apple fruit. *Horticultural Reviews* 11, 289–355.

FreshAppeal (2016). http://www.fresh-appeal.com/.

Gabioud Rebeaud, S. and Gasser, F. (2015). Fruit quality as affected by 1-MCP treatment and DCA storage – a comparison of the two methods. *European Journal of Horticultural Science* 80 (1), 18-24 | DOI: 10.17660/eJHS.2015/80.1.3.

Gago, C. M. L., Guerreiro, A. C., Miguel, G., Panagopoulos, T., Sanchez, C. and Antunes, M. D. C. (2015). Effect of harvest date and 1-MCP (SmartFresh™) treatment on 'Golden Delicious' apple cold storage physiological disorders. *Postharvest Biology and Technology* 110, 77–85.

Gapper, N. E., Rudell, D. R., Giovannoni, J. J. and Watkins, C. B. (2013). Biomarker development for external CO_2 injury prediction in apples through exploration of both transcriptome and DNA methylation changes. AoB Plants 2013; doi: 10.1093/aobpla/plt021.

Gasser, F., Eppler, T., Naunheim, W., Gabioud, S. and Hoehn, E. (2008). Control of the critical oxygen level during dynamic CA storage of apples. *Agrarforschung* 15, 98–103.

Greene, D. W. and Schupp, J. R. (2004). Effect of aminoethoxyvinylglycine (AVG) on preharvest drop, fruit quality, and maturation of 'McIntosh' Apples. II. Effect of timing and concentration relationships and spray volume. *HortScience* 39, 1036–41.

Harker, F. R., Kupferman, E. M., Marin, A. B., Gunson, F. A. and Triggs, C. M. (2008). Eating quality standards for apples based on consumer preferences. *Postharvest Biology and Technology* 50, 70–8.

Harker, F. R., Maindonald, J. H. and Jackson, P. J. (1996). Penetrometer measurement of apple and kiwifruit firmness: Operator and instrument differences. *Journal of the American Society for Horticultural Science* 121, 927–36.

Harker, F. R., Marsh, K. B., Young, H., Murray, S. H., Gunson, F. A. and Walker, S. B. (2002). Sensory interpretation of instrumental measurements 2: sweet and acid taste of apple fruit. *Postharvest Biology and Technology* 24, 241–50.

Harker, F. R., Watkins, C. B., Brookfield, P. L., Miller, M. J., Reid, S., Jackson, P. J., Bieleski, R. L. and Bartley, T. (1999). Maturity and regional influences on watercore development and its postharvest disappearance in 'Fuji' apples. *Journal of the American Society for Horticultural Science* 124, 166–72.

Hatoum, D., Annaratone, C., Hertog, M. L. A. T. M., Geeraerd, A. H. and Nicolai, B. M. (2014a). Targeted metabolomics study of 'Braeburn' apples during long-term storage. *Postharvest Biology and Technology* 96, 33–41.

Hatoum, D., Buts, K., Hertog, M., Geeraerd, A. H., Schenk, A., Vercammen, J. and Nicolai, B. M. (2014b). Effects of pre- and postharvest factors on browning in Braeburn. *Horticultural Science* 41, 19–26.

Hatoum, D., Hertog, M. L. A. T. M., Geeraerd, A. H. and Nicolai, B. M. (2016). Effect of browning related pre- and postharvest factors on the 'Braeburn' apple metabolome during CA storage. *Postharvest Biology and Technology* 111, 106–16.

Hoehn, E., Prange, R. K. and Vigneault, C. (2009). Storage technology and applications, In Yahia, E. M. (Ed.), *Modified and Controlled Atmospheres for the Storage, Transportation, and Packaging of Horticultural Commodities.* CRC Press, Baca Raton.

Jaeger, S. R., Andani, Z., Wakeling, I. N. and MacFie, H. J. H. (1998). Consumer preferences for fresh and aged apples: A cross-cultural comparison. *Food Quality and Preference* 9, 355–66.

Jaeger, S. R., Antunez, L., Ares, G., Johnston, J. W., Hall, M. and Harker, F. R. (2016). Consumers' visual attention to fruit defects and disorders: A case study with apple images. *Postharvest Biology and Technology* 116, 36–44.

Johnston, J. W., Gunaseelan, K., Pidakala, P., Wang, M. and Schaffer, R. J. (2009). Co-ordination of early and late ripening events in apples is regulated through differential sensitivities to ethylene. *Journal of Experimental Botany* 60, 2689–99.

Jung, S. -K., Nock, J. F. and Watkins, C. B. (2006). Internal ethylene concentration and harvest date effects on responses of apple fruit to 1-MCP. *HortScience* 41, 1051–.

Jung, S.-K. and Watkins, C. B. (2011). Involvement of ethylene in browning development of controlled atmosphere stored 'Empire' apple fruit. *Postharvest Biology and Technology* 59, 219–26.

Karlsen, A. M., Aaby, K., Sivertsen, H., Baardseth, P. and Ellekjaer, M. R. (1999). Instrumental and sensory analysis of fresh Norwegian and imported apples. *Food Quality and Preference* 10, 305–14.

Khan, M. A., Olsen, K. M., Sovero, V., Kushad, M. M. and Korban, S. S. (2014). Fruit quality traits have played critical roles in domestication of the apple. *Plant Genome* 7, unpaginated.

Kittemann, D., McCormick, R. and Neuwald, D. A. (2015b). Effect of high temperature and 1-MCP application or dynamic controlled atmosphere on energy savings during apple storage. *European Journal of Horticultural Science* 80, (1) 33-38 | DOI: 10.17660/eJHS.2015/80.1.5.

Kittemann, D., McCormick, R. and Streif, J. (2008). Influence of 1-MCP application timing on the ripening behaviour of 'Elstar' apples after storage. *Acta Horticulturae* 796, 155–9.

Knee, M. (1972). Anthocyanin, carotenoid, and chlorophyll changes in peel of Cox's Orange Pippin apples during ripening on and off tree. *Journal of Experimental Botany* 23, 184-96.

Krawitzky, M., Orera, I., Lopez-Millan, A. F., Oria, R. and Val, J. (2016). Identification of bitter pit protein markers in Malus domestica using differential in-gel electrophoresis (DIGE) and LC-MS/MS. *Postharvest Biology and Technology* 111, 224–39.

Kumar, S., McGlone, A., Whitworth, C. and Volz, R. (2015a). Postharvest performance of apple phenotypes predicted by near-infrared (NIR) spectral analysis. *Postharvest Biology and Technology* 100, 16–22.

Kumar, S., Rowan, D., Hunt, M., Chagne, D., Whitworth, C. and Souleyre, E. (2015b). Genome-wide scans reveal genetic architecture of apple flavour volatiles. *Molecular Breeding* 35.

Lachapelle, M., Bourgeois, G., DeEll, J., Stewart, K. A. and Seguin, P. (2013). Modeling the effect of preharvest weather conditions on the incidence of soft scald in 'Honeycrisp' apples. *Postharvest Biology and Technology* 85, 57–66.

Larrigaudiere, C., Vilaplana, R., Recasens, I., Soria, Y. and Dupille, E. (2010). 'Diffuse skin browning' in 1-MCP-treated apples: etiology and systems of control. *Journal of the Science of Food and Agriculture* 90, 2379–85.

Lee, J., Cheng, L., Rudell, D. R. and Watkins, C. B. (2012a). Antioxidant metabolism of 1-methylcyclopropene (1-MCP) treated 'Empire' apples during controlled atmosphere storage. *Postharvest Biology and Technology* 65, 79–91.

Lee, J., Mattheis, J. P. and Rudell, D. R. (2012b). Antioxidant treatment alters metabolism associated with internal browning in 'Braeburn' apples during controlled atmosphere storage. *Postharvest Biology and Technology* 68, 32–42.

Lee, J., Mattheis, J. P. and Rudell, D. R. (2013). Fruit size affects physiological attributes and storage disorders in cold-stored 'Royal Gala' apples. *HortScience* 48, 1518–24.

Lee, J., Mattheis, J. P. and Rudell, D. R. (2016). Storage temperature and 1-methylcyclopropene treatment affect storage disorders and physiological attributes of 'Royal Gala' apples. *HortScience* 51, 84–93.

Lee, J., Rudell, D. R., Davies, P. J. and Watkins, C. B. (2012c). Metabolic changes in 1-methylcyclopropene (1-MCP)-treated 'Empire' apple fruit during storage. *Metabolomics* 8, 742–53.

LehmanSalada, L. (1996). Instrument and operator effects on apple firmness readings. *HortScience* 31, 994–7.

Leisso, R., Buchanan, D., Lee, J., Mattheis, J. and Rudell, D. (2013). Cell wall, cell membrane, and volatile metabolism are altered by antioxidant treatment, temperature shifts, and peel necrosis during apple fruit storage. *Journal of Agricultural and Food Chemistry* 61, 1373–87.

Leisso, R. S., Buchanan, D. A., Lee, J., Mattheis, J. P., Sater, C., Hanrahan, I., Watkins, C. B., Gapper, N., Johnston, J. W., Schaffer, R. J., Hertog, M. L. A. T. M., Nicolai, B. M. and Rudell, D. R. (2015). Chilling-related cell damage of apple (*Malus x domestica* Borkh.) fruit cortical tissue impacts antioxidant, lipid and phenolic metabolism. *Physiologia Plantarum* 153, 204–20.

Li, J. and Yuan, R. (2008). NAA and ethylene regulate expression of genes related to ethylene biosynthesis, perception, and cell wall degradation during fruit abscission and ripening in 'Delicious' apples. *Journal of Plant Growth Regulation* 27, 283–95.

Little, C. R. and Holmes, R. J. (2000). Storage technology for apples and pears: a guide to production, postharvest treatment and storage of pome fruit in Australia. Department of Natural Resources and Environment, Victoria, Australia, Melbourne.

Lu, R. F. (2004). Multispectral imaging for predicting firmness and soluble solids content of apple fruit. *Postharvest Biology and Technology* 31, 147–57.

Lu, X., Nock, J. F., Ma, Y., Liu, X. and Watkins, C. B. (2013). Effects of repeated 1-methylcyclopropene (1-MCP) treatments on ripening and superficial scald of 'Cortland' and 'Delicious' apples. *Postharvest Biology and Technology* 78, 48–54.

Lum, G. B., Brikis, C. J., Deyman, K. L., Subedi, S., DeEll, J. R., Shelp, B. J. and Bozzo, G. G. (2016). Pre-storage conditioning ameliorates the negative impact of 1-methylcyclopropene on physiological injury and modifies the response of antioxidants and y-aminobutyrate in Honeycrisp' apples exposed to controlled-atmosphere conditions. *Postharvest Biology and Technology* 116, 115–28.

Lurie, S. (1998). Postharvest heat treatments. *Postharvest Biology and Technology* 14, 257–69.

Lurie, S., Klein, J. and BenArie, R. (1989). Physiological-changes in diphenylamine treated Granny Smith apples. *Israel Journal of Botany* 38, 199–207.

Lurie, S. and Watkins, C. B. (2012). Superficial scald, its etiology and control. *Postharvest Biology and Technology* 65, 44–60.

Marin, A. B., Colonna, A. E., Kudo, K., Kupferman, E. M. and Mattheis, J. R. (2009). Measuring consumer response to 'Gala' apples treated with 1-methylcyclopropene (1-MCP). *Postharvest Biology and Technology* 51, 73–9.

Marlow, G. C. and Loescher, W. H. (1984). Watercore. *Horticultural Reviews* 6, 189–251.

Mattheis, J. P. (2008). How 1-methylcyclopropene has altered the Washington state apple industry. *HortScience* 43, 99–101.

Mattheis, J. P. and Rudell, D. R. (2008). Diphenylamine metabolism in 'Braeburn' apples stored under conditions conducive to the development of internal browning. *Journal of Agricultural and Food Chemistry* 56, 3381–5.

Maxin, P., Williams, M. and Webe, R. W. S. (2014). Control of fungal storage rots of apples by hot-water treatments: a Northern European perspective. *Erwerbs-Obstbau* 56, 25–34.

Mazzurana, E. R., Argenta, L. C., Do Amarante, C. V. T. and Stefens, C. A. (2016). Potential benefits of temperature increase during storage under controlled atmosphere of 'Gala' apples treated with 1-MCP. *Revista Brasileira de Fruticultura* 38, 43–52.

McArtney, S. J., Obermiller, J. D., Schupp, J. R., Parker, M. L. and Edgington, T. B. (2008). Preharvest 1-methylcyclopropene delays fruit maturity and reduces softening and superficial scald of apples during long-term storage. *HortScience* 43, 366–71.

McCormick, R., Neuwald, D. A. and Streif, J. (2010). A case study: potential energy savings using 1-MCP with 'Gala' apples in commercial CA storage. *Acta Horticulturae*, pp. 323–6.

McGlone, V. A., Jordan, R. B., Seelye, R. and Clark, C. J. (2003). Dry-matter – a better predictor of the post-storage soluble solids in apples? *Postharvest Biology and Technology* 28, 431–5.

Mditshwa, A., Fawole, O. A., Vries, F., Van Der Merwe, K., Crouch, E. and Opara, U. L. (2016). Classification of 'Granny Smith' apples with different levels of superficial scald severity based on targeted metabolites and discriminant analysis. *Journal of Applied Botany and Food Quality* 89, 49–55.

Mellidou, I., Buts, K., Hatoum, D., Quang Tri, H., Johnston, J. W., Watkins, C. B., Schaffer, R. J., Gapper, N. E., Giovannoni, J. J., Rudell, D. R., Hertog, M. L. A. T. M. and Nicolai, B. M. (2014). Transcriptomic events associated with internal browning of apple during postharvest storage. *BMC Plant Biology* 14.

Merzlyak, M. N., Solovchenko, A. E. and Gitelson, A. A. (2003). Reflectance spectral features and non-destructive estimation of chlorophyll, carotenoid and anthocyanin content in apple fruit. *Postharvest Biology and Technology* 27, 197–211.

Mir, M. A., Beigh, M. A. and Shafi, F. (2013). Effect of 1-MCP on quality attributes of Red Delicious apples under non-refrigerated and refrigerated conditions of storage. *Research on Crops* 14, 522–9.

Mir, N. A., Curell, E., Khan, N., Whitaker, M. and Beaudry, R. M. (2001). Harvest maturity, storage temperature, and 1-MCP application frequency alter firmness retention and chlorophyll fluorescence of 'Redchief Delicious' apples. *Journal of the American Society for Horticultural Science* 126, 618–24.

Moran, R. E., DeEll, J. R. and Halteman, W. (2009). Effects of preharvest precipitation, air temperature, and humidity on the occurrence of soft scald in 'Honeycrisp' apples. *HortScience* 44, 1645–7.

Moran, R. E. and McManus, P. (2005). Firmness retention, and prevention of coreline browning and senescence in. 'Macoun' apples with 1-methylcyclopropene. *HortScience* 40, 161–3.

Nicolai, B. M., Beullens, K., Bobelyn, E., Peirs, A., Saeys, W., Theron, K. I. and Lammertyn, J. (2007). Nondestructive measurement of fruit and vegetable quality by means of NIR spectroscopy: A review. *Postharvest Biology and Technology* 46, 99–118.

Nicolai, B. M., Defraeye, T., De Ketelaere, B., Herremans, E., Hertog, M., Saeys, W., Torricelli, A., Vandendriessche, T. and Verboven, P. (2014). Nondestructive measurement of fruit and vegetable quality. *Annual Review of Food Science and Technology* 5, 285–312.

Nock, J. F. and Watkins, C. B. (2013). Repeated treatment of apple fruit with 1-methylcyclopropene (1-MCP) prior to controlled atmosphere storage. *Postharvest Biology and Technology* 79, 73–9.

Noiton, D. A. M. and Alspach, P. A. (1996). Founding clones, inbreeding, coancestry, and status number of modern apple cultivars. *Journal of the American Society for Horticultural Science* 121, 773–82.

Nyasordzi, J., Friedman, H., Schmilovitch, Z., Ignat, T., Weksler, A., Rot, I. and Lurie, S. (2013). Utilizing the I_{AD} index to determine internal quality attributes of apples at harvest and after storage. *Postharvest Biology and Technology* 77, 80–6.

Opara, L. U. (2007). Bruise susceptibilities of 'Gala' apples as affected by orchard management practices and harvest date. *Postharvest Biology and Technology* 43, 47–54.

PACE (2016). http://www.paceint.com/.

Palmer, J., Diack, R., Johnston, J. and Boldingh, H. (2013). Manipulation of fruit dry matter accumulation and fruit size in 'Scifresh' apple through alteration of the carbon supply, and its relationship with apoplastic sugar composition. *Journal of Horticultural Science & Biotechnology* 88, 483–9.

Palmer, J. W., Harker, F. R., Tustin, D. S. and Johnston, J. (2010). Fruit dry matter concentration: a new quality metric for apples. *Journal of the Science of Food and Agriculture* 90, 2586–94.

Peirs, A., Scheerlinck, N., Perez, A. B., Jancsok, P. and Nicolai, B. M. (2002). Uncertainty analysis and modelling of the starch index during apple fruit maturation. *Postharvest Biology and Technology* 26, 199–207.

Pesis, E., Ibanez, A. M., Phu, M. L., Mitcham, E. J., Ebeler, S. E. and Dandekar, A. M. (2009). Superficial scald and bitter pit development in cold-stored transgenic apples suppressed for ethylene biosynthesis. *Journal of Agricultural and Food Chemistry* 57, 2786–92.

Prange, R. K., DeLong, J. M., Harrison, P. A., Leyte, J. C. and McLean, S. D. (2003). Oxygen concentration affects chlorophyll fluorescence in chlorophyll-containing fruit and vegetables. *Journal of the American Society for Horticultural Science* 128, 603–7.

Prange, R. K., Wright, A. H., DeLong, J. M. and Zanella, A. (2013). History, current situation and future prospects for dynamic controlled atmosphere (DCA) storage of fruits and vegetables, using chlorophyll fluorescence. *Acta Horticulturae* 1012, 905–15.

Prange, R. K., Wright, A. H., DeLong, J. M. and Zanella, A. (2015). A review on the successful adoption of dynamic controlled atmosphere (DCA) storage as a replacement for diphenylamine (DPA), the chemical used for control of superficial scald in apples and pears. *Acta Horticulturae* 1071, 389–96.

Pre-Aymard, C., Fallik, E., Weksler, A. and Lurie, S. (2005). Sensory analysis and instrumental measurements of 'Anna' apples treated with 1-methylcyclopropene. *Postharvest Biology and Technology* 36, 135–42.

Raffo, A., Kelderer, M., Paoletti, F. and Zanella, A. (2009). Impact of innovative controlled atmosphere storage technologies and postharvest treatments on volatile compound production in cv. Pinova apples. *Journal of Agricultural and Food Chemistry* 57, 915–23.

Rebeaud, S. G. and Gasser, F. (2015). Fruit quality as affected by 1-MCP treatment and DCA storage – a comparison of the two methods. *European Journal of Horticultural Science* 80.

Regenstein, J. M., Chaudry, M. M. and Regenstein, C. E. (2003). The kosher and halal food laws. *Comprehensive Reviews in Food Science and Food Safety* 2, 111–27.

Reid, M. S., Padfield, C. A. S., Watkins, C. B. and Harman, J. E. (1982). Starch iodine pattern as a maturity index for cultivar Granny Smith apples (*Malus sylvestris*) 1. Comparison with flesh firmness and soluble solids content. *New Zealand Journal of Agricultural Research* 25, 239–44.

Robatscher, P., Eisenstecken, D., Sacco, F., Poehl, H., Berger, J., Zanella, A. and Oberhuber, M. (2012). Diphenylamine residues in apples caused by contamination in fruit storage facilities. *Journal of Agricultural and Food Chemistry* 60, 2205–11.

Rodriguez-Perez, L. C., Harte, B., Auras, R., Burgess, G. and Beaudry, R. M. (2009). Measurement and prediction of the concentration of 1-methylcyclopropene in treatment chambers containing different packaging materials. *Journal of the Science of Food and Agriculture* 89, 2581–7.

Rudell, D. R., Buchanan, D. A., Leisso, R. S., Whitaker, B. D., Mattheis, J. P., Zhu, Y. and Varanasi, V. (2011). Ripening, storage temperature, ethylene action, and oxidative stress alter apple peel phytosterol metabolism. *Phytochemistry* 72, 1328–40.

Rudell, D. R., Mattheis, J. P. and Fellman, J. K. (2006). Influence of ethylene action, storage atmosphere, and storage duration on diphenylamine and diphenylamine derivative content of Granny Smith apple peel. *Journal of Agricultural and Food Chemistry* 54, 2365–71.

Rudell, D. R., Mattheis, J. P. and Hertog, M. L. A. T. M. (2009). Metabolomic change precedes apple superficial scald symptoms. *Journal of Agricultural and Food Chemistry* 57, 8459–66.

Saei, A., Tustin, D. S., Zamani, Z., Talaie, A. and Hall, A. J. (2011). Cropping effects on the loss of apple fruit firmness during storage: The relationship between texture retention and fruit dry matter concentration. *Scientia Horticulturae* 130, 256–65.

Schaffer, R. J., Friel, E. N., Souleyre, E. J. F., Bolitho, K., Thodey, K., Ledger, S., Bowen, J. H., Ma, J. -H., Nain, B., Cohen, D., Gleave, A. P., Crowhurst, R. N., Janssen, B. J., Yao, J. -L. and Newcomb, R. D. (2007). A genomics approach reveals that aroma production in apple is controlled by ethylene predominantly at the final step in each biosynthetic pathway. *Plant Physiology* 144, 1899–912.

Schouten, S. P., Prange, R. K., Verschoor, J., Lammers, T. R. and Oosterhaven, J. (1997). Improvement of quality of Elstar apples by dynamic control of ULO conditions. In: Mitcham, E. J. (Ed.) CA'97. Proc. 7th Intl. Controlled Atmosphere Research Conf., Univ. Calif., Davis, California, USA. Vol. 2, pp. 71–8.

Schupp, J. R. and Greene, D. W. (2004). Effect of aminoethoxyvinylglycine (AVG) on preharvest drop, fruit quality, and maturation of 'McIntosh' apples. I. Concentration and timing of dilute applications of AVG. *HortScience* 39, 1030–5.

Serra, S., Leisso, R., Giordani, L., Kalcsits, L. and Musacchi, S. (2016). Crop load influences fruit quality, nutritional balance, and return bloom in 'Honeycrisp' apple. *HortScience* 51, 236–44.

Serrano, M., Martinez-Romero, D., Zapata, P. J., Guillen, F., Valverde, J. M., Diaz-Mula, H. M., Castillo, S. and Valero, D. (2015). Advances in Edible Coatings. In Wills, R. B. H. and Golding, J. (Eds), *Advances in Postharvest Fruit and Vegetable Technology*. CRC Press, pp. 147–66.

Sisler, E. C. and Blankenship, S. M. (1996). *Method of Counteracting An Ethylene Response In Plants*. North Carolina State University.

Smock, R. M. (1961). Methods of scald control on the apple. *Bull. Cornell Agric. Exp. Stn.* 970, 55, 55.

Smock, R. M. (1977). Nomenclature for internal storage disorders of apples. *HortScience* 12, 306–8.

Snowdon, A. L. (2008). *Post-Harvest Diseases and Disorders of Fruits and Vegetables: Volume 1: General Introduction and Fruits*. CRC Press.

Solovchenko, A. E., Chivkunova, O. B., Merzlyak, M. N. and Gudkovsky, V. A. (2005). Relationships between chlorophyll and carotenoid pigments during on- and off-tree ripening of apple fruit as revealed non-destructively with reflectance spectroscopy. *Postharvest Biology and Technology* 38, 9–17.

Stover, E., Fargione, M. J., Watkins, C. B. and Iungerman, K. A. (2003). Harvest management of Marshall 'Mcintosh' apples: Effects of AVG, NAA, ethephon, and summer pruning on preharvest drop and fruit quality. *HortScience* 38, 1093–9.

Talamini do Amarante, C. V., Argenta, L. C., Vieira, M. J. and Steffens, C. A. (2010). Changes of 1-MCP efficiency by delaying its postharvest application on 'Fuji Suprema' apples. *Revista Brasileira De Fruticultura* 32, 984–92.

Tatsuki, M., Endo, A. and Ohkawa, H. (2007). Influence of time from harvest to 1-MCP treatment on apple fruit quality and expression of genes for ethylene biosynthesis enzymes and ethylene receptors. *Postharvest Biology and Technology* 43, 28–35.

Thompson, J. F. (2016). Precooling and storage facilities. In Gross, K. C., Wang, C. Y. and Saltveit, M. E. (Eds), *The Commercial Storage of Fruits, Vegetables, and Florist and Nursery Stocks*, http://www.ba.ars.usda.gov/hb66/precooling.pdf.

Toivonen, P. M. A. and Hampson, C. R. (2014). Relationship of I_{AD} index to internal quality attributes of apples treated with 1-methylcyclopropene and stored in air or controlled atmospheres. *Postharvest Biology and Technology* 91, 90–5.

Toivonen, P. M. A. and Lu, C. W. (2005). Studies on elevated temperature, short-term storage of 'Sunrise' Summer apples using 1-MCP to maintain quality. *Journal of Horticultural Science & Biotechnology* 80, 439–46.

Tomic, N., Radivojevic, D., Milivojevic, J., Djekic, I. and Smigic, N. (2016). Effects of 1-methylcyclopropene and diphenylamine on changes in sensory properties of 'Granny Smith' apples during postharvest storage. *Postharvest Biology and Technology* 112, 233–40.

Tong, C. B. S., Bedford, D. S., Luby, J. J., Propsom, F. M., Beaudry, R. M., Mattheis, J. P., Watkins, C. B. and Weis, S. A. (2003). Location and temperature effects on soft scald in 'Honeycrisp' apples. *HortScience* 38, 1153–5.

Travers, I., Jacquet, A., Brisset, A. and Maite, C. (2002). Relationship between the enzymatic determination of starch and the starch iodine index in two varieties of cider apple. *Journal of the Science of Food and Agriculture* 82, 983–9.

Vallejo, F. and Beaudry, R. (2006). Depletion of 1-MCP by 'non-target' materials from fruit storage facilities. *Postharvest Biology and Technology* 40, 177–82.

Vandendriessche, T., Schaefer, H., Verlinden, B. E., Humpfer, E., Hertog, M. L. A. T. M. and Nicolai, B. M. (2013). High-throughput NMR based metabolic profiling of Braeburn apple in relation to internal browning. *Postharvest Biology and Technology* 80, 18–24.

Varanasi, V., Shin, S., Johnson, F., Mattheis, J. P., and Zhu, Y. (2013). Differential suppression of ethylene biosynthesis and receptor genes in 'Golden Delicious' apple by preharvest and postharvest 1-MCP treatments. *Journal of Plant Growth Regulation* 32, 585–95.

Veltman, R. H., Verschoor, J. A. and van Dugteren, J. H. R. (2003). Dynamic control system (DCS) for apples (*Malus domestica* Borkh. cv 'Elstar'): optimal quality through storage based on product response. *Postharvest Biology and Technology* 27, 79–86.

Volz, R. K., Alspach, P. A., Fletcher, D. J. and Ferguson, I. B. (2006). Genetic variation in bitter pit and fruit calcium concentrations within a diverse apple germplasm collection. *Euphytica* 149, 1–10.

Wang, H., Peng, J., Xie, C., Bao, Y. and He, Y. (2015). Fruit quality evaluation using spectroscopy technology: a review. *Sensors* 15, 11889–927.

Wang, Z. Y. and Dilley, D. R. (2000). Initial low oxygen stress controls superficial scald of apples. *Postharvest Biology and Technology* 18, 201–13.

Wargo, J. M. and Watkins, C. B. (2004). Maturity and storage quality of 'Honeycrisp' apples. *HortTechnology* 14, 496–9.

Watkins, C. B. (2003). Principles and practices of postharvest handling and stress. In Ferree, D. C. and Warrington, I. J. (Eds), *Apples: Botany, Production and uses*. CAB International, pp. 585–613.

Watkins, C. B. (2008). Overview of 1-methylcyclopropene trials and uses for edible horticultural crops. *HortScience* 43, 86–94.

Watkins, C. B. (2015). Advances in the use of 1-MCP. In Wills, R. B. H. and Golding, J. (Eds), *Advances in Postharvest Fruit and Vegetable Technology*. CRC Press, pp. 117–46.

Watkins, C. B., Bramlage, W. J. and Cregoe, B. A. (1995). Superficial scald of Granny Smith apples is expressed as a typical chilling injury. *Journal of the American Society for Horticultural Science* 120, 88–94.

Watkins, C. B., James, H., Nock, J. F., Reed, N. and Oakes, R. L. (2010). Preharvest application of 1-methylcyclopropene (1-MCP) to control fruit drop of apples, and its effects on postharvest quality. *Acta Horticulturae* 877, 365–73.

Watkins, C. B. and Liu, F. W. (2010). Temperature and carbon dioxide interactions on quality of controlled atmosphere stored 'Empire' apples. *HortScience* 45, 1708–12.

Watkins, C. B., and Nock, J. F. (2004). Effects of postharvest delay before application on responses of apple to 1-MCP. *HortScience* 39, 846–.

Watkins, C. B. and Nock, J. F. (2005). Effects of delays between harvest and 1-methylcyclopropene treatment, and temperature during treatment, on ripening of air-stored and controlled-atmosphere-stored apples. *HortScience* 40, 2096–101.

Watkins, C. B. and Nock, J. F. (2012). Rapid 1-methylcyclopropene (1-MCP) treatment and delayed controlled atmosphere storage of apples. *Postharvest Biology and Technology* 69, 24–31.

Watkins, C. B., Nock, J. F., Weis, S. A., Jayanty, S. and Beaudry, R. M. (2004). Storage temperature, diphenylamine, and pre-storage delay effects on soft scald, soggy breakdown and bitter pit of 'Honeycrisp' apples. *Postharvest Biology and Technology* 32, 213–21.

Watkins, C. B., Nock, J. F. and Whitaker, B. D. (2000). Responses of early, mid and late season apple cultivars to postharvest application of 1-methylcyclopropene (1-MCP) under air and controlled atmosphere storage conditions. *Postharvest Biology and Technology* 19, 17–32.

Watkins, C. B., Silsby, K. J. and Goffinet, M. C. (1997). Controlled atmosphere and antioxidant effects on external CO_2 injury of 'Empire' apples. *HortScience* 32, 1242–6.

Wilkinson, B. G. and Fidler, J. C. (1973). Physiological disorders. In Fidler, J. C., Wilkinson, B. G., Edney, K. L. and Sharples, R. O. (Eds), *The Biology of Apple and Pear Storage* Commonwealth Agricultural Bureaux, East Malling, Kent, England, pp. 67–80

Wills, R. B. H. and Scott, K. J. (1982). Use of dips containing diphenylamine and edible oils to reduce soft scald of apples. *HortScience* 17, 964–5.

Wunsche, J. N. and Heyn, C. S. (2015). Consumer responses to fruit quality of 'Jonagold' apples treated with postharvest application of 1-methylcyclopropene (1-MCP) under air and controlled atmosphere storage conditions. *European Journal of Horticultural Science* 80, (1) 3-10 | DOI: 10.17660/eJHS.2015/80.1.1

Yang, X., Song, J., Du, L., Forney, C., Leslie, C.-P., Sherry, F., Wismer, P. and Zhang, Z. (2016). Ethylene and 1-MCP regulate major volatile biosynthetic pathways in apple fruit. *Food Chemistry* 194, 325–36.

Yost, G. E. (1984). Energy saving through the use of fan and refrigeration cycling in apple cold storage. *Transactions of the ASAE* 27, 497–501.

Yuan, R. and Carbaugh, D. H. (2007). Effects of NAA, AVG, and 1-MCP on ethylene biosynthesis, preharvest fruit drop, fruit maturity, and quality of 'Golden supreme' and 'Golden delicious' apples. *HortScience* 42, 101–5.

Zanella, A. (2003). Control of apple superficial scald and ripening – a comparison between 1-methylcyclopropene and diphenylamine postharvest treatments, initial low oxygen stress and ultra low oxygen storage. *Postharvest Biology and Technology* 27, 69–78.

Zanella, A. and Rossi, O. (2015). Post-harvest retention of apple fruit firmness by 1-methylcyclopropene (1-MCP) treatment or dynamic CA storage with chlorophyll fluorescence (DCA-CF). *European Journal of Horticultural Science* 80, 11–17.

Zanella, A., Vanoli, M., Rizzolo, A., Grassi, M., Zerbini, P. E., Cubeddu, R., Torricelli, A. and Spinelli, L. (2013). Correlating optical maturity indices and firmness in stored 'Braeburn' and 'Cripps Pink' apples. *Acta Horticulturae* 1012, 1173–80.

Ziosi, V., Noferini, M., Fiori, G., Tadiello, A., Trainotti, L., Casadoro, G. and Costa, G. (2008). A new index based on vis spectroscopy to characterize the progression of ripening in peach fruit. *Postharvest Biology and Technology* 49, 319–29.

Zou, X. and Zhao, J. (2015). Machine vision online measurements. *Nondestructive meaurement in food and agro-products* http://link.springer.com/chapter/10.1007/978-94-017-9676-7_2/fulltext.html.

Zude-Sasse, M., Truppel, I. and Herold, B. (2002). An approach to non-destructive apple fruit chlorophyll determination. *Postharvest Biology and Technology* 25, 123–33.

Part 3

Diseases and pests

Chapter 14

Pre- and postharvest fungal apple diseases

Wayne M. Jurick II, USDA-ARS, USA; and Kerik D. Cox, Cornell University, USA

1 Introduction: apple cultivation, production and storage and the impact of fungal apple diseases
2 Pre- and postharvest apple diseases
3 Chemical, biological and cultural forms of disease management during apple production and storage
4 Emerging pathogens affecting apple production in the United States
5 Conclusion
6 Where to look for further information
7 References

1 Introduction: apple cultivation, production and storage and the impact of fungal apple diseases

1.1 Cultivation

The World Apple and Pear Association reports that global apple production reached 64.3 million tons in 2013. China led with 31.7 million tons, while the United States placed a distant second with 3.2 million tons of apples produced. The farm gate value for US apple production in 2013 was valued at $2.7 billion and exports to Mexico, Canada and Taiwan resulted in $1.1 billion in revenue in 2012 (US Apple Association). There are about 7500 apple producers growing fruit on approximately 363 000 acres located in Washington State, New York, Michigan, Pennsylvania and California according the US Apple Association. Approximately two-thirds of the apples produced in the United States are consumed fresh, while the remaining third are processed into juices, sauces and butters. There are hundreds of different kinds of apples grown in the United States, including heirloom varieties. However, according to the US Apple Association, the top ten commercial cultivars include Red Delicious, Gala, Golden Delicious, Fuji, Granny Smith, McIntosh, Honeycrisp, Rome, Empire and Cripps Pink.

1.2 Production

Commercial production of apple fruit has constantly evolved over the past few decades, with the overall trend to maximize crop output and increase productivity on a global

http://dx.doi.org/10.19103/AS.2016.0017.17

Figure 1 Apple-planting archetypes showing (A) modern, high-density planting with trellis system and (B) low-density traditional planting.

scale (Fig. 1A). Gone are the days of standard or semi-dwarf plantings, which are largely abandoned or relegated to agri-tourism operations (Fig. 1B). In modern, commercial operations, apples are grafted to fully dwarfing rootstocks of the Malling, Geneva or Budagovsky series (e.g. M.9, G.935, G.41 and B.9), and planted in tightly spaced trellis rows at densities often exceeding 3000 trees per hectare (Robinson et al. 2013; Robinson 2008a,b). In addition, such plantings are kept at a high state of vigour in the first three years to accelerate growth to maximize production. However, the planting paradigm of tightly packed, highly vigorous, small trees leads to increased concerns for pathogen management. Diseases such as apple scab, powdery mildew and Glomerella leaf spot in the United States have greater impact on fruit production as smaller stature trees have less foliage to balance the photosynthetic impact/load caused by fungal foliar pathogens. While pathogens vary with climate and location, apple foliar and fruit pathogens are a significant concern in these higher-density plantings. Aside from tree architecture, consumer preference and disease resistance are the primary foci of contemporary apple-breeding efforts.

1.3 Storage

Apples are picked into bags and placed directly into wooden or plastic bins in the field and moved directly to the storage facility. There are many factors that govern the treatment regime and length of storage for apples based on their end use, market demand and sophistication of the storage facility. For example, high-quality fruit are typically sold for fresh consumption and are sorted for size and colour, graded and treated with food-grade wax and/or a postharvest fungicide prior to storage in the United States (Fig. 2). By contrast, low-grade fruit may be sold directly to processors or consumers for a much lower

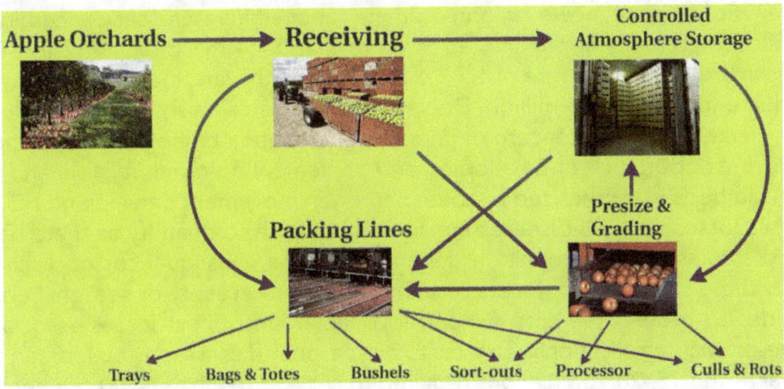

Figure 2 Potential routes for apples from harvest to storage at a typical large-scale commercial-scale packing facility.

price and are typically devoid of treatment. In European countries where postharvest fungicides are not permitted, fruit go unprotected in storage, regardless of their grade or length of storage. In general, apples are stored for up to six months at 1°C in air and can be kept as long as one year in controlled atmosphere cold storage where temperature, oxygen and carbon dioxide are maintained at specific levels (Sutton et al. 2014). Many large-scale producers in the United States and Europe, with the intent to maintain fruit quality, treat storage rooms with ethylene inhibitor 1-MCP (1-methylcyclopropene) to delay ripening. Antioxidants like diphenylamine (DPA) to control superficial scald in susceptible apple varieties like 'Law Rome' and 'Granny Smith' are used in the United States, banned in Europe and have been shown to provide additional control against blue mould (Errampalli et al. 2006). Extended storage times provide the consumer with fruit all-year round, give the producer the opportunity to sell on the basis of market demand for optimal profitability and have the unintended consequence of increased risk for decay over time (Jurick II, pers. obs.).

1.4 Impact of fungal apple diseases

Decay caused by filamentous fungal plant pathogens is a globally significant problem, especially in areas where postharvest fungicides are not used (i.e. Europe), or in developing countries where storage technologies and tools are either cost prohibitive or unavailable. Rot reduces profit margins by necessitating additional expenditures for labour, fungicide applications, energy and costs for sanitization, rejection from processors, overall product losses and ultimately a reduction in fruit quality. Processors in the United States reject loads at very low thresholds (i.e. 1% decay or less), thus eliminating any chance for the producer to recover costs (Lee Showalter – Grower Services at Rice Fruit Company in Carlisle Pennsylvania, pers. comm.). Losses attributed to postharvest apple decay in the United States typically range from 1% to 15% for untreated fruit and/or where decay failures on fungicide-treated apples have occurred (Jurick II, pers. obs.). However, in countries like South Africa, which export 46.6% of their fruit, or in Europe where postharvest fungicides are not utilized, losses can be upwards of 50% or greater (Louw and Korsten 2014). Specific dollar amounts associated with crop losses are very limited for agricultural commodities in the scientific literature; however, one estimate for postharvest losses concerning apples in the United States was approximated at more than $4.4 million in 1992 (Rosenberger 1997). Assuming an inflation rate of 1.7% over the past 23 years, that loss in today's dollars would equate to nearly $6.5 million. Decreased revenue can also result from the closure of export markets with the concern of importing quarantine pathogens on asymptomatic fruit. Thus, the importance of developing and implementing rapid, reliable and accurate detection methods is emphasized to abate pathogen movement between global markets.

Economic losses for field diseases are similarly difficult to quantify as there have been little to no studies on the subject. In general, the fiscal consequences of disease in the field are twofold, with indirect losses resulting from the necessity of seasonal application of pesticides to protect the crop (where applicable), and direct losses associated with the failure of such applications. The costs of seasonal disease protection can be fairly substantial as the 15–20 applications required may be as much as $550 per hectare (Cox 2015). In the current apple production climate, principal direct losses are associated with blemishes, in the case of fungal diseases like apple scab. Aberrations to the fruit finish, will downgrade the crop from fresh market to 'utility' grade, which are used exclusively for processing and juice. Moreover, the revenue that producers receive from 'utility'-grade apples often does not exceed production costs. This is primarily due to the inexpensive juice concentrates imported from abroad, a trend that has substantially increased over the past two decades.

2 Pre- and postharvest apple diseases

2.1 Preharvest diseases

There are a number of preharvest diseases around the world with which apple producers must contend (Fig. 3). For the US pome fruit growing areas, the most common pathogens are *Alternaria mali* (Alternaria leaf spot), *Colletotrichum* species (bitter rot), *Glomerella cingulata* (Glomerella leaf spot), *Gymnosporangium juniperi-virginianae* (cedar apple rust), *Podosphaera leucotricha* (powdery mildew), *Botryosphaeria dothidea* (white rot), *Schizathyrium pomi* (flyspeck) and *Venturia inaequalis* (apple scab) (Li et al. 2002; Xu et al. 2002). Of these diseases, apple scab is the most prevalent and globally persistent problem, particularly in Europe and the United States. There has been considerable study for over a century on the topic, and it remains at the forefront of fungal disease management (MacHardy 1996). Despite research efforts to develop cultural and biological management practices for apple scab, this disease requires considerable applications of fungicides to manage losses, maintain fruit quality and ensure grower profitability.

Apple scab, caused by the fungus *V. inaequalis* (Cke) Wint is most problematic in production regions where cool rainy weather conditions in the spring are commonplace (MacHardy 1996). The disease is one of the more economically important fungal diseases of apple, given that pathogen can cause lesions on both the leaves and fruit (Fig. 3A and B). If left unchecked or poorly managed, scab can cause complete fresh market loss. Even a single lesion will downgrade an apple to 'Utility' grade, which affords little to no profitability to producers. Foliar infections can lead to premature defoliation, which can impact cold hardiness and reduce flowering in the subsequent season (MacHardy et al. 2001). To complicate management, the pathogen undergoes sexual reproduction

Figure 3 Various apple fruit and foliar diseases that occur in the field. (A) scab on fruit (*Venturia inequalis*), (B) scab on leaves (*V. inequalis*), (C) cedar apple rust on leaves (*Gymnosporangium juniperi-virginianae*), (D) powdery mildew on leaves (*Podosphaera leucotricha*), (E) Glomerella leaf spot (*Glomerella cingulata*) and (F) fly speck on apple (*Schizathyrium pomi*).

providing *V. inaequalis* with the propensity to overcome single-gene resistance and develop tolerance to single-site mode-of-action fungicides (MacHardy 1996). Epidemics occur during the spring and early summer months when weather remains cool (< 30°C) and infections coincide with wetting events in excess of 6–12 hours. Such infections are initiated by the release of ascospores from pseudothecia on overwintering leaf lesions produced on the orchard floor. The ascospore inoculum infects the leaves around flower clusters, giving rise to olive-coloured raised lesions that subsequently produce secondary conidia as inoculum. The conidia infect developing fruit and terminal leaves later in the season. Infections will continue until mid-summer when the primary ascospore inoculum has been discharged and several days of average daily temperatures in excess of 30°C render secondary inoculum non-viable.

2.2 Postharvest diseases

Packers, processors and storage managers have to contend with an array of fungal species that cause storage rots on apples. While disease incidence varies by region, the following pathogens are of global concern, and these include *Botrytis cinerea* (grey mould), *Colletotrichum* spp. (bitter rot), *Mucor piriformis* (Mucor rot) and *Penicillium expansum* (blue mould) (Kou et al. 2013; Li et al. 2014). Less common pathogens include *Fusarium avenaceum* (Fusarium rot), *Alternaria alternata* (dry core rot and mouldy core), *Phacidiopycnis washingtonensis* (speck rot), *Monilinia fructicola* (brown rot), *Sphaeropsis pyriputrescens* (Sphaeropsis rot) and *V. inaequalis* (storage scab) (Jurick et al. 2103; Kou et al. 2013; Peter et al. 2015) (Fig. 4A–F). By far, the most ubiquitous and economically significant postharvest disease of apple in the world is blue mould caused by *Penicillium* spp. (Sommer et al. 2002).

Figure 4 Postharvest fungal diseases of apple fruit from commercial storage. (A) blue mould (*Penicillium expansum*), (B) grey mould (*Botrytis cinerea*), (C) Mucor rot (*Mucor pyriformis*) (D) mouldy core (*Alternaria alternata*), (E) storage scab (*Venturia inequalis*) and (F) bitter rot (*Colletotrichum acutatum*).

Penicillium expansum is the cause of blue mould on pome fruits, but several other *Penicillium* species (*P. aurantiogriseum, P. commune, P. glabrum, P. roqueforti, P. rugulosum, P. solitum* and *P. verrucosum*) are also implicated to cause the disease (Sanderson 1995). Infected fruit develop mushy, tan-to-brown decayed areas with a sharp margin between diseased and healthy tissues. During advanced stages of decay, bluish conidia develop on the fruit surface, giving the disease its descriptive common name (Fig. 4A). *Penicillium* spp. not only cause decay but also produce a variety of mycotoxins that impact human health, most notably patulin, citrinin and penicillic acid (McKinley and Carlton 1991; Andersen et al. 2004). Due to its potentially carcinogenic effects, the European Union has set limits on the maximum amount of patulin to 50 µg/L for fruit juices and derived products, 25 µg/L for solid apple products and 10 µg/L for juices and foods for babies and infants. The US Food and Drug Administration (FDA), as well as other countries, including the European Union, have imposed strict limits on the permissible amounts of patulin (at 50 µg/L) in fruit juices and processed pome fruit products for human consumption (Puel et al. 2010).

3 Chemical, biological and cultural forms of disease management during apple production and storage

3.1 Chemical

With regard to preharvest fungicide use, the multi-site, broad-spectrum protectant fungicides, captan and mancozeb, form the basis for devising apple scab fungicide programmes in the United States since the 1940s. The introduction of single-site fungicides with curative or 'post-infection' properties, including dodine and the methyl benzimidazole carbamates (MBCs) in the 1950s, the sterol biosynthesis inhibitors (SBIs) in the 1980s, anilinopyrimidines (APs) in the 1990s and the quinone-outside inhibitors (QoIs) in the 1990s, allowed growers to continue to pursue post-infection application paradigms for apple scab management. Unfortunately, frequent reliance on single-site fungicides has led to a succession of resistance to each newly introduced fungicide class in the United States from the 1950s to the present (Frederick et al. 2014; Jones 1991; Lesniak et al. 2011; Wicks 1974; Villani et al. 2015). The AP fungicides, the DMI fungicide difenoconazole released in 2010, and succinate dehydrogenase inhibitor (SDHI) fungicides released after 2012 (i.e. penthiopyrad, fluopyram, fluxapyroxad and benzovindiflupyr), are the only fungicides to which resistance has not been reported at the time of writing this chapter (Villani et al. 2016). In the face of resistance, apple growers are relying on applications of the multi-site, broad-spectrum protectant fungicides. On susceptible cultivars, fungicide applications are made at five-to seven-day intervals from bud break to early summer, which coincides with the presence of ascospore and conidial inoculum (MacHardy 1996; MacHardy et al. 2001). For apple scab alone, the number of fungicide applications needed may be a few as 8 or as many as 12, depending on which ones are used, the level of cultivar susceptibility and the number of seasonal wetting events (Jones 1991; MacHardy et al. 2001).

Currently, there are several fungicides labelled for preharvest application in the United States that have efficacy against postharvest pathogens including, Merivon (pyraclostrobin + fluxpyroxad), Luna sensation (fluopyram + trifloxystrobin) and Topsin M (thiophanate methyl) (Dr. Kari A. Peter, pers. comm.). While these preharvest fungicides are helpful at combating diseases in the field, their postharvest efficacy is generally not the principal method of control and should be viewed as supplemental for the United States where postharvest fungicides are utilized. Conversely, a limited number of chemicals (six) are currently labelled to control postharvest decay in the United States on apple fruit: Academy (fludioxonil and difenaconazole), Captan (*N*-trichloromethylthio-4-cyclohexene-1.2-dicarboximide), Judge (fenhexamid), Mertect (thiabendazole), Penbotec (pyrimethanil) and Scholar (fludioxonil). Depending on the label, postharvest fungicides can be applied in the form of drenches, dips and line sprays and/or incorporated into waxes during the sorting and packing process prior to long-term storage. However, postharvest fungicides are not used in European countries, thus necessitating the development of alternative methods of control that do not involve genetic modification of the host but targeting the pathogen directly.

3.2 Biological

Utilization of biological controls has been scarce for pre- and postharvest diseases for a multitude of reasons. In general, cost, ease of use, efficacy and labelling have all impacted implementation of biocontrols for both pre- and postharvest apple diseases. Field control of apple scab using organic-approved approaches, including biocontrol agents, is not only costly, but does not offer adequate control of the pathogen and has non-target impacts (Cromwell et al. 2011). However, because of the controlled nature of the packinghouse environment, advances using organic-approved biocontrol agents have been made to combat postharvest rots, specifically blue and grey mould. BioSave™ is by far the most successful and widely used product in the market that contains a saprophytic strain of *Pseudomonas syringae* originally isolated from the surface of apple fruit. It was introduced in 1996 and labelled in the United States to control postharvest diseases of fruits and vegetables, including blue and grey mould on pome and stone fruits (Janisiewicz and

Jeffers 1997). Since that time, the product has had its registration expanded for use on cherries, potatoes and sweet potatoes (Janisiewicz 2010). The commercial implementation of BioSave™ has been steadily increasing ever since it was released, and during the 2000–2001 storage season, it was applied to over 34 million bushel boxes of produce (Janisiewicz and Korsten 2002).

3.3 Cultural

Postharvest decay caused by fungal plant pathogens (i.e. blue mould) are also managed by careful harvest and handling practices that decrease fungal inoculum (spores, conidia and other survival/resting structures) levels. Good cultural practices in the orchard should focus on the removal of organic debris and decayed fruits. After harvest, bin sterilization, stringent packinghouse sanitization, packing dry product, culling bruised and infected fruit and the maintenance of proper storage temperature are all essential for managing postharvest decay (Hansen et al. 2010; Rosenberger 2012). Specific sanitization methods for packing houses in the United States include treating bins with steam, cleaning packing house walls and floors with quaternary amines or peroxyacetic acids and maintaining proper chlorine levels in sizing flumes (Spotts 1994; Rosenberger 2012). In addition to being effective on their own merits, cultural practices function synergistically when integrated with either chemical and/or biological controls, which increases their efficacy. Thus, they should be practised as part of an integrated control strategy to achieve the highest level of postharvest decay management to ensure treatment efficacy, while reducing the risk of developing fungicide-resistant pathogen populations (Rosenberger 2012).

Cultural management of field diseases (i.e. apple scab) in the United States begins with the reduction of overwintering inoculum in late fall and/or earlier spring. This is accomplished by shredding leaves with specialized equipment or by applying 5% urea to the orchard floor in the spring or fall to degrade leaf litter (MacHardy 1996). Additionally, copper is also effective at reducing overwintering apple scab inoculum (Frederick et al. 2014; Montag et al. 2006). With the development of durable apple scab-resistant cultivars being the goal of apple-breeding programmes for decades (MacHardy 1996), fruit quality and taste become a matter of primary concern while deciding consumer preferences. Despite this trend, there are many cultivars with single-gene resistance to apple scab that have been commercially available for more than half a century, but grower and consumer acceptance of resistant cultivars has been limited (Gessler and Pertot 2012; Gallardo et al. 2012; Yue et al. 2013). Fortunately, single-gene resistance remains effective in the United States, given that the development of virulent races of *V. inaequalis* has been curtailed by seasonal fungicide applications necessary to manage other fungal diseases (e.g. powdery mildew, bitter rot and flyspeck and sooty blotch).

4 Emerging pathogens affecting apple production in the United States

Washington State is the largest producer of apples in the United States, according to the Washington Apple Commission, as they grow approximately 2.5 million tons of apples per year. Both conventional and organic exports in Washington State lead the country. Therefore, monitoring fruit destined for export via the development of rapid and sensitive detection methods is critical to maintain exports. Two fungal pathogens have emerged in Washington State and these are speck rot caused by *Phacidiopycnis washingtonensis* and Sphaeropsis rot caused by *Sphaeropsis pyriputrescens*. Hence, nucleic acid-based detection methods via application of the polymerase chain reaction (PCR) have been developed to discover these quarantine fungal plant pathogens on apple fruit (Sikdar et al. 2014). While these methods are a good first step to prevent global movement of these pathogens, more user-friendly methods that can be implemented in the field (i.e. packing houses, orchards, loading docks) by growers and inspectors are still needed.

According to the US Apple Association, New York is the second-largest producer of apples in the United States. In the New York and Northeast apple-producing areas, there have been little or no occurrences of new or emerging field diseases for apples. However, it should be said that the curtailment of single-site fungicides for apple scab management, in combination with global climate change, has led to increased occurrences of summer fruit rots (i.e. bitter rot and black rot), Glomerella leaf spot and Botryosphaeria and Nectria shoot and canker blights in warmer production regions. Therefore, producers in other regions should be aware of these issues and be watchful of them in their production areas as the effects of climate change may impact their incidence and severity.

Pennsylvania is the main producer of pome fruit for the east coast (US Apple Association). Several new postharvest pathogens have been observed in Pennsylvania by Dr. Wayne M. Jurick II and colleagues. An outbreak of Mucor rot has been observed for D-grade 'Gala' fruit in Pennsylvania storage (Li et al. 2014). This pathogen is particularly troubling as there are no pre- or postharvest fungicides labelled for its control in the United States, and its management is relegated to sanitization and cultural practices. Previously undocumented species of postharvest pathogens have been found causing postharvest decay on apples, and these are *Alternaria tenuissima* (dry core rot), *Colletotrichum fioriniae* (bitter rot) and *Fusarium avenaceum* (Fusarium rot) on commercially stored fruit in Pennsylvania (Kou et al. 2013). However, their incidence is low, and continued monitoring is required to determine their economic importance and potential impact on export markets.

5 Conclusion

Global apple production will continue to increase with consumer demand and rising population. It is envisioned that additional countries and niche markets will continue to emerge and contribute to overall apple production to meet increasing consumer demands. Apples are a popular food and convenient snack that has many health benefits and new varieties are being constantly developed and improved upon. However, sustainable apple production is at risk if society does not continually reinvest in agricultural research. This is particularly important in the area of pre- and postharvest diseases concerning fundamental, applied and translational studies that can provide innovative solutions to move towards permanence with regard to sustainability. Breakthroughs in control methods, storage technologies and disease resistance have been achieved, but there is much more work ahead to keep growers profitable while supplying consumers with high-quality fruit and safe-processed apple products devoid of mycotoxins. Scientists of the current and future generation must combine molecular, genetic, genomic, breeding and classical scientific knowledge of these pathogens and the host to develop futuristic detection methods, new control strategies and hardy disease-resistant varieties to deliver quality fruit that will be appealing to consumers today and in the future.

6 Where to look for further information

Good sources of information are mostly found, via a simple Google search, on the internet concerning apple production, diseases, storage conditions and disease management. Many Land-grant Universities have research faculty with websites solely dedicated to pre- and postharvest disease management which include Washington State University, Cornell University, Penn State University, and West Virginia University to name a few. American Phytopathological Society (APS) webpage (APSnet) can be a very good starting point for specific information concerning plant disease research. In particular, the *Compendium of Apple and Pear Diseases and Pests*, 2nd edition is an excellent reference along with the *Tree Fruit Production Guide* (2014–15) offered by Penn State Extension and Handbook 66 *The Commercial Storage of Fruits, Vegetables, and Florist and Nursery Stocks* published by the USDA-ARS Food Quality Laboratory. Finally, web-based resources of public interest include the U.S. Apple Association and the World Apple and Pear Association for varietal information, production numbers and various factoids.

7 References

Andersen, B., Smedsgaard, J. and Frisvad, J. C. (2004), '*Penicillium expansum*: consistent production of patulin, chaetoglobosins, and other secondary metabolites in culture and their natural occurrence in fruit products', *J. Agric. Food Chem.*, 52, 2421–8.

Cox, K. D. (2015), 'Fungicide resistance in *Venturia inequalis*, the causal agent of apple scab in the United States', in H. Ishii and D. W. Hollowman (eds), *Fungicide Resistance in Plant Pathogens*, pp. 433–7. Springer Japan, Tokyo.

Cromwell, M. L., Berkett, L. P., Darby, H. M. and Ashikaga, T. (2011), 'Alternative organic fungicides for apple scab management and their non-target effects', *HortScience*, 46, 1254–9.

Errampalli, D., Brubacher, N. R. and DeEll, J. R. (2006),'Sensitivity of *Penicillium expansum* to diphenylamine and thiabendazole and postharvest control of blue mold with fludioxonil in "McIntosh" apples', *Postharvest Biol. Technol.*, 39, 101–7.

Frederick, Z. A., Villani, S. M., Cooley, D. R., Biggs, A. R., Raes, J. J. and Cox, K. D. (2014), 'Prevalence of quinone outside inhibitor resistance and the stability of qualitative QoI resistance in Venturia inaequalis', *Plant Dis.*, 98, 1122–30.

Gallardo, K., Nguyne, D., McCracken, V., Yue, C., Luby, J. and McFerson, J. (2012), 'An investigation of trait prioritization in Rosaceous fruit breeding programs', *HortScience*, 47, 771–6.

Gessler, C. and Pertot, I. (2012), 'Vf scab resistance of Malus', *Trees – Structure and Function*, 26(1), 95–108.

Hansen, J., Xiao, C. L. and Kupferman, G. (2010), 'Bin Sanitation: an effective was to reduce codling moth and fungal decay spores', *WSU Publication*, 1–3.

Janisiewicz, W. J. (2010), 'Quo vadis of biological control of postharvest diseases', *Postharvest Pathology*, 2, 137–48.

Janisiewicz, W. J. and Jeffers, S. N. (1997), 'Efficacy of commercial formulation of two biofungicides for control of blue mold and gray mold of apples in cold storage', *Crop Prot.*, 16, 629–33.

Janisiewicz, W. J. and Korsten, L. (2002), 'Biological control of postharvest diseases of fruits', *Annu. Rev. Phytopathol.*, 40, 411–41.

Jones, A. (1991), 'Fungicide resistance: Past experience with benomyl and dodine and future concerns with sterol biosynthesis inhibitors', *Plant Dis.*, 65, 990–992.

Jurick II, W. M., Kou, L., Gaskins, V. L. and Luo Y. (2013), 'First report of *Alternaria alternata* causing postharvest decay on apple fruit from cold storage in Pennsylvania', *Plant Dis.*, 98, 960.

Kou, L., Gaskins, V. L., Luo, Y. and Jurick II, W. M. (2013), 'First report of *Alternaria tenuissima* causing postharvest decay on apple fruit from cold storage in the United States', *Plant Dis.*, 98, 690.

Kou, L., Gaskins, V. L., Luo, Y. and Jurick II, W. M. (2013), 'First report of *Fusarium avenaceum* causing postharvest decay of 'Gala' apple fruit in the United States', *Plant Dis.*, 98, 690.

Kou, L., Gaskins, V. L., Luo, Y. and Jurick II, W. M. (2013), 'First report of *Colletotrichum fioriniae* causing postharvest decay on "Nittany" apple fruit in the United States', *Plant Dis.*, 98, 7.

Lesniak, K. E., Proffer, T. J., Beckerman, J. L. and Sundin, G. W. (2011), 'Occurrence of QoI resistance and detection of the G143A mutation in Michigan populations of Venturia inaequalis', *Plant Dis.*, 95, 927–34.

Li, B. H. and Xu, X. M. (2002), 'Infection and development of apple scab (Venturia inaequalis) on old leaves', *J. Phytopathol.*, 150, 687–91.

Li, J., Gaskins, V. L., Yan, H., Lou, Y. and Jurick II, W. M. (2014), 'First report of Mucor rot on stored 'Gala' apple fruit caused by *Mucor piriformis* in Pennsylvania', *Plant Dis.*, 98, 8.

Louw, J. P. and Korsten, L. (2014), 'Pathogenic *Penicillium* spp. on apple and pear', *Plant Dis.*, 98, 590–8.

MacHardy, W. E. (1996), 'Apple Scab: Biology, Epidemiology, and Management', The American Phytopathological Society, St. Paul, MN.

MacHardy, W. E., Gadoury, D. M. and Gessler, C. (2001), 'Parasitic and biological fitness of *Venturia inaequalis*: Relationship to disease management strategies', *Plant Dis.*, 85, 1036–51.

McKinley, E. R. and Carlton, W. W. (1991), 'Patulin', in R. P. Sharma and D. K. Salunkhe (eds), *Mycotoxins and Phytoalexins*, pp. 191–236, CRC Press, Boca Raton.

Montag, J., Schreiber, L. and Schönherr, J. (2006), 'An in vitro study of the nature of protective activities of copper sulphate, copper hydroxide and copper oxide against conidia of *Venturia inaequalis*', *J. Phytopathol.*, 154, 474–81.

Peter, K. A., Gaskins, V. L., Lehman, B. and Jurick II, W. M. (2015), 'First report of brown rot on apple fruit caused by *Monilinia fructicola* in the United States', *Plant Dis.*, 99, 1179.

Puel, O., Galtier, P. and Oswald, I. P. (2010), 'Biosynthesis and toxilogical effects of patulin', *Toxins*, 2, 613.

Robinson, T. L. (2008a), 'The evolution towards more competitive apple orchard systems in the USA', *Acta Hortic*, 772, 491–500.

Robinson, T. L. (2008b), 'Crop load management of new high-density apple orchards', *N.Y. Fruit Q*, 16, 3–7.

Robinson, T. L., Hoying, S. A., Miranda-Sazo, M., DeMarree, A. and Dominguez, L. (2013), 'A vision for apple orchard systems of the future', *N.Y. Fruit Q*, 21, 11–17

Rosenberger, D. A. (1997), 'Recent research and changing options for controlling postharvest decays of apples', in *Proc. Harvesting, Handling, and Storage Workshop*, Cornell University, pp. 42–53.

Rosenberger, D. A. (2012), 'Sanitize apple storage rooms to minimize postharvest decays', *Scaffolds Fruit J.*, 21, 4–5.

Sanderson, P. G. and Spotts, R. A. (1995), 'Postharvest decay of winter pear and apple fruit caused by species of *Penicillium*', *Phytopathology*, 85, 103–10.

Sikdar, P., Okubara, P., Mazzola, M. and Xiao, C. L. (2014), 'Development of PCR assays for diagnosis and detection of the pathogens *Phacidiopycnis washingtonensis* and *Sphaeropsis pyriputrescens* in apple fruit', *Plant Dis.*, 98, 241–6.

Sommer, N. F., Fortlage, R. J. and Edwards, D. C. (2002), 'Postharvest diseases of selected commodities', *Postharvest Technology of Horticultural Crops*, 3rd ed., 197–249.

Spotts, R. A. and Cervantes, L. A. (1994), 'Contamination of harvest bins with pear decay fungi and evaluation of disinfectants on plastic and wood bin material', *Acta Hort*, 367, 419–25.

Sutton, T. B., Aldwinckle, H. S., Agnello, A. M. and Walgenbach, J. F. (2014), Harvest and Handling in 'Compendium of apple and pear diseases and pests', APS Press, pp. 6–7.

Villani, S. M., Ayer, K. and Cox, K. D. (2016), 'Molecular characterization of the *sdhB* gene and baseline sensitivity to penthiopyrad, fluopyram, and benzovindiflupyr in *Venturia inaequalis*', *Plant Dis.*, (In Press).

Villani, S. M., Biggs, A. R., Cooley, D. R., Raes, J. J. and Cox, K. D. (2015), 'Prevalence of myclobutanil resistance and difenoconazole insensitivity in populations of Venturia inaequalis', *Plant Dis.*, 99, 1526–36.

Wicks, T. (1974), 'Tolerance of the apple scab fungus to benzimidazole fungicides', *Plant Dis. Rep.*, 58, 886–9.

Xu, X. M. and Madden, L. V. (2002), 'Incidence and density relationships of powdery mildew on apple', *Phytopathology*, 92, 1005–14.

Yue, C., Gallardo, R. K., Luby, J., Rihn, A., McFerson, J., McCracken, V., Bedford, D., Brown, S., Evans, K., Weebadde, C., Sebolt, A. and Iezzoni, A. (2013), 'An investigation of U.S. apple producers' trait prioritization – evidence from audience surveys', *Hortscience*, 48, 1378–84.

Chapter 15

Management of viruses and virus-like agents affecting apple production

Kenneth C. Eastwell, Washington State University, USA

1 Introduction

In the context of this chapter, the term 'virus' will include virus-like agents that share certain characteristics such as persistence in the tree and graft transmissibility. Therefore, the comments to follow apply to viruses, viroids and phytoplasma, as well as agents of graft-transmissible diseases that have yet to be identified. Many viruses and virus-like agents that infect apple do not elicit acutely obvious symptoms. Consequently, they are frequently and inadvertently propagated and dispersed, creating a negative impact on production and the economic sustainability of the industry from the nursery to the consumer. Unlike many insect or microbial infections, viruses persist within infected tissues, and thus, exert their effects throughout the life of an orchard tree, which may span 20 years or more. Most viruses encountered in apple production do not have insect vectors known to transmit the pathogen – perhaps the long history of vegetative propagation has allowed the viruses to evolve without the need to associate with insect or other vectors other than humankind. For this reason, many diseases caused by virus-like agents can be managed most effectively by planting trees that are tested and found free of pathogens. The effectiveness of this strategy was demonstrated in the 1950s with fruit tree and grape programmes and is currently reinforced by a system of clean plant programmes operating throughout the world (Gergerich et al., 2015). Planting stock that is tested and found free of virus-like agents based on prescribed protocols is referred to by one of several terms, including foundation, nuclear or G1 stock. The strategy of building the industry on nuclear stock is particularly critical in the face of rapid globalization of fruit tree production, increased global competitiveness and the exchange of large volumes of planting and breeding material in the search for new and desirable product characteristics.

http://dx.doi.org/10.19103/AS.2016.0017.18

The quest to obtain new genetic material from other growing regions and countries is tempered by the desire to prevent the introduction of disease agents, or potentially more damaging isolates of existing agents, into a particular growing region. A substantial barrier to rapid access to virus-tested material is the absence of knowledge of aetiological agents associated with many graft-transmissible diseases. The North American Plant Protection Organization currently lists 39 graft-transmissible apple diseases for which the associated pathogen is not known (NAPPO, 2009). To strengthen assurances that these agents and other pathogens are absent from propagation material, a complex system of testing has developed. These procedures are used in whole or in part (tailored to regional disease pressures) in foundation and certification programmes in many countries. In the past 20 years, molecular and serological assays have been increasingly incorporated into these programmes because of their rapidity and dependability. However, for the diseases with no known pathogen(s), biological assay procedures remain the current standard. The introduction of high-throughput sequencing has the potential to dramatically impact this model, and the technology is being introduced into several testing facilities. From the research perspective, this technology provides a tool that will allow scientists to determine the identity of aetiological agents associated with many of these diseases. However, this powerful technique is also revealing previously unrecognized viruses in plants that are not expressing overt symptoms, thus raising the question as to whether these agents should be considered pathogens.

2 Reducing the economic impact of virus-like agents

Quantifying crop losses attributable to infection by virus-like agents is one of the most difficult parameters to assess in crop production (Posnette, 1989; Waterworth and Hadidi, 1998; Hadidi and Barba, 2011). Yet, the presence of virus-like agents can be a determining factor in the economical sustainability of the apple industry. The impacts of viruses begin in the nursery where infection significantly reduced tree growth over their two-year residency in the nursery (Maxim et al., 2004). The impact carries forward through the production of trees and fruit, and into the marketplace. One analysis reported that the use of virus-free trees for apple production would avoid losses to growers in the United States of $63M annually (Cembali et al., 2003). The same study suggested that uniform adoption of virus-free planting stock would be reflected in substantial cost benefits to consumers in the United States.

The economic impact of viruses on production can be viewed at the individual grower level. A 14-year study determined the impact on the production of 'Golden Delicious' apples from trees infected with three common 'latent viruses' of apple: *Apple chlorotic leafspot virus* (ACLSV) (genus *Trichovirus*), *Apple stem pitting virus* (ASPV) (genus *Foveavirus*) and *Apple stem grooving virus* (ASGV) (genus *Capillovirus*) (van Oosten et al., 1983). The combination of these viruses resulted in a 17% reduction in yield over the course of the study. This reduced growth triggered a reduction in pruning costs and harvest costs that only partially compensated for reduced yield. The application of these reported yield and cost reductions to a balance sheet for the establishment and maintenance of a 'Red Delicious' apple orchard in Washington State (planted at 900 trees per acre) (Gallardo and Galinato, 2012) provides a gauge for the value of anticipated losses. It would require a price of $421.50 per 850-pound bin for the orchard planted with infected trees to match the economic return from an orchard planted with virus-free trees receiving only $400.00 per bin. This disparity is imposed throughout the 30-year life expectancy of the orchard and dramatically reduces the economic sustainability of the orchard operation. Moreover, the study of the 'Golden Delicious' block (van Oosten et al., 1983) also noted an increased rate of blemished fruit that would further reduce the profit margin from the virus-infected orchard. A similar 12-year study on 'McIntosh' grown on a variety of rootstocks confirmed these basic trends (Warner et al., 1984). Virus-infected trees not only produced lower crop yield, but the grade of the fruit produced was also reduced, resulting in lower profitability of orchards. In spite of the apparent economic impact of these 'latent viruses', the absence of easily recognizable symptoms has led to their widespread distribution in the industry.

The production and distribution of virus-tested propagation material is key to reducing the burden of virus diseases in commercial production (Gergerich et al., 2015). Clean plant programmes provide propagation material that is free of known economically important virus and virus-like agents, and are essential elements in any holistic approach to manage the diseases caused by these pathogens, and to improve economic sustainability. The constant flow of virus-tested material into commercial production displaces virus-infected trees and, over time, improves the overall phytosanitary status of the industry. Many of the diseases reported in the mid-twentieth century and caused by graft-transmissible agents can no longer be found in commercial production orchards. Also, clean plant programmes provide a means to safely access foreign cultivars for both new market development and new cultivar development, critically important to future viability of the industry.

Producers who are able access new varieties within the first three years of availability gain higher returns from their planting than growers who enter the market at a later time in the cycle of new variety development. Therefore, a central issue in the management of viral diseases in sustainable apple production is the balance between the need to establish new apple clones in orchards of commercial growers as quickly as possible, and the need for biosecurity (Foster and Hadidi, 1998). Since trees that are tested in the various quarantine and certification programmes will be expanded to thousands or millions of trees, these programmes are circumspect in their efforts to ensure that trees are free of harmful pathogens that could impact the broader industry.

The heart of clean plant programmes is the production of foundation planting material. Clean plant programmes represent substantial commitment and economic investment because of extensive testing and quarantine requirements, particularly for material from foreign sources. These programmes are the first line of defence in the effort to exclude new pathogens or genetic variants of pathogens from an area. For example, the Clean Plant Center Northwest located at Washington State University – IAREC in Prosser, Washington, reports that 40% of the propagation material submitted to the programme is infected with one or more virus-like agents. Similar programmes dedicated to the exchange of foundation level material exist throughout the world, and they facilitate the domestic and foreign exchange of commercial cultivars and germplasm (Gergerich et al., 2015). The nuclear stock sourced from these centres becomes the foundation from which large numbers of trees are expanded by nurseries and growers. In the United States, state-operated certification programmes offer a mechanism by which the released foundation material is propagated in such a manner as to minimize the risk of reintroduction of virus-like agents. The regulations for certification are generally based on the risk of re-infection imposed by regional conditions; consequently, testing requirements can vary between states.

3 Viruses and virus-like agents of apple

3.1 ACLSV, ASPV and ASGV

There are three '*latent viruses*' of apple, so called because most isolates of these viruses induce no readily visible symptoms on most commercial cultivars. The viruses are ACLSV, ASPV and ASGV. Through decades of vegetative propagation of presumably non-symptomatic apple trees, the strains that induce no or mild symptoms on most commercial cultivars have been inadvertently preferentially selected. This repeated propagation, coupled with the high mutation rate associated with RNA-containing viruses (Roossinck, 1997; Garcia-Arenal et al., 2001), may have also contributed to the absence of recognized natural vectors of these viruses. Man's propagation and dispersal of infected trees has become the main mechanism by which these viruses are transmitted and perpetuated.

There are some apple clones that express symptoms in response to infection by the 'latent viruses', and these clones have become the basis of biological assays used for virus detection. ACLSV induces chlorotic rings and necrotic spots on leaves of *Malus sylvestris* (L.) Mill. cv. R12740-7 ('Russian seedling'); this clone is the preferred indicator host for detecting this virus in most clean plant programmes (Jelkmann, 2004). ACLSV is also a contributing agent to 'topworking disease' in which a hypersensitive reaction occurs at the graft union on rootstock of *M. pumila* (Willd.) Borkh. Affected trees ultimately decline as the impairment of nutrient and water to the scion becomes more severe. This host specificity is paralleled by ASPV. There is a lack of acute symptom expression on most commercial cultivars, but some clones are selected as indicators because of a pronounced and distinct reaction that can be readily observed. *M.* × *domestica* Borkh. 'Spy 227' develops severe epinasty of young developing shoots. On this indicator, and on *M. sylvestris* (L.) Mill. 'Virginia Crab' and 'Ottawa 524' rootstocks, ASPV induces severe pitting under the cambium layer, impacting normal tree development. On these sensitive scions and rootstocks, there is a clear reduction in tree growth and consequently, in fruit production. ASGV was only detected once testing programmes were established for the other common viruses. The 'Virginia Crab' used for detecting ASPV develops deep grooves in the wood cylinder when infected with ASGV. Over time, ASGV will induce a hypersensitive reaction at the graft union of the scion with 'Virginia Crab' rootstock, and the scion ultimately declines and fails. In the nursery, on less sensitive rootstocks, ASGV reduced tree height by 23% and girth by 14% after the two-year growth cycle in nursery rows (Maxim et al., 2004). This degree of reduced growth was cultivar dependent, and some cultivars also exhibited symptoms of chlorosis.

Despite the absence of clearly defined symptoms on many commercial cultivars, the 'latent viruses' still extract a cost. There are few detailed reports of viral impacts, although yield reductions of 12% to 30% have been reported for mixed infections of these viruses (Meijnske et al., 1975; van Oosten et al., 1983; Penrose et al., 1988). These viruses are subject to genetic variation typical of most RNA-containing viruses, and ASPV is an extreme example of this variability (Yoshikawa et al., 2001; James et al., 2013). Through the rapid expansion of virus variants, there is enhanced potential for a unique isolate to develop and rise to prominence; these variants may induce symptoms on commercial cultivars. It has been suggested that a variant of ASPV may be responsible for apple green crinkle disease (see discussion below) (James et al., 2013). Adding to the complexity of host responses to infection, fruit tree breeding in general is broadening the scope of its efforts to identify desirable traits from different sources, often involving interspecific hybrids. This creates the potential to introduce sensitivity to viruses that did not occur in existing commercial cultivars.

3.2 Apple mosaic virus

Apple mosaic virus (ApMV) (genus *Ilarvirus*) is widespread wherever apples are cultivated. It induces symptoms on several commercial cultivars, but the intensity of these symptoms depends on cultivar and virus isolate (Posnette and Cropley, 1956). On sensitive varieties, distinct cream-coloured patterns appear on leaves early in the annual growth cycle, particularly in seasons with cool springtime weather (Fig. 1). Despite these transient foliar symptoms, ApMV is frequently regarded as a latent virus. Contributing to this perception is the fact that severely symptomatic leaves are frequently cast early in the season and so infections are overlooked. However, yield reductions of up to 30% have been reported for some cultivars sensitive to ApMV (Posnette and Cropley, 1959). ApMV belongs to a genus of viruses that are pollen transmitted, but there is no evidence of pollen or insect vector transmission of ApMV in commercial orchards. Root grafting has been identified as a potential source of transmission (Hunter et al., 1958; Dhingra, 1972).

Figure 1 *Apple mosaic virus* causes pale patterns on leaves of many apple cultivars such as 'Fuji'. The symptomatic leaves develop early in the season and are typically dropped.

3.3 Flat apple disease

Flat apple disease is caused by the virus *Cherry rasp leaf virus* (CRLV) (genus *Cheravirus*). Incidence of this disease is limited to the north-western region of the United States, and is believed to have originated in the native vegetation (Zawadska and Millikan, 1989). In addition to graft transmission, CRLV is transmitted through the nematode *Xiphinema americanum* Cobb (*sensu lato*). Both apple and cherry are hosts of CRLV, and fruit production of both is severely impacted. Additionally, the wide host range of CRLV includes many broadleaf weed species that are commonly found in orchards. This combination of weedy hosts and nematode transmission creates a circumstance in which CRLV is very difficult to control once it is established in an orchard. Production of both apple and cherry is seriously impacted in areas experiencing infections. Pre-planting treatments to reduce nematode populations are only partially successful. Repeated disking and fumigation may provide a period of relief from disease development, but complete eradication of the virus and viruliferous nematodes is unlikely, so the disease ultimately returns. Some horticultural practices such as disking to provide a clean orchard floor and trenching for rill irrigation contribute to the spread of virus-carrying nematodes along the rows within infected orchards. The development and availability of resistant rootstocks will provide an opportunity to regenerate orchards lost to production because of this virus.

3.4 Apple union necrosis and decline

Another nematode-transmitted virus of apple is *Tomato ringspot virus* (ToRSV) (genus *Nepovoirus*) that causes *apple union necrosis and decline*. Affected trees develop a necrotic layer of cells between the rootstock and scion resulting in decline of the scion. Trees on Malling-Merton 106 (MM106) rootstock frequently exhibit decline associated with ToRSV. Over a period of 4–5 years, leaves of the scion become chlorotic, and fruit size and overall yield decreases. ToRSV shares many biological properties with CRLV in that it is nematode transmitted, and has a wide host range. However, in contrast to CRLV, apple rootstocks are available that can be used to manage apple union necrosis and decline. Several available rootstock selections are either tolerant or resistant to ToRSV and permit rehabilitation of affected orchards (Gonzalves and Eastwell, 2014).

3.5 Viroids

Viroids are sub-viral particles that do not code for any proteins; thus, they are totally reliant on the host to provide the proteins needed for replication. There are several viroids that infect apple and induce symptoms, including *Apple scar skin viroid* (ASSVd), *Apple dimple fruit viroid* (ADFVd), and *Apple fruit crinkle viroid* (AFCVd); these are all members of the viroid genus *Apscaviroid*. These viroids directly affect fruit quality, resulting in surface defects of the fruit that range from dappling on cultivars with red skin to deep fissures and cracks. The fruit is unmarketable and the crop loss from affected trees is often 100%. The global distribution of these viroids is sporadic and so they are subject to intense regulation to minimize the introduction of the pathogens into new growing areas. There are no known insect vectors associated with these viroids, and graft transmission is the major means of transmission. It has been demonstrated that ASSVd, AFDVd and AFCVd can be transmitted mechanically by razor cuts (Koganezawa, 1985; DiSerio et al., 2001; Ito et al., 1993, respectively). Therefore, the use of contaminated implements in the nursery or orchard is a potential cause for spread of these agents. Natural spread has been observed, but the relative importance of natural root grafting and mechanical transmission through orchard maintenance operations is not known.

3.6 Apple proliferation

Although most virus-like agents that infect *Malus* are not known to be transmitted by insect vectors, an important exception is *apple proliferation* caused by the phytoplasma *Candidatus* Phytoplasma mali. This serious pathogen has been successfully excluded from North America by aggressive quarantine and certification programmes (Kunze, 1989; Mink, 1998); but in Europe, apple proliferation is regarded as the most destructive virus-like agent of *Malus* (Smith, 2004). The pathogen causes significant damage to affected trees (i.e. severe disease symptoms that include overall poor tree growth, poor fruit flavour and marked reduction in fruit size) and is vectored by the psyllids *Cacopsylla melanoneura* and *C. picta* (Tedeschi and Alma, 2004; Frisinghelli et al., 2000, respectively). In addition to exclusion programmes that provide pathogen-tested planting material, active vector control programmes are necessary in regions where the disease is found (Kunze, 1989). Although appropriate insecticide sprays are key elements to containing the spread of the pathogen (Kunze, 1989), testing and roguing are not universally recommended (Schmidt, 1975). From an economic management perspective, roguing is recommended for young productive orchards, but not in the case of older orchards. Many old trees will enter a recovery phase where the symptoms are subdued and the infected trees can continue to produce an economic crop until the orchard is renovated by complete removal (Schmidt, 1975). It remains to be determined if the current economic environment and the use of many new rootstocks and scions support this strategy.

As with most diseases caused by virus-like agents, removing a symptomatic limb does not eliminate the pathogen. By the time symptoms develop, the pathogen has typically invaded the root system and will initiate infection of new portions of the scion the following season. This is particularly true of apple proliferation which overwinters in active sieve tubes associated with the phloem. Since apple proliferation is an obligate parasite with a very specific environmental niche, it essentially re-infects the scion from the root system each season. It has been reported that recolonization of the aerial portion of the scion may not result in symptoms every year, but the potential persists once the tree has become infected (Kunze, 1989). An important consequence of the seasonal fluctuation in phytoplasma populations is that the efficiency of graft transmission from dormant budwood is relatively low (Cornaggia et al., 2012).

3.7 Apple green crinkle disease

Apple green crinkle disease highlights the challenges confronting the control of virus-like agents through quarantine and certification programmes. Over the last two decades, there has been a dramatic intensification in the search for new and exciting apple cultivars that tree brokers and apple producers hope will captivate the consumer market. In the new era of globalization, it is essential that producers have access to these new varieties early in the production cycle in order to remain competitive. This need for rapid accessibility to new cultivars is in direct conflict with phytosanitary programmes that ensure that each country or region remains free of harmful agents, or at least ensures that new introductions do not contribute to the virus load in that production area. Apple green crinkle disease directly affects the marketability of affected fruit as it results in severe cracks and blemishes that eliminate the fruit from commerce (Fig. 2). Affected trees of some cultivars have reduced fruit size, and the storage life of the fruit is also reduced (Fridlund and Drake, 1987). The disease is one of the most common fruit marking diseases of apple in North America, but occurs worldwide (Thomsen, 1989). No reliable leaf symptoms are associated with the disease. Moreover, since the aetiology of this disease is unknown, the only diagnostic method is observation of fruit symptoms on varieties known to express the disease. To further complicate the diagnostic procedure, fruit symptoms do not develop reliably each year. The development of fruit symptoms depends on temperature during the post-bloom period (Johnstone and Sampson, 1986), although exact parameters necessary for symptom expression are still under investigation. Inconsistent symptom development has two major consequences. In practice, healthy appearing trees are often used as a source of propagation material. In any particular year, if symptoms are absent, an infected tree may be inadvertently selected as a budwood source. In the context of clean plant programmes, to ensure that this disease is not present in plant material that will be disseminated through the industry, quarantine programmes rely on a bioassay whereby fruiting trees of a sensitive cultivar such as 'Golden Delicious' are inoculated with budwood from the suspect tree and then observed for three cropping years. This test alone delays the release of material for a minimum of three years.

The absence of rapid and inexpensive testing methods hampers the ability of growers to correctly identify the cause of fruit deformation in their orchards. Feeding by rosy apple aphid on developing fruit often leads to fruit deformation (Andris et al., 2015) that is difficult to distinguish from that caused by apple green crinkle disease. Aphid-induced symptoms can be managed in future cropping years, whereas apple green crinkle disease has the potential to impact production year after year. However, existing diagnostic capability is rapidly changing with new technology. Expanding access to high-throughput sequencing is accelerating the rate at which aetiological agents are being identified. In the case of apple green crinkle disease, it has been demonstrated that this disease is caused by a

Figure 2 *Apple green crinkle disease* is characterized by deformation of sensitive apple cultivars. The creases often become russeted. The early season symptoms are shown on 'Gala'. The depressions can become more pronounced and necrotic as the season progresses.

virus that is very closely related to ASPV (James et al., 2013). Further research is needed to determine if this agent represents a variant of ASPV or a new, very closely related virus. However, the large accumulation of information and molecular testing protocols related to ASPV can now be applied to this newly described pathogen. ASPV is tested and eliminated routinely in plant protection programmes internationally, suggesting that protocols for the detection and elimination of the agent of apple green crinkle disease can soon be implemented. The result of this research is that testing for the disease can be conducted in a matter of days instead of three years, and new apple cultivars will be made available more rapidly. However, the ability to differentiate the agent of green crinkle disease from closely related strains of ASPV that are not associated with acute disease symptoms in established orchard trees is a lingering problem.

3.8 Apple decline

Apple decline is a disease with very limited geographical distribution (Parish, 1989b). It is believed to be limited to areas of North America west of the continental divide, and has been reported in several orchards with different apple cultivars in the north-western United States (Parish, 1989b) and western Canada (Eastwell, unpublished). Although the regions affected by this disease appear to be quite isolated, once trees in an orchard are affected, control has not been accomplished, even with aggressive roguing operations. A resolution of this situation will depend on identification of the pathogen(s) involved. This will enable further research on the mechanism of transmission and reveal critical control points in the disease cycle that could be addressed by management strategies. Trees affected with this disease exhibit a dramatic crop loss; the undersized fruit is not commercially marketable, and the annual extension of new growth is dramatically reduced (Eastwell, 2014). The symptoms will first appear on a single limb but will spread to the remainder of the tree. The aetiological agent associated with this disease has not been identified, but it was demonstrated that the injection of oxytetracycline can temporarily ameliorate symptoms. This implicates the involvement of a phytoplasma or similar agent in the disease, but many attempts to verify this association have been unsuccessful. There are no known vectors of apple decline, but radial spread has been observed, and root grafting may play an important role in the local spread of the disease (Parish, 1989a). Since there are no appropriate detection methods, apple decline does not appear on most international quarantine regulations. It is assumed that material selected for propagation will be from healthy appearing trees not afflicted with this condition.

3.9 Apple rubbery wood disease and apple flat limb disease

Apple rubbery wood disease is likely distributed throughout the world wherever apples are grown commercially (Waterworth and Fridlund, 1989), but through the efforts of quarantine and certification programmes, severe forms of the disease are becoming rare. The disease is manifested by poor lignification of limbs leading to branches that are very supple; consequently, branches droop under the weight of fruit (Fig. 3) (Németh, 1986; Waterworth and Fridlund, 1989). On sensitive cultivars, there is a concomitant loss of yield (Németh, 1986) with degraded fruit quality (van Oosten, 1983). *Apple flat limb disease* is believed to be a manifestation of the same aberrant lignification caused by the

Figure 3 Rubbery wood causes depressed lignification of limbs. In addition to reduced branch and fruit growth, the weight of the fruit causes the branches to droop dramatically. This is an extreme example of rubbery wood on the indicator 'Lord Lambourne'.

apple rubbery wood agent. Attempts to determine the pathogen associated with these diseases have resulted in ambiguous or contradictory results (Poggi Pollini et al., 1995; Bertaccini et al., 1998). Consequently, biological indexing is the only accepted method at the moment to detect the disease-causing agent. Because of the potential impact on apple production globally, a reliable and rapid assay method to detect the aetiological agent would dramatically accelerate research into determining the mechanism of disease, the potential economic impact and key factors in the management of the disease in commercial production.

4 Advancing diagnostic technology

Among many formats of molecular-based assays, the application of high-throughput sequencing to virus detection in quarantine programmes is becoming increasingly contemplated as an alternative to the current mixture of biological, serological and molecular assays that constitute virus testing programmes (Ho and Tzanetakis, 2014; Massart et al., 2014; Al Rwahnih et al., 2015). The primary benefit of this strategy is that no prior knowledge of a particular pathogen is needed before it can be detected in the pool of nucleic acids extracted from the subject sample. However, a protocol to distinguish the presence of virus-like sequences from actual viruses remains a subject of investigation. The detection of sequence fragments that are similar to known virus sequence does not necessarily indicate an active virus infection as segments of virus-like sequence can be detected in the genomes of plants; this complication is being addressed (e.g. Ho and Tzanetakis, 2014). Other molecular methods revealed this potential, and the sensitivity and broad spectrum of high-throughput sequencing increases the risk of repeating these incidents. The screening parameters necessary to process raw data from high-throughput sequencing to identify active infections of novel viruses may be situational, thus hampering the development of universal guidelines that can be applied across multiple crops and/or situations. However, the detection of known, characterized viruses by high-throughput sequencing is less prone to the consequences of mutations than other molecular testing formats. Many plant viruses are RNA-containing viruses, and through the idiosyncrasy of virus replication, many viruses generate an abundance of mutations. This is the basis of the quasi-species concept of plant viruses (Roossinck, 1997; Garcia-Arenal et al., 2001). The spectre of a high mutation rate raises concerns of the reliability of many sequence-specific assays, such as RT-PCR, that may fail if a sequence variant arises that eludes detection. In grapevines, it was demonstrated that high-throughput sequencing exceeds the ability of current programmes to detect even known pathogens (Al Rwahnih et al., 2015). The increasing reliance on high-throughput sequencing is gaining acceptance in the area of disease management worldwide, but is not without unintended consequences.

High-throughput sequencing has been a valuable tool in the determination of aetiology of many fruit tree diseases. Apple green crinkle disease is but one example. Other diseases of uncertain aetiology are awaiting the application of new technology; two important examples include apple rubbery wood and apple decline. This technology has been used in *Prunus* to identify several diseases with previously unknown aetiology (Bag et al., 2015; Marais et al., 2015; Villamor et al., 2015). However, this technology has revealed the presence, in various crops, of virus-like agents that are not necessarily pathogenic. In *Malus*, this was recently demonstrated by the description of a geminivirus infecting apple trees (Liang et al., 2015). This genus of virus that represents DNA-containing viruses was not known to occur in apple trees prior to this discovery. It was verified that the sequence is a graft-transmissible agent. Moreover, it was demonstrated that the virus was present in 7.2% of the small number of trees tested. This raises the question of whether this virus is of quarantine significance. The trees that were sampled and found to be infected did not express any acute symptoms. This does not necessarily mean that the virus does not impact yield or fruit quality. It will take many years to address these latter questions. In the meantime, the regulatory community is in limbo in its efforts to develop an appropriate response to this discovery in the absence of critical facts pertaining to the biology and economic impact of the virus. Because the agent was unknown prior to this revelation, it is presumed that the dissemination of the virus was not halted by the existing quarantine or certification systems in place around the world. Is it present outside of the country where it was originally detected? What is the impact on fruit production? Even though no symptoms were observed on the sample trees, the rapidly expanding array of new cultivars and hybrids permeating the market place make it difficult to designate the virus as latent across the spectrum of cultivars and rootstocks currently in production or that may be introduced in the future. It will take many years to evaluate this situation. However, based on experiences with other asymptomatic or latent viruses, this virus will likely take a toll on production even in the absence of acute symptoms. As the application of high-throughput sequencing expands, this uncertainty will be confronted on a reoccurring basis.

An additional issue that challenges efforts to exchange planting stock that is free of deleterious agents is the abundance, in the literature, of many disease names for which identified voucher samples are no longer available. Many of the disease names may reflect the impact of a known agent that exerts a unique spectrum of symptoms on a specific apple cultivar, or that of a specific variant of a known agent that causes symptoms different from the type isolates of the same agent. Many of the listed diseases have disappeared from commercial production through the advent of clean plant programmes, but their legacy remains in the literature. This hampers the ability of agencies, and those concerned with the impact of diseases, to provide material designated free of these agents. How can you verify that an agent is not present if you do not have a control sample to develop appropriate assays? The answer to this question must be reached by a consensus of regulatory agencies. In part, high-throughput sequencing will facilitate identification of some of these agents, but this strategy will only advance identification if voucher sample material is available. A mechanism that allows diseases that have not been reported for a specified period of years to be removed from the list of agents of concern would provide a balance between allowing free movement of propagation material and biosecurity. An important proviso is that, if an incidence of the disease is subsequently confirmed, the disease will be returned to the 'active' list. To be effective, this strategy requires the cooperation of surveillance programmes and regulators at all levels.

5 Remaining challenges

The adoption of high-throughput sequencing as a tool in the development and distribution of apple propagation material and germplasm is in its infancy. Validation and expansion of this technology globally to increase confidence and acceptance for virus detection is critical. Concurrently, improvements in the application of technology to recognize viral sequences and differentiate them from host sequences are imperative. The acquisition of more data will aid in the refinement of data analysis parameters and increase the reliability of this technology as a diagnostic tool for vegetatively propagated perennial crops. Additional studies will increase confidence in these methods and perhaps, in time, they will replace all other existing diagnostic protocols for graft-transmissible agents.

High-throughput sequencing provides an approach that will allow the aetiological agent of described diseases to be determined. As promising as this may appear, the effort still requires rigorous and often very time-consuming analyses. Many diseased trees are infected with multiple viral agents. Determining which one or combination of these agents is responsible for eliciting specific disease symptoms is a complicated task. Although high-throughput sequencing greatly enhances the pace at which sequence data can be accumulated, the sequence data must ultimately be coupled with biological data to confirm the aetiological relationship. The reward is knowledge of the specific pathogen, and the ability to address many of the questions regarding disease epidemiology.

Similarly, the acquisition of biological data related to newly discovered virus-like agents is a critical task that must be addressed before appropriate action plans can be formulated. Presumably, the pace at which new virus-like agents are discovered by high-throughput sequencing and reported will diminish over time as application of the technology finds greater use, at least until the next technology that reveals another group of virus-like agents is developed. The regulatory implications of each new discovery should be harmonized. A globally accepted response protocol will help reduce interruptions in trade that frequently occur when a virus is reported from a certain area. Fear of trade repercussions currently slows the reporting of these findings. Clear guidelines for criteria that constitute a new virus report are important. Is this a new pathogen? What are the production impacts? These are the first questions to be asked when a new virus is described and response guidelines must be in place until these questions can be answered.

As previously mentioned, the reliance on field studies to reveal impacts on production is very expensive and very lengthy. While field plots yield the most reliable data to expose economic impacts, the development of molecular markers of the host that can help predict eventual impacts will be beneficial in addressing this question. The strategy could be based on the comparison of expression levels of key genes in infected young trees relative to healthy plants as a predictor of eventual impacts on yield. However, this strategy will require extensive research to gain confidence in its predictive value. Identification of the 'key genes' whose expression levels should be monitored becomes more plausible with the advent of high-throughput sequencing.

Once a virus has been detected, the next challenge is the elimination of the agent from the propagation material. There are many strategies that have been investigated, including thermal therapy, chemotherapy, meristem culture, cryotherapy or combinations of these. However, one basic concern remains: this process presents a genetic bottleneck. All plants propagated from that source after the virus-elimination process will be derived from that bud. As apples pass through apical meristem culture, physiological changes can be induced. Examples include delayed flowering or delayed fruiting (A. I. Campbell, 1974 as cited in Walkey, 1978). The source of this variation is not always clear. The origin may be elimination of an undetected infectious agent, a genetic mutation that was captured through the meristem process, periclinal chimeric starting material (Einset and Blaser, 1948; Pratt et al., 1967; Embree and McRae, 1991) or epigenetic modification. *Malus* species have a tendency to produce sports, and this has resulted in the development of many desirable apple clones (Granhall, 1954). However, point mutations in a single gene can also result in serious deleterious effects such as the pale green lethal disorder (Orcheski et al., 2015); this single change apparently arose as a unique mutation that was fixed in the commercial apple population and perpetuated (Orcheski et al., 2015). Somaclonal variation is low in apple (Skirvin et al., 1994; McMeans et al., 1998), which limits the use of this strategy to generate new apple cultivars. The expression of a different physiological characteristic is expected in 1–3% of regenerated explants (Skirvin et al., 1994). However, even a low rate of variation in a foundation programme can have dire consequences, and the physiological stress imposed by virus-elimination techniques may exacerbate this potential (Galhardo et al., 2007). Thus, after production of material free of the designated pathogen, it must be verified that the resulting buds have the same genetic constitution as the starting material. Currently, this is achieved through gross horticultural trait evaluation. It may take many years before subtle differences in fruit yield or quality are noted. Regardless of whether virus-like agents are eliminated through meristem culture, chemotherapy, thermal therapy, cryotherapy or a combination of methods, assurance that the product will display the correct phenotype (true-to-type) is critical. In reality, this consideration is more important to most producers than the virus status. The development of genetic keys for new apple variety identification will greatly help to alleviate this concern. The application of high-throughput sequencing may play a significant role in this process. Many current molecular methods based on RAPD, AFLP (Goulão et al., 2001), microsatellite (Nybom et al., 1990) and SSR (Goulão and Oliveira, 2001; Galli et al., 2005) technology that are used for developing a genetic fingerprint of an apple clone (Nybom et al., 2014) are not adequately refined to routinely differentiate clones whose horticultural differences may be the consequence of minor genetic variations resulting from somatic mutation. The power of these techniques has been demonstrated for some established clones (Chagné et al., 2012) but is insufficient for routine application. Comparison of high-throughput sequencing data of the entire plant genome will become the next tool for plant identification. In a different application of this strategy, genome sequencing of apple cultivars has been used to identify genetic differences that may be responsible for certain preferred horticultural traits (Zhang et al., 2014). This demonstrates the power of high-throughput sequencing and its application for clone identification among three closely related but genetically distinct clones. High-throughput sequencing has been used to identify single nucleotide polymorphisms (SNPs) (Khan et al., 2012) and this strategy may be used for apple clone identification in the future.

The uncertainties that remain in the epidemiology of the diseases mentioned above highlight the need for fundamental biological research to reveal key control points that allow for disease management. The many gaps in understanding of how these diseases may behave in the orchard emphasize the need to exert the first line of defence for disease control: exclusion. Nationally, this is governed by federal quarantine regulations (Gergerich et al., 2015). However, domestic nursery certification programmes are critical components in an industry-wide disease management strategy. State-operated certification programmes actively monitor the intermediate steps in the expansion of the very limited number of trees produced by foundation programmes to the millions of trees planted in commercial orchards.

Understanding modes of transmission once a virus-like agent is introduced into an orchard is critical for disease management, and horticultural management practices of individual orchards can dramatically impact disease potential. Transmission through root grafting has been clearly demonstrated for several pathogens, including ApMV and apple proliferation phytoplasma (Dhingra, 1972; Lešnik et al., 2008, respectively). Many, if not all, of the virus-like agents that are known to infect apple are potentially transmitted via root grafting. A thorough understanding of this potential as influenced by different orchard architecture may help provide guidance for minimizing the impact of virus-like agents on production. The migration of the industry to extremely high-density plantings will only facilitate root grafting and pathogen transmission unless specific rootstocks are developed that reduce the frequency of root grafting.

Unlike most insects and other pathogens that impact apple production, there are no practical means to treat or eliminate viruses from a commercial planting once a tree has become infected. This led to the emphasis on exclusion programmes as the primary defence against diseases caused by these agents. Since *Malus* germplasm has been very limited in terms of providing natural resistance to viruses, the development of transgenic plant lines offers the promise of providing resistance to select viruses. However, concerns about trade barriers and public acceptance have limited the commercial development of genetically modified apples. This landscape is rapidly changing and genetically modified Arctic® apples that resist browning have been approved for release (Lehnert, 2015). The use of transgenics to develop resistance to virus infection has been of interest for several decades (Register and Beachy, 1988; Prins et al., 2008). As the understanding of the mechanisms of induced resistance becomes more clear (Waterhouse et al., 1998; Duan et al., 2012), the possibilities of achieving this goal increase. The many strategies for the development of transgenics for virus resistance are beyond the scope of this chapter, but there is little doubt that they will eventually become a large part of crop production. Transgenic versions of other fruit tree crops that provide resistance to viruses have been developed and released for commercial production (Ferreira et al., 2002; Scorza et al., 2013).

Only with increased speed and responsiveness of the clean plant programmes will the use of clean planting stock be more widely accepted. With the exception of plant movement between defined quarantine zones, the use of virus-tested planting material largely remains voluntary. The presence or absence of virus-like agents is frequently the last consideration in the development of an orchard business plan, if it is considered at all. There is little patience for the protracted testing protocols required to ensure freedom from graft-transmissible agents. Propagative material that is free of harmful viruses and virus-like agents must be made available as quickly as possible at a reasonable cost. Only then will the adoption of an 'all clean' planting system be embraced. Currently, in the eyes of the industry, the short-term encumbrances imposed by the clean plant process outweigh the long-term benefits. The universal adoption of a clean plant approach depends on an improved cost/benefit ratio as viewed by the industry. With the advent of many new cultivars and armed with the knowledge that has been gained over the recent decades, developing and maintaining a tree free of pathogens from inception in a breeding programme until it is planted in a commercial orchard is possible. This would negate the need for the virus-elimination process and the attendant risk of sport induction.

6 Where to look for further information

Much of the innovation for the detection of apple pathogens is driven by the needs of the industry components to remain globally competitive. Various international quarantine agencies are responsive to this need, and implement the most appropriate technologies to maintain the phytosanitary standards of each country. The European Plant Protection Organization (OEPP/EPPO, 1999) and the North American Plant Protection Organization (NAPPO, 2009) maintain lists of pathogens and their status in each growing region, as well as general descriptions of the overall disease management strategies. In the United States, the National Clean Plant Network (NCPN) (http://nationalcleanplantnetwork.org) initiative is committed to the propagation and distribution of virus-tested vegetatively propagated specialty crops including apple. NCPN centres and participants are actively engaged in all aspects of clean plant production, but the primary focus is on the development of nuclear clean stock (Gergerich et al., 2015). Research aspects of fruit tree virus management are coordinated through the Multistate Research Coordinating Committee and Information Exchange Group of USDA-NIFA in the WERA-020 Virus and Virus-Like Diseases of Fruit Trees, Small Fruits, and Grapevines (http://www.nimss.org/projects/view/mrp/outline/13496). This group of technical and regulatory experts convenes annually to share current challenges and to facilitate collaborative research efforts that address emerging disease issues. The International Council for the Study of Virus and other Graft Transmissible Diseases of Fruit Crops (ICVF) (http://www.icvf.net/) is a scientific organization that promotes and coordinates research for the characterization and management of graft-transmissible disease. Active since 1954, its membership convenes every three years to share information on current and emerging research topics.

The American Phytopathological Society (APS) and its many committees are major coordinating bodies. In addition to annual meetings, APS Press associated with the society facilitates publication of current journals (*Plant Disease*, and *Phytopathology*) and books devoted to the subject of disease management. Recent publications addressing specific virus-like agents include

Hadidi, A., Barba, M., Candresse, T. and Jelkmann, W., eds. 2011. Virus and Virus-Like Diseases of Pome and Stone Fruits, American Phytopathological Society Press, St. Paul, MN, p. 429.

Sutton, T., Aldwinckle, H. S., Agnello, A. M. and Walgenbach, J. F., eds. 2014. Compendium of Apple and Pear Diseases and Pests, Second Edition, American Phytopathological Society Press, St. Paul, MN, p. 218.

A general description to the principles of disease control that includes many sections on perennial crops can be found in:

Hadidi, A., Khetarpal, R. K. and Koganezawa, H., eds. 1998. Plant Virus Disease Control, American Phytopathological Society Press, St. Paul, MN, p. 684.

An in-depth description that incorporates most of the critical elements of viroids as pathogens of many horticultural crops can be found in:

Hadidi, A., Flores, R., Randles, J. W. and Semancik, J. S., eds. 2003. Viroids, CSIRO Publishing, Collingwood, VIC, p. 370.

7 References

Al Rwahnih, M., Daubert, S., Golino, D., Islas, C. and Rowhani, A. 2015. Comparison of next-generation sequencing versus biological indexing for the optimal detection of virus pathogens in grapevine. *Phytopathology* 105: 758–63.

Andris, H. L., Bentley, W. J., Coates, W. W. and Pickel, C. 2015. Rosy apple aphid. UC IPM Pest Management Guidelines: Apple. UC ANR Publication 3432. http://www.ipm.ucdavis.edu/PMG/r4301511.html.

Bag, S., Al Rwahnih, M., Li, A., Gonzalez, A., Rowhani, A., Uyemoto, J. K. and Sudarshana, M. R. 2015. Detection of a new Luteovirus in imported nectarine trees: a case study to propose adoption of metagenomics in post-entry quarantine. *Phytopathology* 105: 840–6.

Baric, S., Berger, J., Cainelli, C., Kerschbamer, C., Letschka, T. and Dalla Via, J. 2010. Seasonal colonisation of apple trees by '*Candidatus* Phytoplasma mali' revealed by a new quantitative TaqMan real-time PCR approach. *European Journal of Plant Pathology* (5 November 2010), pp. 1–13. doi: 10.1007/s10658-010-9706-x.

Bertaccini, A., Vibio, M., Janeková, M. and Fránová-Honetšlegrová, J. 1998. Molecular detection of phytoplasmas in apple with rubbery wood symptoms. *Acta Horticulturae* 472: 693–700.

Cembali, T., Folwell, R. J., Wandschneider, P., Eastwell, K. C. and Howell, W. E. 2003. Economic implications of a virus prevention program in deciduous tree fruits in the US. *Crop Protection* 22: 1149–56.

Chagné, D., Crowhurst, R. N., Troggio, M., Davey, M. W., Gilmore, B., Lawley, C., Vanderzande, S., Hellens, R. P., Kumar, S., Cestaro, A., Velasco, R., Main, D., Rees, J. D., Iezzoni, A., Mockler, T., Wilhelm, L., Van de Weg, E., Gardiner, S. E., Bassil, N. and Peace, C. 2012. Genome-wide SNP detection, validation, and development of an 8 K SNP array for apple. *PLoS ONE* 7: e31745.

Cornaggia, D., Grasseau, N. and Gentit, P. 2012. Efficacy of apple proliferation phytoplasma transmission by vegetative propagation techniques and impact on its epidemiology. *Petria* 22: 183.

Dhingra, K. L. 1972. Transmission of apple mosaic virus by natural root grafting. *Indian Journal of Horticulture* 29: 348–50.

Di Serio, F., Malfitano, M., Alioto, D., Ragozzino, A., Desvignes, J. C. and Flores, R. 2001. Apple dimple fruit viroid: fulfillment of Koch postulates and symptom characteristics. *Plant Disease* 85: 179–82.

Duan, C.-G., Wang, C.-H. and Guo, H.-S. 2012. Application of RNA silencing to plant disease resistance. *Silence* 3: 5.

Eastwell, K. C. 2014. Viruslike or graft-transmissible diseases with no known causal agents, in T. Sutton, H. S. Aldwinckle, A. M. Agnello and J. F. Walgenbach (eds), *Compendium of Apple and Pear Diseases and Pests*, Second Edition, American Phytopathological Society Press, St. Paul, MN, p. 106.

Einset, J. and Blaser, H. W. 1948. Chromosomal periclinal chimeras in leaves and flowers of six apple sports. *Genetics* 33: 103–4.

Embree, C. G. and McRae, K. B. 1991. Variation in growth, spur density, and yield of 'MacSpur McIntosh'. *HortScience* 26: 188–90.

Ferreira, S. A., Pitz, K. Y., Manshardt, R., Zee, F., Fitch, M. and Gonsalves, D. 2002. Virus coat protein transgenic papaya provides practical control of Papaya ringspot virus in Hawaii. *Plant Disease* 86: 101–5

Foster, J. A. and Hadidi, A. 1998. Exclusion of plant viruses, in A. Hadidi, R. K. Khetarpal and H. Koganezawa (eds), *Plant Virus Disease Control*, American Phytopathological Society Press, St. Paul, MN, pp. 208–29.

Fridlund, P. R. and Drake, S. R. 1987. Influence of apple green crinkle disease on the quality of Granny Smith apples. *Plant Disease* 71: 585–7.

Frisinghelli, C., Delatti, L., Grando, M. S., Forti, D. and Vindimian, M. E. 2000. *Cacopsylla costalis* (Flor 1861), as a vector of apple proliferation in Trentino. *Journal of Phytopathology* 148: 425–31.

Galhardo, R. S., Hastings, P. J. and Rosenberg, S. M. 2007. Mutation as a stress response and the regulation of evolvability. *Critical Reviews in Biochemistry and Molecular Biology* 42: 399–435.

Gallardo, R. K. and Galinato, S. P. 2012. Cost estimates of establishing, producing and packing Red Delicious apples in Washington. Washington State University Enterprise Budgets for Crop Commodities (FS099E).

Galli, Z., Halász, G., Kiss, E., Heszky, L. and Dobránszki, J. 2005. Molecular identification of commercial apple cultivars with microsatellite markers. *HortScience* 40: 1974–7.

Garcia-Arenal, F., Fralle, A. and Malpica, J. M. 2001. Variability and genetic structure of plant virus populations. *Annual Review of Phytopathology* 39: 157–86.

Gergerich, R. C., Welliver, R. A., Gettys, S., Osterbauer, N. K., Kamenidou, S., Martin, R. R., Golino, D. A., Eastwell, K., Fuchs, M., Vidalakis, G. and Tzanetakis, I. E. 2015. Safeguarding fruit crops in the age of agricultural globalization. *Plant Disease* 99: 176–87.

Gonzalves, D. and Eastwell, K. C. 2014. Apple union necrosis, in T. B. Sutton, H. S. Aldwinckle, A. M. Agnello and J. F. Walgenbach (eds), *Compendium of Apple and Pear Diseases and Pests*, Second Edition, American Phytopathological Society Press, St. Paul, MN, pp. 99–100.

Goulão, L. and Oliveira, C. M. 2001. Molecular characterisation of cultivars of apple (Malus × domestica Borkh.) using microsatellite (SSR and ISSR) markers. *Euphytica* 122: 81–9.

Goulão, L., Cabrita, L., Oliveira, C. M. and Leitão, J. M. 2001. Comparing RAPD and AFLP analysis in discrimination and estimation of genetic similarities among apple (Malus × domestica Borkh.) cultivars. *Euphytica* 119: 259–70.

Granhall, I. 1954. Spontaneous and induced bud mutations in fruit trees. *Acta Agriculturae Scandinavica* 4: 594–400.

Hadidi, A. and Barba, M. 2011. Economic impact of pome and stone fruit viruses and viroids, in A. Hadidi, M. Barba, T. Candresse and W. Jelkmann (eds), *Virus and Virus-like Diseases of Pome and Stone Fruits*, American Phytopathological Society Press, St. Paul, MN, pp.1–7.

Ho, T. and Tzanetakis, I. E. 2014. Development of a virus detection and discovery pipeline using next generation sequencing. *Virology* 471–3: 54–60.

Hunter, J. A., Chamberlain, E. E. and Atkinson, J. D. 1958. Note on the transmission of apple mosaic by natural root grafting. *New Zealand Journal of Agricultural Research* 1: 80–2.

Ito, T., Kanematsu, S., Koganezawa, H., Tsuchizaki, T. and Yoshida, Y. 1993. Detection of a viroid associated with apple fruit crinkle disease. *Annals of the Phytopathological Society of Japan* 59: 520–7.

James, D., Varga, A., Jesperson, G. D., Navratil, M., Safarova, D., Constable, F., Horner, M., Eastwell, K. and Jelkmann, W. 2013. Identification and complete genome analysis of a virus variant or putative new foveavirus associated with apple green crinkle disease. *Archives of Virology* 158: 1877–87.

Jelkmann, W. 2004. Detection of virus and virus-like diseases of fruit trees. *Acta Horticulturae* 657: 575–96.

Johnstone, G. R. and Sampson, P. J. 1986. The influence of temperature on the expression of green crinkle in Granny Smith apple fruit. *Acta Horticulturae* 193: 113–18.

Khan, M. A., Han, Y., Zhao, Y. F. and Korban, S. S. 2012. A high-throughput apple SNP genotyping platform using the GoldenGate™ assay. *Gene* 494: 196–201.

Koganezawa, H. 1985. Transmission to apple seedlings of low molecular weight RNA from apple scar skin diseased trees. *Annals of the Phytopathological Society of Japan* 51: 176–82.

Kunze, L. 1989. Apple proliferation, in P. Fridlund (ed.), *Virus and Viruslike Diseases of Pome Fruits and Simulating Noninfectious Disorders*. Washington State University Cooperative Extension Publications SP0003, pp. 99–113.

Lehnert, R. 2015. USDA clears non-browning GMO apples. http://www.goodfruit.com/usda-clears-non-bowning-gmo-arctic-granny-and-golden-apples/. Accessed 31 March 2016.

Lešnik, M., Brzin, J., Mehle, N. and Ravnikar, M. 2008. Transmission of '*Candidatus phytoplasma mali*' by natural formation of root bridges in M9 apple rootstock. *Agricultura* 5: 43–6.

Liang, P., Navarro, B., Zhang, Z., Wang, H., Lu, M., Xiao, H., Wu, Q., Zhou, X., Di Serio, F. and Li, S. 2015. Identification and characterization of a novel geminivirus with a monopartite genome infecting apple trees. *Journal of General Virology* 96: 2411–20.

Marais, A., Faure, C., Mustafayev, E., Barone, M., Alioto, D. and Candresse, T. 2015. Characterization by deep sequencing of Prunus virus T, a novel *Tepovirus* infecting *Prunus* species. *Phytopathology* 105: 135–40.

Massart, S., Olmos, A., Jijakli, H. and Candresse, T. 2014. Current impact and future directions of high throughput sequencing in plant virus diagnostics. *Virus Research* 188: 90–6.

Maxim, A., Zagrai, L., Zagrai, I. and Isac, M. 2004. Studies on the influence of Apple stem grooving virus on tree growth of various cultivars in the nursery. *Acta Horticulturae* 657: 41–9.

McMeans, O., Skirvin, R. M., Otterbacher, A. and Mitiku, G. 1998. Assessment of tissue culture-derived 'Gala' and 'Royal Gala' apples (Malus × domestica Borkh.) for somaclonal variation. *Euphytica* 103: 251–7.

Meijnske, C. A. R., van Oosten, H. J. and Peerbooms, H. 1975. Growth, yield and fruit quality of virus-infected and virus-free 'golden delicious' apple trees. *Acta Horticulturae* 44: 209–12.

Mink, G. I. 1998. Virus certification of deciduous fruit trees in the United States and Canada, Chapter 23, in A. Hadidi, R. K. Khetarpal and H. Koganezawa (eds), *Plant Virus Control*, American Phytopathological Society Press, St. Paul, MN, pp. 294–300.

NAPPO. 2009. Guidelines for the movement of stone and pome fruit trees and grapevines into a NAPPO member country, RSPM No. 35.

Németh, M. 1986. *Virus, Mycoplasma and Rickettsia Diseases of Fruit Trees*, Martinus Nijhoff, The Netherlands, p. 840.

Nybom, H., Rogstad, S. H. and Schaal, B. A. 1990. Genetic variation detected by use of the M13 'DNA fingerprint' probe in Malus, Prunus, and Rubus (Rosaceae). *Theoretical and Applied Genetics* 79: 153–6.

Nybom, H., Weising, K. and Rotter, B. 2014. DNA fingerprinting in botany: past, present, future. *Investigative Genetics* 5: 1.

OEPP/EPPO. 1999. Pathogen-tested material of Malus, Pyrus and Cydonia. Bulletin OEPP/EPPO Bulletin 29, 239–52. https://www.eppo.int/STANDARDS/council2015.htm.

Orcheski, B., Parker, R. and Brown, S. 2015. Pale green lethal disorder in apple (*Malus*) is caused by a mutation in the *PHYLLO* gene which is essential for phylloquinone (vitamin K_1) biosynthesis. *Tree Genetics and Genomes* 11: 131. doi: 10.1007/s11295-015-0956-9.

Parish, C. L. 1989a. Apple decline. *Washington State Horticultural Association Proceedings* 85: 225–7.

Parish, C. L. 1989b. Apple decline: characterization, cause and cure. *Acta Horticulturae* 235: 93–97.

Penrose, L. J., Davis, K. C. and Valentine, B. J. 1988. Comparative performance of three apple clones derived from a virus-tested scheme, with clones infected with latent viruses and a mycoplasma. *Scientia Horticulturae* 36: 55–65.

Poggi Polini, C., Giunchedi, L., Seemüller, E., Filippini, G. and Vindimian, E. 1995. Etiological studies of apple rubbery wood disease. *Acta Horticulturae* 386: 503–5.

Posnette, A. F. 1989. Losses caused by virus and viruslike diseases, in P. Fridlund (ed.), *Virus and Viruslike Diseases of Pome Fruits and Simulating Noninfectious Disorders*, Washington State University Cooperative Extension Publications SP0003, pp. 3–6.

Posnette, A. F. and Cropley, R. 1956. Apple mosaic viruses. Host reactions and strain differences. *Journal of Horticultural Science* 31: 119–33.

Posnette, A. F. and Cropley, R. 1959. The reduction in yield caused by apple mosaic. *Report East Malling Research Station*, 1958: 89–90.

Pratt, C., Ourecky, D. K. and Einset, J. 1967. Variation in apple cytochimeras. *American Journal of Botany* 54: 1295–301.

Prins, M., Laimer, M., Noris, E., Schubert, J., Wassenegger, M. and Tepfer, M. 2008. Strategies for antiviral resistance in transgenic plants. *Molecular Plant Pathology* 9: 73–83.

Register, J. C. and Beachy, R. N. 1988. Resistance to TMV in transgenic plants results from interference with an early event in infection. *Virology* 166: 524–32.

Roossinck, M. J. 1997. Mechanisms of plant virus evolution. *Annual Review of Phytopathology* 35: 191–209.

Schmid, G. 1975. Prolonged observations on spread and behaviour of proliferation disease in apple orchards. *Acta Horticulturae* 44: 183–91.

Scorza, R., Callahan, A., Dardick, C., Ravelonandro, M., Polak, J., Malinowski, T., Zagrai, I., Cambra, M. and Kamenova, I. 2013. Genetic engineering of Plum pox virus resistance: 'HoneySweet' plum-from concept to product. *Plant Cell Tissue and Organ Culture* 115: 1–12.

Skirvin, R. M., McPheeters, K. D. and Norton, M. 1994. Sources and frequency of somaclonal variation. *Hortscience* 29: 1232–7.

Smith, I. M. 2004. Apple proliferation MLO. CABI/EPPO www.eppo.int/QUARANTINE/data_sheets/bacteria/PHYPMA_ds.pdf. Accessed 21 April 2016.

Tedeschi, R. and Alma, A. 2004. Transmission of apple proliferation phytoplasma by *Cacopsylla melanoneura* (Homoptera:Psyllidae). *Journal of Economic Entomology* 97: 8–13.

Thomsen, C. 1989. Green crinkle, in P. Fridlund (ed.), *Virus and Viruslike Diseases of Pome Fruits and Simulating Noninfectious Disorders*, Washington State University Cooperative Extension Publications SP0003, pp. 55–7.

van Oosten, H. J., Meijneke, C. A. R. and Peerbooms, H. 1983. Growth, yield and fruit quality of virus-infected and virus-free Golden Delicious apple trees, 1968–82. *Acta Horticulturae* 130: 213–20.

Villamor, D. E. V., Susaimuthu, J. and Eastwell, K. C. 2015. Genomic analyses of cherry rusty mottle group and cherry twisted leaf-associated viruses reveal a possible new genus within the family Betaflexiviridae. *Phytopathology* 105: 399–408.

Walkey, D. G. A. 1978. In vitro methods for virus elimination, chapter 25, in T. A. Thorpe (ed.), *Frontiers of Plant Tissue Culture 1978*, International Association for Plant Tissue Culture 1978, CalgaryAB, pp. 245–54.

Warner, J., Heeney, H. B., Leuty, S. J. and Potter, C. L. 1984. Effect of virus infection on the performance of McIntosh apple trees on selected seedling and clonal rootstocks. *Canadian Journal of Plant Science* 64: 361–8.

Waterhouse, P. M., Graham, M. W. and Wang, M.-B. 1998. Virus resistance and gene silencing in plants can be induced by simultaneous expression of sense and antisense RNA. *Proceedings of the National Academy of Sciences, U.S.A.* 95: 13959–64.

Waterworth, H. E. and Fridlund, P. R. 1989. Apple rubbery wood, in P. Fridlund (ed.), *Virus and Viruslike Diseases of Pome Fruits and Simulating Noninfectious Disorders*, Washington State University Cooperative Extension Publications SP0003, pp. 118–26.

Waterworth, H. E. and Hadidi, A. 1998. Economic losses due to plant viruses, in A. Hadidi, R. K. Khetarpal and H. Koganezawa (eds), *Plant Virus Disease Control*, American Phytopathological Society Press, St. Paul, MN, pp. 1–13.

Yoshikawa, N., Matsuda, H., Oda, Y., Isogai, M. and Takahashi, T. 2001. Genome heterogeneity of *Apple stem pitting virus* in apple. *Acta Horticulturae* 550: 285–90.

Zawadska, B. and Millikan, D. F. 1989. Other infectious disease of apple, in P. Fridlund (ed.), *Virus and Viruslike Diseases of Pome Fruits and Simulating Noninfectious Disorders*, Washington State University Cooperative Extension Publications SP0003, pp. 157–61.

Zhang, S., Chen, W., Xin, L., Gao, Z., Hou, Y., Yu, X., Zhang, Z. and Qu, S. 2014. Genomic variants of genes associated with three horticultural traits in apple revealed by genome re-sequencing. *Horticulture Research* 1: 14045. doi:10.1038/hortres.2014.45.

Chapter 16

Bacterial diseases affecting apples

John Norelli, USDA-ARS, USA

1 Introduction

1.1 General features of bacterial plant disease

Bacteria are simple, microscopic organisms consisting of a single prokaryotic cell lacking an organized or bound nucleus or other cellular organelles. In general, they are not capable of physically penetrating the plant cuticle and require a wound, the aid of insect vectors, or a natural opening, such as stomata, hydathode or nectathode, to infect plants. Bacteria can grow rapidly with generation times of less than an hour, therefore bacterial diseases can develop very rapidly under optimal environmental conditions. Bacteria can actively move through liquids by means of flagella; however, some are non-motile and move passively. In general, bacterial plant pathogens are dispersed by water, wind-driven rain, insects, contaminated plant material and humans.

1.2 Management of bacterial diseases in apple in comparison to annual crop plants

Bacterial diseases of plants are usually difficult to control and often require a combination of control measures to be successfully managed. Chemical control of bacterial plant diseases is generally much less effective than chemical control of fungal diseases and

http://dx.doi.org/10.19103/AS.2016.0017.19

is not widely used in either annual or perennial crops. There are often stark differences between the means available to control bacterial diseases in annual crops versus a woody tree crop, such as apple. In annual crops, the most common and effective means of control for bacterial plant diseases are sanitation, host resistance and the use of pathogen free seed and transplants. In apple, pruning to remove infected plant tissue and orchard sanitation are important but often less effective than with annual crops. Host resistance is a common means of controlling bacterial diseases in annual crops; however, in apple, resistant cultivars with required market quality are often not available. Control of bacterial diseases in apple has relied heavily on the use of antibiotics.

1.3 Chapter overview

Fire blight is a serious economic threat to apple production in regions of the world where it occurs. Due to the economic importance of fire blight to the pome fruit industry, both fire blight and its causative agent *Erwinia amylovora* have been extensively studied since the late 1800s with over 5000 literature citations (van der Zwet et al., 2012). It will therefore be used as an example for the key issues and challenges presented by bacterial diseases for apple production with sections on its history and spread around the globe, economic impact, symptoms, disease biology, and management. There has been less research conducted on other bacterial diseases of apple and these, including blister spot, crown gall and hairy root, bacterial blossom blast, and bacterial blister bark, will be discussed in less detail. Apple proliferation, which is caused by a phytoplasma, is also of economic importance in regions of the world where it occurs and will be discussed in a separate section on diseases caused by phytoplasma. A case study will be presented on current efforts to develop apple cultivars that are resistant to fire blight disease.

2 Bacterial diseases: the example of fire blight

Fire blight caused by the bacterium *E. amylovora* (Burr.) Winslow et al. is a destructive disease of apple, pear and woody ornamentals. The name 'fire blight' was derived from the characteristic blackening of vegetative tissue caused by the disease on pear, often making trees appear as if they were swept by fire. Fire blight occurs only within the rose family (Rosaceae) with species of the Maleae, or apple tribe, and the genera *Rubus* being most susceptible to the disease (van der Zwet and Keil, 1979). The disease has been reported on several *Prunus* and rose species including Japanese plum (*Prunus salicina*) and apricot (*P. armeniaca*); however, the disease cycle cannot be completed on these rosaceous hosts (Mohan and Thomson, 1966; Végh and Palkovics, 2013). Asian pear blight (also known as bacterial shoot blight of pear) is a fire blight-like disease of similar symptomatology and biology that occurs in Korea and Japan, and is caused by the closely related pathogen *E. pyrifoliae* (Kim et al., 2001).

2.1 History and spread

Fire blight is indigenous to North America. It was first reported on pear and apple in the Hudson Valley of New York in 1780 (Denning, 1793). Research to determine the causal agent of fire blight conducted between 1877 and 1886 by Thomas J. Burrill and

Joseph C. Arthur played an important role in the genesis of the concept that bacteria could be pathogenic to plants (Griffith et al., 2003). By the late 1800s the disease was reported in the newly established pome fruit industry in California. Until recently it was believed that fire blight was indigenous to the east of the Mississippi River and the disease was brought to California by settlers transporting apple and pear trees (van der Zwet and Keil 1979). More recent research studying the diversity of clustered regularly interspaced short palindromic repeats (CRISPRs) within the *E. amylovora* genome indicates that most east and west coast strains are of distinct lineage (McGhee and Sundin, 2012), and that the pathogen may have been established on native west coast hosts prior to European colonization and the introduction of apple. Fire blight spread to New Zealand in the 1910s and then to England in the 1950s (van der Zwet et al., 2012). From England it spread first to Germany and then to France, and was also introduced into Egypt. Analysis of *E. amylovora* strains from Europe and the Middle East by pulsed field gel electrophoresis (PFGE) found limited diversity and suggested only 3–4 distinct introductions (Jock et al., 2002). Currently, the disease is widespread throughout North America, in most of Europe, the Middle East, Central Asia and Korea. It has not yet been reported in India, China, South America or South Africa.

Attempts are often made to eradicate fire blight after its introduction into a new region by removing and destroying infected trees. These measures are costly, damaging to the local fruit industry and almost always ineffective. Fire blight was successfully eradicated from Australia after its introduction into the Royal Botanic Gardens, Melbourne in 1997 at an estimated cost of $20 million (Rodoni et al., 2006). However, the circumstances in Australia were unique in that the introduction was not in a fruit production area and the infested area was separated from production areas by a large distance with environmental conditions that were not conducive to fire blight development.

2.2 Economic impact

Since the mid-nineteenth century, fire blight has caused serious economic losses for apple and pear growers. In 1914 in Illinois, fire blight caused losses of apples and pears valued at $1.5 million (van der Zwet and Keil, 1979). From 1900 to 1910 the loss of pear trees in California was catastrophic and the U.S. Census indicated a 28% decline in total pear trees in California during this period (Baker, 1971). More recently, in 2000, a fire blight epidemic that occurred in Michigan was estimated to have caused the death of 350 000–450 000 trees, resulting in a total economic loss of $42 million.

Pear has always been considered highly susceptible to fire blight and the disease has long been the main factor limiting pear production in the eastern United States. In the last century, fundamental changes in the apple industry resulting from the adoption of high-density orchard systems, and planting of susceptible cultivars and rootstocks have increased the danger of fire blight in apple orchards to unprecedented levels. A high-density planting system depends upon dwarfing apple rootstocks to control tree size. The most commonly used dwarfing apple rootstocks, Malling (M.) 9 and M.26, are highly susceptible to *E. amylovora* and infection kills trees by girdling the rootstock in almost all cases (Norelli et al., 2003b). A 10% incidence of rootstock blight in a 4-year-old high-density planting can result in losses up to $8,400 per hectare when the costs of tree replacement, lost investment in tree maintenance, and reduced yields over several years are considered (Momol et al., 1997). Currently, there are no effective cultural practices or chemical treatments available that can protect these rootstocks from infection.

Furthermore, most commercially successful apple cultivars introduced in recent years, including 'Braeburn', 'Fuji', 'Gala', 'Ginger Gold', 'Jonagold', Pink Lady™ and RubyFrost™ are much more susceptible to fire blight than most older cultivars (Brown, 2012). The increased establishment costs of high-density plantings combined with the extreme fire blight susceptibility of new apple rootstocks and scion cultivars has resulted in devastating financial losses for many apple growers due to fire blight. Many of the orchards lost due to fire blight in Michigan in 2000 were 2–5 years old high-density orchards that were just coming into production.

2.3 Symptomatology

The host–pathogen interactions between *E. amylovora* and its hosts are in many ways similar to an incompatible or resistant response of a bacterial pathogen on a non-host plant leading to a hypersensitive response (Venisse et al., 2001, 2003). A hypersensitive response is characterized by the rapid death of cells in the local region surrounding the infection, restricting the growth and spread of the pathogen to other parts of the plant. Unlike a hypersensitive response, the interaction between *E. amylovora* and its hosts is not restricted and the pathogen spreads systemically through its host. Initially, *E. amylovora* will invade parenchyma tissue which is composed of thin walled cells concerned with the vegetative activities of the plant, before eventually moving into vascular tissue. Although *E. amylovora* can infect many different types of tissues, resulting in diverse symptoms, the development of these symptoms shares certain features due to the host–pathogen interactions described above. These include rapid cell death resulting in a water-soaked appearance of infected tissue that rapidly leads to the necrosis, or death, of the tissue which turns either black in some hosts, such as pear, or dark brown in other hosts, such as apple. In addition, due to the rapid growth of the bacteria in parenchyma tissue, an ooze consisting of bacteria and plant sap will often be exuded through stomata. The ooze can be white, orange or red in colour, and will dry to a shiny dark red, brown or black gum-like substance.

E. amylovora can infect blossoms, stems, immature fruits, woody branches, tree trunks and root crowns. Infected blossoms and peduncles first appear water-soaked or grey-green, then shrivel and turn brownish to black (Fig. 1A). Droplets of ooze (plant sap and *E. amylovora*) sometimes exude from the blossom peduncle. Similarly on immature fruit, symptomatic tissues first appear water-soaked and later with the onset of necrosis become brown to black, with droplets of ooze frequently present (Fig. 1B and 1D). Symptoms on shoots generally appear first in young leaves with necrosis along the petiole and leaf midrib. Necrosis will spread through the secondary veins and the entire leaf blade will become necrotic. Blighted shoots often wilt and form a shepherd's crook-like appearance (Fig. 1C). Stem tissue will shrivel and darken, eventually turning dark brown in apple. Droplets of ooze sometimes exude from stem or leaf petioles (Fig. 1D). In addition, aerial bacterial strands that can be dispersed by wind can form on stems (van der Zwet 1994). Necrosis can spread downward from infected blossoms, shoots and fruits through the current season's growth and into older wood. Infected wood will appear shrivelled or sunken and eventually turn dark brown to black. Abundant ooze can sometimes flow along the bark of infected branches, trunk, or the rootstock crown (Fig. 1E).

Figure 1 Fire blight symptoms on apple and pear. A: infected pear blossoms (photo courtesy of Dr Tom van der Zwet); B: infected immature apple fruits with bacterial ooze (photo courtesy of Dr Alan Jones); C: blighted pear shoot with typical form of shepherd's crook (photo courtesy of Dr Ken Hickey); D: bacterial ooze exuding from infected apple shoot (photo courtesy of Dr Alan Jones); and E: infected apple rootstock crown with abundant ooze. From Norelli (2004).

2.4 Disease biology

E. amylovora most commonly overwinters in cankers formed as a result of the previous season's infection of woody tissue (Fig. 2) (Thompson, 2000). In spring, bacteria multiply at the margin between diseased and healthy tissue. Usually only a small proportion of the cankers in an orchard produce inoculum and cankers with smooth, indeterminate margins are more likely to become active (Beer and Norelli, 1977). Although the production of ooze is common, *E. amylovora* can be isolated from active cankers in the absence of ooze. Active cankers can continue to expand during the growing season resulting in the spread of infection to both blossom spurs and shoots. This phase of the disease is often referred to as the canker blight phase (Biggs and Turechek, 2010; Steiner, 1990). In addition to cankers, bacteria have been reported to overwinter in dormant buds and in internal non-symptomatic plant tissue (endophytic bacteria) (Kiel and van der Zwet, 1972; McManus and Jones, 1994). Although contaminated planting material is important in the introduction of the pathogen to new orchards and regions, infested buds and endophytic bacteria are not considered important sources of primary inoculum in orchards with a prior history of fire blight.

During bloom, bacteria can be transferred from active cankers to blossoms by water (rain, irrigation, frost-protection or dilute chemical sprays) or by insects (ants and flies) (Fig. 2) (Thompson, 2000). Bees are important in spreading bacteria among blossoms. In blossoms, *E. amylovora* can multiply as an epiphyte on the surface of the stigma without

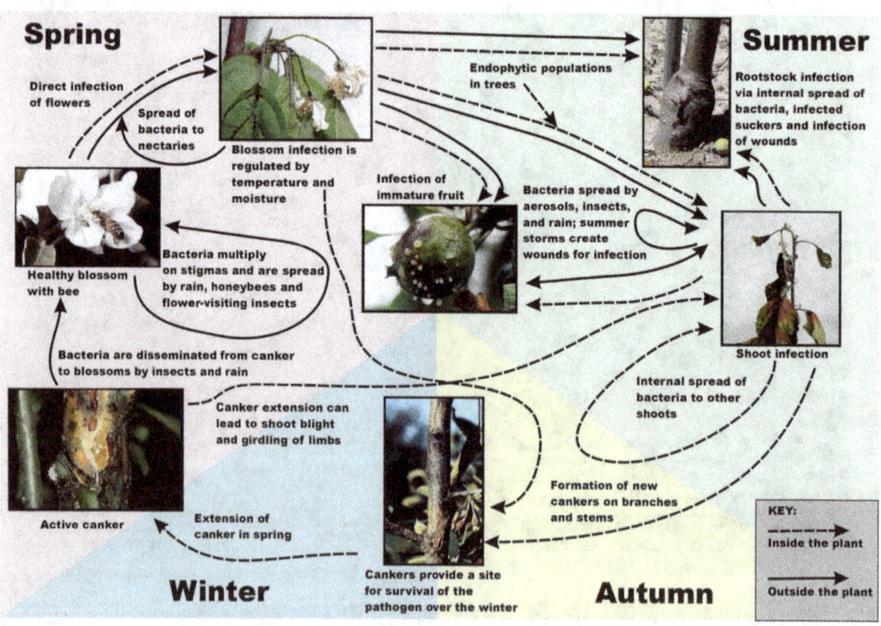

Figure 2 Disease cycle of fire blight caused by *Erwinia amylovora*. Coloured sections represent the season of activity, with blue: winter, pink: spring, green: summer and yellow: autumn. Dashed lines represent movement of *E. amylovora* and spread of the disease within the plant, and solid lines represent movement of bacteria outside the plant. From Norelli et al. (2003).

causing disease (Thomson, 1986). Large epiphytic populations can develop when warm temperatures occur during bloom, enhancing the likelihood of blossom infection. Blossom infection usually occurs when bacteria are washed by rain into nectaries located in the hypanthium or floral cup, although other flower parts can also be invaded (Thompson, 2000). This phase of the disease cycle is often referred to as the blossom blight phase.

Much less is known about the process by which shoots become infected by *E. amylovora*. If inoculum is present, shoot infection can occur following injury caused by hail, high winds or pruning. *E. amylovora* can also move systemically from blossom infections to proximal shoots to initiate infection. Endophytic *E. amylovora* can also be isolated from more distal symptomless shoots; however, the role of endophytic bacteria in initiating new shoot infections is poorly understood. In addition, it is thought that insect feeding in the presence of inoculum can result in shoot infection (Thompson, 2000). However, although almost 100 species of insects have been reported to be associated with the dissemination of fire blight (van der Zwet et al., 2012), there is a lack of information demonstrating the importance of specific species in the shoot blight phase of the disease. Infected shoots resulting from the shoot blight and canker blight phases of the disease are often confused, but can be distinguished as to the origin of the infection. The symptoms of shoot infection from canker blight usually start at the base of the shoot, whereas in shoot blight, symptoms are first observed at the top of the shoot.

Fire blight of apple rootstocks has become a serious economic problem in high-density orchards (Norelli et al., 2003b) that depend on dwarfing rootstocks to control tree size. The commonly used dwarfing rootstocks, Malling (M.) 9 and M.26, are highly susceptible to *E. amylovora,* and infection usually kills trees by girdling the rootstock (Norelli et al., 2003a). Several avenues of rootstock infection have been demonstrated, including infection of rootstock suckers (vegetative shoots developing from the rootstock), internal spread of endophytic bacteria, or direct infection of the rootstock through bark cracks, ruptures associated with burrknots or various injuries (Momol et al., 1998).

2.5 Disease management

Despite the destructive potential of fire blight, it is sporadic in its occurrence. The disease can cause little damage in an area for several years but then can be followed by severe epidemics once environmental conditions are favourable. Effective management of fire blight requires the integration of several practices aimed at reducing the amount of initial inoculum, imposing barriers to successful establishment of the pathogen, and reducing host susceptibility to infection.

Removal of cankers during winter pruning is a critical component of fire blight management. Other strategies include the application of early-season copper sprays to reduce primary inoculum in the orchard and pruning out early-season infections after bloom to reduce inoculum available for shoot infection (van der Zwet and Beer, 1995).

Preventing blossom infection is also a critical component of fire blight management. Antibiotics are effective in preventing blossom infection but must be applied precisely to coincide with fire blight infection periods. Several models including Maryblyt™ (Steiner, 1988), Cougar Blight (Smith, 1993 and 1999) and the Billing Integrated/Revised System (Billings, 1979 and 1992) can predict the occurrence of blossom infection periods. In general, antibiotic applications are recommended during bloom when weather is relatively warm, and there is a high probability of rain and temperatures greater than 18°C in the next 24 hours. Streptomycin has been the most effective antibiotic for fire blight control. Resistance to streptomycin has been reported in the west coast of USA, Michigan, New York, western Canada and in Israel (Coyier and Covey, 1975; Manulis et al., 1998; McGhee et al., 2011; Russo et al., 2008). Kasugamycin has also been registered for fire blight control (McGhee and Sundin, 2011). Although blossom blight control has relied heavily on antibiotic sprays, significant progress has been made in the biological control of blossom infection and commercial products are currently available (Johnson and Stockwell, 1998; Johnson and Temple, 2013; Sundin et al., 2009). In general, biological control agents should be applied during early bloom so that they can become established prior to the occurrence of infection periods. Antibiotics are not recommended for the control of shoot infection due to their limited efficacy and to reduce the chance of antibiotic resistance developing within *E. amylovora* populations; however, antibiotic application is recommended following a hailstorm in orchards that already have fire blight infections.

Management practices that minimize excessive vegetative growth will reduce host susceptibility to *E. amylovora* infection. Prohexadione-calcium is a plant growth regulator that reduces longitudinal shoot growth by inhibiting biosynthesis of gibberellins and results in cell wall thickening of cortical parenchyma (McGrath et al., 2009). Controlling vegetative growth with prohexadione-calcium in apple has also been shown to reduce the incidence and severity of fire blight shoot infection (Yoder et al., 1999; Jones, 2002). Suppression of apple shoot growth and fire blight control by prohexadione-calcium requires application near petal fall as a single spray or as multiple sprays over an extended period (Norelli and Miller, 2004).

Serious fire blight damage can usually be avoided by planting scion varieties and rootstocks with resistance (van der Zwet and Beer, 1995). Fire blight resistant rootstocks have been developed and provide effective control of the rootstock blight phase of the disease (Fazio et al., 2015; Norelli et al., 2003a). These include G.65 with tree size control similar to M.27; G.11 and B.9 similar to M.9-T337; G.41, G.16 and G.213 intermediate between M.9-T337 and M.9-PAJ2, G.214 and B.814 similar to M.9-PAJ2; G.935, G.222 and G.202 similar to M.26; and G.969, G.30, G.210 and G.890 similar to M.7 and MM.106 in tree size control.

3 Development of fire blight tolerant apple scion cultivars

The approach to developing fire blight resistant cultivars in the United States and Europe differs slightly due largely to different perspectives on the disease. In Europe, fire blight is an exotic disease which resulted in devastating losses for European pome fruit industries after its recent introduction and many quarantine restrictions remain in place for fire blight. Hence, most European programmes are seeking to develop cultivars that will be immune to the disease and are using highly resistant wild species as sources of resistance such as *Malus* × *robusta* 5, 'Evereste' crabapple, and *M. fusca* (Durel et al., 2009; Emeriewen et al., 2014; Peil et al., 2007).

In contrast, North American apple growers have had fire blight in their orchards for centuries. They may have zero tolerance limits for fungal diseases that annually reduce the marketability of fruit in fresh markets, such as scab, cedar-apple rust and powdery mildew, but they are prepared to accept low levels of fire blight in their orchards if it does not affect profitability. Although they fear fire blight, they know that devastating losses result from permanent structural damage or outright death of young, highly susceptible trees, not from annual reduction in fruit quality. Therefore, cultivars with tolerance, or quantitative resistance to fire blight would be acceptable to most North American apple growers. For example, fire blight was not considered a disease priority by growers in Washington State 30 years ago when fire blight tolerant 'Delicious' was the primary apple cultivar grown; however, now that highly susceptible cultivars are being grown, it is consistently a disease priority among these growers.

The advantage to breeding for tolerance to fire blight is that adequate resistance can be found in existing domesticated apple cultivars with good to excellent horticultural quality, making it easier and faster to incorporate tolerance into elite breeding line. Genes conferring the highest level of resistance are typically from wild species with undesirable fruit qualities, such as small size, and bitter or astringent flavours. The apple industry is dependent on growing high value scion cultivars prized for their fruit quality and the negative effects on fruit quality of introgressing wild species are time-consuming to overcome with conventional breeding methods.

3.1 Potential sources of tolerance to fire blight

The fire blight resistance of 'Splendour' is comparable to that of 'Delicious'. 'Splendour' has been used as a parent is several breeding programmes for its desirable flavour attributes, and Aurora Golden Gala™ and Pacific Rose™ are both offspring of 'Splendour' (Taylor et al., 2002; Hampson and Sholberg, 2008). When introduced in the 1950s from New Zealand, it was of unknown parentage; however, genetic and pedigree analysis has determined that it resulted from a cross between 'Golden Delicious' and 'Delicious'. Mapping populations derived from crosses of 'Splendour' with 'Gala', 'Cripps Pink' and 'Pinova' have been established for genetic analysis and evaluated for their resistance to fire blight.

Several of the scab-resistant cultivars developed in the Purdue-Rutgers-Illinois breeding programme also have tolerance to fire blight, including 'Enterprise' and 'GoldRush'. The source of scab resistance for these cultivars was *M. floribunda* 821 for which a quantitative trait loci (QTL) for fire blight resistance has been described on linkage group (LG) 12 (Durel et al., 2009). In addition, QTLs for fire blight resistance have been described on LG02, LG06 and LG15 of the 'Coop 17' parent of 'Goldrush' (Khan et al., 2013). Based upon these previously published results, DNA markers are being developed for the QTLs and will be validated in populations segregating for fire blight resistance that resulted from crosses of 'Pinova' with 'CrimsonCrisp', 'Enterprise', and 'GoldRush' as part of the U.S. multi-institutional RosBREED project (USDA-NIFA-SCRI 2014-51181-22378; https://www.rosbreed.org/).

A reference germplasm including pedigreed populations, established as part of the RosBREED project (Peace et al., 2014), has been established to identify and validate QTLs in apple using a Bayesian approach (Bink et al., 2008). The apple reference germplasm is currently being evaluated for its resistance to fire blight which should lead to the identification of several additional QTLs for fire blight tolerance in elite breeding lines and cultivars.

3.2 Using DNA-informed breeding and marker-assisted selection to increase breeding efficiency

The RosBREED project is based on the premise that routine use of DNA information can enhance efficiency, accuracy, creativity and pace of breeding (Peace and Norelli, 2009;

Xu and Crouch, 2008). Efficiency is enhanced when DNA information is used to select parental cross combinations and among the resulting seedlings (marker-assisted selection). Breeding resources are focused on crosses and seedlings predicted to most efficiently lead to commercially viable new cultivars. This strategy is particularly useful when accurate phenotyping costs are very high or the target trait requires several years to phenotype for traits such as fruit quality, productivity and disease resistance (Peace and Norelli, 2009). DNA information can enhance accuracy by revealing genetic potential of breeding germplasm, supplementing or replacing phenotypic data influenced by the test environment or management system. Using DNA tests to incorporate multiple resistance alleles for a disease from different sources (pyramiding) is both efficient and accurate. Identifying individuals with pyramided disease resistance alleles is often not even possible without DNA markers linked to the resistance alleles because it is impossible to select on phenotype alone for the presence of a second resistance allele.

3.3 Improving breeding efficiency by shortening the juvenile phase of apple

Historically, the introgression of disease resistance loci from exotic germplasm into elite breeding lines of fruit tree crops has been a long-term process requiring several decades (Crosby et al., 1992). More recently, high-speed breeding technology using transgenic lines of apple with reduced juvenility has improved breeding efficiency by reducing generation time from 4–5 years to 11–18 months (Flachowsky et al., 2011; Le Roux et al., 2012). To shorten the juvenile stage the *BpMADS4* gene from silver birch (*Betula pendula* Roth.), which is similar to *Arabidopsis FRUITFUL*, was constitutively over-expressed in transgenic 'Pinova' apple (Elo et al., 2001; Flachowsky et al., 2007). Transgenic clones started to initiate flowers within 3–4 months after transfer to the glasshouse. Primary flowers of transgenics were solitary and in a terminal position on the main shoot, in some transgenics subsequent lateral flower clusters consisted of three to five individual flowers. Flowers were viable and produced seed. There were no significant differences between pollen viability and vitality of *BpMADS4* transgenic flowers and pollen of non-transgenic control plants. Transgenic seedling flowered 15 to 50 weeks after planting of stratified seed (Le Roux et al., 2012). The time required for fruit development, seed maturation and seed stratification is not altered in transgenic seedlings.

To develop the high-speed breeding technology, single *BpMADS4* insertions were selected on various chromosomes of 'Pinova' transgenics (Flachowsky et al., 2007). These single, stable *BpMADS4* insertions behave genetically as single dominant heterozygous alleles in subsequent crosses that segregate 1:1 such that 50% of the resulting population will carry *BpMADS4* for reduced juvenility, and the other 50% will be non-transgenic with normal growth and juvenility. Initially, individuals carrying the *BpMADS4* transgene are selected to facilitate high-speed breeding during introgression and/or the combining of desirable alleles at multiple trait loci. After achieving the desired degree of introgression or level of desirable allele accumulation, non-transgenic individuals no longer carrying the *BpMADS4* transgene but retaining the desired trait alleles are selected for use as elite breeding lines. For the high-speed breeding system to work the trait loci of interest and the *BpMADS4* transgene cannot be within the same linkage group (chromosome) because in subsequent generations the desired trait allele and the *BpMADS4* transgene would segregate and could not be simultaneously selected. The availability of multiple single

BpMADS4 insertions on different chromosomes allows the breeder to select the most appropriate *BpMADS4* donor line for a given set of breeding objects. Additionally, the system relies heavily on DNA-marker-assisted selection of seedling because the rapidity of breeding cycles does not allow for extensive phenotypic evaluation.

In Europe, high-speed breeding technology has been used both for the introgression of high levels of fire blight resistance from *M. fusca* Mfu10 (Emeriewen et al., 2014), *M.* × *robusta* 5 (Peil et al., 2007) and 'Evereste' crab apple (Durel et al., 2009) into *M.* × domestica and to pyramid or combine multiple disease resistance alleles into elite breeding parents (Flachowsky et al., 2011; Le Roux et al., 2012). The programme at the Julius Kühn-Institut in Dresden, Germany, has pyramided the fire blight resistance of *M. fusca* Mfu10 with that of 'Fiesta' (Khan et al., 2007) as well as combining it with either pyramided resistance to apple scab (*Rvi2, Rvi4*) or powdery mildew (*Pl-1, Pl-2*) (Flachowsky et al., 2011).

4 Other diseases caused by bacteria

4.1 Blister spot

Blister spot, caused by *Pseudomonas syringae* pv. *papulans*, results in purplish black lesions associated with fruit lenticels that will reduce market value of fruit (Burr, 1990). It is primarily a problem on 'Mutsu' (also known as Crispin™) where it can cause major economic losses. Although it can occur on many other apple cultivars including 'Golden Delicious', many of its progeny, such as 'Gala' and 'Jonagold', and other cultivars, such as 'Braeburn' and 'Fuji' (Sholberg and Bedford, 1997), the disease rarely occurs on these cultivars unless they are planted adjacent to affected 'Mutsu' (OMAFRA, 2009). *P. syringae* pv. *papulans* can also cause a midvein necrosis on 'Mutsu' apple leaves which does not have major economic impact (Bonn, 1986).

P. syringae pv. *papulans* overwinters in leaf scars and buds (Bedford et al., 1981). It is an excellent epiphyte and is probably present in most apple orchards in humid environments (Kerkoud et al., 2000). High inoculum levels are required for disease development which may explain why most apple cultivars do not become infected naturally (Burr and Hurwitz, 1979). The pathogen enters fruit through stomata, with susceptibility beginning approximately 2–2.5 weeks after petal fall and remaining for approximately 6 weeks (Burr and Hurwitz, 1981). This period of susceptibility coincides with the period when stomata develop into lenticels. Lesions are first noticeable 2–3 months after petal fall as small water-soaked, raised blisters associated with lenticels on the fruit surface. Lesions rarely extend more than 1–2 mm deep into the fruit flesh and do not result in fruit decay.

Newly planted blocks of 'Mutsu' obtained from disease-free nursery stock often remain free of blister spot for several years if they are not planted near older 'Mutsu' blocks where blister spot is present (OMAFRA, 2009). A PCR assay to detect and monitor the pathogen has been developed (Kerkoud et al., 2002). The bacteria are spread by water and the disease is favoured by wet weather (Burr, 1990). In 'Mutsu' orchards, overhead irrigation should be avoided during the period of fruit susceptibility (2–8 week after petal fall). The disease can be controlled by streptomycin sprays first applied 10–14 days after petal fall and followed by 2 additional sprays at weekly intervals. However, *P. syringae* pv. *papulans* can rapidly develop resistance to streptomycin (Burr et al., 1988).

4.2 Crown gall and hairy root

Crown gall of apple, caused by *Rhizobium radiobacter* (formerly *Agrobacterium tumefaciens*), results in galls located on the roots and crown of the tree that can vary in diameter from a few millimetres to 5–15 centimetres on older trees (Moore, 1990a). Anatomically, the galls consist of disorganized vascular and parenchyma tissues. Hairy root, caused by *Rhizobium rhizogenes* (formerly *Agrobacterium rhizogenes*), results in an extensive proliferation of adventitious roots from either single foci or clusters at the base of trunk, crown or roots (Moore, 1990b). The morphological and biochemical characteristics of *R. radiobacter and R. rhizogenes* are very similar and they share a high degree of DNA homology. In both species, pathogenicity is dependent upon large, tumour-inducing (Ti) plasmids containing transfer DNA (T-DNA), which is inserted into a plant cell chromosome during infection. Oncogenes present in the T-DNA result in the over-expression of various growth hormones which result in gall and hairy root formation (Moore et al., 1979; White et al., 1982).

The biology and management of both diseases are similar. *Rhizobium* spp. can live in soil and will escape from galls or hairy roots to contaminate healthy roots or soil. The bacteria are disseminated by splashing rain, irrigation water, tools, farm machinery and plant parts used for propagation (Moore, 1990a). The bacteria infect through wounds created by pruning, winter injury or the emergence of new lateral roots. Although both diseases occur worldwide in nurseries and orchards, economic damage is usually associated with nursery production where infected trees are culled prior to sale. However, due to the extremely broad host range of *R. radiobacter and R. rhizogenes* (at least 93 families), damage can be extensive in orchards planted on former nursery sites (Moore, 1990a). To the grower, the disease is usually more damaging due to an overall reduction in growth. In older trees, secondary symptoms may include poor root systems, one-sided root system development and blackening of stem and root tissues near galls.

Good sanitation and cultural practices are important for crown gall and hairy root management. All symptomatic planting stock should be discarded as soon as it is harvested, trees should be budded rather than grafted, rootstocks chosen with low susceptibility, and management practices adopted that minimize wounding (Moore, 1990a). Irrigation water from wells is less likely to contain *Rhizobacterium* than surface water. The susceptibility of the rootstock cultivar is also an important factor in disease severity. Both M.7 and M.9 apple rootstocks are susceptible to *Rhizobacterium*. Rootstock breeding efforts in Japan are attempting to incorporate resistance to *R. radiobacter* from *M. sieboldi* 'Sanashi 63' into 'JM' rootstocks (Moriya et al., 2008 and 2010). Biological control has been effective in crown gall management. Typically rootstocks are dipped in suspensions of nonpathogenic *Rhizobium* spp. prior to planting (Kawaguchi et al., 2012 and 2015; Moore, 1900a) and dipping rootstocks in a solution of copper oxychloride prior to planting has also been reported to reduce disease severity (Mikcinski et al., 2012).

4.3 Bacterial blister bark

Blister bark is caused by *P. syringae* pv. *syringae* and is characterized by raised, tan areas that are visible on the bark of apple trees in spring (Mansvelt and Hattingh, 1990). Beneath the blister is a spongy bark layer of green cells. The blisters dry and flake off to expose the underlying necrotic tissue.

Blister bark may cause serious damage in nurseries, particularly after fall budding. M.9 has been reported to be susceptible to infection, particularly when inoculated after a freezing thaw (Sholberg and Quamme, 1999). The disease has also been reported to be a problem in orchards in Italy on sandy soils (Scortichini and Morone, 1997). No practical control is known other than avoiding fall grafting and attempting to improve soil structure (Mansvelt and Hattingh, 1990; Polverigiani et al., 2016).

5 Diseases caused by phytoplasmas

5.1 General features of plant diseases caused by phytoplasmas

Mycoplasmas are bacteria, or prokaryotes, that lack a true cell wall and the ability to synthesize the substances required to form a cell wall. They are very small, usually measuring 175 to 250 nm in diameter. Wall-less, non-helical prokaryotes that colonize the phloem and phloem-feeding insects are referred to as phytoplasma (Firrao et al., 2004). They are generally present in the sap of phloem cells and are transmitted by leafhoppers and other phloem-feeding insects, such as psyllids and possibly some aphids. The insects acquire the phytoplasma after feeding on infected plants for several hours or days and may then become infective agents several days after acquiring the phytoplasma. Phytoplasmas are dependent upon the metabolism of their plant and insect hosts for their survival and many cannot be cultured *in vitro*. Diseases caused by phytoplasmas are usually managed by using disease-free plant material for propagation, control of the insect vectors that transmit the phytoplasma and the use of resistant cultivars.

5.2 Apple proliferation

Apple proliferation is caused by the unculturable wall-less bacterium '*Candidatus* Phytoplasma mali'. The disease was first described in Italy in 1950 and it is the most important graft-transmissible and vector-borne disease of apple in Europe (Seemüller, 1990). The disease rarely occurs in northern Europe, most likely due to limitation of the insect vector in colder climates. Although the phytoplasma is unculturable it can be observed as highly pleomorphic bodies, 200–800 nm in diameter. The phytoplasma is found only in sieve tubes of the current season's phloem and may severely reduce translocation. Strains of '*Candidatus* Phytoplasma mali' are known to differ in their virulence (Seemüller and Schneider, 2007). The disease causes major economic losses through the reduction of total yield, fruit size and tree vigour.

Apple proliferation induces symptoms in shoots, leaves, fruits and roots. Apple proliferation specific symptoms include witches' broom or leaf rosette at shoot tips in late summer and enlarged stipules (Seemüller, 1990). Non-specific symptoms include foliar reddening or yellowing, growth suppression and undersized fruit. The root systems of affected trees are poorly developed and characterized by dense tufts of short, knotted feeder roots. The incubation period for the disease is dependent upon tree age, with younger trees developing symptoms more rapidly. The population of phytoplasma in the aerial portions of the tree will sharply decline in late fall and winter as sieve tubes cease to function; the phytoplasma survives in the rootstock where functional sieve tubes are present throughout the year. The phytoplasma then re-colonize the aerial portions of the tree in the spring as new phloem is formed. However, in older trees re-colonization may be only partial or variable resulting in a fluctuation in symptom severity. Several PCR assays have been developed to detect and quantify '*Candidatus*

Phytoplasma mali' which have aided in the study of the disease (Aldaghi et al., 2007; Torres et al., 2005).

The primary vectors of the '*Candidatus* Phytoplasma mali' in Europe are three species of psyllid, *Cacopsylla costalis, C. melanoneura* and *C. picta* (Frusunghelli et al., 2000; Jarausch et al., 2003; Tedeschi and Alma, 2004). The phytoplasma is able to overwinter in approximately 10% of overwintering adults of *C. picta* (Jarausch et al., 2011). Several leafhoppers are also capable of vectoring the disease. Insecticide applications greatly reduce the spread of apple proliferation. Disease-free trees should be used for establishing new orchards. Infected young trees should be removed, although the benefit of rougueing older trees is unclear because they will sometimes recover (Seemüller, 1990). '*Candidatus* Phytoplasma mali' is sensitive to tetracycline antibiotics and symptom remission has been obtained by trunk injections of oxytetracycline, with beneficial effects lasting for 1–2 years (Seemüller, 1990). The use of resistant rootstocks could be an effective means of control because the phytoplasma overwinters in the root system of the plant. Resistance to apple proliferation has been found in *M. sieboldii* and resistance to apple proliferation is a goal of some apple rootstock breeding programmes in Europe (Jarausch et al., 2007; Seemüller et al., 2008).

5.3 Other apple diseases associated with phytoplasma

Rubbery wood of apple is caused by a phytoplasma in the aster yellows group (Bertaccini et al., 1998). The most characteristic symptom of rubbery wood is the excessive flexibility of branches up to 3 years old resulting from incomplete lignification (Seemüller et al., 1990b). Affected trees are also stunted, often with strong shoots arising from the base of the trunk. The causal agent is readily transmitted by grafting; however, natural spread does not occur, suggesting that the disease is not transmitted by insect vectors. The causal agent of rubbery wood can be eliminated from infected plants by heat therapy at 30°C for 3 weeks (Seemüller et al., 1990b).

Chat fruit (Beakbane et al., 1971) and apple decline are also suspected of being caused by phytoplasmas. Chat wood symptoms develop mostly in 'Lord Lambourne' and 'Tydeman's Early Worcester' where fruit can be slightly flattened and lopsided. On cultivars that do not develop fruit symptoms, affected trees may be more vigorous and have dull green foliage. The disease is transmitted by budding and grafting. Apple decline has been reported in northern Washington State and Brazil. Initially, apple-decline-affected trees will exhibit small, rolled leaves and reduced terminal growth (Parish, 1989). Eventually the wood will die back. Fruits are small and late maturing. The disease will spread to adjacent trees.

6 Outlook and summary

Due to the susceptibility of domesticated apple to both insect and disease damage, pest management is an ongoing challenge to the sustainability of apple production and bacterial diseases are no exception. Fire blight and apple proliferation are the two bacterial diseases of apple that pose the greatest threat to sustainability and are likely to remain the focus of future research in the coming decades. The use of pathogen-free plant material in the establishment of new orchards is a critical 1st step to sustainable apple production.

Fire blight is endemic to North America, whereas apple proliferation is endemic to Europe. The spread of these diseases to new un-infested regions will continue to pose major threats to sustainability. Infected, non-symptomatic plant material coming from nurseries in infested areas is the primary means of long-distance movement of both diseases.

Plant disease is an interaction between host, pathogen and environment, and in an age of global climate change, alterations in normal temperature, humidity, rainfall or high wind storm events are likely to impact the distribution, occurrence and severity of bacterial plant diseases. For example, insect transmission of '*Candidatus* Phytoplasma mali' will remain an important area of apple proliferation research as global climate change could alter the range of its insect vectors and extend the disease further north in Europe. For fire blight, the development of new biological control products has provided growers more sustainable options for blossom blight management; however, regional climatic differences are known to affect the effectiveness of these material and it is likely that climate change will affect the microbial ecology of apple blossoms (Johnson and Stockwell, 1998; Sundin et al., 2009).

The effective utilization of host resistance could provide more sustainable disease management for apple in the future. Because the root system serves as the overwinter reservoir for '*Candidatus* Phytoplasma mali', the development of apple proliferation resistant rootstocks is likely to provide control for the disease and to be of high impact. Although new apple proliferation resistant rootstocks could take several decades to develop, evaluate and bring into production, they are likely to be readily accepted by growers due to the economic importance of apple proliferation disease. Similarly, the utilization of available fire blight resistant rootstocks is likely to greatly reduce the incidence of the rootstock phase of fire blight. The development of fire blight resistant scion cultivars is a longer-term goal. Nonetheless, the increased efficiency of DNA-informed breeding, the utilization of high-speed breeding systems and the growing trend toward new cultivars in the apple industry indicate that the scion cultivars of the future could, and should, be more resistant to fire blight.

7 Future research trends

7.1 Advances in the development of resistant cultivars

There is a large body of research on the molecular biology of *E. amylovora* and its interactions with both apple and pear which has not been covered here but has been reviewed elsewhere (Khan et al., 2012; Malnoy et al., 2012; Oh and Beer, 2005; Piqué et al., 2015). Although this research has greatly expanded our understanding of fire blight disease, to date it has had relatively little impact on disease management practices in the orchard. The potential application of using new gene editing technologies to alter host–pathogen interactions could radically change the practical impact of this field of research (Bortesi and Fischer, 2015; Jiang et al., 2013). The technology has been successfully used in apple (Nishitani et al., 2016), although it remains unclear if it will be a more acceptable method of genetic engineering for the general public. Regardless of its acceptance by the public and the apple industry, both the identification of the best targets for gene editing and the application of gene editing technology are likely to be an important area of research in the immediate future for both fire blight and apple proliferation.

7.2 The role of biofilm formation in bacterial plant disease

Biofilms are complex bacterial structures which attach to solid structures, have a defined three-dimensional architecture and are surrounded by a self-produced matrix generally composed of exopolysaccharides, proteins, lipids and extracellular DNA (Castiblanco and Sundin, 2016). Biofilm formation allows bacteria to cope with harsh environmental conditions as well as to establish both antagonistic and beneficial interactions with their hosts. *E. amylovora* is capable of forming a biofilm and it has been demonstrated that biofilm formation plays a critical role in its pathogenesis (Koczan et al., 2009). Research on the regulation of biofilm formation and its role and function in fire blight, as well as other bacterial plant diseases, could someday lead to new disease management strategies based upon either preventing biofilm formation or disrupting its integrity.

7.3 Biology of *E. amylovora* on its native North American hosts

Although the biology of fire blight on apple and pear has been thoroughly investigated, there has been very little research conducted on the biology of *E. amylovora* on its native North American rosaceous hosts. Fire blight has many of the characteristics of a plant disease recently adapted to its hosts. At the genome level, *E. amylovora* is more closely related to enteric bacteria found in association with animals and insects than other bacterial plant pathogens (Malnoy et al., 2012). Its genome has clear evidence of gene erosion and has undergone a relatively recent genetic bottleneck, both signs of adaptation to new environments. At the level of host–pathogen interactions, fire blight is a devastating disease that often kills its apple and pear hosts and is very sporadic in its occurrence, both characteristics of diseases newly adapted to their hosts. Even the unusual symptoms and physiology of fire blight development appear more like an out-of-control resistance hypersensitive response than typical disease host responses to bacterial pathogens.

It is often assumed that the host–pathogen characteristics listed above are due to the recent adaptation of *E. amylovora* to its apple and pear hosts after their introduction by European settlers. However, *E. amylovora*'s similarity to animal and insect enteric bacteria suggests it may have adapted relatively recently, in evolutionary time scales, to its native North American hosts as well. Although devastating fire blight epidemics are not observed in natural woodlands or forests, the low density of natural hosts in these environments would not be conducive to epidemic development. Studies on the biology of *E. amylovora* on its native hosts should illuminate our understanding of the evolution of host–pathogen relationships regardless of whether or not *E. amylovora* is recently adapted to its North American hosts, as comparative studies could be conducted between both native and exotic hosts and with other host–pathogen systems.

8 Where to look for further information

The *Compendium of Apple and Pear Diseases and Pests*, 2nd edition by Turner B. Sutton, Herb S. Aldwinckle, Arthur M. Agnello and James F. Walgenbach (2014, APS Press, 224 pages) provides more in-depth information on the symptoms (including colour photos), causal organisms, disease cycle, epidemiology and control of the bacterial disease covered in this chapter. Several texts have been written on fire blight. *Fire Blight/History,*

Biology and Management by Tom van der Zwet, Noemi Orolaza-Halbrendt and Wolfgang Zeller was published in 2012 by the American Phytopatholgoical Society (APS Press) and provides a comprehensive review (460 pages) of the scientific literature on fire blight. Similarly, *Fire Blight/The Disease and its Causative Agent, Erwinia amylovora* edited by Joel Vanneste and published in 2000 by CABI contains 17 separately authored review chapters covering the disease, *E. amylovora* and the control of fire blight. *Fire Blight/The Foundation of Phytobacteriology* edited by Clay S Griffith, Turner B. Sutton and Paul D. Peterson and published by APS Press is a compilation and review of the original publications of Thomas J. Burrill, Joseph C. Arthur and Merton B. Waite which in the nineteenth century led to the genesis of the concept that bacteria can be pathogenic to plants (Baker, 1971). *Fire Blight – Its Nature, Prevention, and Control/A Practical Guide to Integrated Disease Management* by T. van der Zwet and S. V. Beer (USDA, Agriculture Information Bulletin number 631, 91 pages, out of press) was an in-depth extension publication targeted to apple and pear growers with separate sections on integrated management practices to control fire blight in the nursery and the orchard. *Fire Blight/A Bacterial Disease of Rosaceous Plants* by Tom van der Zwet and Harry L. Keil was published in 1979 by the USDA (Agricultural Handbook number 510, 200 pages); although it is currently out of press, its contents are covered in the more recent APS Press 2012 text mentioned above.

In addition to the above texts, three review articles on fire blight have been published in *Annual Review of Phytopathology*: 1) Fire Blight: Applied genomic insights of the pathogen and host (Malnoy, M., Martens, S., Norelli, J. L., Barney, M-A, Sundin, G. W., Smits, T. H. M. and Duffy, B., 2012, 50: 475–94; DOI: 10.1146/annurev-phyto-081211-172931), 2) Management of Fire Blight: A case study in microbial ecology (Johnson, K. B. and Stockwell, V. O.,1998, 36: 227–48; DOI: 10.1146/annurev.phyto.36.1.227), and 3) Epidemiology and control of fire blight (Schroth, M. N., Thomson, S. V., Hildebrand, D. C. and Moller, W. J., 1974, 12: 389–412; DOI: 10.1146/annurev.py.12.090174.002133). There have also been numerous review articles on the molecular biology of the host–pathogen interaction of apple and *E. amylovora* that are published in various journals, some of which are cited above in section 7.1. The American Phytopathological Society also has websites on both fire blight (http://www.apsnet.org/edcenter/intropp/lessons/prokaryotes/Pages/FireBlight.aspx) and crown gall (http://www.apsnet.org/edcenter/intropp/lessons/prokaryotes/Pages/CrownGall.aspx) which are targeted toward K-12 education.

An international working group on fire blight has been meeting once every three years since 1977. The first 13 international workshops were held under the auspices of the International Society for Horticultural Sciences and their proceeding are published in *Acta Horticulturae* (1978, vol 86, Wageningen, Netherlands, 28–30 November 1977; 1981, vol 117, Kiel-Schilksee, Fed. Rep. Germany, 16–19 September 1980; 1984, vol 151, Bordeaux, France, 12–16 September 1983; 1987, vol 217, Ithaca, USA, 22–26 June 1986; 1990 vol 273 Diepenbeek, Belgium, 19–22 June 1989;1993, vol 338, Athens, Greece, 20–23 October 1992; 1996, vol 411, St Catherines, Canada, 7–10 August 1995; 1999, vol.489, Kusadasi, Turkey, 12–15 October 1998; 2002, vol 590, Napier, New Zealand, 8–12 October 2001; 2006, vol 704, Bologna, Italy, 5–9 July 2004; 2008, vol 793, Portland, USA, 12–17 August 2007; 2011, vol 896, Warsaw, Poland, 16–20 August 2010; and 2014, vol 1056, Zürich, Switzerland, 2–5 July 2013). In 2016, the working group reorganized and met as the 1st International Symposium on Fire Blight of Rosaceous Plants in Girona, Spain (http://intea.udg.edu/ISFB2016/Arxius/Abstracts%20book.pdf). In June 2019, the working group is scheduled to meet in East Lansing, USA.

9 References

Agrios, G. N. (1988). *Plant Pathology*. 3rd ed. Cambridge: Academic Press.

Aldaghi, M., Massart, S., Roussel, S. and Jijakli, M. H. (2007). Development of a new probe for specific and sensitive detection of '*Candidatus Phytoplasma* mali' in inoculated apple trees. *Annals Applied Biology* 151 (2): 251–8.

Baker, K. F. (1971). Fire blight of pome fruits, the genesis of the concept that bacteria can be pathogenic to plants. *Hilgardia* 40 (18): 603–33.

Beakbane, A. B., Mishra, M. D., Posnette, A. F. and Slater, C. H. W. (1971). Mycoplasma-like organisms associated with chat fruit and rubbery wood diseases of apple *Malus domestica* Borch., compared with those in strawberry with green petal disease. *Journal of General Microbiology* 66: 55–62.

Bedford, K. E., Macneill, B. H. and Bonn, W. G. (1984). Survival of a genetically marked strain of the blister spot pathogen *Pseudomonas syringae* pv *papulans* in leaf scars and buds of apple. *Canadian Journal of Plant Pathology* 8 (2): 167–9.

Beer, S. V. and Norelli, J. L. (1976). Fire blight epidemiology: factors affecting release of *Erwinia amylovora* by cankers. *Phytopathology* 67 (9): 1119–25.

Bertaccini, A., Vibio, M., Franova-Honetslegrova, J. and Janeckova, M. (1998). Molecular detection of phytoplasmas in apple with rubbery wood symptoms. *Acta Horticulture* 472: 693–700.

Biggs, A. R. and Turechek, W. W. (2010). Fire blight of apples and pears: Epidemiological concepts comprising the Maryblyt forecasting program. Online. *Plant Health Progress*, doi:10.1094/PHP-2010-0315-01-RS.

Billing, E. (1979). Warning systems for fireblight. *EPPO Bulletin* 9: 45–51.

Billing, E. (1992). Billing's revised system (BRS) for fireblight risk assessment. *EPPO Bulletin* 22: 1–102.

Bink, M. C. A. M., Boer, M. P., Ter Braak, C. J. F., Jansen, J., Voorrips, R. E. and Van De Weg, W. E. (2008). Bayesian analysis of complex traits in pedigreed plant populations *Euphytica* 161 (1–2): 85–96.

Bonn, W. G. (1986). Midvein necrosis of Mutsu apple leaves caused by *Pseudomonas syringae* pv *papulans*. *Canadian Journal of Plant Pathology* 6: 17–20.

Bortesi, L. and Fischer, R. (2015). The CRISPR/Cas9 system for plant genome editing and beyond. *Biotechnology Advances* 33 (1): 41–52.

Brown, S. (2012). Apple. In Badenes, M. L. and Byrne, D. H. (eds), *Fruit Breeding*. Springer: Science & Business Media, pp. 329–67.

Burr, T. J. (1990). Blister spot. In Jones, A. L. and Aldwinckle, H. S. (eds), *Compendium of Apple and Pear Diseases*. APS Press, pp. 63–4.

Burr, T. J. and Hurwitz, B. (1979). The etiology of blister spot of 'Mutsu' apple in New York State. *Plant Disease Report* 63: 157–60.

Burr, T. J. and Hurwitz, B. (1981). Seasonal susceptibility of Mutsu apples to *Pseudomonas syringae* pv *papulans*. *Plant Disease* 65: 334–6.

Burr, T. J., Norelli, J. L., Katz, B., Wilcox, W. F. and Hoying, S. A. (1988). Streptomycin resistance of *Pseudomonas syringae* pv. *papulans* in apple orchards and its association with a conjugative plasmid. *Phytopathology* 78: 410–13.

Crosby, J. A., Janick, J., Pecknold, P. C., Korban, S. S., O'Connor, P. A., Ries, S. M., Goffreda, J. and Voordeckers, A. (1992). Breeding apples for scab resistance – 1945–1990. *Fruit Var. Journal* 46: 145–66.

Castiblanco, L. F. and Sundin, G. W. (2016). New insights on molecular regulation of biofilm formation in plant-associated bacteria. *Journal of Integrative Plant Biology* 58 (4): 362–72.

Coyier, D. L. and Covey, R. P. (1975). Tolerance of Erwinia amylovora to streptomycin sulfate in Oregon and Washington. *Plant Disease Report* 59: 849–52.

Durel, C.-E., Denancé, C. and Brisset, M.-N. (2009). Two distinct major QTL for resistance to fire blight co-localize on linkage group 12 in apple genotypes 'Evereste' and Malus floribunda clone 821. *Genome* 52 (2): 139–47.

Emeriewen, O., Richter, K., Kilian, A., Zini, E., Hanke, M.-V., Malnoy, M. and Peil, A. (2014). Identification of a major quantitative trait locus for resistance to fire blight in the wild apple species Malus fusca. *Molecular Breeding* 34 (2): 407–19.

Griffith, S. S., Sutton, T. B. and Peterson, P. D. (2003). *Fire Blight – The Foundation of Phytobacteriology*. APS Press.

Denning, W. (1794). On the decay of apple trees. *New York Society for the Promotion of Agriculture, Arts, and Manufacture Transmission* 2: 219–22.

Elo, A., Lemmetyinen, J., Turunen, M.-L., Tikka, L. and Sopanen, T. (2001). Three MADS-box genes similar to APETALA1 and FRUITFULL from silver birch (*Betula pendula*). *Physiologia Plantarum* 11: 93–103.

Emeriewen, O., Richter, K., Kilian, A., Zini, E., Hanke, M.-V., Malnoy, M. and Peil, A. (2014). Identification of a major quantitative trait locus for resistance to fire blight in the wild apple species *Malus fusca*. *Molecular Breeding* 34 (2): 407–19.

Fazio, G., Robinson, T. L. and Aldwinckle, H. S. (2015). The Geneva Apple Rootstock Breeding Program (Book Chapter). *Plant Breeding Reviews* 39: 379–424.

Firrao, G., Andersen, M., Bertaccini, A., et al. (2004). '*Candidatus* Phytoplasma', a taxon for the wall-less, non-helical prokaryotes that colonize plant phloem and insects. *International Journal of Systematic and Evolutionary Microbiology* 54 (4): 1243–55.

Flachowsky, H., Le Roux, P.-M., Peil, A., Patocchi, A., Richter, K. and Hanke, M.-V. (2011). Application of a high-speed breeding technology to apple (*Malus* × *domestica*) based on transgenic early flowering plants and marker-assisted selection. *New Phytologist* 192: 364–77.

Flachowsky, H., Peil, A., Sopanen, T., Elo, A., Hanke, V. (2007). Overexpression of BpMADS4 from silver birch *(Betula pendula* Roth.) induces early-flowering in apple (*Malus x domestica* Borkh.). *Plant Breeding* 126: 137–45.

Frisinghelli, C., Delaiti, L., Frando, M. S., Forti, D. and Vindimian, M. E. (2000). *Cacopsylla costalis* (Flor, 1861), as a bector of apple proliferation in Trentino. *Journal of Phytopathology* 148: 425–31.

Hampson, C. R. and Sholberg, P. L. (2008). Estimating combining ability for fire blight resistance in apple progenies. *Acta Horticulturae* 793: 337–44.

Jiang, W., Zhou, H., Bi, H., Fromm, M., Yang, B. and Weeks, D. P. (2013). Demonstration of CRISPR/Cas9/sgRNA-mediated targeted gene modification in Arabidopsis, tobacco, sorghum and rice. *Nucleic Acids Research* 41 (20): e188.

Jock, S., Donat, V., López, M. M., Bazzi, C. and Geider, K. (2002). Following spread of fire blight in Western, Central and Southern Europe by molecular differentiation of *Erwinia amylovora* strains with PFGE analysis. *Envinonmental Microbiology* 4 (2): 106–14.

Johnson, K. B. and Stockwell, V. O. (1998). Management of fire blight: A case study in microbial ecology. *Annual Review of Phytopathology* 36: 227–48.

Johnson, K. B. and Temple, T. N. (2013). Evaluation of strategies for fire blight control in organic pome fruit without antibiotics. *Plant Disease* 97 (3): 402–9.

Jarausch, W., Bisognin, C., Schneider, B., Grando, S., Velasco, R. and Seemüller, E. (2007). Breeding of apple rootstocks resistant to '*Candidatus* Phytoplasma mali'. *Bulletin of Insectology* 60 (2): 299–300.

Jarausch, B., Schwind, N., Fuchs, A. and Jarausch, W. (2011). Characteristics of the spread of apple proliferation by its vector *Cacopsylla picta*. *Phytopathology* 101: 1471–80.

Jarausch, B., Schwind, N., Jarausch, W., Krczal, G., Dickler, E. and Seemüller, E. (2003). First report of *Cacopyslla picta* as a vector of apple proliferation phytoplasma in Germany. *Plant Disease* 87: 101.

Johnson, K. B. and Stockwell, V. O. (1998). Management of Fire Blight: A case study in microbial ecology. *Annual Review of Phytopathology* 36: 227–48, doi:10.1146/annurev.phyto.36.1.227.

Jones, A. J. (2002). Fire blight – Challenges and progress on new controls. *Mountaineer Grower* 567: 4–17.

Kawagychi, A., Inoue, K. and Tanina, K. (2015). Evaluation of the nonpathogenic *Agrobacterium vitis* strain ARK-1 for crown gall control in diverse plant species. *Plant Disease* 99 (3): 409–14.

Kawagychi, A., Kondo, K. and Inoue, K. (2012). Biological control of apple crown gall by nonpathogenic *Rhizobium vitis* strain VAR03-1. *Journal of General Plant Pathology* 78 (4): 287–93.

Kerkoud, M., Manceau, C., Gardan, L., Samson, R. and Paulin, J.-P. (2000). Epiphytic occurrence of Pseudomonas syringae pv. papulans (Rose) in France, where blister spot has never been seen. *European Journal of Plant Pathology* 106 (5): 481–5.

Kerkoud, M., Manceau, C. and Paulin, J. P. (2002). Rapid diagnosis of Pseudomonas syringae pv. papulans, the causal agent of blister spot of apple, by polymerase chain reaction using specifically designed hrpL gene primers. *Pytopathology* 92 (10): 1077–83.

Khan, M. A., Duffy, B., Durel, C. E., Denancé, C., Kellerhals, M., Patocchi, A. and Gessler, C. (2007). Development of markers linked to the 'Fiesta' 7 major QTL for fire blight resistance and their application for marker-assisted selection. *Genome* 50: 568–77.

Khan, M. A., Zhao, Y. F. and Korban, S. S. (2012). Molecular Mechanisms of Pathogenesis and Resistance to the Bacterial Pathogen *Erwinia amylovora*, Causal Agent of Fire Blight Disease in Rosaceae. *Plant Molecular Biology Reporter* 30 (2): 247–60.

Khan, M. A., Zhao, Y. F. and Korban, S. S. (2013). Identification of genetic loci associated with fire blight resistance in Malus through combined use of QTL and association mapping. *Physiologia Plantarum* 148 (3): 344–53.

Kim, W.-S., Hildebrand, M., Jock, S. and Geider, K. (2001). Molecular comparison of pathogenic bacteria from pear trees in Japan and the fire blight pathogen *Erwinia amylovora*. *Microbiology* (Reading) 147: 2951–9.

Koczan, J. M., McGrath, M. J., Zhao, Y. and Sundin, G. W. (2009). Contribution of Erwinia amylovora exopolysaccharides amylovoran and levan to biofilm formation: Implications in pathogenicity. *Phytopathology* 99 (11): 1237–44.

Keil, H. L. and van der Zwet, T. (1972). Recovery of *Erwinia amylovora* from sumptomless stems and shoots of Johathan apple and Bartlett pear trees. *Phytopathology* 62: 39–42.

Le Roux, P., Flachowsky, H., Hanke, M.-V., Gessler, C. and Patocchi, A. (2012). Use of a transgenic early flowering approach in apple (*Malus* × *domestica* Borkh.) to introgress fire blight resistance from cultivar Evereste. *Molecular Breeding* 30: 857–74.

Malnoy, M., Martens, S., Norelli, J. L., Barny, M.-A., Sundin, G. W., Smits, T. H. M. and Duffy, B. (2012). Fire blight: Applied genomic insights of the pathogen and host. *Annual Review of Phytopathology* 50: 475–94.

Mansvelt, E. L. and Hattingh, M. J. (1990). Bacterial blister bark. In Jones, A. L. and Aldwinckle, H. S. (ed.), *Compendium of Apple and Pear Diseases*. APS Press, p. 64.

Manulis, S., Zutra, D., Kleitman, F., Dror, O., David, I., Zilberstaine, M. and Shabi, E. (1998). Distribution of streptomycin-resistant strains of *Erwinia amylovora* in Israel and occurrence of blossom blight in the autumn. *Phytoparasitica* 26 (3): 223–30.

McGhee, G. C., Guasco, J., Bellomo, L. M., Blumer-Schuette, S. E., Shane, W. W. Irish-Brown, A. and Sundin, G. W. (2011). Genetic analysis of Streptomycin-resistant (SmR) strains of *Erwinia amylovora* suggests that dissemination of two genotypes is responsible for the current distribution of SmR E. amylovora in Michigan. *Phytopathology* 101 (2): 182–91.

McGhee, G. C. and Sundin, G. W. (2011). Evaluation of kasugamycin for fire blight management, effect on nontarget bacteria, and assessment of kasugamycin resistance potential in *Erwinia amylovora*. *Phytopathology* 101 (2): 192–204.

McGhee, G. C. and Sundin, G. W. (2012). *Erwinia amylovora* CRISPR elements provide new tools for evaluating strain diversity and for microbial source tracking. *PLoS ONE* 7 (7): e41706.

McGrath, M. J., Koczan, J. M., Kennelly, M. M. and Sundin, G. W. (2009). Evidence that prohexadione-calcium induces structural resistance to fire blight infection. *Phytopathology* 99 (5): 591–6.

McManus, P. S. and Jones, A. L. (1994). Role of wind-driven rain, aerosols, and contaminated budwood in incidences and spatial pattern of fire blight in an apple nursery. *Plant Disease* 78: 1059–66.

Mikiciński, A., Sobiczewski, P. and Berczyński, S. (2012). Efficacy of fungicides and essential oils against bacterial diseases of fruit trees. *Journal of Plant Protecion Research* 52 (40): 467–71.

Mohan, S. K. and Thomson, S. V. (1996). An outbreak of fire blight in plums. *Acta Horticulturae* 411: 73–6.

Momol, M. T., Norelli, J. L., Breth, D. I. and Aldwinckle, H. S. (1997). Rootstock blight of apple caused by Erwinia amylovora: Case studies from New York State. (Abstr.) *Phytopathology* 87: S67.

Momol, M. T., Norelli, J. L., Piccioni, D. E., Momol, E. A, Gustafson, H. L., Cummins, J. L and Aldwinckle, H. S. (1998). Internal movement of *Erwinia amylovora* through symptomless apple scion tissues into the rootstock. *Plant Disease* 82 (6): 646–50.

Moore, L. W., Warren, G. and Strobel, G. (1979). The infectivity of *Agrobacterium rhizogenes* is controlled by a plasmid. *Plasmid* 2: 617–26.

Moore, L. W. (1990a). Crown gall. In Jones, A. L. and Aldwinckle, H. S. (eds), *Compendium of Apple and Pear Diseases*. APS Press, pp. 64–5.

Moore, L. W. (1990b). Hairy root. In Jones, A. L. and Aldwinckle, H. S. (eds), *Compendium of Apple and Pear Diseases*. APS Press, p. 66.

Moriya, S., Iwanami, H., Takahashi, S., Kotoda, N., Suzaki, K. and Abe, K. (2008). Evaluation and inheritance of crown gall resistance in apple rootstocks. *Journal of the Japanese Society for Horticultural Science* 77 (3): 236–41.

Moriya, S., Iwanami, H., Takahashi, S., Kotoda, N., Suzaki, K., Yamamoto, T. and Abe, K. (2010). Genetic mapping of the crown gall resistance gene of the wild apple *Malus sieboldii Tree Genet. Genomes* 6 (2): 195–203.

Nishitani, C., Hirai, N., Komori, S., Wada, M., Okada, K., Osakabe, K., Yamamoto, T. and Osakabe, Y. (2016). Efficient genome editing in apple using a CRISPR/Cas9 system. *Science Reports* 6: 31481, doi:10.1038/srep31481.

Norelli, J. L. (2004). Fire blight. In Goodman, R. M. (ed.), *Enclyclopedia of Plant and Crop Science*. Marcel Dekker, Inc., pp. 443–7.

Norelli, J. L., Holleran, H. T., Johnson, W. C., Robinson, T. L. and Aldwinckle, H. S. (2003a). Resistance of geneva and other apple rootstocks to *Erwinia amylovora*. *Plant Disease* 87 (1): 26–32.

Norelli, J. L., Jones, A. L. and Aldwinckle, H. S. (2003b). Fire blight management in the 21st century: using new technologies that enhance host resistance in apple. *Plant Disease* 87: 756–65.

Norelli, J. L. and Miller, S. S. (2004). Effect of prohexadione-calcium dose level on shoot growth and fire blight in young apple trees. *Plant Disease* 88 (10): 1099–1106.

Oh, C.-S. and Beer, S. V. (2005). Molecular genetics of Erwinia amylovora involved in the development of fire blight. *FEMS Microbiology Letters* 253 (2): 185–92.

Ontario Ministry of Agriculture, Food and Rural Affairs (2009). *Integrated Pest Management of Apples*. OMAFRA Publication 310.

Parish, C. L. (1989). Apple decline: Characterization, cause and cure. *Acta Horticulturae* 235: 93–7.

Peace, C. P., Luby, J. J., van de Weg, W. E., Bink, M. C. A. M. and Iezzoni, A. F. (2014). A strategy for developing representative germplasm sets for systematic QTL validation, demonstrated for apple, peach, and sweet cherry. *Tree Genetics Genomes* 10 (6): 1679–94.

Peace, C. P. and Norelli, J. L. (2009). Genomics approaches to crop improvement in the Rosaceae. In Folta, K. and Gardiner, S. (eds), *Genetics and Genomics of Rosaceae*. New York: Springer, pp. 19–53.

Peil, A., Garcia-Libreros, T., Richter, K., Trognitz, F. C., Trognitz, B., Hanke, M.-V. and Flachowsky, H. (2007). Strong evidence for a fire blight resistance gene of Malus robusta located on linkage group 3. *Plant Breeding* 126 (5): 470–5.

Piqué, N., Miñana-Galbis, D., Merin, S. and Tomás, J. M. (2015). Virulence factors of Erwinia amylovora: A review. *International Journal of Molecular Science* 16 (6): 12836–54.

Polverigiani, S., Franzina, M., Salvetti, M., Folini, L., Ferrante, P., Scortichini, M. and Neri, D. (2016). The effect of growth substrate on apple plant status and on the occurrence of blister bark symptoms. *Scientia Horticulturae* 198: 233–41.

Rodoni, B. C., Merriman, P. R. McKirdy, S. J. and Wittwer, G. (2006). Costs associated with fire blight incursion management and predicted costs of future incursions *Acta Horticulturae* 704: 55–61.

Russo, N. L., Burr, T. J., Breth, D. I. and Aldwinckle, H. S. (2008). Isolation of streptomycin-resistant isolates of *Erwinia amylovora* in New York. *Plant Disease* 92 (5): 714–18.

Scortichini, M. and Morone, C. (1997). Bacterial blister bark of apple trees in Italy. *Journal of Phytopathology* 145 (8–9): 401–3.

Seemüller, E. (1990a). Apple proliferation. In Jones, A. L. and Aldwinckle, H. S. (eds), *Compendium of Apple and Pear Diseases*. APS Press, p. 67.

Seemüller, E. (1990b). Rubbery wood. In Jones, A. L. and Aldwinckle, H. S. (eds), *Compendium of Apple and Pear Diseases*. APS Press, p. 69.

Seemüller, E., Moll, E. and Schneider, B. (2008). Apple proliferation resistance of Malus sieboldii-based rootstocks in comparison to rootstocks derived from other Malus species. *European Journal of Plant Pathology* 121 (20): 109–19.

Seemüller, E. and Schneider, B. (2007). Differences in virulence and genomic features of strains of '*Candidatus* Phytoplasma mali', the apple proliferation agent. *Phytopathology* 97 (8): 964–70.

Sholberg, P. L. and Bedford, K. E. (1997). Characterization of blister spot (*Pseudomonas syringae* pv. *papulans*) in British Columbia and its potential for spread to new apple cultivars. *Canadian Journal of Plant Pathology* 19: 347–51.

Sholberg, P. L. and Quamme, H. A. (1999). Dieback of pome fruit rootstocks caused by *Pseudomonas syringae*. *Canadian Journal of Plant Pathology* 79: 387–94.

Smith, T. J. (1993). A predictive model for forecasting fire blight of pear and apple in Washington State. *Acta Horticulture* 338: 153–60.

Smith, T. J. (1999). Report on the development and use of Cougar-blight 98C – A situation-specific fire blight risk assessment model for apple and pear. *Acta Horticulture* 489: 429–36.

Steiner, P. W. (1988). *MARYBLYT*, a predictive model for eastern apple fire blight. Abstr. *Phytopathology* 78: 1571.

Steiner, P. W. (1990). Predicting canker, shoot and trauma blight phases of apple fire blight epidemics using the *MARYBLYT* model. *Acta Horticulture* 273: 149–58.

Sundin, G. W., Werner, N. A., Yoder, K. S. and Aldwinckle, H. S. (2009). Field evaluation of biological control of fire blight in the eastern United States. *Plant Disease* 93: 423–30.

Taylor, R. K., Ranatunga, C., Bus, V. and Alspach, P. A. (2002). Host-pathogen interactions of Erwinia amylovora on apple and pear in New Zealand. *Acta Horticulturae* 590: 363–7.

Tedeschi, R. and Alma, A. (2004). Transmission of apple proliferation phytoplasma by *Cacopsylla melanoneura* (Homopters: Psyllidae). *Journal of Economic Entomology* 97 (1): 8–13.

Thomson, S. V. (1986). The role of the stigma in fire blight infections. *Phytopathology* 76: 476–82.

Thomson, S. V. (2000). Epidemiology of fire blight. In Vanneste, J. L. (ed.), *Fire Blight, the Disease and its Causative Agent, Erwinia Amylovora*. CABI Publishing, pp. 9–36.

Torres, E., Bertolini, E., Cambra, M., Montón, C. and Martín, M. P. (2005). Real-time PCR for simultaneous and quantitative detection of quarantine phytoplasmas from apple proliferation (16SrX) group. *Molecular Cellular Probes* 19 (5): 334–40.

van der Zwet, T. (1994). The various means of dissemination of the fire blight bacterium Erwinia amylovora. *EPPO Bulletin* 24 (1): 209–14.

van der Zwet, T. and Beer, S. V.(1995). *Fire Blight – its Nature, Prevention and Control; a Practical Guide to Integrated Disease Management*. Agric. Information Bulletin #631; US Dept. Agric.

van der Zwet, T. and Keil, H. L. (1979). *Fire Blight – a Bacterial Disease of Rosaceous Plants*, Agric. Hdbk. 510; US Dept. Agric.

van der Zwet, T., Orolaza-Halbrendt, N. and Zeller, W. (2012). *Fire Blight – History, Biology and Management*. APS Press.

Végh, A. and Palkovics, L. (2013). First occurence of fire blight on apricot (*Prunus armeniaca*) in Hungary. *Notulae Botanicae Horti Agrobotanici Cluj-Napoca* 41: 440–3, doi:10.15835/nbha4129183.

Venisse, J.-S., Barny, M.-A., Paulin, J.-P. and Brisset, M.-N. (2003). Involvement of three pathogenicity factors of *Erwinia amylovora* in the oxidative stress associated with compatible interaction in pear. *FEBS Letters* 537: 198–202.

Venisse, J.-S., Gullner, G. and Brisset, M.-N. (2001). Evidence for the involvement of an oxidative stress in the initiation of infection of pear by *Erwinia amylovora*. *Plant Physiology* 125 (4): 2164–72.

White, F. F., Ghidossi, G., Gordon, M. P. and Nester, E. W. (1982). Tumor induction by *Agrobacterium rhizogenes* involves the transfer of plasmid DNA to the plant genome. *Proceedings of the National Academic Science USA* 79: 3193–7.

Xu, Y. and Crouch, J. H. (2008). Marker-assisted selection in plant breeding: From publications to practice. *Crop Science* 48: 391–407.

Yoder, K. S., Miller, S. S. and Byers, R. E. (1999). Suppression of fireblight in apple shoots by prohexadion-calcium following experimental and natural inoculation. *Hortscience* 34: 1202–4.

Chapter 17

Sustainable arthropod management for apples

Elizabeth H. Beers, Washington State University, USA

1 Introduction

The long history of apple cultivation in temperate latitudes of the world has given us ample opportunities to become acquainted with the arthropod pest complex that attacks this crop. Considerable attention has been given to managing these pests because of the numerous challenges they present, and the relatively high value of the crop. The apple tree is large, with significant amounts of biomass invested in roots, aerial woody tissues, leaves and fruits; all of these structures are potential ecological niches for multiple pest species. Temporal factors also play a role in the agroecosystem. The life of an orchard is typically >20 years, allowing the development of permanent orchard residents, uninterrupted by an annual cropping cycle. In fact, the pest complex that has been recorded to attack apple is drawn from nine orders of arthropods, including Lepidoptera, Coleoptera, Diptera, Hymenoptera, Hemiptera, Thysanoptera, Orthoptera, Dermaptera and Acari. However, in a given region at any point in time, only a few dozen pests constitute serious threats to production and require annual management.

http://dx.doi.org/10.19103/AS.2016.0017.20

Apples have historically required relatively high levels of pesticide inputs, particularly insecticides, acaricides and fungicides. While herbicides dominate the pesticide market worldwide, their importance in apple production is considerably less than in row crops. Indeed, other plant species are deliberately introduced and managed as part of the apple orchard ecosystem, in the form of ground cover between the tree rows. This mixture of grass and herbaceous species is tolerated because apple trees are relatively good competitors for water and nutrients, and a narrow plant-free strip beneath the trees suffices for them to thrive. These additional plant species contribute to the orchard ecosystem by the pest or natural enemies they harbour, and provide an opportunity for cultural approaches to pest management.

A major change in approach to insect control came with the articulation of the integrated control concept (Stern et al., 1959). While the original definition focused on integrating chemical and biological control, the concept of integrated pest management (IPM) expanded on this definition to include economic, environmental and societal cost/benefits (Philips et al., 2014). The development of IPM was part of the backlash caused by the explosion of new synthetic pesticides after the Second World War. The use of these materials on a calendar-based schedule leads to outbreaks of secondary pests, insecticide resistance and environmental damage. The principles of IPM sought to bring moderation and reason to pest control, with emphasis on biological control as a foundation and insecticides only when necessary (Philips et al., 2014). This approach brought fundamentally changed pest control in many crops, including apples (Jones et al., 2009). In the decades since the IPM approach gained wide acceptance, there has been a gradual shift to more selective tools and tactics in apple production around the world.

There are numerous ways to classify the pests of apple: taxonomically (order and family), ecologically (orchard resident or invader), by tissue or structure attacked (leaves, shoots, roots or fruits), by damage type (direct or indirect) or by the intrinsic threat level to crop production. From an IPM perspective, the nature of the crop damage is the factor driving the management system, and shapes the opportunities and constraints. The term 'key pest' and 'secondary pest' arise from this mode of thinking; the classification of a pest determines largely what tactics will be used against it. In this chapter, a key pest is defined as a chronic threat to production that will likely never be brought below economic injury levels by biological control. For this reason, key pests are the 'drivers' of an IPM programme. The following is an IPM-centred review of arthropod pests of apple, with emphasis on North America, although the inclusion of insect families or genera will broaden application to many other parts of the world where apple is grown.

2 Key arthropod pests

The Lepidoptera contains a number of the world's more damaging pest species for a wide variety of crops, and the same is true for apple. For most lepidopteran pests, only the larvae cause damage. The family Tortricidae, in particular, is associated with apple worldwide, with different species fulfilling the same niche in different regions. Two active groups within the tortricids that attack apple as internal feeders (generally considered the most harmful) are the genera *Cydia* and *Grapholita*. The second group, often referred to as leafrollers, feed on leaves inside shelters they construct and also feed on the surface of the nearby fruit. In this section we will focus on the case of codling moth in detail. Section 3 will then cover other key arthropod pests.

2.1 Codling moth (*Cydia pomonella*)

This species is arguably the most important apple pest worldwide. The larvae tunnel into the centre of the fruit and feed on the seeds; the resulting decay and deposition of frass renders the fruit inedible (Fig. 1). Even an unsuccessful attack (where the larva chews a small hole in the skin and underlying flesh, known as a 'sting') can cause the fruit to be downgraded during packing. Codling moth is a pest of quarantine concern for a number of countries, creating a zero tolerance for pest presence if export to these countries is to be accomplished. This level of control imposes a severe constraint on IPM tactics for this pest, which ramify throughout the rest of the system. Even where export is not the goal, codling moth is typically managed to produce extremely low damage levels, in part due to the complete unacceptability of a deep (successful) entry and complete larval development. The capacity of codling moth populations to increase rapidly from one generation to the next is another concern. A loss of control would rapidly lead to complete crop loss; after a certain percentage of the crop is infested, it is no longer economically feasible to harvest and grade the fruit, and the entire crop may be abandoned. While this threshold varies by region and crop market destination, 10–20% damage would be considered unusable in most production systems. The general expectation is that good commercial practice will result in <0.5% codling moth infestation.

The high bar imposed by this low infestation level is usually met with a combination of tactics, frequently including use of insecticides. Insecticides have targeted codling moth eggs, larvae, and adults; only the pupal stage has not been a specific target. The adult moth was susceptible at one point in time, but the primary targets today are eggs and newly hatched larvae. It should be noted that the latter stage poses a considerable challenge for an insecticide, in that they rapidly penetrate to the fruit's interior, and only ingest a small amount of the treated fruit surface. Historically, the pesticides used for codling moth have been central to the pest management programme since the advent of the first really effective materials in the late 1800s: the arsenicals. These materials, particularly lead arsenate, were widely used in apple production until after the Second World War, when they were replaced by a synthetic and even more effective group of pesticides, the organochlorines. The adoption of dichlorodiphenyltrichloroethane (DDT) as

Figure 1 Codling moth larva in apple.

a replacement for lead arsenate was rapid due to the high levels of resistance to the latter, and the increasing threat of environmental and human toxicity. While acute mammalian toxicity of DDT was substantially lower than the arsenicals, environmental issues eventually outweighed the benefits of its use. Like the arsenicals, resistance had become widespread by the time its use was cancelled in 1972 (Croft, 1982). The organophosphates (OPs) were developed contemporaneously with the organochlorines, which replaced them for codling moth control by the 1960s. One material in particular (azinphos methyl) was used for decades, and was still considered moderately effective (despite documentation of resistance) in many regions until its cancellation in 2012. Azinphos methyl and other OPs were replaced by a plethora of insecticides, under an expedited Environmental Protection Agency registration programme for OP replacement and reduced-risk insecticides (USEPA, 2013). Members of the neonicotinyl, spinosyn and anthranilic diamide groups were brought into use in the codling moth programme, along with ovicidal compounds (benzoylurea and diacylhydrazine insect growth regulators). Petroleum oils have been used as ovicides for many years, and regained popularity for use in selective programmes. Biorational compounds saw wider use during this period, specifically the codling moth granulosis virus (Beers et al., 2003; NASS, 2008). Improvements in formulations and strains made this a more valuable addition to the IPM toolbox.

More selective insecticide programmes, however, were enabled by the development of a non-insecticidal tactic: mating disruption. Mating disruption is a control technique that slows or inhibits reproduction by permeating the atmosphere around the crop (in this case, the orchard) with a synthetic version of the insect's pheromone. In a non-disrupted orchard, females emit a sex pheromone, which enables the males to find and mate with them; the synthetic pheromone interferes with this vital communication. The pheromones of the tortricids were characterized in the 1960–1970s, but it took several decades of research to bring this tactic into commercial use in tree fruit. The first North American registration for codling moth mating disruption using pheromones took place in 1990 (Fig. 2), and adoption slowly and steadily increased over the next decades. In Washington State, it is estimated that >75% of the apple acreage is treated with mating disruption (Thomson et al., 2009). The severity of the pressure in the arid interior districts and the

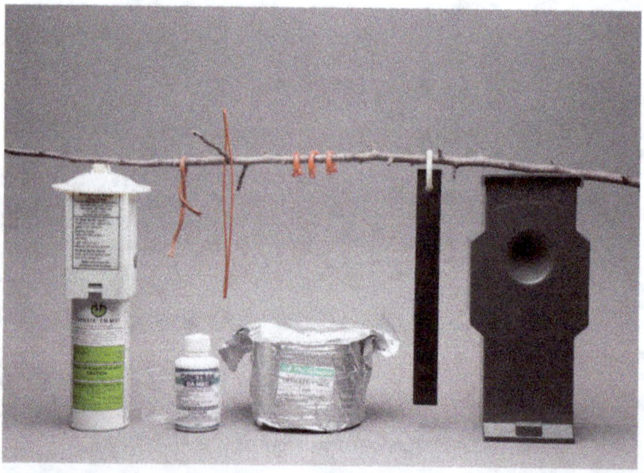

Figure 2 Mating disruption products for codling moth.

absence of other severely problematic direct pests provided the impetus for adoption by Washington apple growers. In other areas of the country, additional direct pests such as plum curculio, apple maggot, oriental fruit moth and leafroller meant that insecticides would be needed despite the use of mating disruption for codling moth, making it economically less attractive.

The success of mating disruption is reliant to a degree on several factors. For many species (including codling moth) it is density-dependent (Cardé and Minks, 1995) and only works at a commercially acceptable level when initial pest densities are low (Witzgall et al., 2008). It also relies on spatial scale to be effective. Small isolated blocks, or those with a high border to interior ratio, have seen less success with mating disruption. For both of these reasons, an 'area-wide' approach is considered optimal, where large contiguous blocks of orchards (regardless of ownership) are managed in concert to control this pest (Brunner et al., 2002). Similar to insecticide-based control, mating disruption also relies on technology for success: the correct pheromone and a long-lasting dispenser system. Considerable research has been devoted to developing an effective dispenser system. The hand-applied dispenser was the first technique to be registered, and still remains in use today. Further developmental research attempted to make the technique more economical. The number of point sources per acre needed for optimum disruption has been investigated (Epstein et al., 2006; Miller et al., 2006), and alternatives to dispensers have taken many forms: the aerosol emitter systems (Knight, 2004; McGhee et al., 2016) employed larger amounts of pheromone with fewer point sources; at the opposite end of the scale, pheromone embedded in fibers or flakes, and applied with specialized sprayers, gave orders of magnitude more point sources of pheromone. The principal mechanism of mating disruption for codling moth is competition between pheromone sources and traps and females for male attention (Miller and Gut, 2015). Surprisingly, competitive disruption operates for both multiple (low-dose) and few (high-dose) point source formulations (Miller et al., 2010; McGhee et al., 2014). Thus, future improvements in mating disruption technologies will entail optimizing release rates and application efficiency.

The other technique that allows suppression of codling moth to low levels without the use of insecticides is the sterile insect technique (SIT). This technique again relies on an area-wide approach, preferably with a degree of isolation from re-infestation. It also requires a production facility for sterilized moths, a substantial initial investment, and an ongoing (usually centralized) programme for distribution. These factors, along with the competing non-insecticidal tactic of mating disruption, have been the major impediments to the adoption of SIT. One of the few extant examples is the Okanagan Valley of British Columbia, which began a SIT programme in 1992 (Wood et al., 2005), and is still active today. Although the initial goal of eradication was not achieved, high levels of suppression mean that growers need few additional inputs for control of this key pest (Bloem et al., 2005).Like mating disruption, SIT is a density-dependent tactic. High initial populations must be reduced (usually with insecticides) to a level where there is a high probability of a female mating with a sterile, rather than a fertile (wild), male. This ratio of sterile to fertile (or overflooding ratio) is considered key to population suppression, and the target set in the British Columbia programme was 40:1 (sterile:wild) (Bloem et al., 2005). Clearly, given the cost of production and distribution of sterile moths, balancing the equation with a low wild population is the most economical approach.

Another recent non-insecticidal development is physical exclusion. This technique (essentially building a netting cage over the orchard) was tested in France in the 1990s, and found to be at least as effective as (if not more than) mating disruption (Romet et al., 2010).

Once the populations inside the cage are eradicated, this method can be used where the ambient populations are high if the cages completely prevent immigration into the block. The high initial installation cost is a barrier to adoption, especially while retrofitting the existing orchards. However, the prevalence of some sort of wooden or steel posts in orchards (for trellising, irrigation lines or anti-hail nets) has changed the perspective on orchard infrastructure. If built into the installation plan of a new planting, multiple purposes can be achieved, justifying the greater cost. In the Po Valley of Italy, the frequency of hail damage provoked the widespread use of anti-hail nets, which in turn promoted the idea of physical exclusion of insects. In hot, arid climates, the use of shade netting to prevent sunburn has provided the initial impetus. The mesh size suitable for sunburn prevention was shown to be effective in codling moth exclusion. Although shade nets need only be placed horizontally over the orchard canopy, exclusion nets must also have vertical sides to achieve a complete enclosure (with attendant problems of equipment access).

It should be noted that exclusion cages are a long-standing technique for entomologists to study the benefits of biological control, thus we can expect some disruption of biological control of some pest species. Indeed, problems have already been noted with increased levels of secondary pests, which are normally regulated by predators and parasitoids. Solving a direct pest problem not infrequently leads to inducing a secondary pest problem, and this will need to be addressed if net enclosures become widely used. Other consequences of altering the microclimate inside a net enclosure (e.g. changes in drying time due to reduced airflow or changes in relative humidity) may include disease pressure, especially in more humid climates.

There has also been some research devoted to autocidal control (Marec et al., 2005; Ferguson et al., 2011; Garczynski, 2017). Like SIT, this method involves rearing codling moth in large numbers for release into orchards. The reared moths carry a conditional lethal gene (usually triggered by temperature) that will be passed on to their progeny. To date, this approach to autocidal control is experimental only.

3 Other key arthropod pests

3.1 Oriental fruit moth and other *Grapholita* spp.

Oriental fruit moth (*Grapholita molesta)* is also a worldwide pest of tree fruits; although its primary hosts are peach and nectarine, it will also attack apple. It was introduced into North America from China in the early 1900s, and its pest status on apple has fluctuated since that time. It differs from codling moth in that it bores into shoots as well as feeds on fruits. The damage is somewhat similar to that of codling moth, but the larvae meander through the flesh instead of going directly to the core, and the characteristic frass pile of codling moth is absent (although frass may be present on the fruit surface). It shares one important characteristic, however: attacked fruits are rendered inedible.

Most of the pesticides that control codling moth will also control oriental fruit moth, although the timing of application may be different. Oriental fruit moth has 3–4 generations in temperate climates, whereas codling moth typically has 2–3 generations. In the US Midwest and Mid-Atlantic States, it is considered an important pest, nearly on par with codling moth; in other production areas, it is rarely a problem in conventionally managed orchards.

Oriental fruit moth has been very successfully controlled with mating disruption (Vickers et al., 1985; Cardé and Minks, 1995), and is one of the easier species to disrupt with this technique. Because of the similarity of the pheromones of the codling moth and oriental fruit moth, a combined dispenser is available commercially, making this technique more attractive where both species are present.

Two other species of *Grapholita* can sometimes cause damage to apple: *G. prunivora*, the lesser appleworm, and *G. packardi*, the cherry fruitworm. While historically they are considered direct pests of apple, their importance today is minor. Because these species are quarantine pests in some markets, and difficult to identify in the larval stage, PCR diagnosis has been developed to aid in export certification (Barcenas et al., 2005).

3.2 Leafrollers

The designation 'leafroller' is based on feeding behaviour, and embraces a number of genera in the Tortricidae. They tend to be more polyphagous than codling moth and the *Grapholita*, which specialize on a few (usually related) host plants. Leafrollers feed on not only the leaves, but also the fruit surface beneath a shelter they construct of a rolled leaf and webbing. Leafrollers include members of the genera *Adoxophyes, Archips, Choristoneura, Argyrotaenia, Epiphyas, Pandemis, Platynota* and *Spilonota*. Unlike codling moth, which has close to worldwide distribution, leafroller species more often occur in regional complexes, although their ecology and damage are similar. Leafrollers are poised between key and secondary pests, depending on the region and point in history, and their relative importance in IPM programmes has waxed and waned. Because they consume larger amounts of leaf tissues, they are more exposed to pesticides than those whose larvae bore directly into the fruit (Fig. 3). The insecticides used overlap, but are not identical to those that control codling moth and *Grapholita* spp. One notable example is the bacteria *Bacillus thuringiensis*, which is very useful against leafrollers but not against codling moth (Brunner, 1994; Brunner et al., 1995).

Mating disruption of the more important leafroller pests has been studied with mixed results (Deland et al., 1994; Agnello et al., 1996; Lawson et al., 1996; Knight et al., 1998; Trimble and Appleby, 2004), but this technique has not been as widely adopted in the United States as codling moth mating disruption (Warner, 1998).

Figure 3 Leafroller damage to fruit surface.

If mating disruption has been less successful with leafrollers than with codling moth, biological control has been relatively more successful. Their feeding behaviour leaves them more exposed to predators and parasitoids than a larva which spends the majority of its development deep inside a fruit. Outbreaks of leafrollers have been associated with broad-spectrum pesticides (especially the OPs), which likely wiped out their natural enemies; the use of more selective products may be responsible for the lessening of leafroller problems. To this end, investigations in conservation biological control (insectary plantings) have been conducted, and met with a degree of success (Stephens et al., 1998; Irvin et al., 1999, 2006; Pfannenstiel and Unruh, 2003; Mullinix et al., 2011). Even without insectary plantings, however, leafroller problems were eventually ameliorated after the withdrawal of OPs from the IPM programme.

3.3 Apple maggot (*Rhagoletis pomonella*)

Apple maggot is a tephritid fruit fly native to eastern North America, and has been recognized as a pest since the 1860s (Bush, 1966). Dipteran pests are relatively uncommon as pests of apple, but this particular species has proven quite problematic in the regions where it has become established. It was detected in western North America (Oregon) in 1977 (Ali Niazee and Penrose, 1981), and has spread slowly since that time. Initially, its distribution was confined to the coastal areas of Washington and Oregon, but has now spread to the primary apple production districts east of the Cascades, an isolated area in Mexico, and more recently, into British Columbia. Other parts of the world remain free of this pest.

Apple maggot is a direct, internal fruit feeder, leaving tunnels throughout the flesh of the fruit. Like the rest of the internal feeders, fruit damaged by apple maggot must be culled (Fig. 4). Because of its limited distribution, it is a quarantine pest – meaning there is essentially zero tolerance for fruit infestation. Thus pesticides are necessarily a primary tactic for control. Historically this function has been fulfilled by the OPs, but other products have largely supplanted them in recent years. Attracticides (a mixture of an attractant with a very low rate of a toxicant, in this case spinosad) have been tested with some success against this species (Pelz et al., 2005), but are typically only used in organic production.

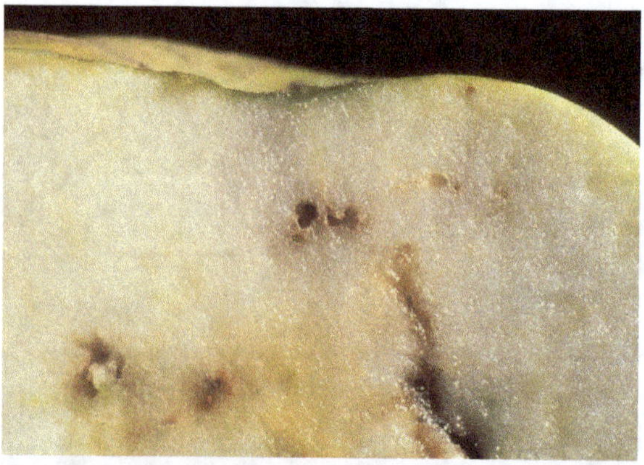

Figure 4 Apple maggot larval tunneling in fruit flesh.

Figure 5 Plum curculio adult.

Overall, the attracticide approach has achieved less success for apple maggot than for other tephritid fruit flies, including the tropical species, and the more closely related *Rhagoletis* species attacking sweet cherry.

3.4 Plum curculio (*Conotrachelus nenuphar*)

Given the size and importance of the Coleoptera as crop pests in general, they are somewhat under-represented in the pests of tree fruits. The plum curculio is one of the few that could conceivably be given key pest status (Glass and Lienk, 1971). Its distribution is limited to North America, and has been detected only as far west as Utah. It is a direct pest in that adult weevils feed and oviposit near the surface of the fruit, but damage is limited to a characteristic crescent-shaped scar (Fig. 5). This is sufficient to cause fruit downgrading, but without the internal contamination caused by the tortricids or apple maggot. The timing of control (early in the post-bloom period) coincides with control timing for other species, and limits the use of more selective options.

4 Secondary arthropod pests: introduction, leaf and bud damage

4.1 Introduction to secondary pests

Broadly speaking, a secondary pest can be any pest not designated as a key pest, regardless of which plant tissues it affects; however, most secondary pests are also indirect feeders (leaves, shoots, roots). Regardless of how they are grouped, they share a common characteristic for IPM purposes: they are the passengers, whereas the key pests are the drivers, shaping the nature and outcomes of the IPM programme. They can, of course, rise to high levels and cause significant damage under the right circumstances, but management tends to respond to their presence on an 'as-needed' basis, rather than forming the backbone of the strategy. The number of potential species included in this designation is very large, so only the more familiar examples will be included in this discussion.

The sporadic pests (occurring with less frequency in time or space) deserve a special note with regard to their impact on pest management. Sporadic pests are those that are known to occur in the region, but not often enough to warrant inclusion in the basic structure of the IPM programme. They can be dangerous in that their densities can be elevated without warning from a low level (with inconsequential damage) to a damaging outbreak. They are frequently orchard invaders, completing the majority of their life cycle outside the orchards, where they are beyond the reach of many management activities.

There is quite a bit of blurring of the lines between the various categories, and no single classification system will fit all situations. Pests that were formerly common secondary pests may become sporadic or rare with the passage of time and the change in production practices or orchards environment. The reverse is also true: a little-known pest can suddenly emerge at outbreak levels, for reasons that are often poorly understood, but usually ascribed to changes in pesticide programmes. This can be caused by the disruption of biological control or the removal of an effective pesticide, leaving a window of opportunity for the pest. Similarly, although grouped by the plant tissue that is attacked, many species can attack several plant tissues, in either the same or different life stages.

4.2 Leaf damage

The most common type of indirect pest feeds on leaves and to some extent compromises the tree's ability to photosynthesize and supply the saleable commodity (the fruit) with the water and nutrients it requires to obtain commercial size and colour. Leaf feeders can be broadly classified as either chewing or sucking pests.

Chewing pests

Chewing insects, which remove leaf tissue in bulk, are generally not major apple pests. Leafrollers (mentioned under key pests) remove some leaf tissue, but not enough to significantly damage the tree; it is the fruit damage that is paramount. Other lepidopterans are more efficient at foliar consumption, including species in the Lymantriidae (e.g. gypsy moth), Lasiocampidae (e.g. tent caterpillars), Arctiidae (e.g. webworms) and Geometridae (cankerworms, loopers, winter moth). The larvae of generalist moths and butterflies (Papilionidae, Nymphalidae, Notodontidae, Saturniidae, Sphingidae, Lycaenidae and Limacodidae) can also denude a branch of foliage, but rarely survive in commercial orchards. Leafminers (Gracillariidae, Lyonetiidae and Bucculatricidae) can also damage leaves by consuming tissue, but do so from inside a mine formed between the epidermal layers. Larvae and adults of some scarabs (e.g. the Japanese beetle, rose chafer) can cause significant defoliation under some circumstances, as can some sawflies (Pamphilidae and Tenthredinidae). Some of the Curculionidae notch leaves, but generally do not remove enough tissue to cause damage.

Sucking pests

Foliar damage to apple is more likely to be caused by arthropods with sucking mouthparts. These pests use stylets to penetrate the leaf, removing chlorophyll and other cell contents, thereby reducing the tree's photosynthetic capacity. Pests in this category can occur in very high densities, and thus can cause extensive leaf damage.

Historically, one of the most problematic groups has been the Acari, especially the Tetranychidae (spider mites). Two species (*Panonychus ulmi*– the European red mite, and *Tetranychus urticae* – the two-spotted spider mite) are cosmopolitan in their distribution and are recognized as chronic pests in most apple-growing regions (Fig. 6). In the absence of disruptive pesticides, biological control will typically hold these in check (Glass and Lienk, 1971; Schmidt-Jeffris and Beers, 2016), but outbreaks occur when non-selective pesticides wipe out predators (Hoyt, 1969). A considerable amount of research in the 1960–1970s was devoted to determining which compounds were detrimental to the most common mite predators, the phytoseiids. OPs, while initially quite toxic, became partially selective as predators were selected for resistance. Other compounds, such as pyrethroids and carbamates, are still more or less toxic. Learning the hard lesson of non-target effect on mite predators has taken decades, and the struggle continues today as older materials, whose properties were well documented, are replaced by newer materials (Bostanian et al., 2009; Lefebvre et al., 2011; Beers and Schmidt, 2014).

Leafhoppers are common pests of apple, and occasionally cause extensive foliar damage. While the species varies with region (e.g., white apple leafhopper – *Typhlocyba pomonella*, and rose leafhopper – *Edwardsiana rosae*, in the western US; Froggatt's apple leafhopper, *Edwardsiana crataegi*, in New Zealand (Shaw et al., 1997)), the damage is similar: stippling of leaves caused by penetration of stylets and removal of chlorophyll. Biological control, unless disrupted by pesticides, may keep leafhoppers at sub-economic levels; indeed, it has been difficult to demonstrate crop loss by *T. pomonella* (Beers et al., 1995; Welker et al., 1995)

Apple leafcurling midge or apple leaf gall midge (*Dasineura mali*) is a cecidomyiid of European origin, but has since spread to New Zealand, Argentina, Canada, and parts of the United States (CDFA, 2015). As its common name implies, larval feeding causes a gall, which distorts growth of the leaf. Larvae are protected from contact with sprays inside a tight curl that they form, thus spray timing and materials are a consideration in their

Figure 6 European red mite adult and eggs.

control. A pheromone was discovered that can be used to monitor this pest (Cross and Hall, 2009; Cross et al., 2009). Low-to-moderate populations are tolerated by most trees; only high levels in nursery trees or young orchards may cause economic damage. A parasitoid (*Platygaster demades*) can exert high levels of parasitism in the apple leafcurling midge if not disrupted by pesticides (Sandanayaka and Charles, 2006). While its damage potential is low, the apple leafcurling midge is a pest of concern largely because of quarantines imposed by several countries and the state of California.

4.3 Bud damage

There are a number of taxa that cause bud damage, which can be serious on small trees, but less of a problem on large trees. Various species in the Noctuidae are bud feeders (*Orthosia, Lithophane, Xestia*), but this family is primarily an opportunist on tree fruits; its primary hosts are herbaceous plants. There are numerous families of microlepidoptera bearing a common name of 'budmoth' or 'budworm' based on their feeding habits: Blastobasidae, Gelechiidae, Pyralidae and Tortricidae (*Archips, Argyrotaenia, Choristoneura, Hedya, Olethreutes, Platynota, Pseudexentera, Sparganothis* and *Spilonota*). A few of the curculionids, chrysomelids, scarabs, tenebrionids and elaterids will also feed on buds, again with serious damage typically only to young trees.

5 Secondary arthropod pests: shoot, branch, trunk, root and fruit damage

5.1 Shoot damage

Damage to the shoots usually falls into two categories: phloem feeders, or borers. The major phloem feeders are the aphids, which can feed on either leaves or shoots. Some aphid species have a complex life history, wherein part of the year (usually overwintering as eggs) is spent on apple and part of the year is spent on an alternate (typically herbaceous) host (viz heterocyclic). Examples are aphids in the genus *Dysaphis* (*D. plantaginea* and *D. devecta*) and *Rhopalosiphum*. However, one of the most common aphid pests of apple worldwide, *Aphis pomi*, completes its life history on apple alone (holocyclic). An intermediate between these two extremes is the woolly apple aphid (*Eriosoma lanigerum* Hausmann), a North American native species, which was originally heteroecious (alternating between *Ulmus* and a Rosaceous host). Woolly apple aphid adapted to the single-host, parthenogenesis-only lifestyle, and it can now be found year round on apple wherever the overwintering host, elm, does not occur. Despite losing its capacity for sexual reproduction in most populations, it is still considered one of the most widespread and troublesome of the aphid pests.

Damage to shoots is in two forms. First is the simple withdrawal of phloem, and associated loss of photosynthates. In large quantities, this loss can devitalize a tree, but this type of economic loss is difficult to quantify (Hamilton et al., 1986). The second involves injection of a salivary toxin, causing a plant deformity of some sort. This occurs with both *Dysaphis* and *Eriosoma*, and the appearance of the deformity is quite distinct between species. *Dysaphis* spp. cause distortion of leaf and shoot growth, resulting in crooked or curling shoots, as well as pygmy fruit in sensitive cultivars. *Eriosoma* forms galls

Figure 7 Woolly apple aphid colony on apple shoot.

at the feeding site, which can result in blind wood (dead buds) or poorly healing pruning wounds. Regardless of the physiological consequences of feeding, all aphids produce honeydew as a result of their feeding activities. When present in large quantities, this will drip on leaves and fruit, providing a substrate for black sooty mould. Although fairly innocuous, sooty mould is considered a cosmetic flaw and commercial washing cannot completely remove the mould from the stem bowl of the apple fruit. The presence of honeydew also makes fruit unpleasant to harvest (Fig. 7).

Aphids vary in the types of plant tissues they can feed upon. *Dysaphis* and *Aphis* are confined mostly to leaves and shoots, and prefer vigorous tender growth as their feeding site. *Eriosoma*, while still found frequently on shoots, will attack other woody tissues, including the trunk, branches and roots. When populations reach outbreak conditions, aphids may also be found on fruits, although this is largely a contamination problem.

All aphids are quite susceptible to biological control, especially generalist predators (lacewings, syrphids, lady beetles, earwigs). Parasitoids also affect the population densities to a greater or lesser degree depending on the aphid species. A specialist parasitoid of *Eriosoma*, *Aphelinus mali*, was distributed worldwide in the early part of the twentieth century to help combat this pest in an early effort of classical biological control (Howard, 1929). Recent research challenges the notion that *A. mali* alone can regulate populations without the aid of some of the generalist predators (Gontijo et al., 2015), but rather these two groups provide complementary suppression.

A related group, the mealybugs, cause injury similar to aphids (phloem feeding in shoots and leaves; production of honeydew and sooty mould). Shoots are a frequent feeding site, but mealybugs are also commonly found on heavy wood, especially in pruning cuts at the top of trees. Whether this is a preferred habitat or a consequence of poor pesticide coverage is uncertain, but the association of mealybugs in older plantings with large canopies is noteworthy.

Most apple-growing regions have one or more mealybug pests. In the western US, grape mealybug (*Pseudococcus maritimus*) and apple mealybug (*Phenacoccus aceris*) attack apple as well as other deciduous fruits. In New York, the Comstock mealybug (*Pseudococcus comstocki*) is the most prevalent species, while in Oceana and Africa, similar damage is caused by the longtailed mealybug (*Pseudococcus longispinus*).

Parasitoids are most often listed as a means of biological control, although generalist predators will also attack mealybugs. Only the male mealybugs are winged, so dispersal is by wind or crawlers (the first instar nymphs). Although spread of an infestation may be slow, mealybugs are difficult to eradicate with pesticides.

Oriental fruit moth (discussed under the direct pests) can also legitimately be considered an indirect pest during the first two generations. The larvae bore into the shoot, causing flagging (wilting) of the distal portion. The later generations are more prone to feed on fruit (Brunner and Rice, 1993).

5.2 Branch/trunk damage

The number of insects attacking the trunk is somewhat limited, but includes several families of insects in the orders Coleoptera and Lepidoptera. Moth larvae in the family Sesiidae that bore in the trunk are the most common of these pests. Apple is attacked by the apple clearwing moth (*Synanthedon myopaeformis*) and the dogwood borer (*Synanthedon scitula*), which both cause similar types of damage. Heavy attacks can be quite devastating to the tree's health, interrupting the transport in xylem and phloem. One of the few tortricids that attack the trunk is the cherry bark tortrix, *Enarmonia formosana*, a relatively recent introduction to North America from Europe. As the common name implies, it is more often found in stone fruits, but will attack apple, causing sesiid-like damage. Larvae of a pyralid moth *Euzophera semifuneralis* damage the trunk in a similar manner, especially where graft unions provide a point of entry.

The family Cossidae has two species whose larvae attack apple. The carpenterworm, *Prionoxystus robiniae*, has apple as a recorded host, but its attacks are more frequently found in cherries or other deciduous trees. While *P. robiniae* is a North American species, Europe has a related species of cossid, *Zeuzera pyrina*, which causes the same type of damage. *Z. pyrina* was introduced into eastern North America in the late 1800s, but is still only known from northeastern and mid-Atlantic areas. The large larvae develop over several years, and can cause significant damage to the lower trunk if left unchecked; the holes created by their tunnelling are up to 10 mm in diameter.

Several families of beetles attack the trunks and major scaffold limbs of apple. These can be devastating to young trees and, given sufficient populations, debilitating to older trees. There are several scolytid beetles (shothole borers and ambrosia beetles) that are a constant, but usually low-level threat to orchards. Sanitation is often the most useful measure, viz, cutting out infested limbs, and burning or removal of infested wood from the vicinity of the orchard.

Several species of armoured scales (Diaspididae) and soft scales (Coccidae) attack apple. The most common in the United States is San Jose scale (*Quadraspidiotus perniciosus*). With modern pesticides, scales are easily controlled, but have been a significant and dangerous pest in past eras. For many decades, petroleum oil plus an OP insecticide in the dormant/delayed dormant period prevented the scale from becoming a problem. Removal of this application from the seasonal programme (with no replacement) will allow the scales to rapidly expand their populations. The scale crawlers settle on the fruit and will mature there; even if they are removed in the packing process, their feeding activities leave a red ring on lighter skinned fruit. Ultimately, the more dangerous damage is to scaffold limbs, which can be killed by a heavy encrustation of scales. If left unchecked, the orchard will become so devitalized that removal is the best recourse.

5.3 Root damage

There are a limited number of insects that attack the apple root system; given its importance in water/nutrient translocation and tree anchoring, they can have serious consequences. One of the aphid species noted above, *E. lanigerum*, has root as well as aerial colonies, and the galling of the root system can devitalize trees, particularly those with a limited or dwarfing root system.

Larvae of several species of scarabs feed on roots of apple trees, and fairly readily kill young trees. The ten-lined June beetle, *Polyphylla decemlineata* (Say), is a pest in western North America, limited to sandy soils. Roots may be chewed back to the main stem, allowing the affected plants to be pulled from the ground by hand. Evidence of attack consists of poor growth and water stress caused by lack of water and nutrients. Several scarab larvae in the genus *Pleocoma* (rain beetles) cause similar types of damage.

Other beetles that attack roots and lower trunk are the cerambycids and buprestids, often referred to as the roundheaded and flatheaded borers, respectively (Agnello, 1999). Most are considered sporadic but capable of causing serious damage if left unchecked. Their tunnelling provides an entry point for decay organisms, exacerbating the damage. They are characterized by long life cycles, often taking several years to complete their development. The roundheaded apple tree borer (*Saperda candida*) is widely distributed in the US and southern Canada, with a wide host range in addition to apple. A group of cerambycids in the genus *Prionus* (*P. laticollis, P. imbricornis, P. californicus*) are large insects (as adults and larvae) and are capable of causing extensive damage where they are present. The buprestid borers (*Chrysobothris* spp.) can also cause extensive damage, especially in nurseries where a single larva may debilitate or kill the tree.

5.4 Fruit damage

There are several examples in the sucking bugs (Hemiptera) that are secondary or sporadic pests, especially the family Miridae. The genus *Lygus* is a key pest of many other field and forage crops, and is usually only incidental on tree fruit. However, when other host resources are compromised (e.g., by harvest of alfalfa), this pest can invade and cause substantial damage to developing fruitlets. Depending on the species, the damage can be in the form of a deep indentation, or corky scarring and deformities; the common name 'dimpling bug' is highly descriptive of the damage. Other genera causing similar damage worldwide include *Lygocoris*, *Heterocordylus*, *Blepharidopterus* and *Atractotomus*. *Campylomma verbasci*, a European native, is now widespread in North America, and can occasionally cause extensive fruit damage when it feeds on fruitlets shortly after bloom. The rest of the growing season is spent as a predator, either in the orchard or on herbaceous weeds (e.g. *Verbascum thapsus*). A related species, *Campylomma liebknechti*, causes similar types of damage in Australia (Bone et al., 2009).

The other family that causes this type of damage is the Pentatomidae, or stink bugs. While the mirids generally attack fruit early in their development, the pentatomids are more of a problem as fruit mature in mid-to-late summer. They increase in numbers on native plants outside the orchard; as these plants shut down growth during the summer heat, a nearby irrigated orchard becomes increasingly attractive. Unlike the mirids, which leave readily visible scars, the stylet penetration by pentatomids near harvest can be nearly invisible from the surface. The flesh surrounding the sheath left by the stylets eventually turns brown so damage becomes more apparent after months of storage when the true level is discovered. Brown marmorated stink bug, *Halyomorpha halys*, can attack apple throughout the main part of the growing season, with feeding and development by both nymphs and adults. Damage occurring early in the season is most likely to result in obvious deformity, while that occurring later may be less visible from the surface. *H. halys* is highly polyphagous, and the use of multiple hosts (especially their fruiting structures) appears to be optimal to complete its development (Acebes-Doria et al., 2016).

A few Orthoptera are sporadic pests of apple, including the grasshoppers (Acrididae), katydids (Tettigoniidae) and tree crickets (Gryllidae). Although the primary hosts of grasshoppers and katydids are rangeland plants, under certain circumstances they can build to surprising numbers, and invade orchards. They are generalist feeders, and can strip foliage and chew holes in fruit.

6 Models, decision aids and monitoring

6.1 Phenology models/decision aids

The notion that temperature could be used to predict the occurrence of various life stages of codling moth was proposed nearly a century ago (Glenn, 1922). But it was not until the widespread availability of computers that phenology models became popularized, allowing the use of such models in an extension mode (Welch et al., 1978). Many of the early models targeted lepidopteran pests, which were well studied because of their importance. However, it was evident that a wide array of pests and diseases could be modelled, along with the host plant itself. The codling moth model was widely adopted because of its ease of use and accuracy; the previous rule of thumb (21 days after full bloom) suffered from frequent, often large errors in predicting the correct spray timing (Beers and Brunner, 1992). Many models began heat unit accumulation from a specified

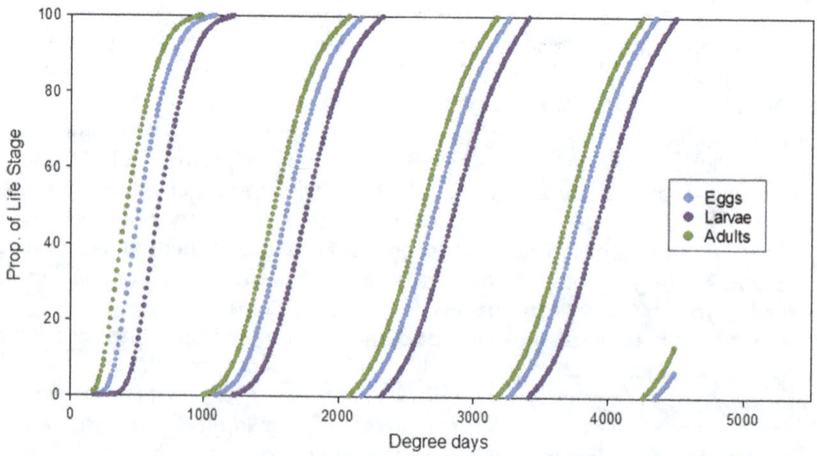

Figure 8 Codling moth phenology model output.

point in the calendar year (e.g. January 1, March 1), but a few used a biological event or 'biofix' to improve their accuracy. For lepidopteran pests, this was the first capture of an adult in a pheromone trap. However, this method may have introduced more error than it corrected, if population densities were below the detection point with pheromones. Multiple years of data resulted in the development of a biofix-free version of the model, which corrected this problem (Jones et al., 2008) (Fig. 8).

The dawn of personal computers profoundly changed the ability to deliver models to the pest management decision-maker. Even more profound is the now-ubiquitous availability of internet/web connections, allowing developers to integrate all the tools and resources a decision-maker might need in a single venue. One such example is the Decision Aid System (DAS) (https://www.decisionaid.systems/) developed by Jones et al. (2010). This system uses a network of weather stations to power temperature-driven models, and presents them to the user in an attractive and informative format. The major leap forward from the more primitive systems is the ability to interface with other databases (pesticide recommendations, biology and identification), and provide real-time site-specific information on a given pest or disease. Sixteen WSU-DAS models are currently available, including insects, diseases and storage disorders.

6.2 Monitoring/thresholds

While phenology models provide information on the best time to spray, they must be linked with two other elements to complete the decision-making process: a monitoring method to assess population densities, and a threshold to which the estimated population density may be compared. Development of the latter two tools has lagged woefully behind the development of phenology models. Population growth/prediction models have been developed but have not been widely adopted by end users; they are usually reserved for hypothesis testing in a research mode. Population growth models require a starting population from which to forecast, and determining this number is problematic. The number of biotic and abiotic factors that modulate the outcome introduces a formidable degree of complexity into such calculations. For example, pesticides applied to the orchard may have both lethal and sublethal effects, and the persistence of the residues may influence the magnitude of the toxic effect.

A wide variety of monitoring methods have been employed in apple (Beers et al., 1994). They can be loosely grouped into two categories: where the sampler goes to the pests, or the pests come to the sampler. The reason for this grouping is the amount of time, labour and money that must be invested in the sampling process.

Most sampling methods, both past and present, involve the sampler walking through an orchard. The most basic method is simply visual inspection and pest counts on a certain number of plant parts (leaves, shoots, fruits) per unit area. One variation on this theme is the presence/absence or binomial sampling scheme, where the percentage of sampling units occupied by one or more individuals is related to the average number per sampling unit. Binomial sampling can be considerably more efficient when arthropods are small and densities are large, and has been used with mites (Jones, 1990) and leafminers (Jones, 1991). A variation of presence/absence is percentage infested, but without any knowledge of how that relates to averages; this method is sometimes used with aphid-infested terminals. A third variation is a timed count; there is no specified number of sampling units, but rather all the pests observed in a specified period of time (e.g. 1–5 minutes) are recorded, giving a relative measure of density. While the loss of correspondence to a specific surface area or sampling unit is problematic, timed counts can give surprisingly informative relative differences between orchards or management regimes. This method is best suited to pests that are highly visible, and can be seen while maintaining a walking pace through the orchards (e.g. woolly apple aphid).

The beating tray or beating sheet is a useful tool for detecting insects that will fall from the tree when a limb is jarred. The efficiency of this method varies considerably with the size and behaviour of the arthropod involved. It is very effective with weevils (curculionid beetles), which are inclined to fall to the ground when disturbed. Conversely, it is very ineffective for eriophyid mites, which are too small to be easily jarred and detected on the surface of the tray. Winged adults of some species may be poor candidates for this method, as they fly rather than drop; this can be modulated to an extent by sampling at cooler temperatures, when they are less likely to fly when disturbed. If they can be dislodged, larvae may be more susceptible to beating tray sampling, such as those of generalist predators (syrphids, lacewings, lady beetles). If sampling is done during the daylight hours, then nocturnal feeders such as earwigs may be missed altogether.

While the previous methods described are *in situ*, orchard sampling often entails gathering a specified number of units and returning them to the laboratory to perform the counts. Leaves are usually the sampling unit collected because of their portability. The primary reason for laboratory inspection is when arthropods are very small or numerous, or more detailed information is required. Mites are often sampled in this manner; they can be brushed from the leaves with a machine designed for this purpose to concentrate their numbers on a single glass plate (Henderson and McBurnie, 1943). Similarly, the information sought may require dissection, for example, of a leafminer mine, to determine life stage distribution and parasitism.

While some level of visual inspection is still widely used by decision-makers, traps (where the insect is drawn to a bait, lure, light wavelength or habitat) are increasingly popular. Traps usually retain or kill the insects, accumulating them over the deployment period; thus, instead of single snapshot in time, the sampler retrieves several weeks' accumulation of the target arthropod(s). A larger area can be sampled in considerably less time, making such methods more prone to adoption. Bait traps (fermented mixtures usually containing molasses and yeast) were some of the first traps used for codling moth. While effective, they were also attractive to a wide range of pests and natural enemies, creating a problem with sorting to obtain the target species. Non-target species captures or 'by-catch' is still a problem with bait traps.

An evolutionary step in trapping came with the discovery of insect pheromones of the tortricids (Roelofs et al., 1971). Unlike bait traps, these compounds provided nearly species-specific traps, and their advantage as a monitoring tool was immediately apparent. The term pheromone is most often associated with a compound emitted by a female, whereby she attracts a male for mating; however, pheromones emitted by male are also known (e.g. iridodial for a lacewing, *Chrysopa nigricornis*). The implementation of pheromones for monitoring was made possible by advances in synthetic chemistry; in a few decades, attraction went from the use of tethered virgin females or gland extracts to synthetic blends in dispenser systems that release over a period of weeks. Concurrently, trap designs and adhesives were optimized for capture, retention and ease of use.

The success of sex pheromones led to exploration of other types of semiochemical cues. Aggregation pheromones (e.g. those of stink bugs) attract both sexes and immature stages, overcoming the limitations of single-sex traps. Combinations of compounds (Light et al., 2001) were also used to capture both sexes, with the notion that females were the more dangerous sex, thus potentially providing more pertinent information to the decision-maker.

While most trapping efforts have focused on pests, traps for monitoring natural enemies with plant volatiles are a recent innovation (James, 2003; Jones et al., 2011, 2016). These compounds were discovered by chemical ecologists, who found that damaged plants emitted a different chemical profile than undamaged plants, signalling the presence of a herbivore. Natural enemies could use this cue to their advantage to find their prey. Monitoring natural enemies is a double-edged sword: it allows the sampler to assess the presence (and hopefully impact) of natural enemies in the orchard, but, if an adhesive coated panel is used, it also kills them in the process.

The latest evolutionary step in monitoring is automated or semi-automated traps. One such approach (Spensa Z-trap) uses a pheromone-baited trap to attract the target species. Upon entering the trap, it is killed with an electric current. The large receptacle at the bottom of the trap means these traps are essentially non-saturating; only the lure has to be changed at intervals. Insect count data are sent via cellular communication to a web portal; the trapper can (in effect) monitor insects without leaving the office. A mobile app allows the user to integrate visual inspection with trap data in order to guide decision-making. A second approach (TrapView, Semios) also uses a pheromone-baited trap; in addition, a small camera mounted in the roof of the trap photographs (usually daily) the sticky panel to which the target insects are adhered. This system also uses cellular communication system to upload the information to a web portal; however, the user can examine the photographs and judge which species has been captured. The trap liners only need to be changed when they are saturated with the target species or by-catch. The advantage of such systems is in saving time and labour, and the ability to check traps at more frequent intervals than is normally used (typically 1 week). The disadvantage is the initial expense and maintenance of the systems.

7 Tools and tactics: pesticides, biological control, mating disruption and host plant resistance

7.1 Pesticides

The advent of highly effective insecticides and acaricides was both bane and blessing to the IPM practitioner. They are density-independent means of control, and are most useful when pests reach outbreak densities. However, because of their efficacy, there is a temptation to rely on them exclusively. The knowledge that there is a remedy for almost every pest that attacks apple is a disincentive to expend time and effort in developing a management programme that uses multiple tactics. While both the materials and the application involve expense for the producer, they are widely regarded as the simplest solution to a pest problem.

The issues most often cited concerning the use of insecticides are environmental contamination (soil and water), worker safety (applicators and orchard labour), and human health (residues on foods). The registration process has eliminated older materials that triggered one or more of these requirements, and the vetting process for new materials is increasingly stringent. More recently, pollinator safety has become an additional requirement, either through the intrinsic properties of the materials or through the use restrictions on the label. Ensuring that a new material can meet all safety requirements (via pre-registration testing) is a considerable expense for the registrant, and typically much of the patent life is spent in this phase. The cost must then be passed on to the purchaser, hence the increase in cost for newer materials.

7.2 Biological control

Biological control, or pest control via predators, parasitoids and pathogens, is the most popular IPM tactic among entomologists, if not producers. The study of biological control occupies a considerable amount of the entomological literature, in part because of the complexity of promoting or preserving it in a highly managed agro-ecosystem. In general, biological control is less used for direct, quarantine or disease-vectoring pests because of the low economic threshold. By far the greatest opportunity is for control of secondary pests (mites, aphids, leafminers and leafhoppers) (Fig. 9) where a certain amount of damage is tolerated. One of the precepts of biological control is that if the natural enemy exterminates its prey/host, its population will also crash. This can be addressed to a certain extent by the presence of an alternate prey/host, but the latter introduces greater complexity in the system. Biological control can be divided into several different types, described below.

Conservation biological control

Biological control in apple has emphasized conservation methods. The first aspect of conservation biological control is the minimization of non-target effects of pesticides through choice of material, rate or timing. The success of this method was embodied by integrated mite control, where changing some insecticide choices and reducing the rate of the material for key pests resulted in near-complete biological control of spider mites (Hoyt, 1969). Unfortunately, few pest problems since then have lent themselves to this simple solution. While largely undocumented, it is probable that outbreaks of leafminers

Figure 9 Lacewing larva attacking an aphid.

and leafhoppers were due to disruption of natural enemies, and that withdrawal of a few key pesticides from the programme effectively solved the problem.

A second approach to conservation biological control is providing essential resources for natural enemies, through insectary plantings on the orchard floor (ground cover) or periphery (hedgerows, extra-orchard patches) (Altieri and Letourneau, 1982). Results from these studies in apple are mixed, with some showing potential for improving biological control of pests, while others demonstrating only an increase in natural enemy abundance or diversity (Altieri and Schmidt, 1985; Smith et al., 1989; Haley and Hogue, 1990; Pfannenstiel et al., 2010; Gontijo et al., 2013). An additional consideration in orchard production is the habitat preference of the natural enemy. Generalist predators that are abundant in herbaceous ground cover may not readily attack arboreal pests (Horton et al., 2009). The conceptual framework of this approach remains intriguing, but implementation in commercial production has been limited.

Classical biological control

Classical biological control (the importation of a natural enemy from the pest's region of origin) is less used in apple production than conservation biological control. To some extent, classical biological control presupposes that the pest is exotic, and that the natural enemy provided adequate suppression of the pest in its native habitat. Classical biological control typically entails foreign exploration, along with collection and rearing of candidate species (Caltagirone, 1981). Recent experiences in non-target effects of the introduced natural enemies has made entomologists and regulatory agencies more cautious (Follett et al., 2000; Louda et al., 2003). It is not surprising, therefore, that the most notable success in classical biological control in apple is from the 1920s, before regulatory procedures were in place. This was the dissemination of *Aphelinus mali* (Fig. 10), a specialist parasitoid of woolly apple aphid native to eastern North America (Howard, 1929), to other apple-growing regions of the world infested with woolly apple aphid. While *A. mali* is hailed as one of the success stories of classical biological control (DeBach, 1962), it should be noted that woolly apple aphid remains a pest of importance in many areas despite

Figure 10 *Aphelinus mali* , a parasitoid wasp, ovipositing in a woolly apple aphid.

the presence of *A. mali*. Its effect may be reduced by pesticide use, as parasitoids are often more susceptible to pesticides than predators (Croft, 1990). However, there is also evidence that *A. mali* alone cannot regulate woolly apple aphid below damage levels and that contribution by predators (lacewings, syrphids, coccinellids and earwigs) is essential (Walker, 1985; Nicholas et al., 2005; Gontijo et al., 2015).

Innoculative, augmentative and innundative biological control

Biological control methods where natural enemies are added to the system have seen very little study or use in apple. Inoculations of a natural enemy are likely to be too slow-acting for an imminent pest problem. The need for inoculation is most likely due to the use of non-selective pesticides, and without solving this underlying problem, the technique is unlikely to succeed. However, the release of pesticide-resistant strains of predators may pay long-term dividends (Croft and Hoying, 1975; Hoy et al., 1983; Croft and MacRae, 1992). Inundation with natural enemies (where they are reared and released in large numbers as a 'biological pesticide') has been similarly unsuccessful in apple systems. Release of predatory mites to control spider mites is the one of the more recognized systems, and multiple species of phytoseiids are available for purchase. The primary difficulty is the large volume per acre occupied by the plant and the predator–prey ratio required to rapidly reduce spider mite populations. A case was made for predatory mite releases on apple seedlings in a nursery, where leaf area is greatly reduced in comparison to a mature apple orchard (Croft et al., 2004). A recent study (Schmidt et al., 2013) of release rates of *Galendromus occidentalis* for control of European red mite in a mature orchard estimated that ca. 460 000 predators/acre (1 136 200/ha) would be necessary to effect control, at a cost of ca. $6,000/acre ($14 820/ha). In a similar study on control of green apple aphid (Grasswitz and Burts, 1996), *Chrysoperla rufilabris* and *Aphidoletes aphidimyza* were released at weekly intervals at high rates, but still failed to control aphids. Given a moderate tree planting density (768 trees/ha), the combination of both predators at the rates reported in this study would have cost >US$35 000/acre ($86 450/ha). To put this in perspective, pesticide application costs are typically < $20/acre ($50/ha); the cost of an inexpensive insecticide for aphids is about $10/acre ($25/ha) or the cost of a more expensive acaricide is $50/acre ($124/ha). The relative cost of pesticides versus released predators illustrates why this technique is little used in conventional production, although in organic production, where no effective materials are available, it may be more feasible, albeit likely at release rates that will have little immediate effect. It should be noted that predator releases may be more effective inside cages, where they cannot readily disperse from the target site.

7.3 Mating disruption and the sterile insect technique (SIT)

Mating disruption and SIT (also known as SIR, or sterile insect release) are discussed under management of specific lepidopteran pests (see above 1.1–2) where they have been successfully implemented. The strengths of mating disruption are its species specificity, absence of non-target effects, poor potential for resistance development, and potential for use in combatting insecticide resistance (Welter et al., 2005). In some areas, the use of mating disruption has revolutionized apple IPM, allowing the use of selective programmes that rely on biological control to a much greater extent (Jones et al., 2009). The initial hope that it would be a stand-alone tactic has not been realized in the case of some pests, and a more practical approach of supplementing mating disruption with insecticides is more prevalent. Similarly, although the initial goal of codling moth eradication (Dyck et al., 1993) has not been realized with SIT in the British Columbia programme, it has provided the target benefits of high levels of control and reduction of pesticide inputs (Bloem et al., 2005).

While new pheromones are discovered regularly, not all of them can or will lead to successful pheromonal control. One example is *C. verbasci*, where the discovery of the pheromone (Thistlewood et al., 1989) was followed by attempts to use it for mating disruption (McBrien et al., 1997). While the experiment had limited success, both the inconsistency of the results, and the sporadic nature of this pest have limited implementation. Similarly, to date attempts to control apple clearwing moth via mating disruption have failed, although the seriousness of the pest and lack of organic control alternatives provides the impetus to continue research on this method (Judd et al., 2015). Mating disruption of apple leafcurling midge has been investigated (Cross et al., 2009), but not widely adopted as yet.

7.4 Host plant resistance

Host plant resistance is one of the least used tactics in apple IPM, especially in comparison to annual crops. The reasons are fairly evident: the long breeding cycle and the longevity of orchard plantings mean rapid response is unlikely. While pest and disease resistance have a place in breeding programmes, the top priorities are horticultural characteristics (precocity, annual bearing, high yields); eating qualities (taste and texture); and freedom from storage disorders (scald, bitter pit). There are a few examples, however (see Fazio, Chapter 6 and Kellerhals, Chapter 22), with the most notable example being that of woolly apple aphid resistance incorporated into rootstocks. While this has been helpful in suppressing aphid populations, susceptible scions help perpetuate the cycle. Some cultivars are poorer/less preferred hosts than others for certain pests, or vary in their ability to tolerate damage (Ghate and Howitt, 1965; Downing and Moilliet, 1967; Bengston, 1970; Goonewardene et al., 1976, 1980; Paiva and Janick, 1980; Hogmire and Miller, 2005). The ability to slow the population growth rate is greatly under-rated, but may be particularly useful for pests with many generations per year and a well-established natural enemy complex.

8 Challenges to the sustainability of integrated pest management (IPM)

8.1 Pesticide resistance

Since A. L. Melander famously asked the question 'Can insects become resistant to sprays?' (Melander, 1914), increasingly the answer is 'Yes'. It is appropriate that this first query concerned an apple pest, although the phenomenon has since occurred in many crops and pests. A basic tenet of ecology is co-evolution, where organisms evolve in response to a stimulus from biotic or abiotic factors. After climate and natural enemies, pesticides are the predominant (abiotic) factor in many agro-ecosystems. Apple culture has been associated with a relatively intensive use of pesticides, and predictably, with multiple instances of resistance (Croft, 1982). The speed of resistance development depends on a number of factors, including the number of generations per year, the biology of the pest, and the type of insecticide. Increasingly, prior exposure to various modes of action also influences the speed of resistance development through cross-resistances or multiple resistances (Dennehy et al., 1987; Fergusson-Kolmes et al., 1991; Flexner et al., 1995).

Spider mites are often the poster child for resistance development; with 10–11 generations/year, they have more opportunities to be exposed to insecticides and acaricides, and their populations selected for resistance. Despite the successful implementation of integrated control, the United States has suffered from a history of resistance development. Outbreaks of mites were exacerbated by the use of DDT (Baker, 1952) during the same era when many specific acaricides were being developed. OPs were highly toxic to mites when first introduced, although resistance was selected for fairly rapidly (Hoyt, 1962; Forsythe, 1963). A series of organochlorine acaricides were introduced; in some cases, field failures occurred in just a few seasons' use (Hoyt, 1962). The implementation of integrated mite control in the 1970s (Hoyt, 1969) slowed the rate of resistance development (Croft and Hoyt, 1983), but the trend continued with the organotins (Edge and Phimister, 1986; Croft et al., 1987; Pree and Wagner, 1987; Welty et al., 1987, 1989; Hoy et al., 1988) and the ovicides (Rathman et al., 1990; Reissig and Hull, 1991; Thwaite, 1991; Grosscurt et al., 1994). For the first time, reversion was seriously explored as a means to prolong the life of a miticide (Flexner et al., 1988, 1989; Welty et al., 1989; Beers et al., 1998), but a new mode of action was used whenever it was available.

Codling moth has been subject to ca. 125 years of intensive pesticide management, and unsurprisingly, instances of insecticide resistance have occurred. Codling moth became resistant to lead arsenate (Hough, 1928), DDT (Barnes, 1958; Barnes and Moffitt, 1963), and (to a greater or lesser extent) the OPs (Knight et al., 1994). There are also documented cases of cross-resistance to various modes of action used against this pest (Sauphanor and Bouvier, 1995; Dunley and Welter, 2000; Bouvier et al., 2002). However, resistance development for the most part has occurred more slowly than in spider mites, with a time course of decades rather than years. All things considered, azinphos methyl was a remarkably long-lived product. It was used in Washington from the 1950s until its registration was cancelled in 2012, and was still moderately effective in many areas.

There is a general sentiment that 'soft' control materials and tactics will be exempt from resistance. Petroleum oil has been used for over a century, and is still effective today on apple pests. Unexpectedly, there are documented cases of resistance to the biorational compound, codling moth granulosis virus (Lacey et al., 2008), when it was used as the primary tactic over many years. The potential for resistance to mating disruption has been speculated upon (Welter et al., 2005), but there is no evidence in tree fruit pests to date.

8.2 Climate change

It is difficult to generalize about the effects of warmer temperatures without some information on when (summer/winter, daytime/night-time) the temperature changes occur. Many insects' diapause is triggered by a combination of temperature and photoperiod, thus an opportunity is created for an increase in the number of generations where temperature is currently a limiting factor. Specifically, warmer temperatures in the spring could accelerate insect development in the early part of the growing season, allowing time for additional generations before photoperiod becomes critical. There is already evidence that codling moth in Washington State is shifting from two generations (plus a partial third) per year to three full generations (V. Jones, pers. comm.). This only affects the later-maturing cultivars of apples (October–November harvest), which were typically less impacted by the partial third generation. Early maturing cultivars such as 'Gala', which are harvested in August, need not be managed differently. Less severe winters in the higher latitudes could increase winter survival of some pests; others are hardy to about the same extent as the apple tree and buds, and would not be affected except by a freeze that also damaged the trees. Extremely high summer temperatures (in excess of 35–40 °C) for sustained periods of time could theoretically slow developmental rate. Higher night-time temperatures may enable faster degree day accumulation (and thus speed insect development). Mating may be limited by temperatures only during a specific part of the diurnal cycle (dawn and dusk), thus more favourable temperatures during this period might enhance mating success. The same may be said of rain events, which also limit mating; a dry, moderate climate would promote mating more than a cold, wet one. Pests such as mites and aphids turn over multiple generations each year, and with earlier springs they could presumably have even more generations. However, their population growth is more likely to be limited by host plant quality or biological control.

8.3 Invasive species

It should be noted at the outset that one region's invasive species is another region's native species, and vice versa. The apple is not native to North America, so many apple pests in many growing regions are by definition invasive – it is merely a question of how long ago the invasion happened. The pace of invasion has increased with globalization of trade (Meyerson and Mooney, 2007), but colonization or trade have been consistent factors (Hulme, 2009). Quarantines help slow the impact of trade, but have not eliminated it (Mack et al., 2000). The reverse situation is when a crop is imported into a new region and a native insect switches host (e.g. woolly apple aphid, apple maggot) (Wellhouse, 1920). Regardless of whether the insect is brought to the crop, or the crop is brought to the insect, a new pest may be created.

Several cases illustrate the process of invasion. Codling moth originated in Asia Minor and was likely transported in infested fruits to new areas, including Europe, South America, Australia and New Zealand. Codling moth was thought to be introduced into North America from Europe with settlers in the mid-1700s, and spread along with apple cultivation from the east coast to the west (Barnes, 1991). Much of the invasion was accomplished long before quarantines were conceived of, let alone set in place.

A second case is that of a native species that adapts to using apple as a host. Woolly apple aphid is native to North America, with American elm (*Ulmus*) and hawthorn (*Crataegus*) or mountain ash (*Sorbus*) as native (alternate) hosts. Sexual reproduction occurred on the overwintering *Ulmus* host (Patch, 1912), while reproduction on the rosaceous host was parthenogenetic. When cultivated apple (another rosaceous species) was imported, woolly apple aphid adapted to use it as the alternate host in place of *Crataegus* and then further adapted to living on apple year round, eliminating the sexual reproductive phase (Hoyt and Madsen, 1960). This species has been remarkably successful despite the loss of the opportunity provided by genetic recombination via sexual reproduction. Woolly apple aphid spread throughout the world wherever apples were grown (Theobald, 1921) and is an intermittently serious pest in many of those regions. Movement was likely with infested trees; aphid colonies can survive on both roots and shoots.

Similarly, apple maggot is native to North America, and it also used native hawthorns as its primary host prior to the arrival of cultivated apples (Ali Niazee and Penrose, 1981). Spread of apple maggot has been much slower, but can also occur via infested fruit. In contrast to woolly apple aphid, which had spread to much of the world by the early part of the twentieth century, apple maggot has been spreading at glacial speed within North America alone, apparently only adapting to *Malus* after 200 years of exposure (Bush, 1966). This species most likely moves with infested fruit in the larval stage, with a slow extension of range by the adults.

The term 'invasive species' became popularized relatively recently, underscoring the urgency related to dealing with the increasing number of new pests (Mack et al., 2000). A recent example is brown marmorated stink bug (*H. halys*), a pentatomid bug native to Asia (Leskey et al., 2012). Pome fruits, including apples, are among the high-risk crops, and its spread across North America and throughout Europe has occurred in a relatively short time span. *H. halys* adults are the long-distance dispersive stage; they attach themselves to cargo or vehicles and get transported for hundreds or thousands of kilometres. Transport across the breadth of North America (from the first detection in Pennsylvania in 1996 to Oregon in 2004, a linear distance of ca. 3,800 km) took only eight years. In contrast to *H. halys*, it took the apple maggot almost 30 years to move from Portland, Oregon to Wenatchee, Washington, a linear distance of only 277 km. Clearly, there are great differences between the mobility and invasion potential of these two species. While many biological factors are in play, the extreme polyphagy and 'hitchhiker' behaviour of *H. halys* have contributed to its rapid dissemination, while the near-monophagy and weak adult dispersal of apple maggot has helped slow its spread.

More disturbing at a global scale is the deep reservoir of species that can be transported to areas where apples are grown, where they do not currently occur. Given the propensity of a species to survive with a suitable host in a favourable environment, a greater array of pests appears to await us in the future. While the native environment of some apple-growing areas (e.g. the arid western US) may seem unhospitable to many pest species, the microclimate modification brought about by horticultural operations (irrigation, shading) may tip the balance in favour of otherwise excluded species.

Invasive pests typically disrupt IPM programmes that may have taken decades to develop (Leskey et al., 2012). While it is understood that at all times IPM must adapt to changing conditions, the rate of change may outstrip the resources devoted to their study. One of the challenges to sustainability must therefore be matching the rate of change to the rate of solutions.

8.4 Societal factors

Many of the factors that challenge sustainable apple IPM may be broadly grouped as 'societal', or external to the agricultural production system. In a relatively short amount of time, agriculture has changed from a means of subsistence or a family occupation, to agriculture as a business and an important component of world trade. This has radically changed the standards to which the crop must be produced, in terms of eating quality, cosmetic standards, phytosanitary regulations and pesticide residues. More stringent standards place more constraints on IPM systems. Consumers in developed countries are increasingly concerned about the safety and quality of their food, and the health and safety of agricultural workers. Their concerns are translated into export standards; failure to meet a standard will close that market to a producer wishing to sell there. While increased governmental regulation plays a role, many IPM decisions are market-driven rather than regulatory-driven; in either case, much of the initiative is taken out of the hands of the IPM practitioner. At the same time, we are increasing production to meet the increasing demand for food, which must be accomplished in part by greater production efficiencies. For apples, this entails concentrating production in the areas most favourable for productivity, along with access to human resources (labour) and infrastructure (packing and shipping facilities). This concentration reduces landscape-level diversity in apple-growing regions and tends to further destabilize agro-ecosystems, in direct conflict with the application of ecological principles to IPM.

9 Where to look for further information

Much of the practical information on apple IPM can be found on websites of land-grant universities, including Washington State (http://treefruit.wsu.edu/), California (http://ipm.ucanr.edu/PMG/selectnewpest.apples.html), New York (https://nysipm.cornell.edu/agriculture/fruits/publications), Michigan (http://www.ipm.msu.edu/agriculture/fruit) and the Mid-Atlantic (http://extension.psu.edu/plants/tree-fruit/insects-mites). Many of these states also publish crop protection guides annually (Washington – http://www.tfrec.wsu.edu/pages/cpg/); Pennsylvania – (http://extension.psu.edu/plants/tree-fruit/tfpg). Washington has two additional websites, one with biology, identification and management of pests (http://jenny.tfrec.wsu.edu/opm/), and one with a DAS (subscription only) (https://www.decisionaid.systems/). The International Organization for Biological and Integrated Control (IOBC) West Palearctic Regional Section (WPRS) is a rich source of information on biocontrol-based IPM; while many crops are covered, the tree fruit group is very active. The conference proceedings are published as a series of bulletins (https://www.iobc-wprs.org/). Apple IPM is reviewed in a book chapter (Beers et al., 2003; see references), and many reviews on aspects of IPM pertaining to apple can be found in the Annual Review of Entomology (http://www.annualreviews.org/journal/ento). The Entomological Society of America has several journals that include articles on IPM, viz, the Journal of Economic Entomology and Environmental Entomology. A newer journal, the Journal of Integrated Pest Management (http://jipm.oxfordjournals.org/), contains review articles on the state of IPM practice in various crops. Many European journals also include articles on tree fruit IPM (e.g. Crop Protection, Pest Management Science, Experimental and Applied Entomology). Proceedings of national or regional conferences can also be a good source on current thinking and practice of tree fruit IPM: Orchard Pest and Disease Management Conference (http://www.tfrec.wsu.edu/pages/wopdmc/; North America, with some international participation), and the New Zealand Plant Protection Conference (https://www.nzpps.org/conference.php).

10 Acknowledgements

I thank my colleagues for conversations sharing their experience and perspective on many aspects of apple IPM, including Art Agnello (Cornell University), Vince Jones (Washington State University), Howard Thistlewood (Ag Canada, Summerland, BC), Wee Yee (USDA-ARS Wapato, WA) and David Granatstein (Washington State University). Special thanks to Larry Gut for his thoughtful review of the mating disruption sections.

11 References

Acebes-Doria, A. L., Leskey, T. C. and Bergh, J. C. (2016), 'Host plant effects on *Halyomorpha halys* (Hemiptera: Pentatomidae) nymphal development and survivorship', *Environ. Entomol.*, 45(3), 663–70.

Agnello, A. (1999), 'Apple-boring beetles', Cornell Cooperative Extension, Tree Fruit Fact Sheet 102GFSTF-I26.

Agnello, A. M., Reissig, W. H., Spangler, S. M., Charlton, R. E. and Kain, D. P. (1996), 'Trap response and fruit damage by obliquebanded leafroller (Lepidoptera: Tortricidae) in pheromone-treated apple orchards in New York', *Environ. Entomol.*, 25(2), 268–82.

Ali Niazee, M. T. and Penrose, R. L. (1981), 'Apple maggot in Oregon: a possible new threat to the Northwest apple industry', *Bull. Entomol. Soc. Amer.*, 27(4), 245–6.

Altieri, M. A. and Letourneau, D. K. (1982), 'Vegetation management and biological control in agroecosystems', *Crop Prot.*, 1(4), 405–30.

Altieri, M. A. and Schmidt, L. L. (1985), 'Cover crop manipulation in northern California orchards and vineyards: effects on arthropod communities', *Biol. Agric. Hortic.*, 3, 1–24.

Baker, H. (1952), 'Spider mites, insects, and DDT', *Yearbook of Agriculture*, pp. 562–6.

Barcenas, N. M., Unruh, T. R. and Neven, L. G. (2005), 'DNA diagnostics to identify internal feeders (Lepidoptera: Tortricidae) of pome fruits of quarantine importance', *J. Econ. Entomol.*, 98(2), 299–306.

Barnes, M. M. (1958), 'Technique for testing insecticide deposits with newly hatched codling moth larvae', *J. Econ. Entomol.*, 51(4), 547–9.

Barnes, M. M. and Moffitt, H. R. (1963), 'Resistance to DDT in the adult codling moth and reference curves for Guthion and Carbaryl', *J. Econ. Entomol.*, 56(6), 722–5.

Barnes, M. M. (1991), Codling moth occurrence, host race formation, and damage, In: Van Der Geest, L. P. S. and Evenhuis, H. H. (eds), *Tortricid Pests, their Biology, Natural Enemies and Control*, Elsevier Science, pp. 313–27.

Beers, E. H. and Brunner, J. F. (1992), 'Implementation of the codling moth phenology model on apple in Washington state, USA', *Acta Phytopathol. Entomol. Hung.*, 27, 97–102.

Beers, E. H., Hull, L. A. and Jones, V. P. (1994), Sampling pest and beneficial arthropods of apple, In: Pedigo, L. A. and Buntin, G. D. (eds), *Handbook of Sampling Methods for Arthropods in Agriculture*, CRC Press, pp. 383–416.

Beers, E. H., Elsner, E. A. and Drake, S. R. (1995), 'White apple leafhopper (Homoptera: Cicadellidae) effect on fruit size, quality, and return bloom of apple', *J. Econ. Entomol.*, 88(4), 973–8.

Beers, E. H., Riedl, H. and Dunley, J. E. (1998), 'Resistance to abamectin and reversion to susceptibility to fenbutatin oxide in spider mite (Acari: Tetranychidae) populations in the Pacific Northwest', *J. Econ. Entomol.*, 91(2), 352–60.

Beers, E. H., Suckling, D. M., Prokopy, R. J. and Avilla, J. (2003), Ecology and management of apple arthropod pests, In: Ferree, D. C. and Warrington, I. J. (eds), *Apples: Botany, Production and Uses*, CAB International, pp. 489–519.

Beers, E. H. and Schmidt, R. A. (2014), 'Impacts of orchard pesticides on *Galendromus occidentalis*: Lethal and sublethal effects', *Crop Prot.*, 56, 16–24.

Bengston, M. (1970), 'Effect of different varieties of the apple host on the development of *Tetranychus urticae* (Koch)', *Queensland J. Agr. Animal Sci.*, 530(1), 95–114.

Bloem, K. A., Bloem, S. and Carpenter, J. E. (2005), *Impact of moth suppression/eradication programmes using the sterile insect technique or inherited sterility. Chap. 7.3.*, In: Dyck, V. A. and Robinson, A. S. (eds), *Sterile Insect Technique. Principles and Practice in Area-Wide Integrated Pest Management*, Springer, pp. 677–700.

Bone, N. J., Thomson, L. J., Ridland, P. M., Cole, P. and Hoffmann, A. A. (2009), 'Cover crops in Victorian apple orchards: effects on production, natural enemies and pests across a season', *Crop Prot.*, 28, 675–83.

Bostanian, N. J., Thistlewood, H. M. A., Hardman, J. M., Laurin, M.-C. and Racette, G. (2009), 'Effect of seven new orchard pesticides on *Galendromus occidentalis* in laboratory studies', *Pest Manag. Sci.*, 65, 635–9.

Bouvier, J.-C., Boivin, T., Beslay, D. and Sauphanor, B. (2002), 'Age-dependent response to insecticides and enzymatic variation in susceptible and resistant codling moth larvae', *Arch. Insect Biochem. Physiol.*, 51, 55–66.

Brunner, J. F. and Rice, R. (1993), *Oriental fruit moth*, In: Beers, E. H., Brunner, J. F., Willett, M. J. and Warner, G. M. (eds), *Orchard Pest Management: A Resource Guide for the Pacific Northwest*, The Good Fruit Grower, Yakima, WA, pp. 69–71.

Brunner, J. F. (1994), 'Using Bt products as tools in pest control', *Good Fruit Grower*, 45(15), 34–8.

Brunner, J. F., Doerr, M. and Smith, L. (1995), '*Bacillus thuringiensis* products and leafroller control in Washington apple orchards', *Proceedings, 90th Annual Meeting of the Washington State Horticultural Association*, Yakima, WA, 5–7 December 1994, pp. 102–8.

Brunner, J. F., Welter, S., Calkins, C., Hilton, R., Beers, E. H., Dunley, J. E., Unruh, T., Knight, A., Van Steenwyk, R. and VanBuskirk, P. (2002), 'Mating disruption of codling moth: a perspective from the Western United States', *IOBC WPRS Bulletin*, 25(9), 207–15.

Bush, G. L. (1966), 'The taxonomy, cytology, and evaluation of the genus *Rhagoletis* in North America (Diptera: Tephritidae)', *Bull. Mus. Comp. Zool. Harvard Univ.*, 134, 431–562.

Caltagirone, L. E. (1981), 'Landmark examples in classical biological control', *Annu. Rev. Entomol.*, 26, 213–32.

Cardé, R. T. and Minks, A. K. (1995), 'Control of moth pests by mating disruption: successes and constraints', *Annu. Rev. Entomol.*, 40(1), 559–85.

CDFA (2015),'Apple leaf gall midge', http://blogs.cdfa.ca.gov/Section3162/?tag=apple-leaf-gall-midge. (accessed 12 November 2016).

Croft, B. A. and Hoying, S. A. (1975), 'Carbaryl resistance in native and released populations of *Amblyseius fallacis*', *Environ. Entomol.*, 4(6), 895–8.

Croft, B. A. (1982), 'Arthropod resistance to insecticides: a key to pest control failures and successes in North American apple orchards', *Entom. Exp. Appl.*, 31, 88–110.

Croft, B. A. and Hoyt, S. C. (1983),'Integrated management of insect pests of pome and stone fruits'. John Wiley and Sons, New York, NY.

Croft, B. A., Hoyt, S. C. and Westigard, P. H. (1987), 'Spider mite management on pome fruits, revisited: organotin and acaricide resistance management', *J. Econ. Entomol.*, 80(2), 304–11.

Croft, B. A. (1990), *Arthropod Biological Control Agents and Pesticides*, John Wiley and Sons, New York, NY.

Croft, B. A. and MacRae, I. V. (1992), 'Persistence of *Typhlodromus pyri* and *Metaseiulus occidentalis* (Acari: Phytoseiidae) on apple after inoculative release and competition with *Zetzellia mali* (Acai: Stigmaeidae)', *Environ. Entomol.*, 21(5), 1168–77.

Croft, B. A., Pratt, P. D. and Luh, H.-K. (2004), 'Low-density releases of *Neoseiulus fallacis* provide for rapid dispersal and control of Tetrnaychus urticae (Acari: Phytoseiidae, Tetranychidae) on apple seedlings', *Exp. Appl. Acarol.*, 33(4), 327–39.

Cross, J. V. and Hall, D. R. (2009), 'Exploitation of the sex pheromone of apple leaf midge *Dasineura mali* Kieffer (Diptera: Cecidomyiidae) for pest monitoring: Part 1. Development of lure and trap', *Crop Prot.*, 28(2), 139–44.

Cross, J. V., Hall, D. R., Shaw, P. and Anfora, G. (2009), 'Exploitation of the sex pheromone of apple leaf midge *Dasineura mali* Kieffer (Diptera: Cecidomyiidae): Part 2. Use of sex pheromone traps for pest monitoring', *Crop Prot.*, 28(2), 128–33.

DeBach, P. (1962), 'An analysis of successes in biological control of insects in the Pacific Area', *Proc. Hawaiian Entomol. Soc.*, 18(1), 69–79.

Deland, J. P., Judd, G. J. R. and Roitberg, B. D. (1994), 'Disruption of pheromone communication in three sympatric leafroller (Lepidoptera: Tortricidae) pests of apple in British Columbia', *Environ. Entomol.*, 23(5), 1084–90.

Dennehy, T. J., Granett, J., Leigh, T. F. and Colvin, A. (1987), 'Laboratory and field investigations of spider mite (Acari: Tetranychidae) resistance to the selective acaricide propargite', *J. Econ. Entomol.*, 80(3), 565–74.

Downing, R. S. and Moilliet, T. K. (1967), 'Relative densities of predacious and phytophagous mites on three varieties of apple trees', *Can. Entomol.*, 99, 738–41.

Dunley, J. and Welter, S. C. (2000), 'Correlated insecticide cross-resistance in azinphosmethyl resistant codling moth (Lepidoptera: Tortricidae)', *J. Econ. Entomol.*, 93(3), 955–62.

Dyck, V. A., Graham, S. H. and Bloem, K. A. (1993), 'Implementation of the sterile insect release programme to eradicate the codling moth, *Cydia pomonella* (L.) (Lepidoptera: Olethreutidae), in British Columbia, Canada', *Management of insect pests: nuclear and related molecular and genetic techniques*, Vienna, Austria, 19–23 October 1992, International Atomic Energy Agency, pp. 285–97.

Edge, V. E. and Phimister, J. (1986), 'Organotin resistance in the two-spotted mite in Australia ', *Down to Earth*, 42(1), 18–20.

Epstein, D. L., Stelinski, L. L., Reed, T. P., Miller, J. R. and Gut, L. J. (2006), 'Higher densities of distributed pheromone sources provide disruption of codling moth (Lepidoptera: Tortricidae) superior to that of lower densities of clumped sources', *J. Econ. Entomol.*, 99(4), 1327–33.

Ferguson, H. J., Neven, L. G., Thibault, S. T., Mohammed, A. and Fraser, M. (2011), 'Genetic transformation of the codling moth, *Cydia pomonella* L., with piggyBac EGFP', *Transgenic Res.*, 20(1), 201–14.

Fergusson-Kolmes, L. A., Scott, J. G. and Dennehy, T. J. (1991), 'Dicofol resistance in *Tetranychus urticae* (Acari: Tetranychidae): cross-resistance and pharmacokinetics', *J. Econ. Entomol.*, 84(1), 41–8.

Flexner, J. L., Westigard, P. H. and Croft, B. A. (1988), 'Field reversion of organotin resistance in the twospotted spider mite (Acari: Tetranychidae) following relaxation of selection pressure', *J. Econ. Entomol.*, 81(6), 1516–20.

Flexner, J. L., Theiling, K. M., Croft, B. A. and Westigard, P. H. (1989), 'Fitness and immigration: factors affecting reversion of organotin resistance in the twospotted spider mite (Acari: Tetranychidae)', *J. Econ. Entomol.*, 82(4), 996–1002.

Flexner, J. L., Westigard, P. H., Hilton, R. and Croft, B. A. (1995), 'Experimental evaluation of resistance management for twospotted spider mite (Acari: Tetranychidae) on southern Oregon pear: 1987–93', *J. Econ. Entomol.*, 88(6), 1517–24.

Follett, P. A., Duan, J., Messing, R. H. and Jones, V. P. (2000), 'Parasitoid drift after biological control introductions: re-examining Pandora's box', *Amer. Entomol.*, 46(2), 82–94.

Forsythe, H. Y. (1963), '*European red mite*, Tedion, and resistance', *Pesticide Institute News*, 17(1), 1–2.

Garczynski, S. (2017), 'CRISPR/Cas9 genome editing of a codling moth receptor causes female sterility'. *91st Orchard Pest and Disease Management Conference*, 11–13 January 2017, Portland, OR, 91, 11.

Ghate, A. V. and Howitt, A. J. (1965), 'Mite resistance to organophosphorous compounds and the response of apple varieties to mites in Michigan', *Michigan Ag. Exp. Sta. Quart.*, Bull. No. 47.

Glass, E. H. and Lienk, S. E. (1971), 'Apple insect and mite populations developing after discontinuance of insecticides: 10-year record', *J. Econ. Entomol.*, 64(1), 23–6.

Glenn, P. A. (1922), 'Relation of temperature to development of the codling-moth', *J. Econ. Entomol.*, 15(3), 193–9.

Gontijo, L., Beers, E. H. and Snyder, W. E. (2013), 'Flowers promote aphid suppression in apple orchards', *Biol. Control*, 66(1), 8–15.

Gontijo, L., Beers, E. H. and Snyder, W. E. (2015), 'Complementary suppression of aphids by predators and parasitoids', *Biol. Control*, 90, 83–91.

Goonewardene, H. F., Williams, E. B., Kwolek, W. F. and McCabe, L. D. (1976), 'Resistance to European red mite, *Panonychus ulmi* (Koch), in apple', *J. Amer. Soc. Hortic. Sci.*, 101(5), 532–7.

Goonewardene, H. F., Kwolek, W. F., Dayton, D. F. and Hayden, R. A. (1980), 'Preference of the European red mite for strains of 'Delicious' apple with differences in leaf pubescence', *J. Econ. Entomol.*, 73, 101–3.

Grasswitz, T. and Burts, E. (1996), 'Effect of native natural enemies and augmentative releases of *Chrysoperla rufilabris* Burmeister and *Aphidoletes aphidimyza* (Rondani) on the population dynamics of the green apple aphid, *Aphis pomi* De Geer', *Intl. J. Pest Mgmt.*, 41(3), 176–83.

Grosscurt, A. C., Wixley, R. A. J. and ter Haar, M. (1994), 'Cross-resistance between flucycloxuron, clofentezine and hexythiazox in *Panonychus ulmi* (fruit tree red spider mite)', *Exp. Appl. Acarol.*, 18, 445–58.

Haley, S. and Hogue, E. J. (1990), 'Ground cover influence on apple aphid, *Aphis pomi* DeGeer (Homoptera: Aphididae), and its predators in young apple orchard', *Crop Prot.*, 9, 225–30.

Hamilton, G. C., Swift, F. C. and Marini, R. (1986), 'Effect of *Aphis pomi* (Homoptera: Aphididae) density on apples', *J. Econ. Entomol.*, 79(2), 471–8.

Henderson, C. F. and McBurnie, H. V. (1943), 'Sampling technique for determining populations of the citrus red mite and its predators', United States Department of Agriculture, Circular No. 671.

Hogmire, H. W. and Miller, S. S. (2005), 'Relative susceptibility of new apple cutlivars to arthropod pests', *HortSci.*, 40(7), 2071–5.

Horton, D., Jones, V. P. and Unruh, T. R. (2009), 'Use of a new immunmarking method to assess movement by generalist predators', *Amer. Entomol.*, 55(1), 49–56.

Hough, W. S. (1928), 'Relative resistance to arsenical poisoning of two codling moth strains', *J. Econ. Entomol.*, 21, 325–9.

Howard, L. O. (1929), '*Aphelinus mali* and its travels', *Ann. Entomol. Soc. Amer.*, 22(3), 311–68.

Hoy, M. A., Westigard, P. H. and Hoyt, S. C. (1983), 'Release and evaluation of a laboratory-selected, pyrethroid-resistant strain of the predaceous mite *Metaseiulus occidentalis* (Acari: Phytoseiidae) in southern Oregon pear orchards and a Washington apple orchard', *J. Econ. Entomol.*, 76(2), 383–8.

Hoy, M. A., Conley, J. and Robinson, W. (1988), 'Cyhexatin and fenbutatin-oxide resistance in Pacific spider mite (Acari: Tetranychidae): stability and mode of inheritance', *J. Econ. Entomol.*, 81(1), 57–64.

Hoyt, S. C. and Madsen, H. F. (1960), 'Dispersal behavior of the first instar nymphs of the woolly apple aphid', *Hilgardia*, 30(10), 267–99.

Hoyt, S. C. (1962), 'McDaniel mite control, problems and promise', *Proceedings, 58th Annual Meeting of the Washington State Horticultural Association*, 58, 57–60.

Hoyt, S. C. (1969), 'Integrated chemical control of insects and biological control of mites on apple in Washington', *J. Econ. Entomol.*, 62, 74–86.

Hulme, P. E. (2009), 'Trade, transport and trouble: managing invasive species pathways in an era of globalization', *J. Appl. Ecol.*, 46, 10–18.

Irvin, N., Wratten, S., Chapman, R. and Frampton, C. (1999), 'Effects of floral resources on fitness of the leafroller parasitoid (*Dolichogenidea tasmanica*) in apples', *Proceedings of the New Zealand Plant Protection Conference*, 84–8.

Irvin, N. A., Scarratt, S. L., Wratten, S. D., Frampton, C. M., Chapman, R. B. and Tylianakis, J. M. (2006), 'The effects of floral understoreys on parasitism of leafrollers (Lepidoptera: Tortricidae) on apples in New Zealand', *Agric. Forest Entomol.*, 8(1), 25–34.

James, D. G. (2003), 'Synthetic herbivore-induced plant volatiles as field attractants for beneficial insects', *Environ. Entomol.*, 32(5), 977–82.

Jones, V. P. (1990), 'Sampling and dispersion of the twospotted spider mite (Acari: Tetranychidae) and the western orchard predatory mite (Acari: Phytoseiidae) on tart cherry', *J. Econ. Entomol.*, 83(4), 1376–80.

Jones, V. P. (1991), 'Binomial sampling plans for tentiform leafminer (Lepidoptera: Gracillaridae) on apples in Utah', *J. Econ. Entomol.*, 84(2), 484–8.

Jones, V. P., Doerr, M. D. and Brunner, J. F. (2008), 'Is biofix necessary for predicting codling moth (Lepidoptera: Tortricidae) emergence in Washington State apple orchards?', *J. Econ. Entomol.*, 101(5), 1651–7.

Jones, V. P., Unruh, T. R., Horton, D. R., Mills, N. J., Brunner, J. F., Beers, E. H. and Shearer, P. W. (2009), 'Tree fruit IPM programs in the Western United States: The challenge of enhancing biological control through intensive management', *Pest Manag. Sci.*, 65, 1305–10.

Jones, V. P., Brunner, J. F., Grove, G. G., Petit, B., Tangren, G. V. and Jones, W. E. (2010), 'A web-based decision support system to enhance IPM programs in Washington tree fruit', *Pest Manag. Sci.*, 66(6), 587–95.

Jones, V. P., Steffan, S., Wiman, N. G., Horton, D. R., Miliczky, E. R., Zhang, Q.-H. and Baker, C. C. (2011), 'Evaluation of herbivore-induced plant volatiles for monitoring green lacewings in Washington apple orchards', *Biol. Control*, 56(1), 98–105.

Jones, V. P., Horton, D. R., Mills, N. J., Unruh, T. R., Baker, A. C., Melton, T. D., Miliczky, E. R., Steffan, S., Shearer, P. W. and Amarasekare, K. G. (2016), 'Evaluating plant volatiles for monitoring natural enemies in apple, pear and walnut orchards', *Biol. Control*, 102, 53–65.

Judd, G. J. R., Aurelian, M., Eby, C., Evenden, M. L., Kwon, J., Gries, R. and Gries, G. (2015), 'Towards semiochemical-based management of apple clearwing moth: decade of experimentation'. *89th Orchard Pest and Disease Management Conference*. Portland, OR, p. 33.

Knight, A. (2004), 'Managing codling moth (Lepidoptera: Tortricidae) with an internal grid of either aerosol puffers or dispenser clusters plus border applications of individual dispensers ', *J. Entomol. Soc. B. C.*, 101, 69–77.

Knight, A. L., Brunner, J. F. and Alston, D. (1994), 'Survey of azinphosmethyl resistance in codling moth (Lepidoptera: Tortricidae) in Washington and Utah', *J. Econ. Entomol.*, 87(2), 285–92.

Knight, A. L., Thomson, D. R. and Cockfield, S. D. (1998), 'Developing mating disruption of obliquebanded leafroller (Lepidoptera: Tortricidae) in Washington state', *Environ. Entomol.*, 27(5), 1080–8.

Lacey, L. A., Thomson, D. R., Vincent, C. and Arthurs, S. P. (2008), 'Codling moth granulovirus: a comprehensive review', *Biocontr. Sci. Tech.*, 18(7), 639–63.

Lawson, D. S., Reissig, W. H., Agnello, A. M., Nyrop, J. P. and Roelofs, W. L. (1996), 'Interference with the mate-finding communication system of the obliquebanded leafroller (Lepidoptera: Tortricidae) using synthetic sex pheromones', *Environ. Entomol.*, 25(5), 895–905.

Lefebvre, M., Bostanian, N. J., Thistlewood, H. M. A., Mauffette, Y. and Racette, G. (2011), 'A laboratory assessment of the toxic attributes of six 'reduced risk insecticides' on *Galendromus occidentalis* (Acari: Phytoseiidae)', *Chemosphere*, 84(1), 25–30.

Leskey, T. C., Hamilton, G. C., Nielsen, A. L., Polk, D. F., Rodriguez-Saona, C., Bergh, J. C., Herbert, D. A., Kuhar, T. P., Pfeiffer, D., Dively, G. P., Hooks, C. R. R., Raupp, M. J., Shrewsbury, P. M., Krawczyk, G., Shearer, P. W., Whalen, J., Koplinka-Loehr, C., Myers, E., Inkley, D., Hoelmer, K. A., Lee, D.-H. and Wright, S. E. (2012), 'Pest status of the brown marmorated stink bug, *Halyomorpha halys* in the USA', *Outlooks on Pest Management*, 23(5), 218–26.

Leskey, T. C., Short, B. D., Butler, B. R. and Wright, S. E. (2012), 'Impact of the invasive brown marmorated stink bug, *Halyomorpha halys* (Stål), in mid-Atlantic tree fruit orchards in the United States: case studies of commercial management', *Psyche*, 1–14.

Light, D. M., Knight, A. L., Henrick, C. A., Rajapaska, D., Lingren, B., Dickens, J. C., Reynolds, K. M., Buttery, R. G., Merrill, G., Roitman, J. and Campbell, B. C. (2001), 'A pear-derived kairomone with pheromonal potency that attracts male and female codling moth, *Cydia pomonella* (L.)', *Naturwissenschaften*, 88(8), 333–8.

Louda, S. M., Pemberton, R. W., Johnson, M. T. and Follett, P. A. (2003), 'Nontarget effects – the Achilles' heel of biological control? Retrospective analyses to reduce risk associated with biocontrol introductions', *Annu. Rev. Entomol.*, 48, 365–96.

Mack, R. N., Simberloff, D., Lonsdale, W. M., Evans, H., Clout, M. and Bazzaz, F. (2000), 'Biotic invasions: causes, epidemiology, global consequences and control', *Issues in Ecology*, (Issue #5, pp. 1-20). https://www.esa.org/esa/science/issues/.

Marec, F., Neven, L. G., Robinson, A. S., Vreysen, M., Goldsmith, M. R., Nagaraju, J. and Franz, G. (2005), 'Development of genetic sexing strains in Lepidoptera: from traditional to transgenic approaches', *J. Econ. Entomol.*, 98(2), 248–59.

McBrien, H. L., Judd, G. J. R. and Borden, J. H. (1997), 'Population suppression of *Campylomma verbasci* (Heteroptera: Miridae) by atmospheric permeation with synthetic sex pheromone', *J. Econ. Entomol.*, 90(3), 801–8.

McGhee, P., Gut, L. J. and Miller, J. R. (2014), 'Aerosol emitters disrupt codling moth, *Cydia pomonella*, competitively', *Pest Manag. Sci.*, 70, 1859–62.

McGhee, P. S., Miller, J. R., Thomson, D. R. and Gut, L. J. (2016), 'Optimizing aerosol emitters for mating disruption of codling moth, *Cydia pomonella* L. ', *J. Chem. Ecol.*, 42, 612–16.

Melander, A. L. (1914), 'Can insects become resistant to sprays?', *J. Econ. Entomol.*, 7(2), 167–73.

Meyerson, L. A. and Mooney, H. A. (2007), 'Invasive alien species in an era of globalization', *Frontiers Ecol. Environ.*, 5(4), 199–208.

Miller, J. R., Gut, L. J., de Lame, F. M. and Stelinski, L. L. (2006), 'Differentiation of competitive vs. non-competitive mechanisms mediating disruption of moth sexual communication by point sources of sex pheromone (part I): theory', *J. Chem. Ecol.*, 32, 2089.

Miller, J. R., McGhee, P. S., Siegert, P. Y., Adams, C. G., Huang, J., Grieshop, M. J. and Gut, L. J. (2010), 'General principles of attraction and competitive attraction as revealed by large-cage studies of moths responding to sex pheromone', *Proc.Nat. Acad. Sci.*, 107, 22–7.

Miller, J. R. and Gut, L. J. (2015), 'Mating disruption for the 21st century: Matching technology with mechanism', *Environ. Entomol.*, 44, 427–53.

Mullinix, K., Brunner, J. F. and Isman, M. B. (2011), 'Apple leafroller (Lepidoptera: Tortricidae) populations and parasitism in an orchard managed with either a grass or alfalfa cover and without insecticides over four growing seasons', *International Journal of Fruit Science*, 11(2), 99–110.

NASS(2008),'Washington'sagriculturalfruitchemicalusage,2007',NationalAgriculturalStatisticsService, http://usda01.library.cornell.edu/usda/nass/AgriChemUsFruits//2000s/2008/AgriChemUsFruits-05-21-2008.pdf

Nicholas, A. H., Spooner-Hart, R. N. and Vickers, R. A. (2005), 'Abundance and natural control of the woolly aphid *Eriosoma lanigerum* in an Australian apple orchard IPM program', *Biocontrol*, 50, 270–91.

Paiva, M. and Janick, J. (1980), 'Relationship between leaf pubescence and resistance to European red mite in apple', *HortSci.*, 15(4), 511–12.

Patch, E. (1912), 'Woolly aphid migration from elm to mountain ash', *J. Econ. Entomol.*, 5(5), 395–8.

Pelz, K. S., Isaacs, R., Wise, J. C. and Gut, L. J. (2005), 'Protection of fruit against infestation by apple maggot and blueberry maggot (Diptera: Tephritidae) using compounds containing spinosad', *J. Econ. Entomol.*, 98, 432–7.

Pfannenstiel, R. and Unruh, T. (2003), 'Conservation of leafroller parasitoids through provision of alternate hosts in near-orchard habitats', *Proceeding 1st International Symposium on the Biological Control of Arthropods. USDA Forest Service Publ. FHTET* -03-5, 256-262.

Pfannenstiel, R., Unruh, T. R. and Brunner, J. F. (2010), 'Overwintering hosts for the exotic leafroller parasitoid, *Colpoclypeus florus*: Implications for habitat manipulation to augment biological control of leafrollers in pome fruits', *J. Insect Sci.*, 10(75).

Philips, C. R., Kuhar, T. P., Hoffmann, M. P., Zalom, F. G., Hallberg, R., Herbert, D. A., Gonzales, C. and Elliott, S. (2014), Integrated Pest Management, John Wiley and Sons, Ltd, .

Pree, D. J. and Wagner, H. W. (1987), 'Occurrence of cyhexatin and dicofol resistance in the European red mite, *Panonychus ulmi* (Koch) (Acari: Tetranychidae), in southern Ontario', *Can. Entomol.*, 119, 287–90.

Rathman, R. J., Beers, E. H., Flexner, J. L., Riedl, H., Hoyt, S. C., Westigard, P. H. and Knight, A. L. (1990), 'Baseline bioassays with hexythiazox and clofentezine of three mite species (Acari: Tetranychidae) occurring on Washington and Oregon tree fruits', *J. Econ. Entomol.*, 83(5), 1711–14.

Reissig, W. H. and Hull, L. A. (1991), 'Hexythiazox resistance in a field population of European red mite (Acari: Tetranychidae) on apples', *J. Econ. Entomol.*, 84(3), 727–35.

Roelofs, W., Comeau, A., Hill, A. and Milicevic, G. (1971), 'Sex attractant of the codling moth: characterization with electroantennogram technique', *Science*, 174(4006), 297–9.

Romet, L., Severac, G. and Warlop, F. (2010),'Overview of "ALT'CARPO" concept and its development in France.' , http://www.ecofruit.net/proceedings-2010.html#3. Accessed Nov. 2016.

Sandanayaka, W. R. M. and Charles, J. G. (2006), 'Host location and ovipositional behavior of *Platygaster demades* Walker (Hymenoptera: Platygastridae), an egg parasitoid of apple and pear leaf curling midges', *J. Ins. Behav.*, 19(1), 99–113.

Sauphanor, B. and Bouvier, J. C. (1995), 'Cross-resistance between benzoylureas and benzoylhydrazines in the codling moth, *Cydia pomonella* L.', *Pesticide Science*, 45(4), 369–75.

Schmidt-Jeffris, R. A. and Beers, E. H. (2016), 'Phenology and structure of a phytoseiid community in an insecticide-free apple orchard', *Exp. Appl. Acarol.*, 68(2), 173–82.

Schmidt, R. A., Beers, E. H., Unruh, T. R. and Horton, D. R. (2013), 'Releases of insectary-reared *Galendromus occidentalis* (Acari: Phytoseiidae) in commercial apple orchards', *J. Econ. Entomol.*, 106(5), 1996–2005.

Shaw, P. W., Bradley, S. J. and Walker, J. T. S. (1997), 'The impact of early season insecticides in an integrated fruit production programme on apple', *Proceedings of the 50th New Zealand Plant Protection Conference*, 50, 283–7.

Smith, G. D., Stiles, W. C. and Weires, R. W. (1989), 'The effcts of ground cover manipulations on pest and predator mite populations on apple in eastern New York', *NY Food Life Sci. Bull.*, 128, 1–4.

Stephens, M. J., France, C. M., Wratten, S. D. and Frampton, C. (1998), 'Enhancing biological control of leafrollers (Lepidoptera: Tortricidae) by sowing buckwheat (*Fagopyrum esculentum*) in an orchard', *Biocontr. Sci. Tech.*, 8(4), 547–58.

Stern, V. M., Smith, R. F., van den Bosch, R. and Hagen, K. S. (1959), 'The integrated control concept.', *Hilgardia*, 29(2), 81–101.

Theobald, F. V. (1921), 'The woolly aphid of the apple and elm', *J. Pomol.*, 2, 73–94.

Thistlewood, H. M. A., Borden, J. H., Smith, R. F., Pierce, H. D., Jr. and McMullen, R. D. (1989), 'Evidence for a sex pheromone in the mullein bug, *Campylomma verbasci* (Meyer) (Heteroptera: Miridae)', *Can. Entomol.*, 121(9), 737–44.

Thomson, D., Brunner, J. F., Jenkins, J. and Gut, L. J. (2009), 'Commercial use of codling moth mating disruption: A success story despite the limitations', *IOBC WPRS Bulletin*, 41, 53–60.

Thwaite, W. G. (1991), 'Resistance to clofentezine and hexythiazox in *Panonychus ulmi* from apples in Australia', *Exp. Appl. Acarol.*, 11, 73–80.

Trimble, R. M. and Appleby, M. E. (2004), 'Comparison of efficacy of programs using insecticide and insecticide plus mating disruption for controlling the obliquebanded leafroller in apple (Lepidoptera: Tortricidae)', *J. Econ. Entomol.*, 97(2), 518–24.

USEPA (2013),'Reduced risk/organophosphate alternative decisions for conventional pesticides', http://www2.epa.gov/sites/production/files/2014–02/documents/reduced-risk-op-decisions.pdf. (accessed November 2016).

Vickers, R. A., Rothschild, G. H. L. and Jones, E. L. (1985), 'Control of the oriental fruit moth, *Cydia molesta* (Busek) (Lepidoptera: Tortricidae), at a district level by mating disruption with synthetic female pheromone', *Bull. Entomol. Res.*, 75(4), 625–34.

Walker, J. T. S. (1985), 'The influence of temperature and natural enemies on population development of woolly apple aphid, *Eriosoma lanigerum* (Hausmann)', Ph.D. Thesis, Dept. Entomol., Washington State University, Pullman, WA.

Warner, G. (1998), 'Leafrollers tough to control with mating disruption', *Good Fruit Grower*, 49(6), 14.

Welch, S. M., Croft, B. A., Brunner, J. F. and Michels, M. F. (1978), 'PETE: an extension phenology modeling system for management of multi-species pest complex', *Environ. Entomol.*, 7(4), 487–94.

Welker, R. M., Marini, R. P. and Pfeiffer, D. G. (1995), 'Influence of first-generation white apple leafhopper (Homoptera: Cicadellidae) and leaf-to-fruit ratio on apple fruit size and quality', *J. Econ. Entomol.*, 88(4), 959–64.

Wellhouse, W. H. (1920), 'Wild hawthorns as hosts of apple, pear, and quince pests', *J. Econ. Entomol.*, 13, 388–91.

Welter, S. C., Pickel, C., Millar, J. G., Cave, F., Van Steenwyk, R. and Dunley, J. E. (2005), 'Pheromone mating disruption offers selective management options for key pests', *Calif. Agric.*, 59(1), 16–22.

Welty, C., Reissig, W. H., Dennehy, T. J. and Weires, R. W. (1987), 'Cyhexatin resistance in New York populations of European red mite (Acari: Tetranychidae)', *J. Econ. Entomol.*, 80(1), 230–6.
Welty, C., Reissig, W. H., Dennehy, T. J. and Weires, R. W. (1989), 'Stability of cyhexatin resistance in field populations of European red mite (Acari: Tetranychidae)', *J. Econ. Entomol.*, 82(3), 692–7.
Witzgall, P., Stelinski, L., Gut, L. J. and Thomson, D. (2008), 'Codling moth management and chemical ecology', *Annu. Rev. Entomol.*, 53, 503–32.
Wood, S., Tomlin, L., Fugger, R., Bloem, S. and Carpenter, J. (2005), 'Area-wide codling moth SIT programme (1994–2004) in British Columbia, *Canada*', International Atomic Energy Agency (IAEA), (IAEA-CN-131), pp. 143–4.

Chapter 18

Advances in pest- and disease-resistant apple varieties

Markus Kellerhals, Agroscope, Switzerland

1 Introduction: the development of disease-resistant apple varieties

Disease- and pest-resistant cultivars can significantly contribute to sustainable and resilient cultivation of apples. Developing pest-resistant and especially disease-resistant varieties has been a major objective in most apple breeding programmes worldwide for several decades. However, specific breeding aims differ for different production areas and their climatic conditions. For example in Washington State (the United States), owing to little rainfall and very dry climatic conditions, apple scab (*Venturia inaequalis*) is rarely present and therefore scab-resistant cultivars would not bring an added value. However, breeding for resistance to scab is a major target for breeders elsewhere. A survey in 1996 of 42 apple breeders from 29 countries representing four continents reported a common principle objective of combining high fruit quality with disease and pest resistance in new cultivars (Laurens, 1999). Typically, pest resistances are less frequently considered as breeding objectives. In apple breeding, it is necessary to focus on a limited number of objectives as the breeding cycle is long due to the juvenile phase and a high level of heterozygosity.

The development of pest- and disease-resistant apple varieties has been considerable in recent decades. Fruit quality has been improved and a range of flavour, texture and appearance features considered for different markets. Resistances are being combined

http://dx.doi.org/10.19103/AS.2016.0017.22

and pyramided in order to allow for more durable resistance. However, despite the huge efforts in breeding and the underpinning breeding research, the impact of pest- and disease-resistant varieties in commercial apple growing and consumption is still very small.

Breeding efforts for disease resistance in apple started in the twentieth century and were based on close collaborations between pathologists and breeders. The first attempts to breed scab-resistant apples were undertaken in Germany and in the United States (Kellerhals, 1989). As sources of resistance, mainly wild apple species were considered. As they are small-fruited and often with low fruit quality, this approach required a long-term breeding approach with several pseudo-backcross generations. As these wild apple resistances are mainly qualitatively inherited and therefore vulnerable to resistance breakdown, alternative sources for resistances were also considered, among them traditional local varieties (Kellerhals et al., 2012; Lefrancq, 2004).

We are indeed grateful for the pioneers of disease resistance breeding as we still rely on the selections and varieties they developed and made publically available.

2 Resistance to apple scab, powdery mildew and fire blight

2.1 Apple scab (*Venturia inaequalis*)

An important gene for scab resistance used in the US programme was identified from a collection of *Malus* species and hybrids at the University of Illinois in the early 1900s. In 1914, Crandall made a cross between 'Rome Beauty' and '*Malus floribunda* 821' in a study on the inheritance of fruit size in apple and in 1926, presumably a full sib cross between two F1 selections (Crandall, 1926). When a severe apple scab epidemic occurred in 1942, Hough noticed that one progeny from Crandall's F1 cross showed an almost perfect 1:1 segregation for scab resistance. Further crosses with two scab-resistant selections from this cross (F2 26829-2-2 and F2 26830-2) indicated that resistance was determined by a single dominant gene subsequently named *Vf* (*Venturia* resistant from *floribunda*). The assumption was put forward as resistance conferred by '*Vf*' showed no evidence of being violated by new fungal races (Crosby et al., 1992) and was not linked to any undesirable effects and acted as a true dominant in the diploid, triploid or tetraploid condition. *Vf* was found in two selections, F226829-2-2 and F226830-2, that were horticulturally advanced over the original source *M. floribunda* 821, which has fruit of about 1 cm in diameter. These two selections formed the basis of the present scab resistance programme in apples and all *Vf*-resistant apple cultivars can be traced back to this cross (Fig. 1).

The PRI apple breeding programme for disease resistance was a cooperation among Purdue University, Rutgers; The State University of New Jersey; and the University of Illinois. The programme started in 1945 to breed scab-resistant apples. A complete history of the programme can be found in the work of Crosby et al. (1992). They described the state of the art at that time as follows: 'The effort that continues today in at least 17 breeding programmes throughout the world was based on a modified backcross programme to combine genes for resistance to apple scab from *M. floribunda* 821, and other species with commercially-accepted traits'.

Figure 1 Cross-pollination.

Selection for scab resistance is based on phenotypic early selection of seedlings inoculated with scab conidia in the glasshouse (Fig. 2). The seedlings are then scored using the scale of Chevalier et al. (1991), which is a further development of the scale developed by Shay and Hough (1952). Depending on the progenies, varying proportions can be discarded as susceptible to scab. In the case of one parent carrying a major scab resistance gene in a heterozygous state crossed with a susceptible one, a 1:1 segregation can be expected. The broad use of a single resistance gene (*R*-gene) resulted in high selection pressure on the pathogen and the emergence of *V. inaequalis* strains able to overcome *Vf* (later named *Rvi6*, Bus et al., 2011), rendering these *Vf* cultivars highly susceptible to the disease (Parisi et al., 1993; Roberts and Crute, 1994). The risk of resistance breakdown can be reduced by combining (pyramiding) several functionally different resistances in a cultivar (Baumgartner et al., 2015), which can be achieved using DNA-based selection. Molecular selection for genotypes with such pyramided genetic resistance against scab and powdery mildew is applied in the Swiss apple breeding programme at Agroscope (Baumgartner et al., 2016) and in several breeding programmes worldwide.

In a joint effort of the University of Bologna and the Swiss Federal Institute of Technology in Zürich, *Vf* was cloned. The position of *Vf* was first clarified and the physical distance of the markers (AL07 and M18) bracketing the gene was estimated (Patocchi et al., 1999a). Upon construction of a bacterial artificial chromosome library of the cv 'Florina' (Vinatzer et al., 1998), the positional cloning of the gene was conducted (Patocchi et al., 1999b). Finally, one of the candidate resistance genes found by Vinatzer et al. (2001) was shown to confer *Vf* resistance by Belfanti et al. (2004). Cloning of *Vf* was the basis for developing cisgenic scab-resistant apple lines.

Schouten et al. (2006) defined a cisgenic plant as a plant that does not contain any transgenes and that has been genetically modified with one or more genes isolated from a crossable donor plant. Recently, cisgenic *Rvi6* plants of the apple cv 'Gala Galaxy' were created (Vanblaere et al., 2013). The vector used for genetic modification allowed a post-selection marker gene elimination to achieve cisgenesis. For two of these cisgenic lines developed, a single insertion was observed. Despite a low gene expression compared to 'Florina', this insertion was sufficient to induce plant reaction and reduce fungal growth by 80% compared with the scab-susceptible 'Gala'. Currently, in Europe cisgenic crops are still under GMO regulation and therefore not yet appropriate for commercial use.

Figure 2 (Top) Resistance symptoms after artificial scab inoculation; (Bottom) Scab symptoms after artificial inoculation in the greenhouse.

Partial resistance has been mapped as quantitative trait loci (QTLs) or quantitative resistance loci (QRLs) in the apple genome (Gessler et al., 2006; Calenge et al., 2005; Durel et al., 2003; Liebhard et al., 2003). Laloi et al. (2016) examined the efficiency of pyramiding three QRLs to apple scab and came to the conclusion that pyramiding remains a promising strategy to increase the efficiency of resistance. They showed a gain in resistance efficiency after pyramiding QRLs that individually were not effective against aggressive strains of the pathogen.

Since 1970, a range of scab-resistant cultivars have been released worldwide, of which most carry the *Vf* resistance from *M. floribunda* 821. Some cultivars such as 'Freedom' carry additional polygenic resistance from 'Antonovka'. One variety ('Imrus') carries *Vf* from *M. atrosanguinea* 804. Some cultivars carry other resistance genes such as 'Rouville' which carries the *Vm* gene from *M. atrosanguinea* 804. The 'Nova Easygro' variety, which was supposed to carry the *Vr* gene from a Russian apple seedling from the Caucasus Mountains (R#12740-7A), actually carries *Vf* (Vinatzer et al., 2004). Other examples are

'Murray' which carries *Vm* from *M. micromalus*, and 'Romus 2', 'Gavin' and 'Generos' which have only polygenic resistance.

2.2 Powdery mildew (*Podosphaera leucotricha*)

Breeders and pathologists initially proposed that the cultivated apple provides a complete range of reaction from full susceptibility to 'immunity' to the fungus powdery mildew (Brown, 1975). However, examination of some 2000 cultivars at East Malling and the UK National Fruit Trials near Faversham, England, suggested that it is improbable that true immunity exists in the cultivated apple (Alston, 1969). Brown (1959) suggested that mildew resistance in the cultivated apple is invariably under polygenic control. Knight and Alston (1968) conducted a search for wild *Malus* species as sources that could provide immunity or near immunity to mildew as part of the apple breeding programme at East Malling (UK). *M. robusta* and *M. zumi* both proved to be reliable sources of resistance. It was assumed that this resistance might be determined by single dominant genes, *Pl1* and *Pl2*, respectively. Later, this hypothesis was revised to account for two different genes plus modifiers (Alston, 1977). The genetics of these mildew resistances has been further elucidated by molecular approaches. The position of the *Pl1* locus at the bottom part of apple linkage group (LG) 12 indicates that it might be part of an important resistance gene cluster located in this region of the apple genome (Dunemann et al., 2007). *Pl2* was mapped on LG11 (Seglias and Gessler, 1997) and cloned by Rikkerink et al. (2007). Unfortunately, it is reported that strains of apple powdery mildew were able to overcome *Pl2* (Caffier and Laurens, 2005). Recently, it was shown that the silencing of MdMLO19, a homolog of the more famous *Mlo* resistance gene of barley, lead to an increased resistance to apple powdery mildew (Pessina et al., 2016).

In greenhouse tests for mildew resistance, Dayton (1977) identified a single plant grown from open-pollinated seeds of 'Starking Delicious' as highly resistant. One parent appears to have been a *Malus* species. It was named 'Mildew immune selection' (MIS). Progeny tests indicated a single gene from MIS conferring immunity to *P. leucotricha* in both the greenhouse and the field, distinct from those reported by Knight and Alston (1968). Lespinasse (1983) reported a virulent biotype of mildew, which seemed to overcome the resistance of MIS in Dayton's plantation. Other sources for mildew resistance were found in the Netherlands (D-Series), in *M. hupehensis* and the ornamental cultivar 'White Angel', in 'David', *M. robusta* 'Robusta 5', *M. robusta* 'Korea' and *M. robusta* ('24-7-7,8') (Gallot et al., 1985).

2.3 Fire blight (*Erwinia amylovora*)

Efforts to breed fire blight-resistant apple and pear cultivars started in the United States at the end of the nineteenth century. The disease was first observed around 1793 in the United States. In 1880, Burrill discovered the causal agents (bacteria) for this disease and Waite examined the way these pathogens enter the host and started a pear breeding programme in 1908. Many pear and apple cultivars from Europe were introduced to the United States at that time and showed differential susceptibility to the disease. Fire blight-resistant material, especially crab apples and pears, were also introduced from China and Japan; therefore, a range of resistance sources was available for breeding. As the disease spread to Europe and other parts of the world, breeding activities for host resistance were intensified. Fire blight spread from North America after 1950, first

Figure 3 Evaluation of apple hybrids in the orchard.

to the United Kingdom and subsequently to continental Europe (Peil et al., 2007). A combination of phenotypic testing and molecular selection is currently underpinning the development of fire blight-resistant cultivars carrying commercial fruit quality in several breeding programmes worldwide (Fig 3). For phenotypic resistance testing, artificial needle inoculation of shoot tips (Khan et al., 2006) or infection by cutting leaves with scissors dipped in a bacterial solution (Peil et al., 2009) are common practice in glasshouse tests. Additionally, artificial flower inoculation tests in protected orchards are performed (Kellerhals et al., 2014).

Towards the end of the last century a range of apple cultivars with resistance to scab and some with additional resistance or tolerance to fire blight were released, especially from breeding programmes in the United States, Germany and France. However, only a few are available commercially (see list at the end of this chapter).

In order to obtain useful resistance sources for breeding, QTLs were described in wild apple species accessions including *M. x robusta* 5, *M. fusca*, *M. floribunda* 821 and the ornamental cultivar 'Evereste' (Fig. 4) (Peil et al., 2007; Emeriewen, 2014; Durel et al., 2009). Peil et al. (2007) found a resistance gene for fire blight on LG 3 of *Malus x robusta* 5 (*FB_MR5*). This strong QTL explained 67–83% of the phenotypic variation, and in each family tested, about 20–25% of the seedlings remained disease-free (Peil et al., 2008). Another strong QTL explaining 50–70% of phenotypic variation was identified in 'Evereste' (*FB_E*, Durel et al., 2009). The QTLs from 'Robusta 5' and 'Evereste' in contrast to the QTL from 'Fiesta' (Calenge et al., 2005; Khan et al., 2006) are supposed to confer strong resistance. As they were found in small-fruited crab apples with low fruit quality, several pseudo-backcross generations are required before a new commercial cultivar carrying this resistance can be expected.

SSR and in part SNP markers closely linked to these fire blight resistance loci are nowadays available to breeders for marker-assisted breeding (in addition to the above-mentioned references: Parravicini et al., 2011; Jänsch et al., 2015). This is especially valuable as the phenotypic screening for fire blight resistance or susceptibility is far more laborious and expensive compared to screening for scab or mildew resistance.

Figure 4 Wild apple 'Evereste' carrying fire blight resistance.

Transgenic pear and apple lines expressing *attacin* genes and therefore enhanced fire blight resistance were developed in Geneva, New York, in the late 1990s (Borejsza-Wysocka, 2010).

Recently, the fire blight resistance gene *FB_MR5* has been cloned (Fahrentrapp et al., 2013; Broggini et al., 2014) and those of *FB_E* (Parravicini et al., 2011) and *FB_Mfu10* (Emeriewen et al., 2014) are in progress. Kost et al. (2015) reported the development of the first cisgenic apple with increased resistance to fire blight. The fire blight susceptible apple cultivar 'Gala Galaxy' was genetically engineered to carry the fire blight resistance gene *FB_MR5*. A previously established method involving *Agrobacterium tumefaciens*-mediated transformation was used to develop marker-free genetically modified plants according to the definition of cisgenesis. Following regeneration of transgenic lines a heat-induced recombinase-mediated excision was used to remove all foreign helper genes. Thirteen cisgenic lines were produced and out of these, a marker-free line could be regenerated. Inoculation experiments with an *MR5* avirulent strain of *Erwinia amylovora* were performed using both scissors and syringe shoot inoculation methods. Significantly less disease symptoms were detected on shoots of the cisgenic line compared to untransformed Gala (Kost et al., 2015). However, Vogt et al. (2013) reported that strains of *E. amylovora* able to overcome the *FB_MR5* resistance exist. A single point mutation is sufficient to transform an avirulent strain into a virulent one. Therefore, *FB_MR5* resistance must be deployed in pyramids of resistance genes and not alone.

3 Resistance to *Nectria* canker, *Marssonina* apple blotch and other diseases

3.1 *Nectria* canker (*Neonectria ditissima*)

European canker, caused by the fungus *Neonectria ditissima* (formerly *Neonectria galligena*), is a severe disease in apple production in many European countries, especially in the northern more humid areas. Apple canker, *Neonectria* canker and Europaen canker are common names of fruit tree canker (Ghasemkhani, 2015). A study using RAPD and ribosomal DNA polymorphisms has shown that this pathogen originated in North America (Plante et al., 2002). Besides Europe and North America, fruit tree canker has also been reported from Australia, Chile, South Africa, New Zealand and Japan (Weber, 2014; Beresford and Kim, 2011; Xu and Robinson, 2010; Grove, 1990). In some years, devastating outbreaks can destroy numerous trees. Infection of young trees can even happen in the nursery. After a latent period of 3–5 years, systemic infections suddenly break out on these trees and damage the stem. In some cases, epidemics have necessitated the removal of entire orchards (Jones and Aldwinckle, 1990). Eventually the cankers girdle the tree, resulting in a dieback of infected limbs or trees. The cankers are sometimes visible on the trunks of nursery trees at the time of planting, although latent infections can appear later. *Neonectria ditissima* can also cause rotting of fruits in storage. According to Garkava-Gustavsson (2016, pers. comm.) the resistance is a highly quantitative trait and there is no complete resistance. She compared the transcriptomes of partially resistant and highly susceptible apple cultivars in response to *Neonectria ditissima* infection and found cultivar differences and temporal changes in the host–pathogen interaction.

Complete resistance to *N. ditissima* has not yet been demonstrated in apple (Gomez-Cortecero et al., 2016). *Malus* species and apple cultivars show variation in susceptibility to *N. ditissima* (Alston, 1970; Van de Weg, 1987; Van de Weg, 1989; Ghasemkhani et al., 2015). For further research and breeding, heritable variation in quantitative resistance needs to be investigated by phenotyping large sets of cultivars. Garkava-Gustavsson et al. (2013) inoculated one-year-old shoots from mature trees in the greenhouse with a standardized volume and concentration of conidia suspension using different inoculation methods. Two-year-old trees of five cultivars were inoculated in the field. Length of the occurring cankers was measured at regular intervals throughout a period of up to three months. The investigated cultivars showed considerable differences in colonization rate. In cultivars known to be highly resistant, such as 'Santana', lesions progressed much slower compared to susceptible cultivars such as 'Cox's Orange Pippin' and 'James Grieve'. As the inoculation-based phenotyping is expensive, qPCR-based assessment of fungal biomass at early stages of infection was explored as an alternative or complementary approach for phenotyping (Garkava-Gustavsson et al., 2013).

3.2 *Marssonina* apple blotch (*Diplocarpon mali*)

Marssonina apple blotch, caused by the fungus *Diplocarpon mali* [Y. Harada and K. Sawamura (anamorph *Marssonina coronaria* (Ell. and J. J. Davis) J. J. Davis, syn. *Marssonina mali* (Henn.) S. Ito)], is a severe apple disease (Harada et al., 1974; Lee et al., 2011) leading to defoliation during the growing season, thereby weakening tree vigour and diminishing fruit yield and quality (Donghoon et al., 2011). The fungus mainly infects leaves but can also attack twigs and fruits (Sharma et al., 2004). The disease is first manifested by brown spots and dark green circular patches on the leaf upper surfaces in mid-summer. As it progresses, spots coalesce and black asexual fruiting bodies (acervuli) develop on the affected tissue. Severe infections result in leaves turning yellow, followed by premature defoliation. The fungus also attacks the fruit by causing circular dark brown spots of varying size (3–5 mm diameter) on all the commercial cultivars. The disease is favoured by high rainfall and moderate temperature ranging from 20–22°C during the fruit development stages of apple. *Marssonina* defoliation in apple was first noticed in 1992 in some of the orchards in the Kotkhai area in Himachal Pradesh (northern India) and developed to an epidemic in the subsequent years. The disease has been recorded in Canada, the United States, Japan, Brazil, China, Korea, India, Italy, Germany and Switzerland. Usually occurring in consecutive years, it has become a serious problem within major regions of apple production especially in organic orchards and standard tree orchards with only few or no treatments against fungal diseases.

Suppression subtractive hybridization was used to identify differentially expressed resistance genes in the 'Qinguan' apple in response to challenge with *M. coronaria* (Zhou et al., 2012). In conclusion, responses to *M. coronaria* in apples are complex, whilst some common genes in apple have been found showing resistance to fungal infections such as *V. inaequalis*.

3.3 Storage diseases

With the increasing restriction of pesticide use in fruit production, the relevance or resistance to storage diseases will further increase. For example, blue mould primarily caused by the fungus *Penicillium expansum* is a common postharvest disease in apple. Worldwide, blue mould of apples is considered as the most important postharvest rot of apple fruit (Capellini et al., 1987; Jurick et al., 2011) and is of great concern to the fruit processing industry due to the production of the mycotoxin patulin (Wouters and Speijers, 1996; McCallum et al., 2002). A source of blue mould resistance was found in *M. sieversii*, a wild apple native in Central Asia. As *M. sieversii* is a genetically diverse species, numerous sources of resistance to blue mould may be present and the resistance genes may have diverse functions. In a collaborative project of the USDA-ARS at Kearneysville (West Virginia) and the Volcani Institute in Israel, QTL(s) for blue mould resistance in a mapping population of 'Royal Gala' × *M. sieversii* PI613981 were identified (Norelli et al., 2014).

3.4 Other diseases

Other diseases of importance in apple include crown rot (*Phytophthora cactorum*), ceder apple rust (CAR; *Gymnosporangium juniperi-virginianae*) and quince rust *Gymnosporangium clavipes* as well as various storage diseases, sooty blotch and fly speck. More information about these diseases is available in Peil et al. (2011).

4 Resistance to pests

Apple can be attacked by both fruit- and leaf-damaging lepidopteran pest insects, such as the carpophagous codling moth, *Cydia pomonella*, or the phyllophagous apple leaf miner, *Lyonetia clerkella* (Stoeckli et al., 2009a). This requires regular control. Codling moth is the most widely reported apple pest in both conventional and organic orchards. Insect resistance is partly based on morphological plant traits such as pubescence, fruit colour or fruit shape, and surface waxes (Smith, 2005).

Monogenic resistances to rosy apple aphid (*Dysaphis plantaginea*), rosy leaf-curling aphid (*Dysaphis devecta*) and woolly apple aphid (*Eriosoma lanigerum*) have been found in either cultivated or wild apples (Korban and Tartarini, 2009). Alston and Briggs (1970) reported on a hypersensitive response of '*Malus robusta* 5' to rosy apple aphid. For the rosy apple aphid resistance from 'Florina' named *Dp-fl*, located at the distal end of LG 8, SNP markers linked to *Dp-fl* have been identified and validated with orchard data (Pagliarani et al., 2016). Resistance to three biotypes of the rosy leaf-curling aphid has been identified in different apple cultivars (Alston and Briggs, 1970) and later the Sd1 resistance gene locus was identified (Korban and Tartarini, 2009). Three genes for woolly apple aphid resistance were also mapped (Bus et al., 2008).

The apple rust mite (*Aculus schlechtendali* Nalepa) is a serious pest in many apple growing regions of the world (Easterbrook and Palmer, 1996). *A. schlechtendali* infestation of apple trees increased with changes in fungicide application, namely the replacement of acaricidal lime sulphur by non-acaricidal compounds, and the application of pesticides that reduce predators such as gall midges or predatory mites (Spieser et al., 1998). The potential influence of apple cultivar on *A. schlechtendali* resistance has been discussed by Stöeckli et al. (2009b). The resistance of cultivars was tested by phenotypic evaluation but no information on the genetic basis of this resistance is yet available (Stoeckli et al., 2009b). The use of such cultivars for resistance breeding in apple is limited, due to the unreliable phenotypic evaluation and the additional time and money needed for complex field trials (Brown and Maloney, 2003; Francia et al., 2005).

5 Selection techniques and resistance mechanisms

5.1 DNA-based selection techniques for pest and disease resistance

From a breeding point of view, it is extremely challenging to develop commercial varieties with a broad range of durable resistances against a range of pests and diseases as well as high productivity, fruit quality and storage. To date, most resistant varieties have only a single resistance against one disease, mainly apple scab, and a few have scab and mildew resistance and a certain level of fire blight resistance. There is scope that modern molecular screening methods will improve the efficiency of producing multi-resistant high-quality varieties.

Marker-assisted selection in apple has been mainly applied in such areas as pyramiding different scab resistance genes (Peil et al., 2007; Bus et al., 2009; Kellerhals et al., 2009). This is most likely because most markers were developed for resistance to scab (Gardiner et al., 2007) and that the scab resistance inherited from '*Malus floribunda* 821' has already been overcome (Parisi et al., 1993; Bénaouf and Parisi, 1997).

Pyramiding of (major) resistance genes is supposed to achieve more durable resistance to a pathogen, which is especially important in a perennial crop such as apple (MacHardy et al 2001; Gessler et al., 2006; Kellerhals et al., 2009; Patocchi et al., 2009). The presence of natural pyramids in apple consisting of major genes has shown to be durable, for example, scab resistance of the 'Russian seedling' (Bus, 2006). However, examples in other plant pathosystems have shown that pyramiding of major genes can be risky and not durable as, for example, in *Bremia* (Lebeda, 1992). Scab studies in apple demonstrated that at least race 6 is able to overcome the pyramided genes *Rvi1/Rvi6* (Parisi and Lespinasse, 1996; Laurens et al., 2004). Bus et al. (2009) stated that *V. inaequalis* isolates can be expected to overcome several resistance genes and QTLs as an increasing number of them are identified and used in apple breeding. Nevertheless, the combination of major genes with polygenic quantitative resistance may improve durability of resistance (Fischer and Fischer, 1999; Soufflet-Fresslon et al., 2008) as polygenic quantitative resistance is considered to be more durable (Parlevliet, 1993, 2002; Lindhout, 2002).

Pyramiding of resistance genes requires the application of molecular markers as the detection by phenotyping alone usually cannot prove the presence of pyramided resistance genes, but phenotyping can be performed to reduce the number of progeny plants directed to marker-assisted seedling selection.

5.2 Resistance mechanisms

Plants have developed sophisticated mechanisms to protect themselves against pathogenic infections. Besides physical constraints hindering pathogens from entering the host, the plant immune system comprises a range of proteins and signalling components that either perceive the presence of a microbe in general, or the imminent threat of a specific hostile organism. So-called pattern-recognition receptors (PRRs) that are usually embedded in the plant's plasma membrane recognize the presence of pathogens through conserved molecular pattern such as flagellin (Felix et al., 1999) in the case of bacteria or constituents of chitin (Kaku et al., 2006) in the case of fungi. These patterns are thus called 'pathogen-associated molecular patterns' (PAMPs) or also 'microbe-associated molecular patterns' (MAMPs) (Jones and Dangl, 2006). They are commonly shared within large pathogen classes, allowing wide multiplicity in the recognition of pathogens by PRRs, thereby increasing the efficiency of detection (Medzhitov and Janeway, 1997). The recognition of a PAMP/MAMP by a PRR results in a complex immune response involving several measures that finally help the plant warding off the attacker – a state called PAMP-triggered immunity (PTI) (Jones and Dangl, 2006). To counteract PTI, pathogens may use their type three secretion systems (T3SS) to inject special molecules called 'effectors' into their host cells. These effectors may inhibit the host's immune reaction or increase the pathogen's chance of success in general, resulting in so-called 'effector-triggered susceptibility' (ETS) (Boller and He, 2009; Jones and Dangl, 2006). Plants, in turn, may carry *R*-genes that recognize the presence of a pathogen effector either directly or indirectly. Upon perception of the corresponding effector by the *R*-gene product, a signalling cascade often elicits apoptosis in the affected cell, thereby preventing the spread of the invading pathogen – this is called 'effector-triggered immunity' (ETI) (Jones and Dangl, 2006). The interaction between *R*-gene product and pathogen effector is very specific. For this reason, it is frequently referred to as a 'gene-for-gene' relationship (Flor, 1971).

In apple, this host–pathogen interaction can be exemplified with the *E. amylovora*–apple host pathosystem. As mentioned above, the wild apple *Malus x robusta* 5 was found to be fire blight resistant. The genetic basis for this resistance is *FB_MR5*. and the corresponding pathogen effector candidate *avrRPT2ea* has been determined (Vogt et al., 2013).

The evolutionary arms race between pathogen and plant is an ongoing process and has been described as a 'zigzag' model: after initial PTI, the pathogen evolves to reach the state of ETS; following this, the host acquires an appropriate *R*-gene to achieve ETI and so on (Jones and Dangl, 2006; Ingle et al., 2006). The rapid evolution of pathogen populations in the context of the large-scale use of host plants with a single *R*-gene has led to so-called 'boom-and-bust' cycles of plant resistance and susceptibility in agriculture (McDonald and Linde, 2002). A single change in an effector that results in the molecule no longer being recognised by the corresponding host R-protein may allow the pathogen to evade the host's immune response, thereby rendering the plant susceptible again (Jones and Dangl, 2006).

6 Current disease-resistant apple varieties

Current disease-resistant apple varieties (adapted and updated from Susan Brown and Kevin Maloney, www.nyshs.org, Brown and Maloney, 2008) are as follows:

'Ariane': This *Rvi6*-resistant apple was bred at INRA Angers, France, from a cross of a hybrid of 'Florina' × 'Prima' as the pollen parent and an open-pollinated selection from 'Golden Delicious' (Laurens et al., 2005). Around 500 ha of Ariane are currently grown in France. The quality of the fruit is excellent as is the storage life, but thinning must be aggressive to obtain good fruit size. 'Ariane' is slightly susceptible to powdery mildew and to fire blight and woolly apple aphids. The scab resistance of 'Ariane' has proven not to be very strong and resistance breakdown is quite frequent.

'Ariwa': This *Rvi6* scab-resistant apple (Golden Delicious x A 849-5) from the Swiss Agroscope breeding programme also has good resistance to powdery mildew in European trials (Kellerhals et al., 2003). The apple has a good storage life but acidity decreases during that period. Correct fruit thinning is required to obtain regular yields. Resistance breakdown has occasionally been observed.

Bonita: A recent release with *Rvi6* scab resistance bred by the Czech breeding programme at Střížovice in collaboration with KSB in South Tyrol. It is a cross between Topaz and Cripps Pink. A bi-coloured red apple, slightly tart, with uniform shape.

'CrimsonCrisp'™ (Co-op 39) carries the scab resistance (*Rvi6*) but it is susceptible to CAR, powdery mildew and to fire blight. This red apple with fine texture and good crunch and well-balanced flavour ripens in mid-September (Janick et al., 2006).

Dalinette (Choupette®) a red, rustic *Rvi6* apple mainly for organic growers and marketing. Bred by INRA Angers with the cross X4598 × X3174 in collaboration with the current company Dalival. The fruit is firm and juicy.

'Ecolette': A hybrid of 'Elstar' × 'Prima' with *Rvi6* resistance that was developed in the Netherlands. Fruit size and yield are moderate; fruits are somewhat tart with good firmness and keeping quality. Low susceptibility to mildew. No more impact in commercial growing.

'**Enterprise**' (Co-op 30) carries the *Rvi6* scab resistance and a low susceptibility to fire blight (Crosby et al., 1994). The fruit quality as table fruit is not outstanding, perhaps limiting it to juice production or other processing.

Galant is a scab-resistant apple that originates from Switzerland from the cross of 'Resi' × 'Delbard Jubilé'. The apple is more tart than Gala and matures shortly before 'Golden Delicious'. It is a bi-coloured red apple.

'Galarina' (X 4982): This variety with *Rvi6* scab resistance was originally bred by INRA Angers, France. It is a hybrid of 'Florina' × 'Gala' and was released jointly by Agriculture and Agri-Food Canada and INRA France (Khanizadeh et al., 2003).

'Goldrush' (Co-op 38): A slightly acid, green/gold-skinned cultivar that has excellent firmness, crispness, storage and shelf life. 'Goldrush' carries the *Rvi6* scab resistance. It is moderately resistant to fire blight but very susceptible to powdery mildew and also susceptible to CAR. The late maturing restricts the cultivation to sites with a long growing season. Surprisingly, this 'tart' cultivar (in the United States) is classified as 'mild to sweet' when tested in Switzerland (Weibel et al., 2003).

'Goldstar': This *Rvi6* scab-resistant apple was developed in the Czech Republic, has 'Golden Delicious' and 'Lord Lambourne' in its pedigree and has looked promising in some tests. It was rated as worth a trial by Lind et al. (2003) but has not gained more extensive cultivation.

Inored – Story® was released from the French cooperative breeding programme of Novadi with INRA Angers. A variety that is adapted to warmer climatic conditions where it gets good colouration. A rather sweet and firm apple with regular shape. Under cooler climatic conditions, internal browning can be observed.

Juliet (Korban et al., 2003) (Co-op 43): Another release from the PRI programme. It carries the *Rvi6* scab resistance. It ripens two to three weeks after 'Delicious'. 'Juliet' is susceptible to CAR. In France it is grown under organic production and licensed as an exclusive brand (www.juliet.eu). Late ripening around Fuji, it is very productive. A bi-coloured apple with slight bloom on the skin, sweet taste and good storage. Slight tendency to bitter pit. The multiplication and growing of this variety is restricted.

Ladina was recently released by Agroscope, Switzerland, from a cross between Topaz and Fuji. It carries the *Rvi6* scab resistance and has proved to be little susceptible to fire blight. The extremely juicy and rather sweet fruit is dark red and develops an exotic flavour resembling peach and litchi (Fig. 5).

'Liberty': One of earlier and well-known of the *Rvi6*-resistant apples and also tolerant to fire blight (Lamb et al., 1979). Ripens with 'Empire' but storage life is limited and core browning as well as odd over-ripe flavour is frequent. Adequate thinning is essential. Trees are productive and winter hardy. Liberty is also suitable for apple juice production.

'Modi' (CIV G198): This *Rvi6* scab-resistant apple hybrid of 'Gala' × 'Liberty' was bred by the Consorzio Italiano Vivaisti in Italy. It is a solid red apple that ripens with 'Golden Delicious'. It has a well-balanced taste. Storage is limited as there is a tendency for softening. Access to this variety is restricted.

Natyra® was bred in the Netherlands. An *Rvi6* scab-resistant apple cultivar with aromatic, slightly sub-acid to acid fruit with excellent aroma and good storage. Tree vigour is rather

Figure 5 Ladina variety with resistance to scab and fire blight.

low and requires a more vigorous rootstock. Currently the variety is mainly produced in Germany by organic growers.

'Otava': This is a product of a breeding programme in the Czech Republic. This 'Sampion' × 'Jolana' *Rvi6* hybrid was rated worth a trial by Lind et al. (2003). 'Otava' has a good growth habit, yield and multiple resistances. Concerns included susceptibility to sooty blotch and acidic taste that might be too acidic given its resemblance to 'Golden Delicious' (Weibel and Häseli, 2003).

'Pixie Crunch'™ (Co-op 33): It is named for its small-to-moderate fruit size and its crisp texture (Janick et al., 2004b). The open calyx at the base of the fruit is a concern for the development of mouldy core. Suggested mainly for home gardeners but not as a commercial variety to place in cold storage due to its relative short storage life. 'Pixie Crunch' is scab resistant but susceptible to powdery mildew. It ripens in mid-September but has the ability to hang on the tree.

'Pristine' (Co-op 32): This yellow scab-resistant apple was from PRI (Janick et al., 1995). It ripens in mid-August and may be site selective, with some sites unable to produce enough sugar in the fruit to balance its high acidity. Fire blight has been observed in some test sites and 'Pristine' is susceptible to CAR.

'Redfree' (Co-op 13): A slightly sweet and relatively early ripening scab-resistant apple that has been successful as a niche variety. The tree habit is weepy and prone to bare wood. 'Redfree' is highly susceptible to rosy apple aphid infection and to fire blight (Minarro and Dapena, 2008).

'Rubinola' is scab resistant, low susceptible to powdery mildew and matures 10 days to 2 weeks before 'Golden Delicious' originating from the Czech Republic (http://home.ueb.cas.cz/eng/varieties-rubinola.htm). The trees are vigorous with bare wood. The cultivar was the result of a cross between 'Prima' and 'Rubin'. The fruits are medium to large, flat, globose and with a skin that is bright red over most of the surface, although some russeting can occur. The flesh is yellow, firm, fine textured, juicy and has an aromatic flavour.

'Santana': This hybrid of 'Elstar' × 'Priscilla' was developed in the Netherlands. This *Rvi6* variety produces good-sized fruits with good colour and a sweet, yet sharp flavour. Trees are productive annual producers. 'Santana' is susceptible to powdery mildew and may defoliate (Lind et al., 2003). 'Santana' has also been determined to be low in apple allergens.

'Scarlet O'Hara' (Co-op 25): (Janick et al., 2000). This *Rvi6*-resistant cultivar has good firmness and storage life, but the susceptibility to fire blight and to mouldy core have limited its commercialization.

'Sundance'™ (Co-op 29): A yellow-skinned variety with little blush (Janick et al., 2004a). This variety ripens in mid-October. Fruits of 'Sundance' are prone to russet in the stem cavity and this russet can extend over the side of the fruit. Prone to biennial bearing, fruit clustering and bare wood.

'Topaz': This *Rvi6*-resistant hybrid of 'Rubin' × 'Vanda' was developed and named in the Czech Republic by the team of Jaroslav Tupy in Střížovice. It ripens towards the end of September and has good fruit quality. Fruits are prone to bitter pit and may soften after long storage. Trees are also prone to collar rot and very susceptible to fire blight (noted in the United States and Europe). It is extensively grown in Europe as one of the better-quality scab-resistant cultivars especially in organic fruit farms.

A range of scab- and multiple resistant cultivars was developed in Dresden-Pillnitz, Germany (Fischer, 1993). They include among others the cultivars 'Rewena' (*Rvi6*), 'Reanda' (*Rvi6*), 'Regine' (*Rvi6*), 'Reglindis' (*VA*), 'Relinda' (*Rvi6*), 'Remo' (*Rvi6*), 'Resi' (*Rvi6*), 'Retina'

(*Rvi6*), 'Regia' (*Rvi2, Rvi4*) and 'Reka' (*Rvi2*). It noteworthy that some of these varieties carry alternative scab resistances than *Rvi6*, which is by far the most common scab resistance gene. 'Reanda' and 'Reglindis' are suitable for processing and 'Relinda', 'Rewena' and 'Remo' for juice and cider. 'Resi' is appreciated for home gardening.

This list does not give a complete overview of all commercially available disease-resistant apple varieties, but it shows a major part of the current range.

7 Summary

Advances in disease- and pest-resistant varieties are considerable and encouraging in the last years. Scientific knowledge about the disease and pest resistance genetics and genomics has dramatically increased in the last decades and new breeding technologies open new perspectives for the future. Nevertheless, pest- and disease-resistant apples have to fulfil the high-quality expectation of the market and the consumers. Commercial impact of disease-resistant apple varieties is still low. In Switzerland the overall share of scab-resistant varieties in apple production is around 5% and in South Tyrol it is even lower (Walter Guerra, pers. comm.). However, with the increasing concern of consumers related to pesticide use and residues on the product, there will be an urgency to look for production systems requiring less input of plant protection agents.

8 Where to look for further information

Many breeders worldwide continue their efforts to develop disease- and pest-resistant apple varieties with commercial fruit quality for sustainable production systems. Often there is also a close collaboration with molecular biologists, bioinformaticians, genetecists, phytopathologists, entomologists and other relevant experts. Networks such as RosBREED (www.rosbreed.org) in the USA are stimulating these collaborative developments. The Eucarpia (European Association of Plant Breeding, www.eucarpia.org) with the Fruit Breeding and Genetics Section is bringing together breeders and scientists worldwide in meetings every four years.

9 References

Alston F. H. (1969), Response of apple cultivars to mildew, *Podosphaera. Reports East Malling Research Station for 1968*, 133–5.

Alston, F. H. (1970), Response of apple cultivars to canker, *Nectria galligena. Annu. Rep. East Malling Res. Stn.* 1969 A53, 143–5.

Alston, F. H. (1977), Practical aspects of breeding for mildew (*Podosphaera*) resistance in apples. *Proc. Eucarpia Fruit Sec. Symp. Top Fruit Breed., 6th, Wageningen* 1976: 4–17.

Alston, F. H. and Briggs, J. B. (1970), Inheritance of hypersensitivity to rosy apple aphid *Dysaphis plantaginea* in apple. *Can. J. Genetic. Cytol.* 12, 257–8.

Baumgartner, I. O., Patocchi, A., Frey, J. E., Peil, A. and Kellerhals, M. (2015), Breeding Elite Lines of Apple Carrying Pyramided Homozygous Resistance Genes Against Apple Scab and Resistance Against Powdery Mildew and Fire Blight, *Plant Molecular Biology Reporter* 33(5) 2015· DOI:10.1007/s11105-015-0858-x

Baumgartner, I. O., Kellerhals, M., Costa, F., Dondini, L., Pagliarani, G., Gregori, R., Tartarini, S., Leumann, L., Laurens, F. and Patocchi, A. (2016), Development of SNP-based assays for disease resistance and fruit quality traits in apple (*Malus × domestica Borkh.*) and validation in breeding pilot studies, *Tree Genetics & Genomes* (2016) 12: 35. doi:10.1007/s11295-016-0994-y.

Bénaouf, G., Parisi, L. and Laurens, F. (1997), Inheritance of *Malus floribunda clone 821* resistance to *Venturia inaequalis*, IOBC/WPRS Bulletin, 20, 1–7.

Belfanti, E., Silfverberg-Dilworth, E., Tartarini, S., Patocchi, A., Barbieri, M., Zhu, J., Vinatzer, B. A., Gianfranceschi, L., Gessler, C. and Sansavini, S. (2004), The HcrVf2 gene from a wild apple confers scab resistance to a transgenic cultivated variety, *Proc. Natl. Acad. Sci. U.S.A.* 101: 886–90.

Beresford, R. M. and Kim, K. S. (2011), Identification of regional climatic conditions favorable for development of European canker of apple, *Phytopathology*, 101: 135–46.

Boller, T. and He, S. Y. (2009), Innate immunity in plants: an arms race between pattern recognition receptors in plants and effectors in microbial pathogens, *Science*, 324(5928): 742–4.

Borejsza-Wysocka, E., Norelli, J. L. Aldwinckle, H. S. and Malnoy, M. (2010), Stable expression and phenotypic impact of attacin E transgene in orchard grown apple, *BMC Biotechnol.* 10: 41. doi:10.1186/1472-6750-10-41

Broggini, G. A. L., Wöhner, T., Fahrentrapp, J., Kost, T. D., Flachowsky, H., Peil, A., Hanke, M. V., Richter, K., Patocchi, A. and Gessler, C. (2014), Engineering fire blight resistance into the apple cultivar 'Gala' using the FB_MR5 CC-NBS-LRR resistance gene of Malus x robusta 5, *Plant Biotech. J.* 12: 728–33.

Brown, A. G. (1959), The inheritance of mildew resistance in progenies of the cultivated apple, *Euphytica* 8, 81–8.

Brown, S. K. and Maloney, K. E. (2003), Genetic improvement of apple: breeding, markers, mapping and biotechnology. In Ferree, D. C. and Warrington, I. J. (Eds), *Apples: Botany, Production And Uses.* CABI Publishing, Wallingford, United Kingdom, pp. 31–61.

Brown, S. K. and Maloney, K. E. (2008), Scab-resistant cultivars (varieties), *NY Fruit Quarterly* 16 (4): 3–6.

Bus, V. G. M. (2006), Gene-for-gene relationships and durable resistance to apple scab. In Mercer, C. F. (Ed.), *Proceedings of the 13th Australasian Plant Breeding Conference*. Breeding for Success: Diversity in Action, Christchurch, NZ, pp. 1159–69.

Bus, V. G. M., Basset, H. C. M., Bowatte, D., Calenge, F., Celton, J.-M., Durel, C.-E., Malone, M. T., Patocchi, A., Ranatunga, A. C., Rikkerink, E. H. A., Tustin, D. S., Zhou, J. and Gardiner, S. (2008), Genome mapping of three major resistance genes to woolly apple aphid (*Eriosoma lanigerum* Hausm.). *Tree Genetics & Genomes* 4: 233–6. doi 10.1007/s11295-007-0103-3

Bus, V. G. M., Esmenjaud, D., Buck, E. and Laurens, F. (2009), Application of genetic markers in rosaceous crops. In Folta, K. M. and Gardiner, S. E. (Eds), *Genetics and Genomics of Rosaceae.* Springer-Verlag, Berlin Heidelberg, pp. 563–600.

Bus, V. G. M., Rikkerink, E. H. A., Caffier, V., Durel, C. E. and Plummer, K. M. (2011), Revision of the nomenclature of the differential host-pathogen interactions of *Venturia inaequalis* and *Malus*, *Ann. Rev. Phytopathol.* 49: 391–413.

Caffier, V. and Laurens, F. (2005), Breakdown of *Pl2*, a major gene of resistance to apple powdery mildew, in a French experimental orchard, *Plant Pathol.* 54: 116—24.

Calenge, F., van der Linden, C. G., van de Weg, W. E., Schouten, H. J., van Arkel, G., Denancé, C. and Durel, C.-E. (2005), Resistance genes analogues identified through the NBS-profiling method map close to major genes and QTL for disease resistance in apple, *Theo. Appl. Genet.* 110: 660–8.

Cappellini, R. A., Ceponis, M. J. and Lightner, G. (1987), Disorders in apple and pear shipments to the New York market, 1972–1984, *Plant Dis.* 71: 852–6.

Chevalier, M., Lespinasse, Y. and Renaudin, S. (1991), A microscopic study of the different classes of symptoms coded by the *Vf* gene in apple for resistance to scab (*Venturia inaequalis*), *Plant Pathol.* 40: 249–56.

Crandall, C. S. (1926). Apple breeding at the University of Illinois. *Illinois Agric. Exp. Stn. Bull.* 275: 341–600.

Crosby, J. A., Janick, J., Pecknold, P. C., Korban, S. S., O'Conner, P. A., Ries, S. M., Goffreda, J. and Voodeckers, A. (1992), Breeding apples for scab resistance: 1945–1990, *Fruit Varieties J.* 46: 145–66.

Crosby, J. A., Janick, J., Pecknold, P. C., Goffreda, J. C. and Korban, S. S. (1994), 'Enterprise' apple, *HortSci.* 29(7): 825–6.

Donghoon, S., Hunjoohg, K. and Yangyik, S., et al. (2011), Influence of defoliation by *Marssonina* blotch on vegetative growth and fruit quality in'Fuji'/M.9 apple tree, *Korean J. Hortic. Sci.* 29: 531–8.

Dayton, D. F. (1977), Genetic Immunity to Apple Mildew Incited by *Podosphaera, Hort. Sci.* 12: 225–6.

Dunemann, F., Peil, A., Urbanietz, A. and Garcia-Libreros, T. (2007), Mapping of the apple powdery mildew resistance gene *Pl1* and its genetic association with an NBS-LRR candidate resistance gene, *Plant Breed.* 126: 476–81

Durel, C.-E., Parisi, L., Laurens, F., van de Weg, W. E., Liebhard, R. and Jourjon, M. F. (2003), Genetic dissection of partial resistance to race 6 of *Venturia inaequalis* in apple. *Genome* 46: 224–34.

Durel, C.-E., Denancé, C. and Brisset, M.-N. (2009), Two distinct major QTL for resistance to fire blight co-localize on linkage group 12 in apple genotypes 'Evereste' and *Malus floribunda* clone 821, *Genome* 52: 1–9.

Easterbrook, M. A. and Palmer, J. W. (1996), The relationship between early-season leaf feeding by apple rust mite, *Aculus schlechtendali* (Nal.), and fruit set and photosynthesis of apple. *J. Hortic. Sci.* 71: 939–44.

Emeriewen, O., Richter, K., Kilian, A., Zini, E., Hanke, M.-V., Malnoy, M. and Peil, A. (2014), Identification of a major quantitative trait locus for resistance to fire blight in the wild apple species *Malus fusca*. *Mol. Breed.* 34: 407–19.

Fahrentrapp, J., Broggini, G. A., Kellerhals, M., Peil, A., Richter, K., Zini, E. and Gessler, C. (2013), A candidate gene for fire blight resistance in *Malus × robusta* 5 is coding for a CC–NBS–LRR. *Tree Genet. Genom.* 9(1): 237–51.

Felix, G., Duran, J. D., Volko, S. and Boller, T. (1999), Plants have a sensitive perception system for the most conserved domain of bacterial flagellin. *The Plant J.* 18(3): 265–76.

Flor, H. H. (1971), Current status of the gene-for-gene concept. *Ann. Rev. Phytopathol.* 9(1): 275–96.

Fischer, C. (1993). Results in the resistance breeding of apple in Dresden-Pillnitz, *Acta Horticult.* 347: 163–8

Fischer, M. and Fischer, C. (1999), Evaluation of *Malus* species and cultivars at the Fruit Genebank Dresden-Pillnitz and its use for apple resistance breeding. *Genetic Res. Crop Evol.* 46: 235–41.

Francia, E., Tacconi, G., Crosatti, C., Barabaschi,. D., Bulgarelli, D., Dall'Aglio, E. and Valè, G. (2005), Marker assisted selection in crop plants, *Plant Cell, Tissue Organ Cult.* 82: 317–42.

Gallot, J. C., Lamb, R. C. and Aldwinckle, H. S. (1985), Resistance to powdery mildew from some small-fruited *Malus* cultivars, *HortSci.* 20: 1085–7.

Gardiner, S. E., Bus, V. G. M., Rusholme, R. L., Chagné, D. and Rikkerink, E. H. A. (2007), Apple. In Kole, C. (Ed.), *Genome Mapping and Molecular Breeding in Plants, Fruits and Nuts*, Vol. 4. Springer-Verlag Berlin Heidelberg, pp. 1–62.

Garkava-Gustavsson, L., Zborowska, A., Sehic, J., Rur, M., Nybom, H., Englund, J.-E., Lateur, M., Van de Weg, E. and Holefors, A. (2013), Screening of apple cultivars for resistance to European canker, *Neonectria ditissima, Acta Hort.* 976: 529–36.

Ghasemkhani, M. (2015), *Resistance Against Fruit Tree Canker In Apple*. Doctoral Thesis, Swedish University of Agricultural Sciences, Alnarp, p. 64.

Ghasemkhani, M., Liljeroth, E., Sehic, J., Zborowska, A. and Nybom, H. (2015), Cut-off shoots method for estimation of partial resistance in apple cultivars to fruit tree canker caused by *Neonectria ditissima, Acta Agric. Scand. Sect. BSoil PlantSci.* 65: 412–21. doi:10.1080/09064710.2015.10 16101

Gessler, C., Patocchi, A., Sansavini, S., Tartarini, S., Gianfranceschi, L. (2006), *Venturia inaequalis* resistance in apple. Crit. Rev. Plant Sci. 25: 473–503.

Gómez-Cortecero, A., Saville, R. J., Scheper, R. W. A., Bowen, J. K., Agripino De Medeiros, H., Kingsnorth, J., Xu, X. and Harrison, R. J. (2016), Variation in Host and Pathogen in the *Neonectria/Malus* Interaction; toward an Understanding of the Genetic Basis of Resistance to European Canker. *Front. Plant Sci.* 7: 1365. doi:10.3389/fpls.2016.01365

Grove, G. G. (1990), *Nectria* canker. In Jones, A. L. and Aldwinckle, H. S. (Eds), *Compendium of Apple And Pear Diseases*. St. Paul, APS, pp. 34–5.

Harada, Y., Sawamura, K. and Konno, K. (1974), *Diplocarpon mali* sp. nov., the perfect state of apple blotch fungus *Marssonina coronaria*, *Ann. Phytopathol. Soc. Jpn*. 40: 412–18.

Ingle, R. A., Carstens, M. and Denby, K. J. (2006), PAMP recognition and the plant–pathogen arms race. *Bioessays*, 28(9): 880–9.

Janick, J., Crosby, J. A., Pecknold, P. C., Goffreda, J. C. and Korban, S. S. (1995), 'Co-op 32' (Pristine™) apple. *HortSci*. 30: 1312–13.

Janick, J., Goffreda, J. C. and Korban, S. S. (2000), 'Co-op 25' (Scarlet O'Hara™) Apple, *HortSci*. 35(1): 150–1.

Janick, J., Goffreda, J. C. and Korban, S. S. (2004a), 'Co-op 29' (Sundance™) apple, *HortSci*. 39: 450–1.

Janick, J., Goffreda, J. C. and Korban, S. S. (2004b), 'Co-op 33' (PixieCrunch™) apple, *HortSci*. 39: 452–3.

Janick, J., Goffreda, J. C. and Korban, S. S. (2006), 'Co-op 39' (CrimsonCrisp™) apple, *HortSci*. 41: 465–6.

Jänsch, M., Broggini, A. L., Weger, J., Bus, V. G. M., Gardiner, S. E., Bassett, H. and Patocchi, A. (2015), Identification of SNPs linked to eight apple disease resistance loci, *Mol Breed*. 35: 45. doi 10.1007/s11032-015-0242-4

Jones, A. L. and Aldwinckle, H. S. (1990), *Compendium of Apple and Pear Diseases*, Am. Phytopath. Soc. Press, St. Paul, MN.

Jones, J. D. and Dangl, J. L. (2006), The plant immune system, *Nature*, 444(7117): 323–9.

Jurick, W. M., II, Janisiewicz, W. J., Saftner, R. A., Vico, I., Gaskins, V. L., Park, E., Forsline, P. L., Fazio, G. and Conway, W. S. (2011), Identification of wild apple germplasm (Malus sp.) accessions with resistance to the postharvest decay pathogens Penicillium expansum and Colletotrichum acutatum. *Plant Breed*. 130: 481–6.

Kaku, H., Nishizawa, Y., Ishii-Minami, N., Akimoto-Tomiyama, C., Dohmae, N., Takio, K., Minami, E. and Shibuya, N. (2006), Plant cells recognize chitin fragments for defense signaling through a plasma membrane receptor. *Proceed. Nat. Acad. Sci*. 103(29): 11086–91.

Kellerhals, M. (1989), Breeding of pome fruits with stable resistance to diseases. Development of disease resistant pome fruit varieties, *OILB Working Group 'Integrated Control of Pome Fruit Diseases Vol. II, WPRS Bulletin XII/6*, pp. 116–29.

Kellerhals, M., Angstl, J. and Weibel, F. (2003), Portrait schorfresistenter Apfelsorten, *Schweiz. Z. Obst- und Weinbau* 19: 7–14.

Kellerhals, M., Szekely, T., Sauer, C., Frey, J. E. and Patocchi, A. (2009), Pyramiding scab resistances in apple breeding. *Erwerbs-Obstbau* 51: 21–8.

Kellerhals, M., Szalatnay, D., Hunziker, K., Duffy, B., Nybom, H., Ahmadi-Afzadi, M., Höfer, M., Richter, K., and Lateur, M. (2012), European pome fruit genetic resources evaluated for disease resistance. *Trees* 26: 179–89.

Kellerhals, M., Baumgartner, I. O., Leumann, L., Lussi, L., Schütz, S. and Patocchi, A. (2014), Breeding high quality apples with fire blight resistance, *Acta Hort*. 1056: 225–30.

Khan, M. A., Duffy, B., Gessler, C. and Patocchi, A (2006), QTL mapping of fire blight resistance in apple, *Mol Breed*. 17: 299–306

Khan M. A., Durel C.-E., Duffy B., Drouet D., Kellerhals M., Gessler C. and Patocchi A (2007), Development of molecular markers linked to the 'Fiesta' linkage group 7 major QTL for fire blight resistance and their application for marker-assisted selection, *Genome* 50: 568–77.

Khanizadeh, S., Laurens, F., Lespinasse, Y., Groleau, Y., Cousineau, J. and Carisse, O. (2003), 'Galarina' apple. *HortSci.* 38: 477–8.

Knight, R. L. and Alston, F. H. (1968), Sources of field immunity to mildew (*Podosphaera*) in apple. *Ca. J. Genet. And Cytol.* 10: 294–8.

Kost, T. D., Gessler, C., Jänsch, M., Flachowsky, H., Patocchi, A. and Broggini, G. A. L. (2015), Development of the First Cisgenic Apple with Increased Resistance to Fire Blight. *PLoS ONE* 10(12): e0143980. doi:10.1371/journal.pone.0143980

Korban, S. S., Goffreda, J. C. and Janick, J. (2003), 'Co-op 43' (Juliet™), *HortSci.* 38: 144–5.

Korban, S. S. and Tartarini, S., (2009). Apple Structural Genomics. In Folta, K. M. and Gardiner, E. (Eds), *Genetic and Genomics of Rosaceae*, Springer, Plant Genetics and Genomics: Crops and Models 6, doi 10.1007/978-0-387-77491-6 1

Laloi, G., Vergne, E., Durel, C. E., LeCam, B. and Caffier, V. (2016), Efficiency of pyramiding of three quantitative resistance loci to apple scab. *Plant Pathol.* doi:10.1111/ppa.12581.

Lamb, R. C., Aldwinckle, H. S. Way, R. D. and Terry, D. E. (1979), 'Liberty' apple, *HortSci.* 14: 757–8.

Laurens, F. (1999), Review of the current apple breeding programmes in the world: objectives for scion cultivar improvement, *Acta Hort.* 484: 163–8. https://doi.org/10.17660/actahortic.1998.484.26

Laurens, F., Chevalier, M., Dolega, E., Gennari, F., Goerre, M., Fischer, C., Kellerhals, M., Lateur, M., Lefrancq, B., Parisi, L., Schouten, H. J. and Tartarini, S. (2004), Local European cultivars as sources of durable scab resistance in apple, *Acta Hort.* 663: 115–21.

Laurens, F., Lespinasse, Y. and Fouillet, A. (2005), A new scab resistant apple: 'Ariane'. *HortSci.* 40: 484–5.

Lebeda, A. (1992), The level of field resistance to *Bremia lactucae* in lettuce (*Lactuca sativa*) cultivars carrying the resistance gene DM11, *Plant Breed.* 108: 126–31.

Lee, D. H., Back, C. G., Win, N. K. K., et al., (2011), Biological characterization of *Marssonina coronaria* associated with apple blotch disease, *Mycobiology*, 39: 200–5.

Lefrancq, B., Lateur, M. and Rondia, A. (2004), Screening method for polygenic scab resistance within an apple breeding programme: relationship between early greenhouse screening test on young seedlings and their susceptibility in natural field conditions, *Acta Hort.* 663, doi:10.17660/ActaHortic.2004.663.143

Lespinasse, Y. (1983), Amélioration du pommier pour la résistance à l'oidium (*Podosphaera*). Premiers résultats concernant la virulence du champignon. *Proceedings IOBC Meeting: Disease Resistance As Component of Integrated Control In Orchards*, Angers, pp. 96–110.

Liebhard, R., Koller, B., Gianfranchesci, L. and Gessler, C. (2003a), Creating a saturated reference map for the apple (Malus x domestica Borkh) genome, *Theo. Appl. Genet.* 106: 1497–508.

Lind, K., Lafer, G., Schloffer, K., Imerhofer, G. and Meister, H. (2003), *Organic Fruit Growing*, CABI Publishing, Cambridge, MA.

Lindhout, P. (2002), The perspectives of polygenic resistance in breeding for durable disease resistance, *Euphytica* 124: 217–26.

McCallum, J. L., Tsao, R. and Zhou, T. (2002), Factors affecting patulin production by Penicillium expansum, *J. Food Prot.* 65: 1937.

MacHardy, W. E., Gadoury, D. M. and Gessler, C. (2001), Parasitic and biological fitness of *Venturia inaequalis*: relationship to disease management strategies. *Plant Dis.* 85: 1036–51.

McDonald, B. A. and Linde, C. (2002), The population genetics of plant pathogens and breeding strategies for durable resistance, *Euphytica*, 124(2): 163–80.

Medzhitov, R. and Janeway, C. A. (1997), Innate immunity: the virtues of a nonclonal system of recognition, *Cell*, 91(3): 295–8.

Minarro, M. and Dapena, E. (2008), Tolerance of some scab resistant apple cultivars to the rosy apple aphid, *Dysaphis plantaginea, Crop Protection* 27: 391–5.

Norelli, J. L., Wisniewski, M.. and Droby, S. (2014), Identification of a QTL for Postharvest Disease Resistance to *Penicillium expansum* in *Malus sieversii*. *Acta Hort.* 1053: 199–203.

Patocchi, A., Gianfranceschi, L. and Gessler, C. (1999a). Towards the map-based cloning of *Vf:* Fine and physical mapping of the *Vf*-region, *Theor. Appl. Genet.* 99: 1012–17.

Patocchi, A., Vinatzer, B. A., Gianfranceschi, L., Tartarini, S., Zhang, H. B., Sansavini, S. and Gessler, C. (1999b). Construction of a 550 kb BAC contig spanning the genomic region containing the apple resistance gene Vf, *Mol. Gen. Genet.* 262: 884–91

Patocchi, A., Frei, A., Frey, J. E. and Kellerhals, M. (2009), Towards improvement of marker assisted selection of apple scab resistant cultivars: Venturia inaequalis virulence surveys and standardization of of molecular marker alleles associated with resistance genes. *Mol Breeding* 24: 337. doi:10.1007/s11032-009-9295-6

Pagliarani, G., Dapena, E., Miñarro, M., Denancé, C., Lespinasse, Y., Rat-Morris, E., Troggio, M., Durel, C.-E., and Tartarini, S. (2016), Fine mapping of the rosy apple aphid resistance Dp-fl locus on Linkage Group 8 of the apple cultivar'Florina'. *Tree Genet. Genom.* 12: 56. doi:10.1007/s11295-016-1015-x

Parravicini, G., Gessler, C., Denance, C., Lasserre-Zuber, P., Vergne, E., Brisset, M. N., Patocchi, A., Durel, C. E., Broggini, G. A. L. (2011), Identification of serine/threonine kinase and nucleotide-binding site-leucine-rich repeat (NBS-LRR) genes in the fire blight resistance quantitative trait locus of apple cultivar 'Evereste', *Mol. Plant Pathol.* 12: 493–505.

Parisi, L., Lespinasse, Y., Guillaumes, J. and Krüger, J. (1993), A new race of *Venturia inaequalis* virulent to apples with resistance due to the *Vf* gene. *Phytopathology* 83: 533–7.

Parisi, L. and Lespinasse, Y. (1996), Pathogenicity of *Venturia inaequalis* strains of race 6 on apple clones (*Malus* sp), *Plant Dis.* 80: 1179–83.

Parlevliet, J. E. (1993), What is durable resistance, a general outline, In Jacobs, T. and Parlevliet, J. E. (Eds), *Durability of Disease Resistance*, Kluwer Academic Publishers, Dordrecht, The Netherlands, pp. 23–9.

Parlevliet, J. E. (2002), Durability of disease resistance against fungal, bacterial and viral pathogens; present situation, *Euphytica* 124: 147–56.

Peil, A., Garcia-Libreros, T., Richter, K., Trognitz, F. C., Trognitz, B., Hanke, M. V. and Flachowsky, H. (2007), Strong evidence for a fire blight resistance gene of *Malus robusta* located on linkage group 3, *Plant Breed.* 126: 470–5.

Peil, A., Hanke, M. V., Flachowsky, H., Richter, K., Garcia-Libreros, T., Celton, J. M., Gardiner, S., Horner, M. and Bus, V. (2008), Confirmation of the fire blight QTL of *Malus x robusta* 5 on linkage group 3, *Acta Hortic.* 793: 297–303.

Peil, A., Bus, V. G. M., Geider, K., Richter, K., Flachowsky, H. and Hanke, M. V. (2009), Improvement of fire blight resistance in apple and pear, *Int. J. Plant Breed.* 3: 1–27.

Peil, A., Kellerhals, M., Höfer, M. and Flachowsky, H. (2011), Apple Breeding – From the origin to Genetic Engineering, In Flachowsky, H. and Hanke, M.-V. (Eds), *Methods in Temperate Fruit Breeding, Fruit, Vegetable and Cereal Science and Biotechnology* (Spec. Iss. 1): 118–38.

Pessina, S., Angeli, D., Martens, S., Visser, R. G. F., Bai, Y., Salamini, F., Velasco, R., Schouten, H. and Malnoy, M. (2016), The knock-down of the expression of *MdMLO19* reduces susceptibility to powdery mildew (*Podosphaera leucotricha*) in apple (*Malus domestica*), *Plant Biotechnol. J.* 14: 2033–44. doi:10.1111/pbi.12562

Plante, F., Hamelin, R. C. and Bernier, L. (2002), A comparative study of genetic diversity of populations of *Nectria galligena* and *N. coccinea var. faginata* in North America, *Mycological Res.* 106: 183–93.

Rikkerink, H. A., Hilario-Andrade, E. M., Dare, A. P., Gardiner, S. E., Yoon, M. and Bus, V. G. M. (2007), Resistance gene and uses thereof. New Zealand Patent Application 563032, filed 31 Oct 2007.

Roberts, A. L. and Crute ,I. R. (1994), Apple scab resistance from *Malus floribunda 821* (Vf) is rendered ineffective by isolates of *Venturia inaequalis* from *Malus floribunda*, *Norw. J. Agr. Sci. Suppl.* 17: 403–6.

Schouten, H. J., Krens, F. A. and Jacobsen, E. (2006), Cisgenic plants are similar to traditionally bred plants—International regulations for genetically modified organisms should be altered to exempt cisgenesis, *Embo Rep.* 7(8): 750–3. doi:10.1038/sj.embor.7400769 PMID: 16880817

Seglias, N. P. and Gessler, C. (1997), Genetics of apple powdery mildew resistance from *Malus zumi* (PI2). *IOBC/WPRS Bulletin* 20: 195–208.

Shay, J. R. and Hough, L. F. (1952), Evaluation of apple scab resistance in selections of *Malus*, *Amer. J. Bot.* 39: 288–97.

Smith, C. M. (2005), *Plant Resistance to Arthropods: Molecular and Conventional Approaches*. Springer Science & Business Media, Dordrecht, The Netherlands

Sharma, J. N., Sharma, A. and Sharma, P. (2004), Out-break of *Marssonina* blotch in warmer climates causing premature leaf fall problem of apple and its management, *Acta Hort.* 662: 405–9. doi: 0.17660/ActaHortic.2004.662.61

Soufflet-Freslon, V., Gianfranceschi, L., Patocchi, A. and Durel, C. E. (2008), Inheritance studies of apple scab resistance and identification of *Rvi14*, a new major gene that acts together with other broad-spectrum QTL, *Genome* 51: 657–67.

Spieser, F., Graf, B., Walther, P. and Noesberger, J. (1998), Impact of apple rust mite (Acari: Eriophyiidae) feeding on apple leaf gas exchange and leaf color associated with changes in leaf tissue, *Environ. Entomol* 27: 1149–56.

Stoeckli, S., Mody, K., Gessler, C., Christen, D. and Dorn, S. (2009a), Quantitative trait locus mapping of resistance in apple to *Cydia pomonella* and *Lyonetia clerkella* and of two selected fruit traits, *Ann. Appl. Biol.* 154: 377–87.

Stoeckli, S., Mody, K., Patocchi, A., Kellerhals, M. and Dorn, S. (2009b), Rust mite resistance in apple assessed by quantitative trait loci analysis, *Tree Genet. Genom.* 5: 257–67.

Vanblaere, T., Flachowsky, H., Gessler, C. and Broggini, G. A. L. (2014), Molecular characterization of cisgenic apple lines of the cultivar `Gala` carrying the Rvi6 scab resistance gene, *Plant Biotechnol. J.* 12(1): 2–9.

Vinatzer, B. A., Zhang, H-B. and Sansavini, S. (1998), Construction and characterization of a bacterial artificial chromosome library of apple. *Theor. Appl. Genet.* 97: 1183–90.

Vinatzer, B. A., Patocchi, A., Gianfranceschi, L., Tartarini, S., Zhang, H. B., Gessler, C. and Sansavini, S. (2001), Apple contains receptor-like genes homologous to the *Cladosporium fulvum* resistance gene family of tomato with a cluster of genes cosegregating with *Vf* apple scab resistance. *Mol. Plant Microb. Interact.* 14: 508–15.

Vinatzer, B. A., Patocchi, A., Tartarini, S., Gianfranceschi, L., Sansavini, S. and Gessler, C. (2004), Isolation of two microsatellite markers from BAC clones of the Vf scab resistance region and molecular characterization of scab-resistant accessions in Malus germplasm. *Plant Breed.* 123: 321–6.

Van De Weg, W. E. (1987), Note on an inoculation method to infect young apple seedlings with *Nectria galligena* Bres, *Euphytica* 36: 853–4. doi:10.1007/BF00051869

Van de Weg, W. E. (1989), Screening for resistance to *Nectria galligena* Bres. In cut shoots of apple. *Euphytica* 42: 233–40. doi:10.1007/BF00034459

Vogt, I., Wöhner, T., Richter, K., Flachowsky, H., Sundin, G. W., Wensing, A., Savory, E. A., Geider, K., Day, B., Hanke, M.-V. and Peil, A. (2013), Gene-for-gene relationship in the host–pathogen system *Malus× robusta* 5–*Erwinia amylovora*. *New Phytologist*, 197(4): 1262–75.

Weber, R. W. S. (2014), Biology and control of the apple canker fungus *Neonectria ditissima* (syn. N. *galligena*) from a Northwestern European perspective. *Erwerbs-Obstbau* 56: 95–107.

Weibel, F. and Häseli, A. (2003), Organic apple production – with emphasis on European experiences. In Ferree, D. C. and Warrington, I. J. (Eds), *Apples Botany, Production and Use*, pp. 551–83. CABI Publishing, Cambridge, MA.

Wouters, M. F. A. and Speijers, G. J. A. (1996), Patulin Food Additives Series 35 Toxicological Evaluation of Certain Food Additives and Contaminants, vol. 35. World Health Organization, Geneva, pp. 377–402, http://www.inchem.org/documents/jecfa/jecmono/v35je16.htm

Xu, X.-M. and Robinson, J. D. (2010), Effects of fruit maturity and wetness on the infection of apple fruit by *Neonectria galligena*. *Plant Pathol.* 59: 542–7.

Zhou, Q., Gao, H., Wanga, M., Xue, Y., Guob, Y. Z., Wana, Y. Z. and Zhaoa, Z. Y. (2012), Characterization of defense-related genes in the 'Qinguan' apple in response to *Marssonina coronaria*. doi:10.1016/j.sajb.2012.01.005

Part 4

Sustainability

Chapter 19

The economics of apple production

R. Karina Gallardo, Washington State University, USA; and Hildegard Garming, Thünen Institute of Farm Economics, Germany

1 Introduction

Typically there are three dimensions of sustainability identified: environmental, social and economic (United Nations Environmental Program, 2015; Goodland and Daly, 1993). These three dimensions of sustainability must be integrated in a 'coherent, holistic, comprehensive and balanced' manner to accomplish the ultimate goal of enhancing human and planet welfare (United Nations Environmental Program, 2015). These three aspects are intrinsically interlinked and overlapped when dealing with the agricultural industry, in particular specialty crops (e.g. fruits, vegetables, tree nuts, dried fruits, horticultural and nursery crops).

Apple production, as a perennial crop production system with a time span of at least 20 years, is a good example to illustrate the concept of sustainability.

1.1 Environmental sustainability

As with all agricultural production, apple production depends on the natural environment, and the availability of water, fertile soils and appropriate climatic conditions. Decision-makers in apple operations must, and an increasing number of them have already started, search for alternative production methods that will be less disruptive to the natural ecosystems hosting their apple operation (Jones et al., 2009). Any production practices that compromise these basic functions of the agro-ecosystem would also impact the economic sustainability of operations. Management of pests and diseases is often seen as conflicting with environmental goals; however, studies show that investing in pest management programmes that are less disruptive to the ecosystem could lead to diminishing input

http://dx.doi.org/10.19103/AS.2016.0017.27

(pesticide) costs, among other benefits (Jones et al., 2016). For example, in the U.S. Pacific Northwest (PNW), using integrated pest management based on conservation biological control could lead to reducing pesticide applications to control secondary pests because beneficial insect populations are sufficient to control secondary pests through predation (Gallardo et al., 2016).

1.2 Social sustainability

The human aspect of an apple operation is of primal importance, given that apple production is heavily labour dependent (Gallardo and Brady, 2015). Hence to guarantee sustainable apple production, enterprise decision-makers must invest in their human capital by guaranteeing a healthy working environment, access to health services, advancing workers by making training opportunities accessible and so on. Investing in human and environmentally oriented agricultural practices translates into enhancing profit margins for agricultural operations. Investing in human capital could ensure the availability of worker pools during periods of peak demand for labour. This is key to guarantee the economic sustainability of the agricultural enterprise in times when labour pools are perceived to be in less abundance and mechanization alternatives are not yet at their optimum (Taylor et al., 2012; Charlton and Taylor, 2016).

Social and environmental sustainability are not only conditions for producing apples, but also are increasingly a factor for marketing them.

1.3 Economic sustainability

Profitability of apple production is the main incentive for farmers to maintain their operations and hence could be considered the pillar of the human and the environmental dimensions. In order to ensure the economic sustainability of agricultural operations, it is necessary to make business- and production-related decisions based on reliable information. This is where production costs become a preponderant factor because these studies provide such necessary information. A production cost study is a general guide for evaluating the feasibility of producing a crop in a given location over a given period. This type of study provides information considered crucial in assessing the potential profitability of the agricultural operation and is essential to farm and market planning. In addition, these studies are helpful in evaluating profits and alternative strategies to maximize returns from land, capital and labour (Missouri Extension, 2011; Kay et al., 2008; Garrison and Noreen, 1994).

In the next section, basic approaches to the analysis of production costs are presented and their application to apple production is described. Data on apple production costs in the United States and other countries are provided in the third section.

2 Cost analysis of apple production

Production cost studies are highly important for specialty crops such as apples. This is because specialty crops, different from arable crops, require intensive crop management.

Hence, the value-added and the general production costs for specialty crops are higher compared to other crops.

The establishment of an apple orchard is a major investment and can be expected to pay off only after a number of years. In most instances, the production life cycle of an apple orchard is at least 15 years. Within that time span, the initial years are considered establishment years when the tree has not reached its full maturation; hence, it is not in full production or full bearing. Only after four to five years is the tree in its full production of fruits. Yields across full production years are also highly variable, which induces uncertainty about yield levels and revenues.

2.1 Classification of production costs

When analysing production costs, it is useful to follow a common classification of costs, which enables a comparison and assessment of cost structures of different enterprises or production systems. For a comprehensive analysis, production costs include both cash and non-cash (imputed) costs.

Cash costs include direct or variable costs such as expenses for seeds/trees, fertilizers, plant protection, wages for seasonal and permanent labour, and overhead costs such as fuel, energy, water, farm office, advisory fees and insurances, as well as the costs for renting land and capital (interests) (Fig. 1). *Imputed costs* refer to depreciation and opportunity

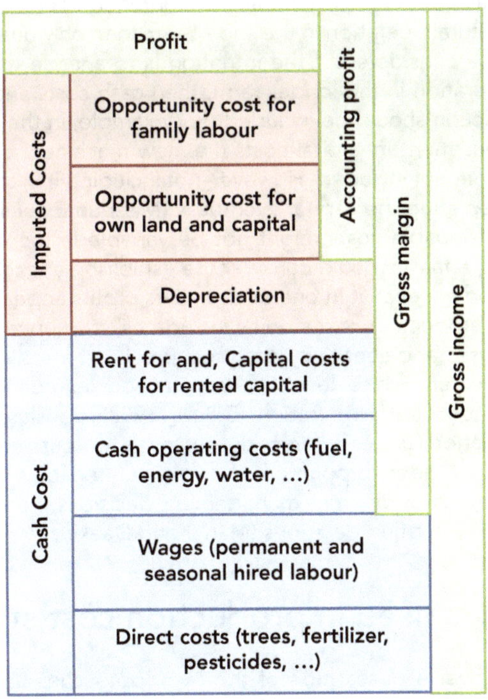

Figure 1 Classification of costs and revenues of apple production.

costs. Opportunity costs are the forgone income that could have been realized if the funds would have been invested in an alternative activity, or if an input was sold or rented. Examples of opportunity costs are unpaid family and operator labour, pre-owned machinery and pre-owned land (Langemeier, 2015; Gallardo et al., 2010).

Other complementary cost categories can be used, according to the objective of the analysis: The distinction between variable versus fixed costs is a common approach for farm management. *Variable costs* would vary depending on the expected yield per unit of production. They include all production cost centres or field activities, the inputs for each activity and labour associated with each. For apple production, this includes labour costs associated with winter pruning, flower thinning, green fruit thinning, application of fertilizers, pesticides, plant growth regulators, among others. *Fixed costs* would not vary with the expected yield per unit of production. These costs will generally be calculated for the whole farm enterprise and be allocated for the unit of production. Examples of these costs are depreciation rates, cost of opportunity interest rates and management costs.

To determine profit, first, the overall revenue or gross income is calculated, which is the physical output multiplied by the market price. Key figures usually used to assess profitability of an enterprise are the gross margin (gross income minus direct costs and seasonal labour costs), accounting profit (gross income minus cash costs and depreciation) and profit (gross income minus total cost of production), illustrated in Fig. 1.

It should be noted that agricultural operations do not always have profits above zero. Owing to varying yield levels as well as output prices, full cost recovery is not achieved in all years. Therefore, different levels of production costs are analysed, depending on the intention of the assessment. In general if the intention is to analyse the economic situation of the agricultural operation in the short term, then only direct costs and seasonal labour costs should be considered. If the intention is to analyse the economic situation of the agricultural operation in the longer term, then cash costs and imputed costs, that is, total cost of production should be included. For example, in the short term, it is often considered that reinvestment in capital costs (i.e. new machinery and equipment) is not needed, and is therefore not included. However, not including it in the long-term analyses involves the risk of depleting the capital eventually (Missouri Extension, 2011). Note that for apple production, variable costs might not be variable in the strict sense; once the orchard is established, a farmer could consider the establishment costs as fixed costs and would continue to produce even if in one or two years profits appear as negative.

In some instances, production cost studies include risk and uncertainty considerations with break-even analyses and sensitivity analyses. Break-even analyses offer information on the minimal output and price levels to be attained in order to ensure economic sustainability of the agricultural operation (Dillon, 1993). Sensitivity analysis generally consists of profit simulations under different output levels, output and input prices (Ludena et al., 2003). In general, the incorporation of uncertainty enables further exploration of the economic implications of different management decisions, hence enabling the farm operator to make more informed decisions (Maher and Deakin, 1994).

2.2 Approaches used in production cost studies

This subsection will cover a description of the two approaches used when developing production cost studies, analyses of financial statements and accounts, and the engineering approach.

2.2.1 Analysis of financial statement and accounts

In many countries, regular statistical analysis of farm accounts or financial statements based on large representative samples of farms is conducted as part of public agricultural information systems. In the United States, the Economic Research Service (ERS), a branch of the U.S. Department of Agriculture, publishes indicators of economic performance for farm businesses by resource region and commodity specialization (USDA, 2016). Financial performance information includes farms' receipts and expenses, gross and net value added, net cash farm income and net farm income (USDA, 2016). The ERS publishes estimates of farm income and wealth based on annual responses to USDA's Agricultural Resource Management Survey (ARMS), as well as information provided by the National Agricultural Statistics Service (NASS), Farm Service Agency (FSA), Risk Management Agency (RMA) and various other sources of sector-level information (USDA, 2016).

In the European Union, the Farm Accounting Data Network (FADN) provides similar data on the economic situation of farms. The major objective is to create a basis to assess the impact of agricultural policies on the income and development of farms. The database is a representative sample of farms, stratified by farm type and size in each member state of the European Union (EU COM, 2016; Marongui and Cesare, 2013). For this purpose, a harmonized approach of data collection and a common set of variables are used in the analysis. The studies provide an assessment of economic performance with the possibility of a comparison of different farm types on a national level as well as across European countries.

With regard to apple production and specialty crop production in general, a major drawback is that these specialized farms are not adequately represented in the samples. Compared to other arable crops or animal production, specialized fruit and vegetable farms are few and cultivated land is less. The statistical analyses generally focus on those crops with major production area.

Analysis of financial statements and accounts is targeted at the sector level and aims to provide a basis for policy-making. Although data can be disaggregated for different farm types and locations, the results are not appropriate for crop-specific cost analysis for farm management purposes.

2.2.2 Engineering approach

In order to generate crop-specific cost information that can be used for decision-making in farm management, models of production systems are often used. Such engineering approaches to economic analysis start from the physical description of the production process (see Isermeyer, 2012; Marongui and Cesaro, 2013; Sav, 1984). Based on specific assumptions on physical quantities and prices of all inputs, that is, a defined set of operations to be considered with all corresponding inputs, including labour and contractor costs, as well as average output, a detailed model of a production system or farm enterprise is constructed. Establishment and analysis of production system models therefore requires a detailed technical and economic understanding of the production process.

This type of model can be used either to construct an enterprise budget or to generate cost information for a typical farm. The enterprise budget provides a detailed analysis of average costs and benefits for a crop based on a comprehensive survey among growers. The objective is to assess the profitability of the investment in the apple orchard. For this purpose, an individual farm needs to adjust the underlying assumptions to their own specific conditions. In apple production, an enterprise budget usually refers to a specific variety, a composite of varieties, or a generic variety and includes the estimation of all expected costs and benefits from orchard establishment to the end of the utilization period (Gallardo, 2010).

A typical or representative farm represents the prevailing farm type for a specific commodity and region based on expert assessment and focus group discussions with farmers and technical advisers (Deblitz and Zimmer, 2005). Usually, the costs and revenues of production, referring to production area or to metric ton (hereafter ton) of output are computed for a specific year. Every year, prices and quantities of inputs and outputs are updated. The aim is to assess competitiveness of different typical farms or producing regions over time. A typical apple farm will include a set of varieties that are grown in the respective producing region (Garming et al., 2015). The whole farm is considered with land, capital (machinery and equipment) and labour resources. Resources are then allocated to the typical combination of farm enterprises or production systems. The allocation of overhead costs is done according to revenue shares. The annual updates of yields and prices generate a time series, which enables profitability to be assessed annually and sheds light on variability and risk.

The typical farm approach is the methodological base of the international network *agri benchmark* (Isermeyer, 2012), where production systems of various commodities around the globe are compared. Network members are agricultural economists in different organizations, for example, national research institutes, universities or farmer organizations, aiming to assess international competitiveness of the production systems in their countries and/or specific production regions. Using harmonized methods for data collection and analysis established in the standard operating procedure (Deblitz and Zimmer, 2005), a common database is created, where costs of production can be analysed and compared across countries and even commodities. One advantage of the typical farm approach is the rather inexpensive implementation, based on available statistics, local experts and focus group discussions with farmers, rather than the more expensive farm surveys with large samples. The *agri benchmark* network is coordinated by the Thünen Institute of Farm Economics in Germany and is organized into a number of networks focusing on different commodities, such as cash crops, meat production and horticulture. Under the *agri benchmark* 'Horticulture', so far only wine grapes and apple production are covered (Dirksmeyer et al., 2014).

Other approaches include the Farm Level Income and Policy Simulation Model (FLIPSIM) in the U.S. developed by Texas A&M University (Texas A&M University, Agricultural and Policy Research Center, 2016). This model uses accounting equations and probability distributions to simulate annual economic activities of a representative farm over a multiple year planning horizon. Another approach is the specialty crop representative farm models in California (Paggi, 2007). These are stochastic simulation models used to analyse the impacts of current and changing market conditions and government policies on farms yearly net income, cash flow position, financial ratios, and net present values of net income (Paggi, 2007).

2.3 Categories included in cost-of-production studies

In general, cost-of-production studies (enterprise budgets) are not intended to be a definitive guide to production practices, but they could be helpful in estimating the physical and financial requirements of comparable plantings (Gallardo et al., 2010). Because cost centres and costs change from operation to operation, specific assumptions are considered to formulate these studies. These assumptions may not fit every situation since production costs and returns vary across agricultural operations, depending on factors such as level of investment, production technology, efficiency in space and labour, and quality of management (Ludena et al., 2003). Cost estimations in the enterprise budget also vary depending on its intended use. It is often advised that to avoid drawing unwarranted conclusions for any particular agricultural operation, readers must closely examine the assumptions made in the production cost studies, and then adjust the costs and/or returns as appropriate for their own operation (Gallardo et al., 2010).

Typically, enterprise budgets summary tables for fruits have five sections. The first section is a summary of the projected total revenue from the agricultural operation, including expected yields and market prices per unit of production. The second section is an outline of all expected variable costs (e.g. operating costs), during a production period. It includes all production field activities, the inputs for each activity and labour associated with each. These costs would vary depending on the expected yield per unit of production. For example, labour costs associated with winter pruning, flower thinning, green fruit thinning, application of fertilizers, pesticides, plant growth regulators, among others. The third section is an outline of all fixed costs; these costs would not vary with the expected yield per unit of production but will generally be calculated for the whole farm enterprise and be allocated for the unit of production. Examples of these costs are depreciation rates, opportunity costs, interest rates and management costs. The fourth section is the estimation of the total cost of production per unit of production; generally it is the sum of variable and fixed costs. The fifth section includes the estimation of the net returns or profits, which are the total returns minus the total costs (University of Missouri Extension, 2011; Kay et al., 2008).

In the apple production areas of the western region of the United States, enterprise budgets could refer to a specific variety produced (e.g. budgets from Washington State University) or to a composite of varieties (e.g. budgets from University of California – Davis) or just refer to the plantation density considering a generic variety (e.g. budgets from Oregon State University). Enterprise budgets can either include establishment or full production years (e.g. Oregon State University) or the same study can include both establishment and full production (or full bearing) years (e.g. Washington State University) (Gallardo et al., 2010; Galinato and Gallardo, 2012; Gallardo and Galinato, 2012; West et al., 2012; Klonsky and Stewart, 2014). The difference between establishment and full production years relies on the fact that apple trees – and in general tree fruits – start producing fruit when the tree has reached some degree of maturity, which is not the year the tree was planted. Generally, apple trees start yielding significant amounts of fruit in the third year, and yield per tree increases gradually until the tree achieves full maturity or full production, which is in most instances the 6th or 7th year. Hence apple enterprise budgets could focus on either establishment (the years before trees reached full production) and full production years, or both.

The apple industry in the United States generally encompasses growers, packers, shippers and processors. Apples destined for the fresh market are generally shipped from the orchard either to a packing operation or channelled to direct marketing (United States International Trade Commission, 2010). Packing costs are considered in some enterprise budgets (e.g. Washington State University) as part of the operating costs as it is seen as

Table 1 Operating cost categories considered in establishment, maintenance and full production years of apples

Cost category	Establishment year (1st year)	Maintenance years (2nd to 4th year)	Full production years (5th to 20th year)
Soil preparation	X		
Set up/maintain trellising system	X	X	X
Set up irrigation system			X
Planting	X		
Binding/trellising	X	X	X
Run irrigation	X	X	X
Fertilization	X	X	X
Plant protection	X	X	X
Winter pruning		X	X
Summer pruning		X	X
Fruit thinning		X	X
Weed management	X	X	X
Harvest		X	X
Packing costs			X

Note: The 'X' in each cell means that the cost category related to each operation belongs to the establishment (1st year), maintenance year (2nd to 4th year) and full production year (5th to 20th year).

a service provided to and paid by the grower. See Table 1 for an illustration of the cost categories considered across production years.

3 Apple production costs around the world

This subsection presents data on the cost of production of apples. First, enterprise budgets for four apple varieties in Washington State are compared. Then results from typical apple farms in the *agri benchmark* Horticulture network are presented.

3.1 Case study: Washington State

During 2004–14, Washington State produced on average 60% of all apples in the United States. Total utilized production in Washington during the same period increased by 15%

(2.8–3.2 million tons). The increase in production volume in Washington is higher than the average increase in production volume in the United States at 8% (4.7 million tons in 2004 to 5.1 million tons in 2014).

On average from 2004–14, 81% of total utilized Washington apple production was sold in the fresh market. For the same time, packinghouse door prices for Washington fresh market apples ranged from $0.46–0.99/kg. This mirrors the United States range of $0.49–0.99/kg. From 2004–14, 33% of all Washington fresh apple shipments were exported to international markets, including Mexico, Canada, United Arab Emirates, Taiwan, India, Indonesia, Vietnam, Hong Kong, Saudi Arabia and Thailand. Average per capita consumption of fresh apples in the United States for 2004–14 remained stagnant at 7.6 ± 0.5 kg.

Over the last ten years, Washington State apple production moved from low-density standard rootstocks with abundant canopy plantings to high-density fruiting wall systems, due in part to the proliferation of dwarfing rootstocks.[1] These improved planting systems result in higher productivity levels per unit area, and more rapid yield increase in establishment years, but require increased upfront investments (e.g. trellis support system), and more intensive management than under traditional planting systems. Apple productivity in Washington increased by 19% from 2004 to 2014 (from 44.5 in 2004 to 55.3 t/ha in 2014), with yearly fluctuations mainly due to weather factors. The average productivity in Washington State for the last ten years was at 42.5 t/ha, which was higher than the average productivity in the United States at 31.4 t/ha.

Washington apple growers have diversified their range of apple varieties produced. Washington apple shipments 10 years ago (2003–4) show 'Red Delicious' represented 37%, 'Gala' and 'Golden Delicious' 14% and 'Granny Smith' and 'Fuji' 13% of total apple shipments. By 2013–14, 'Red Delicious' continued to be the dominant apple variety grown in Washington, comprising 30% of total apple shipments, followed by 'Gala' with 21%, 'Granny Smith' with 14%, 'Fuji' with 13%, 'Golden Delicious' with 8% and 'Honeycrisp' with 4%. It is expected that production of traditional apple varieties will continue to decrease in volume, and varieties with enhanced eating quality characteristics such as 'Honeycrisp' will expand their market share.

Results from an enterprise budget for 'Red Delicious' apples for 2014 concluded that the total production costs during full production (6th year) are estimated at $60792/ha. The break-even price for 'Red Delicious' apples is estimated at $984/t (considering 85% pack-out). Assuming a price of $989/t and a discount rate of 3%, the net present value (NPV) of the investment over the lifetime (30 years) of the orchard block (62 ha) is $675733. The estimated discounted payback period for the orchard investment, which gives the number of years to recover an investment while accounting for the time value of money, including all production costs, was estimated at 16.3 years (Galinato and Gallardo, 2016).

Results from a production cost study for 'Gala' apples in 2014 concluded that total production costs on a full production year (6th year) were at $60923/ha. The break-even price for 'Gala' apples was estimated at $1130/t (considering 80% pack-out). The NPV over the lifetime (20 years) of the orchard block (99 ha) was $1.4 million given a price of

1. Apple trees are not grown on their own roots, but they are propagated on rootstocks. Rootstocks are divided into three groups: dwarf, semi-dwarf and standard (Washington State Department of Agriculture, 2011). Dwarf rootstocks control the amount of wood in the tree, directing its energy into fruit production. The more dwarfing of the rootstock, the better suited it is for high-density plantings (>2471 trees/ha). Dwarf rootstocks require some type of support for the trees given the narrow canopy, thus they are associated with fruiting walls or planar systems. See chapter in this book by Musacchi and Greene.

Table 2 Basic assumptions for enterprise budgets of apples grown in Washington State

	Gala	Fuji	Honeycrisp	Red Delicious
Block size (ha)	104	52	27	64
Productive block size (ha)	99	49	25	62
FOB price ($/t)	1,263	1,230	1,549	989
Average pack-out (%)	80	80	72	85
In-row spacing (m)	1.2	1.2	0.9	1.2
Between row spacing (m)	3.0	3.0	3.0	3.7
Root stock	M9	M9	M9	M106
Life of planting (years)	20	20	15	30
Tree density (trees/ha)	2,691	2,691	3,588	2,241
Trellis system	Spindle trellis system			
Block architecture	Randomly trained w/18 radius from tree centre			Three-dimensional central leader

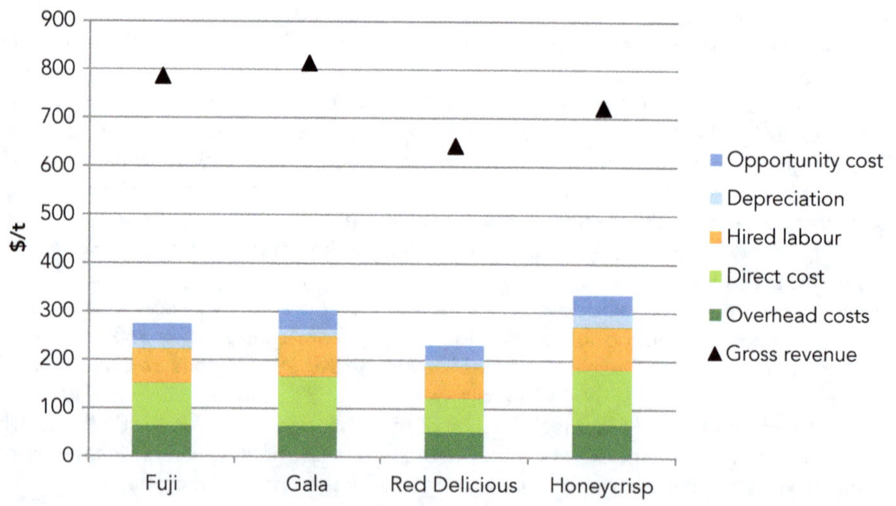

Figure 2 Average annualized production costs for 'Fuji', 'Gala', 'Red Delicious' and 'Honeycrisp' varieties in Washington State. Source: Based on Galinato and Gallardo, 2012 and 2016.

$1263/t and a discount rate of 4%. The estimated discounted payback period, including all production costs, was 10.7 years (Galinato et al., 2016).

'Fuji' apple production costs on a full production year (6th year) were at $68 034/ha for 2014. Assuming an 80% pack-out, the break-even price was estimated at $1094/t when grown under a spindle system. Assuming a price of $1230/t and a discount rate of 5%, the NPV over the lifetime (20 years) of the orchard block (49 ha) is $0.67 million. The estimated discounted payback period, including all production costs, was 10.8 years (Galinato and Gallardo, forthcoming).

'Honeycrisp' production costs on a full production year (6th year) were at $53094/ha for 2011. Assuming a 72% pack-out, the break-even price was at $1189/t. The NPV over the lifetime (15 years) of the orchard block (25 ha) is $439 796. The discounted payback period, including all production costs, is at 7.8 years (Galinato and Gallardo, 2012).

All enterprise budgets for apple varieties grown in Washington State are available free of charge at http://ses.wsu.edu/enterprise_budgets/. Each study presents a PDF summary and an Excel spreadsheet in which users can input their own information to tailor the study to their own operations.

3.2 International comparison by typical farm

Apples are traded on international markets in large quantities (see also Section 5), implying international competition for apple producers as well. Hence, it is useful to compare costs around major producing countries in the world. The *agri benchmark* 'Horticulture' network provides an annual benchmarking of apple production costs and farm-level profitability. Based on standardized methods to establish and update typical apple farms (see Section 2.3), detailed analyses of cost structures are available to the network members and, over time, trends in production and technologies can be identified (Dirksmeyer et al., 2014). So far, major apple production regions in Chile, Germany, Italy, South Africa and Switzerland are covered within the network. Production structures differ widely between the countries and regions; we present a description of production costs and returns for major apple-producing countries in the world.

Chile produces between 1.6 and 1.8 million tons of apples from about 37000 ha (ODEPA, 2015). Chile's apple industry is strongly export-oriented. In 2014, 800000 tons were exported as fresh apples. Production concentrates in the El Maule region, with 60% of apple production area and the O'Higgins region with another 27%. For Chile, one typical farm, medium sized (25 ha) and an average yield of 67 t/ha in the El Maule region, is included in the benchmarking (Table 3).

Germany has an apple production area of about 31500 ha, producing on average 950000 tons of apples (average 2002–12; Destatis, 2002, 2012). Variation in yields between years is high and can exceed 15% of the average yield. Apples are mainly consumed within the country; an additional 600000 tons are imported, 40% of which come from Italy. Per capita consumption of apples has decreased over the past few decades but is still high compared to other countries, with per capita consumption of more than 20 kg/year. In Germany, major production regions for apples are the Lake Constance and the Lower Elbe regions, with about one-third of national production each. These regions differ in farm structures and climate condition. In the South of Germany, near Lake Constance, farms tend to be smaller than in the Lower Elbe, which is the Northern area of the country, close to Hamburg. The size of the typical apple operation varies from 15 ha in the South to 40 ha in the North. Adding to the differences, Northern Germany appears to have a climatic disadvantage, with significantly less sunshine than Southern Germany. This has an effect on yield levels. On average, only 32 t/ha were harvested in 2012, compared to above 40 t/ha in the Lake Constance region.

Italy is the second largest apple-producing country in Europe with more than 2 million tons per year from 55000 ha (FAOstat, 2016). The Alpine regions of Trentino and Alto

Table 3 Typical apple farms in the *agri benchmark* 'Horticulture'

Country	Region	Farm apple area	Utilization period	Ø Yield in 2012	Planting density	Main varieties	Ø Pack out ratio	Mechanization of harvest
		ha	*years*	*t/ha*	*Trees/ha*		%	
Germany	Lake Constance (B)	15	20	46	2,500	Jonagold, Elstar, Gala	85	Self-propelled platform
	Lower Elbe Region (AL)	41	18	32	2,500	Elstar, Jonagold, Braeburn	92	Towed platform
Italy	Emilia Romagna (ER)	5	25	49	2,500	Fuji, Pink Lady, Modi, Gala	92	Self-propelled platform
	Trentino (T)	2.5	18	60	3,000	Red/Golden Delicious, Gala, Renetta	94	Pluko-Trak or Self-propelled platform
Switzerland	Thurgau (TH)	6	18	47	2,500	Golden Delicious, Gala, Jonagold	87	Self-propelled platform
South Africa	EGVV	80	25	53	1,667	Golden Delicious, Granny Smith, Gala	65	Ladder
	Ceres (C)	120	25–30	54	1,770	Red and Golden Delicious, Pink Lady	75	Ladder
Chile	El Maule	25	25	67	990	Granny Smith, Gala, Fuji	80	Ladder

Source: agri benchmark 2016.

Adige are the major apple-producing clusters in Italy with 70% of total output. Italy exports approximately half of its apple production, with Germany being the most important destination. Italian apple farms are rather small: the typical Trentino farm in the benchmark is a family business with 2.5 ha apple orchards. As the second region for production, Emilia Romagna is included with a 5 ha typical farm. Good climatic conditions in both regions are the main reason for high yields with more than 50 t/ha on average.

South African apple production of approximately 725000 tons on 22000 ha is concentrated in the Cape region where the temperate climate favours the crop. Typically, farms are larger compared to Europe; hence, two typical farms in the regions of Elgin, Grabouw, Villiersdorp and Vyeboom (EGVV) as well as in Ceres were established with 80 and 120 ha apple production area.

In **Switzerland**, apples are mainly produced for local markets. Production is between 200000 and 300000 tons from about 4000 ha (FAOstat, 2016). The typical apple farm is located in Thurgau and has 6 ha of apple orchards.

The comparison of production costs between typical farms is based on the cost categories explained above. For a quick overview, the costs are presented in a bar chart (see Fig. 3); the total height of the bar represents total costs as the sum of direct costs (inputs, pollination), overhead costs (farm management, insurances and fees), hired labour costs (both permanent and seasonal labour), depreciation costs (for buildings and machinery) and opportunity costs (for family labour, own land and own capital). For calculation of annual costs, the establishment costs are classified in the same way and considered according to the utilization period assuming a continuous replanting. As an example, for a utilization period of an orchard of 20 years, annual costs include 5% of the

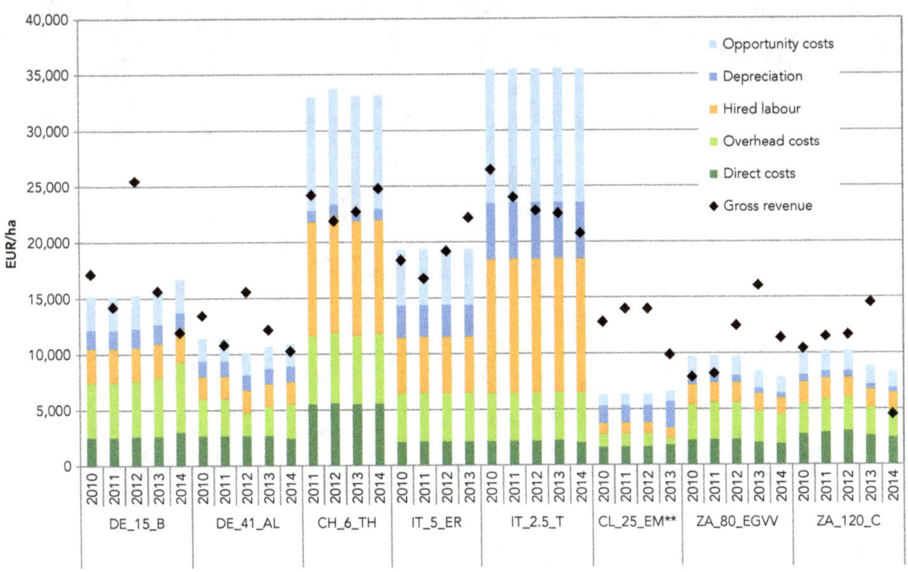

Figure 3 Total production costs and gross revenues Euros per hectare, 2010–14. Source: agri benchmark, 2016. ** typical farm update based on ODEPA, 2013.

establishment costs. Revenues are calculated based on yields and the weighted output price for the variety mix, and refer to farm gate prices.

Figure 3 illustrates the cost and revenue structure of the typical apple farms over a period of five years. The cost structure shows basic differences in production expenses between countries. In South Africa (ZA_80_C and ZA_120_EGVV) and Chile (CH_25_EM), per ha production costs are below 10000 Euro, substantially lower than in the European farms where total production costs can reach 35000 Euro, in the case of the Italian region Trentino (IT_2.5_T).

A closer look at the cost structure reveals that direct costs are similar in all countries, with one exception, the Swiss typical farm (CH_6_TH). Overhead costs can be expected to be lower for larger farms, as is the case for the South African farms, compared to the much smaller typical farms in Europe. The same pattern can be observed by comparing the larger German farm (DE_41_AL) to the smaller farms in Germany and Italy (DE_15_B and IT_5_ER, IT_2.5_T). Hired labour costs are low in South Africa, which is a major cost advantage. However, the larger German farm (DE_41_AL) also has relatively low costs for hired labour. This is related to mechanization – and corresponding higher depreciation costs – but also to the lower yields and respective costs for harvesting. The picture changes therefore when analysing costs per ton of apples (Fig. 4). The very small Italian farms are characterized by particularly high depreciation costs and high opportunity costs. Both items are directly linked to the size of the operation. A basic set of equipment and machinery is needed for apple production, but is most likely not utilized at full capacity, thereby driving depreciation costs. Also, these farms rely heavily on family labour, which is a major part of the opportunity costs.

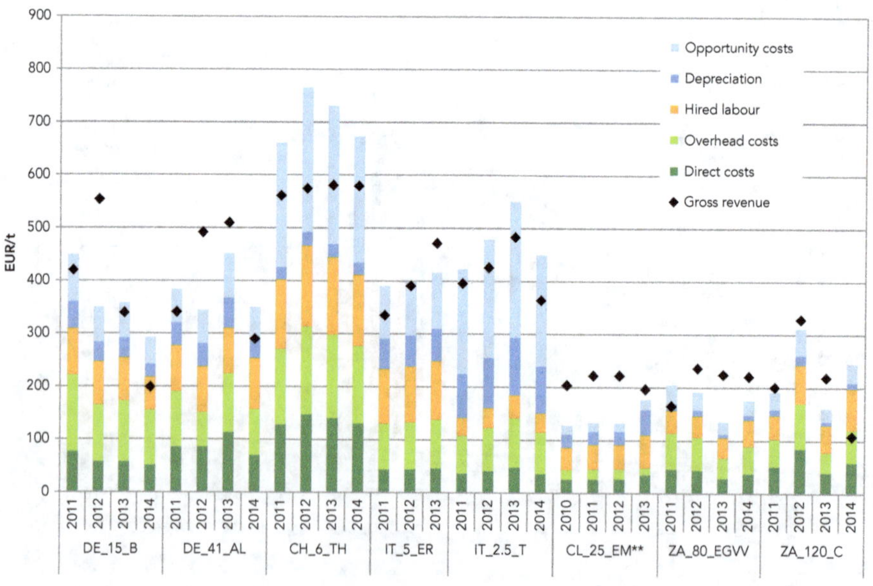

Figure 4 Total production costs and gross revenues per ton, 2010–14. Source: agri benchmark 2016. **typical farm update based on ODEPA, 2013.

In all countries, hired labour is the most significant variable cost in apple production. This confirms findings of Zahniser et al. (2012): for 2006 through 2010, hired labour represented 17% of the total average variable cost of production for agricultural crops in the United States and a much larger proportion for fruit (48%) and vegetable (35%) production. A cost-of-production study (Gallardo et al., 2010) of Washington-grown 'Gala' apples revealed that labour represented 46% of the variable production cost for a full production year.

The comparison of gross revenue to the costs indicates profitability of apple production. The first observation to note is the high variability of gross revenue between years. In the South African and German farms (DE_15_B and DE_41_AL) gross revenue is higher than the full economic costs in most years. For the Italian and Swiss farms, at least the cash costs (direct, overhead and hired labour costs) are covered by revenues in most years. However, the returns on use of family labour, own land and capital are below expectations, that is, the opportunity costs are not covered. This means no real profits are achieved over the analysed years, with the exception of IT_5_ER in 2013.

For an assessment of competitiveness, the cost per unit of output is more relevant; Fig. 4 shows the total costs and revenues per ton. Owing to high average yields, the costs per ton of apples for the Italian farms are approximately the same (around 400 €/t) as in Germany, while South Africa and Chile still have a significant advantage of only half of that cost. However, gross returns at the farm level are also substantially lower in South Africa than in Europe. While European farmers find their major markets within the same regions, the majority of the South African apple harvest is exported and therefore subject to higher transportation costs. As a result, variation of profitability is similar to Europe, and there are also years where full cost recovery is not achieved.

Gross revenue per ton indicates the price level at farm gate, excluding grading and packhouse charges. These prices are extremely variable over time. In particular, 2014 was a year of very low prices in Germany and Italy, but also in South Africa. Variation in total costs per ton for one farm over time can be explained by variation of yields.

3.3 International comparison by variety

Choice of variety is a major factor for the profitability of apple production. For the decision on the best variety or mix of varieties to be planted in a specific apple orchard, market aspects followed by production aspects have to be considered. Large differences between apple varieties can be observed with respect to yield, optimum climate conditions and market demand. Output prices vary greatly between varieties. Data from Germany and Italy show that producer prices for modern varieties such as 'Fuji', 'Braeburn' and 'Kanzi' are generally higher than the more traditional varieties 'Golden Delicious', 'Red Delicious' or 'Granny Smith' (Garming et al., 2015; AMI, 2015). The proportion of new varieties, for example, 'Cripps Pink', 'Pinova', 'Honeycrunch' (or 'Honeycrisp'), 'Jazz' and 'Junami', is an indicator of international competitiveness (Belrose, 2013).

However, with respect to production practices, varieties differ in their specific requirements of crop management and hence in production costs. The price of trees and

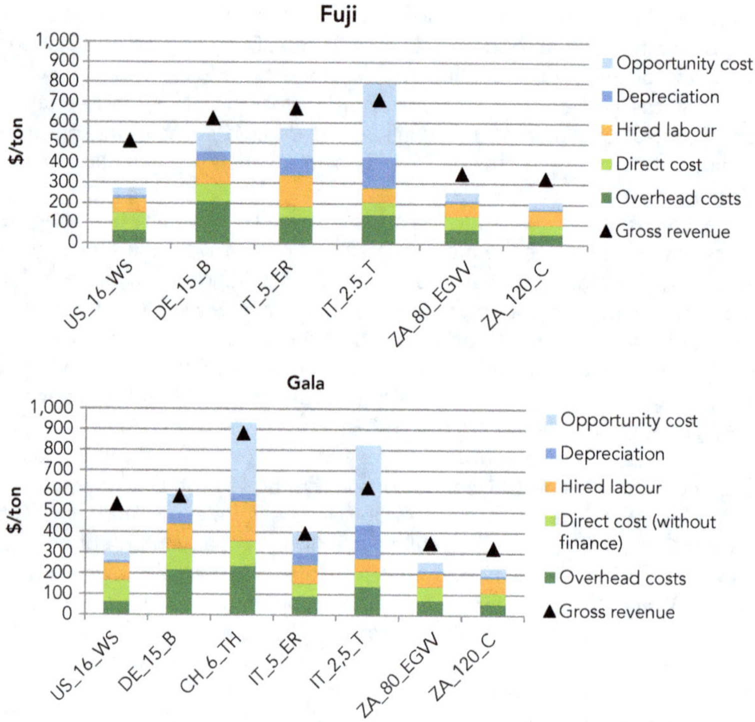

Figure 5 Production costs and gross revenue for the varieties 'Fuji' and 'Gala', 2013, $/t. Source: agri benchmark 2016 and Gallardo, 2016.

amount of pesticide required, as well as the use of labour for pruning and fruit thinning, can vary largely between varieties. These specific costs are collected and modelled in both typical farm data and enterprise budgets, allowing a comparison of production costs between regions that produce the same apple variety.

Variety-specific production costs of 'Gala' and 'Fuji' apples (Fig. 5) differ from the overall farm average (Fig. 4). The German typical farm (DE_15_B)[2] produces apples at a cost of approximately 350 €/t in the overall farm average, which is much lower than the Italian farm in Emilia Romagna (IT_5_ER) with about 420 €/t. However, the 'Fuji' production costs of the two typical farms are similar at about 550 $/t (converting euros to dollars). 'Gala' is less profitable for the German and Italian typical farms than 'Fuji', mainly due to high production costs compared to other regions.

This variety-specific cost comparison confirms the general observation that South Africa has the lowest production costs (around 200 $/t), followed by the United States, with approximately 300 $/t and the much higher level of costs in Europe. Gross returns vary across countries; for 'Fuji', higher gross returns are observed in Germany and Italy. However, these are offset with higher production costs net; therefore, profits are lower

2. Farms are coded having the 2 digit ISO 3166 code for countries followed by underscore, size of the apple operation it has and the region where apple was produced represented by the initials of the region that could range from 1 to 4 capital letters.

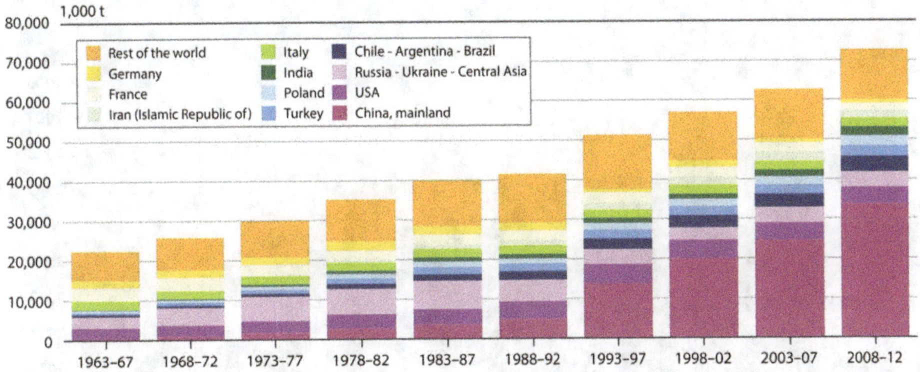

Figure 6 Global apple production by country and region – five-year averages from 1963 to 2012. Source: FAOstat, 2013.

compared to other regions. In general for both varieties, the United States exhibits the highest net profits across all countries included in this comparison.

4 Outlook and challenges for the global apple industry

In this subsection we present trends in production and international trade, a discussion of the structural change in the apple industry, trends in production systems and technologies, and labour supply as a major stressor for apple producers worldwide.

4.1 Global trends in production and international trade

Between the 1960s and 2010, apple production worldwide has increased threefold. While production in the United States and Italy has been relatively stable, other countries have strongly expanded production, such as China, Turkey, Poland, Chile, Argentina and Brazil.

Advances in orchard systems and horticultural technology have enabled production to reach unprecedented levels. While in most countries around the globe, domestic consumption of apples has been stagnant for the past 20 years; the increased surplus has made industries in major exporting countries depend on international markets to guarantee their economic sustainability (U.S. Apple Association, 2015). International trade data show an increasing trend in apple exports from the major producing countries over the past ten years (UN Comtrade, 2016). Italy and the United States have been leading exporters for many years and have both increased export volumes recently (Fig. 7). A particularly dynamic apple-exporting country is Poland, which has more than doubled its apple exports since 2010 and has been ranked first in export volume in 2014. In 2006, China was the world's largest exporter and further increased to reach a peak in 2009. In the Southern Hemisphere, Chile, New Zealand and South Africa are the leading apple exporters, with relatively stable volumes.

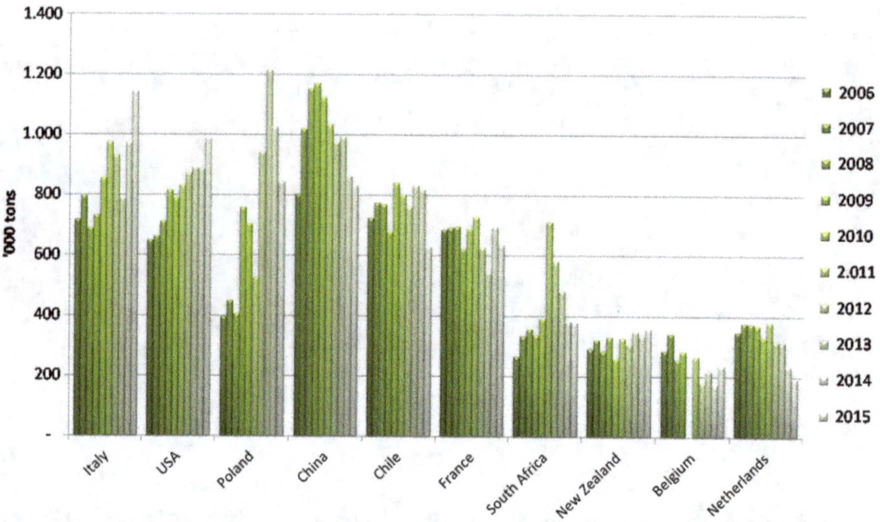

Figure 7 Apple exports from top 10 exporting countries, 2006–15. Source: UN Comtrade, 2016.

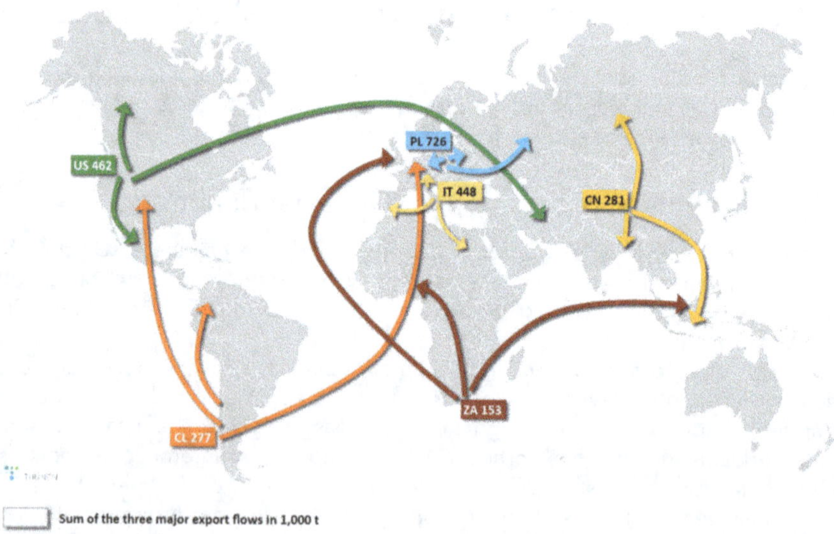

Figure 8 Major apple export destinations – in volume, 2016.

Figure 8 and Table 4 illustrate the main destinations for major apple exporters. One can observe as expected that principal export destinations are neighbouring countries. Since 2000, trade data show increasing apple imports in to new markets in South and Central America, Asia and Africa. Between 2001 and 2015, import volumes significantly

Table 4 Top five export destinations for apple exports from major exporting countries, 2015

Exporter	Destination	Export (t)	Exporter	Destination	Export (t)
Italy	*Total*	1,143,883	*USA*	*Total*	988,539
	Germany	304,561		Mexico	305,521
	Egypt	129,173		Canada	145,994
	Spain	81,391		India	104,408
	Algeria	60,888		Other Asia, nes	61,122
	Saudi Arabia	48,014		Un. Arab Emirates	55,918
Poland	*Total*	845,104	*Chile*	*Total*	629,045
	Belarus	256,682		USA	79,075
	Latvia	69,216		Colombia	75,593
	Lithuania	64,975		Peru	43,715
	Germany	63,935		Netherlands	43,409
	Ukraine	55,708		Ecuador	41,589
China	*Total*	833,355	*South Africa*	*Total*	381,051
	Bangladesh	129,128		Nigeria	75,398
	Thailand	103,443		United Kingdom	67,926
	Viet Nam	86,459		Malaysia	35,910
	Russian Federation	85,660		Angola	14,583
	Philippines	69,229		Un. Arab Emirates	13,236

Source: UN Comtrade, 2016.

increased in Bolivia (55%), Brazil (58%), Colombia (98%) and Peru (54%) (UN Comtrade, 2016), which meant growing market opportunities for the nearby producing countries such as Chile and Argentina. Kenya, Ghana, Benin, Zambia and Zimbabwe have also shown large increases in apple imports since 2001. In Asia, apple imports have augmented in countries such as the Philippines (41%), Malaysia (34%) and Indonesia (58%). In Thailand, apple imports have tripled over the same period, and India increased from only 12000 tons in 2001 to 190000 tons in 2015 (UN Comtrade, 2016). The Middle Eastern countries have become increasingly important export markets of apples and are found among the top five destinations for apples from the United States, Italy and South Africa (Table 4).

Although there are large differences in production volumes and export orientation, there are several common trends in global apple industry. Most are undergoing consolidation, with a decrease in number of farms and an increase in average farm size. In addition, because of the high importance of labour costs in apple production and the increasing availability of new technologies, interest in mechanization to reduce dependence on manual labour has increased in the past few years, even in countries with relatively low wages. The following sections address these two major trends in apple production.

4.2 Structural change: consolidation of apple production

In the United States, in 2013, there were approximately 7500 apple producers growing nearly 200 varieties of apples in approximately 132742 ha (U.S. Apple, 2013). In Washington and New York, two of the largest producing states in terms of volume, apple agricultural operations have consolidated during the 2000s. This was mainly driven by industry-wide financial losses caused by low apple prices in the late 1990s and early 2000s. As a result, a number of small and less efficient agricultural operations were forced out of the industry. In Washington, the number of apple operations declined by more than 20% between 1996 and 2001, while the average size expanded from 45 to 55 acres (18–22 ha).

The packinghouse sector also consolidated; the number declined from 44 in the late 1990s to 24 in 2008 in Washington State. The U.S. apple industry, in general, had become more vertically integrated, as large-scale packinghouses typically owned and operated their own apple orchards. In addition, the retail sector in the United States had consolidated; large-sized retailers preferred to buy fruit from large operations who can supply apples consistently throughout the whole year, pushing smaller packers and producers out of business (United States International Trade Commission, 2010).

In Europe, a similar process of consolidation of the fruit industry can be observed, reflecting a common trend in agriculture. In Germany, the fruit tree census in 2002 counted 10228 apple producers with 0.5 ha or more. According to the most recent census in 2012, the number of operations has declined by about 40% to only 6070, while the total apple area in Germany remained stable with about 32000 ha (Destatis, 2002, 2012). On average, apple production area per farm increased from 3.8 to 5.2 ha. Even though there is still a large number of very small apple farms, in 2012, 43% of apple production area accrued to farms of 10 ha or more.

In Italy, the decline in the number of farms during 2000–15 has been even more dramatic. Between 2000 and 2010, the number of apple-producing operations fell by 56% from 113000 to about 50000 farms. At the same time, total apple production area declined by 15% from 64400 to 54700 ha (Istat, 2000, 2010), leading to an increase of average apple area per farm from 0.5 to 1.1 ha.

4.3 Trends in production systems and technologies

The planting of apple trees on dwarfing rootstocks has become the industry standard. Compared to standard-sized apple trees, lower vigour trees can be planted at higher tree density, reach full production faster and achieve greater yields per acre; trees are also faster and easier to prune, thin and harvest (United States International Trade Commission, 2010). However, trees on dwarfing rootstocks often require intensive capital investment (e.g. trellis system to help support the canopy structure) and labour (e.g. more trees to manage on a per unit of production basis). In the United States, traditional varieties (e.g. 'Red Delicious' and 'Golden Delicious') are often still grown under traditional systems using standard rootstocks. However, the tendency is that relatively new more profitable varieties (e.g. 'Honeycrisp', 'Gala', 'Fuji' etc.) are planted in modern, high-density systems.

In Europe, the use of dwarfing apple rootstocks and Spindle canopy structures started from the 1970s (Weber, 2002; Bradlwarter, 2013). The aim was to reduce labour need for harvest, increase yields and reduce the period from planting to full production. Decreasing tree sizes and increasing planting density led to a significant increase in productivities. Bradlwarter (2013) estimated that yields in South Tyrol apple production doubled between the 1970s and the 1990s. Today, these modern planting systems can be considered the industry standard for all apple varieties for commercial fresh fruit production in Germany and Italy. Tree height and planting densities increase from North to South, corresponding to climate conditions and, in particular, radiation. In Northern Germany, about 2500 trees per ha are planted with a height of about 2.5 m, while in Southern Germany and Italy, planting density can reach 4000 trees per ha and heights of up to 3.5 m, with corresponding higher yields. In these planting systems, production starts in the second year. Apple farmers in Germany expect their orchards to reach full production in the fourth year after establishment (Dirksmeyer et al., 2014).

4.4 Labour use in apple production

Labour constitutes a major share of worldwide apple production costs. Pruning, manual fruit thinning and especially harvest require a particularly high labour input. Reducing labour input, that is, increase in labour productivity, has been a major driver for innovation in apple production systems, such as dwarfing rootstocks and orchard design that allows for mechanization of pruning and harvesting with platforms.

In Europe, despite ongoing structural changes, apple farms are still rather small and the farmer and family members usually cover a major share of labour needs. Yet, availability and cost of labour have become important components of competitiveness in specialty crop production in general, and migrant labour plays a key role for the viability of the production of fruits and vegetables in many countries (see Gertel and Sippel, 2014, for a comprehensive overview of seasonal worker dynamics in the Mediterranean countries).

In Europe, the use of migrant labour in fruit production has evolved within the context of the European Union, but was also determined by historical trends based on geography and colonial ties between countries. For example, for many years, Spain has attracted North African migrants, particularly from Morocco, as well as from Latin America (Lindner and Kathmann, 2014). From 2004 onwards, when ten Eastern European countries joined the EU, the free labour mobility for the citizens of the European Union member states led to significant increases in labour migration from these countries, motivated by differences in wages and labour demand (Koikkalainen, 2011). This trend increased again with the entry of Bulgaria and Romania to the European Union in 2007. Italy and Spain were major destinations for migrant workers from these countries who found opportunities in fruit and vegetable production (Portes, 2015).

In Germany, Polish workers had increasingly been hired for harvest in specialty crop production from the 1990s, based on new regulations enabling seasonal contracts (Dietz, 2004). After the entry of Poland to the European Union in 2004, access to seasonal workers for agriculture became much easier. Today, an estimated 300000 seasonal workers mainly from Poland, Romania and Bulgaria are employed during the harvest of fruits and vegetables in Germany every year (Darpeix et al., 2012). Agreements on exemptions for payment of social insurance for short-term contracts or other tax exemptions lowered the costs of seasonal labour compared to permanent labour in several European countries, encouraging the proliferation of migrant seasonal workers in Germany (Darpeix et al., 2012).

In the United States, the entire specialty crop industry, including apples, has long depended on migrant labour (Martin, 2009). For example, the 1870s Chinese and the 1900s Japanese immigration to California, the 1920s Mexican immigration to Southwestern states, the World War I Bracero Program targeting Mexican farmworkers, the 1930s Dust Bowl migration from Oklahoma and Arkansas to Western States and the World War II Mexican Bracero Program. Since the end of the Bracero Program, in 1963, undocumented migrant pools have supplemented the farm labour demand, in special Mexican nationals (Martin, 2009). By 2010, Mexico accounted for 75% of hired farm workers in the United States (Calvin and Martin, 2010).

However, in the wake of the 2008 U.S. economic recession, fewer migrant workers have been available to harvest fruit and vegetable crops. This, coupled with increased spending on border enforcement has raised the cost of migration for potential farm workers (Taylor et al., 2012). The number of unauthorized Mexican immigrants estimated to be living in the United States by 2012, has decreased 12.9% since its peak of 12.2 million in 2007 (Passel et al., 2012). In addition, fertility rates in Mexico dropped around 1980 in response to numerous cultural and economic factors. This demographic shift is now showing up in the working-age population. Furthermore, Mexico's economic growth and productivity in both farm and non-farm sectors have accelerated plus the government has been displaying unprecedented efforts to guarantee farm worker wages (Charlton and Taylor, 2016). All the aforementioned factors would likely contract the U.S. farm labour supply (Zahniser et al., 2012). While, the supply of U.S. farm labour appears to be shrinking, the demand for it has remained relatively constant at an annual average of 1 million workers since 2007. However, by 2012, labour economists found no evidence of a widespread shortage (Zahniser et al., 2012; Hertz and Zahniser, 2013). In this scenario, reforms to immigration legislation, like an expansion to the H2A guest worker programme or investment in research and developing for labour-saving or labour-enhancing technologies, appear as alternatives. The H2A programme is a temporal solution, considering that the tendency around the world is that as countries grow the population is likely to move away from farm jobs, which are typically more physically demanding, lower in intellectual skills, and not as well paid as non-farm jobs (Charlton and Taylor, 2016). Developing of mechanization technologies to diminish the high dependency on labour seems to be a promising long-run solution. However, there is an insufficient widespread adoption of labour-saving technologies by the U.S. fresh market apple industry. If mechanization technologies were massively adopted, there would likely be numerous effects on farm structures such as increased farm size and geographic transitions to different terrains. Besides, there are significant differences in the production systems across labour-intensive crops that would impact the ability to mechanize, as opposed to switching crops (Calvin and Martin, 2010). Notable advances have been made in developing mechanization and automation technologies in recent years; however, widespread adoption is yet to happen.

5 Conclusion

This chapter discusses a dimension of sustainability that is not often considered: the economic dimension of sustainability. We present information on the different approaches existing to the analyses of production costs. Among the different approaches, we cite the analysis of financial statements and accounts, and the models of production systems based on an engineering approach. The chapter illustrates a description of the different categories included in enterprise budgets, including imputed costs and cash costs.

We include a description of apple production costs for representative apple operations around the globe. When comparing costs and returns for 'Fuji' apple, we found that prices tend to be higher in European countries (e.g. Italy and Germany) compared to the United States and South Africa. However, the costs of production, particularly the opportunity and the overhead costs, are notoriously higher in European countries compared to their U.S. and South African counterparts. Given that prices in the United States are higher than those in South Africa, net profits are the highest for the United States compared to Europe and South Africa.

Challenges facing apple production economic sustainability include the consolidation phenomenon, the high dependence on manual labour and shrinking labour pools in the United States and the competition to maintain and expand market share in international markets. In general, one can conclude that the substitution of labour by capital investment given the technological advancements is irreversible. This phenomenon has led to industry consolidation. The increase in production volumes has augmented the dependence on international markets, and competition to maintain international market share has become one major stressor for growers in major (volume) apple-producing countries in the world.

The recent proliferation of international trade agreements such as the Transpacific Partnership (TPP) or the – in negotiation – Transatlantic Trade and Investment Partnership (T-TIP) would surely increase the volume traded of fresh apples. In that scenario, competition to gain access to strategic markets will surely increase, thus it is critical for major apple producers to remain competitive in order to sustain global apple production.

When talking about 'sustainability', concepts related to the environment or social fairness often come to mind initially. It is usual that the economic component is left behind. The aim of this chapter is to present the importance of the economic dimension to ensure sustainable apple production. Agricultural operations not meeting their bottom line are likely not adopting sustainable practices. Likewise, sustainable practices are showing to be an investment rather than a philosophy for agricultural management. It is proven that investing in environmental and social sustainable practices would benefit overall profits for the agricultural enterprise and society as a whole.

6 Where to look for further information

Updated information on apple production and market statistics for the United States can be found at the website of the U.S. Department of Agriculture Economic Research Service: http://www.ers.usda.gov/data-products/fruit-and-tree-nut-data

Information on global production can be found at the website of the U.N. Food and Agriculture Organization: http://faostat.fao.org

Information on fruit production in the European Community are published at the website of Eurostat: http://ec.europa.eu/eurostat/de

Information on international trade statistics can be found at the website of the U.N. Comtrade: http://comtrade.un.org

Further information on the international network agri benchmark Horticulture is available at the website: www.agribenchmark.org

7 References

Belrose, 2013. 'World Apple Review'. 2013 Edition. Pullman. USA, p. 146.

Bradlwarter, M. 2013. 'Organization of the apple industry in South Tyrol'. Variety Innovation Consortium South Tyrol. Presentation, July 2013. Teran. Italy.

Calvin, L. and Martin, P. 2010. 'Labor-intensive U.S. fruit and vegetable industry competes in a global market'. Economic Research Service, U.S. Department of Agriculture, Amber Waves, Vol. 8, No. 4, December.

Charlton, D. and Taylor, E. 2016. A declining farm workforce: Analysis of panel data from rural Mexico. *American Journal of Agricultural Economics*, 98 (4): 1158–80.

Darpeix, A., Bignebat, C. and Perrier-Cornet, P. 2013. Demand for seasonal wage labor in agriculture: What does family farming hide? *Journal of Agricultural Economics*, 65 (1): 257–72.

Deblitz, C. and Zimmer, Y. 2005. *IFCN Cash Crops: A Standard Operating Procedure to Define Typical Farms.* International Farm Comparison Network (IFCN), FAL, Braunschweig, Germany.

Destatis, 2002. 'Baumobstanbauerhebung 2002' Fruit tree census 2002. Statistisches Bundesamt Series 3.1.4, Landwirtschaftliche Bodennutzung – Baumobstflächen, Wiesbaden, Germany.

Destatis, 2012. 'Baumobstanbauerhebung 2012' Fruit tree census 2012. Statistisches Bundesamt Series 3.1.4, Landwirtschaftliche Bodennutzung – Baumobstflächen, Wiesbaden, Germany.

Dietz, B. 2004. Is there an alternative to the employment of seasonal workers from Poland in Germany? in German: Gibt es eine Alternative? – Zur Beschäftigung polnischer Saisonarbeitnehmer in Deutschland. Osteuropa-Institut. Working Papers Nr. 253. Munich, Germany.

Dillon, C. R. 1993. Advanced breakeven analysis of agricultural enterprise budgets. *Agricultural Economics*, 9: 127–43.

Dirksmeyer, W., Strohm, K. and Garming, H. (eds). 2014. Horticulture Report. *Agri Benchmark.* Thünen-Institute, Braunschweig, Germany.

European Commission (EU COM), 2016. Farm Accountancy Data Network. http://ec.europa.eu/agriculture/rica/ (Accessed 20 June 2016).

Galinato, S., Gallardo, R. K. and Hong, Y. 2016. 2014 Cost Estimates of Establishing, Producing, and Packing Red Delicious Apples in Washington. Washington State University Extension Factsheet TB12E. http://ses.wsu.edu/enterprise_budgets/ (Accessed 1 August 2016).

Galinato, S. and Gallardo, R. K. 2014. Cost Estimates of Establishing, Producing, and Packing Gala Apples in Washington. Washington State University Extension Factsheet TB18E. http://ses.wsu.edu/enterprise_budgets/ (Accessed 1 August 2016).

Galinato, S. and Gallardo, R. K. 2015. Cost Estimates of Establishing, Producing, and Packing Fuji Apples in Washington. Washington State University Extension Factsheet (Forthcoming).

Galinato, S. and Gallardo, R. K. 2012. 2011 Cost Estimates of Establishing, Producing, and Packing Honeycrisp Apples in Washington. Washington State University Extension Factsheet FS062E. http://ses.wsu.edu/enterprise_budgets/ (Accessed 1 February 2016).

Gallardo, R. K. and Galinato, S. 2012. 2012 Cost Estimates of Establishing, Producing, and Packing Red Delicious Apples in Washington. Washington State University Extension Factsheet FS099E, December 2012. http://ses.wsu.edu/enterprise_budgets/ (Accessed 1 February 2016).

Gallardo, R. K. and Brady, M. 2015. Adoption of labor-enhancing technologies by the specialty crop agriculture: The case of the washington apple industry. *Agricultural Finance Review*, 75 (4): 514–32.

Gallardo, R.K., Brunner, J.F. and Castagnoli, S. 2016. Capturing the Economic Value of Biological Control in Western Tree Fruit. *Biological Control*. http://www.sciencedirect.com/science/article/pii/S1049964416300846.

Gallardo, R.K., Taylor, M. and Hinman, H. 2010. 2009 Cost Estimates of Establishing and Producing Gala Apples in Washington. Washington State University Extension Factsheet FS005E. http://ses.wsu.edu/enterprise_budgets/ (Accessed 1 February 2016).

Garming, H., Dirksmeyer, W. and Strohm, K. 2015. Profitability of apple production in Germany and Italy. In German: Wirtschaftlichkeit der Apfelproduktion in Deutschland und Italien. *Thünen Rep.*, 22: 129–47. Braunschweig, Germany.

Garrison, R.H. and Noreen, E.W. 1982. *Managerial Accounting: Concepts for Planning Control, Decision Making*. 7th Ed. Richard D. Irwing, Inc. 912p.

Gertel J. Sippel S. R. 2014. Seasonality and temporality in intensive agriculture. In J. Gertel and S. Sippel (eds), *Seasonal Workers in Mediterranean Agriculture – The Social Costs of Eating Fresh*, Earth Scan from Routledge, Oxon and New York, pp. 3–22.

Goodland, R. and Daly, H. 1996. Environmental sustainability: Universal and non-negotiable. *Ecological Applications*, 6 (4): 1002–17.

Hertz, T. and Zahniser, S. 2013. Is there a farm labor shortage? *American Journal of Agricultural Economics*, 95 (2): 476–81.

Isermeyer, F. 2012. Methodologies and Comparisons of Production Costs – a Global Overview. In S. Langrell et al. (eds), *Sustainability and Production Costs in the Global Framing Sector: Comparative Analysis and Methodologies*, JRC Scientific and Policy Reports, Luxembourg, pp. 25–50.

Instituto nazionale di statistica, Censimiento agricole (Istat), 2010. Agricultural Census. http://dati-censimentoagricoltura.istat.it/?lang=en.

Jones, V.P., Unruh, T.R., Horton, D.R., Mills, N.J., Brunner, J.F., Beers, E.H. and Shearer, P.W. 2009. Tree fruit IPM programs in the Western United States; the challenge of enhancing biological control through intensive management. *Pest Management Science*, 65 (12): 1305–10.

Jones, V., Mills, N., Brunner, J., Horton, D., Beers, E., Unruh, T., Shearer, P., Goldberger, J., Milickzy, E., Steffan, S., Amarasekare, K., Chambers, U., Gadino, A., Gallardo, R.K. and Jones, W. 2016. From Planning to Execution to the Future: An Overview of a Concerted Effort to Enhance Biological Control in Apple, Pear and Walnut Orchards in the Western U.S. Biological Control. *Biological Control*. http://www.sciencedirect.com/science/article/pii/S1049964416300421.

Kay, R.D., Edwards, W.M. and Duffy, P.A. 2008. *Farm Management*, 6th Ed. McGraw Hill, New York, NY, p. 160.

Klonsky, K.M. and D. Stewart. '2014 Sample Costs to Produce Organic Processing Apples. Various Varieties. Central Coast Santa Cruz County', University of California Cooperative Extension.

Koikkalainen, S. 2011. 'Free Movement in Europe: Past and Present. Migration Policy Institute. Feature', 10p. http://www.migrationpolicy.org/article/free-movement-europe-past-and-present

Langemeier, M. Using Enterprise Budgets to Compute Crop Breakeven Prices. *Farmdoc Daily*, (5): 195. Department of Agricultural and Consumer Economics, University of Illinois at Urbana-Champaign, 21 October 2015.

Linder and Kathmann. 2014. Mobility partnerships and circular migration: managing seasonal migration to Spain. In J. Gertel and S. Sippel (eds), *Seasonal Workers in Mediterranean Agriculture – The Social Costs of Eating Fresh*. Earth Scan from Routledge, Oxon and New York, pp. 121–9.

Ludena, C.E., McNamara, K., Allen Hammer, P. and Foster, K. Development of a Stochastic Model to Evaluate Plant Growers' Enterprise Budgets. Applied Agricultural Economics Association Annual Meeting, Montreal, Canada, 27–30 July 2003.

Maher, M. and Deaking, E.B. 1994. *Cost Accounting*, 4th Ed. McGraw-Hill, New York, NY, 984p.

Marongui, S. and Cesaro, L., 2013. Estimation and calculation of the cost of production in agriculture. In L. Cesaro and S. Marongui (eds), *The Use of Rica to Estimate the Cost of Production in Agriculture. Application of Econometric and Mathematical Programming Methodologies*. Chapter 1, Instituto Nazionale di Economia Agraria (INEA) Rome, Italy, pp. 13–49.

Martin, P.L. (2009), *Importing Poverty? Immigration and the Changing Face of Rural America*, Yale University Press, New Haven, CT.

Missouri Extension. 2011. Know your cost of production – Develop and enterprise budget. http://extension.missouri.edu/henry/documents/newsreleases/Enterprise%20Budgets.pdf.

ODEPA Oficina de Estudios y Políticas Agrarias, Ministerio de Agricultura, Chile. 2015. Boletín frutícola. Febrero 2015. http://www.odepa.cl/wp-content/files_mf/1425495729boletinFruticolaFeb2015.pdf

ODEPA Oficina de Estudios y Políticas Agrarias, Ministerio de Agricultura, Chile. 2013. Ficha téchnica económica Manzano. Region del Maule. Diciembre 2013. http://www.odepa.cl/costo-de-produccion-manzano/

Paggi, M., Yamazaki, F. and Qiao, F. 2007. 'Final Report: Representative Farm Model Specialty Crop Policy Study', 117p. California State University. Fresno. USA.

Passel, J., Cohn, D. and Gonzalez-Barrera, A. 2012. 'Net migration from Mexico falls to zero – and perhaps less', Pew Hispanic Center, Washington DC. http://www.pewhispanic.org/2012/04/23/net-migration-from-mexico-falls-to-zero-and-perhaps-less (Accessed 17 September 2014).

Portes, J. 2015. Labor mobility in the European Union. In S. N. Durlauf and L. E. Blume (eds), *The New Palgrave Dictionary of Economics.* Online Edition.

Sav, G. T. 1984. The engineering approach to economic production functions revisited: An application to solar processes. *The Journal of Industrial Economics*, 33 (1): 21–35.

Taylor, J.E., Charlton, D. and Yúnez-Naude, A. 2012. The end of farm labor abundance. *Applied Economic Perspectives and Policy*, 34 (4): 587–98.

Texas A&M University. Agricultural and Food Policy Center. 2016. Description of the FLIPSIM: The Farm Level Income and Policy Simulation Model. https://www.afpc.tamu.edu/models/flipsim/ (Access 25 February 2016).

U.S. Apple Association. 2015. Production and Utilization Analysis. www.usapple.org (Accessed 1 June 2016).

U.S. Apple Industry. 2013. Industry Statistics An Overview of the U.S. Apple Industry. http://www.usapple.org/index.php?option=com_content&view=article&id=179&Itemid=285 (Accessed 24 February 2016).

United Nations Environmental Programs. 2015. The Three Dimensions of Sustainable Development. http://web.unep.org/ourplanet/march-2015/unep-work/three-dimensions-sustainable-development (Accessed 10 March 2016).

United Nations Food and Agriculture Organization (FAOstat). 2012. 'Food and Agricultural Commodities Production. http://faostat.fao.org/site/339/default.aspx (Accessed 4 March 2015).

United Nations Food and Agriculture Organization (FAOstat). 2013. 'Food and Agricultural Commodities Production'. http://faostat3.fao.org/download/ (Accessed 20 June 2013).

United States International Trade Commission. 2010. Apple Industry and Trade Summary. Office of Industries Publication ITS-04. February 2010.

United States Department of Agriculture Economic Research Service (USDA – ERS). 2016. http://www.ers.usda.gov/data-products/arms-farm-financial-and-crop-production-practices/arms-data.aspx (Accessed 20 June 2016).

United Nations Statistical Division: Commodity Trade Statistics Database (UN Comtrade). 2016. http://comtrade.un.org/ (Accessed 21 July 2016).

Weber, M. 2002. Anbausysteme für Baumobst. In H. Link (ed.), *Lucas' Anleitung zum Obstbau.* Chapter 7, Eugen Ulmer, Stuttgart, Germany, pp. 171–3.

West, T., Sullivan, R., Seavert, C. and Castagnoli, S. 2012. 'Orchard Economics: Establishing and Producing Medium-Density Apples in Hood River County'. AEB 0028. Oregon State University Extension Service. May 2012.

Zahniser, S., Hertz, T., Dixon, P. and Rimmer, M. 2012. 'The potential impact of changes in immigration policy on U.S. agriculture and the market for hired farm labor: A simulation analysis', Economic Research Report 135, U.S. Department of Agriculture Economic Research Service, Washington, DC.

Chapter 20

Consumer trends in apple sales

Desmond O'Rourke, Washington State University and Belrose Inc., USA

1 Introduction

Any analysis of consumer trends in apple sales needs to take account of the fact that different participants in the global apple market have different goals and different time frames for making decisions. For example, consumers are concerned primarily about the immediate satisfaction for themselves and their families. Retailers want current information so they can serve their apple customers more profitably. Marketers want rapid feedback so that they can service their retailer customers more effectively. On the other hand, breeders, nurseries, producers and packers must make investments whose success will be affected by consumer trends one or two decades into the future. Since the share of fresh apples entering international trade continues to rise, decision makers also need to be concerned about consumer trends in both the domestic market and numerous foreign markets. Analysing consumer trends in apple sales across many countries and years into the future is a very challenging intellectual exercise.

Consumer trends are also influenced by numerous supply and demand factors. On the supply side, decisions are being made every day by industry participants on which varieties to retain, remove or replace in their orchards. In aggregate, these decisions determine what mix of apple varieties will be offered to retail buyers. In turn, numerous retailers every day make decisions about what varieties they will stock or promote. These constraints limit the selection of apples that are available to consumers. On the demand side, demographic and socio-economic changes have the greatest influence. Demographic changes include trends in total population, geographic dispersion, rural–urban mix, distribution by age group, household size and household composition. The major economic factor is income, but the related socio-economic factors include education, culture, ethnicity, religion and

http://dx.doi.org/10.19103/AS.2016.0017.28

influences that contribute to attitudes and behaviour. For almost a century, economic and business researchers have been systematically assessing the impact of demographic and socio-economic factors on consumer preferences for many products, including apples and competing fruits and snacks. It is the complex interaction between such supply and demand factors which determines the future trajectory of apple sales.

2 The influence of suppliers and retailers on apple sales

Many of the entities involved in developing or commercializing fresh apples have a major influence (one they often underestimate) in framing the choices that retailers and consumers will face in the future. In the long process of developing new cultivars and commercial strains, in addition to the obvious eating quality traits, heavy emphasis is placed on intrinsic attributes such as size, colour, yield potential, disease resistance and other physiological measures. There is also a heavy emphasis on selecting cultivars that are deemed to be 'grower friendly' or 'packer friendly'. Before commercialization of a new cultivar begins, there is little systematic evaluation about the market segments where that cultivar might be most acceptable, either in domestic or in foreign markets.

The choice of parent cultivars for breeding programmes appears to have an element of fashion. Golden Delicious, Gala and now Honeycrisp all have had their periods of prominence. There is also a long history of chance seedlings, such as Granny Smith, Golden Delicious, Braeburn and Ambrosia, emerging outside organized breeding programmes. In either case, systematic evaluations of cultivars at the consumer level begin seriously only at a late stage in the development process. Even then, current test methods, such as taste panels or in-store demonstrations, are a very weak guide to how a new cultivar will perform under commercial market conditions. Even in the case of club varieties, sponsors have little objective evidence on the consumer appeal of their chosen variety until large volumes have already been produced and marketed.

Major food retailers also play a critical role in framing the choices of apple varieties that will be offered to consumers. Most apples are bought in the course of broader food shopping, so trends in food shopping are of vital concern to apple marketers. Food shopping systems continue to change. In the traditional food retailing system, with very large numbers of very small stores, the individual store owner or manager had discretion in allocating produce space between alternative fruits and vegetables and could purchase apples to match the expressed preferences of known customers. In the modern retailing system, most retail outlets are controlled by very large chains whose competitive strength derives from their ability to buy in bulk through central buying offices and to dictate terms to fruit suppliers for price, quality and other certifications (such as traceability and food safety). For decades prior to the onset of the Great Recession in 2008, chain outlets became increasingly larger, and the number of stock-keeping units (SKUs) handled expanded. However, individual store managers or produce managers had limited discretion in deviating from the SKUs selected by central buying offices. Thus, in modern food retailing, the selection of apple varieties offered was almost identical across all outlets within a chain. Since a number of different chain outlets competed in close proximity at the local level, they tended to mimic each other's selection of produce items, including fresh apples.

Modern retailing formats now dominate food distribution in most developed countries. They have also been gradually replacing traditional formats, such as Mom-and-Pop stores and open air markets, in most emerging markets. In the last two decades, even within the modern retailing system, retail food chains have become fewer and larger through mergers or acquisitions. A small number of chain buyers have gained increasing influence over the varieties of fresh apples that will be offered to consumers across individual countries and even across world regions. That, in turn, has led to increased consolidation among apple suppliers who needed to get bigger to provide chain customers with the volume of apples they demanded 12 months a year and to meet rigorous quality and other certifications. Most of these larger suppliers of fresh apples are now integrated grower–packer–marketers.

However, since 2008, the balance has again begun to shift towards greater fractionation of food retailing. The large mass marketers, such as Walmart and Tesco, have begun to rethink their 'larger is better' philosophy, in terms of the number of countries they serve, the number of outlets they build, the size of each outlet and the number of SKUs stocked. Limited assortment discount chains, like Aldi and Lidl, have gained market share among economy shoppers with no-frills, smaller stores, private label goods and very competitive prices. Warehouse club stores, like Costco and Meier's, have gained market share among middle and upper income shoppers with limited assortments of leading branded or private label products, sold in large packages.

Upscale stores, like Whole Foods, Trader Joe's and Waitrose, have appealed to more affluent shoppers with selections of natural, organic, specialty and luxury items. Other store formats, such as drug stores, convenience stores, large discount stores and small dollar stores, have been adding more food items in order to attract shoppers for their main lines of merchandise. Each of these different types of food outlets has different expectations for the number and type of apple varieties they will stock. In turn, shoppers are continually reassessing their patronage of different forms of food outlets. This affects the apple varieties to which they are exposed.

There has also been a dramatic increase in the availability of food in alternative channels. Amazon now offers about 500 000 food items that can be ordered online for home delivery. Another online supplier, Fresh Direct, focuses almost entirely on produce items. Television programmes, like the Home Shopping Network, often feature food specials. These channels avoid the expense of bricks and mortar outlets and can supply customers either from a central warehouse or directly from the producers. Farmers' markets of various diversity and seasonality have sprung up partly to meet the demand for local food. Mainline retailers have responded both by setting up their own online markets and by incorporating features of farmers' markets within their own stores.

Another factor affecting the apple availability is that about half of the consumer's food dollar in developed countries is now spent on prepared foods, either away from home, or delivered to the home. One can eat out at fast-food restaurants, white table cloth restaurants, upscale restaurants or institutional cafeterias, buy food to take home, or have food (from ready-made pizza to complete gourmet meals) delivered to the home. In general, apples, either whole or sliced, have made limited inroads into this huge segment of the food business and remain heavily dependent on grocery retailers for sales.

In response to changing competition, mainline food retailers have begun to rethink the number of fresh apple SKUs that they stock. Using advanced analytical techniques, they can now estimate the contribution of any individual apple variety to frequency of shopping trips, dollar sales, purchases of related items and overall contribution to store profitability. Apple varieties that perform poorly on these analytical measures can be summarily removed from the produce section. On the other hand, some smaller retail chains and direct marketers have attempted to distinguish themselves by stocking apple varieties that are not readily available elsewhere. For example, there has been increased interest in 'heirloom' varieties, apple varieties that were once popular, but they are now rarely available in mainstream outlets.

Amid all this retail churning, a steady stream of attractive, new apple varieties has been offered to retailers, most under some form of managed supply. A small number, like Pink Lady™ and Honeycrisp, have become permanent fixtures in the produce department, where they are gradually displacing old favourites like Braeburn, Cameo, Jonagold and Cox's Orange. However, the largest integrated marketers now have at least one new, unique apple variety that they can offer to selected retailers on an exclusive basis. Retailers have welcomed such exclusive varieties because they can be used to help differentiate their produce departments from those of other retailers and because they can generally be sold at higher prices than mainstream apple varieties. These developments have meant that the apple category is becoming as fractionated as the overall food retailing business. This process of fractionation in apple varieties will continue as long as it serves the competitive needs of major retailers, both bricks and mortar and online, and of their key suppliers.

3 Global forces affecting apple demand

Whether selling into a local market, or to a foreign market, demand for apples is most influenced by three factors, (a) the size and make-up of the resident population, (b) their level of incomes and resulting expenditure patterns and (c) the availability of competing fruits and snacks. Since the supply of apples in many countries can be increased much more rapidly than demand, countries with surpluses must continually seek export markets for that product. Even producers who are currently comfortable supplying just their local market must be able to compete in quality, price and other services against potential surges in imports. Thus, demand for fresh apples everywhere is being increasingly affected by global demographic and economic forces.

Demographic changes influence food consumption in many ways. The absolute size and rate of growth of the population establishes the limits of the market for any product. The most recent (2015) update of United Nations' population forecasts suggests that the total world population will grow from less than 7 billion in 2010 to over 9.7 billion in 2050, an absolute increase of almost 2.8 billion and a 40% increase from 2010. However, increases in world population will not be uniformly shared. Over 50% of the increase is forecast to occur in Asia, almost 40% in Africa and about 10% in the Americas. In Asia, most of the increases will occur in Southern Asia; in Africa, it will be in tropical Africa, and in the Americas, in South America. Only 5% of the increase will occur in high-income, developed countries. Europe is forecast to be exceptional in experiencing a modest population decline.

Some notable changes are also forecast among the top 20 countries with the largest populations in 2010. Most of these countries will experience some increases through 2050. India's population will pass that of China shortly after 2020. In contrast, China's population will peak around 2030 and then begin to slowly decline. The population declines that are already evident in Russia, Japan and Germany will continue. By 2050, three other countries, Nigeria, Indonesia and Pakistan, will join the United States in having more than 300 million people. By then, Brazil and Bangladesh are forecast to have more than 200 million people.

The make-up of the relevant populations also has a major influence on market potential. The number of children in the world aged 14 or under will grow by only 12% between 2010 and 2050, while the number aged 65 or over will triple, as birth rates slow down and longevity increases. The share of world population in the working age group, 15–64, will gradually shrink from 65.7% in 2010 to 62.7% in 2050. Fewer workers will have to support more younger and older dependents. The ratio of younger or older dependents to the working age population will rise from 52.2% in 2010 to 59.5% in 2050. The increases in the elderly population and in the dependency ratio will be more severe in higher income countries that have traditionally been the main export markets for fresh apples. In those countries, the dependency ratio is forecast to reach 71.2% by 2050. The effects of shifting age groups on apple sales are discussed further in Section 5.

The ethnic, religious or cultural make-up of populations also have a major influence on consumption patterns for many products, but reliable information of use to marketers is less readily available. Many countries collect data on their foreign-born populations. For example, the American Community Survey for 2014 estimated that 42.4 million people (13.3% of the total) in the United States in 2014 were foreign born. Hispanics made up 45.7% of the foreign born and 17.3% of the total U.S. population. Asians accounted for 26.2% of the foreign born, but only 5.2% of the total. In Europe, for millennia, many different peoples with different languages and ethnicities have migrated and settled, and assumed different national identities, as borders of states and countries have changed. In the early twenty-first century, many European countries included both majority and minority ethnic groups within their borders. Since World War II, many millions of newer immigrants from different racial and religious backgrounds have migrated to the developed world of Europe, North America and Oceania from Asia, Africa and the Caribbean, adding further to the diversity of their populations.

EUROSTAT estimated that as on 1 January 2014 there were 33.5 million people born outside the EU-28 living in a member state, while there were 17.9 million persons who had been born in a different member state from the one where they were resident. This represented 6.6% and 3.5%, respectively, of the total EU-28 population of 506.9 million. A new wave of immigrants arrived in 2015 as refugees from the civil strife in the Middle East and North Africa. In 2012, the Organization for Economic Cooperation and Development (OECD) estimated that the share of the total population that was foreign born was less than 1% in major countries like China, India, Indonesia and Brazil, exceeded 10% for most countries in Western Europe and exceeded 20% in Canada, Australia and New Zealand. The two countries with the biggest share of foreign-born residents were Israel (31.2%) and Luxembourg (42.4%). Thus, the population melting pot varies widely between major countries.

The level of per capita incomes in different countries, income distribution within countries and the rate of change of per capita incomes over time also have a major influence on demand for many products. World Bank data for the last two decades show that growth rates of real per capita incomes of developing countries like China (7.6%), India (5.2%), Indonesia (4.3%) and Brazil (1.2%) far outdistanced the rate of growth in such major developed countries as the United States (1.3%), Japan (0.9%), Germany (1.7%) and the United Kingdom (1.0%). In general, the response of apple consumption to rising incomes is much higher in developing countries than in developed countries. Thus, the best prospects for rapid growth in fresh apple consumption will tend to be seen in developing countries, not in developed countries where apples have traditionally been important.

4 Recent trends in apple consumption

Recent trends in per capita consumption are a good indicator of how receptive the populations in different countries have been to purchase and consume fresh apples. They are also a good indicator of the challenges that may be faced in any country or region in further increasing per capita consumption. In general, per capita consumption of fresh apples has been high, but stagnant, in the high-income countries of Europe and North America, but has been growing fastest in countries where per capita incomes have been growing most rapidly. Some of these countries, like China and Brazil, have been able to expand their indigenous apple industries to meet growing domestic demand. In others, like India, Indonesia and Russia, increases in demand have been filled primarily from increased imports.

The *World Apple Review* 2015 presents three-year average estimates of per capita consumption of fresh apples for numerous major apple producing and non-apple producing countries and regions for the three-year periods, 1992–4, 2002–4 and 2012–14. Use of three-year averages is intended to reduce distortions due to crop failures or other unusual market circumstances in any single year. Table 1 shows the results for major apple producing regions and selected countries. All three developed regions shown, the EU-11 (the 11 major apple producing countries in Western Europe), Other Europe, and North America, have experienced substantial declines in per capita consumption of fresh apples

Table 1 Estimated per capita consumption of fresh apples in producing regions and countries, 1992–4, 2002–4 and 2012–14 (kg)

Location	1992–4 (kg)	2002–4 (kg)	2012–14 (kg)	1992–4 to 2012–14 (per cent change)
Selected regions				
EU-11	19.54	16.81	15.14	−22.5
Other Europe	16.29	14.26	14.79	−9.2
North America	8.26	7.88	7.08	−14.3
Southern Hemisphere	5.07	5.08	5.63	+11.0
Selected countries				
Germany	24.31	18.92	18.95	−22.0
Poland	12.23	13.37	17.92	+46.5
Russia	4.92	6.40	13.66	+177.6
United States	8.72	7.80	6.69	−23.3
China	4.78	12.64	24.22	+406.7
Turkey	34.65	32.24	33.85	−2.3

in the last two decades. Only in the Southern Hemisphere region, which includes six major apple producing countries, has per capita consumption risen from relatively low levels.

Also shown in Table 1 are data for six of the largest individual countries. Per capita consumption in Germany and the United States followed a similar declining pattern to that of their region. Turkey's per capita consumption also declined modestly from a very high level. However, dramatic rates of growth were achieved in China, Russia, and to a lesser extent, Poland. Among these countries, only in Russia was increased demand met by increased imports.

Table 2 shows three-year average estimates of per capita fresh apple consumption in selected non-apple producing countries and regions for 1995–7, 2003–5 and 2011–13. Per capita fresh consumption is measured as net imports divided by current population in each period. Table 2 again shows a wide disparity between regions and countries. It also shows that per capita fresh consumption in most non-producing countries remains considerably below that in producing countries. The Middle East, the highest-income, non-producing region in 1995–7, had the highest per capita consumption in that period, but the slowest rate of growth in the next two decades. The two highest-income, non-producing countries listed, Singapore and Saudi Arabia, experienced similar trends. They had the highest per capita consumption in the earliest period, 1995–7, but suffered declines in the next two decades.

In contrast, all the lower-income, developing countries listed, Indonesia, Egypt, Costa Rica and Colombia, had substantial growth in per capita consumption in the next two decades. Table 2 also shows UN Food and Agriculture Organization (UN FAO) population estimates for each region and country. Because of its relatively large population, East Asia consumed the greatest total volume of fresh apples in 2011–13. In terms of individual countries, Indonesia, Saudi Arabia and Egypt consumed similar total volumes of fresh apples in the same period.

Table 2 Estimated per capita consumption of fresh apples in non-producing regions and countries, 1995–7, 2003–5 and 2011–13 (kg)

Location	Population, 2013 (millions)	1995–7 (kg)	2003–5 (kg)	2011–13 (kg)	1995–7 to 2011–13 (per cent change)
Selected regions					
East Asia	480.9	1.16	1.40	1.54	+32.8
Middle East	134.8	2.90	2.93	3.42	+17.9
Latin America	148.8	1.03	1.27	1.40	+35.9
Selected countries					
Singapore	5.4	11.68	8.47	7.55	−35.4
Indonesia	249.9	0.26	0.58	0.71	+173.1
Egypt	82.1	0.37	0.63	2.01	+443.2
Saudi Arabia	28.8	6.87	5.37	6.41	−6.7
Costa Rica	4.9	1.98	2.38	3.26	+64.6
Colombia	48.3	1.53	1.10	1.99	+30.1

Many factors other than income and population growth have affected past trends in per capita consumption of fresh apples in different countries and regions. For example, trade to Asia has flourished in a generally peaceful environment. In contrast, many other regions and countries have suffered trade disruptions due to civil strife. The buying power in Middle Eastern countries remains heavily influenced by fluctuations in the export price of petroleum. Other countries have seen their purchasing power affected, both positively and negatively, by fluctuations in the value of their national currencies compared to those of exporting countries. It is very difficult to forecast future trends in these extraneous, but powerful, forces. Any individual country may have a large, rapidly growing population, and a strong propensity to import apples, but its demand can be easily disrupted by such extraneous forces. That is why, it is so important for the global apple industry to seek expansion of many different markets simultaneously.

5 Factors affecting consumer demand for apples

In the developing world, the average levels and rates of growth of per capita incomes are good indicators of demand for apples, which generally must be imported. In rapidly growing markets, the challenge has been to expand the number of consumers that will make consumption of fresh apples a regular part of their diet. This correlates strongly with the number of households that were able to adopt middle-class lifestyles, shop at modern retail outlets and afford imported products. For this reason, where income data are not available, a frequently used proxy is the proportion of the population that is considered to have entered the middle classes. The emphasis in these markets has been on supplying mass market varieties like Red Delicious or Fuji that can be sold in bulk at highly competitive prices. As more consumers in developing countries achieve higher incomes, they too, like their counterparts in the developed world, become more willing to try newer, more expensive, apple varieties.

In the developed world, where average incomes are relatively high and income growth is slow, researchers have had to probe more deeply to explain changes in apple demand. In the past, when a few varieties dominated the apple market, analyses of consumer influences on apple purchases focused primarily on the socio-economic make-up of shoppers and on their attitudes towards such intrinsic apple qualities as firmness and sweetness. In countries such as the United Kingdom, predefined social classes, from A (higher managerial, professional, etc.) to E (retired, unemployed or widowed), were used to differentiate incomes and lifestyles. In other countries, like the United States, income categories were usually defined in numerical terms.

In addition to income levels, demand for mainstream varieties like Red Delicious, Golden Delicious, Granny Smith and McIntosh in the United States was compared on the basis of region, the size of the community, the homemaker's education and age, and the size and composition of the household. A United States Department of Agriculture SRS (USDA SRS) study in 1966 found 94% of U.S. households had purchased fresh apples in the previous 12 months. The heaviest purchasers of all varieties were upper income families, in the middle age group, with larger households and more children, both those under six and those aged 6–17. McIntosh was distinguished from other mainstream varieties in being purchased primarily in the Northeastern United States and in having an older customer profile.

In the 1960s, large, intact families were the core purchasers of fresh apples. However, in the intervening decades, the number of such families has become a smaller share of the total. Average family size has continued to shrink. Over half of all households now have only one or two persons. At the same time, the number of apple varieties available to consumers has exploded. Apple producers and marketers have been forced to adopt different strategies for increasing sales in mature markets and in rapidly growing markets.

In mature markets, the emphasis has shifted increasingly to targeting existing market niches with new varieties, new tastes, new packaging and innovative marketing strategies. Market researchers have become increasingly more discriminating in exploring potential market niches. For example, surveys began to add stage-of-life descriptors such as 'childless younger couples' and 'empty nesters'. A recent development has been to give age cohorts fancy names and to try to attribute different preferences and purchasing patterns to these different groups. The age categories include 'Baby Boomers', born between 1946 and 1964, 'Generation X', born between 1965 and 1980, 'Millennials', born between 1981 and 1995, and 'Generation Y', born since 1996. However, further research is indicating great diversity even among these different generations, and so any generalizations about their consumption patterns can be misleading.

Psychographic profiles of different types of consumers were developed by using computer techniques like cluster analysis of differences in values or attitudes, or involvement in sports or leisure activities. Motivations behind food purchasing decisions were used to characterize shoppers as 'health shoppers' or 'price sensitive' or 'premium spending shoppers'. A *Wall Street Journal* article of 14 August 2015 noted that consumers could also be grouped by their food-texture preferences into 'chewers 43%', 'crunchers 33%', 'smooshers 16%' or 'suckers 8%'. In addition to intrinsic qualities, such as firmness or sweetness, researchers found an increasing number of shoppers being influenced by extrinsic factors, such as the sustainability of the growing and processing methods used and the social responsibility of producers and marketers. Simultaneously, a new 'industry' has developed to provide agribusinesses with certifications on different extrinsic factors.

Retailers have also gained new tools for evaluating the performance of different apple varieties in terms of their volume of sales, retail revenue generated, contribution to store gross margins or relationship to purchases of other items. Many now have huge databases on consumer purchasing patterns, drawn from loyalty programmes, smart phone applications, in-store tracking devices and other sources. They can use these analyses in trying to optimize decisions about the size of orders, positioning on store shelves and merchandising support for hundreds of individual fruit items. Online retailers, such as Amazon, have begun to use their data trove to anticipate what products consumers will wish to buy and to target sales messages accordingly.

As incomes have increased around the world, an increasingly important influence on the demand for fresh apples has been the explosion in availability of a wide array of competing fruits. These include other deciduous fruits, citrus fruits, tropical fruits and exotic fruits, many that cannot be grown successfully in countries where apples have traditionally been produced. Aid agencies, governments and private investors in the Southern Hemisphere and in many poorer, tropical countries have recognized the economic potential of this trend and have expanded plantations of competing fruits to tap into the growing markets in the Northern Hemisphere. In general, Southern Hemisphere countries have concentrated on providing off-season supplies of familiar fruits, while tropical countries have focused on expanding sales of less familiar fruits.

Younger, more affluent, consumers have tended to be leaders in trying newer fruits and in incorporating them into their diets. However, over time, this practice has permeated other major consumer segments. Many fruits that were once novelties have now become mainstream and are available all year round. Food influencers, such as food columns in major newspapers and magazines, food shows on television and radio, celebrity chefs and food bloggers continue to promote the use of alternative fruits in new and exciting ways. Retailers have steadily increased the shelf space available for these competing fruits since they can generally be sold at higher prices and contribute more to store profitability. For these reasons, consumer interest in buying alternative fruits is likely to continue to grow, and the number and volume of alternatives to fresh apples will continue to expand.

6 Challenges in marketing apples

The new world of apple marketing presents many complex challenges. Consumers have become increasingly disparate and selective in what they expect from the products they buy. They remain concerned about the intrinsic qualities of apples, such as size, colour, taste and texture, but continue to add new concerns such as social responsibility, environmental sustainability, human health and worker safety. They have also become more disparate in where they shop, how often they shop and when and why they prefer different retail outlets. They have also become very assertive through mobile devices and social media in providing feedback to suppliers and retailers and in exchanging comments with their online friends on the products they buy.

Retailers are under intense competitive pressure because there is overcapacity in many trading areas, and a proliferation of different outlets, from farmers markets to drug stores to supercentres, selling the same foods. More and more retailers are using powerful, new analytical tools in measuring the performance of individual items and in monitoring the reactions of consumers in helping them choose which items they will stock. Breeders are continuing to introduce new apple varieties that they believe have superior characteristics to those already in the market, but they continue to be focused primarily on supply side attributes, rather than on potential market demand. Marketers are increasingly adopting new varieties to add excitement or exclusivity to their portfolios and are further differentiating varieties through the use of selective standards, unique branding or distinctive packaging.

However, the fruit industry in general, and the apple industry in particular, needs to greatly increase its understanding of the combinations of factors that currently drive demand for any variety in each individual market segment. These factors include consumer preferences, retailer business needs, unique product attributes and potential contribution of each product to the profitability of affiliated growers and packers. They then need to use that knowledge in determining what new varieties to adopt and what old varieties to gradually phase out.

No single trend will dominate consumer demand for apples, but multiple, interwoven trends will play out for different varieties, in different countries and regions. To prosper in this environment, the apple industry will need greatly enhanced ability to evaluate the wide array of variety choices it will face. Currently, it is seriously underinvesting in acquiring that ability to understand and adapt to changing markets. Unless it acquires that ability, the apple industry will continue to waste enormous amounts of public and private funds in developing varieties that consumers may not want and will be at increased risk of making unprofitable choices.

7 Where to look for further information

Since the apple market is truly a world market involving over 200 countries and territories and many billions of consumers, only a small sample of potential sources can be included here. The annual *World Apple Review*, published by Belrose, Inc., attempts to summarize the current situation and future outlook for the world apple market in the major producing and consuming countries. A number of annual conferences selectively cover similar topics. Two of these that draw international audiences are Prognosfruit, hosted by the World Apple and Pear Association (www.wapa-association.org/) every August in a different European country, and the Apple Crop Outlook and Marketing Conference, hosted by the U.S. Apple Association (www.usapple.org/) every August in Chicago. Many major apple producing regions host annual conferences that generate published proceedings.

A number of agencies within the U.S. Department of Agriculture provide information relevant to apple markets. These include daily, weekly and annual price and shipment reports from the Agricultural Marketing Service's Market News (https://www.marketnews.usda.gov/mnp/fv-home), the quarterly and annual Fruit Tree and Nuts Outlook and special reports from the Economic Research Service (www.ers.usda.gov/), and annual commodity and retail trade reports on major countries from the Foreign Agriculture Service's Global Agricultural Information Network (gain.fas.usda.gov/Lists/Advanced Search). Departments of Agriculture within the European Commission (ec.europa.eu/agriculture/statistics/index_en.htm) and in other countries provide statistical information and occasional reports on apple marketing and consumption.

The Economist magazine (www.economist.com/), published weekly, covers current issues and long-term trends on many of the issues related to population, income and demand discussed in this chapter. While the level of economic sophistication is high, articles are written to be understood by the intelligent lay person. Professional economic or marketing journals rarely have articles dealing with commodity issues. The exception is the *Journal of Food Distribution Research* (https://.fdrsinc.org/journal), which covers more applied marketing and consumption topics. Consumption data are also available for a fee from numerous retail auditing and marketing research firms in developed countries.

Finally, both government and non-government organizations in different countries, which are committed to increasing consumption of fresh fruits and vegetables, occasionally issue status reports on the progress of efforts to expand consumption. For example, the Produce for Better Health Foundation (www.pbhfoundation.org/) in the United States publishes a periodic study on consumption of fruits and vegetables.

8 References

Belrose, Inc. The World Apple Review – 2015. Pullman, Washington, 2015 (http://www.e-belrose.com/).

European Commission. Eurostat Statistics Explained, 'Migration and Migrant Population Statistics', 2015 (http://ec.europa.eu), accessed 8 December 2015.

OECD. OECD Factbook, 2013, 'Economic, Environmental and Social Statistics', 2012 (http://www.oecd.library.org), accessed 5 December 2015.

Produce for Better Health Foundation. 'State of the Plate. 2015 Study on America's Consumption of Fruit & Vegetables' (http://www.pbhfoundation.org/pdfs/about/res/pbh_res/State_of_the_Plate_WEB_Bookmarked.pdf).

United Nations, FAOSTAT database. FAO population estimates for each region and country (http://faostat3.fao.org/home), accessed 10 December 2015.

United Nations Population Division. Department of Economic and Social Affairs. World Population Prospects: 2015 Revision (http://www.un.org/en/development/desa/population), accessed 10 December 2015.

United States Census Bureau. American Community Survey for 2014, 2015 (http://www.census.gov/programs-surveys/acs/), accessed 10 December 2015.

USDA SRS. August 1966. 'Homemakers', Use of and Opinions About Selected Fruits and Fruit Products, MRR 765.

Wall Street Journal. 'Why People Love to Crunch', 14 August 2015, page D1.

Chapter 21

Assessing the environmental impact and sustainability of apple cultivation

D. Granatstein, Washington State University, USA; and G. Peck, Cornell University, USA

1 Introduction

Our modern apple production systems have substantially increased in productivity and sophistication over the last several decades. Much of the rest of this book focuses on the science and technology by which those increases have occurred. However, like many other modern farming systems, increased productivity often results in undesirable environmental impacts, which lower the overall sustainability of the enterprise (Merwin and Pritts, 1993) (Fig. 1). Focusing on increasing yields and fruit quality at the sacrifice of maintaining agricultural resource bases such as land, water and biodiversity can significantly threaten the long-term viability of orchards, particularly if resources become so degraded or contaminated that future production becomes limited or impossible.

Minimizing negative environmental impacts, such as pesticide residues, greenhouse gas emissions and habitat loss of farming systems, is embodied in the concept of 'sustainable agriculture'. In this chapter, we explore how this concept that became institutionalized in the 1980s applies to modern apple production systems. The authors, both based in the United States, draw on their familiarity with examples in that country in order to illustrate ideas and experiences, rather than provide an exhaustive review.

http://dx.doi.org/10.19103/AS.2016.0017.23

Figure 1 The interaction of apple orchards with the environmental elements (water, resource inputs, native vegetation and biodiversity) are evident in Yakima Valley, Washington, USA. Photo: Greg Peck.

2 Sustainability: definitions and trends

'Sustain' as related to agriculture means 'to endure, to last indefinitely' according to *Webster's Dictionary*. Simply put, this implies that orchards should be managed in a way that allows the land to continue producing at the same or increased level indefinitely. One of the most cited definitions of sustainability was developed by the United Nations Brundtland Commission, which stated that the goal was to 'meet the needs of today without compromising the needs of future generations to meet their needs' (Brundtland, 1987). Sustainability is often described as having three main goals: economically viable, environmentally sound and socially acceptable. The objective is to maximize all three goals without serious trade-offs with any of the others (Francis and Youngberg, 1990). A three-legged stool analogy is often used: if two legs are long (strong) such as economic and social, but one leg is short (weak), such as environmental, the stool is unstable and falls over. In business management, these tripartite goals are referred to as the 'triple bottom line' (Anon., 2009). More recent definitions of sustainability take into consideration the multiple interactions within the broader tripartite goals with visual representations resembling a network or spider web.

Sustainability may also be described as a path or journey rather than a destination (Granatstein and Kupferman, 2008). It is generally much easier to say what is not sustainable than what is, as hindsight is more accurate than prediction. A line is not crossed to achieve sustainability; rather, it is a goal to pursue that is unlikely to be achieved. However, definable progress towards the goal can be and is being achieved. Shifting from reliance on broad-spectrum insecticides to cultural or biological pest control approaches is a major goal of apple integrated pest management (IPM) programmes and, as described below and elsewhere in this book, a step towards reducing negative environmental impacts of pesticides and thus also towards achieving greater sustainability. For example, the shift from organophosphate insecticides to pheromone mating disruption (PMD) control for codling moth (*Cydia pomonella*) has been accomplished by apple producers worldwide.

Sustainability is a relative concept dependent upon the orchard's location and historical context, the regional environmental, soil and climatic features, and many sociopolitical factors. A small-scale, direct market apple grower in New York will have different criteria for achieving sustainability than a large-scale export-oriented apple producer in Washington State or a grower producing apples for processing in Virginia. Apple producers in China, Europe and elsewhere in the world face different challenges than their counterparts in North America. Therefore, any discussion of sustainability must take scale, market destinations and geographic location into account.

Concerns about agricultural sustainability are not new. The Roman writers Cato, Columella and Varro all described soil, fertility, crop rotation and erosion issues in their treatises (Columella and Ash, 1941; Hooper and Ash, 1934). With the advent of the Industrial Revolution and its impacts on farming, a number of writers in Europe, North America, India and Japan discussed environmental and social issues related to emerging and past agricultural practices (King, 2004; Lowdermilk, 1953; Howard, 2006). The Dust Bowl in the United States during the 1920s brought the soil erosion and farm management issues to a national level (Bennett, 1947). Rachel Carson's book *Silent Spring* was important in founding the modern environmental movement with its indictment of the organochlorine insecticide dichlorodiphenyltrichloroethane (DDT). The chemical was widely used in US agriculture after World War II and contributed to increased crop productivity, but the off-target impacts nearly drove many birds of prey species to extinction (Carson, 1962). At that time, various insect pests had already developed resistance to the pesticides being used, a trend that continues to this day (Schooley et al., 2008). Energy use in agriculture became an issue during the 1970s oil embargo and the obvious dependency of food production on non-renewable petroleum resources (Lockeretz, 1977), which has become amplified with concerns over global climate change (Smith et al., 2014).

On a global scale, agricultural sustainability is being driven by population growth, a reduction in the number of people living in poverty, and continuing or accelerating resource depletion and degradation. Additional environmental issues that are faced within the realm of apple production include ground and surface water quality and availability, air quality in terms of dust generation and pollution, pesticide use, loss of biodiversity and alterations to the biogeochemical cycle for nitrogen, carbon and other minerals that are used as fertilizers. In addition, the social and economic sustainability of agriculture and the food system are now widely discussed in many societies, touching on topics such as food security, food safety, food justice, nutritional density, farmworker treatment, labour availability versus mechanization, the need to train the next generation of farmers, increasing urbanization and the loss of farmland, and conflicts between apple producers and new neighbours who are impacted by noise, smell and spray operations.

Many of these issues are embodied in the recent focus on the food–energy–water nexus, which is receiving policy attention at national and global levels (FAO, 2014). Beyond the farm, the development of corporate sustainability consciousness has spurred activity at other levels of the food system, which may require very specific definitions of sustainability in order to codify and then certify that orchards are using sustainable practices (Dyllick and Hockerts, 2002).

3 Environmental impacts of apple production

Producing apples requires managing an agroecosystem and as such achieving zero environmental impact or footprint is unlikely. As a perennial crop, apple production has certain inherent advantages over annual cropping systems for achieving greater sustainability. For example, there is little to no soil tillage after the orchard has been planted, replanting is only required every 20–30 years and plant cover exists year round in the row middles, which account for up to 75% of the land area. Conversely, there is no crop rotation, and soil organic matter additions through cover crops, composts and mulches are typically not practiced by larger-scale conventional growers. Additionally, genetic diversity is low with about a dozen cultivars dominating the world apple trade, which increases potential for pest and disease problems. In order to achieve high yields and to meet consumer expectations for well-coloured, large and unblemished fruit, apple orchards require frequent (from 10 or fewer to 25 or more) pesticide applications. These applications have direct environmental impacts on the ecosystem (e.g. on pollinators and other beneficial organisms, water and soil quality, and farm worker and consumer health), as well as indirect effects from the manufacturing process, transportation and application of the materials, and represent arguably the main environmental impact from apple production.

3.1 Pesticides

Long-term apple production in many sites has resulted in a legacy of pesticide contamination. This is most notable in the elevated levels of lead and arsenic in soils of US apple orchards from the late 1800s to mid-1900s due to the use of lead arsenate insecticide for control of codling moth and other economically important apple pests (Schooley et al., 2008). Even though lead arsenic applications in Washington State ceased in the late 1940s and in the Eastern US in the 1960s, lead and arsenic levels in orchards can still exceed regulatory environmental clean-up thresholds (Peryea, 1998). For example, soil from a former orchard in Washington State was found to have a lead concentration of 1300 ppm (D. Granatstein, unpublished data). For relative comparison, the US Environmental Protection Agency has an action level of 250 ppm of lead for child play areas and 500 ppm for residential yards (EPA, 2016). Similarly, residues of DDT and its metabolites can be still found in streams, rivers and fish some 30–40 years after its use ceased (Johnson et al., 1988; Cuffney et al., 2000).

3.2 Resource use

A second major impact area is use of non-renewable resources, including energy, nutrients and feedstocks for various products. Modern orchard operations require numerous machine operations that are often performed down every drive alley. These operations can comprise the majority of the carbon footprint of apple production from both the fuel consumed and the energy embedded in manufacturing the equipment (Milà i Canals et al., 2006; see Section 4 below). Current harvest operations rely on continual tractor use to place empty bins and to retrieve full ones and account for a large portion of the machinery use and resulting carbon footprint. The advent of solid-set water systems in the canopy to deliver pesticides, automated platform systems, mechanical harvesting and robotics may reduce energy consumption as they become standard practices in the apple industry. Additionally, replacing petroleum-based diesel with biodiesel may also lead to a reduced

Figure 2 Sun/hail netting covers over 200 hectares of apple in May 2015 in an orchard in southcentral Washington State, USA. Photo: D. Granatstein.

carbon footprint (Milà i Canals et al., 2006). As orchard systems more frequently involve greater than 2500 trees per hectare, trellises, irrigation and protected cultivation such as hail nets and rain covers, each practice creates new resource demands and environmental impacts. Trellises require posts, which are often chemically treated wood, and galvanized steel wire. Irrigation systems are made from plastic pipes and fittings, with petroleum as the feedstock. Protected cultivation entails various netting and rain cover products largely made from plastic and requiring additional post and cable resources (Fig. 2). Many of these inputs end up as waste that is burned or buried in landfills at the end of the orchard lifespan. Developing orchard systems that utilize existing trellis and protection systems for sequential plantings would help to reduce the impact per unit of production by amortizing the environmental footprint over many more years.

3.3 Water

Water availability and quality is a third important resource that affects orchard sustainability. Increasingly, apple orchards have irrigation systems, even in humid climates, due to a combination of dwarfing rootstocks that have less drought tolerance and more variable precipitation events. The impact of irrigation systems on water quantity is most acute in arid or semi-arid regions where there are greater evapotranspiration rates and limited water supplies both in surface waters and in terms of groundwater recharge. The majority of US apple production is located in the arid regions of central Washington State, which is reliant upon a complex system of irrigation canals fed by mountain snowmelt. Currently, this system is sustainable because the resource is renewed each year. However, under a changing climate, the winter snowpack is decreasing, thus shifting the balance towards less summer water availability and making the apple production in Washington State less sustainable (Bauman et al., 2006).

Irrigation is more sustainable in regions with access to rivers that rely on annual recharge from snow or rain if irrigation withdrawals are regulated to ensure minimum stream flows that satisfy other uses such as fish, municipal water or recreation. Orchards that rely on 'fossil' groundwater that is not recharged, or on aquifers that are rapidly being depleted, do not have a long-term sustainable source. For water quality, orchard pesticides have been found in water, as in the case of azinphos-methyl, the primary codling moth insecticide used in Washington State prior to 2012. This was detected in numerous surface and ground water samples taken in the Columbia Basin region of Washington from 1992 to 1995. Concentrations were mostly between 0.01 and 0.1 ppm, with a 0.01 ppm concentration considered as a threshold for chronic risk to aquatic life (Williamson et al., 1998). Improved irrigation efficiency through the use of irrigation scheduling utilizing apple-specific evapotranspiration models (Dragoni and Lakso, 2011) and new drip and microirrigation technology is being more widely adopted and can address both the water quantity and quality impacts of apple production (Ferguson, 2015).

Multipurpose irrigation systems for frost and sunburn control are also becoming standard practice among many apple growers, thus increasing the use and eventual disposal of tonnes of plastic tubing.

3.4 Crop nutrients

Orchard nutrition and fertilization practices are another impact category. In comparison with grains and other crops, apple orchards export relatively little nitrogen (N) typically in the range of 0.4 kg N per metric tonne fresh weight of apples (Neilsen and Neilsen, 2003). Nitrogen can come from soil organic matter, organic sources such as manures or legumes, or from synthetic sources that rely on the Haber-Bosch nitrogen fixation process that requires high fossil energy inputs. Both organic and inorganic nitrogen sources have the potential to off-gas nitrous oxide, a potent greenhouse gas, and urea and ammonia-based products can volatilize ammonia. Once in the nitrate form, nitrogen can also be lost via leaching. Management strategies can be used to minimize these losses, including soil incorporation, proper application timing, water management and soil pH adjustment. Potassium is the other macronutrient exported in the fruit, at quantities about double the nitrogen rates. Potassium fertilizers are available from multiple sources that are relatively plentiful, compared with phosphorus, another important macronutrient needed for some soils that is derived from mined mineral deposits of limited extent. Phosphorous exports in harvested fruit are relatively small, but phosphorus can be leached from the soil profile when applied in excess.

3.5 Biodiversity

A more difficult environmental aspect to evaluate is biodiversity. The domesticated apples that are grown for human consumption have their genetic centre of origin in central Asia (Dzhangaliev et al., 2003) and are thus an 'exotic' species in most places where they are commercially grown. As apple trees are clonally propagated, their genetic diversity within an orchard is extremely low. To some extent an apple orchard displaces the flora and fauna of a native ecosystem, which likely represents a loss of biodiversity. However, most orchards have vegetative understories that add biodiversity (Fig. 3). The perennial nature of an apple orchard makes it more amenable to a variety of fauna than an annual crop field that is tilled with all habitat destroyed each year. Multi-species tree plantings that include apple trees have been attempted (Logsdon, 1981), but it is difficult to achieve the high fruit quality and yields that have become standard for commercial apple production.

Figure 3 Plant biodiversity was increased in the row middles of this Swiss high-density experimental apple orchard in an attempt to increase predatory and parasitoid insect populations. Photo: D. Granatstein.

4 Studies of sustainable apple production

Environmental impact and sustainability can be measured and rated in a number of ways. These include empirical field studies quantifying specific parameters and comparing treatments under controlled conditions, surveys of grower practices – such as pesticide or fertilizer use – that can illustrate trends over time, Life Cycle Assessment (LCA) and similar environmental impact metrics, and various assessment schemes that are more qualitative in nature as described in Section 5. Apple orchard systems studies often focus on the pesticide aspect of production, as it can impact water and air quality, ecosystem biodiversity, human health and soil biological properties.

4.1 Pesticide impacts

With the wide adoption of Integrated Fruit Production (IFP) in Europe and countries who export apples to Europe, measures of total use of pesticides (usually kg active ingredient) in IFP versus non-IFP apples have been compared (e.g. VanLierde and VanDenBossche, 2004; Peck et al., 2010). Apples grown under IFP production practices often use less total active ingredient than conventional pest control management; however, this does not account for the different toxicity impacts among pesticides. Kovach et al. (1992) developed the environmental impact quotient (EIQ) to address this problem. Using existing data on the environmental effects of specific pesticides to quantify their relative impacts, the EIQ accounts for farmworkers, consumers and ecological effects that cover various toxicities (dermal, chronic, fish, bird, bee and beneficial arthropod), water contamination (leaching and surface loss potential), soil and plant surface half-life, and systemicity. A case study calculation was done for 'Red Delicious' apples grown in New York, under conventional, IFP and organic pest management using the field-use EIQ equation (EIQ value × per cent

Figure 4 The use of kaolin clay (Surround WP) aided organic insect control in this New York orchard, but also significantly increased the environmental impact of the pest control program. Photo: Greg Peck.

active ingredient × application rate per land area) to generate an overall impact score (lower EIQ scores equate to lower environmental impact). In this theoretical case study, the field-use EIQ was 938 for conventional, 167 for IFP and 1799 for organic. Under organic management, sulphur accounted for 95% of the EIQ score. IFP had much lower impact for both insecticides and fungicides than conventional. Similar results were obtained from a 4-year field study in New York where organic production had 4.2–7.9 times more negative impact from pesticides than IFP (Peck et al., 2010) (Fig. 4). In this study, the impact was largely due to the volume of kaolin clay (Surround WP) used for pest control and lime sulphur used for chemical fruit thinning (Fig. 5).

Using a modified EIQ field-use rating adapted to Washington State conditions (Reed, 1995), a similar comparison was done for 'Golden Delicious' apples from 1994 to 1999

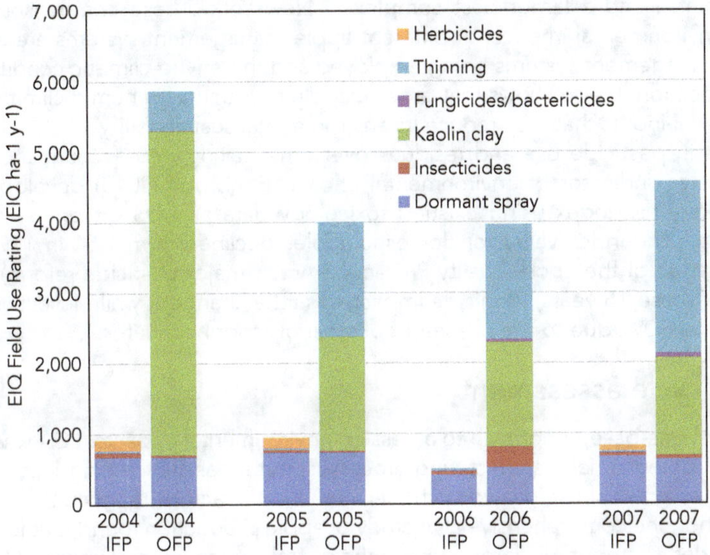

Figure 5 Field Use Environmental Impact Quotient (EIQ) rating for integrated (IFP) and organic fruit production (OFP) systems during four years, New York, USA. The Field Use rating equation multiplies the EIQ value × the percent a.i. × the rate per hectare for each material used (Peck et al., 2010).

Table 1 Effects of apple orchard management system on three environmental impacts, Washington State (Reganold et al., 2001).

	Conventional*	Integrated	Organic
Environmental impact rating (pesticides) over 6 yr	2,893	2,211	466
Energy input (MJ ha^{-1})	516,489	488,661	445,328
Energy input (MJ/MT fruit)	2,108	1,955	1,855
Soil organic matter % after 6 yr	2.2 a	3.7 b	3.9 b

*With pheromone mating disruption

for organic, integrated and conventional with pheromone mating disruption (PMD), as well as a conventional without PMD (Reganold et al., 2001). The rating for organic apple production was far lower than for the other systems (Table 1). The conventional PMD and integrated impacts were 6.2 and 4.7 times higher than organic, respectively, while the conventional no PMD system was 7.7 times greater than the organic orchard system. The lower impact for organic in Washington than in New York reflects the dry summer climate and fewer number of arthropod pests and diseases in the former compared with frequent summer rainfall and a larger pest complex in New York. These comparisons illustrate how the environmental impacts of different apple management systems are affected by both the management systems being employed and the enviro-climatic conditions of the orchard's location. It also shows that organic apple production in humid climates requires a high level of inputs that may reduce its environmental sustainability.

Changes in pesticide use and residues over time, taking into account their risk, can be used as an indicator of environmental impact. Benbrook (2013) developed a food supply dietary risk index using existing toxicology data and US government pesticide residue data. The index value for domestic apples declined from 1.51 in 1995 (prior to implementation of the Food Quality Protection Act, a major pesticide reform) to 0.43 in 2010. Over those 15 years, fungicide impact was little changed, while insecticide impact dropped over 90%, due to the phase-out of most organophosphates.

4.2 Life cycle assessment

LCA is a systems-based approach to assessing environmental impacts. LCA was originally developed for industrial manufacturing processes, but it has also been adopted to assess energy, water, greenhouse gas and other environmental factors in agricultural systems. In New Zealand, one comprehensive LCA study of apple production found that fuel, fertilizer and pesticides contributed a large share to the overall environmental impact (Milà i Canals et al., 2006). Pesticides had the largest impact, in the human toxicity soil category, and this category was one to two orders of magnitude higher than others such as global warming potential, human toxicity water and acidification potential. Field machinery use accounted for 64–71% of total energy use, with machinery manufacture at 7–12%, pesticide manufacture at 10–20% and fertilizer manufacture at 5–11% (Fig. 6). Harvest was the main contributor to machine energy use, and differences among producers for the same operation ranged from 40 to 80%, indicating real opportunities to increase efficiency and reduce energy impacts. The total amount of energy needed to produce apples ranged from 420 to 720 MJ per metric tonne (fresh), slightly lower than reported by Stadig (1997) for apple production in Sweden and France. Energy-related CO_2 emissions accounted for 34–50% of the global warming potential impact, along with 25–51% from fertilizer use. The total global warming potential was estimated to be 40–98 kg CO_2e per metric ton of fruit, which compares to the emissions from driving a typical passenger car

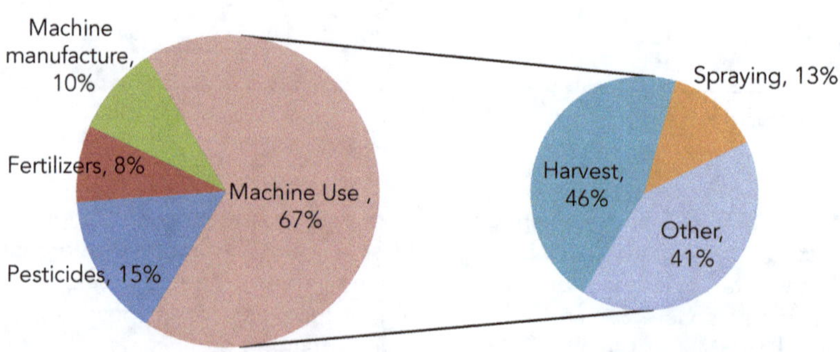

Figure 6 Energy consumption by use (% of total) in apple production, New Zealand. Adapted from Milà i Canals et al., 2006.

46–113 km. Estimated nitrate leaching of 18–53 kg N ha^{-1} yr^{-1} accounted for 83–91% of the eutrophication potential. The study authors suggest that apple growers have many opportunities to reduce their environmental impacts, particularly related to energy, given the large variation in the magnitude of environmental impact with similar end results for the targeted management. Energy use could be reduced by using renewable energy and biofuels, changing to lighter weight machinery and its use pattern, and altering fertilizer type and timing to reduce emissions.

A study of Swiss apple orchards used LCA to look at the variability among farms with similar conditions that all followed Swiss IFP guidelines, seeking to identify opportunities to improve environmental performance (Mouron et al., 2006). Five impact categories (energy use, global warming potential, ozone formation, terrestrial eutrophication and acidification) were strongly related to energy inputs, and thus a reduction in energy use would result in a reduction in the other four impacts. Aquatic eutrophication was almost entirely related to phosphorus fertilizer use, and aquatic ecotoxicity was largely related to insecticide and fungicide inputs. Farms varied 5–10 fold in the range of input use. Farms with the lowest energy use per unit apple sales also had the lowest value for the two aquatic impact categories, indicating no trade-offs among impacts (if one went down, another did not go up), or with economic performance.

LCA was used to compare conventional, low-input and organic apple production in southern France among apple orchards with trees that had varying levels of scab susceptibility (Alaphilippe et al., 2013). All treatments had similar amounts of global warming potential (GWP, kg CO_2e ha^{-1} yr^{-1}), with fertilization accounting for the largest

Table 2 Comparison of energy use and CO_2 emissions between apples grown and stored in the UK versus grown in New Zealand and shipped to the UK (Saunders et al., 2006).

	Energy (MJ/MT apple)		Emissions (kg CO_2/MT apple)	
	NZ	UK	NZ	UK
Farm	950	2,961	60.1	186.0
Direct energy	573	2,337	29.8	152.1
Indirect energy (fertilizer, pesticides, etc.)	300	624	24.7	33.8
N fertilizer	104	362	4.8	18.1
Equipment, buildings	78	n.a.	5.6	n.a.
Post-harvest	2,030	2,069	124.9	85.8
Cold storage UK 6 mo.	–	2,069	–	85.8
Ocean ship (17.8K km)	2,030	–	124.9	–
	2,980	**5,030**	**185.0**	**271.8**

n.a. Data not available. There is no entry for NZ cold storage and UK ocean ship as these were not part of those respective systems.

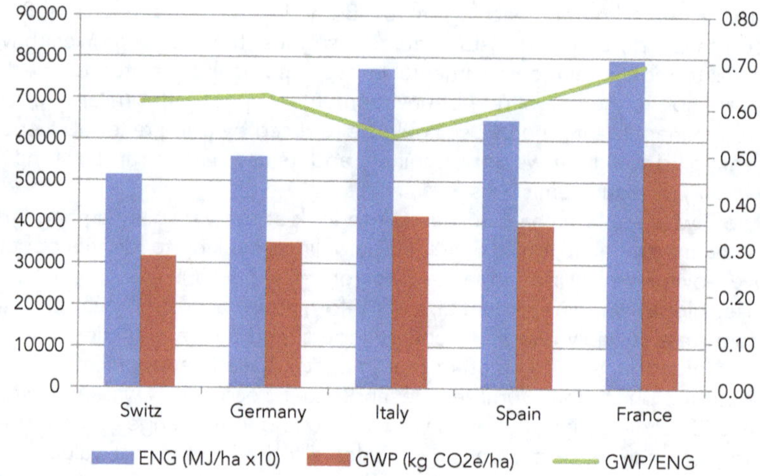

Figure 7 Comparison of energy use (ENG) and global warming potential (GWP) for integrated apple production in five European countries using Life Cycle Assessment. Adapted from Kägi et al., 2008.

share of this impact in all systems. Values were similar to those found in the New Zealand analysis above (Milà i Canals et al., 2006). The organic system had greater GWP per kg of fruit due to lower yields. Machinery use and manufacture accounted for the majority of the energy used in the production of fruit from these orchards.

LCA has also been used to assess the relative impacts of apple production in different regions. Saunders et al. (2006) compared apples grown, stored and consumed in the United Kingdom (UK) to apples grown in New Zealand and then shipped to the UK. Energy use and CO_2 emissions were three times higher in the UK for farm production, while in the post-harvest phase (i.e. storage and transport), energy use was similar but emissions were lower for the UK, with overall substantially lower energy use and emissions for the New Zealand fruit sold in the UK (Table 2). It is important to note that the energy use estimate for New Zealand is slightly higher than that of Milà i Canals et al. (2006) above, illustrating the variation in estimates that come from different data inputs, system assumptions and methods. Kägi et al. (2008) found large regional differences in an LCA comparison of apple production among five European countries for organic and IFP production (Fig. 7). For example, both energy use and GWP were lowest per hectare for Swiss apples, with values 50–74% greater for Italy and France. Lower yields in Switzerland explained much of the difference. The same was true for organic production, which tended to have lower impacts for most categories on a per hectare basis, but higher impacts on a yield output basis due to typically lower yields from organic orchards. Interpreting results from LCA studies must be done with care to avoid misrepresenting the findings.

4.3 Apple systems study

A comprehensive study of organic, IFP and conventional apple production in Washington State looked at several environmental and economic indicators of sustainability over a 10-year period in replicated, field-scale plots (Reganold et al., 2001). The pesticide environmental impacts were mentioned above. Energy inputs over the initial six years were calculated based on actual production practices. Compared with conventional production, energy use was 5–7% lower under IFP, and 12–13% lower with organic production (Table 1). This was true when calculated under either a land area or yield output basis. The energy values (yield output basis) were more than double the New Zealand values presented above and similar to those for the UK. Soil organic matter, which sequesters carbon and aids in soil health and tree nutrition, was significantly higher in organic and IFP plots than in the conventional plots due to compost and mulch additions (Table 1). But once tillage for weed control was implemented in year 9 for the organic treatment, soil organic matter levels dropped to the same as conventional, foregoing this environmental benefit. Overall, the authors concluded that the organic and IFP systems had higher soil quality and lower potential for negative environmental impacts, all systems had similar yields, and organic had higher profitability; thus organic was assessed to be economically and environmentally more sustainable. Similar soil organic matter reductions were found in a comparison between mulched IFP and mechanically cultivated organic orchard soils in New York, along with reductions in soil pH, nutrient availability and microbial biomass (Peck et al., 2011).

4.4 Nutrient loss

Nutrient loss, particularly nitrate leaching and phosphorus runoff, is an environmental concern for many annual cropping systems. Orchards are less prone to phosphorus runoff due to minimal soil erosion from the perennial crop and groundcovers typically used in apples. Nitrate leaching, however, does occur and is often related to amount and timing of N fertilization as well as management of irrigation, if used. Leaching of both nitrate and benomyl fungicide tended to be greater where herbicides were used to keep the tree row bare, compared with a mowed sod or wood chip mulch system, and appeared to be associated with preferential flow in soil macropores (Merwin et al., 1996). Nitrogen inputs and outputs, including leaching losses, were monitored 13 and 15 years after orchard establishment with different groundcover management in a New York apple orchard (Atucha et al., 2011). There was some increase in surface runoff in the herbicide plots due to loss of water infiltration and hydraulic conductivity over time, but nitrate-N levels ranged from 0.67 to 5.83 mg/L, less than the 10 mg/L USEPA drinking water guidance. More nitrate was lost through leaching, which equalled 30% of N mineral fertilizer inputs. Nitrogen leaching also occurred in a series of trials in apple orchards that were less than six years old in the Okanagan Valley, British Columbia, Canada, on coarse, low organic matter soils (Neilsen and Neilsen, 2002). The nitrogen use efficiency (NUE; nitrogen removed in leaves and fruit/nitrogen in fertilizer added) varied from 18% with unscheduled irrigation to 38% with irrigation scheduling based on evapotranspiration demand. Carefully managing drip irrigation with low nitrogen fertigation rates during periods of peak tree uptake led to the highest NUE and lowest leaching loss in the Okanagan region where precipitation is often too low to induce leaching by itself. Nitrogen can also be lost as volatile nitrous oxide (N_2O), a potent greenhouse gas. The Intergovernmental Panel on Climate Change (IPCC) provides a default value for agricultural systems of 1.25% of applied fertilizer N lost as N_2O (Smith et al., 2014). One field study on loess soils in a semi-arid climate of China found emission of 3.22 kg N_2O ha^{-1} y^{-1} from an apple orchard, which equalled 0.66% of applied fertilizer N (Pang et al., 2009). In this environment, apple production had half the rate of N_2O emission as the IPCC default. In Washington State, N_2O emissions

Figure 8 Mulches can be used to conserve soil moisture and also help suppress weeds and improve soil. In this Argentine apple orchard, custom mowers were used to cut the vegetation in the row middles and place it under the trees to form a mulch layer. Photo: David Granatstein.

were found to be similar among organic, IFP and conventional apple orchard systems at ~325 g N_2O-N ha^{-1} over a one month period after fertilization (Kramer et al., 2006). These amounted to <1% of the 112 kg N ha^{-1} of fertilizer N applied. Annual nitrate leaching was found to be 4.4–5.6 times greater in conventionally managed soils than in organically managed soils, with IFP management falling in between.

4.5 Water conservation

Reducing irrigation water consumption is possible using more efficient hardware, altered water scheduling and mulches (Fig. 8). Drip irrigation can reduce water use by 70% or more compared to micro-sprinklers (Yin et al., 2011). Partial root zone drying led to water savings of 45–50% without a significant impact on fruit yield or size in a Washington State apple orchard, whereas deficit irrigation provided similar water savings but with some reduction in yield (Leib et al., 2006). Wood chip mulch on the tree row led to a 20–30% water savings relative to the unmulched rows, with the entire block using irrigation scheduling based on continual soil moisture monitoring (Granatstein and Mullinix, 2008). Thus, apple growers with irrigation can significantly reduce their impact on water supplies by using one or more of these approaches.

4.6 Effects of apple cultivation on biodiversity

Apple orchards are generally less diverse than the native plant community they replace. There are fewer species of both flora and fauna in the orchard, and all commercially sold apples are clonally propagated. Various strategies have been tested to increase diversity; for example, peach was interplanted with apple which led to measurable increase in biocontrol (Brown et al., 2010). Manipulating ground cover by using mulches, planting different species and letting certain weeds dominate can also enhance biocontrol as well as general insect faunal diversity (Yan et al., 1997; Brown and Tworkoski, 2004). This indicates that some practices may positively influence orchard sustainability at multiple trophic levels. In a New Zealand study, management system (i.e. organic, conventional and green) was associated with different densities of native birds in orchards, but larger landscape level factors such as habitat composition and pesticide use were better predictors of population differences (MacLeod et al., 2012). This suggests that protecting the nearby vegetation can improve biodiversity more so than the use of organic apple management. It is not uncommon for these biodiversity strategies to attract more beneficial organisms as well as some pests (Horton et al., 2003), and thus finding an acceptable balance for commercial orchards is necessary.

5 Regulatory and marketing schemes for sustainable orchard systems

With the large number of factors that encompass sustainability, the precise definition of a sustainable orchard system is by nature site and situationally dependent. Thus, there is a real need for consistent and objective assessment systems whereby scientists, growers, politicians and regulators, businesses, consumers and others can understand the impact orchard systems and specific practices have on sustainability. However, converting qualitative definitions of sustainability into quantitative metrics that are universally relevant to all systems in all regions is challenging.

In order for a sustainability assessment system to be effective, it must have a well-defined set of objectives and systematic methods for evaluating the impact of management practices (Levitan et al., 1995). Under the three-legged stool definition of sustainability, economic and environmental impacts are somewhat more quantifiable than social impacts. Additionally, many assessment systems focus on short-term and local impacts, such as the effects on beneficial insects or the concentration of pesticide residues, rather than longer-term impacts such as biodiversity or epidemiology of pesticide-related cancer. The systems described below often rely on a qualitative standard that is rooted in quantitative research data. For example, it is well established that careful irrigation management can prevent nitrate leaching (Neilsen and Neilsen, 2002). Thus, a grower can demonstrate low risk of nitrate leaching via records of soil moisture or water management. Similarly, by using pesticides with low toxicity, a grower demonstrates compliance with minimizing environmental impacts. More difficult is a parameter such as soil organic matter, where a change can be readily quantified and related to desirable environmental outcomes, but for which there is no threshold to use in judging level of compliance.

With the increased scrutiny of the environmental impacts of modern agriculture, including apple orchards, public policies have been enacted that require certain changes in practice or performance in order to be in compliance. The regulatory framework put in place for pesticide registration and use since the 1970s is a prime example that has impacted apple production. Water quality protection is another example. Sustainability claims have also entered the marketplace, where they need to be validated in order to minimize consumer fraud, particularly when the product garners a price premium or other financial advantage. Thus, numerous programmes have been developed over time to evaluate sustainability claims using various qualitative and quantitative metrics. Many programmes are voluntary, some are legally regulated and all face the challenge of characterizing a biological system in binary terms when reality is more nuanced. Some IPM programmes have attempted to develop very quantitative approaches to measuring pest management performance, such as the EIQ (Kovach et al., 1992; Carroll and Robinson, 2006). However, while consumers are clearly concerned about the pesticides used on their fruit, IPM programmes that must communicate the complexities of pest management have not become effective marketing labels.

5.1 Integrated fruit production

IFP was one of the earliest evaluation programmes. It originated in the scientific body International Organization for Biological and Integrated Control of Noxious Animals and Plants (IOBC, 1992) during the late 1960s and 1970s to help increase understanding and adoption of IPM, which emerged in response to burgeoning pesticide use and problems in Europe. IFP and IPM programmes have been developed for specific crops in specific locations with evaluation criteria to determine whether a product meets the standard and can bear the label. For example, Cornell University developed a set of IFP protocols for New York apple growers (Carroll and Robinson, 2006). Some programmes were oriented towards regulatory compliance while others carried the IFP identity to the market and the consumer. By mid-1990s, an estimated 42% of apple and pear area in Europe was grown under an IFP programme, with 31 formal programmes operating in 10 countries (Reed, 1995). Although programmes were started in many countries, a number of them have not endured, in part due to wider recognition of programmes such as 'organic'. This is also the case in the United States, where very few apple growers market their fruit as 'IFP-grown', but US-grown organic apples can be found in almost every retail market. In some European countries, IFP has become the new 'conventional' production method, and although growers follow the protocols, this identity is often not communicated to consumers.

IFP guidelines for pome fruits in Europe specify a number of environmental performance measures that are much more intensive than IPM practices (Cross and Dickler, 1994). Growers must implement at least two options to enhance biodiversity, such as nesting boxes for birds, refuges for predators, host plants for beneficial organisms or new wildlife habitats. Pre-plant soil fumigation is not permitted. Soil or plant sampling is required to inform fertilization, and a maximum nitrogen input level per year is typically set to minimize nitrate leaching. Soil residual herbicides are not allowed, and orchard alleys must be vegetated. Where irrigation is used, water monitoring is required to determine irrigation schedule and amount. IPM is required, giving priority to natural, cultural, genetic and biological controls over chemical controls. Populations of key natural enemies must be preserved. Pesticides are categorized into red (not permitted), yellow (use with restrictions when needed) and green (permitted, use based on monitoring) lists, taking into account environmental and human health risks and potential to disrupt biological control. Pests, diseases and weeds must be regularly monitored with records kept. Pesticide applications should then be based on the monitoring and action thresholds where established. Quantitative monitoring of environmental impact, such as nitrate leaching or chemical drift, is not normally required. However, with the data recorded by growers and the prohibition of certain materials, quantitative impacts of IFP programmes over time can be documented. For example, farms in Belgium growing apples under IFP or traditional management ($n = 55$) were compared for total pesticide application in 1998; IFP farms had reduced use by 26% (Van Lierde and Van Den Bossche, 2004).

5.2 Organic farming

Organic farming is arguably the most widely used sustainability programme in agriculture, with over 43 million hectares worldwide (both certified and uncertified) in 2014 (Willer and Lernoud, 2016). There were an estimated 93 219 ha under organic apple management in 2013, approximately 1.8% of all apple hectares worldwide (Granatstein et al., 2015). This area had grown 165% since 2008, with a large share in Europe where EU subsidies greatly encouraged the adoption of organic farming. Concomitantly, global market demand for organic products is increasing 10–12% annually (OTA, 2015). Organic apple returns to growers have been nearly twice that of conventional in both Europe and the United States for several years (Kirby and Granatstein, 2015; Anon., 2015) (Fig. 9). In the United States, apples are the second-highest selling fresh fruit after berries in terms of dollar sales at retail grocery stores for either conventional or organic purchases (Granatstein and Kirby, 2016b). A survey of organic growers in Washington State, the largest apple producing state in the country, indicated a potential 60–70% increase in organic apple area between 2015 and 2018 (Granatstein and Kirby, 2016a). However, in New York, the second largest apple producing state in the US, organic apple production accounts for less than one per cent of the total. Based on comparative research studies among different production systems in

Figure 9 Organic apples are often sold for a significant price premium in US supermarkets, which helps to offset some of the additional costs associated with organic production. Photo: Greg Peck.

Washington and New York, it is clear that climate and pest complexes can greatly impact the ability of growers to successfully grow organic apples (Peck et al., 2006, 2010).

The US Dept. of Agriculture (USDA) considers organic production to be a marketing scheme and they have spent considerable resources harmonizing organic standards among countries and regions to reduce potential barriers to trade. The US organic standard delineates common management guidance and restrictions across all crops in all locations, in contrast to the more tailored approach that IFP takes. Historically, emphasis has been on materials, with natural materials allowed and synthetics prohibited. Lists of exceptions to these principles are continually updated and evaluated (USDA, 2016). For example, in the initial US standard, the antibiotics streptomycin and oxytetracycline were specifically allowed for control of fire blight (*Erwinia amylovora*) on organic apples and pears; but no other uses of antibiotics in US organic agriculture were allowed. As new control options have been developed, that exception was re-examined in 2012 and the National Organic Standards Board (the regulatory body that oversees decision on allowable materials) voted to end the exception in October 2014.

The US organic standard also requires growers to maintain or improve the natural resources of the farm, which pertains to soil, water and biodiversity. Typically, the certification process uses checklists of practices to demonstrate compliance with this, such as the use of crop rotation, cover crops, buffer strips or wildlife plantings to address biodiversity. Direct measurement of the impact is seldom done but given the prohibition on many materials, certain environmental outcomes are essentially guaranteed. For example, in the Midwest US maize production area, groundwater contamination by the widely used herbicide atrazine has been common (Hallberg, 1989). Organic farms are prohibited from using this material, and thus in an area with contamination, the organic hectares would not be a contributor to the problem. For apples, the prohibition of neonicotinoid insecticides in organic production would similarly guarantee that that this class of insecticides was having no adverse effects on pollinators and other beneficial insects.

Organic apple production is expanding in Chile, China and the United States, in response to growing consumer demand (Granatstein and Kirby, 2016b). In the United States, over 90% of hectares and production are in the semi-arid western states, particularly Washington State, where there is little summer rainfall and disease incidence is low (Granatstein and Kirby, 2016a). Organic apple production is more challenging in climates with warm, wet summers and with key pests and diseases such as plum curculio (*Conotrachelus nenuphar*), scab (*Venturia inaequalis*), and Marssonina blotch (*Marssonina coronaria*). However, over 20 years of continual research and governmental subsidies have supported the expansion of organic apple production in many areas of Europe with a climate similar to the northeastern US. Much of Europe's organic apple supply is now produced from within the continent (Weibel et al., 2014). The potential to expand organic apple production in the northeastern US would likely require similar research support and subsidies.

5.3 Other sustainability evaluation protocols

Several other farm-level sustainability evaluation programmes have been launched in the United States. The Food Alliance (www.foodalliance.org) based in Portland, Oregon, is one of the more established programmes. It was formed by a partnership among growers, consumers, scientists and other stakeholders to create market incentives for adoption of more sustainable agricultural practices. The main areas under the programme's evaluation are soil and water conservation; integrated pest, disease and weed management; pesticide risk reduction; wildlife habitat and biodiversity conservation; and safe and fair working conditions. It does not include a strict natural versus synthetic rule like organic, but it does prohibit genetically modified crops and includes a social sustainability element not included in organic. Farms undergo a third-party audit and if they meet the qualifications, they can use the Food Alliance logo for marketing purposes. With the increase in sustainability demands from buyers, participation in Food Alliance has maintained or opened markets for participating growers more than it has garnered an increased price from the consumer. Similar programmes applicable to apples include Protected Harvest, Salmon Safe, Sustainability Index for Specialty Crops and Eco Apple by the non-profit company, Red Tomato. The Eco Apple programme has standards developed by the IPM Institute (2015) in collaboration with stakeholders from the northeastern US, the target area for the programme. Red Tomato distributes and markets the qualified fruit with the Eco Apple label. The evaluation protocol has a series of pest-specific criteria under minimum requirements, with a pass/fail rating and advanced practices that have point values. A 'red, yellow, green' list of pesticides indicates preferred and prohibited materials to use, as per the IFP approach. Orchards undergo an on-site third-party inspection prior to harvest. Standards are modified annually to reflect changing technology and knowledge, and the introduction of new pests. The programme has been operational since 2005 and currently has 12 participating orchards encompassing some 400 ha of apple orchards.

Buyers and retailers have instituted programmes as part of corporate sustainability efforts. Unilever, Wal-Mart, Tesco, Sysco, Wegmans and Cadburys are examples of multinational and regional companies that have sustainability criteria for the food products they buy. Some have developed their own programmes while others accept accreditation by third-party programmes. The challenge to producers in complying with this escalating number of standards has led to efforts for harmonization similar to what has occurred in the organic sector. Some programmes involve a paper self-assessment, while others use on-site third-party auditors. Aggregate data for growers are generally compiled yearly, leading to estimates of environmental benefit for that year as well as tracking trends over time. For example, the Sysco 2014 sustainability report states that 56 billion litres of water were conserved in growing, processing and manufacturing their food products; and their growers avoided using 8 million kg of fertilizer and 2 million kg of pesticides by following IPM protocols (Sysco, 2014).

The differing approaches to measuring environmental and other sustainability impacts among these programmes are illustrated in Table 3. The column on the left lists the criteria that are assessed by the different programmes. All programmes focus on pesticide records and equipment calibration, some look at drift management and resistance management, and most have lists of prohibited materials. Only organic and GlobalGAP have stipulations for maximum residue levels (MRL).

For environmental issues other than pesticides, almost all programmes assess soil resource management, water quality and/or quantity, nutrient management and biodiversity/ wildlife habitat. Less universal are assessments of energy use/conservation, air quality and greenhouse gas emissions, use of genetically modified organisms, recycling and waste management, and use of packaging materials. Programmes that cover 10 or more of the environmental impacts include Sysco, Food Alliance and Protected Harvest; however, how they assess each issue can vary. To help understand some of these differences, the Consumers Union group in the United States has a Greener Choices website (http://www.greenerchoices.org/eco-labels/).

Table 3 Standards for assessing environmental impacts in a range of US food sustainability evaluation programmes. √ = included. Risk rankings denote different approaches or tools used

Pesticides	USDA Organic	Food Alliance	Salmon Safe	Protected Harvest	SYSCO	Eco Apple	Global GAP
IPM; Records	√	√	√	√	√	√	Minor must
Pesticide Mgt; Records	√	√	√	√	√	√	√
Drift Mgt	√	√	√	√	√		
Sprayer Calibration		√	√	√	√	√	√
Resistance Mgt		√			√		√
MRLs	√						√
Prohibited Pesticide List	√	√		MeBr		√	EU
Restricted List	√	Tox I-II	High risk, limited use			√	
Risk Rankings		Tox class	√	EIQ	Tox class	√	
Soil and Water							
Soil Quality & Conservation	√	√	√	√	√	√	Rec.
Water Quality	√	√	√	√	√		√
Water Use		√	√	√	√	√	√
Nutrient Mgt	√	√	√	√	√		√
Biodiversity							
Biodiversity	√	√	√		√		Rec.
Wildlife Habitat	√	√	√	√			Rec.
Pollinators		√					
Beneficials Mgt	√	√				√	
Sensitive area Mgt		√	√		√	√	Rec.
Air Quality	√			√	√		
GHG emission							
Non-GMO	√	√			√		
Energy use, farm				√	√		Rec.
Recycling, farm					√	√	Rec.
Packaging, farm					√		√

USDA = US Dept. of Agriculture; MeBr = methyl bromide prohibited; EU = European Union pesticide lists; EIQ = environmental impact quotient; Rec.=recommended; Protected Harvest for stone fruit; GHG = greenhouse gas; GMO = genetically modified organism; Tox class = US Environmental Protection Agency toxicity ratings of specific pesticides

There have also been programmes developed for specific commodities at the regional, national and global levels. The Roundtable on Sustainable Palm Oil and the US Dairy Sustainability Commitment are two examples where industry has initiated the programme in partnership with other key stakeholders. For palm oil, producers undergo a third-party audit every five years for compliance with the standards in order to use the programme trademark in any market claim. The US dairy effort produces an annual sustainability report that documents progress in meeting goals set under the following environmental categories: greenhouse gas emissions, energy, water, resource recovery, landscape stewardship, soil health and air quality. Progress is documented by implementation of specific projects both on the farm and off. No such commodity-wide sustainability programme exists on a global scale for apples other than IFP.

LCA as discussed in the previous section is increasingly being used to evaluate sustainability in food production. This assessment can be conducted by a properly trained consultant or business, and various software programmes (e.g. EcoInvent) and databases are available to help with the data-intensive process. An LCA can be tailored to the specific product (e.g. apple) and be used to develop a snapshot of impact under current management as well as to contrast how changing management might alter the impact (Baumann and Tillman, 2004). Impact category results can be difficult to interpret in order to come up with an overall rating for a system.

6 Future trends and conclusion

Decreasing the environmental impact of apple production is going to become a continual management goal, along with increasing productivity and profitability. This change will be driven by consumers, governmental regulations and producers gaining a greater awareness of the impacts of their management practices. For the past several decades, apple producers have responded to both social and biological pressures to test and

Figure 10 Growing legumes in the orchard drive alley may offer a more sustainable source of nitrogen. Current research is focusing on developing more precise management strategies to make cover cropping more widely implemented in apple orchards. Photo: D. Granatstein.

adopt more sustainable pest management strategies, and significant progress is evident. Pesticide use will likely continue to be a focal area for apple sustainability. With the shift to input-intensive high-density orchards, there will be increased emphasis on reducing the environmental footprint of apple production, whether from energy, greenhouse gas emissions, material inputs or generation of waste at orchard renewal (e.g. plastic pipe and trellis wire). As apple production has concentrated in semi-arid climates in some regions due to the lower number of pests and diseases, issues around water will become more critical and producers will need to maximize efficiency of water use while also protecting water quality. This shift to drier regions illustrates the fact that there will be trade-offs that occur during the on-going search for greater sustainability, and managing those trade-offs will be an on-going challenge. In a broader context, the energy used by relatively inefficient cars to transport apples from retail outlets to consumers' homes has been shown to be greater than the energy required to produce the apple (Jones, 2002). This bodes the question as to whether it is more sustainable to produce apples in horticulturally desirable locations, such as Washington State, or to produce them closer to where they will by consumed.

Apple production will be both a contributor to climate change to the degree that it is a net emitter of greenhouse gases, and the recipient of its effects in terms of more variable weather; increased frost, drought and heat stress risk; and extended ranges of insect pests and diseases, as well as new invasive pests. Greenhouse gas emissions can be countered by increasing soil organic matter, recycling woody debris instead of burning, using biofuels and renewable nitrogen sources, and extending the life of farm machinery and orchard infrastructure. Many of these aspects can be measured and may need to be documented to various customers or regulators (Fig. 10). Thus, more monitoring of the orchard and recordkeeping is likely. Some growers will be compensated for their extra efforts and costs by participating in a formal sustainability programme such as IFP or organic. For others, these will simply be additional costs of doing business that are needed to maintain access to customers and markets which have sustainability requirements.

Social sustainability issues that growers may face include worker safety, labour supply, the nutritional and health qualities of the fruit (nutraceuticals, pesticide residues), rural community development, farm inheritance, land issues (urbanization, access for new farmers) and food security.

Regardless, apple producers must address economic sustainability in the short-term or their operation will fail in an unsubsidized system. The economic issues include rising production costs (including materials, labour, land, etc.) with static or declining prices; retail consolidation leading to more sellers than buyers and less economic power for producers; minimal increase in apple consumption with more production than consumption leading to greater competition and ultimately low farm-gate returns. Apples are an expensive commodity to produce. High-density orchards typically cost $35–$50 000 per hectare to install and that does not include land costs. If investment costs for these systems are not recouped within the first 10–12 years after planting, it is often difficult for growers to be economically sustainable. Without economic sustainability, investments are less likely for improving environmental or social sustainability.

7 Where to look for further information

The references used in this chapter provide a good entrée into the literature on the environmental impacts and sustainability of apple production. There is no single volume written on these topics. Several journals do focus on the broader subject of sustainability or environment, such as *Renewable Agriculture and Food Systems, Agriculture Ecosystems and Environment, Organic Agriculture, Agroecology and Sustainable Food Systems* (formerly *Journal of Sustainable Agriculture*) and *International Journal of Agricultural Sustainability*. Most scientific journals will publish articles on environmental impacts of apple production, but there are far fewer for apple than for major agronomic crops. Proceedings from the 17 Ecofruit conferences held in Hohenheim, Germany, provide a longitudinal look at many specific production challenges for more ecological apple culture (http://www.ecofruit.net/). The *Good Fruit Grower* magazine (http://www.goodfruit.com/) from the United States has featured many articles on the practical application of new techniques and tools that address sustainability and environmental impact. Technical handbooks have been produced for specific regions to help with different production regimes, such as *A Grower's Guide to Organic Apples* (Peck and Merwin, 2009; for the northeastern US), *Organic Tree Fruit Management* (Edwards, 1998; for British Columbia, Canada), *Integrated Pest Management for Ontario Apple Orchards* (Solymar, 1999; for eastern Canada) and *Management Guide for Low-Input Sustainable Apple Production* (USDA Northeast LISA Apple Production Project, 1990), but are often not updated regularly. Conference proceedings papers in the *Acta Horticulturae* (http://www.actahort.org/) from the International Society of Horticultural Science address the topic based on specific conferences and symposia, such as 'Sustainability of Horticultural Systems in the 21st Century' (Bertschinger and Anderson, 2004) and the '2nd International Organic Fruit Symposium' (Granatstein et al., 2013).

8 References

Alaphilippe, A., Simon, S., Brun, L., Hayer, F. and Gaillard, G. (2013), 'Life cycle analysis reveals higher agroecological benefits of organic and low-input apple production', *Agron. Sust. Devel.* 33, 581–92. https://hal.archives-ouvertes.fr/hal-01201368/document.

Anon. (2009), 'Triple bottom line', *The Economist*. 17 November 2009. Retrieved 25 May 2016.

Anon. (2015), 'Bio-Kernobst Saison 2014/15', *AMI*, unpublished market data.

Atucha, A., Merwin, I. A., Purohit, C. K. and Brown, M. G. (2011), 'Nitrogen dynamics and nutrient budgets in four orchard groundcover management systems', *HortSci.* 46, 1184–93.

Bauman, Y., Doppelt, R., Mazze, S. and Wolf, E. C. (2006), 'Impact of climate change on Washington's economy', Report to Wash. State. Dept. Ecology, Institute for a Sustainable Environment, University of Oregon, Eugene, OR.

Baumann, H. and Tillman, A.-M. (2004), *The Hitch-hiker's Guide to LCA*. Studentlitteratur, Lund, Sweden, 543pp.

Benbrook, C. M. (2013), 'Impacts of changing pest management systems and organic production on tree fruit pesticide residues and risk', *Acta Hortic.* 1001, 91–101.

Bennett, H. H. (1947), *Elements of Soil Conservation*. McGraw-Hill.

Bertschinger, L. and Anderson, J. D. (eds) (2004), 'Sustainability of horticultural systems in the 21st century', *Acta Hort.* 638, 1–512.

Brown, M. W., Mathews, C. R. and Krawczyk, G. (2010), 'Extrafloral nectar in an apple ecosystem to enhance biological control', *J. Econ. Entom.* 103, 1657–64.

Brown, M. W. and Tworkoski, T. (2004), 'Pest management benefits of compost mulch in apple orchards', *Agric. Ecosys. Environ.* 103, 465–72.

Brundtland, G. H. (chair) (1987), 'Our common future', World Commission on Environment and Development, United Nations, New York, p. 318.

Carroll, J. E. and Robinson, T. L. (eds) (2006), 'New York Integrated Fruit Production Protocol for Apples', *NYS Food and Life Sciences Bulletin* 158, 31pp. ISSN 0362-0069. https://ecommons.cornell.edu/handle/1813/5219.

Carson, R. (1962), *Silent Spring*. Houghton-Mifflin.

Columella, L. and Ash, H. B. (ed.) (1941), *On Agriculture*. Loeb Classical Library, Vol. 361, W. Heinemann, London.

Cross, J. V. and Dickler, E. (1994), 'Guidelines for integrated production of pome fruits in Europe. Technical guideline III', 2nd edition, *IOBC wprs Bulletin* 17(9), 40pp.

Cuffney, T. F., Meador, M. R., Porter, S. D. and Gurtz, M. E. (2000), 'Responses of physical, chemical, and biological indicators of water quality to a gradient of agricultural land use in the Yakima River Basin, Washington', *Environ. Monitoring Assessment* 64, 259–70.

Dragoni, D. and Lakso, A. N. (2011), 'An apple-specific ET model', *Acta Hort.* 903, 1175–80.

Dyllick, T. and Hockerts, K. (2002), 'Beyond the business case for corporate sustainability', *Business Strat. Environ.* 11, 130–41.

Dzhangaliev, A. D., Salova, T. N. and Turekhanova, P. M. (2003), 'The wild fruit and nut plants of Kazakhstan', *Hort. Rev.* 29, 305–404.

Edwards, L. (1998), *Organic Tree Fruit Management*. COABC, Keremeos, BC, Canada, 240pp..

Environmental Protection Agency (EPA) (2016). Hazard Standards for Lead in Paint, Dust and Soil (TSCA Section 403). Available at: https://www.epa.gov/lead/hazard-standards-lead-paint-dust-and-soil-tsca-section-403.

Ferguson, H. (2015), 'Optimizing water use with Irrigation Scheduler', *Good Fruit Grower*, 4 June 2015. http://www.goodfruit.com/optimize-water-use-with-irrigation-scheduler/

Food and Agriculture Organization (FAO) (2014), 'Water-Energy-Food Nexus: A new approach in support of food security and sustainable agriculture', U. N. FAO, Rome, 28pp. http://www.fao.org/nr/water/docs/FAO_nexus_concept.pdf.

Francis, C. A. and Youngberg, G. A. (1990), 'Sustainable agriculture: an overview', in C. A. Francis, C. Butler Flora and L. D. King (eds), *Sustainable Agriculture in Temperate Zones*, pp. 1–23. J. Wiley, New York.

Granatstein, D. and Kupferman, E. (2008), 'Sustainable horticulture in fruit production', *Acta Hort.* 767, 295–308.

Granatstein, D. and Mullinix, K. (2008), 'Mulching options for Northwest organic and conventional orchards', *HortSci.* 43, 45–50.

Granatstein, D., Andrews, P. K., Bishop, S. D. and Janisiewicz, W. (eds) (2013), 'Proc. 2nd International Organic Fruit symposium', *Acta Hort.* 1001, 1–409.

Granatstein, D., Kirby, E., Ostenson, H. and Willer, H. (2015), 'Global situation for organic tree fruits', *Scientia Hort.* Available at: http://dx.doi.org/10.1016/j.scienta.2015.12.008.

Granatstein, D. and Kirby, E. (2016a), 'Status of organic production of temperate fruits in the USA', *Proc. 17th Intl. Conf. on Organic Fruit-Growing*, University of Hohenheim, Germany, pp. 128–34. http://www.ecofruit.net/2016/17_Granatstein_128bis134.pdf.

Granatstein, D. and Kirby, E. (2016b), Increasing organic production: where are we headed? Organic Trend Series, Center for Sustaining Agriculture and Natural Resources, Washington State University, Wenatchee, WA. http://www.tfrec.wsu.edu/pdfs/P2979.pdf.

Hallberg, G. (1989), 'Pesticides pollution of groundwater in the humid United States', *Agric. Ecosys. Environ.* 26, 299–367.

Hooper, W. D. and Ash, H. B. (1934), *Translation of On Agriculture by Marcus Terentius Varro*. Harvard University Press.

Horton, D. R., Broers, D. A., Lewis, R. R., Granatstein, D., Zack, R. S., Unruh, T. R., Moldenke, A. R., and Brown, J. J. (2003), 'Effect of mowing frequency on densities of natural enemies in three Pacific Northwest pear orchards', *Entomologia Experimentalis et Applicata* 106, 135–45.

Howard, A. (2006), *The Soil and Health: A Study of Organic Agriculture*. University Press of Kentucky, Lexington, KY, 352pp. Original edition 1947, Devin-Adair.

IOBC. (1992), 'Integrated production, principles and technical guidelines', *IOBC/WPRS Bulletin*, 16(1), 1-97. Intl. Org. Biological Integrated Control of Noxious Animals and Plants, Montfavet, FR. 97pp.

IPM Institute (2015), 'Eco Apple Supplement. Complete Core Eco Protocol. 2015 growing season, ver. 1.6'. http://www.ipminstitute.org/ecoapple.htm.

Johnson, A., Norton, D. and Yake, B. (1988), 'Persistence of DDT in the Yakima River Drainage, Washington', *Arch. Environ. Contam. Toxicol.* 17, 289–97.

Jones, A. (2002), 'An environmental assessment of food supply chains: a case study on dessert apples', *Environ. Manag.* 30, 560–76.

Kägi, T., Casado, D., Delval, P., Gaillard, G., Otto, S. and Strassemeyer, J. (2008), 'Life Cycle Assessment of Integrated and organic apple production systems in Europe'. *Proc. 6th Intl. Conf. on Life Cycle Assessment in the Agri-Food Sector*, Zurich, CH, 12–14 November 2008, pp. 95–6. http://www.agroscope.admin.ch/aktuell/02720/02722/03985/index.html.

King, F. H. (2004), *Farmers of Forty Centuries: Organic Farming in China, Korea, and Japan*. Dover Publ, 464pp.

Kirby, E. and Granatstein, D. (2015), 'Recent trends in certified organic tree fruit: Washington State 2014', Organic Trend Series, Center for Sustaining Agriculture and Natural Resources, Washington State University, Wenatchee, WA. http://csanr.wsu.edu/wp-content/uploads/2015/01/WA_OrgTreeFr2014_annotate_final.pdf.

Kovach, J., Petzoldt, C., Degni, J. and Tette, J. (1992), 'A method to measure the environmental impact of pesticides', NYS Food and Life Sciences Bulletin No. 139, St. Agr. Expt. Station, Geneva, New York, USA, 8p. https://ecommons.cornell.edu/handle/1813/5203.

Kramer, S. B., Reganold, J. P., Glover, J. D., Bohannan, B. J. and Mooney, H. A. (2006), 'Reduced nitrate leaching and enhanced denitrifier activity in organically fertilized soils', *PNAS* 103, 4522–7.

Leib, B. G., Caspari, H. W., Redulla, C. A., Andrews, P. K. and Jabro, J. J. (2006), 'Partial rootzone drying and deficit irrigation of 'Fuji' apples in a semi-arid climate', *Irr. Sci.* 24, 85–99.

Levitan, L., Merwin, I. and Kovach, J. (1995), 'Assessing the relative environmental impacts of agricultural pesticides: the quest for a holistic method', *Agric. Ecosys. Environ.* 55:153–68.

Lockeretz, W. (ed.) (1977), *Agriculture and Energy*. Academic Press, 782pp.

Logsdon, G. (1981), *Organic Orcharding: A Grove of Trees to Live In*. Rodale Press, Emmaus, PA, 415pp.

Lowdermilk, W. C. (1953), 'Conquest of the Land Through 7,000 Years', Agriculture Information Bulletin No. 99, Reprint Washington, DC, Government Printing Office, 1975, 13pp.

MacLeod, C. J., Blackwell, G. and Benge, J. (2012), 'Reduced pesticide toxicity and increased woody vegetation cover account for enhanced native bird densities in organic orchards', *J. Appl. Ecology* 49, 652–60.

Merwin, I. A., Ray, J. A., Steenhuis, T. and Boll, J. (1996), 'Groundcover management systems influence fungicide and nitrate-N concentrations in leachate and runoff from a New York apple orchard', *J. Amer. Soc. Hort. Sci.* 121, 249–57.

Merwin I. A. and M. P. Pritts. (1993), 'Are modern fruit production systems sustainable?', *HortTechnol.* 3, 128–36.

Milà i Canals, L., Burnip, G. M. and Cowell, S. J. (2006), 'Evaluation of the environmental impacts of apple production using Life Cycle Assessment (LCA): Case study in New Zealand', *Agric. Ecosys. Environ.* 114, 226–38.

Mouron, P., Nemecek, T., Scholz, R. W. and Weber, O. (2006), 'Management influence on environmental impacts in an apple production system on Swiss fruit farms: Combining life cycle assessment with statistical risk assessment', *Agric. Ecosys. Environ.* 114, 311–22.

Neilsen, D. R. and Neilsen, G. H. (2002), 'Efficient use of nitrogen and water in high-density apple orchards', *HortTech.* 12, 19–25.

Neilsen, G. H. and Neilsen, D. (2003), 'Nutritional requirements of apple', in Ferree, D. C. and Warrington, I. J. (eds), *Apples: Botany, Production, and Uses*, CABI, pp. 267–302.

OTA. (2015), *Organic Industry Survey*. Organic Trade Assoc., Brattleboro, VT, 139pp.

Pang, J., Xiaoke, W., Yujing, M., Zhiyun, O. and Wenzhao, L. (2009), 'Nitrous oxide emissions from an apple orchard soil in the semi-arid Loess Plateau of China', *Biol. Fertility Soils* 46, 37–44.

Peck, G. M., Andrews, P. K., Reganold, J. P. and Fellman, J. K. (2006), 'Apple orchard productivity and fruit quality under organic, conventional, and integrated management', *HortSci.* 41, 99–107.

Peck, G. M. and Merwin, I. A. (2009), *A Grower's Guide to Organic Apples*. New York State IPM Publication 223, Geneva, NY, 64pp. https://ecommons.cornell.edu/handle/1813/42886.

Peck, G. M., Merwin, I. A. and Brown, M. G. (2010), 'Integrated and organic fruit production systems for 'Liberty' apple in the Northeast United States: A systems-based evaluation', *HortSci.* 45, 1038–48.

Peck, G. M., Merwin, I. A., Theis, J. E., Schindelbeck, R. R. and Brown, M. G. (2011), 'Soil properties change during the transition to integrated and organic apple production in a New York orchard', *Applied Soil Ecology* 48, 18–30.

Peryea F. J. (1998), Historical use of lead arsenate insecticides, resulting in soil contamination and implications for soil remediation. *Proceedings 16th World Congress of Soil Science*, Montpellier, France, pp. 20–6. Available at: http://soils.tfrec.wsu.edu/leadhistory.htm.

Reganold, J. P., Glover, J. D., Andrews, P. K. and Hinman, H. R. (2001), 'Sustainbility of three apple production systems', *Nature* 410, 926–30.

Reed, N. (1995), 'Responsible Choice: A system approach to growing, packing and marketing fruit', *Proc. Mich. St. Hort. Soc.*, 125th Annual Report, E. Lansing, MI. p. 68–78.

Saunders, C., Barber, A. and Taylor, G. (2006), *Food Miles – Comparative Energy/Emissions Performance of New Zealand's Agriculture Industry*. Research Report 285, AERU, Lincoln University, Lincoln, NZ. https://researcharchive.lincoln.ac.nz/handle/10182/125.

Schooley, T., Weaver, M. J., Mullins, D. and Eick, M. (2008), 'The history of lead arsenate use in apple production: Comparison of its impact in Virginia with other states', *J. Pesticide Safety Education* 10, 22–53.

Smith P., Bustamante, M., Ahammad, H., Clark, H., Dong, H., Elsiddig, E. A., Haberl, H., Harper, R., House, J., Jafari, M., Masera, O., Mbow, C., Ravindranath, N. H., Rice, C. W., Robledo Abad, C., Romanovskaya, A., Sperling, F. and Tubiello, F. (2014), 'Agriculture, Forestry and Other Land Use (AFOLU)', in Edenhofer, O., Pichs-Madruga, R., Sokona, Y., Farahani, E., Kadner, S., Seyboth, K., Adler, A., Baum, I., Brunner, S., Eickemeier, P., Kriemann, B., Savolainen, J., Schlömer, S., von Stechow, C., Zwickel, T. and Minx, J. C. (eds), *Climate Change 2014: Mitigation of Climate Change. Contribution of Working Group III to the Fifth Assessment Report of the Intergovernmental Panel on Climate Change*. Cambridge University Press, Cambridge, pp. 811–922. http://www.ipcc.ch/pdf/assessment-report/ar5/wg3/ipcc_wg3_ar5_chapter11.pdf.

Solymar, B. (1999), *Integrated Pest Management for Ontario Apple Orchards*. Publ. 310, Ontario Ministry of Agriculture, Food and Rural Affairs, Toronto, Canada, 230pp.

Stadig, M. (1997), *Livscykelanalys av a¨ppelproduktion-fallstudier fo¨r Sverige, Nya Zeeland och Frankrike*. SIK Report No. 630. Gothenburg, Sweden.

Sysco. (2014), *Sustain: 2014 Sustainability Report*. Sysco Corp., Houston, TX. http://sustainability.sysco.com/supplying-food-responsibly/sourcing-food-responsibly/sustainable-agriculture.php.

USDA Northeast LISA Apple Production Project (1990), Management Guide for Low-Input Sustainable Apple Production: A Publication of the USDA Northeast LISA Apple Production Project and these Institutions: Cornell University, Rodale Research Center, Rutgers University, University of Massachusetts, and University of Vermont, Washington, DC.

USDA (2016). 'The National List of Allowed and Prohibited Substances'. Electronic Code of Federal Regulations. U. S. Govt. Publ. Office, Washington, DC. http://www.ecfr.gov/cgi-bin/text-idx?c=ecfr&SID=9874504b6f1025eb0e6b67cadf9d3b40&rgn=div6&view=text&node=7:3.1.1.9.32.7&idno=7.

VanLierde, D. and VanDenBossche, A. (2004), 'Economical and environmental aspects of Integrated Fruit Production in Belgium', *Acta Hort.* 638, 411–17.

Weibel, F. P., Daniel, C., Tamm, L., Willer, H. and Schwartau, H. (2014), 'Development of organic fruit in Europe', *Acta Hort.* 1001, 19–34.

Willer, H. and Lernoud, J. (eds) (2016), *The World of Organic Agriculture. Statistics and Emerging Trends 2016*. Research Institute of Organic Agriculture (FiBL), Frick, and IFOAM – Organics International, Bonn. 340 pp. http://www.organic-world.net/index.html.

Williamson, A. K., Munn, M. D., Ryker, S. J., Wagner, R. J., Ebbert, J. C. and Vanderpool, A. M. (1998), *Water quality in the central Columbia Plateau, Washington and Idaho, 1992–1995*. U. S. Geological Survey Circular 1144, Washington, DC, 35pp. https://pubs.er.usgs.gov/publication/cir1144.

Yan, Y. -H., Yu, Y., Du, X. -G. and Zhao, B. -G. (1997), 'Conservation and augmentation of natural enemies in pest management of Chinese apple orchards', *Agric. Ecosys. Environ.* 62, 253–60.

Yin, X., Seavert, C. F. and le Roux, J. (2011), 'Responses of irrigation water use and productivity of sweet cherry to single-lateral drip irrigation and ground covers', *Soil Sci.* 176, 39–47.

Chapter 22

Growing organic apples in Europe

Jutta Kienzle, Independent Researcher, Germany; and Markus Kelderer, Research Station of Laimburg, Italy

1 Introduction

1.1 Organic fruit production

Organic agriculture (OA) or organic farming (OF) can be defined as a sustainable holistic farming system that includes the whole food processing chain from farmer to consumer. Private organic associations (under the international umbrella organization IFOAM www.ifoam.bio) and also the legislation for OF in Europe (Council Regulation (EC) No 834/2007) cover the production and processing sector, and there is a specialized and well-labelled market segment for organic products. Four principles best define the ethical values of OA expressed by IFOAM: health, ecology, fairness and care (for more information, see http://www.ifoam.bio/en/organic-landmarks/principles-organic-agriculture). The main principle of OF systems in EC Regulation 834/2007 is specified as the appropriate design and management of biological processes based on ecological systems using natural resources that are internal to the system. Where 'off-farm' inputs are really needed, these are restricted to inputs from organic production, natural or naturally derived substances, and low-solubility mineral fertilizers. Genetically modified organisms (GMOs) and products produced from or by GMOs are incompatible with the concept of organic production.

http://dx.doi.org/10.19103/AS.2016.0017.26

Organic fruit production was one of the later branches of organic cultivation in Europe. The first few pioneer fruit growers in several European regions started OF with little or no scientific support in the 1970s. They experienced losses and took risks that nowadays are difficult to imagine. The first specific research activities were started in close collaboration with local producers in the 1980s. Since then, the sector has continuously increased and now at least 10% of the fruit-growing area in several important fruit-growing regions in Europe, such as Northern Italy, Germany and Austria, is under organic cultivation. This was mainly due to innovation in the plant health care strategy, including development of appropriate equipment for tree row tillage and the introduction of new varieties with lower susceptibility to scab (*Venturia inaequalis*). In fruit-growing regions where OF is more important, specialists are now employed in the regional research stations who collaborate closely with growers. Where local institutes for OA exist (e.g. GRAB in France, FiBL in Switzerland and LBI in the Netherlands), they have established specialised working groups for fruit growing. However, there is still a strong need for research activities to create more resilient systems that provide a high number of ecosystem services. The key challenges for research can be summarized as follows:

- Development of varieties and rootstocks suited to OF systems
- Improvement of the management of soil fertility in organic fruit growing
- Improvement of the strategies for plant health care in organic fruit growing
- Improvement of the ecosystem services of organic orchards

1.2 The participatory and holistic approach to researching in organic apple production

OF was developed by pioneering fruit growers in close cooperation with consumers and retailers who were interested in fruit obtained through organic production. These growers made the initial steps with little research support, and experienced risks and economic failures hardly imaginable today. The first key innovations were on-farm innovations; for example, a fruit grower invented and constructed the first effective machine for tillaging rows of trees (see Fig. 1). These pioneers subsequently organized working groups and national networks of fruit growers, researchers and advisers with organic associations. One such network is the 'working net' of the German FOEKO e.V. (Kienzle et al., 2010, 2012), which also involved the coordination of the international ecofruit conferences. Organic fruit growers not only express their needs for new or improved management tools but actively engage in their development and that of their farming system and of their standards. Participatory research approaches mean a close collaboration between researchers and farmers. The combination of defined trials and on-farm tests where farmers' ideas and demands are included have led to the successful development and implementation of several measures in OF.

However, participatory approach is not limited to the collaboration of farmers and researchers. For many issues, to achieve a real impact on OF practice, the whole production chain must participate. This means that consumers, retailers, farmers and their associations and producers of inputs as fertilizers or biocontrol products must be involved in the generation of solutions.

Figure 1 The first prototype of the equipment for successful tillage of the tree row with the pioneer fruit grower/designer. Author J. Kienzle.

2 Developing varieties and rootstocks suited to organic farming (OF) systems

The availability of suitable varieties and rootstocks is a key factor for the development of a more resilient fruit-growing system in OF.

Varieties should present a durable level of low susceptibility towards the most important diseases, pests and to abiotic damage such as frost or sunburn. Growth and fruit set should be relatively easy to regulate with the measures available in OF and the yield should be reliable and high. Fruits should be tasty, salubrious, low in allergens and with a high dietetic value ingredient content and present a good shelf-life and storage stability. Furthermore, the range of varieties should ensure the availability of organic fruit in good quality over the whole marketing period of the apples in Europe.

In the 1990s, many new varieties from breeding programmes around the world were tested and advertised to be tolerant or to carry resistance to one or more diseases. For a few promising varieties, concerted actions to introduce them into the market were started by the growers. More recently, it has become evident that these varieties did not really meet the long-term needs of sustainability for OF. Breeding techniques, based on genetic modification, are not accepted by the organic agricultural movement regardless of whether genes of the own species (*cis*-) or of other species (*trans*-) (BOELW, 2015; Eurobarometer 1991–2010) are used. Thus, special rules for organic breeding have been developed (Ristel and Sattler, 2014), and the first steps for new breeding programmes dedicated to produce varieties through OF can be observed in different countries, for example, Germany and Switzerland. These programmes particularly reflect the organic ideal of close cooperations between farmers and research stations, aiming to create quantitative resistance using a broad genetic basis. The varieties derived from these programmes are not protected in Variety Clubs but freely available for all interested organic farmers (Haug, 2014, 2015). Cooperation with interested research stations is crucial for these participatory projects and may well lead to innovative breeding programmes.

In addition to the breeding – which may produce long-term solutions – this collaboration is of high importance as many new varieties are released with little information about their susceptibility to pests and diseases in an OF system. Recently, several research stations in the EU have collaborated with organic farmers to establish special test systems under organic conditions, or at least without plant protection measures, to ensure that relevant information about newly introduced varieties that are actually on the market is available to fruit growers (Warlop et al., 2014).

The master plan for varieties suited to OF must cover all steps from breeding to testing to the introduction to the market (Haug, 2014). Some apple varieties ('Topaz', 'Santana' and 'Natyra') were introduced to food retailers and consumers by groups of growers as part of concerted actions. For such an introduction, the availability of sufficient quantity of good quality fruit of the new variety is crucial. To facilitate the marketing of small quantities of new varieties in food retailers, Weibel et al. (2004) developed a 'Flavour Group Concept' (FGC) for Switzerland. The FGC provides additional information to the buyers on the specific flavour direction of that 'unknown' apple cultivar on the display in front of them (mild to sweet, spicy–tart, predominantly tart). To further develop the FGC, in 2002, a consortium of the supermarket chain Coop-Switzerland, organic apple growers, retailers and FiBL experts joined together in a 'Variety Team'. In contrast to Variety Clubs, the focus was not a particular variety but the active creation and promotion of an attractive assortment of scab-resistant apple varieties in a well-co-ordinated and efficient way for both growers and retailers (Weibel et al., 2004). This concept is focused very much on Swiss conditions that are rather different from the market situation in the rest of Europe.

The long-term OF challenge, however, will not be limited to the breeding, testing and market introduction of a couple of suitable varieties. The high susceptibility to pests and diseases of most current commercial varieties is not only due to a genetical predisposition. Some of them, for example, 'Golden Delicious', were initially reported as having a rather low susceptibility; however, if a clonally propagated variety (which means genetically identical plants) is cultivated in large areas, it has to be expected that pathogens adapt to its specific defence mechanisms. Moreover, due to a lack of diversity in parents used in breeding programmes over the last twenty years, many of the limited number of varieties of importance in commercial fruit production are closely related, resulting in a really narrow genetic pool (Bannier, 2011).

Thus, for a truly resilient sustainable system, it will be indispensable to broaden the genetic basis of the tolerance to pathogens not only by breeding but also by increasing the number of more distantly related varieties that are cultivated in large fruit-growing areas. In this context, local varieties that are very well adapted to specific regions should also become more important.

To meet this challenge, close collaboration of researchers of different disciplines (e.g. breeders, plant protection specialists and marketing experts), fruit growers, retailers, consumers, organic associations and NGOs is of great importance to create sustainable variety concepts accepted by all marketing partners based on high genetic biodiversity and good adaptation to local conditions. In this context, biodiversity must become part of innovative quality schemes – a special challenge for participative, interdisciplinary research programmes.

In the following, the specific situation and the main challenges in the variety sector for apples are described.

Over the past 30 years, the organic fruit-growing scene was very much engaged in the introduction of new scab-resistant varieties into production. In the 1990s, fruit quality of these varieties was insufficient for a successful market introduction. The next generation of scab-resistant varieties with higher fruit quality, such as 'Topaz', 'Santana', 'Ariwa' and 'Natyra' were successfully introduced in the market with a concerted action of fruit growers' groups and marketing organizations. However, when growers reduced the applications of fungicides for scab control, other diseases became prominent. In several European regions, sooty blotch quickly advanced to become a serious problem (Weber et al., 2016). In the last couple of years, premature leaf fall from apple blotch disease (*Marssonina coronaria*) caused considerable losses in orchards with scab-resistant varieties where fungicide applications were more extensive. Moreover, all these varieties carry only single-gene resistance (*Rvi6, Vf*) against apple scab, so as a consequence, the resistance breakdown is more and more frequently observed in European orchards. The actual potential of these varieties for more resilience in the German system of OF is shown by Kienzle et al. (2016) using data from a survey on organic farms in 2014. In several German regions, approximately 50% of the area of organic apples (in total about 3300 ha) is covered by scab-resistant varieties (predominantly 'Topaz' and'Santana'). Even if the 'resistance' is overcome in all German regions, the scab-'resistant' varieties

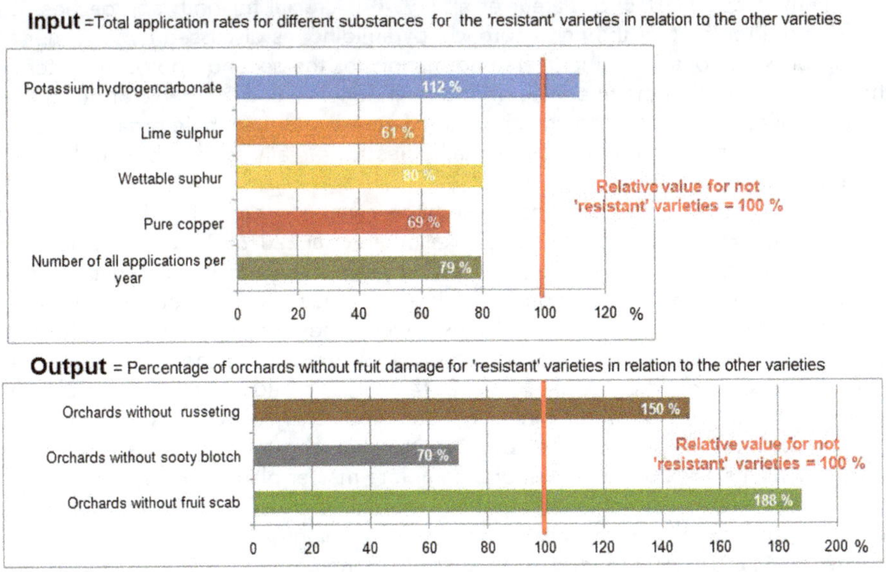

Figure 2 Input and output of the scab-'resistant' varieties in comparison to non-'resistant' varieties in the same farms based on a survey in Germany in 2014 (Kienzle et al., 2016).

still produce a higher output (no fruit damage by scab) with a lower input (defined as the number of applications and the amounts of copper, sulphur and lime sulphur used) in comparison with the usual commercial varieties (Fig. 2). Control of sooty blotch in regions with high pressure seems to be the limiting factor. With an intelligent resistance management strategy, the use of *Vf*-'resistant' varieties still contributes to a general reduction of input and an improvement of the resilience of the growing system. Fruit growers obviously believe this, as data from the survey show the percentage of these 'scab-resistant' varieties in new plantations (from 2010 to 2014) was about 50% or more depending on the region. In Northern European countries, however, these varieties are no longer recommended by the extension services (Korsgard, oral communication, 2016). In other member states, such as Germany, these varieties are considered as a short-term strategy to reduce the inputs and increase the output. Research in these cases is focused on the development of intelligent management strategies for these varieties that preserve the resistance for a reasonable period of time.

The medium- and long-term strategies for varieties are based on two new breeding approaches:

1 Pyramiding of several scab resistances with powdery mildew (*Podosphaera leucotricha*) resistance and fire blight (*Erwinia amylovora*) tolerance (Kellerhals et al., 2016)
2 Breeding for low susceptibility to several diseases based on a quantitative tolerance on a genetical basis as broad as possible from varieties that are not closely related to large-scale commercial varieties. Avoidance of inbreeding is also an important issue (Ristel et al., 2016).

The evaluation of traditional varieties for susceptibility to different diseases as described by Kellerhals et al. (2016) or Lateur et al. (1994) is crucial for both approaches. It can broaden the toolbox for the first approach (pyramiding) as discussed by Woehner et al. (2016) for *M. coronaria* and it offers information for the second approach, in terms of the genetic control of the tolerance of potential donors (i.e. monogenic or quantitative). A close collaboration of researchers, organic farmers and the whole organic food chain in this field is one of the main future challenges for a really sustainable apple-growing system.

The first varieties with pyramided resistance can be expected to be introduced in practice in a few years (Vavra et al., 2014; Kellerhals et al., 2016). This introduction should be a solution for more resilience in the medium term. However, in a perennial crop such as apple, the pathogen goes through multiple generations in the lifetime of an orchard. Thus, it is not surprising that from the 18 resistance genes actually known, 11 have already broken down. In respect to gene-for-gene relationships, it is expected that others will follow (Patocchi and Bus, 2016; Haug, 2014). Thus, breeding for a true quantitative tolerance in combination with a higher genetic biodiversity of varieties in the main fruit-growing areas is the approach needed for sustainable long-term solutions.

The first steps towards initiatives for a complete master plan for varieties suited to OF are underway; however, activities regarding *rootstocks* are still rather limited. Initial test results for interesting new rootstocks for apple under organic cultivation are available (Pfeiffer et al., 2014, 2016; Spornberger et al., 2016; Kelderer et al., 2016). There is a great demand for rootstocks more suited to OF than the rootstocks actually in commerce.

3 Improving soil fertility management

Maintenance and development of soil fertility is a very important part of OF. An 'appropriate design and management of biological processes based on ecological systems using natural resources which are internal to the system' is part of the organic principles (see, http://www.ifoam.bio/en/organic-landmarks/principles-organic-agriculture). Improvement of nutrient flow on the farm and minimization of 'off-farm inputs' are very important for OA. As most fruit growers are specialized farmers without livestock, the use of organic commercial fertilizers and soil conditioners is widespread in OF. The availability of organic fertilizers is limited and this practice is not completely compatible with the organic idea of a closed nutrient flow in the farm. One solution could be: in some regions the organization of local farm clusters of specialized fruit farms and specialized livestock farms or of horse owners where the fruit farmers use the surplus of organic manure from these livestock farms for fertilization. Another solution could be the on-farm production of fertilizers. Buchleither et al. (2014) tested a dense sowing of peas as an alternative nitrogen source to commercial organic fertilizers. He concluded that the success of nitrogen fertilization with pea dense sowing is highly dependent on weather and soil conditions and, thus, fertilization with commercial nitrogen fertilizers like 'Bioilsa' pellets was simpler and more reliable. Furthermore, an early sowing date is necessary to enable an adequate period for germination, growth and mineralization after incorporation. Sufficient availability of nitrogen is only enabled at bloom of the trees if these factors are considered. The idea of producing fertilizer on farm is also discussed by Kienzle et al. (2016) within the framework of vegetation management in the alleys to improve biodiversity. Granatstein (2016) showed that cover crops other than grass in the alley can provide various functions in OF including nitrogen fixation, biomass for mulch, soil improvement and flowers to attract beneficial insects. Further research is required to elaborate such strategies into practice in different regions to ensure the availability of nitrogen when needed and to avoid mobilization in late summer.

For these reasons, fertilization strategies in OF have to always consider a combination of direct fertilization and the vegetation management both in the alley and the tree rows. Timing and modality of tillage in the tree row is especially crucial for accurate timing of nitrogen mobilization. This approach is very clearly specified by Kelderer et al. (2014) who showed the effect of different combinations of tillage and fertilization with different techniques and different application dates in Northern Italy. Kelderer et al. (2014) observed a significant increase in yield due to fertilization. In northern Europe, where soil temperatures in spring are low, the challenge for nitrogen mobilization at the right time is even higher. Weibel et al. (2014) compared different fertilizers and stated that tree performance (yield and growth) in an organic orchard system under Swiss conditions cannot be 'boosted' with organic soil N fertilizer or organic foliar N fertilizers. In each case, he concluded that organic fertilizers can hardly compensate for suboptimal soil conditions. The importance of soil sanitation before planting was highlighted. Different organic fertilizers were compared in a long-term trial in Northern Germany (Ristel et al., 2016); no significant differences in yield, fruit size or colouring between the fertilized and the control plots could be seen from the 5th until the 17th year after planting the orchard.

Some farmers do, however, increase fertilization as yields are 15–30% lower than in conventional fruit growing, a fact they partly attribute to lower fertilization levels (Weibel et al., 2014). Evaluation of the effect of fertilization should consider not only the yield but also the fruit quality and the shelf-life of the fruits. In this context, the approach of Malusa et al. (2014) to use formulations based on rhizosphere microorganisms (e.g. mycorrhizal fungi, plant growth promoter bacteria and rhizobia) seems very interesting. Effects reported can be improved plant nutrition and/or the increase of fruit quality or enhanced tolerance to biotic or abiotic stress. The formulation of inoculum with a reliable and consistent effect under field conditions is still a bottleneck for their wider use. Choice of the technology for inoculum production and for the carrier for the formulation is the key to their successful application (Malusa et al., 2014).

Exploration of locally available organic fertilizers and examination of their quality and potential for application in OF are important regional research in areas close to OF (Kelderer et al., 2008). Malusa et al. (2014) evaluated the feasibility of different organic sources available in Poland as fertilizers for OF. The use of compost of brown coal in combination with organic material increased the frequency of mycorrhizal occurrence in strawberry roots. Moeller (2014) expressed the need to use organic fertilizers in OF that have a nutrient composition close to the needs of the fruit trees, or over time, the balance of the individual nutrients could be compromised. The status quo in orchards under long-term organic cultivation should be investigated to observe the effect of different fertilization practices and to optimize strategies for the future.

Soil management in OF is not limited to the question of nutrient supply. Improvement of the soil structure, and enhancement of organic matter content and of the capacity for water retention have to be considered. For these parameters, preparation of the soil before planting is very important. One possible reason for the lower yields in OF may be that measures to reduce the replant disease are not yet fully developed for OF. Within the framework of a European project, Kelderer et al. (2016) tested different tools for mitigation of replant. The effect of a range of microbial-based preparations and composts tested on potted trees on shoot length, shoot dry weight and leaf colour of plantlets was limited (Kelderer et al., 2016). However, a combination of steaming and certain plant extracts and cover crops showed interesting results in initial field studies on young trees (Kelderer, 2016). These treatments should be tested under different climatic and soil conditions under organic cultivation, especially if new rootstocks with tolerance of replant disease can also be part of the strategy.

The increased use of permanent netting structures for hail protection and resulting inflexibility of spatial arrangement of tree rows raises the importance of intensifying research on replant disease for both conventional and organic production.

4 Strategies for improving plant health care: pest and disease control, regulation of crop set and tree growth

4.1 General aspects

The definition of integrated pest management (IPM) in the thematic strategy (EC directive 2009/128) on the sustainable use of pesticides (p. 74, Art. 2) is:

> IPM means careful consideration of all available plant protection methods and subsequent integration of appropriate measures that discourage the development of populations of harmful organisms and keep the use of plant protection products and other forms of intervention to levels that are economically and ecologically justified

> and reduce or minimise risks to human health and the environment. 'Integrated pest management' emphasises the growth of a healthy crop with the least possible disruption to agro-ecosystems and encourages natural pest control mechanisms.

This definition is also valid for OF. For OF, however, there is a different economic sustainability compared to IPM: OF must respect its principles to maintain its credibility. Due to these principles, it can achieve a higher price with a different level of accepted external qualities that can balance a higher risk of losses and lower yields. However, there is a limit to the feasible and maintainable price level; thus, OF must also balance the organic approach with what is economically realistic within the margins of the accepted prices, accepted qualities and feasible public support.

Biodiversity in plant health care strategies plays a central part of OF principles, including the genetic biodiversity as well as the species and structural biodiversity in the orchards. High biodiversity and a high diversity of measures within the frame of intelligent combination strategies are the backbone of resistance management in OF. Cultivation measures available to reduce the infestation of pests and diseases in practice aim to have a high resilience in the farming system. However, fruit growing is one of the few branches of OF where, in combination with these measures, the direct input of plant protection products and other substances by spray applications is common practice. Within the frame of intelligent combination strategies with the above-mentioned practice, a high efficiency is not an imperative for these substances. Efficiency of the whole management strategy is important. Many substances traditionally used in OF are 'multi-actors' with properties that allow simultaneous use for direct or indirect pathogen control, bio-stimulant and/or fertilization purposes.

OF principles restrict the use of these inputs to natural or naturally derived substances. For substances where there is an already existing exposure in natural systems, the likelihood of unpredictable risks is considerably reduced. Given our lack of complete understanding of the processes in ecosystems, this restriction is due to a precautionary and preventive approach to avoid such unpredictable risks. This holistic system approach is not part of the actual risk evaluation system for pesticides. In these systems, usually only criteria derived from the hazard criteria assessed during the currently established registration process are considered. The unpredictable risk, the probability that there are effects of newly designed substances that are not covered by these assessments, is not considered in these evaluation systems.

For OF in the future, it will be more difficult to draw the line for products that should not be included in organic standards as the number of 'ecopesticides' will increase and probably also include 'borderline cases'. It will then be important to not only consider if a substance occurs in nature but also include quantitative and qualitative (i.e. if the system of occurrence is very different from agroecosystems) aspects regarding how it occurs in nature as part of the comprehensive survey of estimation. It will be an important challenge for researchers to support the responsible organic players in suggesting indicators for this evaluation.

The natural occurrence in the ecosystem and a long experience of historical uses may be advantageous, considering the probability of the occurrence of unpredictable risks. Companies, however, are often reluctant to invest in the registration of such substances as it is nearly impossible to protect the intellectual properties and, thus, the return on investment. Furthermore, the registration of naturally occurring substances is still difficult as the process was designed initially for synthetic substances despite the fact that the special requirements of 'organic' substances are now considered in Europe. Thus, researchers must not only focus on the technical issues of the development of new or on the improvement of traditional substances for plant health care but also attend to and support registration of the substance. Unless these substances can be registered, there will be no impact on OF practice.

4.2 Weed control

Most OF systems use rootstocks that require weed control in the tree row during the whole life cycle of the orchard or at least during the first years of tree growth. Weed control in the row is not only important for the reduction of competition for nutrients and water, tree row management is also responsible for mobilization of nitrogen, water retention and, last but not least, control of voles and field mice.

Establishment of annual crops within the tree row gave better yields than permanent cover with grass or clover in northern Europe (Kuehn and Pedersen, 2009). Benduhn et al. (2006) tested different tillage and mulching types for OF. Mulching at first sight could appear to be very attractive for OF. In practice, however, it is expensive and work-intensive, suppression of weeds is incomplete and there are frequently serious problems with voles and field mice. A vegetation-free soil strip subject to repeated tillage does not meet the organic ideal of sustainable soil management. The Swiss sandwich system, a combination of resident vegetation and tilled strips, was developed in Switzerland (Schmid et al., 2004; Zoppolo et al., 2011) where a small strip of vegetation remains in the middle of the tree row during tillage. This method improves the structure of the nematode web in the soil, an important sustainability parameter (Zoppolo et al., 2011). The sandwich system was not implemented in OF practice in most European countries due to several practical difficulties.

Frequent tillage is usually discussed for negative effects on the soil as well as consumption of energy. Actually, depending on the regional climatic conditions, combination strategies of tillage in spring for nitrogen mobilization followed by a light cover of the tree row with the vegetation cut from the alley in combination with a new machinery that removes weed by brushing is practised in southern Germany and in Italy. The use of annual crops in summer is also in discussion.

Further research is needed to elaborate optimal strategies combining different measures that reduce energy consumption and best meet the requirements for the tree row management mentioned previously. In this context, monitoring of the various available equipment for in-row tillage in different ambient conditions is needed.

Preparation of the soil before planting is very important to determine the infestation pressure of weeds and the quality of the soil in the future. A green manure period before planting is usually advised for all crops.

4.3 Disease control

The strategy for a sustainable disease control in organic apples must target not only the main diseases such as scab and powdery mildew but also the minor diseases such as sooty blotch (Weber et al., 2016) or *Marssonina coronaria* that occur when fungicide applications are reduced (Bohr et al., 2016). Indirect methods are an important part of the OF strategy. As a matter of course, the choice of less susceptible varieties is the first and most crucial point. However, even if 'established' organic fruit growers in regions with a climate favourable for the major diseases usually hold a high proportion of the less susceptible varieties available, newly converted farms present the usual spectrum of more susceptible commercial varieties. For this reason, two strategies are developed: (1) sustainable long-term strategies based on less susceptible varieties and (2) short-term transition strategies based on mainstream varieties typically in use in IPM. However, there is a limit to the feasibility of these strategies depending on the region; it is usually advised that some varieties very susceptible to certain diseases (e.g. 'Golden Delicious' to scab or 'Kanzi' to fruit tree canker) should not cover a large proportion of the farm area in regions where the respective diseases are of high prevalence.

These strategies rely on the combination of several measures. Reduction of infection potential is considered highly relevant. For apple scab, strategies have been developed for the reduction of ascospore concentration before the beginning of the season by mechanical removal of leaves (Benduhn et al., 2016) and different autumnal treatments to accelerate their decomposition (Buchleither et al., 2016; Porsche et al., 2016). Korsgard (2016) tested an irrigation strategy where the orchard floor is irrigated 24 hours before forecast rain to elicit the release of ascospores, which will then not cause infections. A reduction of scab infections could be observed in some years and with some varieties. Further improvement and an intelligent combination of such measures is essential for future research.

Direct control measures in OF are predominantly based on substances of mineral origin, most common are copper, sulphur, lime sulphur and various carbonates. 'Scab stop applications' based on treatments during the period of germination of the spores were developed to reduce the number of applications (Kunz et al., 2014; Kelderer et al., 2012).

Unfortunately, although a useful control measure, copper accumulates in the soil. For this reason, the use of copper in OF is limited within the EU, for example, 6 kg/ha per year in Italy and France, and 3 kg/ha per year in Germany and Austria. In Switzerland, the limit for pome fruit is only 1.5 kg/ha per year. In several northern countries such as the Netherlands, Denmark and Sweden, no product based on copper is registered for plant protection purposes. These upper limits are rarely reached in practice. Real input data of copper in practice are available for Germany for the period 2011–13 (von Mering et al., 2016) where the medium input for organic apple orchards shifts between 1.3 and 1.59 kg pure copper per ha per year (Table 1) and for Switzerland where 0.9 kg pure copper is reported (Speiser et al., 2015) as an average input in apples in 2009–12. Strumpf et al. (2013) showed that in most German orchards the copper content of the soil is comparable in terms of magnitude with the natural copper content of the soils in the region. Copper contents in the soil that are considerably higher than the natural contents are usually a result of the high applications of copper common in conventional farming in the last century (Strumpf et al., 2013). However, a certain accumulation must be expected in the long term. Copper is also toxic to some aquatic organisms so that the drift of applications to water bodies

Table 1 Average application rates of copper (pure copper per ha) in organic apple production in Germany in the years 2010–13 (von Mering et al., 2016)

Year	2010	2011	2012	2013
Total area (ha)	3400	3700	3900	3900
Area treated with copper and analysed (ha)	359.1	697.8	747.1	1.302.9
Copper application rate (kg/ha)	1.59	1.3	1.31	1.47

must also be considered. Thus, strategies must also aim to reduce the amount of copper per single application. Usually, in OF very low amounts of copper per application are used with frequent applications (splitting). Higher applications are used mainly before blossom. Research should aim to reduce the amount of copper needed during this period to reduce the risk for aquatic organisms in adjacent water bodies.

In Germany, organic associations formulated a strategy paper where four pillars for the strategy to reduce the copper amount necessary for the disease control in apples are cited (Roehrig et al., 2010): The first pillar, which is the choice of less susceptible varieties and the increase of the genetic biodiversity of the tolerance to main diseases, is discussed in Section 2. The second pillar, the improvement of the toolbox of measures to reduce the infestation potential, is already mentioned in this chapter. The third pillar is the use of forecasting models for precise timing of applications. Models such as RIMpro (Jamar et al., 2006) were developed specifically for OF, and provide information required for timing of scab 'stop applications'. Improvement of knowledge about the biology of minor pathogens such as sooty blotch and premature leaf fall (*M. coronaria*) can contribute to decision-making models that facilitate their control. The last pillar is the development of new products and/or the improvement of use of known products (Kunz et al., 2014, 2016). In this context, the potential of 'soft products' traditionally used in OF such as *Equisetum arvense* or rock powders should also be explored. No introduction or registration of suitable new products is expected in the short term; however, some reduction was achieved recently by the improvement of copper formulations (Nannen et al., 2010).

As copper as well as sulphur products present side effects of phytotoxicity on leaves and fruits under unfavourable climatic conditions, there is great interest from practitioners of alternative products or measures that respect the principles of OF. Initial results with antagonists are presented by Koehl et al. (2014) and appear promising. More research, but also more political incentives for the investment of small and medium business enterprises in this sector, is needed.

For control of storage diseases, the hot water treatments described by Trierweiler (2003) have been developed further by Maxin et al. (2014). Research is needed to adapt the method to the requirements of both small farmers and big retailers, and to improve efficacy.

Diseases of more regional importance should also be considered in the improvement of the management strategies for control. One of these is fruit canker (*Nectria galligena*) where frequent applications of calcium hydroxide in winter show good results (Heijne et al., 1998). Another is fire blight (*E. amylovora*) where a combination strategy based on antagonists and aluminium sulphate was prepared by Kunz et al. (2010).

Since prevalence of many diseases is related to rainfall in regions with high humidity, and as hail nets are becoming common practice, the idea of rain roofs seems very attractive. For apples, first trials in research stations have recently started (Bertelsen et al., 2014; Buchleither, oral communication, 2014). For some stone fruit species (e.g. sweet cherries) and for small fruit rain roofs are already rather common practice in several regions.

4.4 Pest control

The first and fundamental measure in the strategy for sustainable pest control in OF is the conservation of a high level of functional biodiversity in the orchard. For several pests, there is nothing available for a direct management in the toolbox of OF. Most pests, however, can be effectively controlled with the available tools for direct control in combination with measures to reduce the infestation pressure such as the occurrence of natural predators. Thus, populations of natural predators and a 'certain balance' of the arthropod fauna in the orchard has a high economic relevance that needs to be considered in decision-making for an intervention by farmers.

Several measures for the enhancement of beneficials such as the provision of refugia or overwintering places are common practice. The establishment of flowering plant strips in the alleys to enhance beneficials was started in 1986 by a farmer. Some years later, the first trials were published with weed strips that could be mulched (Kienzle and Straub, 1991), have natural vegetation and extensive mulching (Kienzle et al., 1997), and with unmulched weed strips (Wyss, 1994). However, the enhancement of these predatory arthropods failed to give complete control of key pests such as rosy apple aphid (*Dysaphis plantaginea*); flowering plants in the weed strips failed to establish in the long-term and problems with vole control occurred. For these reasons, weed strips were never introduced as a common practice (Kienzle et al., 1997). Recently, new initiatives were started with participatory and interdisciplinary cooperation of farmers and researchers with expertise in entomology, OF, vole control and plant sociology. When the first feasible management strategies of the alley with weed strips, including vole control, were drafted, a significant enhancement of the abundance of aphid predators in summer could be observed (Kienzle et al., 2014, 2016b). Fernique et al. (2016) provides an overview based on interviews with the farmers and experts involved.

Improvement of practical management strategies for enhancement of functional biodiversity, combined with measures for enhancement of different species important for nature conservation, is certainly one of the key trends in research in OF for the next years. Design of these strategies must also include disease control as some products based on sulphur can partly (e.g. predatory mites) or completely (e.g. *Trichogramma* species) reduce the population of predators sensitive to sulphur. Considerations about the genetic biodiversity of disease tolerance are also valid for susceptibility against pests. Furthermore, there are still considerable gaps to fill in the knowledge about occurrence and biology of many natural predators, even of key pests, and the understanding of the complex interactions of arthropod populations and the effects of climatic changes on their performance in the orchard is still very limited.

For some pest species, functional biodiversity can give sufficient control. The only apple aphid that requires additional direct control measures is the rosy apple aphid. The combination of extract of neem (*Azadirachta indica*) with plant oils was developed especially for OF. For woolly apple aphid (*Eriosoma lanigerum*) control, combination strategies involving various pruning methods, conventional and novel rootstocks, release and enhancement of key predators, mechanical removal (brush) and application of different plant protection products based on different oils were tested (Toups et al., 2010; Kelderer et al.,2016). As coverage with nets and even rain roofs are of increasing importance, and these measures usually enhance infestation with woolly apple aphid, research for additional biocontrol measures and information regarding the tolerance of old and new rootstocks is needed.

Apple sawfly (*Hoplocampa testudinea*) is a key pest in some European regions in OF. Direct control is successfully managed with *Quassia amara* extracts possibly in combination with neem extract (Kienzle et al., 2002, 2004, 2006). The window of application is narrow and exact timing is very important for success (Kienzle et al., 2004). Trapman (2016) developed a pre-dynamic simulation model for the biology of apple sawfly, which has been added to the RIMpro platform; however, there is still need for more biocontrol measures in the toolbox. Happe et al. (2016) reports the first interesting results for a reduction of the sawfly population by application of entomophagous nematodes in spring during hatching of adults. The incidence of the parasitoid *Lathrolestes ensator* (Brauns) was studied in the Netherlands by Zijp and Blommers (2002). In Switzerland, Babendreier (2000) observed high parasitation by *Aptesis nigrocincta*. Grav.

Codling moth (*Cydia pomonella*) is mainly controlled by a combination of mating disruption and several measures to keep the population at low levels. The most important of these measures is application of *C. pomonella* granulosis virus (CpGV), a very selective product. Since emergence of the first cases of resistance to commercial CpGV preparations (Fritsch et al., 2005), new resistance-breaking CpGV products have been developed (Jehle et al., 2016, see also 6.3).There is a clear need to develop strategies for the most effective management of virulence of CpGV. The population of diapausing larvae can be successfully reduced with application of entomophagous nematodes in autumn (Curto et al., 2008; Kienzle et al., 2008, 2010). The use of softwood stakes and bamboo sticks, which offer excellent hiding places for the diapausing larvae, should be avoided in the orchard (Kienzle et al., 2010). Application of *Trichogramma* could be valuable in late autumn when sulphur applications are terminated. Kienzle et al. (2012) tried to develop a practical method for release of this parasitoid using new species and new methods of application; however, due to the short lifespan of all *Trichogramma* species tested, this method was not introduced into practice. Even if the codling moth population is sufficiently reduced by the successful application of netting (Warlop et al., 2014) or by some newly published fructose application methods (Warlop et al., 2016), the development of additional biocontrol measures for codling moth is one of the most important future research challenges in OF.

With climate change and as fruit moves around the world, new exotic pests can appear in both conventional and organic orchards. One example is stinkbugs, which have caused a high level of damage in pear orchards and are now also observed more frequently in apple orchards. The challenge for OF will be to develop feasible control strategies for such pests that are selective enough not to compromise the existing management strategies.

Another key challenge for researchers is the improvement of the direct and indirect measures to prevent damages by voles and field mice. No baits suited to OF are available and the only effective direct control measures are traps.

4.5 Regulation of crop set and tree growth

The use of growth regulators is not allowed in OF. Initially, this is a challenge to the ability of the fruit grower in pruning and tree training. The balance of tree vigour/canopy size and fruit set is one of the most important parameters in the management strategies of different pests (such as aphids or leaf rollers) and diseases (scab). If mechanical pruning is introduced for reasons of labour, cost and time saving, the effects on the development of pests and diseases will have to be studied very carefully (Zimmer et al., 2016).

The regulation of crop set by hand thinning is rather expensive, thus several researchers have tried to reduce or replace this practice by mechanical and/or chemical measures suited to OF. Mechanical measures such as the rope thinner were successfully tested in several countries (Strimmer et al., 1997; Weibel et al., 2008; Sinatsch et al., 2010); their use is now common practice in some countries such as Germany.

Products for application during bloom were examined, including lime sulphur and several different carbonates that gave good results (Strimmer et al, 1997). The utilization of the effects of thinning of lime sulphur applications is widespread mainly in Southern

Europe. The effects of Vinasse (supplier Bioorga) and a product based on potassium carbonate gave interesting results in some trials (Weibel et al., 2008, 2012) whereas in other trials it was not so effective and showed some phytotoxicity (Kelderer et al., 2004).

Tree shading (Byers et al., 1985; McArtney et al., 2004; Stadler et al., 2005; Kelderer et al., 2008) is an interesting trend that is also practised in several countries to induce fruitlet drop. Another method to increase fruitlet drop is the application of transpiration inhibitors (Lardschneider et al., 2016).

The action of thinning treatments may vary depending on factors such as climatic conditions, crop growth stage, applied product(s) or product mixtures, application rates and finally crop selectivity of the applied product(s) on different apple varieties (Kelderer et al., 2008, 2012, 2014). In order to support growers in selecting the most appropriate thinning strategy for their specific farming system, it is important to investigate the influence of these different factors on the most commonly grown apple cultivars, and to develop specific application strategies for each variety and each climatic region.

5 Strategies for improving ecosystem services

This chapter does not aim to give a complete analysis of the ecosystem services of organic orchards and the challenges for their improvement. Instead, it will briefly highlight three important topics that are very much the focus of improvement of the OF system.

Biodiversity as a tool in plant health care strategies in OF offers a chance to extend these concepts into a combined management of functional biodiversity, and general issues of species biodiversity and nature conservation which offer great additional value to OF.

Figure 3 Weed strips with flowering plants in the middle of the alleys in an organic orchard in Southern Germany. Author J. Kienzle.

Figure 4 Single shrubs at the end of the tree rows can provide nectar and food for birds in winter. Author J. Kienzle.

There are a large number of recommendations to practice for such measures; however, implementation in practice is very low. The risks of negative side effects on the general orchard management are well known such as the increase of problems with voles and field mice or the resurgence of minor insect pests such as *Ceresa bubalus* or *Amestategia glabrata* and also diseases such as sooty blotch. Furthermore, the recommended measures often fail to reach the goals they are recommended for in terms of species and nature protection.

There is a strong demand for the development of a toolbox of measures that can be integrated successfully in the general orchard management strategy and be proved to successfully enhance functional or general biodiversity and/or the survival of certain important species. Weed strips with flowering plants in the middle of the alleys (Fig. 3) and at the border of the orchard, hedges or single scrubs at the end of tree rows (Fig. 4) and nesting aids for birds and insects form a key part of these measures.

These approaches must be tested and optimized in a participatory approach in a large number of environments before final recommendation can be made. Based upon such a toolbox of measures to enhance biodiversity, in close collaboration of OF-specialized advisers, nature and species conservation advisers with experience in OF and regional farmer groups, the full potential of organic orchards for species conservation in the agricultural landscape can be explored.

The *reduction of the carbon footprint and the energy consumption* is an important aim for the improvement of the production system. For this second topic, it would be important to have available calculation systems that allow a fast check of the effect of intended changes in the production system on carbon footprint and energy consumption.

The third goal, where there is a growing demand for more research, is the stability and the amount of *yields* from OF. Weibel et al. (2014) state that the yields in OF are about 15–30% lower than in comparable IP orchards that cannot completely be explained by losses due to pathogens or weed competition. There is a strong demand for systematic monitoring of all factors not correlated with plant protection issues which might be involved in this effect. However, the elaboration of strategies to improve yields must consider not only the effect on the amount of yield but also on fruit quality, shelf life and general plant health.

6 Case studies

6.1 Introduction of apple scab-'resistant' varieties in practice

In the 1990s, OF was highly interested in scab-resistant varieties. However, the varieties that were tested were not convincing in terms of fruit quality and shelf life. The situation changed with the variety 'Topaz' bred in 1984 by Professor Jaroslav Tupy (Strizovice, CZ), which seemed very promising. German organic fruit growers founded a society called 'Malus bunda' and bought the licence for 'Topaz'. 'Topaz' was planted in a concerted action by many organic farmers, mainly in Germany and Austria, so that it could be introduced with sufficient volume as a new variety to retailers. However, the fast introduction also had side effects: Only when many orchards were already planted did it become evident that this variety was highly susceptible to apple collar rot (*Phytophthora cactorum*). Grafting of the variety on an interstem was recommended (Kelderer and Lardschneider, 2010), which proved to be a good cultural measure to reduce this problem. Today, 'Topaz' is only sold with an interstem for OF. Losses of more than 50% of the trees by apple collar rot can be seen in the first orchards that were planted. As copper applications are used to control this disease, these first 'Topaz' orchards are sometimes treated with more copper per year than the commercial scab-susceptible varieties (Kienzle et al., 2016a). Fruit growers in this case paid a high price for their pioneering role.

Some years later, the variety 'Santana', bred in the Netherlands was introduced to retailers in a similar concerted action by a group of pioneers. More professional structures exist to launch new disease-resistant varieties. The variety 'Natyra', also bred in the Netherlands, was selected as suitable for OF within a newly established participatory system of collaboration between researchers and organic fruit growers. A concerted action for growing sufficient quantity for a successful market introduction started some years ago (Haug, 2014).

6.2 Improving traditional substances suited to OF

Development of neem extract for the control of rosy apple aphid

In the early 1990s, spring conditions in Europe became more and more favourable for the build-up of populations of rosy apple aphid, resulting in considerable damage in OF. As a consequence, a screening of possible substances suitable to OF for aphid control was started (Schulz et al., 1993). A combination of plant oil and neem (*A. indica*), which was known from traditional use in India for more than 1000 years, showed interesting results

(Kienzle et al., 1992). In a participatory research project with close collaboration of applied and basic researchers, a small business enterprise and farmers, a product based on neem extract for the control of rosy apple aphid in OF was developed and contemporaneously introduced into practice (Schulz et al., 1995; Kienzle et al., 1995). Hermann et al. (1995) studied the side effects on beneficial insects to ensure the optimal integration of this measure in the management strategy.

Registering this product for plant protection was a huge effort by the company involved, but since its registration, this product has been used successfully in most organic farms in Europe as an important part of the management of rosy apple aphid.

Inclusion of lime sulphur and carbonates in scab management strategy

Use of lime sulphur and different carbonates in apple production dates back many years. When OF started to increase in production area and more suitable solutions for disease control were required, these substances were 'rediscovered' and their potential was explored by intensive research activities (Kelderer et al., 1997; Zemmer et al., 2002; Trapman and Drechsler-Elias, 2002; Hinze and Kunz, 2010; Kunz and Hinze, 2016; Kelderer et al., 2006; Tamm et al., 2006). 'Targeted application during infection' supported by the forecasting model RIMpro proved a successful strategy in many European countries (Jamar et al., 2010) and is nowadays common practice in most organic orchards in Europe. These strategies contribute to reduction of the input of copper but also to the reduction of the number of applications needed for disease control in OF.

6.3 Development and improvement of new substances suited to OF: CpGV

The CpGV was first isolated from infected larvae in Mexico and described by Tanada in 1964. All commercial products used in Europe since 2005 were based on this CpGV isolate ('Mexican strain'). Field studies showed its potential for the biocontrol of codling moth (Dickler and Huber, 1977). After an intensive collaboration of researchers and the plant protection industry, in 1988, a Swiss startup registered the first product in Europe based on CpGV: Madex. Within a very short time, CpGV proved to be successful in OF across Europe. Together with mating disruption, CpGV application is the cornerstone of codling moth control (Jehle et al., 2010) in OF.

In 2004, local codling moth populations with significantly reduced susceptibility towards CpGV-M were first observed in orchards in south-west Germany (Fritsch et al., 2005) and later in France (Sauphanor et al., 2006). Since then, further cases of resistance were reported from commercial apple orchards in Germany, France, Italy, Switzerland, the Netherlands and Austria (Asser-Kaiser et al., 2007; Schmitt et al., 2013) and an experimental orchard in the Czech Republic (Zichová et al., 2013). In a concerted action of researchers, producers and organic fruit growers, it was possible to very quickly explore the background to this resistance and to develop solutions: The resistance gene could be located on the Z-chromosome of CM, and methods for resistance monitoring were developed. Furthermore, it could be shown that alternative strains of CpGV with differing virulence to the Mexican CpGV strain could overcome the resistance (Jehle et al., 2006, 2010). New products, based on such isolates, for example, MadexPlus and MadexMAX (Andermatt Biocontrol) or Carpovirusine Evo2 (NPP-Arysta), have been developed and registered. These efforts have ensured the availability of CpGV as a tool in the strategy for codling moth control in OF. However, recently there is evidence for availability of a second type of CpGV resistance from a very few cases in Germany (Jehle et al., 2016). This resistance is targeted against a part of the available new CpGV strains. Thus, for a sustainable virulence management of CpGV, it will be important to find intelligent management strategies that use the full natural biodiversity of CpGV.

7 Future trends in research

The principles of OA (POA; see http://www.ifoam.bio/en/organic-landmarks/principles-organic-agriculture) provide the framework for sustainable development in OF. At first, the four principles of the POA define the sustainability concept including social, environmental and economic aspects. Organic products are 'credency' products that are based on consumers' confidence in the added values of the product. The economic success of OF relies on the acceptance of higher prices for organic products by consumers in return to these added values. Thus, there is a great economic sustainability for OF to improve the sustainability of the production system following the POA. For this reason, organic agricultural systems must continuously try to improve their orientation towards the organic principles that define the sustainability concept and are part of the organic identity. The systems approach involves the whole production chain but the basic unit is always the single farm unit. However, this farm unit is part of the whole agricultural system including consumers, retailers and also the public sector in terms of providing incentives and regulations. Often, these framework conditions are an important part of the bottlenecks that compromise approaches to improving the system. Thus, research that aims for a real practical impact must not only focus on technical issues but also consider the impact of these framework conditions and involve farmers and the appropriate players in a participatory approach.

Part of the organic identity is also that it offers opportunities for a high diversity of farm types. Therefore, researchers should not focus too much on the design of a 'best practice gold standard' for an organic orchard but should develop a toolbox of measures for more resilience in close collaboration with a high number of different organic farm types in their region. This will allow each farm unit a specific adaptation of measures for more resilience with respect to the special situation of the farm and of the farmer family. Key trends for research can be summarized as follows:

- Breeding, testing under organic conditions and introduction into the market of varieties with durable tolerance to pathogens within the framework of variety concepts that are based on high genetic diversity and therefore possibly include also a high number of varieties. The diversity of varieties should be part of the development of quality concepts that include also 'organic' standards for external fruit quality.
- Improvement of the toolbox for soil and vegetation management with the aim of enhancing the total amount of ecosystem services (e.g. yield and fruit quality, biodiversity, soil fertility and others) provided by the farm unit.
- Improvement of the existing toolbox and development of new measures and management strategies suited to OF for plant health care. The challenge in the case of product development includes not only the technical development but also the support of the registration of the substance since otherwise no impact of this research on agricultural practice is to be expected.

8 Where to look for further information

8.1 Important conferences and publications about organic fruit growing in Europe

www.ecofruit.net

'Ecofruit' is a conference that takes place each second year. It aims to bring together European researchers and consultants working on topics related to organic fruit growing. The proceedings of the 17 conferences from 1989 to 2016 are freely available for download on the homepage.

http://orgprints.org/

Organic Eprints is an international open access archive for papers and projects related to research in organic food and farming. The archive contains full-text papers in electronic form together with bibliographic information, abstracts and other metadata. It also offers information on organizations, projects and facilities in the context of OF research.

http://docplayer.pl/9595095-Lidia-sas-paszt-and-eligio-malusa.html

Book of abstracts of the international conference 'Innovative Technologies in Organic Horticultural Production', 22–24 October 2014 in Skierniewice, Poland

http://www.tfrec.wsu.edu/pages/organicfruit2012/Home

Programme and abstract book of the 2nd International Organic Fruit Research Symposium, 2012, Leavenworth, Washington, USA

http://www.ishs.org/symposium/108

Symposium on Organic Fruit Growing in Modena in 2008

8.2 Journals with a focus on OF

Biological Agriculture & Horticulture

http://www.tandfonline.com/toc/tbah20/current

Oeko-Obstbau

Journal in German language, specialized to organic fruit growing;
www.foeko.de/publikationen/zeitschrift-oeko-obstbau/

8.3 Research stations and extension services with a focus on OF

www.laimburg.it
https://www.uni-hohenheim.de/organisation/einrichtung/fg-angewandte-entomologie
www.saat-gut.org/projekte.cfm
Information about the breeding project Apfel:gut
http://www.kob-bavendorf.de
http://www.oeon.de/Versuchswesen.htm
http://www.oekoobstbau.de/F_Versuch.htm
www.fibl.ch
www.grab.fr

http://www.itab.asso.fr/itab/contacts.php
www.louisbolk.org
https://forschung.boku.ac.at/fis/suchen.person_publikationen?sprache_in=en&menue_id_in=102&id_in=253&publikation_typ_id_in=&sortierung_in=reprint
http://www.gartneriraadgivningen.dk
http://www.inhort.pl
http://www.biofruitadvies.nl
http://csanr.wsu.edu/author/granats/

8.4 Important organic umbrella associations and initiatives

IFOAM umbrella organization
www.ifoam.bio
http://www.ifoam-eu.org
Information regarding principles, standards and positions.

A fruit growers' specialist group exists within the frame of the IFOAM EU group. Members are researchers and farmers. Regular meetings take place within the ecofruit conferences.

www.foeko.de
German umbrella organization for fruit growers.

www.bioobst-forum.eu/de
Umbrella organization of fruit grower working groups and networks in western Europe

9 References

Asser-Kaiser, S., Fritsch, E., Undorf-Spahn, K., Kienzle, J., Eberle, K. E., Gund, N. A., Reineke, A., Zebitz, C. P., Heckel, D. G., Huber, J. and Jehle, J. A. (2007), Rapid emergence of baculovirus resistance in codling moth due to dominant, sex-linked inheritance. *Science* 317, 1916–8.

Asser-Kaiser, S., Heckel, D. G. and Jehle, J. A. (2010), Sex linkage of CpGV resistance in a heterogeneous field strain of the codling moth *Cydia pomonella* (L.). *Journal of Invertebrate Pathology* 103, 59–64.

Babendreier, D. (2002), Life history of *Aptesis nigrocincta* (Hymenoptera: Ichneumonidae) a cocoon parasitoid of the apple sawfly, *Hoplocampa testudinea* (Hymenoptera: Tenthredinidae). *Bulletin of Entomological Research* 90(4), 291–7.

Bannier, H.-J. (2011), Moderne Apfelzüchtung: Genetische Verarmung und Tendenzen zur Inzucht. *Erwerbs-Obstbau* 52, 85–110.

Baumgartner, I. O., Patocchi, A., Frey, J. E., Peil, A. and Kellerhals, M. (2015), Breeding elite lines of apple carrying pyramided homozygous resistance genes against apple scab and resistance against powdery mildew and fire blight. *Plant Molecular Biology Reporter* 121 (3), 647–56.

Benduhn, B., Zimmer, J. and Buchleither, S. (2014), Effect of mechanically removing of leaf litter on apple scab (*Venturia inaequalis*) infestation in organic apple production. In Foeko (Ed.), *Proceedings of the 16th International conference on Organic Fruit-GrowingEcofruit*, 40–4

Benduhn, B., Renner, U., Rank, H. and Zimmer, J. (2006), *Pflanzenschonende mechanische Bearbeitung des Pflanzstreifens bei Kernobst und Alternativen: Optimierung der bestehenden Verfahren unter arbeitswirtschaftlichen Gesichtspunkten auf verschiedenen Standorten und Bodentypen.* [Mechanical soil treatment in pipfruit growing: Improvement of established methods on different locations and soil-types under consideration of economic aspects.] Final Report.

Berling, M., Blachère-López, C., Soubabère, O., Lery, X., Bonhomme, A., Sauphanor, B. and López-Ferber, M. (2009), *Cydia pomonella* granulovirus genotypes overcome virus resistance in the codling moth and improve virus efficiency by selection against resistant hosts. *Applied and Environmental Microbiology* 75, 925–30.

Bertelsen, M. and Lindhard Pedersen, H. (2014), Preliminary results show rain roofs to have remarkable effect on diseases of apples. *Proceedings of the 16th International conference on Organic Fruit-GrowingEcofruit* Ed Foeko, 242–3

Bohr, A., Buchleither, S., Späth, S. and Mayr, U. (2016), First observations on disease development of *Marssonina coronaria* on apple in Lake Constance area. In Foeko (Ed.), *Proceedings of the 17th International conference on Organic Fruit-GrowingEcofruit*, 172–5.

Buchleither, S., Bohr, A., Späth, S. and Mayr, U. (2014), Effect of several beetroot Vinasse products on foliage decomposition and ascospore formation of Venturia inaequalis on remaining leaf material. In Foeko (Ed.), *Proceedings of the 16th International conference on Organic Fruit-Growing Ecofruit*, 45–51

Buchleither, S., Mayr, U. and Brandt, M. (2014), Legumes dense sowing with peas as an alternative method for nitrogen fertilization in organic fruit growing. In Foeko (Ed.), *Proceedings of the 16th International conference on Organic Fruit-Growing Ecofruit*, 207–13

Byers, R. E., Lyons, C. G., Yoder, K. S., Barden, J. A. and Young, R. W. (1985), Peach and apple thinning by shading and photosynthetic inhibition. *Journal of Horticultural Science* 60 (4), 465–72

Eurobarometer (1991 (35.1)–2010 (73.1)), Europeans and Biotechnology. Online: http://ec.europa.eu/COMMFrontOffice/PublicOpinion/index.cfm/Archive/index (15.1.2016).

Fankhauser, F., Schumacher, R. and Stadler, W. (1979), Ausdünnung mit unterschiedlichen Brühmengen und Konzentrationen. *Schweizerische Zeitschrift für Obst- Weinbau* 115 (6), 205–13

Fritsch, E., Undorf-Spahn, K., Kienzle, J., Zebitz, C. P. W. and Huber, J. (2005), Apfelwickler-Granulovirus: Erste Hinweise auf Unterschiede in der Empfindlichkeit lokaler Apfelwickler-Populationen. *Nachrichtenblatt des Deutschen Pflanzenschutzdienstes* 57, 29–34.

Gebhardt, M., Eberle, K. E., Radtke, P. and Jehle, J. A. (2014), Baculovirus resistance in codling moth is virus-isolate dependent and the consequence of a mutation in viral gene pe38. *Proceedings of the National Academy of Sciences of the United States of America* 111, 15711–16.

Graillot, B., Berling, M., Blachère-López, C., Siegwart, M., Besse, S. and López-Ferber, M. (2015), Progressive adaptation of a CpGV Isolate to codling moth populations resistant to CpGV-M. *Viruses* 6, 5135–44.

Granatstein, D., Pavek, P. and Kriby, E. (2016), Integrating Cover Crops into organic Orchard Systems, In Foeko (Ed.), *Proceedings of the 17th International conference on Organic Fruit-Growing Ecofruit*, 273–4.

Happe,S., Njezic, B. and Ehlers, R. U. (2016), Control of apple sawfly (*Hoplocampa testudinea* Klug) and plum sawflies (*H. flava* L. and *H. minuata* Christ.) with entomopathogenic nematodes. In Foeko (Ed.), *Proceedings of the 16th International conference on Organic Fruit-Growing Ecofruit*, 258–60.

Haug, P. (2014), Resistente'Sorten Teil 3. *Oeko-obstbau* 1, 21–2.

Haug, P. (2015), AK Sorten und Züchtung. *Oeko-obstbau* 4, 34.

Heijne, B., Anbergen, R. H. N., Jansonius, P. J. and Bloksma, J. (1998), Calcium hydroxide offers opportunities of environmentally friendly canker control. *Fruitteelt (Den Haag)* 88 (37), 14–15.

Hermann, P., Zebitz, C. P. W. and Kienzle, J. (1995), Wirkung verschiedener NeemAzal-Formulierungen auf die Larven der Florfliege *Chrysoperla carnea* in Labor und Halbfreiland. *7. Intern. Erfahrungsaustausch ueber Forschungsergebnisse zum Ökol. Obstbau, Beitr. zur Tagung vom 14.-15.12.1995, Weinsberg*, 114–18.

Hinze, M. and Kunz, S. (2010), Screening of biocontrol agents for their efficacy against apple scab, *Proceedings of the 14th International conference on Organic Fruit-Growing Ecofruit* Ed Foeko, 38–44.

Huber, J. and Dickler, E. (1977), Codling moth granulosis virus: its efficiency in the field in comparison with organophosphorus insecticides. *Journal of Economic Entomology* 70 (3), 557–61.

Huber, J. (1998), Western Europe. In Hunter-Fujita, F. R., Entwistle, P. F., Evans, H. F. and Crook, N. E. (Eds), *Insect Viruses and Pest Management*, 201–15. Wiley and Sons.

Jamar, L., Cavelier, M. and Lateur, M. (2010), Primary scab control using a 'during-infection' spray timing and the effect on fruit quality and yield in organic apple production, *Biotechnology Agronomy Society and Environment* 14 (3), 423–39.

Jehle, J. A., Sayed, S. M. and Wahl-Ermel, B. (2006a), What do we (need to) know about low-susceptibility of codling moth against *Cydia pomonella* granulovirus (CpGV)! In Foeko (Ed.), *Ecofruit – Proceedings to the 12th International conference on cultivation Technique and Phytopathological Problems in Organic Fruit-Growing*, 14–18.

Jehle, J. A., Eberle, K. E., Asser-Kaiser, S., Schulze-Bopp, S. and Schmitt, A. (2010), Resistance of codling moth against *Cydia pomonella* granulovirus (CpGV): State of knowledge. In Foeko (Ed.), *Proceedings to the 14th International conference on cultivation Technique and Phytopathological Problems in Organic Fruit-Growing Ecofruit*, 133–6.

Jehle, J. A., Eberle, K. E., Asser-Kaiser, S., Schulze-Bopp, S. and Schmitt, A. (2010), Resistance of codling moth against *Cydia pomonella* granulovirus (CpGV): State of knowledge. In Foeko (Ed.), *Proceedings of Ecofruit, 14th International conference on organic fruit growing*, 133–6.

Jehle, J. A., Gueli Aletti, G., Fritsch, E., Kienzle, J., Sauer, A. and Undorf-Spahn, K. (2016), Towards a successful resistance and virulence management strategy for CpGV. In Foeko (Ed.), *Proceedings of Ecofruit, 16th International conference on organic fruit growing*, 261–3.

Kelderer, M., Casera, C. and Lardschneider, E. (1997), Schorfregulierung: Verschiedene Kupferformulierungen – Alternativen zu Kupfer – gezielte Behandlung. In Foeko (Ed.), *Proceedings 8. Internationaler Erfahrungsaustausch über Forschungsergebnisse im Ökologischen Obstbau*, 9–14.

Kelderer, M., Casera, C. and Lardschneider, E. (2000), Zwei Jahre Erfahrungen mit der gezielten Schorfbekämpfung durch die Oberkronenberegnung. In Foeko (Ed.), *Proceedings 9. Internationaler Erfahrungsaustausch über Forschungsergebnisse im Ökologischen Obstbau*, 5–11.

Kelderer, M., Casera, C. and Lardschneider, E. (2006), First resuits of the use of potassium bicarbonate against scab in South Tyrol. In Foeko (Ed.), *Proceedings of the 12th International Conference on Cultivation Technique and Phytopathological Problems in Organic Fruit-Growing*, 93–7.

Kelderer, M. Lardschneider, E. and Casera, C. (2008), Tree shading: an efficient method to control alternate bearing? In Foeko (Ed.), *Proceedings of 13th International Conference on Cultivation Technique and Phytopathological Problems in Organic Fruit-Growing Ecofruit*, 310–13.

Kelderer, M. and Lardschneider, E. (2010), Effect of different treatments with oil cakes, plant protection agents and potassium phosphite to control coJJar rot on Topaz. In Foeko (Ed.), *Proceedings of ecofruit – 14th International Conference on Organic Fruit-Growing*, 106–9.

Kelderer, M., Lardschneider, E. and Topp, A. (2010), Effect of transpiration inhibitors on June fruit drop of apple trees. *Ecofruit – 14th International Conference on Organic Fruit-Growing*, Fördergemeinschaft Ökologischer Obstbau e. V. Weinsberg 14, 206–11.

Kelderer, M., Lardschneider, E. and Rainer, A. (2012), Crop regulation on different apple cultivars with transpiration inhibitors. In Foeko (Ed.), Proceedings of *Ecofruit – 15th International Conference on Organic Fruit-Growing*, 131–9.

Kelderer, M. Lardschneider, E. and Telfser, J. (2014), Interaction between varierties, lime sulphur and hailnet on the thinning effect and on side effects using paraffin oil as a June drop thinner. In Foeko (Ed.), *Proceedings of ecofruit – 16th International Conference on Organic Fruit-Growing*, 132–41.

Kelderer. M., Topp, A., Lardschneider, E., Rainer, E. and Matteazzi, A. (2014), Organic apple tree nutrition: Comparison of different organic fertilizers, application timing and rate, and soil management techniques: results of a 5 year field study. In Foeko (Ed.), *Proceedings of the Ecofruit 16th International conference on organic fruit growing*, 108–15.

Kellerhals, M., Baumgartner, I. O., Schütz, S., Lussi, L., Andreoli, R. Gassmann, J. and Patocchi, A. (2016), Approaches in breeding high quality apples with durable disease resistance. In

Foerdergemeinschaft Okologischer Obstbau e. V. (Ed.), *17th International Conference in Organic Fruit-Growing - Ecofruit*, 12–17.

Kienzle, J. and Straub, M. (1991), Nützlingsförderung und Begrünung im Ökologischen Obstbau. In AK Forsch. d. Obstbaufachgr. d. anerk. ökol. Verb. (Hrsg.): 4. Internationaler Erfahrungsaustausch über Forschungsergebnisse zum Ökologischen Obstbau, Weinsberg.

Kienzle, J., Schulz, C., Straub, M., Schmitt, A. and Weil, B. (1992), Einsatz von Neemprodukten zur Regulierung der Mehligen Apfelblattlaus *Dysaphis plantaginea. 5. Internationaler Erfahrungsaustausch über Forschungsergebnisse zum Ökologischen Obstbau*, 19.-20.11.1992, Weinsberg, 21–7.

Kienzle, J., Schulz, C. and Zebitz, C. P. W. (1995), Zweijährige Erfahrungen mit dem Einsatz von NeemAzal in Ökologisch wirtschaftenden Obstbaubetrieben. *7. Intern. Erfahrungsaustausch ueber Forschungsergebnisse zum Oekol. Obstbau, Beitr. zur Tagung vom 14.-15.12.1995, Weinsberg*, 128–32.

Kienzle, J., Zebitz, C. P. W., Brass, S. and Athanassov, A. (1995), Populationsdynamik von Schalenwicklern und ihrer Parasitoide in ökologisch bewirtschafteten Apfelanlagen. Fördergem. Ökol. Obstbau e.V. (Hrsg.): 7. Internationaler Erfahrungsaustausch über Forschungsergebnisse zum Ökologischen Obstbau, Weinsberg. S. 109–13.

Kienzle, J. and Zebitz, C. P. W. (1997), Extensives Mulchsystem – ein Beitrag zur Erhöhung der Arthropodenvielfalt in Obstanlagen? In: Köpke, Ulrich und Jons, A. Eisele (Hrsg.), 1997:Beiträge zur 4. Wissenschaftstagung zum Ökologischen Landbau 3. - 4. März 1997 an der Rheinischen Friedrich-Wilhelms-Universität, Bonn, Verlag Dr. Köster, Berlin

Kienzle, J., Zebitz, C. P. W. and Brass, S. (1997), Floral and Faunal Species Diversity and Abundance of Aphid Predators in Ecological Apple Orchards, *Entomological Research in Organic Agriculture. Biological Agriculture and Horticulture* 15, 233–40.

Kienzle, J., Zebitz, C. P. W., Brass, S. and Athanassov, A. (1997), Abundance of different tortricid species and their parasitoid antagonists in ecological apple orchards,*Entomological Research in Organic Agriculture Biological Agriculture and Horticulture* 15, 211–21.

Kienzle, J., Kopp, B. and Schulz, C. (2002), Control of the apple sawfly (Hoplocampa testudinea Klug) with extracts from *Quassia amara* L.: Quality and combination. In Foeko (Ed.), *Proceedings of Ecofruit – 10th International Conference on Cultivation Technique and Phytopathological Problems in Organic Fruit-Growing*, 28–34.

Kienzle, J., Zimmer, J., Klopp, K., Maxin, P., Yamada, K., Bathon, H., Zebitz, C. P. W.,Ternes, P. and Vogt, H. (2004), Control of the apple sawfly Hoplocampa testudinea Klug in organic fruit growing and possible side effects of control strategies on Aphelinus mali Haldeman and other beneficial insects. In Foeko (Ed.), *Proceedings of Ecofruit – 10th International Conference on Cultivation Technique and Phytopathological Problems in Organic Fruit-Growing*, 14–17.

Kienzle, J., Zimmer, J., Maxin, P., Rank, H., Bathon, H. and Zebitz, C. P. W. (2006), Control of Apple sawfly *Hoplocampa testudinea* Klug in organic fruit growing. In Foeko (Ed.), *Proceedings of the 12th International Conference on cultivation Technique and Phytopathological Problems in Organic Fruit Growing*, 25–9.

Kienzle, J.; Glocker, N. and Haug, P. (2012), POSEIDON – a tool to support the orientation towards a sustainable improvement of organic fruit growing in participative working groups. In Foeko (Ed.), *Proceedings of the 15th International Conference in Organic Fruit-Growing – Ecofruit*, 106–13.

Kienzle, J., Foell, M., Karrer, E., Krismann, A. and C. P. W. Zebitz (2014), Establishment of permanent weed strips with autochthonous nectar plants and their effect on the occurrence of aphid predators, In Foeko (Ed.), *Proceedings of the 16th International Conference in Organic Fruit-Growing – Ecofruit*, 31–9.

Kienzle, J. Rolker, P. and Glocker, N. (2016a), First successful steps of german fruitgrowers towards a more resilient system of organic apple growing. In Foeko (Ed.), *Proceedings of the 17th International Conference in Organic Fruit-Growing – Ecofruit*, 249–51.

Kienzle, J., Krismann, A., Walther,B., Karrer, E., Foell, M. and Zebitz, C. P. W. (2016b), Towards a holistic and participatory approach to integrate flower strips in apple orchards. In Foeko (Ed.), *Proceedings of the 17th International Conference in Organic Fruit-Growing – Ecofruit*, 270–2.

Koehl, J., Scheer, C., Holb, I. J., Masny, S. and Molhoek, W. (2014), Towards an Integrated Use of Biological Control by *Cladosporium cladosporioides* H39 in Apple Scab (*Venturiainaequalis*) Management, *Plant Disease Journal* 99, 535–43.

Korsgaard, M. (2016), Effect of Strategic Irrigation on infection of Apple scab (*Venturia inaequalis*). In Foeko (Ed.), *Proceedings of the 17th International Conference in Organic Fruit-Growing – Ecofruit*, 45–53.

Kuehn, B. F. and Lindhard Pedersen, H. (2009), Cover crop and mulching effects on yield and fruit quality in unsprayed organic apple production. *EJHS* 74 (6), S. 247–53.

Kunz, S., Schmitt, A. and Haug, P. (2010), Fire blight control strategies in organic fruit growing. In Foeko (Ed.), *Proceedings of the 14th International conference on Organic Fruit-Growing Ecofruit*, 118–25.

Kunz, S. and Hinze, M. (2016), Efficacy of biocontrol agents against apple scab in greenhouse trials, In Foeko (Ed.), *Proceedings of the 17th International Conference in Organic Fruit-Growing – Ecofruit, 25–31*.

Kunz, S. and Hinze, M. (2014), Assessment of biocontrol agents for their efficacy against apple scab, In Foeko (Ed.), *Proceedings of the 16th International Conference in Organic Fruit-Growing – Ecofruit*, 65–71.

Lardschneider, E., Schütz, R. and Kelderer, M. (2016), Cultivar-specific adaptation of crop load regulation with transpiration inhibitors on the cultivars Gala, Braeburn and Kanzi®. In Foeko (Ed.), *Proceedings of the 17th International Conference in Organic Fruit-Growing – Ecofruit*, 116–22.

Lateur, M. and Populer, C. (1994), Screening fruit tree genetic resources in Belgium for disease resistance and other desirable characters. *Euphytica* 77, 147–53.

Malusà, E., Sas-Paszt, L., Ciesielska, J. and Stępień, W. (2014), A new generation of fertilizers and soil amendments for organic horticultural crops, *Book of abstracts of the international conference 'Innovative Technologies in Organic Horticultural Production'*, 22–24 October 2014, Skierniewice, Poland, 11–12.

Maxin, P., Williams, M. and Weber, R. W. S. (2014), Control of fungal storage rots of apples by hot-water treatments: a northern european perspective. *Erwerbs-Obstbau* 56 (1), 25–34.

McArtney, S., White, M., Latter, I. and Campbell, J. (2004), Individual and combined effects of shading and thinning chemicals on abscission and dry-matter accumulation of 'Royal Gala' apple fruit. *Journal of Horticultural Science & Biotechnology* 79(3), 441–8.

Moeller, K. (2014), Auf eine ausgewogene Nährstoffversorgung kommt es an! *Oeko-Obstbau* 3, 9–14.

Nannen, D. U., Riecken, I. and Lehne, J. (2010), Technical copper minimising strategies to keep an old friend going. In Foeko (Ed.), *Proceedings of the 14th International conference on Organic Fruit-Growing Ecofruit*, 40–52

Patocchi, A. and Bus, V. (2016), Vinquest. Online: http://www.vinquest.ch (13.1.2016).

Pfeiffer, B. (2014), Comparison of Rootstocks Geneva 16, M9 and CG11 under organic cultivation at the LVWO Weinsberg 2009–2013. In Foeko (Ed.), *Proceedings of the 16th International Conference on Organic Fruit Growing Ecofruit*, 10–14.

Ristel, M. and Sattler, I. (2014), Apfel:gut – Participatory, organic fruit breeding. In Foeko (Ed.), *Proceedings of the Ecofruit 16th International Conference On Organic Fruit Growing Hohenheim*, 158–61.

Ristel, M., Sattler, I. and Bannier, J. (2016), Apfelgut – More vitality, genetic diversity and less susceptibility as an organic fruit breeding strategy. In Foeko (Ed.), *Proceedings of the 17th international Conference on Organic Fruit Growing Ecofruit*, 136–9.

Roehrig, et al. (2010), Strategiepapier zu Kupfer als Pflanzenschutzmittel unter besonderer Berücksichtigung des Ökologischen Landbaus, http://kupfer.jki.bund.de/dokumente/upload/80547_kupfer_strategiepapier_juli_2010.pdf

Sauphanor, B., Berling, M., Toubon, J.-F., Reyes, M. and Delnatte, J. (2006), Carpocase des pommes cas de Resistance aux virus de la granulose dans le Sud-Est. *Phytoma* 590, 24–7.

Schmid, A., Weibel, F., Allemann, P. and Santini, D. (2004), Sandwich-System – die Sache wird konkret, Ed. Häseli, A. *Tagungsband zur FiBL Bioobstbautagung 28.01.2004 in Frick*. Forschungsinstitut für biologischen Landbau, Frick, S. 41–2.

Schmitt, A., Bisutti, I., Ladurner, E., Benuzzi, M., Sauphanor, B., Kienzle, J., Zingg, D., Undorf-Spahn, K., Fritsch, E., Huber, J. and Jehle, J. A. (2013), The occurrence and distribution of resistance of codling moth to *Cydia pomonella* granulovirus in Europe. *Journal of Applied Entomology* 137, 641–9.

Schulz, C., Kienzle, J. and Zebitz, C. P. W. (1997), Optimierung der Anwendung von NeemAzal-T/S zur Regulierung der Mehligen Apfelblattlaus. *8. Intern. Erfahrungsaust. Über Forschungsergebnisse zum Ökol. Obstbau, Beitr. zur Tagung vom 13.-14.11.1997, Weinsberg*, 60–4.

Schwender, C. and Zimmer, J. (2016), Mechanical pruning in organic apple production – especially with regard to pathogens and pests. In Foeko (Ed.), *Proceedings of the 17th international Conference on Organic Fruit Growing Ecofruit*, 224–7.

Speiser, B., Mieves, E. and Tamm, L. (2015), Kupfereinsatz von Schweizer Biobauern in v erschiedenen Kulturen. *Agrarforschung Schweiz* 6 (4), 160–5.

Spornberger, A., Brunmayer, R., Fischer, G., Kaufmann, C. and Osterc, G. (2008), Testing of pear trees on their own roots in comparison with important used rootstocks under organic farming conditions with special regard to fire blight. In Foeko (Ed.), *Proceedings of the 13th International Conference on Cultivation Technique and Phytopathological Problems Ecofruit*, 216–19.

Spornberger, A. and Schueller, E. (2016), Results from a long term trial with pear rootstocks under organic production conditions in Eastern Austria. In Foeko (Ed.), *Proceedings of the 17th International Conference on Organic Fruit-Growing Ecofruit*, 18–24.

Stadler, W., Widmer, A., Dolega, E., Schaffner, M. and Bertschinger, L. (2005), Fruchtausdünnung durch Beschattung der Apfelbäume – eine Methode mit Zukunft? Schweizerische Zeitschrift für Obst- und Weinbau (SZOW) Wädenswil 10, 10–13

Strumpf, T., Strassemeyer, J., Kienzle, J., Palm, G., Klopp, K., Engel, A., Rövekamp, H. L., Müller, S., Balmer, M., Zimmer, J. and Felgentreu, D. (2012), Erhebung von Kupfergesamtgehalten in ökologisch und konventionell bewirtschafteten Böden. Teil 4: Gesamtgehalte in Böden deutscher Baumobstbaugebiete. *Journal für Kulturpflanzen* 64(12), 439–51.

Tamm, L., Thomas, D., Schaerer, H. and Refardt, M. (2006), Efficacy of Armicarb (potassium bicarbonate) against scab and sooty blotch on apples. *Joint Organic Congress, Odense*, May 30–21.

Tanada, Y. (1964), A granulosis virus of the codling moth, *Corpocapsa pomonella* (Linnaeus) (*Olethreutidae*, Lepidoptera). *Journal of Insect Pathology* 6, 378–80.

Trapman, M. C. (2016), The development of a dynamic simulation model for the biology of the Apple Sawfly *Hoplocampa testudinea* and the implementation as Decision Support System. In Foeko (Ed.), *Proceedings of the 17th international Conference on Organic Fruit Growing Ecofruit*, 60–8.

Trapman, M. C. (2016), Validation of the RIMpro Decision Support System for Apple Sawfly (*Hoplocampa testudinea*) with field observations in The Netherlands, Belgium, Denmark and Austria 2010–15. In Foeko (Ed.), *Proceedings of the 17th international Conference on Organic Fruit Growing Ecofruit*, 69–77.

Trapman, M. C. and Drechsler-Elias, E. (2000), Die kurative Wirkung von Schwefelkalk gegen Apfelschorf, *Obstbau* 10, 559–61.

Trierweiler, B., Gräf, V., Schirmer, H. and Tauscher, B. (2003), Thermo-Behandlung ökologisch produzierter Äpfel zur Verbesserung der Lagerfähigkeit. *Frischelogistik* 1, 34–6.

Vavra, R., Litschmann, T. and Falta, V. (2014), Which genes of apple scab resistance keep control against pathogen *V. inaequalis* in future? In Foeko (Ed.), *Proceedings of the 16th international Conference on Organic Fruit Growing Ecofruit*, 162–6.

Von Mering, F., Kienzle, J., Kanthak, S., Reiners, E., Patzwahl, W., Weihrauch, F. and Rückrich, K. (2016), Strategiepapier zu Kupfer als Pflanzenschutzmittel unter besonderer Berücksichtigung des Ökologischen Landbaus – Aktueller Stand der Aktivitäten und weiterer Handlungsbedarf,

Eds. BOELW e. V., FOEKO e. V., ECOVIN e. V., Bioland e. V., Naturland e. V., Demeter e. V., Gäa e. V., Deutscher Weinbauverband e. V.

Warlop, F., Timmermans, B., Brouwer, G., Nybom, H., Lateur, M., Kelderer, M., Weibel, F., Pedersen, HL, Garcin, A., Haug, P. and Dapena, E. (2014), How to optimize fruit and berry cultivar selection for organic farmers? A comparison of European approaches. In Foeko (Ed.), *Proceedings of the 16th international Conference on Organic Fruit Growing Ecofruit*, 167–71.

Weber, R., Späth, S., Buchleither, S., et al. (2016), A review of sooty blotch and flyspeck disease in german organic apple production. *Erwerbs-Obstbau* 58, 63.

Weibel, F. and Leder, A. (2004), Consumer reaction to the 'Flavour Group Concept' to introduce scab resistant apple varieties into the market. 'Variety-Teams' as a further development of the concept. In Foeko (Ed.), *Proceedings of the11th Conference on Cultivation Technique und Phytopathological Problems in Organic Fruit Growing*. 3–5 February 2004, Weinsberg/ Germany, 196–201.

Weibel, F. P., Tamm, L., Wyss, E., Daniel, C., Häseli, A. and Suter, F. (2007), Organic Fruit Production in Europe: Successes in Production and Marketing in the Last Decade, Perspectives and Challenges for the Future Development. *ISHS Acta Horticulturae 737, 1st International Symposium on Organic Apple and Pear* (737), S. 163–72.

Weibel, Franco, P., Chevillat, Véronique, S., Rois, E., Tschabold, J-L. and Stadler, W. (2008), Fruit Thinning in Organic Apple Growing with optimized Strategies including natural Spray Products and Rope-Devices. *European Journal of Horticultural Science* 73.

Weibel, F. P., Lemcke, B., Monzelio, U., Giordano, I. and Kloss, B. (2012), Successful Blossom Thinning and Crop Load Regulation for Organic Apple Growing with Potassium-bi-carbonate (Armicarb®): Results of Field Experiments over 3 Years with 11 Cultivars. *European Journal of Horticultural Science* 77 (4), S. 145–53.

Weibel, F. P., Suter, F., Buchleither, S. and Bantleon, G. (2014), Comparison of different tree nutrition strategies for organic apple growing: long-term and additive effects of compost, N-fertilizer, foliar fertilizer and bio-dynamic preparations on tree growth, yield, fruit quality and soil fertility. In Foeko (Ed.), *Proceedings of the 16th international Conference on Organic Fruit Growing Ecofruit*, 108–15.

Woehner, T., Radatz, S. and Hanke, M. V. (2016), Evaluation of apple cultivars for their resistance to premature leaf fall (*Marssonina coronaria*). In Foeko (Ed.), *Proceedings of the 17th international Conference on Organic Fruit Growing Ecofruit*, 140–4.

Wyss, E. (1994), Biocontrol of apple aphids by weed strip management in apple orchards. Diss. Universtity of Bern.

Zemmer, F., Marschall,K.,Kelderer, M. and Zelger, R. (2002), Inibition of the germination of conidia form *Venturia inaequalis* using lime sulphur, sulphur, copper and Ulmasud in comparison to Dithianon, *10. Internationaler Erfahrungsaustausch über Forschungsergebnisse im Ökologischen Obstbau, ed. Fördergemeinschaft Okologischer Obstbau e. V. Weinsberg*, 96–7.

Zichová, T., Stará, J., Kundu, J. K., Eberle, K. E. and Jehle, J. A. (2013), Resistance to *Cydia pomonella* granulovirus follows a geographically widely distributed inheritance type within Europe. *Biocontrol*58, 525–34.

Zijp, J. P. and Blommers, L. H. M. (2002), Apple sawfly *Hoplocampa testudinea* (Hym., Tenthredinidae) and its parasitoid *Lathrolestes ensator* in Dutch apple orchards (Hym., Ichneumonidae, Ctenopelmatinae). *Journal of Applied Entomology* 126, 265–74.

Zingg, D. (2008), Madex Plus and Madex I12 overcome virus resistance of codling moth. In Foeko (Ed.), *Proceedings of the 13th International Conference on Cultivation Technique and Phytopathological Problems in Organic Fruit-Growing*.

Zoppolo, R. J.,Stefanelli, D., Bird, G. W., Perry, R. L. and Weibel, F. (2011), Soil properties under different orchard floor management systems for organic apple production. *Organic Agriculture* 3, S. 11–18.

Index

www.ingramcontent.com/pod-product-compliance
Lightning Source LLC
Jackson TN
JSHW071327100626
103674JS00003B/23
* 9 7 8 1 7 8 6 7 6 0 3 2 6 *